ESSENTIAL ALGEBRA

THE JOHNSTON/WILLIS
DEVELOPMENTAL MATHEMATICS SERIES

Essential Arithmetic, Fifth Edition (paperbound, 1988)
Johnston/Willis/Lazaris

Essential Algebra, Fifth Edition (paperbound, 1988)
Johnston/Willis/Lazaris

Developmental Mathematics, Second Edition (paperbound, 1987)
Johnston/Willis/Hughes

Elementary Algebra, Second Edition (hardbound, 1987)
Willis/Johnston/Steig

Intermediate Algebra, Second Edition (hardbound, 1987)
Willis/Johnston/Steig

Intermediate Algebra, Fourth Edition (paperbound, 1988)
Johnston/Willis/Lazaris

ESSENTIAL ALGEBRA
FIFTH EDITION

C. L. JOHNSTON
ALDEN T. WILLIS
Formerly of East Los Angeles College

JEANNE LAZARIS
East Los Angeles College

Wadsworth Publishing Company
Belmont, California
A Division of Wadsworth, Inc.

The following problems are used with permission from J. Michael Shaughnessy, Brian M. McCay, Carla A. Randall, Gary L. Winckler, *Application Clusters for Intermediate Algebra* (Boston: Prindle, Weber & Schmidt, 1978): Exercises 5.5, Set I, #14; Exercises 5.5, Set II, #13; Exercises 5.6, Set I, #31, 32; Exercises 8.11, Set I, #13, 14, 15, 16; Exercises 8.11, Set II, #15, 16.

Mathematics Publisher: Kevin J. Howat
Mathematics Development Editor: Anne Scanlan-Rohrer
Assistant Mathematics Editor: Barbara Holland
Editorial Assistant: Sally S. Uchizono
Production Editor: Gary Mcdonald
Managing Designer: James Chadwick
Print Buyer: Karen Hunt
Text and Cover Designer: Julia Scannell
Copy Editor: Mary Roybal
Technical Illustrator: Reese Thornton
Compositor: Graphic Typesetting Service, Los Angeles
Cover Illustration: Frank Miller
Signing Representative: Ken King

Printed in the United States of America   14

2 3 4 5 6 7 8 9 10---92 91 90 89 88

Library of Congress Cataloging-in-Publication Data

Johnston, C. L. (Carol Lee), 1911–
   Essential algebra / C. L. Johnston, Alden T. Willis, Jeanne Lazaris.—
5th ed.
   P. cm.—(The Johnston/Willis developmental mathematics series)
   Includes index.
   ISBN 0-534-08916-X (pbk.)
   1. Algebra. I. Willis, Alden T. II. Lazaris, Jeanne, 1932–
III. Title. IV. Series: Johnston, C. L. (Carol Lee), 1911–
Johnston/Willis developmental mathematics series.
QA152.2.J63 1988
512—dc19

                                                    87-29422
                                                    CIP

This book is dedicated to our students,
who inspired us to do our best
to produce a book worthy of their time.

Contents

4 EQUATIONS AND INEQUALITIES

5 WORD PROBLEMS

6 POLYNOMIALS

7 FACTORING

8 FRACTIONS

9 GRAPHING

10 SYSTEMS OF EQUATIONS

11 RADICALS

12 QUADRATIC EQUATIONS

APPENDIXES

A SETS 553

B BRIEF REVIEW OF ARITHMETIC 563

C FUNCTIONS 571

Preface

Essential Algebra, Fifth Edition can be used in an elementary algebra course in any community college or four-year college, in either a lecture format or a learning laboratory setting, or it can be used for self-study. This book will prepare the student well for intermediate algebra and for any other course that requires the knowledge of elementary algebra.

Features of This Book

The major features of this book include:

1. The book uses a one-step, one-concept-at-a-time approach. The topics are divided into small sections, each with its own examples and exercises. This approach allows students to master each step before proceeding confidently to the next section.

2. Many concrete, annotated examples illustrate the general algebraic principles covered in each section. To prevent confusion, each example ends with this symbol: ■

3. Important concepts and algorithms are enclosed in boxes for easy identification and reference.

4. The approach to solving word problems includes a detailed method for translating a word statement into an algebraic equation or inequality.

5. Visual aids, such as shading and color, and use of annotations guide students through worked-out problems.

6. In special "Words of Caution," students are warned against common algebraic errors.

7. The importance of checking solutions is stressed throughout the book.

8. A review section with Set I and Set II exercise sets appears at the end of each chapter, and some chapters also have a midchapter review that includes Set I and Set II exercise sets. The Set II review exercises allow space for working problems and for answers; they can be removed from the text for grading. (Removal of these pages will not interrupt the continuity of the text.)

9. This book contains more than 6,000 exercises.
   Set I Exercises. The complete solutions for odd-numbered Set I exercises are included in the back of the book, together with the answers for all of the Set I exercises. In most cases (except in the review exercises), the even-numbered exercises are matched with the odd-numbered exercises. Thus, students can use the solutions for odd-numbered exercises as study aids for doing the even-numbered Set I exercises.
   Set II Exercises. Answers to all Set II exercises are included in the Instructor's Manual. No answers for Set II exercises are given in the text. The odd-numbered exercises of Set II are, for the most part (except in the review exercises), matched to the odd-numbered exercises of Set I, while the even-numbered exercises of Set II are *not* so matched. Thus, while students can use the odd-numbered exercises of Set I as study aids for doing the odd-

numbered exercises of Set II, they are on their own in doing the even-numbered exercises of Set II.

10. A diagnostic test at the end of each chapter can be used for study and review or as a pre-test. Complete solutions to all problems in these diagnostic tests, together with section references, appear in the answer section of the book.

11. A set of Cumulative Review Exercises is included at the end of each chapter except Chapter 1; the answers are in the answer section.

Using This Book

Essential Algebra, Fifth Edition can be used in three types of instructional programs: lecture, laboratory, and self-study.

The conventional lecture course. This book has been class-tested and used successfully in conventional lecture courses by the authors and by many other instructors. It is not a workbook, and therefore it contains enough material to stimulate classroom discussion. Examinations for each chapter are provided in the Instructor's Manual, and two different kinds of computer software enable instructors to create their own tests. One software program utilizes a test bank with full graphics capability, and the other is a random-access test generator. Tutorial software is available to help students who require extra assistance.

The learning laboratory class. This text has also been used successfully in many learning labs. The format of explanation, example, and exercises in each section of the book and the tutorial software make the book easy to use in laboratories. Students can use the diagnostic test at the end of each chapter as a pretest or for review and diagnostic purposes. Because several forms of each chapter test are available in the Instructor's Manual, and because of the test generators that are available, a student who does not pass a test can review the material covered on that test and can then take a different form of the test.

Self-study. This book lends itself to self-study because each new topic is short enough to be mastered before continuing, and because more than 900 examples and over 1,500 completely solved exercises show students exactly how to proceed. Students can use the diagnostic test at the end of each chapter to determine which parts of that chapter need to be studied and can thus concentrate on those areas in which they have weaknesses. The tutorial software and the random test generator, which provides answers and cross-references to the text and permits the creation of individualized work sheets, extend the usefulness of the new edition in laboratory and self-study settings.

Changes in the Fifth Edition

The Fifth Edition includes changes that resulted from many helpful comments from users of the first four editions as well as the authors' own classroom experience in teaching from the book. The major changes in the Fifth Edition include the following:

1. The level of the text has been raised. Most of the exercise sets include problems that require the use of techniques learned in an earlier section as well as techniques learned in the current section.

2. Topics such as prime numbers, prime factorization, products of the form $(a + b)(a - b)$, squares of binomials, and so on, are introduced early, along with the topics to which they are related.

3. Chapter 3 in this edition includes most of the topics from Chapters 3 (Simplifying Algebraic Expressions) and 6 (Exponents) of the Fourth Edition. Some repetition in coverage was thus eliminated. In this chapter, problems *already* in simplest form were added to exercise sets where the student is asked to "simplify."

4. A section on scientific notation has been added.

5. The section on solving and graphing inequalities in one variable now appears early in the book; it is included at the end of Chapter 4 (Equations and Inequalities). In that chapter, graphing the solutions of *equations* in one variable is also included.

6. The word problems have been updated and now include some word problems leading to inequalities. The students are cautioned repeatedly to state what their variables represent when they solve word problems. Word problems are included in nearly every chapter and in all Cumulative Review Exercises after Chapter 5.

7. A new section (7.1B) on factoring out a common binomial factor has been added to the chapter on factoring.

8. There are now more problems in the sections on adding and subtracting fractions with unlike denominators, and there are more problems in which finding the least common denominator is challenging. The difference between performing operations on fractions and solving equations involving fractions has been emphasized.

9. A discussion of the slopes of parallel and perpendicular lines is now included in Chapter 9 on graphing, and the discussion of the slope of a line has been expanded.

10. An introduction to functions and functional notation is given in Appendix C.

11. The number of problems in the Set II exercises has been increased; there is now the same number of problems in each Set II exercise set as in the corresponding Set I exercise set.

Ancillaries

The following ancillaries are available with this text:

1. The Instructor's Manual contains five different tests for each chapter, two forms of three midterm examinations, and two final examinations that can be easily removed and duplicated for class use. These tests are prepared with adequate space for students to work the problems. Answer keys for these tests are provided in the manual. The manual also contains the answers to the Set II exercises and to the Critical Thinking exercises.

2. The test bank for *Essential Algebra*, Fifth Edition is also available from the publisher in a computerized format entitled *Micro-Pac© Genie*, for use on the IBM PC or 100 percent compatible machines. This software program allows instructors to arrange items in a variety of ways and print them quickly and easily. Since *Genie* combines word processing and graphics with database management, it also permits instructors to create their own questions—even questions with mathematical notation or geometric figures—as well as to edit the questions provided in the test bank.

3. In addition, Wadsworth offers the *Johnston/Willis/Lazaris Computerized Test Generator* (JeWeL TEST) software for Apple II and IBM PC or compatible machines. This software, written by Ron Staszkow of Ohlone College, allows instructors to produce many different forms of the same test for quizzes, work sheets, practice tests, and so on. Answers and cross-references to the text provide additional instructional support.

4. An "intelligent" tutoring software system is also available for the IBM PC and compatibles. *Expert Algebra Tutor*, written by Sergei Ovchinnikov of San

Francisco State University, uses a highly interactive format and sophisticated techniques to tailor lessons to the specific algebra and prealgebra learning problems of students. The result is individualized tutoring strategies with specific page references to problems, examples, and explanations in the textbook.

5. A set of videotapes and a set of audiocassettes, covering major algebraic concepts, are also available.

To obtain additional information about these supplements, contact your Wadsworth-Brooks/ Cole representative.

Acknowledgments

We wish to thank the members of the editorial staff at Wadsworth Publishing Company for their help with this edition. Special thanks go to Anne Scanlan-Rohrer, Kevin Howat, Gary Mcdonald, Julia Scannell, James Chadwick, Barbara Holland, Sally Uchizono, and Mary Roybal.

We also wish to thank our many friends for their valuable suggestions. In particular, we are deeply grateful to Gale Hughes for preparing the Instructor's Manual, to Mary Jo Steig for proofreading and checking all examples and exercises, to Roland Misajon for solving the Set II exercises, and to the following reviewers: Wm. Christopher Allgyer, Mountain Empire Community College; Carol P. Battle, Erie Community College; Sadie C. Bragg, Borough of Manhattan Community College, CUNY; Bob C. Denton, Orange Coast College; Louise Ettline, Trident Technical College; Philip A. Glynn, Mattatuck Community College; Daniel W. Henry, Los Medanos College; Pauline Jenness, William Rainey Harper College; Adele LeGere, Oakton Community College; Rosalyn Merzer, College of Staten Island; Jack D. Murphy, Williamsport Area Community College; Joyce H. Saxon, Morehead State University; Herbert Stein, Merritt College; Ann T. Thorne, College of DuPage; Susan Townsend, Kalamazoo Valley Community College; Judith Willoughby, Minneapolis Community College; and Ray Wilson, Central Piedmont Community College.

ESSENTIAL ALGEBRA

1 Operations on Signed Numbers

Arithmetic is calculation with numbers using fundamental operations such as addition, subtraction, multiplication, and division. (See Appendix B for a brief review.) **Algebra** deals with the same fundamental operations with numbers, but uses letters to represent some of the numbers. Before beginning the study of algebra, we review a few basic definitions relating to sets and numbers. (See Appendix A for a more complete discussion of sets.)

1.1 Basic Definitions

Set   A **set** is a collection of objects or things.

Example 1   Examples of sets:

   a. A 48-piece set of dishes

   b. A basket of birthday presents

   c. A basket containing an apple, a pillow, and a cat

   d. The number of students attending college in the United States   ■

Element of a Set   The objects or things that make up a set are called its **elements** (or **members**). Sets are usually represented by listing their elements within braces { }. We never use parentheses for sets. Thus, (2, 1, 7) does *not* name a set. The order in which the elements of a set are listed is not important.

Example 2   Here are a few examples showing the elements of sets.

   a. Set {1, 2, 3} has elements 1, 2, and 3.

   b. Set {5, 7, 9} has elements 5, 7, and 9.

   c. Set $\{a, d, f, h, k\}$ has elements $a, d, f, h,$ and k.

   d. Set {Ben, Kay, Frank, Albert} has elements Ben, Kay, Frank, and Albert.

   e. Set {□, Σ, △, ⊙} has elements □, Σ, △, and ⊙.   ■

It is sometimes helpful to think of the braces as a basket or box that contains the elements of the set. This basket, { }, may contain just a few elements, many elements, or no elements at all.

The Meaning of the Equal Sign   The equal sign (=) in a statement means that the expression on the left side of the equal sign *has the same value or values* as the expression on the right side of the equal sign.

Naming a Set   A set is usually named by a capital letter, such as $A, N, W,$ and so on. The expression "$A = \{1, 5, 7\}$" is read "A is the set whose elements are 1, 5, and 7."

Roster Method of Representing a Set   A **roster** is a list of members of a group. When we represent a set by {3, 8, 9, 11}, we are representing the set by a roster (or list) of its members. This method of representing a set is called the **roster method**.

Important Sets of Numbers

Natural Numbers   The numbers

$$1, 2, 3, 4, 5, 6, 7, 8, 9, 10, 11, 12, \text{ and so on}$$

are called the **natural numbers** (or **counting numbers**). These were probably the first numbers invented to enable people to count their possessions, such as sheep or goats. The largest natural number can never be found because no matter how far we count there are always larger natural numbers. Since it is impossible to write all the natural numbers, it is customary to represent them as follows:

$$\{1, 2, 3, 4, \ldots \}$$

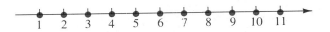

Read "and so on"

The three dots to the right of the number 4 indicate that the remaining numbers are to be found by counting in the same way we have begun: namely, by adding 1 to each number to find the next number. We call the set of natural numbers N; that is,

$$N = \{1, 2, 3, 4, \ldots\}$$

Number Line   Natural numbers can be represented by numbered points equally spaced along a straight line, as in Figure 1.1.1. Such a line is called a **number line.**

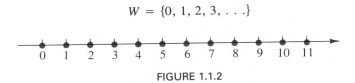

FIGURE 1.1.1   NUMBER LINE

The arrowhead at the right shows the direction in which numbers get larger. Later we will discuss the other kinds of numbers, such as fractions and decimal numbers, that can also be placed on the number line.

Whole Numbers   When 0 is included with the natural numbers, we have the set of **whole numbers** which we call W (see Figure 1.1.2). Therefore,

$$W = \{0, 1, 2, 3, \ldots\}$$

FIGURE 1.1.2

Set of Digits   One important set of numbers is the set of **digits**. This set contains the numbers 0, 1, 2, 3, 4, 5, 6, 7, 8, and 9. These symbols make up our entire number system; *any* number can be written by using some combination of these numerals.

Numbers are often referred to as *one-digit* numbers, *two-digit* numbers, *three-digit* numbers, and so on. Also, we sometimes wish to refer to the *first, second,* or *third* digit of a number; when we do this, we count from left to right.

Example 3   Examples to show the use of digits:

a. An example of a two-digit number is 35.

b. An example of a one-digit number is 7.

c. Three-digit numbers include 275.

d. The first digit of 785 is 7.

e. The second digit of 785 is 8.

f. The third digit of 785 is 5.   ■

Fractions   A **fraction** is an indicated division; the fraction $\dfrac{a}{b}$ is equivalent to the division $a \div b$. (In text, fractions are often denoted a/b.) We call a and b the **terms** of the fraction. In this section, we consider only those fractions in which the *numerator a* is a whole number and the *denominator b* is a natural number. The denominator of a fraction can never equal zero.

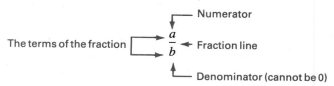

When the numerator of a fraction is less than the denominator, we can think of the fraction as being part of a whole. In this case, the denominator tells us how many equal parts the whole has been divided into, and the numerator tells us how many of those equal parts are being considered.

Example 4   An example of the meaning of a fraction. The fraction $\frac{3}{4}$ is equivalent to $3 \div 4$. The numerator is 3 and the denominator is 4.

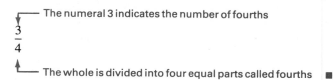

Mixed Numbers   A **mixed number** is made up of both a whole number part and a fraction part.

Example 5   Examples of mixed numbers:

$$2\frac{1}{2} \qquad 3\frac{5}{8} \qquad 5\frac{1}{4} \qquad 12\frac{3}{16} \quad \blacksquare$$

Decimal Fractions   A **decimal fraction** is a fraction whose denominator is 10, or 100, or 1,000, and so on.

Example 6   Examples of decimal fractions:

a. $\dfrac{4}{10} = 0.4$ is read "four tenths"

b. $\dfrac{5}{100} = 0.05$ is read "five-hundredths"

c. $\dfrac{6}{1,000} = 0.006$ is read "six-thousandths"

d. $\dfrac{23}{10} = 2.3$ is read "two and three-tenths"

               is read "twenty-three tenths" $\blacksquare$

Decimal Places   The number of decimal places in a number is the number of digits written to the right of the decimal point.

Example 7   Examples of decimal places:

a. 75.14 (two decimal places)

b. 1.086 (three decimal places)

c. 2.5000 (four decimal places) $\blacksquare$

Real Numbers   Natural numbers, whole numbers, fractions, decimals, and mixed numbers are all elements of a set we call the set of **real numbers**, and they can all be represented by points on the number line. Points representing some fractions, decimals, and mixed numbers are shown in Figure 1.1.3.

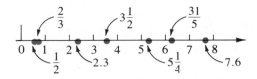

FIGURE 1.1.3

Other kinds of real numbers are introduced in Sections 1.2 and 1.10. It is shown in higher-level mathematics courses that any number that can be represented by a point on the number line is a real number.

"Greater-Than" and "Less-Than" Symbols   The symbol $>$ is read "greater than," and the symbol $<$ is read "less than." Numbers get larger as we move to the right on the number line and smaller as we move to the left.

Example 8 $6 > 3$ is read "6 is greater than 3."

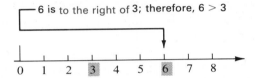

Example 9 $2 < 5$ is read "2 is less than 5."

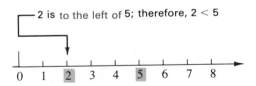

An easy way to remember the meaning of the symbol is to notice that the wide part of the symbol is next to the larger number. On the other hand, some people like to think of the symbols $>$ and $<$ as arrowheads that point toward the smaller number.

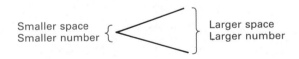

Example 10   Inequalities are written and read as follows:

a. $7 > 6$ is read "7 is greater than 6."

b. $7 > 1$ is read "7 is greater than 1."

c. $5 < 10$ is read "5 is less than 10."

d. $3 < 4 < 5$ is read "3 is less than 4 and 4 is less than 5."   ∎

Note that $7 > 6$ and $6 < 7$ give the same information even though they are read differently.

Another inequality symbol is $\neq$. A slash line drawn through a symbol puts a *not* in the meaning of the symbol.

Example 11   Examples of the use of the slash line:

a. = is read "is equal to."
    ≠ is read "is *not* equal to."

b. < is read "is less than."
    ≮ is read "is *not* less than."

c. > is read "is greater than."
    ≯ is read "is *not* greater than."

d. 4 ≠ 5 is read "4 is *not* equal to 5."

e. 3 ≮ 2 is read "3 is *not* less than 2."

f. 5 ≯ 6 is read "5 is *not* greater than 6." ∎

EXERCISES 1.1

Set I
1. What is the second digit of the number 159?

2. What is the fourth digit of the number 1975?

3. What is the smallest natural number?

4. What is the smallest digit?

5. What is the largest one-digit natural number?

6. What is the largest two-digit whole number?

7. What is the smallest two-digit natural number?

8. What is the smallest three-digit whole number?

9. Which symbol, < or >, should be used to make the statement 10 _?_ 0 true?

10. Which symbol, < or >, should be used to make the statement 3 _?_ 15 true?

11. Which symbol, < or >, should be used to make the statement 8 _?_ 7 true?

12. Which symbol, < or >, should be used to make the statement 0 _?_ 1 true?

13. Is 2.3 a real number?

14. Is $\frac{7}{8}$ a real number?

15. Is 1.8 a real number?

16. Is $\frac{9}{5}$ a real number?

17. Is 1.8 a natural number?

18. Is $\frac{3}{4}$ a natural number?

19. Is 15 a digit?

20. Is 15 a natural number?

21. How many decimal places are there in the number 7.010?

22. How many decimal places are there in the number 41.0005?

Set II
1. What is the third digit of the number 3,187?

2. Is 12 a natural number?

3. What is the largest natural number?

4. What is the smallest whole number?

5. What is the largest digit?

6. What is the smallest three-digit natural number?

7. Is 58.4 a real number?

8. What is the smallest one-digit natural number?

9. Which symbol, $<$ or $>$, should be used to make the statement 18 __?__ 5 true?

10. Which symbol, $<$ or $>$, should be used to make the statement 8 __?__ 17 true?

11. Which symbol, $<$ or $>$, should be used to make the statement 11 __?__ 6 true?

12. Write all the whole numbers < 4.

13. Is $5\frac{1}{2}$ a real number?

14. Is $5\frac{1}{2}$ a natural number?

15. Is 0 a real number?

16. Is 58.4 a natural number?

17. Is 3,628 a natural number?

18. Is $\frac{2}{5}$ a real number?

19. Is 12 a digit?

20. Is $\frac{2}{3}$ a digit?

21. How many decimal places are there in the number 50.602?

22. How many decimal places are there in the number 23.0?

1.2 Negative Numbers and Rational Numbers

In Section 1.1, we showed how whole numbers could be represented by equally spaced points along the number line. We now extend the number line to the left and continue with the set of equally spaced points.

Numbers used to name the points to the left of 0 on the number line are called **negative numbers**. Numbers used to name the points to the right of 0 on the number line are called **positive numbers**. Zero itself is neither positive nor negative. The positive and negative numbers are referred to as **signed numbers** (see Figure 1.2.1).

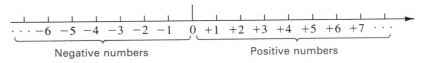

FIGURE 1.2.1

Integers The set of **integers** includes the set of whole numbers as well as the numbers -1, -2, -3, and so on; it can be represented in the following way:

$$\{\ldots, -3, -2, -1, 0, +1, +2, +3, \ldots\}$$

Example 1 Examples of reading positive and negative integers:

a. −1 is read "negative one."

b. −575 is read "negative five hundred seventy-five."

c. 25 is read "twenty-five" or "positive twenty-five." ■

When reading or writing positive numbers, we usually omit the word "positive" and the + sign. Therefore, when there is no sign in front of a number, it is understood to be positive.

Example 2 On an unusually cold day in Minnesota, the temperature was −40° F. This means that the temperature was 40° F below 0° F. ■

Example 3 The altitudes of some unusual places on earth are as follows:

a. Mt. Everest   29,028 ft
  This means that the peak of Mt. Everest is 29,028 feet *above* sea level.

b. Mt. Whitney (California)   14,494 ft

c. Lowest point in Death Valley (California)   −282 ft
  This means that the lowest point in Death Valley is 282 ft *below* sea level.

d. Dead Sea (Jordan)   −1,299 ft

e. Mariana Trench (Pacific Ocean)   −36,198 ft ■

Using Inequality Symbols with Integers Recall that numbers get larger as we move to the right on the number line and smaller as we move to the left.

Example 4 −3 > −5 is read "−3 is greater than −5."

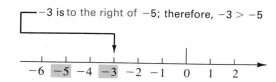

■

Example 5 −5 < −1 is read "−5 is less than −1."

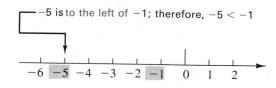

■

Example 6 Verify the following inequalities by noting whether the first number of each pair is to the right or left of the second number (of that pair) on the number line:

a. 6 > 4

b. 0 > −1

c. −2 > −5

d. −20 < −10

e. −5 < 3 ■

Rational Numbers   A **rational number** is any number that can be expressed in the form a/b, where a and b are integers and $b \neq 0$. All integers and all fractions are rational numbers, and all rational numbers are real numbers. When a rational number is expressed in decimal form, the decimal always terminates (see Example 8) or repeats (see Example 9). See Appendix B, if necessary, for a brief review of converting fractions to decimal form.

Example 7   Examples of rational numbers:

a. $\dfrac{3}{4}$, because 3 and 4 are integers and $4 \neq 0$.

b. -5, because $-5 = -\dfrac{5}{1}$.

c. $-1\dfrac{8}{9}$, because $-1\dfrac{8}{9} = -\dfrac{17}{9}$.

d. -3.7, because $-3.7 = -\dfrac{37}{10}$.

e. 0, because $0 = \dfrac{0}{1}$ or $\dfrac{0}{5}$ or $\dfrac{0}{23}$, and so on. ∎

Example 8   Examples of rational numbers whose decimals terminate:

a. $\dfrac{3}{4} = 0.75$ The decimal terminates

b. $\dfrac{1}{16} = 0.0625$ The decimal terminates ∎

Example 9   Examples of rational numbers whose decimals repeat:

a. $\dfrac{1}{3} = 0.33333333 \ldots$ The digit "3" repeats

b. $\dfrac{4}{33} = 0.12121212 \ldots$ The *pair* of digits "12" repeats ∎

Negative Numbers and the Number Line   In Section 1.1, we mentioned that natural numbers, whole numbers, fractions, decimals, and mixed numbers are all real numbers. Negative integers and negative rational numbers are also real numbers. They can be represented by points on the real-number line; some are shown in Figure 1.2.2.

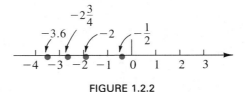

FIGURE 1.2.2

In Section 1.1, we stated that a largest natural number could never be found, because no matter how far we count, there are always larger natural numbers. Similarly, no matter how far we count along the number line to the left of 0, we never reach a smallest negative number.

Other sets of numbers that are often referred to in word problems are the sets of

Consecutive Numbers   Integers that follow one another (without interruption) are called **consecutive numbers**. Thus, 5, 6, 7, and 8 are consecutive numbers, because 6 follows 5, 7 follows 6, and 8 follows 7.

Even Integers   Integers that are *exactly* divisible by 2 are called **even integers**. Therefore, 8, -6, and 0 are even integers.

Odd Integers   Integers that are *not* exactly divisible by 2 are called **odd integers**. Therefore, 1, -7, and 5 are odd integers.

EXERCISES 1.2

Set I   **1.** Write -75 in words.

2. Write -49 in words.

3. Use digits to write negative fifty-four.

4. Use digits to write negative one hundred nine.

5. Which is larger, -2 or -4?

6. Which is larger, 0 or -10?

7. Which is larger, -5 or -10?

8. Which is smaller, -1 or -15?

9. What is the largest negative integer?

10. Is 18 a rational number?

11. Is -3 a rational number?

12. Is 18 a real number?

13. Is -3 a real number?

14. What is the smallest negative integer?

15. Write, in consecutive order, the natural numbers < 5.

16. Write, in consecutive order, the digits > 6.

17. Write, in consecutive order, the even digits that are < 6.

18. Write, in consecutive order, the odd natural numbers that are < 9.

In Exercises 19–24, determine which symbol, $<$ or $>$, should be used to make each statement true.

19. $0 \underline{\ ?\ } -3$ **20.** $-2 \underline{\ ?\ } -6$ **21.** $-5 \underline{\ ?\ } 2$

22. $-7 \underline{\ ?\ } -4$ **23.** $-2 \underline{\ ?\ } -10$ **24.** $-8 \underline{\ ?\ } -3$

25. A scuba diver descends to a depth of sixty-two feet. Represent this number by an integer.

26. The temperature in Fairbanks, Alaska, was forty-five degrees Fahrenheit below zero. Represent this number by an integer.

Set II   **1.** Write -17 in words.

2. Write, in consecutive order, the negative integers > -5.

3. Use digits to write negative two hundred four.

4. Write, in consecutive order, the even natural numbers < 12.

5. Which is larger, −6 or −3?

6. Which is smaller, −5 or 0?

7. Which is larger, −8 or 5?

8. Which is smaller, −10 or 2?

9. What is the largest real number?

10. What is the smallest positive integer?

11. Is $\frac{2}{7}$ a rational number?

12. Is $-\frac{2}{15}$ a rational number?

13. Is $\frac{2}{7}$ a real number?

14. Is $-\frac{2}{15}$ a real number?

15. Write, in consecutive order, the odd negative integers > −7.

16. Write, in consecutive order, the odd digits < 8.

17. Write, in consecutive order, the even digits that are > 6.

18. Write, in consecutive order, the even natural numbers that are < 15.

In Exercises 19–24, determine which symbol, < or >, should be used to make each statement true.

19. 0 _?_ −5

20. −7 _?_ −3

21. 3 _?_ −5

22. −8 _?_ −2

23. −6 _?_ −3

24. −16 _?_ 0

25. The temperature in Fairbanks, Alaska, was six degrees Fahrenheit below zero. Represent this number by an integer.

26. Nitrogen becomes a liquid at (about) 195 degrees Celsius below zero. Represent this number by an integer.

1.3 Adding Signed Numbers

Sum   The *answer* to an addition problem is called the **sum**.

Adding Signed Numbers on the Number Line   We first discuss representing signed numbers by arrows. We can represent a signed number by an arrow that begins at the point representing 0 and ends at the point representing that particular number.

Example 1   Represent 4 by an arrow.

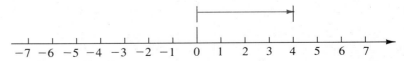

This arrow represents a movement of four units to the *right*.   ■

Any *positive* number is represented by an arrow directed to the *right*. The arrow need not start at zero so long as it has a length equal to the number it represents.

Example 2   Represent −5 by an arrow.

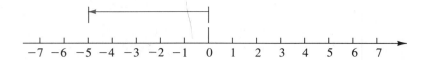

This arrow represents a movement of five units to the *left*. ■

Any *negative* number is represented by an arrow directed to the *left*.
    We can now represent the addition of signed numbers by means of arrows.

Example 3   Add 3 to 2 by means of arrows.

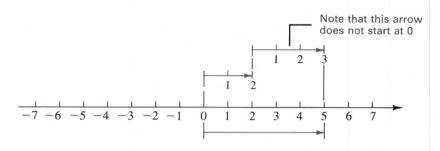

To add 3 to 2 on the number line, begin by drawing the arrow representing 2. Draw the arrow representing 3 starting at the arrowhead end of the arrow representing 2. These two movements represent a net movement to the right of five units. Therefore, 2 + 3 = 5. ■

Example 4   Add −7 to 5 by means of arrows.

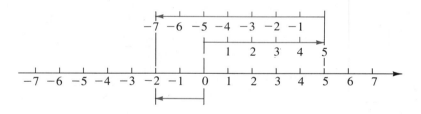

Begin by drawing the arrow representing 5. Draw the arrow representing −7 starting at the arrowhead end of the arrow representing 5. These two movements represent a net movement to the left of two units. Therefore, 5 + (−7) = −2. ■

Example 5   Add −4 to −3 by means of arrows

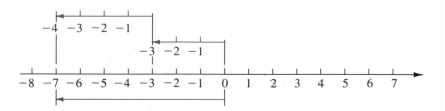

Begin by drawing the arrow representing −3. Draw the arrow representing −4 starting at the arrowhead end of the arrow representing −3. These two movements represent a net movement to the left of seven units. Therefore, −3 + (−4) = −7. ■

Example 6   Add +8 to −5 by means of arrows.

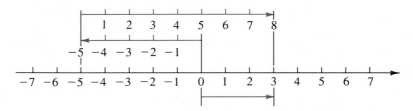

Begin by drawing the arrow representing −5. Draw the arrow representing +8 starting at the arrowhead end of the arrow representing −5. These two movements represent a net movement to the right of three units. Therefore, $(-5) + (+8) = +3$. ∎

Example 7   On the number line, add 0 to −3.

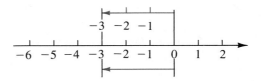

Begin by drawing the arrow representing −3. To add 0, we then need to move zero units from that point. But that leaves us at −3. Therefore, $-3 + 0 = -3$. ∎

Absolute Value

The **absolute value** of a number is the distance between that number and 0 on the number line *with no regard to direction* (see Figure 1.3.1). The symbol for the *absolute value* of a real number x is $|x|$.

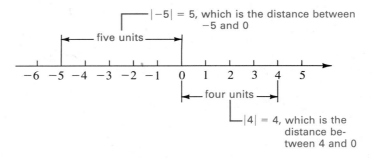

FIGURE 1.3.1   ABSOLUTE VALUE

Example 8   Examples of absolute value of numbers:

a. $|9| = 9$ A positive number
b. $|0| = 0$ Zero
c. $|-4| = 4$ A positive number

Note that the absolute value of a number can never be negative

∎

A signed number has two distinct parts: its absolute value and its sign.

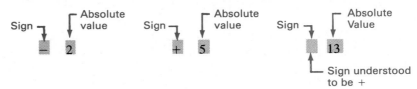

Note that the absolute value of a signed number is the number written without its sign.

The Rules for Adding Signed Numbers   Adding signed numbers by means of arrows is easy to understand, but it is a very slow process. If we needed to add 8,317,005 to $-17,461,037$, it would be impractical to try to do the problem by using the number line. The following rules give an easier, faster, and more accurate method for adding signed numbers. Examples 3, 4, 5, and 6 can be used to verify that the rules are valid.

TO ADD SIGNED NUMBERS

1. When the numbers have the same sign,

   1. Add their absolute values.
     and
   2. Attach (to the left of the sum) the sign of both numbers.

2. When the numbers have different signs,

   1. Subtract the smaller absolute value from the larger absolute value.
     and
   2. Attach (to the left of the sum) the sign of the number with the larger absolute value.

We now show how to add signed numbers by means of the rules.

Example 9   Find $(-29) + (-35)$.

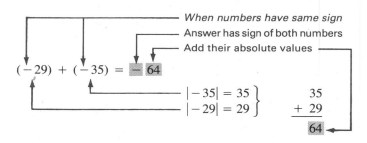

Example 10   Find $(-9) + (+23)$.

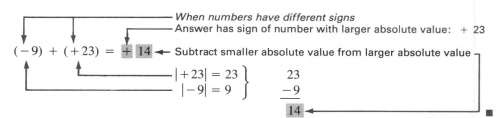

In Example 9, the parentheses around the first number were not necessary, and in Example 10, none of the parentheses were necessary. That is, $(-29) + (-35) = -29 + (-35)$, and $(-9) + (+23) = -9 + 23$.

Example 11   Find $(-7) + (-11)$.
Solution Since -7 and -11 have the same sign, add their absolute values: $7 + 11 = 18$. The sum has the same sign as both numbers: $-$. Therefore, $(-7) + (-11) = -18$. ■

Example 12   Find $(-24) + (17)$.

Solution Since -24 and 17 have different signs, subtract the smaller absolute value from the larger: $24 - 17 = 7$. The sum has the sign of the number with the larger absolute value: $(-)$, since -24 has the larger absolute value. Therefore, $(-24) + (17) = -7$.   ■

Example 13   Find $\left(-2\dfrac{1}{4}\right) + 8\dfrac{1}{8}$.

Solution LCD $= 8$. (The LCD is discussed more fully in Chapter 8.) Since the two numbers have different signs, we subtract the smaller absolute value from the larger: $8\dfrac{1}{8} - 2\dfrac{1}{4} = 5\dfrac{7}{8}$. Vertically:

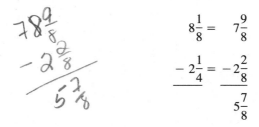

$$8\frac{1}{8} = 7\frac{9}{8}$$
$$-2\frac{1}{4} = -2\frac{2}{8}$$
$$\overline{\phantom{-2\frac{1}{4} = } \ 5\frac{7}{8}}$$

The sum has the sign of the number with the larger absolute value: $(+)$, since $8\dfrac{1}{8}$ has the larger absolute value. Therefore, $\left(-2\dfrac{1}{4}\right) + 8\dfrac{1}{8} = 5\dfrac{7}{8}$.   ■

Example 14   Find $\left(-2\dfrac{1}{2}\right) + \left(-4\dfrac{1}{3}\right)$.

Solution LCD $= 6$

$$-2\frac{1}{2} = -2\frac{3}{6}$$
$$-4\frac{1}{3} = -4\frac{2}{6}$$
$$\overline{\phantom{-4\frac{1}{3} = } -6\frac{5}{6}}$$

Therefore $\qquad \left(-2\dfrac{1}{2}\right) + \left(-4\dfrac{1}{3}\right) = -6\dfrac{5}{6}$ ■

Example 15   At 6 A.M. the temperature in Alamosa, Colorado, was $-15°$ F. By 10 A.M., the temperature had risen $9°$ F. What was the temperature at 10 A.M.?

Solution We must *add* the rise in temperature ($9°$) to the original temperature ($-15°$), and $-15 + (9) = -6$. Therefore, the new temperature is $-6°$ F.   ■

Example 16   Add $\left|-8\right| + (-3)$.

Solution We must remove the absolute-value symbols before adding:

$$\left|-8\right| + (-3) = 8 + (-3) = 5$$ ■

Additive Identity   (Addition involving zero). Because adding zero to a number gives us the identical number we started with (see Example 7), we call zero the **additive identity**.

THE ADDITIVE IDENTITY IS 0

If a is any real number, then

$$a + 0 = a \quad \text{and} \quad 0 + a = a$$

EXERCISES 1.3

Set I In Exercises 1–36, find the sums.

1. (4) + (5)   **2.** (6) + (2) 8   **3.** (−3) + (−4) −7

4. (−7) + (−1)   **5.** (−6) + (5) −1   **6.** (−8) + (3) −5

7. (7) + (−3)   **8.** (9) + (−4) 5   **9.** (−8) + (−4) −12

10. (−5) + (−6)   **11.** (3) + (−9) −6   **12.** (4) + (−8) −4

13. −5 + 0   **14.** 0 + (−17) −17   **15.** (−7) + (9) 2

16. (−5) + (8)   **17.** (−2) + (−11) −13   **18.** (−3) + (−6) −9

19. (5) + (−15)   **20.** (4) + (−12) −8   **21.** (−8) + (9) 1

22. (−7) + (13) 6   **23.** (−4) + (4) 0   **24.** (−9) + (9) 0

25. (−27) + (−13) −40   **26.** (−42) + (−12) −54   **27.** (−80) + (121) 41

28. (−69) + (134) 65   **29.** (105) + (−73) 32   **30.** (218) + (−113) 105

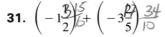

31. $\left(-1\dfrac{3}{2}\right) + \left(-3\dfrac{2}{5}\right)$ **32.** $\left(-2\dfrac{5}{2}\right) + \left(-5\dfrac{1}{4}\right)$ **33.** $\left(4\dfrac{5}{6}\right) + \left(-1\dfrac{4}{3}\right)$

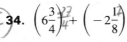

34. $\left(6\dfrac{3}{4}\right) + \left(-2\dfrac{1}{8}\right)$ **35.** (6.075) + (−3.146)   **36.** −4.745 + 93.118

In Exercises 37–40, remove the absolute-value symbols.

37. $|-73|$ **38.** $|-55|$ **39.** $-|48|$ **40.** $-|26|$

In Exercises 41–46, find the sums.

41. $|-6| + (-2)$ **42.** $|-8| + (-5)$ **43.** $-8 + |-17|$

44. $-6 + |-27|$ **45.** $-9 + |23|$ **46.** $-1 + |32|$

47. At 6 A.M. the temperature in Hibbing, Minnesota, was −35° F. If the temperature had risen 53° F by 2 P.M., what was the temperature at that time?

48. At midnight in Billings, Montana, the temperature was −50° F. By noon the temperature had risen 67° F. What was the temperature at noon?

Set II In Exercises 1–36, find the sums.

1. (8) + (7)   **2.** 8 + (−5)   **3.** (−9) + (−3)

4. −19 + (−3)   **5.** (−5) + (9)   **6.** −28 + 28

7. (−4) + (7)   **8.** 0 + (−4,162)   **9.** (−2) + (−5)

10. 18 + (−6)   **11.** (6) + (−10)   **12.** 1,468 + (−1,468)

13. (−16) + 0   **14.** $\left(-\dfrac{2}{3}\right) + \left(-\dfrac{1}{5}\right)$ **15.** (−8) + (5)

Chapter 2 Sect 2
        Pg 77-78

Diagnostic test.

Paula

566
-0494

272 Keela
4422

16. $-146 + (-362)$ **17.** $(-8) + (-9)$ **18.** $-1.724 + 3.6$

19. $(26) + (-35)$ **20.** $-4,728 + (-35)$ **21.** $(-29) + (32)$

22. $-67 + 28$ **23.** $(-6) + 6$ **24.** $147 + (-362)$

25. $(-38) + (-17)$ **26.** $-849 + (-738)$ **27.** $(-132) + (261)$

28. $-536 + (-2)$ **29.** $(872) + (-461)$ **30.** $621 + (-1,417)$

31. $\left(-1\frac{1}{3}\right) + \left(-5\frac{3}{5}\right)$ **32.** $\left(-6\frac{1}{4}\right) + \left(-8\frac{1}{2}\right)$ **33.** $\left(3\frac{2}{3}\right) + \left(-2\frac{2}{9}\right)$

34. $\left(3\frac{1}{3}\right) + \left(-5\frac{1}{6}\right)$ **35.** $-18.0164 + 2.281$ **36.** $-9.6 + (-57.356)$

In Exercises 37–40, remove the absolute-value symbols.

37. $|-81|$ **38.** $|-462|$ **39.** $-|17|$ **40.** $-|-84|$

In Exercises 41–46, find the sums.

41. $(-17) + |-4|$ **42.** $|-5| + |-2|$ **43.** $-18 + |-25|$

44. $|-12| + |-17|$ **45.** $-3 + |-3|$ **46.** $|-18| + |-25|$

47. In Fairbanks, Alaska, the temperature at 2 A.M. was $-35°$ F. By noon the temperature had risen 27° F. What was the temperature at noon?

48. In Duchesne, Utah, the temperature at midnight was $-18°$ F. By 10 A.M. the temperature had risen 12° F. What was the temperature at 10 A.M.?

1.4 Subtracting Signed Numbers

Subtraction is the *inverse* operation of addition; that is, subtraction "undoes" addition.

Difference   The answer to a subtraction problem is called the **difference**.

The Negative of a Number   The idea of *negative* suggests the opposite of something. For example, the negative of taking two steps to the right would be taking two steps to the left. Since $+2$ can be thought of as a movement of two units to the right, then the negative of $+2$ would be thought of as a movement of two units to the left. This means the negative of $+2$ is -2.

Example 1   Examples of the negative of a *positive* number.

a. The negative of 5 is -5.

b. The negative of 12 is -12. ∎

   Since -3 can be thought of as a movement of three units to the left, then the negative of -3 would be thought of as an opposite movement of three units to the right. This means that the negative of -3 is $+3$. This can be written as follows: $-(-3) = +3 = 3$.

Example 2   Examples of the negative of a *negative* number.

a. The negative of -10 is $+10$.   Written $-(-10) = 10$

b. The negative of -14 is $+14$.   Written $-(-14) = 14$ ∎

Examples 1 and 2 lead to the following rules for finding the negative of a number:

TO FIND THE NEGATIVE OF A NUMBER

Change the sign of the number.

The negative of $b = -b$

The negative of $-b = -(-b) = b$

The negative of $0 = 0$

The negative of a number is used in the definition of subtraction.

Additive Inverse   When the sum of two numbers is 0 (which is the additive identity), we say that they are the **additive inverses** of each other. Therefore, the *negative* of a number can also be called the *additive inverse* of that number.

ADDITIVE INVERSE

The additive inverse of a is $-a$.

$$a + (-a) = 0$$

$$-a + a = 0$$

DEFINITION OF SUBTRACTION

$$a - b = a + (-b)$$

In words: To subtract b from a, add the negative of b to a.

This definition leads to the following rule for subtracting signed numbers:

TO SUBTRACT ONE SIGNED NUMBER FROM ANOTHER

1. Change the subtraction symbol to an addition symbol, and change the sign of the number being subtracted.

2. Add the resulting signed numbers as shown in Section 1.3.

Example 3   Find $(-2) - (-5)$.

Solution

Change subtraction to addition
Change the sign of the number being subtracted

$$(-2) \; \underline{-} \; (\; \underline{-} \; 5)$$
$$= (-2) \; \boxed{+} \; (\; \boxed{+} \; 5)$$
$$= 3 \quad \blacksquare$$

Example 4   Find $(-3) - (6)$.

Solution

Change subtraction to addition
Change the sign of the number being subtracted

$$(-3) \; \underline{-} \; (\; \boxed{+} \; 6)$$
$$= (-3) \; \boxed{+} \; (\; \boxed{-} \; 6)$$
$$= -9 \quad \blacksquare$$

Example 5   Find $(+13) - (-14)$.

Solution

$$\begin{array}{lll} (+13) - (-14) & \text{Or} & (+13) \\ = (+13) + (+14) & & \underline{+ \; (+14)} \\ = 27 & & 27 \quad \blacksquare \end{array}$$

Example 6   Find $(+93) - (+59)$.

Solution

$$\begin{array}{lll} (+93) - (+59) & \text{Or} & (+93) \\ = (+93) + (-59) & & \underline{+ \; (-59)} \\ = 34 & & 34 \quad \blacksquare \end{array}$$

Example 7   Subtract 9 from 6.*

Solution Subtract 9 from 6 *means* $6 - 9 = 6 + (-9) = -3$. $\blacksquare$

Example 8   Find $\left(-3\frac{1}{2}\right) - \left(+2\frac{1}{4}\right)$.

Solution

$$\begin{array}{lll} \left(-3\frac{1}{2}\right) - \left(+2\frac{1}{4}\right) & \text{Or} & \left(-3\frac{1}{2}\right) = \left(-3\frac{2}{4}\right) \\ = \left(-3\frac{2}{4}\right) + \left(-2\frac{1}{4}\right) & & \underline{+ \left(-2\frac{1}{4}\right) = \left(-2\frac{1}{4}\right)} \\ = -5\frac{3}{4} & & -5\frac{3}{4} \quad \blacksquare \end{array}$$

---

*"Subtract *a* from *b*" means "*b* − *a*."

Example 9 Find $(-4.56) - (-7.48)$.

Solution

$$(-4.56) - (-7.48) \qquad \text{Or} \qquad 7.48$$
$$= (-4.56) + (+7.48) \qquad\qquad \underline{-4.56}$$
$$= 2.92 \qquad\qquad\qquad +2.92 \quad \blacksquare$$

Example 10 At 4 A.M. the temperature in Missoula, Montana, was $-26°$ F. At noon, the temperature was $-4°$ F. What was the rise in temperature?

Solution We must find the *difference* in temperatures; that is, we must subtract the *lower* temperature $(-26°)$ from the *higher* temperature $(-4°)$. The problem then becomes

$$-4 - (-26)$$
$$= -4 + (+26)$$
$$= 22$$

Therefore, the rise in temperature was $22°$ F. $\blacksquare$

Subtraction Involving Zero Since the subtraction $a - b$ has been defined as $a + (-b)$, the rules for subtractions involving zero are derived from the rules for addition.

IF a IS ANY REAL NUMBER, THEN

1. $a - 0 = a$

2. $0 - a = 0 + (-a) = -a$

EXERCISES 1.4

Set I

1. Find the additive inverse of -6.

2. Find the additive inverse of 2/3.

In Exercises 3–34, find the differences.

3. $(-3) - (-2)$ **4.** $(-4) - (-3)$ **5.** $(-6) - (2)$

6. $(-8) - (5)$ **7.** $(9) - (-5)$ **8.** $(7) - (-3)$

9. $(2) - (-7)$ **10.** $(3) - (-5)$ **11.** $(-5) - (-9)$

12. $(-2) - (-8)$ **13.** $(-4) - (3)$ **14.** $(-7) - (8)$

15. $(6) - (11)$ **16.** $(5) - (9)$ **17.** $(-9) - (-4)$

18. $(-6) - (-3)$ **19.** $(4) - (-7)$ **20.** $(8) - (-9)$

21. $(-15) - (11)$ **22.** $(-24) - (16)$ **23.** $0 - 7$

24. $0 - 15$ **25.** $0 - (-9)$ **26.** $0 - (-25)$

27. $16 - 0$ **28.** $10 - 0$ **29.** $(156) - (-97)$

30. $(284) - (-89)$ **31.** $(-354) - (-286)$ **32.** $(-484) - (-375)$

33. $(-7) - (-2.009)$ **34.** $(-16) - (-7.89)$

35. Subtract (-2) from $(+5)$.

36. Subtract (-10) from (-15).

37. Subtract $\left(2\frac{1}{2}\right)$ from $\left(-5\frac{1}{4}\right)$.

38. Subtract $\left(3\frac{1}{6}\right)$ from $\left(-7\frac{1}{6}\right)$.

39. Mr. Reyes has a balance of $473.29 in his checking account. Find his new balance after he writes a check for $238.43.

40. Ms. Johnson made a $45 deposit on a quadraphonic home music system costing $623.89. What is the balance due?

41. At 5 A.M. the temperature at Mammoth Mountain, California, was $-7°$ F. At noon the temperature was $42°$ F. What was the rise in temperature?

42. At 4 A.M. the temperature in Massena, New York, was $-5.6°$ F. At 1 P.M. the temperature was $37.5°$ F. What was the rise in temperature?

43. A scuba diver descends to a depth of 141 ft below sea level. Her buddy dives 68 ft deeper. What is her buddy's altitude at the deepest point of her dive?

44. When Fred checked his pocket altimeter at the seashore on Friday afternoon, it read -150 ft. Saturday morning it read 9,650 ft when he checked it on the peak of a nearby mountain. Allowing for the obvious error in his altimeter reading, what is the correct height of that peak?

45. Mt. Everest (the highest known point on earth) has an altitude of 29,028 ft. The Mariana Trench in the Pacific Ocean (the lowest known point on earth) has an altitude of $-36,198$ ft. Find the difference in altitude of these two places.

46. An airplane is flying 75 ft above the level of the Dead Sea (elevation $-1,299$ ft). How high must it climb to clear a 2,573-ft peak by 200 ft?

Set II

1. Find the additive inverse of $-\frac{1}{5}$.

2. Find the additive inverse of 0.

In Exercises 3–34, find the differences.

3. $(-8) - (5)$

4. $(7) - (-3)$

5. $(-6) - (-8)$

6. $(-4) - (6)$

7. $(2) - (7)$

8. $(9) - (-7)$

9. $(-14) - (10)$

10. $(184) - (-286)$

11. $(-473) - (389)$

12. $(-784) - (-528)$

13. $-17 - (-12)$

14. $0 - (-12)$

15. $18 - 367$

16. $83 - (-5)$

17. $(-24) - (-15)$

18. $-63 - (-63)$

19. $24 - (-15)$

20. $17 - (+23)$

21. $(-28) - 17$

22. $1 - (26)$

23. $0 - 862$

24. $361 - 0$

25. $0 - (-816)$

26. $19 - (-26)$

27. $83 - 0$

28. $(28) - (362)$

29. $(352) - (-89)$

30. $(-352) - (-89)$

31. $(-563) - (-825)$

32. $563 - (-825)$

33. $(-15) - (-9.794)$

34. $(-25) - (-5.63)$

35. Subtract (-120) from (-285).

36. Subtract (-4) from 8.

37. Subtract $\left(3\frac{1}{5}\right)$ from $\left(-5\frac{1}{10}\right)$.

38. Subtract $\left(4\frac{1}{4}\right)$ from $\left(-9\frac{1}{2}\right)$.

39. John has a balance of $281.42 in his checking account. Find his new balance after he writes a check for $209.57.

40. Sue has a balance of $563.24 in her checking account. Find her new balance after she writes a check for $347.87.

41. At 2 A.M. the temperature in Burlington, Vermont, was $-3°$ F. At 11 A.M. the temperature was 9° F. What was the rise in temperature?

42. At 5 A.M. Don's temperature was 101.8° F. By noon his temperature had risen to 103.2° F. What was the increase in his temperature?

43. At midnight the temperature in Fairbanks, Alaska, was $-5°$ F. At 10 A.M. the temperature was 24° F. What was the rise in temperature?

44. A dune buggy starting from the floor of Death Valley (-282 ft) is driven to the top of a nearby mountain having an elevation of 5,782 ft. What was the change in the dune buggy's altitude?

45. A jeep starting from the shore of the Dead Sea ($-1,299$ ft) is driven to the top of a nearby hill having an elevation of 723 ft. What was the change in the jeep's altitude?

46. At 2 P.M. Cindy's temperature was 103.4° F. By 6 P.M. her temperature had dropped to 99.9° F. What was the drop in her temperature?

1.5 Multiplying Signed Numbers

Product The answer to a multiplication problem is called the **product**.

Factors The numbers that are multiplied together to give a product are called the **factors** of that product.

$$6 \times 2 = 12$$

Factors ⟶ | | ⟵ Product

The numbers 6 and 2 are *factors* of 12; 12 is the *product* of 6 and 2. Similarly, 3 and 4 are factors of 12, because $3 \times 4 = 12$.

Symbols Used in Multiplication Multiplication may be shown in a number of different ways:

1. $3 \times 2 = 6$

2. $3 \cdot 2 = 6$ The multiplication dot "·" is written a little higher than the decimal point.

3. $3(2) = 6$ When there is no operation symbol between a number (or letter) and parentheses, the operation is understood to be multiplication.

4. $(3)(2) = 6$ The operation is understood to be multiplication.

5. ab When two expressions are written next to each other in this way, it is understood that they are to be multiplied. *Exception:* When two *numbers* are written next to each other, they are *not* to be multiplied. For example, 23 does *not* mean $2 \cdot 3 = 6$.

6. $3a$ In this example, it is understood that the value of a is to be *multiplied* by 3. Thus, if a is 7, $3a = 21$.

Multiplication Involving a Positive Integer   Multiplication by a positive integer is a short method for doing repeated addition of the same number.

Example 1 $3 \times 5 =$ three 5s $= 5 + 5 + 5 = 15$ ∎

Example 2 $6 \times 2 =$ six 2s $= 2 + 2 + 2 + 2 + 2 + 2 = 12$ ∎

Therefore, carrying the same idea over into multiplying signed numbers, we have

Example 3 $3 \times (-2) =$ three negative 2s $= (-2) + (-2) + (-2) = -6$ ∎

Example 4 $4 \times (-6) =$ four negative 6s $= (-6) + (-6) + (-6) + (-6) = -24$ ∎

From Examples 3 and 4 we see that *when two numbers having opposite signs are multiplied, their product is negative.*

Example 5 $3 \times 1 = 1 + 1 + 1 = 3$ ∎

Example 6 $1 \times (-8) =$ one negative 8 $= -8$ ∎

Multiplicative Identity   Examples 5 and 6 illustrate the fact that multiplying any real number by 1 gives the *identical* number we started with. Because this is true, we call 1 the **multiplicative identity**.

THE MULTIPLICATIVE IDENTITY IS 1

If a is any real number, then

$$a \cdot 1 = a \quad \text{and} \quad 1 \cdot a = a$$

Multiplication Involving Zero   Since multiplication is a method for doing repeated addition of the same number, multiplying a number by zero gives a product of zero.

Example 7   a. $3 \cdot 0 = 0 + 0 + 0 = 0$

b. $4 \cdot 0 = 0 + 0 + 0 + 0 = 0$

c. $0 \cdot 3 = 0 + 0 + 0 = 0$

d. $0 \cdot 4 = 0 + 0 + 0 + 0 = 0$ ∎

IF a IS ANY REAL NUMBER, THEN

$$a \cdot 0 = 0 \quad \text{and} \quad 0 \cdot a = 0$$

The Product of Two Negative Numbers   *The product of two negative numbers is positive*. We will not prove this statement; however, an examination of the pattern of products as shown below may convince you that it is true:

$$
\begin{array}{rcr}
4 & (-2) = & -8 \\
3 & (-2) = & -6 \\
2 & (-2) = & -4 \\
1 & (-2) = & -2 \\
0 & (-2) = & 0 \\
-1 & (-2) = & 2 \\
-2 & (-2) = & 4 \\
-3 & (-2) = & 6
\end{array}
$$

Decreasing by 1   Increasing by 2

The product of two negative numbers is positive

Multiplying Two Signed Numbers   The rules for multiplying two signed numbers are summarized as follows:

TO MULTIPLY TWO SIGNED NUMBERS

Multiply their absolute values *and* attach the correct sign to the left of the product of the absolute values. That sign is *positive* when the numbers have the same sign, and *negative* when the numbers have different signs.

Example 8   Multiply $(-7)(4)$.
Solution

$$(-7)(4) = -\boxed{28}$$

Product of their absolute values:
$7 \times 4 = 28$

Product negative because the numbers have different signs ■

Example 9   Multiply $(23)(-11)$.
Solution

$$(23)(-11) = -\boxed{253}$$

Product of their absolute values

Product negative because the numbers have different signs ■

Example 10   Multiply $(-14)(-10)$.
Solution

$$(-14)(-10) = +\boxed{140}$$

Product of their absolute values

Product positive because the numbers have the same sign ■

Example 11   Multiply $\left(4\frac{1}{2}\right)\left(-1\frac{1}{3}\right)$.

Solution

$$\left(4\frac{1}{2}\right)\left(-1\frac{1}{3}\right) = \left(\frac{9}{2}\right)\left(-\frac{4}{3}\right) = -\left(\frac{\overset{3}{\cancel{9}}}{\underset{1}{\cancel{2}}} \cdot \frac{\overset{2}{\cancel{4}}}{\underset{1}{\cancel{3}}}\right) = -\frac{6}{1} = -6 \quad \blacksquare$$

Example 12   Multiply $(-2.7)(-4.6)$.

Solution

$$(-2.7)(-4.6) = +(2.7 \times 4.6) = 12.42 \quad \blacksquare$$

We now have an alternate way of finding the *additive inverse* (or *negative*) of a number: The additive inverse of any real number can be found by multiplying the number by -1.

Example 13   Finding the additive inverse of a real number:

a. The additive inverse of 5 is $5(-1)$, or -5.

b. The additive inverse of -2 is $-2(-1)$, or 2. $\blacksquare$

NOTE   We will discuss the *multiplicative inverse* in Section 8.3.   ☑

EXERCISES 1.5

Set I   In Exercises 1–38, find the products.

1. $3(-2)$ **2.** $4(-6)$ **3.** $(-5)(2)$

4. $(-7)(5)$ **5.** $(-8)(-2)$ **6.** $(-6)(-7)$

7. $8(-4)$ **8.** $9(-5)$ **9.** $(-7)(9)$

10. $(-6)(8)$ **11.** $(-10)(-10)$ **12.** $(-9)(-9)$

13. $(8)(-7)$ **14.** $(12)(-6)$ **15.** $(-26)(10)$

16. $(-11)(12)$ **17.** $(-20)(-10)$ **18.** $(-30)(-20)$

19. $(75)(-15)$ **20.** $(86)(-13)$ **21.** $(-30)(+5)$

22. $(-50)(+6)$ **23.** $(-7)(-20)$ **24.** $(-9)(-40)$

25. $\left(-5\frac{1}{2}\right)(0)$ **26.** $\left(-7\frac{2}{7}\right)(0)$ **27.** $\left(6\frac{3}{4}\right)\left(-8\frac{1}{4}\right)$

28. $\left(9\frac{2}{9}\right)\left(-7\frac{1}{3}\right)$ **29.** $\left(-3\frac{3}{5}\right)\left(-8\frac{1}{5}\right)$ **30.** $\left(-2\frac{5}{7}\right)\left(-7\frac{1}{5}\right)$

31. $(-3.5)(-1.4)$ **32.** $(-4.7)(-1.6)$ **33.** $(2.74)(-100)$

34. $(3.04)(-100)$ **35.** $\left(2\frac{1}{3}\right)\left(-3\frac{1}{2}\right)$ **36.** $\left(-5\frac{1}{4}\right)\left(-2\frac{3}{5}\right)$

37. $(0)\left(-\frac{2}{3}\right)$ **38.** $(0)\left(-\frac{7}{8}\right)$

39. Find the additive inverse of -12.

40. Find the additive inverse of 5/7.

Set II In Exercises 1–38, find the products.

1. $(5)(-4)$
2. $(-6)(3)$
3. $(-7)(-3)$
4. $(-4)(8)$
5. $(-9)(-6)$
6. $(+8)(+5)$
7. $(-15)(10)$
8. $(15)(-10)$
9. $(-300)(-10)$
10. $(15)(0)$
11. $(-15)(-15)$
12. $(0)(-28)$
13. $9(-12)$
14. $(-23)(-28)$
15. $(-35)(100)$
16. $(-100)(42)$
17. $(-56)(-100)$
18. $83(0)$
19. $(27)(-81)$
20. $-33(42)$
21. $(-25)(+6)$
22. $(0)(-3,679)$
23. $(-18)(-47)$
24. $(-73)(61)$
25. $\left(-8\frac{1}{3}\right)(0)$
26. $\left(-8\frac{1}{3}\right)\left(-2\frac{1}{5}\right)$
27. $\left(5\frac{1}{7}\right)\left(-3\frac{4}{5}\right)$
28. $\left(2\frac{1}{8}\right)\left(-2\frac{8}{9}\right)$
29. $\left(-12\frac{1}{2}\right)\left(-3\frac{1}{4}\right)$
30. $\left(-18\frac{1}{2}\right)(0)$
31. $(-5.67)(10)$
32. $(-15.1)(32)$
33. $(-0.0032)(-100)$
34. $(6.07)(0)$
35. $\left(-2\frac{1}{5}\right)\left(3\frac{3}{4}\right)$
36. $\left(8\frac{1}{3}\right)\left(-2\frac{3}{5}\right)$
37. $\left(-5\frac{3}{7}\right)(0)$
38. $(0)\left(-9\frac{1}{2}\right)$

39. Find the additive inverse of $-6/11$.

40. What is the multiplicative identity?

1.6 Dividing Signed Numbers

Division is the *inverse* operation of multiplication; that is, division "undoes" multiplication.

Quotient The *answer* to a division problem is called the **quotient**.

Divisor The number we are dividing *by* is called the **divisor**.

Dividend The number we are dividing *into* is called the **dividend**.

Remainder If the divisor does not divide *exactly* into the dividend, the part that is "left over" is called the **remainder**.

When the *remainder* is 0, the *divisor* and the *quotient* are both **factors** of the *dividend*.
Division may be shown in several ways:

$$12 \div 4 = 12/4 = \frac{12}{4} = 4\overline{)12} \qquad \text{The } quotient \text{ in this case is 3}$$

$$\text{Divisor} \rightarrow 4\overline{)\underset{}{12}} \leftarrow \text{Dividend} \qquad \text{Quotient} \qquad \text{Divisor} \rightarrow 6\overline{)23} \leftarrow \text{Dividend}$$

Quotient — 3

Divisor → $4\overline{)12}$ ← Dividend

Quotient — 3

Divisor → $6\overline{)23}$ ← Dividend
  18
  5 ← Remainder

To check the answer in a division problem,

(Divisor × quotient) + remainder = dividend

---

or, if the divisor is a *factor* of the dividend,

Divisor × quotient = dividend

Because of the inverse relation between division and multiplication, the rules for finding the sign of a quotient are the same as the ones used for finding the sign of a product. Therefore,

---

TO DIVIDE ONE SIGNED NUMBER BY ANOTHER

Divide the absolute value of the dividend by the absolute value of the divisor *and* attach the correct sign to the left of the quotient of the absolute values. That sign is *positive* when the numbers have the same sign and *negative* when the numbers have different signs.

Division Involving Zero

Division of zero by a number other than zero is possible, and the quotient is always 0. That is, $0 \div 2 = 0/2 = 0$, *because* $2 \times 0 = 0$ (divisor × quotient = dividend). Therefore, $2\overline{)0}$ with quotient 0.

Division of a nonzero number by zero is impossible. Consider $4 \div 0$ or $4/0$. Suppose the quotient is some unknown number we call q. Then $4 \div 0 = q$ means that we must find a number q such that $0 \times q = 4$. However, there is no such number, since any number times 0 is 0, and so $0 \times q = 0$. Therefore, $4 \div 0$ *has no answer.*

Division of zero by zero cannot be determined. Consider $0/0$ or $0 \div 0$. Suppose that the quotient is 1. Then $0 \div 0 = 1$. This means that 0×1 has to equal 0, and, in fact, it does. This might lead us to assume that the quotient is indeed 1. But now let us suppose that the quotient is 0. (That is, suppose $0 \div 0 = 0$.) This could be true only if 0×0 equals 0, which it does. Furthermore, we *could* say that $0 \div 0 = 7$, because $0 \times 7 = 0$. The quotient could also be -2 or 327 or 2/3, or *any* number! Therefore, we say that $0 \div 0$ *cannot be determined.*

Division involving zero can be summarized as follows:

IF a IS ANY REAL NUMBER *EXCEPT* 0, THEN

1. $\dfrac{0}{a} = 0$

2. $\dfrac{a}{0}$ is not possible

3. $\dfrac{0}{0}$ cannot be determined

Example 1   Divide $(-30) \div (5)$.
Solution

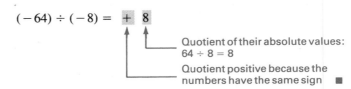

$(-30) \div (5) = \boxed{-}\ \boxed{6}$

Quotient of their absolute values:
$30 \div 5 = 6$
Quotient negative because the
numbers have different signs   ■

Example 2   Divide $(-64) \div (-8)$.
Solution

$(-64) \div (-8) = \boxed{+}\ \boxed{8}$

Quotient of their absolute values:
$64 \div 8 = 8$
Quotient positive because the
numbers have the same sign   ■

Example 3   Divide $\dfrac{35}{-7}$.
Solution

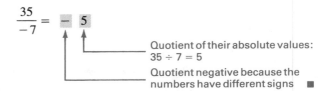

$\dfrac{35}{-7} = \boxed{-}\ \boxed{5}$

Quotient of their absolute values:
$35 \div 7 = 5$
Quotient negative because the
numbers have different signs   ■

Example 4   Divide $\dfrac{-42}{8}$.
Solution

$$\frac{-42}{8} = -\frac{21}{4} \text{ or } -5\frac{1}{4} \text{ or } -5.25 \quad ■$$

Example 5   Divide $0 \div (-3)$.
Solution $0 \div (-3) = 0$.   ■

Example 6   Divide $-7 \div 0$.
Solution $-7 \div 0$ is not possible. (We cannot divide by zero.)   ■

EXERCISES 1.6

Set I   Find the following quotients, or write "Not possible" or "Cannot be determined."

1. $(-10) \div (-5)$ **2.** $(-12) \div (-4)$ **3.** $(-8) \div (2)$

4. $(-6) \div (3)$ **5.** $\dfrac{+10}{-2}$ **6.** $\dfrac{+8}{-4}$

7. $\dfrac{-6}{-3}$ **8.** $\dfrac{-10}{-2}$ **9.** $(-40) \div (8)$

10. $(-60) \div (10)$

11. $16 \div (-4)$

12. $25 \div (-5)$

13. $(-15) \div (-5)$

14. $(-27) \div (-9)$

15. $\dfrac{12}{-4}$

16. $\dfrac{24}{-6}$

17. $\dfrac{-18}{-2}$

18. $\dfrac{-49}{-7}$

19. $\dfrac{-150}{10}$

20. $\dfrac{-250}{100}$

21. $36 \div (-12)$

22. $56 \div (-8)$

23. $(-45) \div 15$

24. $(-39) \div 13$

25. $\dfrac{4}{0}$

26. $\dfrac{8}{0}$

27. $0 \div 0$

28. $\dfrac{0}{0}$

29. $0 \div 7$

30. $0 \div (-12)$

31. $\dfrac{-1}{0}$

32. $\dfrac{-15}{0}$

33. $\dfrac{-15}{6}$

34. $\dfrac{-27}{12}$

35. $\dfrac{7.5}{-0.5}$

36. $\dfrac{1.25}{-0.25}$

37. $\dfrac{-6.3}{-0.9}$

38. $\dfrac{-4.8}{-0.6}$

39. $\dfrac{-367}{100}$

40. $\dfrac{-4860}{1000}$

41. $\dfrac{78.5}{-96.5}$

42. $\dfrac{98.5}{-84.3}$

Set II   Find the following quotients, or write "Not possible" or "Cannot be determined."

1. $(-8) \div (-2)$

2. $(-10) \div (5)$

3. $(12) \div (-6)$

4. $\dfrac{14}{-7}$

5. $\dfrac{-8}{2}$

6. $\dfrac{-50}{-10}$

7. $(36) \div (-12)$

8. $(-18) \div (-9)$

9. $(-16) \div (4)$

10. $\dfrac{-15.6}{10}$

11. $\dfrac{13.8}{-10}$

12. $\dfrac{-6.3}{-0.7}$

13. $(-8.4) \div (0.6)$

14. $(-9.6) \div (-0.8)$

15. $(18.5) \div (-3.7)$

16. $-14 \div 0$

17. $\dfrac{-18}{6}$

18. $-\left(\dfrac{0}{0}\right)$

19. $\dfrac{21}{-14}$

20. $\dfrac{-17}{-34}$

21. $81 \div (-15)$

22. $\dfrac{-51}{17}$

23. $-8 \div 16$

24. $(-35) \div (-7)$

25. $\dfrac{13}{0}$

26. $0 \div (-10)$

27. $-8 \div 0$

28. $\dfrac{-12}{-2}$

29. $0 \div 9$

30. $\dfrac{72}{-18}$

31. $\dfrac{-11}{22}$

32. $\dfrac{16}{-24}$

33. $\dfrac{31}{-62}$

34. $-35 \div 0$ **35.** $\dfrac{-1.2}{0.24}$ **36.** $\dfrac{37}{-10}$

37. $\dfrac{-42}{-36}$ **38.** $\dfrac{-8.1}{2.7}$ **39.** $\dfrac{64}{-12}$

40. $\dfrac{-315}{100}$ **41.** $\dfrac{471}{-10}$ **42.** $\dfrac{-15}{-100}$

1.7 Review: 1.1–1.6

Natural Numbers
1.1
$\{1, 2, 3, \ldots\}$

Whole Numbers
1.1
$\{0, 1, 2, \ldots\}$

Integers
1.2
$\{\ldots, -3, -2, -1, 0, 1, 2, 3, \ldots\}$

Digits
1.1
$\{0, 1, 2, 3, 4, 5, 6, 7, 8, 9\}$

Fractions
1.1
Numbers having the form a/b, where a and b are integers and $b \neq 0$.

Decimal Fractions
1.1
A decimal fraction is a fraction whose denominator is a power of 10: (10, 100, 1000, . . .).

Mixed Numbers
1.1
A mixed number is made up of both a whole-number part and a fraction part.

Equal Sign
1.1
The equal sign ($=$) in a statement means that the expression on the left side of the equal sign *has the same value or values* as the expression on the right side of the equal sign.

"Greater-Than" and "Less-Than" Symbols
1.1
The symbol $>$ is read "greater than," and the symbol $<$ is read "less than." Numbers get larger as we move to the right on the number line and get smaller as we move to the left.

Positive Numbers
1.2
All real numbers greater than 0 are positive numbers.

Negative Numbers
1.2
All real numbers less than 0 are negative numbers.

Real Numbers
1.1 and 1.2
All the numbers that can be represented by points on the number line are called real numbers.

Rational Numbers
1.2
Numbers that can be put in the form a/b, where a and b are integers and $b \neq 0$. (All natural numbers, whole numbers, fractions, decimals, and mixed numbers are rational numbers.) All rational numbers are *real* numbers.
   The *decimal form* of any rational number is always a *terminating* or a *repeating* decimal.

Absolute Value 1.3	The absolute value of a number is the distance between that number and 0 on the number line with no regard to direction. It can never be negative. The absolute value of a real number x is written $|x|$.

Addition of Two Signed Numbers 1.3

1. When the numbers have the same sign,   *First:* Add their absolute values *and*
                     Second: Attach (to the left of the sum) the
                     sign of both numbers.

2. When the numbers have different signs,   *First:* Subtract the smaller absolute value
                     from the larger absolute value *and*
                     Second: Attach (to the left of that answer)
                     the sign of the number that has the larger
                     absolute value.

Additive Identity 1.3

The *additive identity* is 0.

$$a + 0 = a \quad \text{and} \quad 0 + a = a$$

Additive Inverse 1.4

The *additive inverse* (or negative) of a is $-a$.

$$a + (-a) = 0 \quad \text{and} \quad (-a) + a = 0$$

Subtraction of One Signed Number from Another 1.4

1. Change the subtraction symbol to an addition symbol, and change the sign of the number being subtracted.

2. Add the resulting signed numbers.

$$a - b = a + (-b)$$

Subtraction Involving 0 1.4

$$a - 0 = a$$
$$0 - a = -a$$

Multiplication of Two Signed Numbers 1.5

Multiply their absolute values *and* attach the correct sign to the left of the product of the absolute values. That sign is *positive* when the numbers have the same sign, and *negative* when the numbers have different signs.

Multiplicative Identity 1.5

The *multiplicative identity* is 1.

$$a \cdot 1 = a \quad \text{and} \quad 1 \cdot a = a$$

Multiplication Involving 0 1.5

$$a \cdot 0 = 0 \quad \text{and} \quad 0 \cdot a = 0$$

Division of One Signed Number by Another 1.6

Divide the absolute value of the dividend by the absolute value of the divisor *and* attach the correct sign to the left of the quotient of the absolute values. The sign is *positive* when the numbers have the same sign, and *negative* when the numbers have different signs.

Division Involving Zero 1.6

If a is any real number $\neq 0$:

$$\frac{0}{a} = 0$$

$$\frac{a}{0} \qquad \text{is not possible}$$

$$\frac{0}{0} \qquad \text{cannot be determined}$$

Review Exercises 1.7 Set I

1. Write all the digits greater than 7.   *8 9* (handwritten)

2. Write the smallest two-digit natural number.   *11* (handwritten)

3. Write the smallest one-digit integer.   *−9* (handwritten)

4. Write the largest one-digit integer less than zero.   *−1* (handwritten)

5. Which symbol, $<$ or $>$, should be used to make the statement true?

   a. $-3 \underline{<} \ 8$ b. $5 \underline{>} -2$ (handwritten symbols)

6. Which symbol, $<$ or $>$, should be used to make the statement true?

   a. $7 \underline{>} -8$ b. $-3 \underline{>} -9$ (handwritten symbols)

7. What is the multiplicative identity?   *1* (handwritten)

8. What is the additive inverse of 4?   *−4* (handwritten)

In Exercises 9–42, perform the indicated operations, or write "Not possible."

9. $(-2) + (+3)$

10. $(-5) + (+4)$

11. $(-6) \div (-2)$

12. $(-8) \div (-4)$

13. $(-5) - (-3)$

14. $(-7) - (-2)$

15. $(+5) - |-2|$

16. $(+8) - |-3|$

17. $(-3)(-4)$

18. $(-5)(-4)$

19. $(-7) - (3)$

20. $(-2) - (4)$

21. $(4) + (-12)$

22. $(6) + (-8)$

23. $8(-15)$

24. $5(-18)$

25. $(24) \div (-3)$

26. $(42) \div (-7)$

27. $(9) - (-4)$

28. $(4) - (-7)$

29. $\left(-2\frac{2}{3}\right)\left(2\frac{1}{2}\right)$

30. $\left(5\frac{4}{5}\right) + \left(-1\frac{1}{2}\right)$

31. $(-6) \div (2)$

32. $(-12) \div (4)$

33. $(-10) + (-2)$

34. $(-8) + (-3)$

35. $(-4)(6)$

36. $(-5)(7)$

37. $\dfrac{-25}{-5}$

38. $\dfrac{-16}{-2}$

39. $0 \div (-4)$

40. $(0)(-5)$

41. Subtract (-6) from (-10).

42. Subtract (-8) from (-5).

Review Exercises 1.7 Set II

ANSWERS

1. Write all the digits greater than 5.

2. Write all the even whole numbers less than 12.

3. What is the additive identity?

4. Write all the whole numbers less than 4.

5. Write the smallest digit.

6. What is the additive inverse of $-1/6$?

7. Which symbol, $<$ or $>$, should be used to make the statement true?

    a. -5 _?_ -1 b. -2 _?_ 0

8. Which symbol, $<$ or $>$, should be used to make the statement true?

    a. -3 _?_ -8 b. 5 _?_ -2

In Exercises 9–42, perform the indicated operations.

9. $(-5) + (+2)$

10. $(-7) - (2)$

11. $(8) \div (-4)$

12. $(-5)(+2)$

13. $(-8) \div (-2)$

14. $(-8) + (-2)$

15. $|0| - |-5|$

16. $5 - (-12)$

17. $(-8)(-2)$

18. $-8 - 2$

19. $(-9) - (3)$

1. _____

2. _____

3. _____

4. _____

5. _____

6. _____

7a. _____

b. _____

8a. _____

b. _____

9. _____

10. _____

11. _____

12. _____

13. _____

14. _____

15. _____

16. _____

17. _____

18. _____

19. _____

20. $(-9)(-3)$ **21.** $(9) + (-3)$ **22.** $16 \div (-12)$

23. $(41)(-1)$ **24.** $\dfrac{-14}{7}$ **25.** $\dfrac{1}{0}$

26. $\dfrac{3}{-12}$ **27.** $24 \div 0$ **28.** $(-72) \div (-8)$

29. $\dfrac{0}{2}$ **30.** $(-15) + (-23)$ **31.** $-|-4|$

32. $(-437)(0)$ **33.** $0 \div (-15)$ **34.** $0 - 5$

35. $0(-5)$ **36.** $(-20) - 25$ **37.** $(-20)(-25)$

38. $\left(-1\dfrac{7}{8}\right)\left(-3\dfrac{1}{5}\right)$ **39.** $\dfrac{-10}{-2}$ **40.** $0 - (+7)$

41. Subtract -12 from 7.   **42.** Subtract 3 from -15.

20. _____

21. _____

22. _____

23. _____

24. _____

25. _____

26. _____

27. _____

28. _____

29. _____

30. _____

31. _____

32. _____

33. _____

34. _____

35. _____

36. _____

37. _____

38. _____

39. _____

40. _____

41. _____

42. _____

1.8 Commutative and Associative Properties

We have already discussed several properties of the real-number system, namely, that 0 is the additive identity, that 1 is the multiplicative identity, and that the additive inverse of a is $-a$ (where a represents any real number). The *commutative* and *associative properties* are also important properties of the set of real numbers.

Commutative Properties

Addition Is Commutative If we change the order of the two numbers in an addition problem, we get the same sum. This property is called the **commutative property of addition**. We assume that this property is true when *any* two real numbers are added. In Figure 1.8.1, we use the number line to show that $2 + 3 = 3 + 2$.

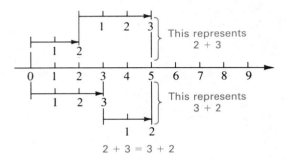

FIGURE 1.8.1

COMMUTATIVE PROPERTY OF ADDITION

If a and b represent any real numbers, then

$$a + b = b + a$$

Example 1   Addition is commutative:

a. $(+7) + (5) = 12$
   $(5) + (+7) = 12$

   Therefore,

   $$(+7) + (5) = (5) + (+7).$$

b. $(-6) + (2) = -4$
   $(2) + (-6) = -4$

   Therefore,

   $$(-6) + (2) = (2) + (-6).$$

c. $(-4) + (-8) = -12$
   $(-8) + (-4) = -12$

   Therefore,

   $$(-4) + (-8) = (-8) + (-4). \quad \blacksquare$$

Subtraction is not commutative, as the next example shows.

Example 2   Subtraction is *not* commutative.

$$3 - 2 = 1$$
$$2 - 3 = -1$$

Therefore,

$$3 - 2 \neq 2 - 3. \quad \blacksquare$$

Multiplication Is Commutative   If we change the order of two numbers in a multiplication problem, we get the same product. This property is called the **commutative property of multiplication**. We *assume* that this property is true when *any* two real numbers are multiplied.

Example 3   Multiplication is commutative:

a. $(4)(5) = 20$
    $(5)(4) = 20$

Therefore, $(4)(5) = (5)(4)$.

b. $(-9)(3) = -27$
    $(3)(-9) = -27$

Therefore,

$$(-9)(3) = (3)(-9). \quad \blacksquare$$

COMMUTATIVE PROPERTY OF MULTIPLICATION

If a and b are real numbers, then

$$a \cdot b = b \cdot a$$

Division is not commutative, as Example 4 will prove.

Example 4   Division is *not* commutative.

$$10 \div 5 = 2$$

$$5 \div 10 = \frac{1}{2}$$

Therefore,

$$10 \div 5 \neq 5 \div 10. \quad \blacksquare$$

Associative Properties

Addition Is Associative   In adding three numbers, the sum is unchanged no matter how we group the numbers. This property is called the **associative property of addition**. We *assume* that this property is true when *any* three real numbers are added. Parentheses or other grouping symbols are used to show which two numbers are to be added first.

> ### ASSOCIATIVE PROPERTY OF ADDITION
>
> If a, b, and c represent any real numbers, then
>
> $$a + b + c = (a + b) + c = a + (b + c)$$

Example 5 Addition is associative.

$(2 + 3) + 4$ The parentheses mean that 2 and 3 are to be added first

$= \quad 5 \quad + 4$

$= 9$

$2 + (3 + 4)$ Here, the parentheses mean that 3 and 4 are to be added first

$= 2 + \quad 7$

$= 9$

Therefore,

$$(2 + 3) + 4 = 2 + (3 + 4). \quad \blacksquare$$

Subtraction is not associative, as the next example will prove.

Example 6 Subtraction is *not* associative.

$(7 - 4) - 8 = 3 - 8 \quad = -5$
$7 - (4 - 8) = 7 - (-4) = 11$

Therefore,

$$(7 - 4) - 8 \neq 7 - (4 - 8). \quad \blacksquare$$

Multiplication Is Associative In multiplying three numbers, the product is unchanged no matter how we group the numbers. This property is called the **associative property of multiplication**. We *assume* that this property is true when *any* three numbers are multiplied.

> ### ASSOCIATIVE PROPERTY OF MULTIPLICATION
>
> If a, b, and c are real numbers, then
>
> $$a \cdot b \cdot c = (a \cdot b) \cdot c = a \cdot (b \cdot c)$$

Example 7 Multiplication is associative.

$(3 \cdot 4) \cdot 2 = 12 \cdot 2 = 24$
$3 \cdot (4 \cdot 2) = 3 \cdot 8 = 24$

Therefore, $(3 \cdot 4) \cdot 2 = 3 \cdot (4 \cdot 2)$. $\blacksquare$

Brackets [ ] and braces { } as well as parentheses ( ) can be used to show which two numbers are to be multiplied first.

Example 8 Use of brackets and parentheses in multiplication.

$$[(-6) \cdot (+2)] \cdot (-5) = [-12] \cdot (-5) = 60$$
$$(-6) \cdot [(+2) \cdot (-5)] = (-6) \cdot [-10] = 60$$

Therefore, $[(-6) \cdot (+2)] \cdot (-5) = (-6) \cdot [(+2) \cdot (-5)]$. ■

Division is not associative, as a single example will prove.

Example 9 Division is *not* associative.

$$[(-16) \div (4)] \div (-2) = [-4] \div (-2) = 2$$
$$(-16) \div [(4) \div (-2)] = (-16) \div [-2] = 8$$

Therefore, $[(-16) \div (4)] \div (-2) \neq (-16) \div [(4) \div (-2)]$. ■

SUMMARY

1. The *commutative property* says that *changing the order* of the numbers in an addition or multiplication problem gives the same answer.

2. The *associative property* says that *changing the grouping* of the numbers in an addition or multiplication problem gives the same answer.

How to Determine Whether Commutativity or Associativity Has Been Used In commutativity, the numbers or letters actually exchange places (commute).

$$a + b = b + a \qquad c \cdot d = d \cdot c$$

The first element occupies the second place and vice versa.

In associativity, the numbers or letters stay in their original places, but the grouping is changed:

$$a + (b + c) = (a + b) + c \qquad d \cdot (e \cdot f) = (d \cdot e) \cdot f$$

Example 10 State whether each of the following is true or false. If the statement is true, give the reason.

a. $(-7) + 5 = 5 + (-7)$ *True* because of the commutative property of addition (*order* of numbers changed)

b. $(+6)(-8) = (-8)(+6)$ *True* because of the commutative property of multiplication (*order* of numbers changed)

c. $[(-3) + 5] + (-2)$
$= (-3) + [5 + (-2)]$ *True*; associative property of addition (*grouping* changed)

d. $-8 + 0 = -8$ — *True*; additive identity property (0 is the additive identity)

e. $-5 + 5 = 0$ — *True*; additive inverse property (additive inverse of -5 is 5)

f. $[(7) \cdot (-4)] \cdot (2) = (7) \cdot [(-4) \cdot (2)]$ — *True*; associative property of multiplication (*grouping* changed)

g. $(+8) - (-7) = (-7) - (+8)$ — *False*

h. $a + (b + c) = (a + b) + c$ — *True*; associative property of addition (*grouping* changed)

i. $y \div z = z \div y$ — *False*

j. $(p \cdot r) \cdot s = p \cdot (r \cdot s)$ — *True*; associative property of multiplication (*grouping* changed)

k. $3 \times 0 = 3$ — *False*

l. $(3 + 5) + 7 = 3 + (7 + 5)$ — *True*: commutative *and* associative properties of addition

m. $8 \cdot (3 \cdot 6) = (8 \cdot 6) \cdot 3$ — *True*; commutative *and* associative properties of multiplication ∎

We can now justify the fact that the product of two negative numbers is positive by using the commutative and associative properties (see Example 11).

Example 11
$(-5)(-4) = (-1)(5)(-4)$ — Because $-5 = (-1)(5)$

$= (-1)(-20)$ — Because $(5)(-4) = -20$

$=$ negative of -20 — Because -1 times a number gives the negative of that number

$= 20$ — Because the negative of a number is found by changing its sign (Section 1.4) ∎

Because the commutative and associative properties of addition and of multiplication hold for all real numbers, we can add more than two numbers in any order, and we can multiply more than two numbers in any order.

Example 12 Multiply $(-2)(-5)(-3)$.
Solution

$$(-2)(-5)(-3)$$
$$= (+10)(-3)$$
$$= -30$$ Notice that an odd number of negative signs gives a product that is negative ∎

Example 13 Multiply $(-2)(-5)(-3)(-4)$.
Solution

$$(-2)(-5)(-3)(-4)$$
$$= (+10)(-3)(-4)$$
$$= (-30)(-4)$$
$$= +120$$ Notice that an even number of negative signs gives a product that is positive ∎

EXERCISES 1.8

Set I   In Exercises 1–32, state whether each of the following is true or false. If the statement is true, give the reason.

1. $7 + 5 = 5 + 7$

2. $9 + 4 = 4 + 9$

3. $(2 + 6) + 3 = 2 + (6 + 3)$

4. $(1 + 8) + 7 = 1 + (8 + 7)$

5. $6 - 2 = 2 - 6$

6. $4 - 7 = 7 - 4$

7. $(a \cdot b) \cdot c = a \cdot (b \cdot c)$

8. $(p \cdot q) \cdot r = p \cdot (q \cdot r)$

9. $8 \div 4 = 4 \div 8$

10. $3 \div 6 = 6 \div 3$

11. $(p)(t) = (t)(p)$

12. $(m)(n) = (n)(m)$

13. $(4) + (-5) = (-5) + (4)$

14. $(-7) + (2) = (2) + (-7)$

15. $5 + (3 + 4) = 5 + (4 + 3)$

16. $6 + (8 + 2) = 6 + (2 + 8)$

17. $e + f = f + e$

18. $j + k = k + j$

19. $-8 \times 1 = -8$

20. $11 + 0 = 11$

21. $15 + (-15) = 0$

22. $-4 + 4 = 0$

23. $3 \times 0 = 3$

24. $0 \times (-4) = -4$

25. $9 + (5 + 6) = (9 + 6) + 5$

26. $3 \cdot (8 \cdot 4) = (3 \cdot 4) \cdot 8$

27. $x - 4 = 4 - x$

28. $5 - y = y - 5$

29. $4(a \cdot 6) = 4a(6)$

30. $m(7 \cdot 5) = (m \cdot 7)5$

31. $H + 8 = 8 + H$

32. $4 + P = P + 4$

In Exercises 33–40, perform the indicated operations.

33. $8 + (-3) + (-7) + (-1)$

34. $-2 + (-5) + 6 + (-11)$

35. $(-5)(-4)(-2)$

36. $(-3)(-2)(-8)$

37. $(2)(-5)(-9)$

38. $(-3)(4)(-2)$

39. $(-2)(-3)(-5)(-4)$

40. $(-4)(-2)(-1)(-7)$

Set II   In Exercises 1–32, state whether each of the following is true or false. If the statement is true, give the reason.

1. $5 + 3 = 3 + 5$

2. $(3 + 1) + 5 = 3 + (1 + 5)$

3. $8 - 2 = 2 - 8$

4. $(x \cdot y) \cdot z = x \cdot (y \cdot z)$

5. $10 \div 2 = 2 \div 10$

6. $(3)(-2) = (-2)(3)$

7. $2 + (3 + 4) = 2 + (4 + 3)$

8. $x + y = y + x$

9. $8 + (2 + 5) = (8 + 5) + 2$

10. $a - 2 = 2 - a$

11. $(4 \cdot c)(3) = 4(3 \cdot c)$

12. $3 \div x = x \div 3$

13. $5(-2) = -2(5)$

14. $8 + (3 + 2) = 8 + (2 + 3)$

15. $12 - (6 - 9) = (12 - 6) - 9$

16. $9 \times (-3) = (-3) \times 9$

17. $3 + (4 + 7) = (3 + 7) + 4$

18. $25 \div 3 = 3 \div 25$

19. $3 + 0 = 3$

20. $0 \times -13 = -13$

21. $\dfrac{1}{2} + \left(-\dfrac{1}{2}\right) = 0$

22. $4 \times (-4) = 0$

23. $-5 \times 0 = -5$

24. $-16 \times 0 = 0$

25. $18 - 3 = 3 - 18$

26. $5 \cdot (12 \cdot 6) = (5 \cdot 6) \cdot 12$

27. $24 + 16 = 16 + 24$

28. $24 \div (12 \div 2) = (24 \div 12) \div 2$

29. $2 \times (3 \times 4) = (2 \times 3) \times 4$

30. $6 - 15 = 15 - 6$

31. $18 + (3 + 5) = (18 + 3) + 5$

32. $9 - (4 - 12) = (9 - 4) - 12$

In Exercises 33–40, perform the indicated operations.

33. $9 + (-5) + 7 + (-12)$

34. $-2 + (-8) + (-6) + 11$

35. $(4)(-5)(-7)$

36. $(-9)(-8)(-1)$

37. $(-5)(-1)(-6)(-2)$

38. $-3(-2)(5)(-4)$

39. $(-5)(-2)(-2)(-2)$

40. $8(-1)(-2)(-3)$

1.9 Powers of Signed Numbers

Now that we have learned to multiply signed numbers, it is possible to consider products in which the same number is repeated as a factor.

The shortened notation for a product such as $3 \cdot 3 \cdot 3 \cdot 3$ is 3^4. That is, by definition, $3^4 = 3 \cdot 3 \cdot 3 \cdot 3$. In the expression 3^4, 3 is called the **base**, and 4 is called the **exponent**. The number 4 (the exponent) indicates that 3 (the base) is to be used as a *factor* four times. The entire symbol 3^4 is called an **exponential expression**, and is commonly read as "three to the fourth power." See Figure 1.9.1.

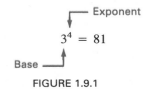

FIGURE 1.9.1

A WORD OF CAUTION

$$3^4 \neq 3 \cdot 4$$

$$3^4 = \underbrace{3 \cdot 3 \cdot 3 \cdot 3}_{\text{Four factors}} = 81$$

☑

A WORD OF CAUTION When you write an exponential number, be sure your exponents look like exponents. For example, be sure that 3^4 doesn't look like 34. ☑

We usually read b^2 as "b squared" rather than as "b to the second power"; likewise, we usually read b^3 as "b cubed" rather than as "b to the third power."

Even Power If a base has an exponent that is an even number, we say that it is an **even power** of the base. For example, 3^2, 5^4, and $(-2)^6$ are even powers.

Odd Power If a base has an exponent that is an odd number, we say that it is an **odd power** of the base. For example, 3^5, 10^3, and $(-4)^5$ are odd powers.

Example 1 Examples of powers of signed numbers.

a. $2^3 = 2 \cdot 2 \cdot 2 = 8$

b. $4^2 = 4 \cdot 4 = 16$

c. $1^4 = 1 \cdot 1 \cdot 1 \cdot 1 = 1$

d. $(-3)^2 = (-3)(-3) = 9$ } Notice that an even power of a negative number is positive

e. $(-1)^4 = (-1)(-1)(-1)(-1) = 1$ }

f. $(-2)^3 = (-2)(-2)(-2) = -8$ } Notice that an odd power of a negative number is negative

g. $(-1)^5 = (-1)(-1)(-1)(-1)(-1) = -1$ }

h. $(-36)^3 = (-36)(-36)(-36) = -46,656$ } ∎

Example 2   Rewrite each of the following with no exponents:

a. $(-6)^2 = (-6)(-6) = 36$

b. $-6^2 = -6 \cdot 6 = -36$ ∎

A WORD OF CAUTION   Students often think that expressions such as $(-6)^2$ and -6^2 are *not* the same. They are *not* the same. The exponent applies only to the symbol immediately preceding it.

$(-6)^2 = (-6)(-6) = 36$ The exponent applies to the ( ), because the immediately preceding symbol is), part of the grouping symbol ( )

$-6^2 = -6 \cdot 6 = -36$ The exponent applies only to the 6   ☑

The Exponent 1

IF a IS ANY REAL NUMBER, THEN

$$a^1 = a$$

Example 3   Examples of 1 as an exponent:

a. $5^1 = 5$

b. $(-8)^1 = -8$ ∎

Powers of Zero

IF a IS ANY POSITIVE REAL NUMBER EXCEPT 0, THEN

$$0^a = 0$$

Example 4   Examples of powers of zero:

a. $0^2 = 0 \cdot 0 = 0$

b. $0^5 = 0 \cdot 0 \cdot 0 \cdot 0 \cdot 0 = 0$ ∎

Cases where 0 appears as an exponent, such as 5^0, or where 0 appears as both exponent and base, 0^0, are discussed in Chapter 3.

You are urged to memorize these powers:

$0^1 = 0,$ $1^1 = 1,$ $2^1 = 2,$ $3^1 = 3$ and so forth,

$0^2 = 0,$ $1^2 = 1,$ $2^2 = 4,$ $3^2 = 9,$ $4^2 = 16,$ $5^2 = 25,$

$6^2 = 36,$ $7^2 = 49,$ $8^2 = 64,$ $9^2 = 81,$ $10^2 = 100,$

$11^2 = 121,$ $12^2 = 144,$ $13^2 = 169,$

$0^3 = 0,$ $1^3 = 1,$ $2^3 = 8,$ $3^3 = 27,$ $4^3 = 64,$ $5^3 = 125,$

$0^4 = 0,$ $1^4 = 1,$ $2^4 = 16,$ $3^4 = 81,$

$0^5 = 0,$ $1^5 = 1,$ $2^5 = 32,$

$0^6 = 0,$ $1^6 = 1,$ and $2^6 = 64.$

EXERCISES 1.9

Set I   Find the value of each of the following expressions.

1. 3^3 27   **2.** 2^4 **3.** $(-5)^2$

4. $(-6)^3$ **5.** 7^2 **6.** 3^4

7. 0^3 **8.** 0^4 **9.** $(-10)^1$

10. $(-10)^2$ **11.** 10^3 **12.** 10^4

13. $(-10)^5$ **14.** $(-10)^6$ 1000000   **15.** 2^1

16. 2^5 **17.** $(-2)^6$ **18.** $(-2)^7$ -128

19. 2^8 **20.** 25^2 **21.** 40^3

22. 0^4 **23.** $(-12)^3$ **24.** $(-15)^2$

25. $(-1)^5$ **26.** $(-1)^7$ **27.** -2^2 -4

28. -3^2 **29.** $(-1)^{99}$ **30.** $(-1)^{98}$

31. $(12.7)^2$ **32.** $(15.4)^2$ **33.** 0^8

34. 0^3

Set II   Find the value of each of the following expressions.

1. $(-2)^3$ **2.** 6^2 **3.** $(-10)^2$

4. -10^2 **5.** 0^5 **6.** $(-1)^{35}$

7. $-(-5)^2$ **8.** $(-1)^{50}$ **9.** 10^5

10. 8^2 **11.** 4^3 **12.** -3^4

13. $(-3)^4$ **14.** 20^3 **15.** 24^2

16. $(-4)^2$ **17.** -4^2 **18.** $(-1)^{43}$

19. 0^{23} **20.** $-(-8)^2$ **21.** $(-1)^{132}$

22. 1^{43} **23.** $(-2)^5$ **24.** 30^2

25. 0^{42} **26.** $(-7)^2$ **27.** $(-3)^3$

28. -3^3 **29.** $-(-3)^3$ **30.** $-(-3^3)$

31. 16^2 **32.** $(-4)^2$ **33.** 0^2

34. 0^1

1.10 Roots of Signed Numbers

1.10A Square Roots

Just as subtraction is the inverse operation of addition, and division is the inverse operation of multiplication, finding *roots* is the inverse operation of raising to powers. Thus, finding the *square root* of a number is the inverse operation of *squaring* a number.

Principal Square Root   Every positive real number has both a positive and a negative square root; the *positive* square root is called the **principal square root**.

Example 1   The number 9 has two square roots: $+3$ and -3.

$+3$ is a square root of 9 because $3^2 = 9$.

-3 is a square root of 9 because $(-3)^2 = 9$.

3 is the *principal* square root of 9 because it is the positive one. ■

The Square Root Symbol   The notation for the principal square root of p is $\sqrt{p}$, which is read, "the square root of p." The entire expression $\sqrt{p}$ is called a **radical expression,** or, more simply, a **radical**. The parts of a square root are shown in Figure 1.10.1.

Radical sign   Radicand

FIGURE 1.10.1

When we are asked to find $\sqrt{p}$, we must find some *positive* number whose *square* is p. For example, "Find $\sqrt{9}$" means we must find a *positive* number whose *square* is 9. The answer, of course, is 3. Therefore, $\sqrt{9} = 3$.

A WORD OF CAUTION   Because $\sqrt{p}$ *always* represents the *principal square root*, when $p \geq 0$, $\sqrt{p}$ is *always* positive or zero. ☑

In this section and in Section 1.10B, the *radicand* will always be the square of some whole number. You will find the problems easier to do if you have memorized the squares of the first thirteen whole numbers (see page 43).

Example 2   Find the square root of 25, written $\sqrt{25}$.

$$\sqrt{25} = 5 \text{ because } 5^2 = 25. ■$$

Example 3   Find the square root of 16, written $\sqrt{16}$.

$$\sqrt{16} = 4 \text{ because } 4^2 = 16. ■$$

Example 4   Square roots by inspection:

a. $\sqrt{4} = 2$ because $2^2 = 4$.

b. $\sqrt{9} = 3$ because $3^2 = 9$.

c. $\sqrt{36} = 6$ because $6^2 = 36$.

d. $\sqrt{0} = 0$ because $0^2 = 0$.

e. $\sqrt{1} = 1$ because $1^2 = 1$. ■

Example 5   Find $-\sqrt{16}$.
Solution We know that $\sqrt{16} = 4$. Therefore, $-\sqrt{16} = -4$.   ■

NOTE   Square roots of negative numbers are *imaginary numbers*, not real numbers. If the problem in Example 5 had been $\sqrt{-16}$, the answer would have been "Not a real number," since *no real number* exists whose square is -16.   ☑

A WORD OF CAUTION   The *square* of 4 is 16. ($4^2 = 16$.) The *square root* of 4 is 2. ($\sqrt{4} = 2$).   ☑

EXERCISES 1.10A

Set I   Find the following square roots.

1. $\sqrt{16}$	**2.** $\sqrt{25}$	**3.** $-\sqrt{4}$
4. $-\sqrt{9}$	**5.** $\sqrt{81}$	**6.** $\sqrt{36}$
7. $\sqrt{100}$	**8.** $\sqrt{144}$	**9.** $-\sqrt{81}$
10. $-\sqrt{121}$	**11.** $\sqrt{64}$	**12.** $\sqrt{169}$

Set II   Find the following square roots.

1. $-\sqrt{100}$	**2.** $-\sqrt{144}$	**3.** $\sqrt{49}$
4. $\sqrt{121}$	**5.** $\sqrt{225}$	**6.** $-\sqrt{36}$
7. $\sqrt{1}$	**8.** $-\sqrt{25}$	**9.** $\sqrt{9}$
10. $\sqrt{400}$	**11.** $-\sqrt{64}$	**12.** $\sqrt{256}$

1.10B   Finding Square Roots by Trial and Error

It is proved in higher mathematics that for all real numbers a and b, if $a > b$, then $\sqrt{a} > \sqrt{b}$, and if $a < b$, then $\sqrt{a} < \sqrt{b}$. We can use these facts in finding square roots.

Example 6   Find $\sqrt{196}$ by trial and error.
Solution We know that $\sqrt{100} = 10$, because $10^2 = 100$. Because $196 > 100$, $\sqrt{196} > 10$.

Try 12; $12^2 = 144$. Therefore, 12 is too small.

Try 13; $13^2 = 169$. Therefore, 13 is too small.

Try 14; $14^2 = 196$. Therefore, $\sqrt{196} = 14$.   ■

This is what we mean by "finding the square root by trial and error."

It is also proved in higher mathematics courses that if b is between a and c, then $\sqrt{b}$ is between $\sqrt{a}$ and $\sqrt{c}$.

Example 7   Find $\sqrt{576}$ by trial and error.

$$\sqrt{400} = 20 \qquad \text{because } 20^2 = 400$$
$$\sqrt{900} = 30 \qquad \text{because } 30^2 = 900$$

Since 576 is between 400 and 900, $\sqrt{576}$ *must* be between 20 and 30. In addition, since the last digit of 57 6 is 6, the last digit of the square root must be 4 or 6 (if the square root is an integer), because $4 \cdot 4 = 1\boxed{6}$ and $6 \cdot 6 = 3\boxed{6}$. Therefore, 24 and 26 are the only possible square roots. Try 24: $24^2 = 576$. Therefore, $\sqrt{576} = 24$. ∎

EXERCISES 1.10B

Set I Find the following square roots by trial and error.

1. $\sqrt{529}$ **2.** $\sqrt{361}$ **3.** $\sqrt{441}$ **4.** $\sqrt{625}$

5. $\sqrt{289}$ **6.** $\sqrt{324}$ **7.** $\sqrt{729}$ **8.** $\sqrt{1{,}296}$

Set II Find the following square roots by trial and error.

1. $\sqrt{400}$ **2.** $\sqrt{484}$ **3.** $\sqrt{676}$ **4.** $\sqrt{1{,}024}$

5. $\sqrt{225}$ **6.** $\sqrt{784}$ **7.** $\sqrt{1{,}444}$ **8.** $\sqrt{2{,}809}$

1.10C   Finding Square Roots by Table or Calculator

Square roots of positive numbers can be found by using a calculator with a square root key $\boxed{\sqrt{}}$ and sometimes by using tables. (See Table I, inside back cover.)

Example 8 Find $\sqrt{710{,}649}$ by using a calculator.
Solution Press the following keys in this order:

$$\boxed{7}\ \boxed{1}\ \boxed{0}\ \boxed{6}\ \boxed{4}\ \boxed{9}\ \boxed{\sqrt{}}$$

The calculator display shows 843. Therefore, $\sqrt{710{,}649} = 843$. (We cannot find $\sqrt{710{,}649}$ by using Table I.) ∎

Irrational Numbers

We often need to find the *approximate* square root of some number that is *not* the square of a whole number. Such a number is called an **irrational** number. The decimal approximation of an irrational number does not terminate and does not repeat. All irrational numbers are real numbers; they can all be graphed on the number line.

Example 9 Find $\sqrt{3}$.
Solution

$$\sqrt{1} = 1 \qquad \text{because } 1^2 = 1$$
$$\sqrt{4} = 2 \qquad \text{because } 2^2 = 4$$

Because 3 is between 1 and 4, $\sqrt{3}$ must be between 1 and 2, and therefore cannot be an integer. We can find the approximate value of $\sqrt{3}$ by referring to Table I, which gives the roots rounded off to three decimal places.

Locate 3 in the column headed *N*.
Read the value of $\sqrt{3}$ to the right of 3 in the column headed $\sqrt{N}$.
We see that $\sqrt{3} \doteq 1.732$*

N	$\sqrt{N}$
1	1.000
2	1.414
3	1.732
4	2.000
5	2.236

*The symbol "$\doteq$" (read "is approximately equal to") is used to show that two numbers are *approximately* equal to each other.

To find $\sqrt{3}$ by using a calculator, press the following keys in order: $\boxed{3}$ $\boxed{\sqrt{}}$. The display probably shows 1.7320508. (Your calculator may not show the same number of digits.) If we round off this answer to three decimal places, we have

$$\sqrt{3} \doteq 1.732 \quad \blacksquare$$

Example 10   Find $\sqrt{94}$.

Solution   To find $\sqrt{94}$ by using a calculator, press the following keys in this order: $\boxed{9}$ $\boxed{4}$ $\boxed{\sqrt{}}$. The display probably shows 9.6953597. When 9.6953597 is rounded off to three decimal places, we get 9.695. To use Table I, proceed as shown below.

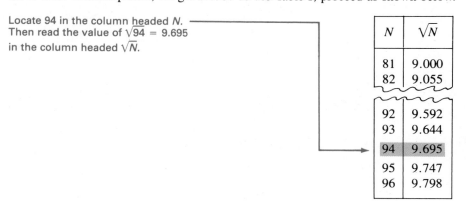

Locate 94 in the column headed *N*.
Then read the value of $\sqrt{94}$ = 9.695
in the column headed $\sqrt{N}$.

N	$\sqrt{N}$
81	9.000
82	9.055
92	9.592
93	9.644
94	9.695
95	9.747
96	9.798

$\blacksquare$

There are methods for calculating square roots. However, calculating square roots is rarely necessary because of the ready availability of tables, calculators, and computers.

EXERCISES 1.10C

Set I   In Exercises 1–8, find each square root by using a calculator and rounding off answers to three decimal places, or by using Table I, inside the back cover.

1. $\sqrt{13}$ **2.** $\sqrt{18}$ **3.** $\sqrt{37}$ **4.** $\sqrt{50}$

5. $\sqrt{79}$ **6.** $\sqrt{60}$ **7.** $\sqrt{86}$ **8.** $\sqrt{92}$

In Exercises 9–12, find each square root by using a calculator.

9. $\sqrt{466,489}$ **10.** $\sqrt{674,041}$ **11.** $\sqrt{272,484}$ **12.** $\sqrt{89,401}$

Set II   In Exercises 1–8, find each square root by using a calculator and rounding off answers to three decimal places, or by using Table I, inside the back cover.

1. $\sqrt{31}$ **2.** $\sqrt{69}$ **3.** $\sqrt{97}$ **4.** $\sqrt{184}$

5. $\sqrt{178}$ **6.** $\sqrt{145}$ **7.** $\sqrt{78}$ **8.** $\sqrt{125}$

In Exercises 9–12, find each square root by using a calculator.

9. $\sqrt{178,929}$ **10.** $\sqrt{373,321}$ **11.** $\sqrt{88,804}$ **12.** $\sqrt{35,344}$

1.10D   Higher Roots

Roots other than square roots are called **higher roots**. The parts of the symbol for higher roots are shown in Figure 1.10.2.

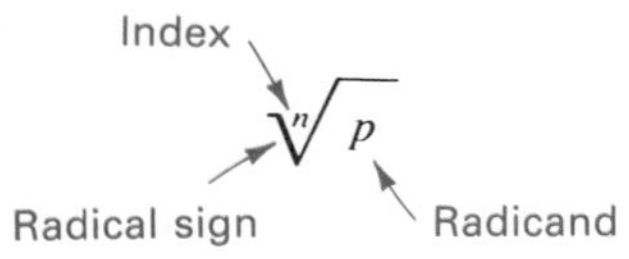

$$\overset{\text{Index}}{\searrow}\; \sqrt[n]{\underset{\nearrow}{\overset{\nwarrow}{p}}}$$

Index · Radical sign · Radicand

FIGURE 1.10.2

When there is no index written, the index is understood to be a 2, and the radical will then be a *square root*. Some examples of higher roots are $\sqrt[3]{8}$, $\sqrt[4]{55}$, and $\sqrt[5]{-32}$.

Principal Roots *Whenever the radical symbol is used, mathematicians agree that it is to stand for the principal root.*

When the index is an *even* number, we call the index an **even index**. When the index is an *odd* number, we call the index an **odd index**.

Principal higher roots are summarized as follows:

PRINCIPAL ROOTS

The symbol $\sqrt[n]{p}$ always represents the *principal* nth root of p. If the *radicand is positive*, the principal root is positive. If the *radicand is negative* and the *index* is

1. odd, the principal root is negative;

2. even, the principal root is *not a real number*.

Some Symbols Used to Indicate Roots The symbol $\sqrt[3]{p}$ indicates the *cubic root* of p. When the *index* of the radical is a 3, we must find a number whose *cube* is p. You will find such problems easier to do if you have memorized the cubes of the first few whole numbers (see page 43).

Example 11 Find the indicated roots.

a. $\sqrt[3]{8}$. We must find a number whose *cube* is 8. That is, we must solve $(?)^3 = 8$. *If* we have memorized that $2^3 = 8$, then we know that the answer is 2. If we *haven't* memorized that the cube of 2 is 8, we must use the "trial-and-error" method." Does $1^3 = 8$? No. Does $2^3 = 8$? Yes. Therefore, $\sqrt[3]{8} = 2$.

b. $\sqrt[3]{-8}$. We must find a number whose *cube* is -8. We know that the principal root will be negative because the index is odd and the radicand is negative. Because $(-2)^3 = -8$, $\sqrt[3]{-8} = -2$. ∎

The symbol $\sqrt[4]{p}$ indicates the *fourth root* of p. When the *index* is a 4, we must find a number whose *fourth power* is p.

Example 12 Find $\sqrt[4]{16}$. We must find a positive number whose *fourth* power is 16. Let's use the "trial-and-error method." Does $1^4 = 16$? No. Does $2^4 = 16$? Yes; $2^4 = 2 \cdot 2 \cdot 2 \cdot 2 = 16$. Therefore, $\sqrt[4]{16} = 2$. ∎

The symbol $\sqrt[5]{p}$ indicates the *fifth* root of p, and so forth.

Example 13 Find $\sqrt[5]{-1}$. We must find a *negative* number whose *fifth* power is -1. Does $(-1)^5 = -1$? Yes. Therefore, $\sqrt[5]{-1} = -1$. ∎

Example 14 Find $\sqrt[4]{-16}$. The radicand is negative and the index is even. Therefore, the answer is "Not a real number." (There is no *real* number whose fourth power is -16.) ∎

Example 15 Examples of roots preceded by a minus sign:

a. $-\sqrt{169} = -(13) = -13$

b. $-\sqrt[3]{8} = -(2) = -2$

c. $-\sqrt[3]{-8} = -(-2) = 2$ ∎

The cubic root of a number that is *not* the cube of some integer is an *irrational* number. Thus, $\sqrt[3]{35}$ and $\sqrt[3]{-17}$ are irrational numbers; their decimal approximations will not terminate and will not repeat.

Example 16 Determine which of the following numbers are rational numbers, which are irrational numbers, which are real numbers, and which are *not* real numbers: $\sqrt{5}$, $\sqrt[3]{17}$, $2.525252\ldots$, 0, $\frac{3}{7}$, -4, $-\sqrt[3]{8}$, $\sqrt{-9}$, and $2.828427125\ldots$.

Solution The rational numbers are $2.525252\ldots$ (the digits "25" repeat), 0, $\frac{3}{7}$, -4, and $-\sqrt[3]{8}$ ($-\sqrt[3]{8} = -2$, which is rational).

The irrational numbers are $\sqrt{5}$ (there is no whole number whose square is 5), $\sqrt[3]{17}$ (there is no integer whose cube is 17), and $2.828427125\ldots$ (there are no repeating digits).

The real numbers are $\sqrt{5}$, $\sqrt[3]{17}$, $2.525252\ldots$, 0, $\frac{3}{7}$, -4, $-\sqrt[3]{8}$, and $2.828427125\ldots$ (all rational numbers and all irrational numbers are real numbers).

The number that is *not* real is $\sqrt{-9}$ (there is no real number whose square is -9). ∎

All roots of positive numbers and zero and all *odd* roots of negative numbers are real numbers and therefore can be represented by points on the number line (see Figure 1.10.3).

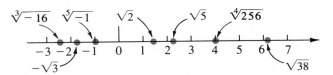

FIGURE 1.10.3

The relationships between the sets of numbers we've discussed above are shown in Figure 1.10.4.

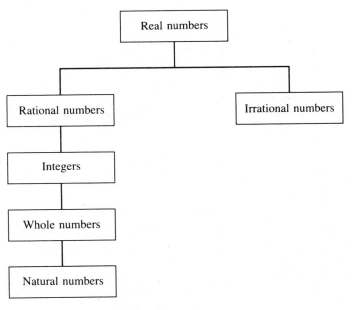

FIGURE 1.10.4

EXERCISES 1.10D

Set I   In Exercises 1–20, find each of the indicated roots by trial and error, or write "Not a real number."

1. $\sqrt[3]{64}$ **2.** $\sqrt[4]{81}$ **3.** $\sqrt[3]{27}$ **4.** $\sqrt[3]{125}$

5. $-\sqrt[3]{27}$ **6.** $-\sqrt{25}$ **7.** $-\sqrt[4]{1}$ **8.** $-\sqrt[5]{-32}$

9. $\sqrt[3]{-125}$ **10.** $\sqrt[3]{-8}$ **11.** $-\sqrt[4]{-16}$ **12.** $-\sqrt{-16}$

13. $\sqrt[3]{-1,000}$ **14.** $\sqrt[3]{-64}$ **15.** $\sqrt[7]{-1}$ **16.** $\sqrt[5]{-32}$

17. $\sqrt[6]{729}$ **18.** $-\sqrt[3]{-216}$ **19.** $\sqrt{-25}$ **20.** $\sqrt{-36}$

In Exercises 21 and 22, determine which of the numbers are rational numbers, which are irrational numbers, which are real numbers, and which are *not* real numbers.

21. $\sqrt[3]{13}, \dfrac{1}{2}, -12, \sqrt{-15}, 0.26262626\ldots, 0.196732468\ldots$

22. $0.67249713\ldots, \sqrt{-27}, 18, \dfrac{11}{32}, 0.37373737\ldots, \sqrt[3]{12}$

Set II   In Exercises 1–20, find each of the indicated roots by trial and error, or write "Not a real number."

1. $-\sqrt[3]{-8}$ **2.** $-\sqrt[3]{-64}$ **3.** $\sqrt[5]{-1}$ **4.** $\sqrt[4]{16}$

5. $\sqrt[3]{1,000}$ **6.** $\sqrt[5]{32}$ **7.** $\sqrt[3]{216}$ **8.** $\sqrt[6]{64}$

9. $\sqrt[4]{256}$ **10.** $\sqrt[5]{243}$ **11.** $\sqrt[4]{-81}$ **12.** $-\sqrt[4]{81}$

13. $\sqrt[5]{-100,000}$ **14.** $-\sqrt[3]{1}$ **15.** $\sqrt[7]{128}$ **16.** $\sqrt[4]{625}$

17. $\sqrt[3]{343}$ **18.** $\sqrt[5]{-243}$ **19.** $\sqrt[6]{-64}$ **20.** $-\sqrt[4]{10,000}$

In Exercises 21 and 22, determine which of the numbers are rational numbers, which are irrational numbers, which are real numbers, and which are *not* real numbers.

21. $-13, -\dfrac{51}{22}, 0.72727272\ldots, \sqrt{-16}, \sqrt[3]{-125}, 0.26925163\ldots$

22. $\sqrt[3]{3}, 1.29715698\ldots, 3.131313131\ldots, \dfrac{23}{52}, 45, \sqrt{-100}$

1.11 Factorization of Positive Integers

Recall that the numbers that are multiplied together to give a product are called the *factors* of that product. The tests for divisibility that follow are quite useful in finding the factors of a number.

Tests for Divisibility

We can use these tests to determine whether a number is divisible by 2, 3, or 5:

Divisibility by 2   A number is divisible by 2 if its last digit is 0, 2, 4, 6, or 8.

Divisibility by 3   A number is divisible by 3 if the sum of its digits is divisible by 3.

Divisibility by 5   A number is divisible by 5 if its last digit is 0 or 5.

While there are tests for divisibility by other numbers, we are not including them.

Example 1   This example shows the use of the tests of divisibility.

  a. 1 2 ; 30 0 ; 2,03 4 ; and 57 8 are divisible by 2, because the last digit of each number is a 0, 2, 4, 6, or 8.

  b. 210 is divisible by 3 because $2 + 1 + 0 = 3$, which is divisible by 3.

  c. 5,162 is not divisible by 3 because $5 + 1 + 6 + 2 = 14$, which is not divisible by 3.

  d. 25 0 and 75 5 are both divisible by 5, because the last digit of each is a 0 or a 5. ∎

Factoring a Positive Integer   We know that $2 \cdot 3 = 6$. There are two ways of looking at this fact. We can think of starting with the $2 \cdot 3$ and *finding the product* 6. Or we can start with the 6 and ask ourselves what *positive integers* multiplied together will give us 6, and then think of $2 \cdot 3$. When we do this we are **finding the factors** of 6, or more simply **factoring** 6.

Finding the product of 2 and 3:

$$2 \cdot 3 = 6 \qquad \text{Starting with } 2 \cdot 3 \text{ and finding the product 6}$$
Product

Factoring 6:

$$6 = 2 \cdot 3 \qquad \text{Starting with 6 and finding the factors 2 and 3}$$
Factors

It is also true, of course, that 1 and 6 are factors of 6.

The "Plus or Minus" Symbol   The symbol "$\pm$" is read "*plus or minus.*" Thus, "± 7" is read "plus or minus 7," and it represents either of the numbers $+7$ or -7.

Finding All the Integral Factors of a Number   We sometimes need to find all of the integers, both positive and negative, that are factors of a number. We do this by first finding all the positive integers that are factors of the number and then attaching a "$\pm$" sign to the left of each of these integers. A systematic method for finding *all* the integral factors of a positive integer will be demonstrated in Example 2.

Example 2   Find all the integral factors (or divisors) of each of the following numbers:

  a. 8

  Factoring
$$\begin{cases} 8 = 2 \cdot 4 \\ 8 = 1 \cdot 8 \end{cases}$$

  By "pairing up" the factors in this manner

$$\{1, \quad 2, \quad 4, \quad 8\}$$

  you can check to see if you have omitted any factors.
  The integral factors of 8 are ± 1, ± 2, ± 4, and ± 8.

  b. 12

$$12 = 1 \cdot 12 = 2 \cdot 6 = 3 \cdot 4$$

$$\{1, \quad 2, \quad 3, \quad 4, \quad 6, \quad 12\}$$

  Therefore, the integral factors of 12 are ± 1, ± 2, ± 3, ± 4, ± 6, and ± 12.

c. 36

$$36 = 1 \cdot 36 = 2 \cdot 18 = 3 \cdot 12 = 4 \cdot 9 = 6 \cdot 6$$

Because 36 is the square of an integer, when we pair up the factors, there will be a single, unpaired number left in the center:

Therefore, the integral factors of 36 are ± 1, ± 2, ± 3, ± 4, ± 6, ± 9, ± 12, ± 18, and ± 36. ∎

Prime and Composite Numbers

Prime Numbers A **prime number** is a natural number greater than 1 that cannot be written as a product of two natural numbers except as the product of itself and 1. That is, a prime number has no natural-number factors other than itself and 1.

A partial list of prime numbers is 2, 3, 5, 7, 11, 13, 17, 19, 23, 29 . . . . There is no largest prime number.

Composite Numbers A **composite number** is a natural number that does have natural-number factors other than itself and 1.

NOTE One (1) is neither prime nor composite. ☑

Example 3 Examples of prime and composite numbers:

a. 9 is a composite number because 1, 3, and 9 are factors of 9, so that 9 has a factor other than itself and 1.

b. 17 is a prime number because 1 and 17 are the only natural-number factors of 17.

c. 45 is a composite number; it has the factors 3, 5, 9, and 15 besides 1 and 45. ∎

Prime Factorization of Natural Numbers The **prime factorization** of a natural number greater than 1 is the indicated product of all of its factors that are themselves prime numbers. The prime factorization is unique except for the order in which the factors are written.

Example 4 Find the prime factorization of 18.

$$\left.\begin{array}{l} 18 = 2 \cdot 9 \\ 18 = 3 \cdot 6 \end{array}\right\} \quad \text{These are } not \text{ prime factorizations because } 9 \text{ and } 6 \text{ are not prime numbers}$$

$$\left.\begin{array}{l} 18 = 2 \cdot 9 = \boxed{2 \cdot 3 \cdot 3} = \boxed{2 \cdot 3^2} \\ 18 = 3 \cdot 6 = \boxed{3 \cdot 2 \cdot 3} = \boxed{2 \cdot 3^2} \end{array}\right\} \quad \text{These } are \text{ prime factorizations because all the factors are prime numbers}$$

Note that the two ways we factored 18 led to the *same* prime factorization $(2 \cdot 3^2)$. This is so because the prime factorization is unique. ∎

A systematic method for finding the prime factorization will be demonstrated in Examples 5 and 6.

Example 5 Find the prime factorization of each of the following numbers:

a. 24

We first try to divide 24 by the smallest prime, 2. This number does divide exactly into 24 and gives a quotient of 12. We again try 2 as a divisor of the quotient, 12. Two *does* divide evenly into 12 and gives a quotient of 6. We *again* try 2 as a divisor of the quotient, 6. Two *does* divide evenly into 6 and gives a quotient of 3, which is itself a prime number, and so the process ends.

The work can be conveniently arranged by placing the quotient *under* the number we're dividing into, as follows:

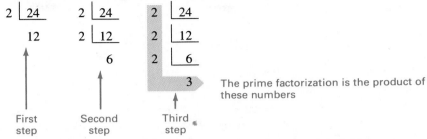

First step   Second step   Third step

The prime factorization is the product of these numbers

Therefore, $24 = 2 \cdot 2 \cdot 2 \cdot 3 = 2^3 \cdot 3$, where $2^3 \cdot 3$ is the prime factorization of 24.

b. 30

```
2 | 30
3 | 15
      5
```
Prime factorization of $30 = 2 \cdot 3 \cdot 5$

c. 20

```
2 | 20
2 | 10
      5
```
Prime factorization of $20 = 2 \cdot 2 \cdot 5 = 2^2 \cdot 5$

d. 36

```
2 | 36
2 | 18
3 | 9
      3
```
Prime factorization of $36 = 2 \cdot 2 \cdot 3 \cdot 3 = 2^2 \cdot 3^2$

e. 315

```
3 | 315
3 | 105
5 | 35
       7
```
Prime factorization of $315 = 3 \cdot 3 \cdot 5 \cdot 7 = 3^2 \cdot 5 \cdot 7$ ∎

When trying to find the prime factors of a number, we do not need to try any prime that has a square greater than that number (see Example 6).

Example 6   Find the prime factorization of 131.

Primes in order of size

2   does not divide 131
3   does not divide 131
5   does not divide 131
7   does not divide 131
11 does not divide 131
13 and larger primes need not be tried because $13^2 = 169$, which is greater than 131

Therefore, the prime factorization of 131 is simply 131, since 131 is a prime number. ∎

EXERCISES 1.11

Set I   In Exercises 1–16, find all the integral factors of each of the numbers.

1. 4	**2.** 9	**3.** 10	**4.** 14
5. 15	**6.** 16	**7.** 18	**8.** 20
9. 21	**10.** 22	**11.** 27	**12.** 28
13. 33	**14.** 34	**15.** 44	**16.** 45

In Exercises 17–32, state whether each of the numbers is prime or composite. To justify your answer, give the set of all positive integral factors for each number.

17. 5	**18.** 8	**19.** 13	**20.** 15
21. 12	**22.** 11	**23.** 21	**24.** 23
25. 55	**26.** 41	**27.** 49	**28.** 31
29. 51	**30.** 42	**31.** 111	**32.** 101

In Exercises 33–48, find the prime factorization of each number.

33. 14	**34.** 15	**35.** 21	**36.** 22
37. 26	**38.** 27	**39.** 29	**40.** 31
41. 32	**42.** 33	**43.** 34	**44.** 35
45. 84	**46.** 75	**47.** 144	**48.** 180

Set II   In Exercises 1–16, find all the integral factors of each of the numbers.

1. 6	**2.** 23	**3.** 24	**4.** 26
5. 30	**6.** 32	**7.** 38	**8.** 46
9. 81	**10.** 105	**11.** 49	**12.** 56
13. 63	**14.** 85	**15.** 111	**16.** 42

In Exercises 17–32, state whether each of the numbers is prime or composite. To justify your answer, give the set of all positive integral factors for each number.

17. 17	**18.** 14	**19.** 18	**20.** 19
21. 61	**22.** 63	**23.** 81	**24.** 73
25. 39	**26.** 83	**27.** 100	**28.** 29
29. 129	**30.** 105	**31.** 89	**32.** 97

In Exercises 33–48, find the prime factorization of each number.

33. 16	**34.** 18	**35.** 28	**36.** 30
37. 65	**38.** 78	**39.** 120	**40.** 112
41. 81	**42.** 49	**43.** 43	**44.** 72
45. 51	**46.** 38	**47.** 111	**48.** 88

1.12 Review: 1.8–1.11

Commutative and Associative Properties 1.8

Addition is commutative

$$a + b = b + a$$

Multiplication is commutative

$$a \cdot b = b \cdot a$$

Subtraction and division are not commutative.

Addition is associative

$$a + (b + c) = (a + b) + c$$

Multiplication is associative

$$a \cdot (b \cdot c) = (a \cdot b) \cdot c$$

Subtraction and division are not associative.

Powers of Signed Numbers 1.9

Exponent

$$3^4 = 3 \cdot 3 \cdot 3 \cdot 3 = 81$$

Base

Roots of Signed Numbers 1.10

$$\sqrt[3]{-8} = -2 \quad \text{because } (-2)^3 = (-2)(-2)(-2) = -8$$

Cubic root of -8

Inverse Operations 1.4, 1.6, and 1.10

Addition and subtraction are inverse operations.

Multiplication and division are inverse operations.

Like powers and roots are inverse operations.

Irrational Numbers 1.10

Numbers such as $\sqrt{7}$ (there is no whole number whose square is 7) and $\sqrt[3]{18}$ (there is no integer whose cube is 18) are irrational numbers. The *decimal form* of any irrational number is always a nonrepeating, nonterminating decimal. All irrational numbers are *real* numbers.

Prime Numbers 1.11

A *prime number* is a natural number greater than 1 that cannot be written as a product of two natural numbers except as the product of itself and 1. That is, a prime number has no natural-number factors other than itself and 1.

Composite Numbers 1.11

A *composite number* is a natural number that does have natural-number factors other than itself and 1.

One (1) is neither prime nor composite.

Prime Factorization of Natural Numbers 1.11

The *prime factorization* of a natural number greater than 1 is the indicated product of all of its factors that are themselves prime numbers. The prime factorization is unique, except for the order in which the factors are written.

Review Exercises 1.12 Set I

1. State whether each of the following is true or false; if the statement is true, give the reason.

   a. $[(-2) \cdot 3] \cdot 4 = (-2)(3 \cdot 4)$

   b. $5 + (-2) = (-2) + 5$

   c. $5 - (-2) = (-2) - 5$

   d. $a + (b + c) = (a + b) + c$

   e. $(c \cdot d) \cdot e = e \cdot (c \cdot d)$

   f. $5 + (x + 7) = (x + 7) + 5$

2. Find the additive inverse of -23.

3. What is the additive identity?

In Exercises 4–13, perform the indicated operations or write "Not a real number."

4. 1^4

5. 0^3

6. $(-4)^2$

7. -5^2

8. $\sqrt{121}$

9. $-\sqrt[4]{16}$

10. $\sqrt{-10}$

11. $\sqrt[3]{-8}$

12. $-\sqrt[5]{-32}$

13. 16^2

14. Determine which of the following numbers are rational numbers, which are irrational numbers, which are real numbers, and which are *not* real numbers:

$$-\sqrt{5}, \quad -31, \quad 5.3892531\ldots, \quad \frac{7}{12}, \quad 4.15151515\ldots, \quad \sqrt[3]{71}$$

15. Use a calculator or Table I to find the decimal approximation to $\sqrt{31}$. (Round off the answer to three decimal places.)

16. a. Find the prime factorization of 270

    b. What are the integral factors of 270?

Review Exercises 1.12 Set II

NAME _____

1. State whether each of the following is true or false. If the statement is true, give the reason.

    a. $(6) - (-2) = (-2) - (6)$

    b. $5 \cdot (2 \cdot 3) = (5 \cdot 2) \cdot 3$

    c. $(-7) + (4 + 3) = [(-7) + (4)] + 3$

    d. $(-6) + (-3) = (-3) + (-6)$

    e. $(-6) \div (-3) = (-3) \div (-6)$.

2. What is the multiplicative identity?

3. What is the additive inverse of $\frac{2}{3}$?

In Exercises 4–13, perform the indicated operations or write "Not a real number."

4. $\sqrt{81}$

5. 4^3

6. 1^3

7. 0^5

8. $\sqrt[3]{-125}$

9. -11^2

10. $(-2)^2$

11. $-\sqrt[3]{1,000}$

12. $\sqrt{-49}$

13. $-\sqrt[7]{-1}$

14. Determine which of the following numbers are rational numbers, which are irrational numbers, which are real numbers, and which are *not* real numbers:

$$\sqrt{-169}, \quad . \quad 0.60606060\ldots, \quad -53, \quad -\frac{5}{17}, \quad 4.35671937\ldots, \quad \sqrt[3]{21}$$

ANSWERS

1a. _____

b. _____

c. _____

d. _____

e. _____

2. _____

3. _____

4. _____

5. _____

6. _____

7. _____

8. _____

9. _____

10. _____

11. _____

12. _____

13. _____

14. **Rational numbers** _____

Irrational numbers _____

Real numbers _____

Numbers that are not real _____

15. Use a calculator or Table I to find the decimal approximation to $\sqrt{53}$. (Round off the answer to three decimal places.)

15. _____

16a. _____

b. _____

16. a. Find the prime factorization of 240.

    b. Determine the integral factors of 240.

Chapter 1 Diagnostic Test

The purpose of this test is to see how well you understand the operations with signed numbers. We recommend that you work this diagnostic test *before* your instructor tests you on this chapter. Allow yourself about 50 minutes to do this test.

Complete solutions for all the problems on this test, together with section references, are given in the Answers section. We suggest that you study the sections referred to for the problems you do incorrectly.

Write the correct symbol, $<$ or $>$, that should be used in Problems 1–3 to make the statement true.

1. $3 \underline{<} 5$ **2.** $0 \underline{>} -4$ **3.** $-5 \underline{<} -4$

4. Write the largest two-digit natural number. 99

5. Write the smallest one-digit integer. 0

6. Write all the digits less than 3. $2\ 1\ 0$

7. What is the additive identity? 0

8. At 5 A.M. the temperature in Denver, Colorado, was $-20°$ F. At noon the temperature was $38°$ F. What was the rise in temperature? $58°$

In Problems 9–11, state whether each statement is true or false. If the statement is true, give the reason.

9. $5 \times (-6 \times 4) = (5 \times [-6]) \times 4$ T

10. $18 \div (-6) = (-6) \div 18$ F

11. $a + b = b + a$ T

In Problems 12–44, perform the indicated operation. If the indicated operation cannot be performed, give a reason.

12. Subtract 24 from 5. -19 **13.** Subtract 8 from -17.

14. $4 + (-18)$ -14 **15.** $5 - 24$ -19 **16.** $3 \div 0$ Not possible

17. $(-10)(-5)$ 50 **18.** $(-10) - 5$ -15 **19.** -7^2 49

20. $(-7)^2$ 49 **21.** $15 + (-8)$ 7 **22.** $-6(0)$ 0

23. $\dfrac{15}{-21}$ $-\dfrac{5}{7}$ **24.** $(-4)^3$ -64 **25.** -4^3 -64

26. $(-24) \div (-4)$ 6 **27.** $\dfrac{-33}{11}$ -3 **28.** $42 \div (-14)$ -3

29. $\dfrac{0}{0}$ cannot be determ. **30.** $-8 - (-17)$ 9 **31.** $-13 + 5$ -8

32. $(-6)(-5)(-2)$ -60 **33.** $-12 + (-9)$ -21 **34.** 0^4 0

35. $-7 - 8$ -15 **36.** $0 \div (-12)$ 0 **37.** $6(-9)$ -54

38. $6 - 9$ -3 **39.** $-7(-8)$ 56 **40.** $-11 - (-4)$ -7

41. $-56 \div (-7)$ 8 **42.** $3 - (-12)$ 15 **43.** $0(-6)$ 0

44. $0 - 6$ -6

In Problems 45 and 46, find the absolute value.

45. $|-4|$ 4　　　　　　　　　　　**46.** $|6|$ 6

In Problems 47–50, find each of the indicated roots by trial and error.

47. $\sqrt{49}$ 7　　　　　　　　　　**48.** $\sqrt{81}$ 9

49. $\sqrt[3]{64}$ 4　　　　　　　　　　**50.** $\sqrt[3]{-8}$ -2

51. Determine which of the numbers $-\sqrt[3]{19}$, -63, $0.38383838\ldots$, $\frac{2}{23}$, $0.25375913\ldots$, and $\sqrt{-121}$ are rational numbers, which are irrational numbers, which are real numbers, and which are *not* real numbers.

52. a. Find the prime factorization of 180.

　　b. Write all the integral factors of 180.

2 Evaluation of Expressions Containing Numbers and Variables

One of the first places a student uses algebra is in the evaluation of formulas in courses taken in science, business, and so on. In this chapter we apply the operations with signed numbers discussed in Chapter 1 to evaluate algebraic expressions and formulas.

2.1 Order of Operations

In Chapter 1, we learned how to perform all six arithmetic operations with signed numbers: addition, subtraction, multiplication, division, raising to powers, and finding roots. However, each problem dealt with only one kind of operation.

Example 1   Problems involving only one kind of operation:

a. $2 + (-4) = -2$ Addition only

b. $(-3) \cdot [4 \cdot (-5)] = (-3)(-20) = 60$ Multiplication only

c. $(-7) - 4 = -11$ Subtraction only   ∎

In this chapter we will evaluate expressions in which more than one kind of operation is used. Consider the problem $5 + 4 \cdot 6$. If we do the addition first, we get one answer, and if we do the multiplication first, we get a different answer. (Try the problem both ways.) So that there will be only one correct answer for problems involving more than one operation, mathematicians have agreed upon the following order of operations:

ORDER OF OPERATIONS

1. If there are any parentheses in the expression, that part of the expression within a pair of parentheses is evaluated first.

2. The evaluation then proceeds in this order:

First: Powers and roots are done.

Second: Multiplication and division are done *in order from left to right.*

Third: Addition and subtraction are done *in order from left to right.*

If we follow these rules in solving the problem $5 + 4 \cdot 6$, we have

$$5 + 4 \cdot 6 = 5 + 24 = 29$$

Therefore, 29 is the only correct answer.

It is important to realize that an expression such as

$$8 - 6 - 4 + 7$$

must be evaluated by doing the addition and subtraction in order from left to right because subtraction is not commutative or associative. If the expression is considered as a sum—that is, as $(8) + (-6) + (-4) + (7)$—then the terms can be added in any order, because addition *is* commutative and associative.

$$\overbrace{\hspace{10cm}}^{\text{Both methods are correct}}$$

Evaluated left to right	*Added in any order*
$8 - 6 - 4 + 7$	$(8) + (-6) + (-4) + (7)$
$=\quad 2\quad -4\ +7$	$= (8) + (7) + (-6) + (-4)$
$=\qquad -2\quad +7$	$=\quad 15\quad +\quad (-10)$
$=\qquad\qquad 5$	$=\qquad\quad 5$

Example 2   Evaluate each of the following expressions, using the correct order of operations:

a. $7 + 3 \cdot 5$ Multiplication is done before addition

$= 7 + \quad 15$

$= 22$

b. $4^2 + \sqrt{25} - 6$ Powers and roots are done first

$= 16 + \quad 5 \ - 6$

$=\qquad 21\qquad - 6$

$= 15$

c. $16 \div 2 \ \cdot \ 4$ Here, division is done first because in reading from left to right the division comes first

$=\quad 8 \quad \cdot \ 4$

$= 32$

d. $(-8) \div 2 - (-4)$ Division is done before subtraction

$=\qquad -4\quad - (-4)$

$=\qquad -4\quad + \ 4$

$= 0$

e. $\sqrt[3]{-8}(-3)^2 - 2(-6)$ Roots and powers are done first

$= -2(9) - 2(-6)$ Multiplication is done before subtraction
$= -18 \ - (-12)$
$= -18 + \quad 12$
$= -6$

f. $12\sqrt{25} + 28 \div 4$ There is an understood multiplication sign between the 12 and the $\sqrt{25}$

$= 12 \cdot 5 + 28 \div 4$ Roots are done first

$=\quad 60 \ + \quad 7$ Multiplications and divisions are done next

$= 67$ Addition is done last

g. $5 \cdot 3^2$

$= 5 \cdot 9$ Powers are done first

$= 45$ Multiplication is done next ■

A WORD OF CAUTION   An exponent applies *only* to the immediately preceding symbol. That is, $5 \cdot 3^2 = 5 \cdot 3 \cdot 3$, *not* $15 \cdot 15$.   ☑

EXERCISES 2.1

Set I   Evaluate each expression. Be sure to perform the operations in the correct order.

1. $12 - 8 - 6$ **2.** $15 - 9 - 4$

3. $17 - 11 + 13 - 9$ **4.** $12 - 8 + 14 - 6$

5. $7 + 2 \cdot 4$ **6.** $10 + 3 \cdot 6$

7. $9 - 3 \cdot 2$ **8.** $14 - 8 \cdot 3$

9. $10 \div 2 \cdot 5$ **10.** $20 \cdot 15 \div 5$

11. $12 \div 6 \div 2$ **12.** $24 \div 12 \div 6$

13. $(-12) \div 2 \cdot (-3)$ **14.** $(-18) \div (-3) \cdot (-6)$

15. $8 \cdot 5^2$ **16.** $6 \cdot 2^4$

17. $(-485)^2 \cdot 0 \cdot (-5)^2$ **18.** $(-589)^2 \cdot 0 \cdot (-3)^2$

19. $12 \cdot 4 + 16 \div 8$ **20.** $4 \cdot 3 + 15 \div 5$

21. $28 \div 4 \cdot 2(6)$ **22.** $48 \div 16 \cdot 2(-8)$

23. $(-2)^2 + (-4)(5) - (-3)^2$ **24.** $(-5)^2 + (-2)(6) - (-4)^2$

25. $2 \cdot 3 + 3^2 - 4 \cdot 2$ **26.** $100 \div 5^2 \cdot 6 + 8 \cdot 75$

27. $(10^2)\sqrt{16} + 5(4) - 80$ **28.** $(5^2)\sqrt{9} + 4(6) - 60$

Set II   Evaluate each expression. Be sure to perform the operations in the correct order.

1. $18 - 10 - 5$ **2.** $13 - 9 + 16 - 8$

3. $2 + 5 \cdot 3$ **4.** $15 \div 3 \cdot 5$

5. $-18 \div (-\sqrt{81})$ **6.** $18 \div 6 \div 3$

7. $10 \cdot 15^2 - 4^3$ **8.** $(-10)^2 \cdot 10 + 0(-20)$

9. $(785)^3(0) + 1^5$ **10.** $(-5)^2 - (2)(-6) + (-2)^2$

11. $2 + 3(100) \div 25 - 10$ **12.** $7^2 + \sqrt{64} - 14$

13. $(-12) \div 6 - (-4)$ **14.** $(10^2)\sqrt{4} - 5 + 36$

15. $15 \cdot 4^2$ **16.** $8\sqrt{4} - 16 \div 4 \cdot 4$

17. $(863)^4 \cdot 0 \cdot (-23)^2$ **18.** $\sqrt{64} + \sqrt{36}$

19. $19 + 5 \cdot 10$ **20.** $\sqrt{64 + 36}$

21. $18 \div 9 \div 3$ **22.** $24 \div 12 \div 3$

23. $2 \cdot 3^2 + 4 \div 4$ **24.** $-5^2 - 6^2$

25. $63 - 12 - 38$ **26.** $35 - 5 \cdot 2$

27. $6\sqrt{9} - 18 \div 6 \cdot 3$ **28.** $8^2 + 12^2$

2.2 Grouping Symbols

Here are the grouping symbols that we have already used:

$$(\quad) \qquad \text{Parentheses}$$

$$[\quad] \qquad \text{Brackets}$$

$$\{ \quad \} \qquad \text{Braces}$$

$$\overline{\qquad} \qquad \text{Bar}$$

$$\text{Fraction line} \longrightarrow \frac{3(-5)+7}{9-4(-2)}$$

Bar in a square root $\longrightarrow$

$$\sqrt{7(+4)-8}$$

Grouping symbols are used to change the normal order of operations. Operations indicated within grouping symbols are carried out before operations outside the grouping symbols. *All grouping symbols have the same meaning:*

$$(8+4)-(9-7) = [8+4]-[9-7]$$
$$= \{8+4\}-\{9-7\} = \overline{8+4} - \overline{9-7}$$
$$= \quad 12 \quad - \quad 2$$
$$= 10$$

Different grouping symbols can be used in the same expression:

$$(8+4)-\{9-7\} = [8+4]-(9-7)$$
$$= \{8+4\}-[9-7]$$
$$= \quad 12 \quad - \quad 2$$
$$= 10$$

In this section we will make more extensive use of these symbols than was done in previous sections.

Example 1 demonstrates how grouping symbols change the normal order of operations.

Example 1   Evaluate each expression:

a. $5 + 4 \cdot 6 = 5 + 24 = 29$ ⎫ Grouping symbols provide a means for changing
b. $(5 + 4) \cdot 6 = 9 \cdot 6 = 54$ ⎭ the normal order of operations ∎

Example 2   Evaluate each of the following expressions:

a. $\quad 5 \cdot (-4) \div 2 \cdot (3 - 8)$ The expression within the parentheses is evaluated first

$= 5 \cdot (-4) \div 2 \cdot (-5)$ The multiplication on the left is done next

$= \quad -20 \div 2 \cdot (-5)$ The division on the left is done next

$= \quad\quad -10 \cdot (-5)$

$= 50$

b. $10 - [3 - (2 - 7)]$

  $= 10 - [3 - \ (-5)]$ When grouping symbols appear within other grouping
                              symbols, evaluate the inner grouping first
  $= 10 - [3 + \quad 5]$ $3 - (-5) = 3 + (+5) = 3 + 5$

  $= 10 - \quad [8]$

  $= 2$

c. $\dfrac{(-4) + (-2)}{8 - 5}$ ← This bar is a grouping symbol for both $\underline{(-4) + (-2)}$ and for $\overline{8 - 5}$.
                              Notice that the bar can be used either above or below the numbers
                              being grouped

  $= \dfrac{-6}{3}$

  $= -2$

d. $20 - 2\{5 - [3 - 5(6 - 2)]\}$

  $= 20 - 2\{5 - [3 - 5(4)]\}$

  $= 20 - 2\{5 - [3 - 20]\}$

  $= 20 - 2\{5 - [-17]\}$

  $= 20 - 2\{5 + 17\}$

  $= 20 - 2\{22\}$

  $= 20 - 44 = -24$

e. $27 \div (-3)^2 - 5\left\{6 - \dfrac{8 - 4}{5}\right\}$

  $= 27 \div (-3)^2 - 5\left\{\boxed{6} - \dfrac{4}{5}\right\}$

  $= 27 \div (-3)^2 - 5\left\{\dfrac{30}{5} - \dfrac{4}{5}\right\}$ Remember that $\boxed{6} = \dfrac{6 \cdot 5}{1 \cdot 5} = \dfrac{30}{5}$

  $= 27 \div (-3)^2 - 5\left\{\dfrac{26}{5}\right\}$

  $= 27 \div \quad 9 \quad - \dfrac{\cancel{5}^{1}}{1}\left\{\dfrac{26}{\cancel{5}_{1}}\right\}$

  $= \quad 3 \qquad - \quad 26 \quad = -23$

f. $(2.5)^2 \div (5.6 - 11.4)$

  $= (2.5)^2 \div (-5.8)$

  $= 6.25 \div (-5.8) = -1.077586207$

  $\doteq -1.078$ Rounded off to three decimal places

g. $\sqrt{13^2 - 12^2}$ ⌐ Since this bar is a grouping symbol, the expression under it is
                           evaluated first. Then the square root is taken

  $= \sqrt{169 - 144}$

  $= \sqrt{25} = 5$ ∎

A WORD OF CAUTION

$$\sqrt{13^2 - 12^2} \neq \sqrt{13^2} - \sqrt{12^2}$$

$$\sqrt{169 - 144} \neq 13 - 12$$

$$\sqrt{25} \neq 1$$

EXERCISES 2.2

Set I   Evaluate each of the following expressions.

1. $2 \cdot (-6) \div 3 \cdot (8 - 4)$ **2.** $5 \cdot (-4) \div 2 \cdot (9 - 4)$

3. $24 - [(-6) + 18]$ **4.** $17 - [(-9) + 15]$

5. $[12 - (-19)] - 16$ **6.** $[21 - (-14)] - 29$

7. $[11 - (5 + 8)] - 24$ **8.** $[16 - (7 + 12)] - 22$

9. $20 - [5 - (7 - 10)]$ **10.** $16 - [8 - (2 - 7)]$

11. $\dfrac{7 + (-12)}{8 - 3}$ **12.** $\dfrac{(-14) + (-2)}{9 - 5}$

13. $15 - \{4 - [2 - 3(6 - 4)]\}$

14. $17 - \{6 - [9 - 2(2 - 7)]\}$

15. $32 \div (-2)^3 - 5\left\{7 - \dfrac{6 - 2}{5}\right\}$

16. $36 \div (-3)^2 - 6\left\{4 - \dfrac{9 - 7}{3}\right\}$

17. $\sqrt{3^2 + 4^2}$ **18.** $\sqrt{13^2 - 5^2}$

19. $\sqrt{16.3^2 - 8.35^2}$ **20.** $\sqrt{23.9^2 + 38.6^2}$

21. $(1.5)^2 \div (-2.5) + \sqrt{35}$ **22.** $(-0.25)^2(-10)^3 + \sqrt{54}$

23. $18.91 - [64.3 - (8.6^2 + 14.2)]$

24. $[\sqrt{101.4} - (73.5 - 19.6^2)] \div 38.2$

Set II   Evaluate each of the following expressions.

1. $4 \cdot (11 - 17) \div 3 \cdot (-2)$ **2.** $[18 - (-15)] - 13$

3. $33 - [(-16) + 11]$ **4.** $[22 - (7 + 12)] - 14$

5. $26 - [8 - (6 - 15)]$ **6.** $\dfrac{(-2) + (-7)}{18 + (-15)}$

7. $23 - \{6 - [5 - 2(8 - 3)]\}$

8. $28 \div (-2)^2 - 6\left\{8 - \dfrac{9 - 4}{3}\right\}$

9. $\sqrt{10^2 - 6^2}$ **10.** $23 - (45 - 73)$

11. $\dfrac{3 + (-15)}{13 - 7}$ **12.** $\dfrac{(-8) - (-12)}{2 - 8}$

13. $37 - \{23 - [8 - 9] - 1\}$ **14.** $13 - \{11 - [1 - 19] - 36\}$

15. $48 \div (-4)^2 - 3\left\{9 - \dfrac{8 - 4}{2}\right\}$

16. $75 \div (-5)^2 - 5\left\{6 - \dfrac{9 - 1}{4}\right\}$

17. $\sqrt{8^2 + 6^2}$ **18.** $\sqrt{3^2 + 4^2}$

19. $\sqrt{2.18^2 + 41.6^2}$ **20.** $\sqrt{1.34^2 + 4.1^2}$

21. $(63.1)^2 \div (-3.8) + \sqrt{89}$ **22.** $\sqrt{34.5^2 - 17.8^2}$

23. $(1.7)^2 \div (-3.2) + \sqrt{43}$

24. $[\sqrt{126.3} - (89.7 - 46.5^2)] \div 52.6$

2.3 Finding the Value of Expressions Having Numbers and Variables

Basic Definitions

Variable   In algebra, we use letters to represent numbers, and we usually call these letters **variables**. A variable is an object or symbol that acts as a placeholder for a number that is unknown. The variable may assume different values in a particular problem or discussion.

Constant   A **constant** is an object or symbol that does *not* change its value in a particular problem or discussion. It is usually represented by a number symbol but it can be represented by one of the first few letters of the alphabet. Thus, in the expression $2x - 7y + 3$, the constants are 2, -7, and 3. In the expression $ax + by + c$, the constants are understood to be a, b, and c.

Algebraic Expression   An **algebraic expression** consists of numbers, variables (letters), signs of operation (such as $+$, $-$, etc.), and signs of grouping. (Not all of these need be present.)

Powers of Variables   In Section 1.9, we discussed bases, exponents, and powers of signed numbers:

$$\overset{\text{Exponent}}{3^4} = 3 \cdot 3 \cdot 3 \cdot 3 = 81$$

Base

The same definitions are carried over to expressions with variables.

$$\overset{\text{Exponent}}{x^4} = x \cdot x \cdot x \cdot x = x^4$$

Base   "Fourth power of x," also read "x to the fourth power"

Example 1   The following are examples of algebraic expressions:

a. $2x - 3y$

b. $5x^3 - 7x^2 + 4$

c. $\dfrac{5a^2 - 2b^3}{\sqrt{3ab}}$

d. $\sqrt{b^2 - 4ac}$

e. $(a - b)^2 + (c - d)^2$ ∎

Example 2 List (a) the constants and (b) the variables for

$$2x + 7y - 3z$$

a. The constants are 2, 7, and -3. (Notice that the negative sign in front of the 3 is considered to be part of the constant.)

b. The variables are x, y, and z. ∎

   In this section, we make use of our knowledge of signed numbers to help us find the value of algebraic expressions containing variables when values of those variables are given.

TO EVALUATE AN EXPRESSION HAVING VARIABLES AND NUMBERS

1. Replace each variable by its number value, usually enclosing the number in parentheses.

2. Carry out all arithmetic operations, using the correct order of operations (see Section 2.1).

Example 3 Find the value of $3x - 5y$ if $x = 10$ and $y = 4$.
Solution Recall from Section 1.5 that $3x$ means 3 times x, and $5y$ means 5 times y.

$$3x - 5y$$
$$= 3(10) - 5(4) \qquad \text{Replace each variable by its number value}$$
$$= 30 - 20 = 10 \qquad \text{Carry out the arithmetic operations} \quad ∎$$

A WORD OF CAUTION When replacing a letter by a number, enclose the number in parentheses to avoid the following common errors:
   Evaluate $3x$ when $x = -2$.

Correct	*Common error*
$3x = 3(-2) = -6$	$3x \neq 3 - 2 = 1$

Evaluate $4x^2$ when $x = -3$.

Correct	*Common errors*
$4x^2 = 4(-3)^2 = 4 \cdot 9 = 36$	$4x^2 \neq 4 - 3^2 = 4 - 9 = -5$
	or $\quad 4x^2 \neq 4 - 3^2 = 4 + 9 = 13$ ☑

Example 4 Evaluate $3x - 5y$ if $x = -4$ and $y = -6$.
Solution

$$3x - 5y$$
$$= 3(-4) - 5(-6)$$
$$= -12 + 30$$
$$= 18 \quad ∎$$

Example 5 Find the value of $\dfrac{2a - b}{10c}$ if $a = -1$, $b = 3$, and $c = -2$.

Solution

Remember this bar is a grouping symbol

$$\frac{2a - b}{10c} = \frac{2(-1) - (3)}{10(-2)} = \frac{-2 - 3}{-20} = \frac{-5}{-20} = \frac{1}{4} = 0.25 \quad \blacksquare$$

Example 6 Evaluate $\dfrac{5hgk}{2m}$ if $h = -2$, $g = 3$, $k = -4$, $m = 6$.

Solution

$$\frac{5hgk}{2m} = \frac{5(-2)(3)(-4)}{2(6)} = \frac{120}{12} = 10 \quad \blacksquare$$

Example 7 Find the value of $2a - [b - (3x - 4y)]$ if $a = -3$, $b = 4$, $x = -5$, and $y = 2$.

Solution

$$2a \quad - [b - \quad (3x \quad - 4y)]$$
$$= 2(-3) - [4 - \{3(-5) - 4(2)\}]$$
$$= 2(-3) - [4 - \{-15 - 8\}]$$
$$= 2(-3) - [4 - \{-23\}]$$
$$= 2(-3) - [4 + 23]$$
$$= \quad -6 \quad - \quad [27]$$
$$= \quad -6 \quad - \quad 27$$
$$= -33 \quad \blacksquare$$

Notice that $\{ \ \}$ were used in place of $(\ )$ to clarify the grouping

Example 8 Evaluate $b - \sqrt{b^2 - 4ac}$ when $a = 3$, $b = -7$, and $c = 2$.

Solution

$$b \quad - \sqrt{b^2 - 4ac} \quad \longleftarrow \text{This bar is a grouping symbol for } b^2 - 4ac$$
$$= (-7) - \sqrt{(-7)^2 - 4(3)(2)}$$
$$= (-7) - \sqrt{49 - 24}$$
$$= (-7) - \sqrt{25}$$
$$= (-7) - \quad 5$$
$$= -12 \quad \blacksquare$$

A WORD OF CAUTION In Example 8, *two* errors are made if you write

$$b - \sqrt{b^2 - 4ac} = -7 - \sqrt{-7^2 - 4(3)(2)} = -7 - \sqrt{49 - 24}$$

It is incorrect to omit the $(\ )$ around -7 here, and $-7^2 = -49$, not 49. ☑

Example 9 Evaluate when $x = 5$:

a. $7 + 3x$. When $x = 5$, $7 + 3x = 7 + 3(5) = 7 + 15 = 22$.

b. $10x$. When $x = 5$, $10x = 10(5) = 50$. ■

NOTE We can see from Example 9 that $7 + 3x \neq 10x$. ☑

EXERCISES 2.3

Set I   In Exercises 1–4, list (a) the different constants and (b) the different variables.

1. $2x + 4y + 2$ **2.** $7a + 3b + 7$

3. $7u - 8v + 2v$ **4.** $3x - 5y - 2x$

In Exercises 5–28, evaluate the expression when $a = 3$, $b = -5$, $c = -1$, $x = 4$, and $y = -7$.

5. $3b$ **6.** $15b$

7. $9 - 6b$ **8.** $8 + 7b$

9. b^2 **10.** $-y^2$

11. $2a - 3b$ **12.** $3x - 2y$

13. $x - y - 2b$ **14.** $a - b - 3y$

15. $3b - ab + xy$ **16.** $4c + ax - by$

17. $x^2 - y^2$ **18.** $b^2 - c^2$

19. $4 + a(x + y)$ **20.** $5 - b(a + c)$

21. $2(a - b) - 3c$ **22.** $3(a - x) + 4b$

23. $3x^2 - 10x + 5$ **24.** $2y^2 - 7y + 9$

25. $a^2 - 2ab + b^2$ **26.** $x^2 - 2xy + y^2$

27. $\dfrac{3x}{y + b}$ **28.** $\dfrac{4a}{c - b}$

In Exercises 29–38, find the value of the expression when $E = -1$, $F = 3$, $G = -5$, $H = -4$, and $K = 0$.

29. $\dfrac{E + F}{EF}$ **30.** $\dfrac{G + H}{GH}$

31. $\dfrac{(1 + G)^2 - 1}{H}$ **32.** $\dfrac{1 - (1 + E)^2}{F}$

33. $2E - [F - (3K - H)]$ **34.** $3H - [K - (4F - E)]$

35. $G - \sqrt{G^2 - 4EH}$ **36.** $H - \sqrt{H^2 - 4EK}$

37. $\dfrac{\sqrt{2H - 5G}}{0.2F^2}$ **38.** $\dfrac{\sqrt{5HG}}{0.5E^2}$

Set II   In Exercises 1–4, list (a) the constants and (b) the variables.

1. $4x - 7y + 4$ **2.** $8u - 5v + 3u$

3. $9s + 3t + 9u$ **4.** $-8x + 2y - 5z$

In Exercises 5–36, evaluate the expression when $a = -2$, $b = 4$, $c = -5$, $x = 3$, and $y = -1$.

5. $8y$ **6.** $-3a$

7. $11 - 3y$ **8.** $9 - 12a$

9. a^2 **10.** $-a^2$

11. $4c - 5y$ **12.** $a - b - 2c$

13. $6x - xy + ab$ **14.** $c^2 - x^2$

15. $7 - x(a + b)$ **16.** $3(x - y) - 4c$

17. $b^2 - 4ac$ **18.** $b^2 - 2bc + c^2$

19. $\dfrac{5b}{x - y}$ **20.** $\dfrac{a - b}{ab}$

21. $\dfrac{(1 - x)^2 - 1}{y}$ **22.** $3a - [b - (5x - y)]$

23. $x - \sqrt{x^2 - 4ay}$ **24.** $\dfrac{b - a}{ab}$

25. $a^2 + 2ab + b^2$ **26.** $(a + b)^2$

27. $a^2 + b^2$ **28.** $x^2 - y^2$

29. $(x - y)^2$ **30.** $x^2 - 2xy + y^2$

31. $\dfrac{a}{b} + \dfrac{c}{x}$ **32.** $\dfrac{a + c}{b + x}$

33. $\dfrac{a + c}{b}$ **34.** $\dfrac{a}{b} + \dfrac{c}{b}$

35. $\dfrac{x}{b} - \dfrac{a}{y}$ **36.** $\dfrac{x - a}{b - y}$

In Exercises 37 and 38, $a = -19.32$, $b = 25.73$, and $c = 47.02$. Evaluate each expression. Round off answers to four decimal places.

37. $\dfrac{\sqrt{3b - 4a}}{0.7c^2}$ **38.** $\dfrac{-b - \sqrt{b^2 - 4ac}}{2a}$

2.4 Evaluating Formulas

In algebra, an **equation** is a statement that two algebraic expressions are equal. The following is an example of an equation:

$$5x - 8 = 3x + 2$$

Left side — Right side — Equal sign

An equation is made up of three parts:

1. The equal sign ($=$).

2. The expression to the left of the equal sign, called the left side (or left member) of the equation.

3. The expression to the right side of the equal sign, called the right side (or right member) of the equation.

The variable x represents an unknown number. Letters other than x may be used in equations to represent unknown numbers.

A **formula** is often an equation that expresses a mathematical or scientific fact. One reason for studying algebra is to prepare us to use formulas. Students will encounter formulas in many courses they take, as well as in real-life situations. In the examples and exercises we have listed the subject areas where the formulas are used.

Formulas are evaluated in the same way any expression having numbers and letters is evaluated. The equal sign is a very important part of the formula, and it must not be dropped or ignored.

Example 1 Given the formula $A = \frac{1}{2}bh$, find A when $b = 17$ and $h = 12$.   (Geometry)
Solution

$$A = \frac{1}{2}bh$$

$$A = \frac{1}{2}(17)(12) \qquad \text{Substitute 17 for } b \text{ and 12 for } h$$

$$A = \frac{(17)(12)}{2}$$

$$A = 102$$

Therefore, $A = 102$ when $b = 17$ and $h = 12$. ∎

Example 2 Given the formula $A = P(1 + rt)$, find A when $P = 1{,}000$, $r = 0.08$, and $t = 1.5$.
(Business)

Solution

$$A = P(1 + rt)$$
$$A = 1{,}000[1 + (0.08)(1.5)] \quad \text{Notice that [ ] were used in place of ( )}$$
$$\text{to clarify the grouping}$$
$$A = 1{,}000[1 + 0.12]$$
$$A = 1{,}000[1.12]$$
$$A = 1{,}120$$

Therefore, $A = 1{,}120$ when $P = 1{,}000$, $r = 0.08$, and $t = 1.5$. ∎

Example 3 Given the formula $s = \frac{1}{2}gt^2$, find s when $g = 32$ and $t = 5\frac{1}{2}$.   (Physics)
Solution

$$s = \frac{1}{2}gt^2$$

$$s = \frac{1}{2}(32)\left(\frac{11}{2}\right)^2$$

$$s = \frac{1}{2}\left(\frac{32}{1}\right)\left(\frac{121}{4}\right)$$

$$s = 484$$

Therefore, $s = 484$ when $g = 32$ and $t = 5\frac{1}{2}$. ∎

Example 4   Given the formula $C = \frac{5}{9}(F - 32)$, find C when $F = -13$.   (Science)

Solution

$$C = \frac{5}{9}(F - 32)$$

$$C = \frac{5}{9}(-13 - 32)$$

$$C = \frac{5}{9}(-45)$$

$$C = -25$$

Therefore, $C = -25$ when $F = -13$.   ■

Example 5   Given the formula $T = \pi\sqrt{\dfrac{L}{g}}$, find T when $\pi \doteq 3.14$, $L = 96$, and $g = 32$.   (Physics)

Solution

$$T = \pi\sqrt{\frac{L}{g}}$$

$$T \doteq (3.14)\sqrt{\frac{96}{32}}$$

$$T \doteq (3.14)\sqrt{3}$$

$$T \doteq (3.14)(1.732)$$

$$T \doteq 5.44 \qquad \text{Rounded off to two decimal places}$$

Therefore, $T \doteq 5.44$ when $\pi \doteq 3.14$, $L = 96$, and $g = 32$.   ■

Example 6   Given the formula

$$S = \frac{a(1 - r^n)}{1 - r}$$

find S when $a = -4$, $r = \frac{1}{2}$, and $n = 3$.   (Mathematics)

Solution

$$S = \frac{a(1 - r^n)}{1 - r}$$

$$S = \frac{(-4)\left[1 - \left(\frac{1}{2}\right)^3\right]]}{1 - \frac{1}{2}} \qquad \left(\frac{1}{2}\right)^3 = \left(\frac{1}{2}\right)\left(\frac{1}{2}\right)\left(\frac{1}{2}\right) = \frac{1}{8}$$

$$S = \frac{(-4)\left[1 - \frac{1}{8}\right]}{1 - \frac{1}{2}}$$

$$S = \frac{(-4)\left[\frac{7}{8}\right]}{\frac{1}{2}}$$

$$S = -7$$

Therefore, $S = -7$ when $a = -4$, $r = \frac{1}{2}$, and $n = 3$.   ■

Formulas will be discussed further in Section 8.10.

EXERCISES 2.4

Set I Evaluate each formula, using the values of the variables given with the formula.

Geometry The area of a triangle is given by the formula

$$A = \frac{1}{2}bh$$

where A is the area, b is the base, and h is the height.

1. Find A when $b = 15$ and $h = 14$.

2. Find A when $b = 27$ and $h = 36$.

Electricity Ohm's law states that

$$I = \frac{E}{R}$$

where I is the current, E is the electromotive force, and R is the resistance.

3. Find I when $E = 110$ and $R = 22$.

4. Find I when $E = 220$ and $R = 33$.

Business The formula for simple interest is

$$I = prt$$

where I is the interest, p is the principal, r is the rate, and t is the time.

5. Find I when $p = 600$, $r = 0.09$, $t = 4.5$.

6. Find I when $p = 700$, $r = 0.08$, $t = 2.5$.

Chemistry The formula to change Celsius to Fahrenheit is

$$F = \frac{9}{5}C + 32$$

where F is degrees Fahrenheit and C is degrees Celsius.

7. Find F when $C = 25$.

8. Find F when $C = -25$.

Geometry The area of a circle is given by the formula

$$A = \pi R^2$$

where A is the area and R is the radius.

9. Find A when $\pi \doteq 3.14$ and $R = 10$.

10. Find A when $\pi \doteq 3.14$ and $R = 20$.

Physics The distance a free-falling object travels is given by the formula

$$s = \frac{1}{2}gt^2$$

where s is the distance, g is the force of gravity, and t is the time.

11. Find s when $g = 32$ and $t = 3$.

12. Find s when $g = 32$ and $t = 5$.

Business   The value of an item that is depreciated each year is given by the formula

$$V = C - Crt$$

where V is the present value, C is the original cost, r is the rate of depreciation, and t is the time.

13. Find V when $C = 500$, $r = 0.1$, $t = 2$.

14. Find V when $C = 1{,}000$, $r = 0.08$, $t = 5$.

Statistics   The standard deviation of a binomial distribution is given by the formula

$$\sigma^* = \sqrt{npq}$$

where σ is the standard deviation, n is the number of trials, p is the probability of success, and q is the probability of a failure.

15. Find σ when $n = 100$, $p = 0.9$, $q = 0.1$.

16. Find σ when $n = 100$, $p = 0.8$, $q = 0.2$.

Chemistry   The formula to change Fahrenheit to Celsius is

$$C = \frac{5}{9}(F - 32)$$

where C is degrees Celsius and F is degrees Fahrenheit.

17. Find C when $F = -4$.

18. Find C when $F = 5$.

Nursing   The formula to determine the dosage for a child is

$$C = \frac{a}{a + 12} \cdot A$$

where C is the child's dosage, a is the age of the child, and A is the adult dosage.

19. Find C when $a = 6$ and $A = 30$.

20. Find C when $a = 4$ and $A = 48$.

Geometry   The volume of a sphere is given by the formula

$$V = \frac{4}{3}\pi r^3$$

where V is the volume and r is the radius.

---

*"σ" is the Greek letter *sigma*; it is used in statistics to represent a quantity called the *standard deviation*.

21. Find V when $\pi \doteq 3.14$ and $r = 3$.

22. Find V when $\pi \doteq 3.14$ and $r = 6$.

Business The formula for compound interest is

$$A = P(1 + i)^n$$

where A is the compound amount, P is the principal, i is the interest rate per period, and n is the number of periods. Round off answers to two decimal places.

23. Find A when $P = 1,000$, $i = 0.06$, $n = 2$.

24. Find A when $P = 2,000$, $i = 0.08$, $n = 3$.

Set II Evaluate each formula, using the values of the variables given with the formula.

Formula		*Problem*
$A = \dfrac{1}{2}bh$	Geometry	**1.** $b = 24$, $h = 19$
		2. $b = 14$, $h = 18$
$I = \dfrac{E}{R}$	Electricity	**3.** $E = 110$, $R = 11$
		4. $E = 220$, $R = 20$
$I = prt$	Business	**5.** $p = 800$, $r = 0.06$, $t = 3.5$
		6. $p = 600$, $r = 0.05$, $t = 5.5$
$s = \dfrac{1}{2}gt^2$	Physics	**7.** $g = 32$, $t = 1\frac{1}{2}$
		8. $g = 32$, $t = 3\frac{1}{2}$
$A = P(1 + i)^n$	Business	**9.** $P = 500$, $i = 0.07$, $n = 2$
		10. $P = 1,000$, $i = 0.06$, $n = 3$
$A = \pi R^2$	Geometry	**11.** $\pi \doteq 3.14$, $R = 50$
		12. $\pi \doteq 3.14$, $R = 15$
$F = \dfrac{9}{5}C + 32$	Chemistry	**13.** $C = -15$
		14. $C = 35$
$A = p(1 + rt)$	Business	**15.** $p = 800$, $r = 0.08$, $t = 2.5$
		16. $p = 1,200$, $r = 0.05$, $t = 6.5$
$C = \dfrac{5}{9}(F - 32)$	Chemistry	**17.** $F = 14$
		18. $F = -22$
$V = \dfrac{4}{3}\pi R^3$	Geometry	**19.** $\pi \doteq 3.14$, $R = 15$
		20. $\pi \doteq 3.14$, $R = 12$
$C = \dfrac{a}{a + 12} \cdot A$	Nursing	**21.** $a = 8$, $A = 20$
		22. $a = 5$, $A = 34$
$V = C - Crt$	Business	**23.** $C = 2,000$, $r = 0.05$, $t = 10$
		24. $C = 1,500$, $r = 0.06$, $t = 8$

2.5 Review 2.1–2.4

Grouping Symbols
2.2

( )   Parentheses

[   ]   Brackets

{   }   Braces

——   Bar

Fraction line $\longrightarrow$ $\dfrac{3x - 2}{5 + 7y}$

Bar in a square root $\longrightarrow$

$\sqrt{5x - 12}$

Order of Operations
2.1

1. If there are any parentheses in the expression, that part of the expression within a pair of parentheses is evaluated first.

2. The evaluation then proceeds in this order:

   First: Powers and roots are done.

   Second: Multiplication and division are done in order from left to right.

   Third: Addition and subtraction are done in order from left to right.

To Evaluate an Expression or Formula
2.3

1. Replace each letter by its number value.

2. Carry out all arithmetic operations using the correct order of operations.

Review Exercises 2.5 Set I

In Exercises 1–10, evaluate each expression.

1. $26 - 14 + 8 - 11$

2. $18 - 22 - 15 + 6$

3. $11 - 7 \cdot 3$

4. $(11 - 7) \cdot 3$

5. $15 \div 5 \cdot 3$

6. $36 - [(-7) - 15]$

7. $6 - [8 - (3 - 4)]$

8. $(35)^2(0) + 18 \div (-6)$

9. $\dfrac{6 + (-14)}{3 - 7}$

10. $\sqrt{10^2 - 8^2}$

In Exercises 11–16, find the value of each expression when $x = -2$, $y = 3$, and $z = -4$.

11. $3x - y + z$

12. $7 - y(x + z)$

13. $5x^2 - 3x + 10$

14. $x^2 + 2xy - y^2$

15. $x - 2[y - x(y + z)]$

16. $\dfrac{(x - y)^2 + z^2}{x - 2z}$

In Exercises 17–22, evaluate each formula using the values of the letters given with the formula.

17. $C = \dfrac{a}{a + 12} \cdot A$ $a = 8, A = 35$

18. $I = Prt$ $P = 400, r = 0.10, t = 2.75$

19. $C = \dfrac{5}{9}(F - 32)$ $F = 15\dfrac{1}{2}$

20. $A = P(1 + i)^n$ $P = 500, n = 3, i = 0.1$

21. $\sigma = \sqrt{npq}$ $n = 100, p = 0.5, q = 0.5$

22. $A = p(1 + rt)$ $p = 500, r = 0.12, t = 10$

Review Exercises 2.5 Set II

NAME _____

In Exercises 1–10, evaluate each expression.

ANSWERS

1. $25 - 8 - 1$

16

2. $48 \div 8 \div 2$

3

3. $8 + 2 \cdot 6$

20

1. _____16_____

2. _____3_____

4. $18 \div 6 \cdot 3$

5. $\dfrac{16 + 8}{4 + 2}$

6

6. $\dfrac{16}{4} + \dfrac{8}{2}$

3. _____20_____

4. _____9_____

7. $5 \cdot 3^2$

9

8. $\sqrt{17^2 - 15^2}$

289 - 225

64

9. $8\sqrt{16} + 8$

4

5. _____4_____

6. _____8_____

7. _____45_____

10. $12 - [14 - (6 - 8)]$

8. _____(8)_____

9. _____32_____

In Exercises 11–16, find the value of each expression when $x = -3$, $y = 5$, and $z = -2$.

10. _____-4_____

11. $4x - y + 2z$

$-12 - 5 - 4$

-21

12. $11 - y(x - z)$

$-5 - 1$

11. _____-21_____

12. _____16_____

13. $3z^2 - 8z + 13$

$12 + 16 + 13$

41

14. $x^2 + 3xy - y^2$

$9 - 45 - 25$

$-36 - 25$

-61

13. _____41_____

14. _____-61_____

15. $z - 5[z(x - y) + 2x]$

-8

16. $\dfrac{x^2 - (y - z)^2}{2z - x}$

15. _____

16. _____

In Exercises 17–22, evaluate each formula, using the values of the letters given with the formula.

17. $A = \pi R^2$ $\qquad R = 30, \pi \doteq 3.14$

17. _____

18. $I = Prt$ $\qquad P = 500, r = 0.09, t = 1.75$

18. _____

19. _____

19. $C = \dfrac{5}{9}(F - 32)$ $\qquad F = 59$

20. $A = P(1 + i)^n$ $P = 2,000, n = 4, i = 0.15$

20. _____

21. _____

21. $A = \dfrac{1}{2}bh$ $b = 4\dfrac{1}{2}, h = 1\dfrac{1}{3}$

22. _____

22. $F = \dfrac{9}{5}C + 32$ $C = -10$

Chapter 2 Diagnostic Test

The purpose of this test is to see how well you understand the evaluation of expressions containing numbers and variables. We recommend that you work this diagnostic test *before* your instructor tests you on this chapter. Allow yourself about 50 minutes.

Complete solutions for all the problems on this test, together with section references, are given in the answer section at the end of the book. For the problems you do incorrectly, study the sections cited.

In Problems 1–12, evaluate the given expression.

1. $23 - 19 - 3 + 11$ $= 12$

2. $8 + 9 \cdot 7$

3. $54 \div 9 \cdot 6$ 36

4. $-5^2 + (-4)^2$ -9

5. $6 \cdot 4^2 - 4$ 92

6. $5\sqrt{36} - 6(-5)$ 60

7. $\dfrac{7 - 15}{-2 + 6}$ -2

8. $(2^3 - 8)(8^2 - 9^2)$ 0

9. $10 - [7 - (4 - 9)]$ -2

10. $\{-10 - [5 + (4 - 7)]\} - 3$ -15

11. $\sqrt{13^2 - 5^2}$ 12

12. $48 \div (-4)^2 - 3\left(10 - \dfrac{10}{-2}\right)$

In Problems 13–16, find the value of each expression when $a = -4$, $b = -7$, $c = -2$, $x = -6$, and $y = 5$.

13. $3c + by + cx$ 41

14. $4x - [a - (3c - b)]$ -19

15. $x^2 + 2xy + y^2$ 1

16. $(x + y)^2$ 1

In Problems 17–20, evaluate each formula, using the values of the variables given with the formula.

17. $C = \dfrac{5}{9}(F - 32)$ $F = 68$ 20

18. $A = \pi R^2$ $\pi \doteq 3.14, R = 3$ 28.26

19. $A = P(1 + rt)$ $P = 500, r = 0.10, t = 3.5$

20. $V = C - Crt$ $C = 800, r = 0.06, t = 10$

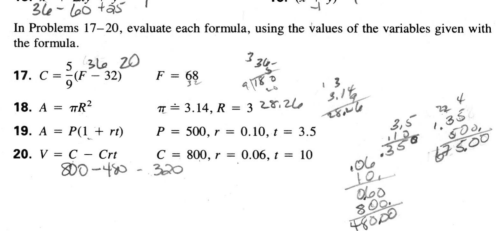

Cumulative Review Exercises
Chapters 1 and 2

In Exercises 1–27, perform the indicated operation. If the operation cannot be performed, give a reason.

1. $(-16)(-2)$ **2.** $(-13) + (-8)$ **3.** $(-5) - (-11)$

4. $(+28) \div (-7)$ **5.** $(-15)(0)(4)$ **6.** $(-2)^3$

7. $(-27) + (10)$ **8.** $\sqrt{16}$ **9.** $\dfrac{-48}{-12}$

10. $(0) \div (-6)$ **11.** $(25)^2$ **12.** $(-15)(3)$

13. -2^4 **14.** Subtract -9 from -13.

15. $\dfrac{-12}{0}$ **16.** $\sqrt[3]{8}$ **17.** $\sqrt{64}$

18. 0^4 **19.** $(8) - (17)$ **20.** $(-20) - (9)$

21. $24 \div 12 \cdot 2$ **22.** $6 \cdot 2^2$ **23.** $8 - 6 \cdot 5$

24. $64 \div 16 \div 4$ **25.** $17 - 9 - 2$ **26.** $4 + 17 \times 2$

27. $0 \div 5$

28. Evaluate the formula $C = \dfrac{5}{9}(F - 32)$ when $F = 21\dfrac{1}{2}$.

In Exercises 29–35, write "True" if the statement is always true. Otherwise, write "False."

29. 1.73 is a rational number.

30. 2/3 is a real number.

31. 1.86235173 . . . is an irrational number.

32. 0 is an irrational number.

33. 0 is a real number.

34. $-\dfrac{3}{5}$ is the additive inverse of $\dfrac{3}{5}$.

35. 0 is the multiplicative identity.

3 Exponents and Simplifying Algebraic Expressions

In this chapter, we discuss the rules of exponents that will be used throughout the remainder of the book. We also discuss the distributive property, combining like terms, simplifying algebraic expressions, and scientific notation.

3.1 Basic Definitions

Terms The plus and minus signs in an algebraic expression break the expression into smaller pieces called **terms.** Each plus and minus sign is part of the term that follows it. An expression within grouping symbols is considered as a single piece even though it may contain plus and minus signs. See Examples 1c, 1d, and 2.

Example 1 a. Consider the algebraic expression $2a + 5c$:

$$2a \quad + \quad 5c$$

First   Second
term   term

The plus sign separates the algebraic expression into two terms

b. In the algebraic expression $3x^2y - 5xy^3 + 7xy$, the minus and plus signs separate the algebraic expression into three terms.

The minus sign is part of the second term

$$3x^2y \quad + \quad -5xy^3 \quad + \quad 7xy$$

First   Second   Third
term   term   term

c. Consider the algebraic expression $3x^2 - 9x(2y + 5z)$:

$$3x^2 \quad + \quad -9x(2y + 5z)$$

First   Second
term   term

d. Consider the algebraic expression $\dfrac{2 - x}{xy} + 5(2x^2 - y)$:

$$\frac{2 - x}{xy} \quad + \quad 5(2x^2 - y)$$

First   Second
term   term ■

Example 2 Examples of the number of terms in an algebraic expression:

a. $3 + 2x - 1$ has three terms.

b. $3 + (2x - 1)$ has two terms.

c. $(3 + 2x - 1)$ has one term. ■

Recall from Section 1.11 that numbers that are multiplied together to give a product are called the *factors* of that product.

Example 3 Examples of factors:

a. $(2)(x) = 2x$

         — Factors of $2x$

b. $(7)(a)(b)(c) = \boxed{7abc}$

         — Product of factors

         — Factors of $7abc$ ■

Coefficients In a term with two factors, the **coefficient** of one factor is the other factor. In a term with more than two factors, the coefficient of each factor is the product of all the other factors in that term.

Example 4 Examples of coefficients:

a. $(3)(5)$

         — 5 is the coefficient of 3

         — 3 is the coefficient of 5

b. $\dfrac{3}{4}y$

         — y is the coefficient of $\dfrac{3}{4}$

         — $\dfrac{3}{4}$ is the coefficient of y

c. $3xy = 3(xy) = (3x)y = (3y)x$

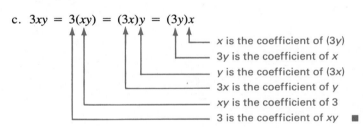

         — x is the coefficient of $(3y)$

         — $3y$ is the coefficient of x

         — y is the coefficient of $(3x)$

         — $3x$ is the coefficient of y

         — xy is the coefficient of 3

         — 3 is the coefficient of xy ■

Numerical Coefficients A **numerical coefficient** is a coefficient that is a number. When we refer to "the coefficient" of a term, it is understood that we mean the *numerical coefficient* of that term.

Example 5 Examples of numerical coefficients:

Term	Numerical coefficient
a. $6w$	6
b. $-12xy^2$	-12
c. $\dfrac{3xy}{4} = \dfrac{3}{4}xy$	$\dfrac{3}{4}$
d. xy	1

         — Even though no number is written, the numerical coefficient can be considered to be 1 because $xy = 1(xy)$; also see Example 5e

e. $-a^2$ -1

         — -1 is the numerical coefficient of a^2

f. $\dfrac{c}{5} = \dfrac{1}{5}c$ $\dfrac{1}{5}$ ■

EXERCISES 3.1

Set I In Exercises 1 and 2, determine whether x is a factor or a term of the expression.

1. a. $3xyz$ b. $3 + x$ c. $4x(y + 2)$ d. $5x$

2. a. $x + 7$ b. $7x$ c. $7xy$ d. $x(2 + y)$

In Exercises 3 and 4, state whether $7x$ is a factor or a term.

3. a. $7x + 3$ b. $y + 7x$ c. $7x(3 + y)$ d. $7x + (3 + y)$

4. a. $7x(y + 6)$ b. $7xy$ c. $1 + 7x$ d. $(y + 6) + 7x$

In Exercises 5–14, (a) determine the number of terms and (b) write the second term if there is one.

5. $7xy$ **6.** $5ab$

7. $E - 5F - 3$ **8.** $R - 2T - 6$

9. $3x^2y + \dfrac{2x + y}{3xy} + 4(3x^2 - y)$ **10.** $5xy^2 + \dfrac{5x - y}{7xy} + 3(x^2 - 4y)$

11. $5u^2 - 6u(2u + v^2)$ **12.** $3E^3 - 2E(8E + F^2)$

13. $[(x + y) - (x - y)]$ **14.** $\{x - [y - (x - y)]\}$

In Exercises 15–22, write the numerical coefficient of the first term.

15. $3x + 7y$ **16.** $4R + 3T$

17. $x^2 - 3xy$ **18.** $x^2 + 5xy$

19. $-b + 4a$ **20.** $-y^2 - x^2$

21. $\dfrac{4xy}{5} + \dfrac{2a}{3}$ **22.** $\dfrac{r}{2} - \dfrac{s}{3} + \dfrac{t}{4}$

Set II In Exercises 1 and 2, determine whether x is a factor or a term of the expression.

1. a. xy b. $x + y$ c. $x + 4y$ d. $x(y - 3)$

2. a. $5y + x$ b. $8x$ c. $x + 8$ d. $9x(y - 1)$

In Exercises 3 and 4, state whether $5x$ is a factor or a term.

3. a. $7 + 5x$ b. $5xy$ c. $5x(5 + y)$ d. $5x + (3 + y)$

4. a. $5x - 2$ b. $5x(y - 1)$ c. $5x + y - 1$ d. $7 + 5x$

In Exercises 5–14, (a) determine the number of terms and (b) write the second term if there is one.

5. $-2x$ **6.** $8(-2x)$

7. $3 - 4y + 2z$ **8.** $3 - (4x + 2z)$

9. $5 - xy$ **10.** $5xy$

11. $3x^3 + 2(y - 3z)$ **12.** $8x^2 - 3(x + y) - z$

13. $[a + (b + c)]$ **14.** $x + y + z$

In Exercises 15–22, write the numerical coefficient of the first term.

15. $18x + 3y$

16. $-18(x + 3y)$

17. $a - b + c$

18. $-x - y$

19. $-x^2 + 5xy$

20. $-y^2 + 3(x + y)$

21. $-2a^2 - 5ab + b^3$

22. $\dfrac{a}{6} + \dfrac{b}{4} - \dfrac{2c}{3}$

3.2 First Basic Rule of Exponents

Consider the product $x^3 \cdot x^2$.

$$x^3 \cdot x^2 = (xxx)(xx) = xxxxx = x^5$$

$$
\begin{aligned}
&3\,\text{factors} \\
+\; &2\,\text{factors} \\
\hline
&5\,\text{factors}
\end{aligned}
$$

Therefore, $x^3 \cdot x^2 = x^{3+2} = x^5$. The general rule for multiplying exponential numbers when the bases are the same is given in Rule 3.1.

RULE 3.1

$$x^a \cdot x^b = x^{a+b}$$

Rule 3.1 can be extended to include more factors:

$$x^a x^b x^c \cdots = x^{a+b+c+\cdots}$$

NOTE   When multiplying powers of the same base, add the exponents but keep the same base. Do not multiply the bases. ☑

Example 1   Rewrite each of the following with the base appearing only once (if possible):

a. $x^5 \cdot x^2 = x^{5+2} = x^7$

b. $x \cdot x^2 = x^{1+2} = x^3$

    $\llcorner x = x^1$

When no exponent is written, the exponent is understood to be 1.

c. $x^3 \cdot x^7 \cdot x^4 = x^{3+7+4} = x^{14}$

d. $10^7 \cdot 10^5 = 10^{7+5} = 10^{12} = \underbrace{1,000,000,000,000}_{12\,\text{zeros}}$

e. $2 \cdot 2^3 \cdot 2^2 = 2^{1+3+2} = 2^6 = 64$

f. $a^4 \cdot a \cdot a^5 = a^{4+1+5} = a^{10}$

g. $a^x \cdot a^y = a^{x+y}$

h. $x^3 \cdot y^2$

Notice that Rule 3.1 does not apply because the bases are different.

i. $x^2 + x^3$

The expression cannot be rewritten with the base appearing only once. Rule 3.1 does not apply because the operation is addition rather than multiplication. ∎

EXERCISES 3.2

Set I Rewrite each of the following with the base appearing only once (if possible).

1. $x^3 \cdot x^4$ 2. $x^2 \cdot x^9$ 3. $y \cdot y^3$

4. $z \cdot z^4$ 5. $m^2 \cdot m$ 6. $a^3 \cdot a$

7. $10^2 \cdot 10^3$ 8. $10^4 \cdot 10^3$ 9. $2 \cdot 2^3 \cdot 2^2$

10. $3 \cdot 3^2 \cdot 3^3$ 11. $x \cdot x^3 \cdot x^4$ 12. $y \cdot y^5 \cdot y^3$

13. $x^2 y^5$ 14. $a^3 b^2$ 15. $3^2 \cdot 5^3$

16. $2^3 \cdot 3^2$ 17. $a^4 + a^2$ 18. $x^3 + x^4$

19. $a^x \cdot a^w$ 20. $x^a \cdot x^b$ 21. $x^y \cdot y^x$

22. $a^b \cdot b^a$ 23. $x^2 y^3 x^5$ 24. $z^3 z^4 w^2$

25. $a^2 b^3 a^5$ 26. $x^8 y x^4$ 27. $s^7 + s^4$

28. $t^3 + t^8$

Set II Rewrite each of the following with the base appearing only once (if possible).

1. $x^2 \cdot x^5$ 2. $y \cdot y^4$ 3. $u^2 \cdot u$

4. $10^3 \cdot 10^4$ 5. $3 \cdot 3^2 \cdot 3^2$ 6. $a \cdot a^3 \cdot a^2$

7. $x^3 y^3$ 8. $x^a \cdot y^b$ 9. $x^2 \cdot x^y \cdot x^3$

10. $3^2 \cdot 4^3$ 11. $a^4 \cdot a^2$ 12. $xy^2 x^3$

13. $x^6 \cdot x^9$ 14. $y^7 \cdot y$ 15. $8^2 \cdot 3^3$

16. $a^4 \cdot b^8$ 17. $y^2 + y^{14}$ 18. $y^2 \cdot y^{14}$

19. $b^x \cdot b^y$ 20. $b^x \cdot b$ 21. $s^x \cdot t^y$

22. $s^2 + s^5$ 23. $a^4 \cdot b^3 \cdot c^2$ 24. $x + x^4$

25. $x \cdot x^7$ 26. $xy^3 x^4$ 27. $x^4 + x^2$

28. $a^5 + a$

3.3 Simplifying Products of Factors

In working with algebraic expressions, you must be able to simplify a product of factors. To remove parentheses in an expression in which a *single term* is inside each set of grouping symbols, use the following rules:

> ### TO SIMPLIFY A PRODUCT OF FACTORS
> ### WHEN EACH FACTOR HAS ONLY ONE TERM
>
> **1.** *Write the sign.* The sign is negative if there is an odd number of negative factors. The sign is positive if there are no negative factors or if there is an even number of negative factors.
>
> **2.** *Write the numerical coefficient.* (Its sign was found in step 1.) The coefficient is the product of the absolute values of the numerical coefficients inside each set of parentheses.
>
> **3.** *Write the variables.* Each variable should appear only once, and the variables are usually written in alphabetical order. Use Rule 3.1 to determine the exponent on each variable.

Example 1   Simplify the following products:

    a. $(6x)(-2x^3)$

$$= -(6 \cdot 2)(xx^3)$$

$$= -(12)(x^4) = -12x^4$$

             Rule 3.1
             12 is the product of the absolute values of the numbers in the factors
             The sign is negative because there is an odd number of negative factors

    b. $(-2a^2b)(5a^3b^3)$

$$= -(2 \cdot 5)(a^2a^3)(bb^3)$$

$$= -(10)(a^5)(b^4) = -10a^5b^4$$

             Rule 3.1
             10 is the product of the absolute values of the numbers in the factors
             The sign is negative because there is an odd number of negative factors

    c. $(-5y^2)(2y)(-4y^3)$

$$= +(5 \cdot 2 \cdot 4)(y^2yy^3)$$

$$= +(40)(y^6) = 40y^6$$

             Positive, because there is an even number of negative factors

    d. $(-4xy^2)(-3x^2y^3)(-2x^3y)$

$$= -(4 \cdot 3 \cdot 2)(xx^2x^3)(y^2y^3y)$$

$$= -(24)(x^6)(y^6) = -24x^6y^6$$

             Negative, because there is an odd number of negative factors

    e. $(3xy^2)^3$

$$= (3xy^2)(3xy^2)(3xy^2)$$

$$= (3 \cdot 3 \cdot 3)(xxx)((y^2y^2y^2)$$

$$= 27x^3y^6 \quad \blacksquare$$

In Example 1, we made use of the associative and commutative properties of multiplication.

EXERCISES 3.3

Set I   Simplify the following products of factors.

1. $(-2a)(4a^2)$ 2. $(-3x)(5x^3)$

3. $(-5h^2)(-6h^3)$ 4. $(-6k^3)(-8k)$

5. $(-5x^3)^2$ 6. $(-7y^4)^2$

7. $(-2a^3)(-4a)(3a^4)$ 8. $(-6b^2)(-2b)(-4b^3)$

9. $(-9m)(m^5)(-2m^2)$ 10. $(4n^4)(-7n^2)(-n)$

11. $(5x^2)(-7y)$ 12. $(-3x^3)(4y)$

13. $(-6m^3n^2)(-4mn^2)$ 14. $(-8h^4k)(5h^2k^3)$

15. $(2x^{10}y^2)(-3x^{12}y^7)$ 16. $(-2a^2b^5)(-5a^{10}b^{10})$

17. $(3xy^2)^2$ 18. $(4x^2y^3)^2$

19. $(5x^4y^5z)(-y^4z^7)$ 20. $(-7E^2F^5G^8)(3F^6G^{10})$

21. $(2^3RS^2)(-2^2R^5T^4)$ 22. $(-3^2xy)(3^3x^8z^5)$

23. $(-c^2d)(5d^1e^3)(-4c^5e^2)$ 24. $(3m^1n^2)(-m^3r^4)(-8n^5r)$

25. $(-2x^2)(-3xy^2z)(-7yz)$ 26. $(14y^2)(-5x^3z^4)(-6xyz)$

27. $(3xy)(x^2y^2)(-5x^3y^3)$ 28. $(xy)(-2xz)(3yz)$

29. $(-2ab)(5bc)(ac)$ 30. $(3xyz)(-2x^2y)(-yz^2)$

31. $(-5x^2y)(2yz^3)(-xz)$ 32. $(2a^2b^3)(-3ab^2)(-a^2b^2)$

33. $(-5x^2y^2z)(x^5y^3z)(7xyz^5)$ 34. $(-4R^2S^3T^4)(-8RS^3T^4)(-R^6S^1T^5)$

35. $(-3h^2k)(7)(-m^3k^5)(-mh^4)$ 36. $(-km^2)(-6n^3)(8)(-k^2m)$

Set II   Simplify the following products of factors.

1. $(-3x^2)(8x^3)$ 2. $(3a^2)(-2a^4)$

3. $(-7x^3)(-12x^4)$ 4. $(-5a^2b)(3ab^2)$

5. $(4x^2)^3$ 6. $(3x^2)^2$

7. $(-4b^2)(-2b)(-6b^5)$ 8. $(-3t^4)(5t^3)(-10t)$

9. $(-4y^2)(y^7)(-3y)$ 10. $(6x^3)(-2x)(7x^4)$

11. $(15x^3)(-3y^2)$ 12. $(-12a^4)(-5b)$

13. $(-4c^2d)(-13cd^3)$ 14. $(-6e^5f^3)(11e^2f^3)$

15. $(2mn^2)(-4m^2n^2)(2m^3n)$ 16. $(5x^2y)(-2xy^2)(-3xy)$

17. $(2m^5n)(-4mn^3)(3m^2n^6)$ 18. $(7ab^4)(2a^2b)(-5a^2b^3)$

19. $(8a^4)(-a^4b^3c)$ 20. $(3x^7)(-y^3)(4z^2)$

21. $(3x^3)(-3^4x^5)$ 22. $(-5^2a^3)(5ab^2)$

23. $(5x^2y)(-3xy^2z)(-2xz^3)$ 24. $(-8st^2)(-s^3t)(2t^4)$

25. $(-3ab^3)(-4a^2b^4)(c^5)$ 26. $(9x^2y^3)(-2x^4y)(-5xy^3)$

27. $(8ab)(-2ab^4)(-2^4ab^3)$ 28. $(-3u^3v^4)(-2^2u^2)(-3^2v^5)$

29. $(-7xy^2)(-3x^2z)(-2xz^3)$ 30. $(8ac)(-9b^2c^3)(abc)$

31. $(3s^4)(-5st^3)(-s^2t^4)$ 32. $(-2xy^2z)^2$

33. $(2H^2K^4M)(-2^2H^4K^5N)(2^3M^6K^7N)$

34. $(-10x^3y^2z)(-10^2xk^{10}z^1)(-10ky)$

35. $(-2e^1f^2)(-3e^3h)(-6fh)(-5e^2h)$

36. $(-4x^2y)(-7y^2z^2)(-5xy^4z)(-2x^3z^5)$

3.4 The Distributive Property

In Section 3.3, we learned how to find products of factors when each factor consists of only one term. If one or more of the factors contains more than one term, we must use the *distributive property*.

RULE 3.2   MULTIPLICATION IS DISTRIBUTIVE OVER ADDITION

If a, b, and c represent any real numbers, then

$$a(b + c) = ab + ac$$

The distributive property is one of the fundamental properties of real numbers that we accept as true without proof. However, examination of the following pairs of arithmetic problems will verify that the distributive property is valid:

$$5(3 + 7) \qquad\qquad 5 \cdot 3 \ + \ 5 \cdot 7$$
$$= 5(10) \qquad\qquad = \ 15 \ + \ 35$$
$$= 50 \qquad\qquad\qquad = \qquad 50$$

Because $50 = 50$, it must be true that $5(3 + 7) = 5 \cdot 3 + 5 \cdot 7$.

$$8(3 + [-10]) \qquad\qquad 8 \cdot 3 + 8 \cdot (-10)$$
$$= 8(-7) \qquad\qquad = \ 24 \ - \ 80$$
$$= -56 \qquad\qquad\qquad = \qquad -56$$

Because $-56 = -56$, it must be true that $8(3 + [-10]) = 8 \cdot 3 + 8 \cdot (-10)$.

When you remove parentheses by using the distributive property, be sure to multiply *each term* inside the parentheses by the other factor; then combine the products, if possible.

Example 1   Use the distributive property to remove the parentheses; then combine the products:

$$8(3 + 5) = \bigcirc\!\!\!8\,(\,3\ +\ 5\,)$$

            First   Second   Each term inside the parentheses is multiplied by
         product   product   the factor outside the parentheses; then these
                            products are added

$$= \quad 24 \ + \ 40 \ = 64$$

Note that $8(3 + 5) = 8(8) = 64$. ■

The distributive property applies to subtraction as well as to addition. That is, $a(b - c) = ab - ac$.

Example 2 Use the distributive property to remove the parentheses; then combine the products:

a. $2(3 - 7) = 2(3) - 2(7)$

$= 6 - 14$

$= -8$

Alternate method:

$2(3 - 7) = 2(3 + [-7])$

$= 2(3) + (2)(-7)$

$= 6 + (-14)$

$= -8$

Note that $2(3 - 7) = 2(-4) = -8$.

b. $-5(2 - 7) = -5(2) - (-5)(7)$

$= -10 + 35$

$= 25$

Alternate method:

$-5(2 - 7) = -5(2 + [-7])$

$= -5(2) + (-5)(-7)$

$= -10 + 35$

$= 25$

Note that $-5(2 - 7) = -5(-5) = 25$. ∎

The distributive property is used to remove parentheses in algebra as well as in arithmetic (see Example 3).

Example 3 Use the distributive property to remove the parentheses (do not try to combine the products):

a. $2(x + y) = (2)(x) + (2)(y)$

$= 2x + 2y$

b. $a(x - y) = (a)(x) - (a)(y)$

$= ax - ay$

c. $x(x^2 + y) = (x)(x^2) + (x)(y)$

$= x^3 + xy$

d. $3x(4x + x^3y) = (3x)(4x) + (3x)(x^3y)$

$= 12x^2 + 3x^4y$ ∎

Extensions of the Distributive Property

The distributive property can be extended to include any number of terms inside the parentheses. That is,

$$a(b + c - d + e + \cdots) = ab + ac - ad + ae + \cdots$$

Example 4 Use the distributive property to remove the parentheses (do not try to combine the products).

$-5a(4a^3 - 2a^2b + b^2) = (-5a)(4a^3) - (-5a)(2a^2b) + (-5a)(b^2)$

$= -20a^4 + 10a^3b - 5ab^2$

Alternate method:

$-5a(4a^3 - 2a^2b + b^2) = -5a(4a^3 + [-2a^2b] + b^2)$

$= (-5a)(4a^3) + (-5a)(-2a^2b) + (-5a)(b^2)$

$= -20a^4 + 10a^3b - 5ab^2$ ∎

The commutative property of multiplication assures us that

$$a(b + c) = (b + c)a$$

It follows that $(b + c)a = a(b + c) = ab + ac = ba + ca$

Therefore, $(b + c)a = ba + ca$

This can be extended to include more terms as follows:

$$(b + c + d + \cdots)a = ba + ca + da + \cdots$$

In Example 5 we show how to use the distributive property when the factor appears on the right.

Example 5 Use the distributive property to remove the parentheses (do not try to combine the products):

a. $(2x - 5)(-4x) = (2x)(-4x) - (5)(-4x)$
$$= -8x^2 + 20x$$

Alternate method:

$(2x - 5)(-4x) = (2x + [-5])(-4x)$
$$= (2x)(-4x) + (-5)(-4x)$$
$$= -8x^2 + 20x$$

b. $(-2x^2 + xy - 5y^2)(-3xy) = (-2x^2)(-3xy) + (xy)(-3xy) - (5y^2)(-3xy)$
$$= 6x^3y - 3x^2y^2 + 15xy^3$$

Alternate method:

$(-2x^2 + xy - 5y^2)(-3xy) = (-2x^2 + xy + [-5y^2])(-3xy)$
$$= (-2x^2)(-3xy) + (xy)(-3xy) + (-5y^2)(-3xy)$$
$$= 6x^3y - 3x^2y^2 + 15xy^3 \qquad \blacksquare$$

A WORD OF CAUTION A common mistake students make is to think that the distributive property applies to expressions like $2(3 \cdot 4)$.

The distributive property applies only when this is an addition

$$2(3 \cdot 4) \neq (2 \cdot 3)(2 \cdot 4)$$
$$2(12) \neq 6 \cdot 8$$
$$24 \neq 48 \qquad \checkmark$$

EXERCISES 3.4

Set I Use the distributive property to remove the parentheses. Do not try to combine the products.

1. $5(a + 6)$ **2.** $4(x + 10)$

3. $7(x + y)$ **4.** $5(m + n)$

5. $3(m - 4)$

6. $3(a - 5)$

7. $4(x - y)$

8. $9(m - n)$

9. $a(6 + x)$

10. $b(7 + y)$

11. $-2(x - 3)$

12. $-3(x - 5)$

13. $-3(2x^2 - 4x + 5)$

14. $-5(3x^2 - 2x - 7)$

15. $4x(3x^2 - 6)$

16. $3x(5x^2 - 10)$

17. $-2x(5x^2 + 3x - 4)$

18. $-4x(2x^2 - 5x + 3)$

19. $(x - 4)(6)$

20. $(3 - 2x)(-5)$

21. $(y^2 - 4y + 3)(7)$

22. $(-9 + z - 2z^2)(-7)$

23. $(2x^2 - 3x + 5)(4x)$

24. $(3w^2 + 2w - 8)(5w)$

25. $x(xy - 3)$

26. $a(ab - 4)$

27. $3a(ab - 2a^2)$

28. $4x(3x - 2y^2)$

29. $(-2x + 4x^2y)(-3y)$

30. $(-3a + 2a^2z)(-2z)$

31. $-2xy(x^2y - y^2x - y - 5)$

32. $-3ab(8 - a^2 - b^2 + ab)$

33. $-3(x - 2y + 2)$

34. $-2(x - 3y + 4)$

35. $(3x^3 - 2x^2y + y^3)(-2xy)$

36. $(4z^3 - z^2y - y^3)(-2yz)$

37. $(2xy^2z - 7x^2z^2)(-5xz^3)$

38. $(3a^2bc^2 - 4ab^3c)(-3ac^2)$

39. $(5x^2y^3z - 2xz^3 + y^4)(-4xz^2)$

40. $(-2xy^2z^2)(6x^2y - 3yz - 4xz^2)$

Set II   Use the distributive property to remove the parentheses. Do not try to combine the products.

1. $3(x + 4)$

2. $2(m - 5)$

3. $x(4 + y)$

4. $-3(x - 4)$

5. $(M + N)(3)$

6. $(x - y)(-4)$

7. $8(a - b)$

8. $(5 - x)(-2x)$

9. $x(y + 2)$

10. $(7 + a)(3b)$

11. $-6(x + 2y)$

12. $(x + 2y)(-6)$

13. $-5(3x^2 + 4x - 7)$

14. $-2(5x^4 - 3x^3 + 4x)$

15. $6x(5x^2 - 4)$

16. $8x^2(7x^3 + 2x^2 - 3)$

17. $-3x(4x^2 - 6x + 8)$

18. $(9 - 5x)(-2)$

19. $(3z^2 - 5z + 4)(-6)$

20. $(2u - 5u^2 + 11)(6)$

21. $x(xy - 6)$

22. $4a(ab - b^2)$

23. $4x(3xy^2 - 2x^2y)$

24. $(6s^2t - 5st)(-3t)$

25. $a(2ab - b)$

26. $-3x^2(-5x^2 + 2x - 1)$

27. $5x(-2xy + y)$

28. $-2xy^2(3xy^2 + 5xy - y)$

29. $(-3x + 2y)(-3x)$

30. $(5x + 7y - z)(-3y)$

31. $-4xy(3x + 5xy - y^2 + 2y)$

32. $-x^2(2xy - 3xy^2 + x^2y + y)$

33. $-8(2a + 3b - 1)$
34. $4x^3(x^3 - 2x^2 + 3x - 1)$
35. $(4x^3 - x^2 + 1)(-4x^2)$
36. $(-3x^2 + x - 1)(4x)$
37. $(x^2 - 4xy + y^2)(-2x)$
38. $3x^2y(x^3 - 3x^2y + 3xy^2 - y^3)$
39. $(-5a^2bc)(2b^2c - 5ac^3 - 3a^2b^3)$
40. $(2a^3bc^2 - 3ac + b^3)(-5ab^2)$

3.5 Removing Grouping Symbols

An algebraic expression is not simplified unless all grouping symbols have been removed. The following rules, which result from applications of the distributive property, can be used to remove grouping symbols that contain more than one term.

REMOVING GROUPING SYMBOLS THAT CONTAIN MORE THAN ONE TERM

1. If a set of grouping symbols containing more than one term is preceded by or followed by a factor, use the distributive property.

2. If a set of grouping symbols containing more than one term is not preceded by and not followed by a factor and is:
   a. preceded by a plus or minus sign, insert a 1 between the sign and the grouping symbol and then use the distributive property.
   b. preceded by no sign at all, drop the grouping symbols.

3. If grouping symbols occur *within* other grouping symbols, remove the innermost grouping symbols first.

Example 1   Remove the grouping symbols. Do not combine the terms.

$$(3x - 5) + 2y$$

The parentheses are not preceded by and not followed by a factor, and they are preceded by no sign at all. Applying part 2b of the above rules, we have

$$(3x - 5) + 2y = 3x - 5 + 2y \quad \blacksquare$$

There is one shortcut we can safely take in using part 2a of the rule. If the grouping symbols are preceded by a $+$ sign and if the sign of the first term inside the grouping symbols is an *understood* $+$ sign, then we can simply drop the parentheses.

Example 2   Remove the grouping symbols (do not combine the terms).

$$3x + (4y + 7)$$

The parentheses are preceded by a plus sign and are not followed by a factor; the sign of the $4y$ is an understood plus sign. Applying the shortcut, we have

$$3x + (4y + 7) = 3x + 4y + 7 \quad \blacksquare$$

Example 3   Remove the grouping symbols (do not combine the terms).

$$-(8 - 6x)$$

The parentheses are neither preceded by nor followed by a factor; they are preceded by a minus sign. Applying part 2a, we have

$$-(8 - 6x) = -1(8 - 6x) = (-1)(8) + (-1)(-6x) = -8 + 6x$$

           └─ Multiplying a number by 1 does not change its value   ∎

Example 4   Remove the grouping symbols (do not combine the terms).

$$-2x(4 - 5x)$$

The parentheses are preceded by a factor. Using the distributive property, we have

$$-2x(4 - 5x) = (-2x)(4) + (-2x)(-5x) = -8x + 10x^2 \quad ∎$$

Example 5   Remove the grouping symbols (do not combine the terms).

a. $(3x - 5) - y$

The parentheses are neither preceded by nor followed by a factor. (The $-y$ is being added.) Applying part 2b, we have

$$(3x - 5) - y = 3x - 5 - y$$

b. $(3x - 5)(-y)$

The parentheses are followed by a factor. Using the distributive property, we have

$$(3x - 5)(-y) = (3x)(-y) + (-5)(-y) = -3xy + 5y$$

NOTE   Notice that in Example 5a, $-y$ is being *added* to $(3x - 5)$, whereas in Example 5b, $-y$ is being *multiplied* by $(3x - 5)$.   ☑

c. $2x - (8 - 6z)$

Applying part 2a, we have

$$2x - (8 - 6z) = 2x - 1(8 - 6z)$$
$$= 2x + (-1)(8) + (-1)(-6z)$$
$$= 2x - 8 + 6z \quad ∎$$

When grouping symbols occur within other grouping symbols, it is usually easier to remove the innermost grouping symbols first (see Example 6).

Example 6   Remove the grouping symbols (do not combine the terms):

a. $x - [y + (a - b)]$

$$x - [y + (a - b)] = x - [y + a - b] \quad \text{Applying the shortcut}$$
$$= x - 1[y + a - b] \quad \text{Applied part 2a}$$
$$= x + (-1)(y) + (-1)(a) + (-1)(-b) \quad \text{Using the distributive property}$$
$$= x - y - a + b$$

b. $\quad 3 + 2[a - 5(x - 4y)]$

$= 3 + 2[a - 5x + 20y]$ Removed ( ) using the distributive property

$= 3 + 2a - 10x + 40y$ Removed [ ] using the distributive property

c. $\quad (3a - b) - 2\{x - [(y - 2) - z]\}$

$= (3a - b) - 2\{x - [y - 2 - z]\}$ Removed ( ) preceded by no sign

$= (3a - b) - 2\{x - 1[y - 2 - z]\}$ Applied part 2a

$= (3a - b) - 2\{x - y + 2 + z\}$ Removed [ ] using the distributive property

$= 3a - b - 2x + 2y - 4 - 2z$ Removed ( ) using the shortcut; removed { } using the distributive property ■

EXERCISES 3.5

Set I   Remove the grouping symbols. Do not try to combine the terms.

1. $8 + (a - b)$ **2.** $7 + (m - n)$

3. $5 - (x - y)$ **4.** $6 - (a - b)$

5. $12 - 3(m - n)$ **6.** $14 - 5(x - y)$

7. $(R - S) - 8$ **8.** $(x - y) - 2$

9. $(R - S)(-8)$ **10.** $(x - y)(-2)$

11. $(3x - 2y)(-5z)$ **12.** $(5a - 7b)(-2c)$

13. $(3x - 2y) - 5z$ **14.** $(5a - 7b) - 2c$

15. $10 - 2(a - b)$ **16.** $12 - 5(x - 2y)$

17. $2(x - y) + 3$ **18.** $4(a - 2b) + 5$

19. $2a - 3(x - y)$ **20.** $5x - 2(a - b)$

21. $a - [x - (b - c)]$ **22.** $a - [x - (-b + c)]$

23. $4 - 2[a - 3(x - y)]$ **24.** $8 - 3[x - 2(3a - b)]$

25. $3(a - 2x) - 2(y - 3b)$ **26.** $2(2x - b) - 3(y - 5a)$

27. $-10[-2(x - 3y) + a] - b$ **28.** $-5[-3(2x - y) + a] - c$

29. $(a - b) - \{[x - (3 - y)] - R\}$ **30.** $(x - y) - \{[a - (c - 5)] - b\}$

Set II   Remove the grouping symbols. Do not try to combine the terms.

1. $5 + (x + y)$ **2.** $-3 + (a - b)$

3. $3 - (x - y)$ **4.** $-6 - (s + t)$

5. $12 - 2(a + b)$ **6.** $8 - 2(x - y)$

7. $(a - b) - c$ **8.** $(s + 3t)(-2u)$

9. $(a - b)(-c)$ **10.** $(x - y)(-2)$

11. $(8u - 5v)(-7w)$ **12.** $(x - y) - 2$

13. $(-4x - 3y) - 9z$ **14.** $(-4x - 3y)(-9z)$

15. $2x - 3(a - b)$ **16.** $5x - (y + z)$

17. $4(x - 2y) - 3a$

18. $a - (b - c)$

19. $-(x - y) + (2 - a)$

20. $(3 + x) - 3(a - b)$

21. $(a + b)(2) - 6$

22. $2 - 6(a + b)$

23. $x - [a + (y - b)]$

24. $y - [x - (a - b)]$

25. $x - [-(y - b) + a]$

26. $[8 + 2(a - b) - x] - y$

27. $P - \{x - [y - (4 - z)]\}$

28. $P - \{x - [y - (z - 4)]\}$

29. $-\{-[-(-2 - x) + y]\} - a$

30. $-(-\{-[x - 3] + y\} - z)$

3.6 Combining Like Terms and Simplifying Algebraic Expressions

Like Terms   Terms that have equal literal parts are called **like terms;** also, all constants are like terms.

Example 1   Examples of like terms:

a. $3x$, $4x$, x, $\frac{1}{2}x$, $0.7x$ are like terms. They are called "x-terms."

b. $2x^2$, $10x^2$, $\frac{3}{4}x^2$, $2.3x^2$ are like terms. They are called "x^2-terms."

c. $5xy$, $2xy$, xy, $\frac{2}{3}yx$, $5.6xy$ are like terms ($yx = xy$). They are called "xy-terms."

d. $4x^2y$, $8xyx$, x^2y, $\frac{1}{5}x^2y$, $2.8yx^2$ are like terms ($xyx = x^2y$, and $yx^2 = x^2y$). They are called "x^2y-terms."

e. 5, -3, $\frac{1}{4}$, 2.6 are like terms. They are called "constant terms." ■

Unlike Terms   Terms that do not have equal literal parts are called **unlike terms.**

Example 2   Examples of unlike terms:

a. $2x$, $3y$ are unlike terms. The variables are different.

b. $5x^2$, $7x$ are unlike terms. The variable x has different exponents.

c. $4x^2y$, $10xy^2$ are unlike terms. ■

Combining Like Terms

$$3 \text{ dollars} + 5 \text{ dollars} = (3 + 5)\text{dollars} = 8 \text{ dollars}$$
$$3 \text{ cars} \quad + 5 \text{ cars} \quad = (3 + 5)\text{cars} \quad = 8 \text{ cars}$$
$$\underline{3x \qquad\quad + 5x \qquad = (3 + 5)x \qquad = 8x}$$

This is an application of the distributive property

When we combine like terms, we usually change the grouping and the order in which the terms appear. The commutative and associative properties of addition guarantee that when we do this the sum remains unchanged.

TO COMBINE LIKE TERMS

1. Identify the like terms.

2. Find the sum of each group of like terms by adding their numerical coefficients; then multiply that sum by the common variables. When no coefficient is written, it is understood to be 1.

Example 3 Examples of combining like terms:

a. $2x + 4x = (2 + 4)x = 6x$

b. $8y - 3y = (8 - 3)y = 5y$

c. $e - 9e = (1 - 9)e = -8e$

⬆ ⬆ — When no coefficient is written, it is understood to be 1

d. $5x^2y - 7x^2y = (5 - 7)x^2y = -2x^2y$

e. $4ab - 3ab + 6ab = (4 - 3 + 6)ab = 7ab$

f. $9x - 3x + 5y = (9 - 3)x + 5y = 6x + 5y$

⬆ — These are the only like terms in the expression; unlike terms cannot be combined

g. $12a - 7b - 9a + 4b$

$= (12a - 9a) + (-7b + 4b)$ Only like terms can be combined

$= (12 - 9)a + (-7 + 4)b$

$= \quad 3a \quad + \quad (-3b) \quad = 3a - 3b$

NOTE It is not necessary to show the step in which the distributive property is used. That is, in an expression such as $3x - 7x + 10x$, we can simply add the numerical coefficients, and then multiply that sum by the variable x. That is, we can add as follows:

— Add these coefficients

$$3\,x + (-7)\,x + 10\,x = 6\,x$$

— Multiply that sum by x ☑

h. $7x - 2y + 9 - 11x + 3 - 4y$

$= (7x - 11x) + (-2y - 4y) + (9 + 3)$

$= \quad -4x \quad + \quad (-6y) \quad + \quad 12 \quad = -4x - 6y + 12$

i. $3x^2 - 5x + 5 - 2x^2 + 7x + 11$

$= (3x^2 - 2x^2) + (-5x + 7x) + (5 + 11)$

$= \quad x^2 \quad + \quad 2x \quad + \quad 16$ ■

NOTE While it is possible to combine (add or subtract) only like terms, it is possible to multiply unlike terms together; for example, $5x^3 + 5x$ cannot be combined (written as a single term), but we can multiply $5x^3$ and $5x$ as follows: $(5x^3)(5x) = 25x^4$. ☑

Simplifying Algebraic Expressions

To simplify an algebraic expression, we remove all grouping symbols, simplify each term (each variable can appear only once in a term), and combine all like terms (see Example 4).

Example 4   Examples of simplifying algebraic expressions:

a. $x + 4(y + 3z + x)$

$= x + 4(y) + 4(3z) + 4(x)$ Using the distributive property

$= x + 4y + 12z + 4x$ Simplifying each term

Like terms

$= x + 4x + 4y + 12z$ Collecting like terms

$= 5x + 4y + 12z$ Combining like terms

b. $7b + 2b(3a + 1 - 5b)$

$= 7b + 2b(3a) + 2b(1) + 2b(-5b)$ Using the distributive property

$= 7b + 6ab + 2b - 10b^2$ Simplifying each term

Like terms

$= 7b + 2b + 6ab - 10b^2$ Collecting like terms

$= 9b + 6ab - 10b^2$ Combining like terms

c. $5x^2(3x^2 - 2x + 1) - 3x(7x^2 + x - 6)$

$= 5x^2(3x^2) + 5x^2(-2x) + 5x^2(1) - 3x(7x^2) - 3x(x) - 3x(-6)$ Using the distributive property

Like terms

$= 15x^4 - 10x^3 + 5x^2 - 21x^3 - 3x^2 + 18x$ Simplifying each term

Like terms

$= 15x^4 - 10x^3 - 21x^3 + 5x^2 - 3x^2 + 18x$ Collecting like terms

$= 15x^4 - 31x^3 + 2x^2 + 18x$ Combining like terms ∎

A WORD OF CAUTION   In Example 4b, it is incorrect to say

$$7b + 2b(3a + 1 - 5b) = 9b(3a + 1 - 5b)$$

Remember: Multiplication must be done *before* addition. ☑

We will discuss simplifying algebraic expressions further in Section 6.3.

EXERCISES 3.6

Set I   In Exercises 1–34, combine like terms.

1. $15x - 3x$

2. $12a - 9a$

3. $5a - 12a$

4. $10x - 24x$

5. $2a - 5a + 6a$

6. $3y - 4y + 5y$

7. $5x - 8x + x$

8. $3a - 5a + a$

9. $3x + 2y - 3x$

10. $4a - 2b + 2b$

11. $4y + y - 10y$

12. $3x - x - 5x$

13. $3mn - 5mn + 2mn$

14. $5cd - 8cd + 3cd$

15. $2xy - 5yx + xy$

16. $8mn - 7nm + 3nm$

17. $8x^2y - 2x^2y$

18. $10ab^2 - 3ab^2$

19. $a^2b - 3a^2b$

20. $x^2y^2 - 5x^2y^2$

21. $5ab + 2c - 2ba$

22. $7xy - 3z - 4yx$

23. $5xyz^2 - 2xyz^2 - 4xyz^2$

24. $7a^2bc - 4a^2bc - a^2bc$

25. $5u - 2u + 10v$

26. $8w - 4w + 5v$

27. $8x - 2y - 4x$

28. $9x - 8y + 2x$

29. $7x^2y - 2xy^2 - 4x^2y$

30. $4xy^2 - 5x^2y - 2xy^2$

31. $5x^2 - 3x + 7 - 2x^2 + 8x - 9$

32. $7y^2 + 4y - 6 - 9y^2 - 2y + 7$

33. $12.67 \text{ sec} + 9.08 \text{ sec} - 6.73 \text{ sec}$

34. $158.7 \text{ ft} + 609.5 \text{ ft} - 421.8 \text{ ft} - 263.4 \text{ ft}$

In Exercises 35–44, simplify each algebraic expression.

35. $a + 5(b + 3c + a)$

36. $x + 9(y + 7z + x)$

37. $7x + 4x(3y - 5 + 8x)$

38. $9y + 8y(4z + 7 - 9y)$

39. $4x - (5 - 2y + 8x)$

40. $8c - (2a - 1 + 3b + 12c)$

41. $6x^2 + 5 - (3x + 4 - x^2)$

42. $9a^2 + 7 - (12a^2 + 10 - 5a)$

43. $8x^2(7x^2 - 3x + 1) - 5x(6x^2 - 8x + 9)$

44. $7a^3(6a^2 + 9a - 1) - 5a^2(9a^2 - 8a + 3)$

Set II   In Exercises 1–34, combine like terms.

1. $12a - 4a$

2. $5x - 10x$

3. $18x - 25x$

4. $-4y - 12y$

5. $9z + 14z - 28z$

6. $8x - x + 3x$

7. $3x - 2x + 10x$

8. $7y - 5y - 2y$

9. $7ab + c - 2ba$

10. $8xy - 2xz - 2yz$

11. $6x + x - 9x$

12. $7x + 3x - 10x$

13. $15xy - 3xy + 6xy$

14. $3ab - 2bc + 4ac$

15. $8ab + 2ba - 4ab$

16. $6x - 6 + 4y - 3$

17. $23ab^2 - 17ab^2$

18. $16xy^2 + 5x^2y$

19. $s^3t^2 - 7s^3t^2$

20. $x^2y^4 + 3x^4y^2$

21. $x - 4xy + 3yx$

22. $5a - 4 + 3b - 3$

23. $7xy^2z - 2y^2xz + 5zy^2x$

24. $2x^3 - 2x^2 + 3x - 5x$

25. $5y^2 - 3y^3 + 2y - 4y$

26. $4x - 3y + 7 - 2x + 4 - 6y$

27. $3b - 5a - 9 - 2a + 4 - 5b$

28. $x - 3x^3 + 2x^2 - 5x - 4x^2 + x^3$

29. $y - 2y^2 - 5y^3 - y + 3y^3 - y^2$

30. $a^2b - 11ab + 12ab^2 - 3a^2b + 4ab$

31. $xy^2 + y - 5x^2y + 3xy^2 + x^2y$

32. $4z^2 - 6z + 5 - z^2 + 9z - 10$

33. $37.9 \text{ cm} - 13.5 \text{ cm} + 24.8 \text{ cm} - 19.3 \text{ cm}$

34. $7.2 \text{ m} - 3.65 \text{ m} + 8.002 \text{ m}$

In Exercises 35–44, simplify each algebraic expression.

35. $x + 6(z + 8y + x)$ **36.** $a - 6(a + 9b - c)$

37. $8c + 6c(7a + 4 - 9b)$ **38.** $15x - 3x(6y - 5x - 7)$

39. $6a - (8a - 3b + 4)$ **40.** $7x - (4w - 6 + 6y + 13x)$

41. $4x^2 + 2 - (6x + 8 - x^2)$ **42.** $7y^2 + 6 - (16y^2 + 15 - 4y)$

43. $4x^2(8x^2 - 7x + 9) - 8x(5x^2 - 9x + 7)$

44. $8x^4(5x^3 + 8x - 1) - 8x^2(7x^5 - 8x^3 + 2x^2)$

3.7 More Rules of Exponents

3.7A Raising Exponential Numbers and Products to a Power

Raising Exponential Numbers to a Power

Consider the expression $(x^4)^2$.

$$(x^4)^2 = (x^4)(x^4) = x^4 \cdot x^4 = x^{4+4} = x^{2 \cdot 4} = x^{4 \cdot 2} = x^8$$

Note that $(x^4)^2 = x^{4 \cdot 2}$. In this case, the exponents are multiplied. The general rule for raising one exponential number to a power is given in Rule 3.3.

RULE 3.3

$$(x^a)^b = x^{ab}$$

Example 1   Remove the parentheses:

a. $(x^5)^4 = x^{5 \cdot 4} = x^{20}$

b. $(x)^4 = (x^1)^4 = x^{1 \cdot 4} = x^4$

c. $(y^4)^3 = y^{4 \cdot 3} = y^{12}$

d. $(10^6)^2 = 10^{6 \cdot 2} = 10^{12} = 1,\underbrace{000,000,000,000}_{12 \text{ zeros}}$

e. $(2^a)^b = 2^{a \cdot b} = 2^{ab}$ ■

Raising a Product of Factors to a Power

Next, consider the expression $(xy)^3$.

$$(xy)^3 = (xy)(xy)(xy) = (xxx)(yyy) = x^3y^3$$

The general rule for raising a product of factors to a power is given in Rule 3.4.

RULE 3.4

$$(xy)^a = x^a y^a$$

Example 2 Remove the parentheses:

a. $(xy)^5 = x^5y^5$

b. $(2x)^3 = 2^3x^3$

Be sure to raise *numbers* in the parentheses to the power also. Remember,

$$(2x)^3 = (2x)(2x)(2x) = (2 \cdot 2 \cdot 2)(xxx) = 2^3x^3 \quad \blacksquare$$

A WORD OF CAUTION $2x^3 \neq (2x)^3$. In the expression $2x^3$, the exponent applies only to the x. That is, $2x^3 = 2xxx$. In the expression $(2x)^3$, the exponent applies to everything inside the parentheses. ☑

Example 3 Evaluate:

a. $5(-2)^2 = 5(4) = 20$

This exponent applies only to the 5

b. $-2 \cdot 5^2 = -2 \cdot 25 = -50$

This exponent applies to everything in the parentheses

c. $(-2 \cdot 5)^2 = (-10)^2 = 100 \quad \blacksquare$

EXERCISES 3.7A

Set I   In Exercises 1–14, remove the parentheses.

1. $(y^2)^5$ **2.** $(N^3)^4$ **3.** $(x^8)^2$

4. $(z^4)^7$ **5.** $(2^4)^2$ **6.** $(3^4)^2$

7. $(xy)^5$ **8.** $(ab)^4$ **9.** $(2c)^6$

10. $(3x)^4$ **11.** $(x^4)^7$ **12.** $(v^3)^8$

13. $(10^2)^3$ **14.** $(10^7)^2$

In Exercises 15–20, evaluate the expression.

15. $(-2 \cdot 3)^2$ **16.** $(-3 \cdot 4)^2$ **17.** $2(-3)^3$

18. $3(-2)^3$ **19.** $-4 \cdot 5^2$ **20.** $-3 \cdot 4^2$

Set II   In Exercises 1–14, remove the parentheses.

1. $(a^3)^4$ **2.** $(2x)^4$ **3.** $(b^9)^4$

4. $(x^3)^2$ **5.** $(2^2)^4$ **6.** $(2x)^5$

7. $(abc)^4$ **8.** $(2xy)^5$ **9.** $(3x)^4$

10. $(st)^8$ **11.** $(u^7)^4$ **12.** $(5x)^2$

13. $(10^3)^3$ **14.** $(x^6)^4$

In Exercises 15–20, evaluate the expression.

15. $(-3 \cdot 5)^2$ **16.** $(-5 \cdot 2)^3$ **17.** $5(-2)^5$

18. $-4 \cdot 3^2$ **19.** $-6 \cdot 2^2$ **20.** $7(-2)^2$

3.7B   Exponential Numbers and Division

Dividing Exponential Numbers When the Bases Are the Same

Consider the expression $\dfrac{x^5}{x^3}$. (Assume $x \neq 0$.)

$$\frac{x^5}{x^3} = \frac{xxxxx}{xxx} = \frac{xxx \cdot xx}{xxx \cdot 1} = \frac{xxx}{xxx} \cdot \frac{xx}{1} = 1 \cdot \frac{xx}{1} = xx = x^2$$

The value of this fraction
is 1 (for $x \neq 0$)

Note that $\dfrac{x^5}{x^3} = x^{5-3} = x^2$. In this case, three of the factors of the denominator canceled with three of the five factors of the numerator, leaving $5 - 3 = 2$ factors of x. The general rule for dividing exponential numbers when the bases are the same is given in Rule 3.5.

RULE 3.5

If $x \neq 0$, $\dfrac{x^a}{x^b} = x^{a-b}$

NOTE   When we use Rule 3.5, the exponent in the denominator is subtracted from the exponent in the numerator. ☑

In Example 4, assume $x \neq 0$, $r \neq 0$, $y \neq 0$, $a \neq 0$, and $b \neq 0$.

Example 4   Rewrite each expression with the base appearing only once (if possible):

a. $\dfrac{x^6}{x^2} = x^{6-2} = x^4$ b. $\dfrac{r^{12}}{r^5} = r^{12-5} = r^7$

c. $\dfrac{y^3}{y} = \dfrac{y^3}{y^1} = y^{3-1} = y^2$

d. $\dfrac{10^7}{10^3} = 10^{7-3} = 10^4 = \underbrace{10{,}000}_{4 \text{ zeros}}$

e. $\dfrac{x^5}{y^2}$

Rule 3.5 does not apply when the bases are different.

f. $x^4 - x^2$

This cannot be rewritten with the base appearing only once. Rule 3.5 does not apply, because the operation is subtraction rather than division.

g. $\dfrac{8x^3}{2x} = \dfrac{\overset{4}{\cancel{8}}}{\underset{1}{\cancel{2}}} \cdot \dfrac{x^3}{x} = 4 \cdot x^{3-1} = 4x^2$

h. $\dfrac{6a^3b^4}{9ab^2} = \dfrac{\overset{2}{\cancel{6}}}{\underset{3}{\cancel{9}}} \cdot \dfrac{a^3}{a} \cdot \dfrac{b^4}{b^2} = \dfrac{2}{3} \cdot a^{3-1} \cdot b^{4-2} = \dfrac{2}{3}a^2b^2 \ \text{ or } \ \dfrac{2a^2b^2}{3}$

i. $\dfrac{2^a}{2^b} = 2^{a-b}$

j. $\dfrac{x^5 - y^3}{x^2}$

The expression is usually left in this form. It could be changed as follows:

$$\frac{x^5 - y^3}{x^2} = \frac{x^5}{x^2} - \frac{y^3}{x^2} = x^3 - \frac{y^3}{x^2}$$

Expressions of this form are discussed in Sections 6.5 and 8.2. ∎

A WORD OF CAUTION We show examples of a common mistake students make in division. (Assume $x \neq 0$.)

Correct method	Incorrect method

a. $\dfrac{2+6}{2} = \dfrac{8}{2} = 4$ $\dfrac{\overset{1}{\cancel{2}}+6}{\underset{1}{\cancel{2}}} = \dfrac{1+6}{1} = 7$ ✗

b. $\dfrac{x^2+y}{x^2} = \dfrac{x^2}{x^2} + \dfrac{y}{x^2}$ $\dfrac{\overset{1}{\cancel{x^2}}+y}{\underset{1}{\cancel{x^2}}} = 1+y$ ✗

$= 1 + \dfrac{y}{x^2}$

This expression is usually left in the form $\dfrac{x^2+y}{x^2}$. ☑

NOTE To see why the restriction $x \neq 0$ is included in Rule 3.5, consider the example $\dfrac{0^5}{0^2}$. $\dfrac{0^5}{0^2} = \dfrac{0 \cdot 0 \cdot 0 \cdot 0 \cdot 0}{0 \cdot 0} = \dfrac{0}{0}$, which cannot be determined. For this reason, x cannot be 0 in Rule 3.5. ☑

In this book, unless otherwise noted, none of the variables has a value that makes a denominator zero.

In Example 4, we were careful to have the exponent in the numerator be larger than the exponent in the denominator, so that the quotient always had a positive exponent. When the exponent in the denominator is larger than the exponent in the numerator, the quotient has a negative exponent. Negative exponents are discussed in Section 3.8.

Example 5   Evaluate each of the following expressions. (Compare parts a and b and note the differences when negative values are raised to an even or an odd power.)

This exponent applies only to the 5

a. $\dfrac{-5^2}{(-5)^2} = \dfrac{-25}{25} = -1$

This exponent applies to everything in the parentheses

b. $\dfrac{-2^3}{(-2)^3} = \dfrac{-8}{-8} = 1$ ∎

Raising Fractions to Powers
Rule 3.6 is used in raising a fraction to a power.

RULE 3.6

If $y \neq 0$, $\left(\dfrac{x}{y}\right)^n = \dfrac{x^n}{y^n}$

Example 6   Remove the parentheses:

a. $\left(\dfrac{a}{b}\right)^5 = \dfrac{a^5}{b^5}$

b. $\left(\dfrac{3}{c}\right)^3 = \dfrac{3^3}{c^3} = \dfrac{27}{c^3}$ Be sure to raise *numbers* in parentheses to the power also ∎

The General Rule of Exponents
Rules 3.3, 3.4, 3.5, and 3.6 can be combined into the following general rule:

RULE 3.7   GENERAL RULE OF EXPONENTS

$$\left(\frac{x^a y^b}{z^c}\right)^n = \frac{x^{an} y^{bn}}{z^{cn}}$$

None of the variables can have a value that makes the denominator zero.

In applying Rule 3.7, notice the following:

1. x, y, and z are *factors* of the expression within the parentheses. They are *not* separated by $+$ or $-$ signs.

2. The exponent of each factor within the parentheses is multiplied by the exponent outside the parentheses.

Example 7 Simplify each of the following expressions:

a. $(x^2y^3)^2 = x^{2\cdot2}y^{3\cdot2} = x^4y^6$

———— The exponents on the 5 and the x are understood to be 1

———— Those 1s must be multiplied by 3

b. $(5xy^2z^5)^3 = (5^1x^1y^2z^5)^3 = 5^{1\cdot3}x^{1\cdot3}y^{2\cdot3}z^{5\cdot3} = 5^3x^3y^6z^{15} = 125x^3y^6z^{15}$

c. $(x^2 + y^3)^4$

Rule 3.7 *cannot* be used here because the $+$ sign means that x^2 and y^3 are terms, *not* factors, of the expression being raised to the fourth power. We cannot simplify $(x^2 + y^3)^4$ at this time. ∎

Example 8 In the following examples, none of the factors appearing in any denominator is zero:

a. $\left(\dfrac{x^2y^5}{z^3}\right)^4 = \dfrac{x^{2\cdot4}y^{5\cdot4}}{z^{3\cdot4}} = \dfrac{x^8y^{20}}{z^{12}}$

——— Simplify the expression within parentheses first whenever possible

b. $\left(\dfrac{x^5y^6}{x^3y^3}\right)^4 = (x^2y^3)^4 = x^{2\cdot4}y^{3\cdot4} = x^8y^{12}$

c. $\left(\dfrac{3b^3}{2c^4}\right)^2 = \left(\dfrac{3^1b^3}{2^1c^4}\right)^2 = \dfrac{3^{1\cdot2}b^{3\cdot2}}{2^{1\cdot2}c^{4\cdot2}} = \dfrac{3^2b^6}{2^2c^8} = \dfrac{9b^6}{4c^8}$ ∎

EXERCISES 3.7B

Set I In Exercises 1–22, rewrite each expression with the base appearing only once, if possible.

1. $\dfrac{x^7}{x^2}$ 2. $\dfrac{y^8}{y^6}$ 3. $x^4 - x^2$ 4. $s^8 - s^3$

5. $\dfrac{a^5}{a}$ 6. $\dfrac{b^7}{b}$ 7. $\dfrac{10^{11}}{10}$ 8. $\dfrac{5^6}{5}$

9. $\dfrac{x^8}{y^4}$ 10. $\dfrac{a^4}{b^3}$ 11. $\dfrac{6x^2}{2x}$ 12. $\dfrac{9y^3}{3y}$

13. $\dfrac{a^3}{b^2}$ 14. $\dfrac{x^5}{y^3}$ 15. $\dfrac{10x^4}{5x^3}$

16. $\dfrac{15y^5}{9y^2}$ 17. $\dfrac{12h^4k^3}{8h^2k}$ 18. $\dfrac{16a^5b^3}{12ab^2}$

19. $\dfrac{x^{5a}}{x^{3a}}$ 20. $\dfrac{M^{6x}}{M^{2x}}$ 21. $\dfrac{a^4 - b^3}{a^2}$

22. $\dfrac{x^6 + y^4}{y^2}$

In Exercises 23–38, remove the parentheses.

23. $\left(\dfrac{s}{t}\right)^7$ **24.** $\left(\dfrac{x}{y}\right)^9$ **25.** $\left(\dfrac{2}{x}\right)^4$

26. $\left(\dfrac{3}{z}\right)^2$ **27.** $\left(\dfrac{x}{2}\right)^6$ **28.** $\left(\dfrac{c}{5}\right)^3$

29. $(a^2b^3)^2$ **30.** $(x^4y^5)^3$ **31.** $(2z^3)^2$

32. $(3w^2)^3$ **33.** $\left(\dfrac{xy^4}{z^2}\right)^2$ **34.** $\left(\dfrac{a^3b}{c^2}\right)^3$

35. $\left(\dfrac{5y^3}{2x^2}\right)^4$ **36.** $\left(\dfrac{6b^4}{7c^2}\right)^3$ **37.** $\left(\dfrac{x^3y^7}{xy^2}\right)^3$

38. $\left(\dfrac{a^6b^8}{a^4b^3}\right)^3$

In Exercises 39 and 40, evaluate each expression.

39. $\dfrac{(-4)^2}{-4^2}$ **40.** $\dfrac{-9^2}{(-9)^2}$

Set II   In Exercises 1–22, rewrite each expression with the base appearing only once, if possible.

1. $\dfrac{x^9}{x^3}$ **2.** $\dfrac{a^4}{b^2}$ **3.** $s^6 - s^4$

4. $\dfrac{s^6}{s^4}$ **5.** $\dfrac{y^7}{y}$ **6.** $\dfrac{3^5}{3}$

7. $\dfrac{8^6}{8^2}$ **8.** $\dfrac{15^9}{15^4}$ **9.** $\dfrac{a^9}{b^4}$

10. $\dfrac{a^6}{a^4}$ **11.** $\dfrac{12b^6}{3b^3}$ **12.** $\dfrac{12x^3}{10x}$

13. $\dfrac{m^4}{n^2}$ **14.** $\dfrac{18z^6}{10z^4}$ **15.** $\dfrac{25a^3b^4}{15ab^3}$

16. $\dfrac{18a^8}{2a^4}$ **17.** $\dfrac{9x^4y^8}{6xy^4}$ **18.** $6x^3y^2 - 4x^2y$

19. $\dfrac{H^{4n}}{H^{2n}}$ **20.** $z^{4a} - z^{2a}$ **21.** $\dfrac{x^3 - y^2}{x^2}$

22. $\dfrac{a^4 - b^5}{b^3}$

In Exercises 23–38, remove the parentheses.

23. $\left(\dfrac{u}{v}\right)^4$ **24.** $\left(\dfrac{8}{c}\right)^2$ **25.** $\left(\dfrac{3}{a}\right)^3$

26. $\left(\dfrac{x}{2}\right)^5$ **27.** $\left(\dfrac{t}{5}\right)^2$ **28.** $\left(\dfrac{a}{c}\right)^7$

29. $(h^3k^4)^2$ **30.** $(2x^4y^2)^3$ **31.** $(5a^3)^2$

32. $(8xy^3)^2$

33. $\left(\dfrac{x^3y}{z^2}\right)^3$

34. $\left(\dfrac{s^4t^2}{u^3}\right)^3$

35. $\left(\dfrac{3a^5}{2x^2}\right)^4$

36. $\left(\dfrac{5}{2t^5}\right)^2$

37. $\left(\dfrac{x^4y^7}{x^3y^3}\right)^5$

38. $\left(\dfrac{6a^9b^3}{2ab}\right)^3$

In Exercises 39 and 40, evaluate each expression.

39. $\dfrac{-2^4}{(-2)^4}$

40. $\dfrac{-3^3}{(-3)^3}$

3.8 Zero and Negative Exponents

The Zero Exponent

When we used Rule 3.5 in Section 3.7, the exponent of the numerator was always greater than the exponent of the denominator. We now consider the case in which the exponents of the numerator and the denominator are the same. Consider the expression $\dfrac{x^4}{x^4}$.

$$\frac{x^4}{x^4} = \frac{xxxx}{xxxx} = 1 \qquad \text{Because a number divided by itself is 1}$$

$$\frac{x^4}{x^4} = x^{4-4} = x^0 \qquad \text{Using Rule 3.5}$$

Therefore, we define x^0 to be 1 (if $x \neq 0$).

RULE 3.8

If $x \neq 0$, $x^0 = 1$.

Example 1 Examples of zero as an exponent:

a. $a^0 = 1$ Provided $a \neq 0$

b. $H^0 = 1$ Provided $H \neq 0$

c. $10^0 = 1$

d. $5^0 = 1$

e. $6x^0 = 6 \cdot 1 = 6$ Provided $x \neq 0$; the 0 exponent applies only to x ■

NOTE $x \neq 0$ in Rule 3.8 because 0^0 can be interpreted as 0^{1-1}, and $0^{1-1} = \dfrac{0^1}{0^1} = \dfrac{0}{0}$, which cannot be determined. ☑

Negative Exponents

Let us now consider the case in which the exponent of the denominator is larger than the exponent of the numerator. Consider the expression $\dfrac{x^3}{x^5}$.

$$\frac{x^3}{x^5} = \frac{xxx}{xxxxx} = \frac{xxx \cdot 1}{xxx \cdot xx} = \frac{xxx}{xxx} \cdot \frac{1}{xx} = 1 \cdot \frac{1}{xx} = \frac{1}{x^2}$$

⎯⎯ The value of this fraction is 1

However, if we use Rule 3.5,

$$\frac{x^3}{x^5} = x^{3-5} = x^{-2}$$

Therefore, we would like the following statement to be true:

$$x^{-2} = \frac{1}{x^2}$$

Because we want x^{-2} to equal $\dfrac{1}{x^2}$, we make the following definition:

RULE 3.9a

If $x \neq 0$, $x^{-n} = \dfrac{1}{x^n}$.

A WORD OF CAUTION x^{-n} is not necessarily a negative number. ☑

When we use Rule 3.9a, the exponent n can be either positive or negative.

Example 2 Remove the negative exponents, using Rule 3.9a:

a. $x^{-5} = \dfrac{1}{x^5}$

b. $a^{-3} = \dfrac{1}{a^3}$

c. $10^{-4} = \dfrac{1}{10^4}$ ∎

The next two rules follow from Rule 3.9a:

RULE 3.9b

If $x \neq 0$, $\dfrac{1}{x^{-n}} = x^n$.

RULE 3.9c

If $x \neq 0$ and $y \neq 0$, $\left(\dfrac{x}{y}\right)^{-n} = \left(\dfrac{y}{x}\right)^{n}$.

Rules 3.9b and 3.9c can both be used to remove negative exponents (see Examples 3 and 4).

Example 3 Remove the negative exponents, using Rule 3.9b:

a. $\dfrac{1}{x^{-4}} = x^4$

b. $\dfrac{1}{w^{-2}} = w^2$ ∎

Example 4 Remove the negative exponent and the parentheses from $\left(\dfrac{a}{b}\right)^{-4}$.

$$\left(\frac{a}{b}\right)^{-4} = \left(\frac{b}{a}\right)^{4} = \frac{b^4}{a^4} \qquad \text{Using Rules 3.9c and 3.6} \quad ∎$$

Rules 3.9a, b, and c can also be used to insert negative exponents (see Example 5).

Example 5 Rewrite $\dfrac{1}{y^3}$ without fractions.

$$\frac{1}{y^3} = y^{-3} \qquad \text{Using Rule 3.9a} \quad ∎$$

NOTE If a single number or variable appears in the numerator or the denominator of a fraction, that number can still be considered a factor of the numerator or denominator. For example, $\dfrac{5}{x} = \dfrac{1 \cdot 5}{1 \cdot x}$. Therefore, we can say that 5 is a factor of the numerator and x is a factor of the denominator of $\dfrac{5}{x}$. ☑

A factor can be moved either from the numerator to the denominator or from the denominator to the numerator simply by so moving it and changing the sign of its exponent. This does not change the sign of the *expression*. If a factor has no exponent, the exponent is understood to equal 1.

Example 6 Remove the negative exponents in each of the following expressions:

a^{-3} is a factor of $a^{-3}b^4$

a. $a^{-3}b^4 = \dfrac{a^{-3}}{1} \cdot \dfrac{b^4}{1} = \dfrac{1}{a^3} \cdot \dfrac{b^4}{1} = \dfrac{b^4}{a^3}$

The factor a^{-3} was moved from the numerator to the denominator by changing the sign of its exponent

y^{-4} and z^{-2} are factors of $y^{-4}w^5z^{-2}$

b. $y^{-4}w^5z^{-2} = \dfrac{y^{-4}}{1} \cdot \dfrac{w^5}{1} \cdot \dfrac{z^{-2}}{1} = \dfrac{1}{y^4} \cdot \dfrac{w^5}{1} \cdot \dfrac{1}{z^2} = \dfrac{w^5}{y^4z^2}$

These steps need not be written; we use them in these examples to show you why this method works

x^{-2} is a factor of x^{-2}

c. $\dfrac{x^{-2}}{y} = \dfrac{x^{-2}}{1} \cdot \dfrac{1}{y} = \dfrac{1}{x^2} \cdot \dfrac{1}{y} = \dfrac{1}{x^2y}$ ∎

Example 7   Rewrite each expression without fractions, using negative exponents when necessary:

a. $\dfrac{h^5}{k^{-4}} = \dfrac{h^5}{1} \cdot \dfrac{1}{k^{-4}} = \dfrac{h^5}{1} \cdot \dfrac{k^4}{1} = h^5k^4$

b. $\dfrac{a^3}{b^2} = \dfrac{a^3}{1} \cdot \dfrac{1}{b^2} = \dfrac{a^3}{1} \cdot \dfrac{b^{-2}}{1} = a^3b^{-2}$

c. $\dfrac{z^3}{x} = \dfrac{z^3}{1} \cdot \dfrac{1}{x} = \dfrac{z^3}{1} \cdot \dfrac{x^{-1}}{1} = z^3x^{-1}$ ∎

Example 8   Write each expression with positive exponents only:

a. $x^4y^{-3}z^{-6} = \dfrac{x^4}{1} \cdot \dfrac{y^{-3}}{1} \cdot \dfrac{z^{-6}}{1} = \dfrac{x^4}{1} \cdot \dfrac{1}{y^3} \cdot \dfrac{1}{z^6} = \dfrac{x^4}{y^3z^6}$

b. $\dfrac{a^{-2}b^4}{c^5d^{-3}} = \dfrac{a^{-2}}{1} \cdot \dfrac{b^4}{1} \cdot \dfrac{1}{c^5} \cdot \dfrac{1}{d^{-3}} = \dfrac{1}{a^2} \cdot \dfrac{b^4}{1} \cdot \dfrac{1}{c^5} \cdot \dfrac{d^3}{1} = \dfrac{b^4d^3}{a^2c^5}$

c. $\dfrac{e^2f}{g^{-1}} = \dfrac{e^2}{1} \cdot \dfrac{f^1}{1} \cdot \dfrac{1}{g^{-1}} = \dfrac{e^2}{1} \cdot \dfrac{f^1}{1} \cdot \dfrac{g^1}{1} = e^2fg$ ∎

Example 9   Write each expression without fractions, using negative exponents when necessary:

a. $\dfrac{m^5}{n^2} = \dfrac{m^5}{1} \cdot \dfrac{1}{n^2} = \dfrac{m^5}{1} \cdot \dfrac{n^{-2}}{1} = m^5n^{-2}$

b. $\dfrac{y^3}{z^{-2}} = \dfrac{y^3}{1} \cdot \dfrac{1}{z^{-2}} = \dfrac{y^3}{1} \cdot \dfrac{z^2}{1} = y^3z^2$

c. $\dfrac{a}{bc^2} = \dfrac{a^1}{1} \cdot \dfrac{1}{b^1} \cdot \dfrac{1}{c^2} = \dfrac{a^1}{1} \cdot \dfrac{b^{-1}}{1} \cdot \dfrac{c^{-2}}{1} = ab^{-1}c^{-2}$ ∎

A WORD OF CAUTION   An expression that is a term of, rather than a factor of, a numerator cannot be moved from the numerator to the denominator of a fraction simply by moving it and changing the sign of its exponents. For example, $\dfrac{a^{-2} + b^5}{c^4} \neq \dfrac{b^5}{a^2c^4}$.

$\dfrac{a^{-2} + b^5}{c^4} = \dfrac{\dfrac{1}{a^2} + b^5}{c^4}$

a^{-2} cannot be moved to the denominator because it is not a factor of the numerator (the plus sign indicates that a^{-2} is a term rather than a factor of the numerator)

Expressions of this kind will be simplified in Section 8.7

Using the Rules of Exponents with Positive, Zero, and Negative Exponents

Rules 3.1, and 3.3–3.7 apply to expressions having zero and negative exponents.

Example 10 Apply the rules of exponents to simplify each of the following expressions:

a. $a^4 \cdot a^{-3} = a^{4+(-3)} = a^1 = a$ Rule 3.1

b. $x^{-5} \cdot x^2 = x^{-5+2} = x^{-3} = \dfrac{1}{x^3}$ Rules 3.1 and 3.9a

c. $(y^{-2})^{-1} = y^{(-2)(-1)} = y^2$ Rule 3.3

d. $(x^2)^{-4} = x^{2(-4)} = x^{-8} = \dfrac{1}{x^8}$ Rules 3.3 and 3.9a

e. $\dfrac{y^{-2}}{y^{-6}} = y^{(-2)-(-6)} = y^{-2+6} = y^4$ Rule 3.5

f. $\dfrac{z^{-4}}{z^{-2}} = z^{(-4)-(-2)} = z^{-4+2} = z^{-2} = \dfrac{1}{z^2}$ Rules 3.5 and 3.9a

g. $h^3 h^0 h^{-2} = h^{3+0+(-2)} = h^1 = h$ Rule 3.1

h. $(7x)^{-2} = 7^{-2}x^{-2} = \dfrac{1}{7^2 x^2} = \dfrac{1}{49x^2}$ Rules 3.4 and 3.9a

i. $\left(\dfrac{5}{d}\right)^{-3} = \left(\dfrac{d}{5}\right)^3 = \dfrac{d^3}{5^3} = \dfrac{d^3}{125}$ Rules 3.9c and 3.6

j. $\left(\dfrac{a}{2bc}\right)^0 = 1$ Rule 3.8 ■

An expression such as $\dfrac{x^3}{x^5}$ can be simplified in either of two ways:

$$\dfrac{x^3}{x^5} = x^{3-5} = x^{-2} = \dfrac{1}{x^2}$$

or

$$\dfrac{x^3}{x^5} = \dfrac{1}{x^5 \cdot x^{-3}} = \dfrac{1}{x^{5-3}} = \dfrac{1}{x^2}$$ Moved factor x^3 to denominator and changed sign of its exponent

Example 11 Simplify the fractions, using the rules of exponents; write results using only positive exponents:

a. $\dfrac{12x^{-2}}{4x^{-3}} = \dfrac{\overset{3}{\cancel{12}}}{\underset{1}{\cancel{4}}} \cdot \dfrac{x^{-2}}{x^{-3}} = \dfrac{3}{1} \cdot \dfrac{x^{-2-(-3)}}{1} = 3x$

b. $\dfrac{5a^4 b^{-3}}{10a^{-2}b^{-4}} = \dfrac{\overset{1}{\cancel{5}}}{\underset{2}{\cancel{10}}} \cdot \dfrac{a^4}{a^{-2}} \cdot \dfrac{b^{-3}}{b^{-4}} = \dfrac{1}{2} \cdot a^{4-(-2)} b^{-3-(-4)} = \dfrac{a^6 b}{2}$ or $\dfrac{1}{2}a^6 b$

c. $\dfrac{9xy^{-3}}{15x^3 y^{-4}} = \dfrac{\overset{3}{\cancel{9}}}{\underset{5}{\cancel{15}}} \cdot \dfrac{x}{x^3} \cdot \dfrac{y^{-3}}{y^{-4}} = \dfrac{3}{5} \cdot \dfrac{x^{1-3}}{1} \cdot \dfrac{y^{-3-(-4)}}{1}$

$= \dfrac{3}{5} \cdot \dfrac{x^{-2}}{1} \cdot \dfrac{y}{1} = \dfrac{3}{5} \cdot \dfrac{1}{x^2} \cdot \dfrac{y}{1} = \dfrac{3y}{5x^2}$ ■

Evaluating Expressions with Numerical Bases
Example 12 demonstrates the evaluation of expressions with numerical bases.

Example 12   Evaluate each expression:

After applying the rules of exponents, it is customary to evaluate the power of a number

a. $10^3 \cdot 10^2 = 10^5 = 10 \cdot 10 \cdot 10 \cdot 10 \cdot 10 = \overbrace{100,000}^{\text{5 zeros}}$

b. $10^{-2} = \dfrac{1}{10^2} = \dfrac{1}{10 \cdot 10} = \dfrac{1}{\underbrace{100}_{\text{2 zeros}}}$

c. $(2^3)^{-1} = 2^{-3} = \dfrac{1}{2^3} = \dfrac{1}{2 \cdot 2 \cdot 2} = \dfrac{1}{8}$

d. $\dfrac{5^0}{5^2} = \dfrac{1}{5 \cdot 5} = \dfrac{1}{25}$

e. $(-2)^{-5} = \dfrac{1}{(-2)^5} = \dfrac{1}{-32}$ or $-\dfrac{1}{32}$

f. $(-3)^{-4} = \dfrac{1}{(-3)^4} = \dfrac{1}{81}$ ∎

A WORD OF CAUTION   A common mistake students make is shown by the following examples:

Correct method	*Incorrect method*
a. $2^3 \cdot 2^2 = 2^{3+2}$	$2^3 \cdot 2^2 \neq (2 \cdot 2)^{3+2} = 4^5$
$= 2^5 = 32$	$= 1,024$
b. $10^2 \cdot 10 = 10^{2+1}$	$10^2 \cdot 10 \neq (10 \cdot 10)^{2+1} = 100^3$
$= 10^3 = 1,000$	$= 1,000,000$

In words, when multiplying powers of the same base, add the exponents; do *not* multiply the bases as well. ☑

Simplified Form of Terms with Exponents

A term with exponents is considered simplified when each different base appears only once and the exponent on each base is a single positive integer.

Example 13   Simplify each expression:

a. $x^2 \cdot x^7 = x^9$

b. $\dfrac{x^5 y^2}{x^3 y} = x^2 y$

c. $(x^2)^3 = x^6$

d. $\dfrac{x^5 y^2}{x^4 y} = \dfrac{x y^2}{y}$

This expression is not considered to be completely simplified because the base y appears twice. Continue the simplification:

$$\frac{x y^2}{y} = x y$$

The expression is now completely simplified. ∎

Example 14 Rule 3.7 also applies to expressions with zero and negative exponents. In the following examples, none of the factors appearing in any denominator is zero:

a. $(x^3y^{-1})^5 = x^{3\cdot5}y^{(-1)5} = x^{15}y^{-5} = \dfrac{x^{15}}{y^5}$

— The same rules of exponents apply to numerical bases as well as literal bases

b. $\left(\dfrac{2a^{-3}b^2}{c^5}\right)^3 = \dfrac{2^{1\cdot3}a^{(-3)3}b^{2\cdot3}}{c^{5\cdot3}} = \dfrac{2^3a^{-9}b^6}{c^{15}} = \dfrac{8b^6}{a^9c^{15}}$

c. $\left(\dfrac{3^2c^{-4}}{d^3}\right)^{-1} = \dfrac{3^{2(-1)}c^{(-4)(-1)}}{d^{3(-1)}} = \dfrac{3^{-2}c^4}{d^{-3}} = \dfrac{c^4d^3}{3^2} = \dfrac{c^4d^3}{9}$

d. $\left(\dfrac{3^{-7}x^{10}}{y^{-4}}\right)^0 = 1$

e. $\left(\dfrac{x^5y^4}{x^3y^7}\right)^2 = (x^2y^{-3})^2 = x^4y^{-6} = \dfrac{x^4}{y^6}$

— Simplify the expression within the parentheses first whenever possible

f. $(5^0h^{-2})^{-3} = (1h^{-2})^{-3} = (h^{-2})^{-3} = h^6$

g. $\left(\dfrac{10^{-2}\cdot10^5}{10^4}\right)^3 = \left(\dfrac{10^3}{10^4}\right)^3 = (10^{-1})^3 = 10^{-3} = \dfrac{1}{10^3}$ ∎

EXERCISES 3.8

Set I In Exercises 1–28, simplify each expression. Write answers using only positive exponents.

1. x^{-4}
2. y^{-7}
3. $\dfrac{1}{a^{-4}}$

4. $\dfrac{1}{b^{-5}}$
5. $r^{-4}st^{-2}$
6. $r^{-5}s^{-3}t$

7. $(xy)^{-2}$
8. $(ab)^{-4}$
9. $\dfrac{h^2}{k^{-4}}$

10. $\dfrac{m^3}{n^{-2}}$
11. $\dfrac{x^{-4}}{y}$
12. $\dfrac{a^{-5}}{b}$

13. $ab^{-2}c^0$
14. $x^{-3}y^0z$
15. $x^{-3}\cdot x^4$

16. $y^6\cdot y^{-2}$
17. $10^3\cdot10^{-2}$
18. $2^{-3}\cdot2^2$

19. $(x^2)^{-4}$
20. $(z^3)^{-2}$
21. $(a^{-2})^3$

22. $(b^{-5})^2$
23. $\dfrac{y^{-2}}{y^5}$
24. $\dfrac{z^{-2}}{z^2}$

25. $\dfrac{10^2}{10^{-5}}$
26. $\dfrac{2^3}{2^{-2}}$
27. $\left(\dfrac{x}{y}\right)^{-3}$

28. $\left(\dfrac{s}{t}\right)^{-2}$

In Exercises 29–34, write each expression without fractions, using negative exponents if necessary.

29. $\dfrac{1}{x^{+2}}$ 　　　　　**30.** $\dfrac{1}{y^3}$ 　　　　　**31.** $\dfrac{h}{k}$

32. $\dfrac{m}{n}$ 　　　　　**33.** $\dfrac{x^2}{yz^5}$ 　　　　　**34.** $\dfrac{a^3}{b^2c}$

In Exercises 35–50, evaluate each expression.

35. $10^4 \cdot 10^{-2}$ 　　　　　**36.** $3^{-2} \cdot 3^3$ 　　　　　**37.** 10^{-4}

38. 2^{-3} 　　　　　**39.** $5^0 \cdot 7^2$ 　　　　　**40.** $4^3 \cdot 2^0$

41. $\dfrac{10^0}{10^2}$ 　　　　　**42.** $\dfrac{5^2}{5^0}$ 　　　　　**43.** $\dfrac{10^{-3} \cdot 10^2}{10^5}$

44. $\dfrac{2^3 \cdot 2^{-4}}{2^2}$ 　　　　　**45.** $(10^2)^{-1}$ 　　　　　**46.** $(2^{-3})^2$

47. $(-5)^{-3}$ 　　　　　**48.** $(-4)^{-3}$ 　　　　　**49.** $(-12)^{-2}$

50. $(-13)^{-2}$

In Exercises 51–82, simplify each expression. Write answers using only positive exponents.

51. $\dfrac{a^3b^0}{c^{-2}}$ 　　　　　**52.** $\dfrac{d^0e^2}{f^{-3}}$ 　　　　　**53.** $\dfrac{p^4r^{-1}}{t^{-2}}$

54. $\dfrac{u^5v^{-2}}{w^{-3}}$ 　　　　　**55.** $\dfrac{8x^{-3}}{12x}$ 　　　　　**56.** $\dfrac{15y^{-2}}{10y}$

57. $\dfrac{20h^{-2}}{35h^{-4}}$ 　　　　　**58.** $\dfrac{35k^{-1}}{28k^{-4}}$ 　　　　　**59.** $\dfrac{15m^0n^{-2}}{5m^{-3}n^4}$

60. $\dfrac{14x^0y^{-3}}{12x^{-2}y^{-4}}$ 　　　　　**61.** $x^{3m} \cdot x^{-m}$ 　　　　　**62.** $y^{-2n} \cdot y^{5n}$

63. $(x^{3b})^{-2}$ 　　　　　**64.** $(y^{2a})^{-3}$ 　　　　　**65.** $\dfrac{x^{2a}}{x^{-5a}}$

66. $\dfrac{a^{3x}}{a^{-5x}}$ 　　　　　**67.** $(m^{-2}n)^4$ 　　　　　**68.** $(p^{-3}r)^5$

69. $(x^{-2}y^3)^{-4}$ 　　　　　**70.** $(w^{-3}z^4)^{-2}$ 　　　　　**71.** $\left(\dfrac{M^{-2}}{N^3}\right)^4$

72. $\left(\dfrac{R^5}{S^{-4}}\right)^3$ 　　　　　**73.** $\left(\dfrac{a^2b^{-4}}{b^{-5}}\right)^2$ 　　　　　**74.** $\left(\dfrac{x^{-2}y^2}{x^{-3}}\right)^3$

75. $\left(\dfrac{mn^{-1}}{m^3}\right)^{-2}$ 　　　　　**76.** $\left(\dfrac{ab^{-2}}{a^2}\right)^{-3}$ 　　　　　**77.** $\left(\dfrac{x^4}{x^{-1}y^{-2}}\right)^{-1}$

78. $\left(\dfrac{x^3}{x^{-2}y^{-4}}\right)^{-1}$ 　　　　　**79.** $(10^0k^{-4})^{-2}$ 　　　　　**80.** $(6^0z^{-5})^{-2}$

81. $\left(\dfrac{r^7s^8}{r^9s^6}\right)^0$ 　　　　　**82.** $\left(\dfrac{t^5u^6}{t^8u^7}\right)^0$

Set II   In Exercises 1–28, simplify each expression. Write answers using only positive exponents.

1. a^{-3}

2. $\dfrac{1}{x^{-2}}$

3. $\dfrac{1}{s^{-3}}$

4. $x^0 y^{-4}$

5. $r^{-2} s t^{-4}$

6. $8^0 x^3 y^{-3}$

7. $h^{-3} k^5$

8. $3x^{-1}$

9. $\dfrac{x}{y^{-3}}$

10. $(x^{-2})^2$

11. $\dfrac{a^{-2}}{b}$

12. $(y^4)^{-2}$

13. $mn^0 p^{-4}$

14. $(y^{-4})^0$

15. $y^{-7} \cdot y^9$

16. $(3^0)^{-4}$

17. $10^{-4} \cdot 10^3$

18. $x^{-8} \cdot x^4$

19. $(n^4)^{-1}$

20. $x^{-8} \cdot y^4$

21. $(z^{-3})^2$

22. $\dfrac{x^0}{y^{-4}}$

23. $\dfrac{y^{-1}}{y^4}$

24. $\dfrac{x^{-2}}{y^{-3}}$

25. $\dfrac{10^3}{10^{-2}}$

26. $\dfrac{5^0 x^{-4}}{x^2}$

27. $p^2 p^0 p^{-3}$

28. $\left(\dfrac{s}{t}\right)^{-6}$

In Exercises 29–34, write each expression without fractions, using negative exponents if necessary.

29. $\dfrac{1}{c^4}$

30. $\dfrac{8}{x^{-3}}$

31. $\dfrac{x}{y}$

32. $\dfrac{5}{ax^2}$

33. $\dfrac{h^4}{kt^3}$

34. $\dfrac{x^{-1}}{y^2}$

In Exercises 35–50, evaluate each expression.

35. $2^{-3} \cdot 2^5$

36. $5^4 \cdot 5^{-6}$

37. 8^{-2}

38. $6^0 \cdot 3^{-4}$

39. $3^0 \cdot 5^2$

40. $(8 \cdot 2^4)^0$

41. $\dfrac{10^0}{10^{-2}}$

42. $\left(\dfrac{1}{2}\right)^{-3}$

43. $\dfrac{10^{-2} \cdot 10^3}{10^4}$

44. $\dfrac{5^0 \cdot 5^3}{5^{-2}}$

45. $(2^{-4})^2$

46. $(3^3)^{-1}$

47. $(-10)^{-3}$

48. $(-3)^{-4}$

49. $(-11)^{-2}$

50. $(-2)^{-3}$

In Exercises 51–82, simplify each expression. Write answers using only positive exponents.

51. $\dfrac{x^0 y^2}{z^{-5}}$

52. $\dfrac{3^4 x^{-4}}{3x^2}$

53. $\dfrac{u^{-1} v^2}{w^{-3}}$

54. $\dfrac{2x^{-3}}{2^3 x}$

55. $\dfrac{16h^{-2}}{10h}$

56. $\left(\dfrac{x}{2}\right)^{-4}$

57. $\dfrac{24m^{-1}}{18m^{-3}}$

58. $(4x)^{-2}$

59. $\dfrac{18w^0 z^{-4}}{16w^{-2} z^2}$

60. $\dfrac{2^{-3}}{x}$

61. $x^{2n} \cdot x^{-n}$

62. $\left(\dfrac{2}{x}\right)^{-3}$

63. $(k^{-2c})^2$

64. $(2y)^{-4}$

65. $\dfrac{y^{4n}}{y^{-3n}}$

66. $2y^{-4}$

67. $(m^{-3}n)^3$

68. $3 \cdot 5^2$

69. $(w^2 z^{-4})^{-3}$

70. $(2x^3)^{-4}$

71. $\left(\dfrac{R^4}{S^{-2}}\right)^3$

72. $\left(\dfrac{x^4}{2y^{-3}}\right)^{-3}$

73. $\left(\dfrac{a^{-3}b^2}{a^{-4}}\right)^3$

74. $\left(\dfrac{2x^2 y^{-3}}{x^{-3}y}\right)^{-4}$

75. $\left(\dfrac{m^{-1}n^3}{m}\right)^{-2}$

76. $\left(\dfrac{a^3 b^{-2}}{b^3 c^{-1}}\right)^{-3}$

77. $\left(\dfrac{x^2}{x^{-3}y^{-2}}\right)^{-1}$

78. $\left(\dfrac{ab^{-3}}{2b^{-2}}\right)^{-3}$

79. $(2^0 h^2)^{-3}$

80. $(5^3 x^{-2} y^{-4})^0$

81. $\left(\dfrac{5t^{-1}u^2}{t^4 u^{-3}}\right)^0$

82. $\left(\dfrac{6^0 a^2 b^{-3}}{2^{-1}ab}\right)^{-2}$

3.9 Scientific Notation

Now that we have discussed zero and negative exponents, we can introduce *scientific notation*, a notation used in many science courses and often seen in calculator displays.

A positive number written in scientific notation is written as the product of some number between 1 and 10 and an integral power of 10. That is, it must be in the form $a \times 10^n$, where a is greater than or equal to 1 but less than 10 and n is an integer. For example, 8.021×10^4 is correctly written in scientific notation, because 8.021 is a number between 1 and 10 and 10^4 is a power of 10.

Any decimal number can be written in scientific notation. We will first discuss how to find the correct power of 10, and then we will discuss how to find a.

FINDING THE EXPONENT OF 10

1. Place a caret (‸) to the right of the first nonzero digit of the number.

2. Draw an arrow from the caret to the actual decimal point.

3. The sign of the exponent of 10 will be positive if the arrow points right and negative if the arrow points left.

4. The number of digits separating the caret and the actual decimal point gives the absolute value of the exponent of 10.

These rules imply that if the number to be converted to scientific notation is greater than or equal to 10, the exponent of 10 will be positive; if the number is less than 1, the exponent of 10 will be negative. If the number is between 1 and 10, the exponent of 10 will be 0.

FINDING a

Place the decimal point so there is exactly one nonzero digit to its left. This means that the decimal point will be where the "caret" is in the instructions for finding the exponent of 10.

A number is then written in scientific notation by writing it in the form $a \times 10^n$.

Example 1   Write the following decimal numbers in scientific notation:

Decimal notation		*Scientific notation*
a.	$0.00641 = 0.006_\wedge 41$	$= 6.41 \times 10^{-3}$
b.	$64{,}100{,}000 = 6_\wedge 4{,}100{,}000.$	$= 6.41 \times 10^7$
c.	$6.41 = 6_\wedge 41$	$= 6.41 \times 10^0$
d.	$6{,}410 = 6_\wedge 410.$	$= 6.41 \times 10^3$
e.	$0.0003015 = 0.0003_\wedge 015$	$= 3.015 \times 10^{-4}$

Number between 1 and 10 ———┘   └—— Power of 10 ■

To change from scientific notation to decimal notation, simply multiply by the power of 10 (see Example 2).

Example 2   Convert each of the following to decimal notation:

a. $8.601 \times 10^4 = 8.601 \times 10{,}000 = 86{,}010$

b. $4.23 \times 10^{-3} = 4.23 \times \dfrac{1}{10^3} = 4.23 \times \dfrac{1}{1{,}000} = 0.00423$ ■

It is sometimes necessary to convert a number such as 672.3×10^2 or 0.0791×10^{-3} to scientific notation or to decimal notation, as shown in Example 3.

Example 3   Convert each number to scientific notation and then to decimal notation:

a. 672.3×10^2 *67230*

$$672.3 = 6.723 \times 10^2$$

Therefore,

$$672.3 \times 10^2 = (6.723 \times 10^2) \times 10^2$$
$$= 6.723 \times (10^2 \times 10^2)$$
$$= 6.723 \times 10^4 \qquad \text{Scientific notation}$$
$$= 67{,}230 \qquad \text{Decimal notation}$$

b. 0.0791×10^{-3}

$$0.0791 = 7.91 \times 10^{-2}$$

Therefore,

$$
\begin{aligned}
0.0791 \times 10^{-3} &= (7.91 \times 10^{-2}) \times 10^{-3} \\
&= 7.91 \times (10^{-2} \times 10^{-3}) \\
&= 7.91 \times 10^{-5} \quad \text{Scientific notation} \\
&= 0.0000791 \quad \text{Decimal notation} \quad \blacksquare
\end{aligned}
$$

When a scientific calculator is used and answers are very large or very small, the calculator display will probably be in scientific notation. On the calculator, however, numbers in scientific notation are displayed in a different (and possibly misleading) way. $5.06\ ^{03}$ does *not* mean 5.06 to the third power. It means 5.06×10^3. (The display $5.06\ \ 03$ also means 5.06×10^3.)

Example 4 Find $80,000,000 \times 300,000$, using a scientific calculator.

The display probably shows $2.4\ ^{13}$ or $2.4\ 13$. Both notations mean 2.4×10^{13}. $\quad \blacksquare$

Example 5 Find $0.0000008 \div 400$, using a scientific calculator.

The display probably shows $2.\ ^{-09}$ or $2.\ -09$. Both notations mean 2.0×10^{-9}. $\quad \blacksquare$

EXERCISES 3.9

Set I In Exercises 1–6, write each number in decimal notation.

1. 8.06×10^3 **2.** 3.14×10^4 **3.** 1.32×10^{-3}

4. 8.2×10^{-4} **5.** 5.26×10^0 **6.** 9.11×10^0

In Exercises 7–18, write each number in scientific notation.

7. 35,300   **8.** 825,000   **9.** 0.00312

10. 0.000145   **11.** 8.97   **12.** 2.497

13. 0.815   **14.** 0.274   **15.** 0.0002

16. 0.006   **17.** 45   **18.** 12

In Exercises 19–22, perform the indicated operations with a scientific calculator, and express each answer correctly in scientific notation.

19. $860,000 \times 630,000$ **20.** $0.0000009 \div 3,000$

21. $\sqrt{0.00000081}$ **22.** $\sqrt{0.00000225}$

Set II In Exercises 1–6, write each number in decimal notation.

1. 5.23×10^4 **2.** 6.34×10^2 **3.** 7.12×10^{-4}

4. 2.3×10^{-3} **5.** 7.32×10^0 **6.** 3.71×10^0

In Exercises 7–18, write each number in scientific notation.

7. 87,600   **8.** 25,000,000   **9.** 0.00631

10. 0.00614   **11.** 3.69   **12.** 3.9

13. 0.153	**14.** 0.456	**15.** 0.0052
16. 0.00003	**17.** 28	**18.** 2.9

In Exercises 19–22, perform the indicated operations with a scientific calculator, and express each answer correctly in scientific notation.

19. $60,000 \times 730,000,000$ **20.** $0.00000012 \div 4,000$

21. $\sqrt{0.00000036}$ **22.** $\sqrt{0.00000289}$

3.10 Review: 3.1–3.9

Terms
3.1

The plus and minus signs in an algebraic expression break it up into smaller pieces called *terms*. A minus sign is part of the term that follows it. *Exception:* An expression within grouping symbols is considered as a single piece even though it may contain plus and minus signs.

Like terms (3.6) Terms having equal literal parts are called *like terms*. Only the numerical coefficients may differ.

Coefficients (3.1) The coefficient of one factor in a term is the product of all the other factors in that term.

Numerical coefficient (3.1) A numerical coefficient is a coefficient that is a number. If we say "the coefficient" of a term, it is understood to mean the *numerical* coefficient of that term.

To Simplify a Product of Factors When Each Factor Has Only One Term
3.3

1. Write the sign. It is negative if there is an odd number of negative factors; otherwise, it is positive.

2. Write the numerical coefficient. It is the product of the absolute values of the numerical coefficients of each factor.

3. Write the variables. Each variable should appear only once.

Rules of Exponents

Rule 3.1	$x^a \cdot x^b = x^{a+b}$	3.2
Rule 3.3	$(x^a)^b = x^{ab}$	3.7
Rule 3.4	$(xy)^a = x^a y^a$	3.7
Rule 3.5	$\dfrac{x^a}{x^b} = x^{a-b}$, if $x \neq 0$	3.7
Rule 3.6	$\left(\dfrac{x}{y}\right)^n = \dfrac{x^n}{y^n}$, if $y \neq 0$	3.7
Rule 3.7	$\left(\dfrac{x^a y^b}{z^c}\right)^n = \dfrac{x^{an} y^{bn}}{z^{cn}}$, $z \neq 0$	3.7
Rule 3.8	$x^0 = 1$, if $x \neq 0$	3.8
Rule 3.9a	$x^{-n} = \dfrac{1}{x^n}$, if $x \neq 0$	3.8
Rule 3.9b	$\dfrac{1}{x^{-n}} = x^n$, if $x \neq 0$	3.8
Rule 3.9c	$\left(\dfrac{x}{y}\right)^{-n} = \left(\dfrac{y}{x}\right)^n$, if $y \neq 0$ and $x \neq 0$	3.8

To Simplify a Product of Factors When One Factor Has More Than One Term
3.4

Use the distributive property:

$$\text{Rule 3.2: } a(b + c) = ab + ac$$

or use the following techniques:

3.5

1. If a set of grouping symbols containing more than one term is preceded by or followed by a factor, use the distributive property.

2. If a set of grouping symbols containing more than one term is not preceded by and not followed by a factor and is:
   a. preceded by a plus or minus sign, insert a 1 between the sign and the grouping symbol and then use the distributive property.
   b. preceded by no sign at all, drop the grouping symbols.

3. When grouping symbols occur within other grouping symbols, it is usually easier to remove the innermost grouping symbols first.

To Combine Like Terms
3.6

1. Identify the like terms.

2. Find the sum of each group of like terms by adding their numerical coefficients; then multiply that sum by the common variables. When no coefficient is written, it is understood to be 1.

Simplified Form of Expressions Having Exponents
3.8

A term having exponents is considered to be simplified when each different base appears only once and when the exponent on that base is a single positive integer.

To Simplify an Algebraic Expression
3.2, 3.3, 3.4, 3.5, 3.6, 3.7 and 3.8

Remove all grouping symbols, simplify each term, and combine all like terms.

Scientific Notation
3.9

A number is correctly expressed in scientific notation if it is in the form $a \times 10^n$, where a is greater than or equal to 1 but less than 10 and n is any integer.

Review Exercises 3.10 Set I

In Exercises 1–4, (a) determine the number of terms and (b) write the second term if there is one.

1. $a^2 + 2ab + b^2$

2. $6x - 2(x^2 + y^2)$

3. $2(m^2 - n^2)$

4. $\dfrac{x + x^2}{4} - 2(x^3 + 1)$

5. Is $2x$ a factor of or a term of the expression $3 + (2x)$?

6. Is $7y$ a factor of or a term of the expression $8(7y)$?

In Exercises 7–12, find the products.

7. $m^2 m^3$

8. yy^7

9. $2 \cdot 2^2$

10. $x^{2y} x^{5y}$

11. $10 \cdot 10^y$

12. $a^5 \cdot a^{-3}$

In Exercises 13–16, combine the like terms.

13. $18x + 23x$

14. $-42xy^2 - 12xy^2$

15. $12ab - 3a + 5b - 21ba$

16. $4 + 2x - y - 7$

In Exercises 17–50, write each expression in simplest form.

17. $8(3x - y)$

18. $-5(a + b)$

19. $(3s - 2t)(-2)$

20. $(3x - 2t) - 2$

21. $5x(x^2 + 7)$

22. $(5x^2y + 3x - 1)(-2x)$

23. $6 + 2(5x^2y + 3x - 1)$

24. $(9x^8y^3)(-7x^4y^5)$

25. $(-5ef^2)(-f^7g^3)(-10e^4g^2)(-2ef)$

26. $(3x)^3$

27. $(x^2y^3)^4$

28. $(n^{-5})^{-3}$

29. $(p^{-3})^5$

30. $(2c)^{-4}$

31. $(k^{-7})^0$

32. $(s^4t^{-1})^{-3}$

33. $(-2x^4)^2$

34. $(-5a^{-4})^3$

35. $(-10)^{-3}$

36. $\dfrac{r^7}{r^5}$

37. $\dfrac{x^{-4}}{x^5}$

38. $\left(\dfrac{c}{d}\right)^4$

39. $\dfrac{m^0}{m^{-3}}$

40. $x^4 + x^2$

41. $\dfrac{n^{-6}}{n^0}$

42. $\left(\dfrac{x^2y^3}{z^4}\right)^5$

43. $\left(\dfrac{a^{-4}}{b^3c^0}\right)^{-5}$

44. $x^{-5} \cdot x^{-3}$

45. $\left(\dfrac{r^{-6}}{s^5t^{-3}}\right)^4$

46. $\left(\dfrac{6x^{-5}y^8}{3x^2y^{-4}}\right)^0$

47. $(5a^3b^{-4})^{-2}$

48. $m + 5 - (m - n)$

49. $2x(3x^2 - x) - (3x^2 - 4)$

50. $2m^2n(3mn^2 - 2n) - 5mn^2(2m - 3m^2n)$

In Exercises 51–53, write each expression without fractions, using negative exponents if necessary.

51. $\dfrac{a^3}{b^2}$

52. $\dfrac{m^2}{n^{-3}}$

53. $\dfrac{u^{-4}v^3}{10^2w^{-5}}$

In Exercises 54–58, evaluate each expression.

54. 4^{-2}

55. $(10^{-2})^2$

56. $\dfrac{2^0}{2^{-3}}$

57. $4^0 \cdot 3^2$

58. $\dfrac{(-8)^2}{-8^2}$

59. Express 45,300 in scientific notation.

60. Express 0.03156 in scientific notation.

Review Exercises 3.10 Set II

NAME _____

In Exercises 1–4, (a) determine the number of terms and (b) write the second term if there is one.

ANSWERS

1. $8x^3 - 2x^2 + 5x + 3$

2. $7x + (3x^4 - 1)$

1a. _____

b. _____

3. $8a - \dfrac{2x + 1}{2} - 4$

4. $\dfrac{p + q}{2} + \dfrac{p}{4} + 3(p^2 - q)$

2a. _____

b. _____

3a. _____

5. Is $8a$ a term of or a factor of the expression $8ab$?

b. _____

6. Is $-3x$ a term of or a factor of the expression $1 - 3x$?

4a. _____

b. _____

In Exercises 7–12, find the products.

7. $x^4y^2x^6$

8. $3 \cdot 3^7$

9. a^2b^7

5. _____

6. _____

7. _____

10. $m^x m^y$

11. $x^{-7} \cdot x^4$

12. $4^2 \cdot 4^4$

8. _____

9. _____

In Exercises 13–16, combine the like terms.

13. $12y - 28y$

14. $3x^2y - 8x^2y$

10. _____

11. _____

15. $14x^2y - 8x^2y + 3xy^2$

16. $7st - 12s + 3t - 15st + s$

12. _____

13. _____

14. _____

15. _____

16. _____

In Exercises 17–50, write each expression in simplest form.

17. $-3(4a + b)$

18. $(7x + y)(-2)$

19. $(7x + y) - 2$

20. $3x(2x - 1)$

21. $9a^2b(ab + 1)$

22. $(3x^4 - x^3 - x^2)(-2x)$

23. $7 + 4(3x^4 - 5 - x^2)$

24. $(-x^2)(3xy^4)(5y^3)$

25. $(-9w^2z^3)(7z^4)(-wx^2z^4)$

26. $(6e^3)^0$

27. $(2xy^2)^5$

28. $(m^{-2})^4$

29. $(m^{-1}n^2)^{-3}$

30. $c^{-5} \cdot d^0$

31. $(-2t^{-3})^2$

32. $(3a^{-2}b)^{-4}$

33. $m^{3x} \cdot m^{-x}$

34. $\dfrac{y^6}{y^2}$

35. $\left(\dfrac{y}{2}\right)^5$

36. $\dfrac{x^{-2}}{x^2}$

37. $x^5 - x^2$

38. $\left(\dfrac{x}{2}\right)^{-4}$

17. _____

18. _____

19. _____

20. _____

21. _____

22. _____

23. _____

24. _____

25. _____

26. _____

27. _____

28. _____

29. _____

30. _____

31. _____

32. _____

33. _____

34. _____

35. _____

36. _____

37. _____

38. _____

39. x^3y^5

40. $\left(\dfrac{r^{-3}}{s^2t^{-2}}\right)^2$

41. $\left(\dfrac{3p}{m^2n^{-1}}\right)^{-3}$

42. $\left(\dfrac{a^2b}{a^{-3}b^4}\right)^{-2}$

43. $\left(\dfrac{18x^{-3}y}{15x^2y^{-2}}\right)^0$

44. $\left(\dfrac{x^8y}{x^5}\right)^2$

45. $(2c^{-4}d^2)^{-2}$

46. $2p - (5 - 6p) + 4$

47. $2 + 3(4x - 5)$

48. $5(2h - 3k) - 2(h - 2k)$

49. $8 - 3(2x^2 + 3x) - (7x + 2x^2 - 1)$

50. $10k(4k^2 - 3k - 5) - 2k(5k^2 - 4k + 10)$

In Exercises 51–53, write each expression without fractions, using negative exponents if necessary.

51. $\dfrac{z}{y^{-2}}$

52. $\dfrac{h^2}{k^2}$

53. $\dfrac{u^{-3}v^2}{5^2w}$

In Exercises 54–58, evaluate each expression.

54. 5^{-2}

55. $(2^{-3})^2$

56. $\dfrac{10^{-2}}{10^0}$

57. $\dfrac{(-11)^2}{-11^2}$

58. $2^4 \cdot 3^0$

59. Express 0.00297 in scientific notation.

60. Express 120,000,000 in scientific notation.

39. _____

40. _____

41. _____

42. _____

43. _____

44. _____

45. _____

46. _____

47. _____

48. _____

49. _____

50. _____

51. _____

52. _____

53. _____

54. _____

55. _____

56. _____

57. _____

58. _____

59. _____

60. _____

Chapter 3 Diagnostic Test

The purpose of this test is to see how well you understand the simplification of algebraic expressions. We recommend that you work this diagnostic test *before* your instructor tests you on this chapter. Allow yourself about 50 minutes.

Complete solutions for all the problems on this test, together with section references, are given in the Answer Section in the back of the book. For the problems you do incorrectly, study the sections referred to.

1. Determine whether $5x$ is a factor or a term.

   a. $5xy$ ~~factor~~   b. $5x + y$ ~~term~~   c. $3 + 5x$ ~~term~~

   d. $(3 + 2y) + 5x$ ~~term~~   e. $(3 + 2y)(+5x)$ ~~factor~~

In Problems 2–5, find the products.

2. $(-3xy)(5x^3y)(-2xy^4)$ **3.** $2(x - 3y)$ $2x - 6y$

4. $(x - 4)(-5)$ $-5x + 20$ **5.** $2xy^2(x^2 - 3y - 4)$ $2x^3y^2 - 6xy^3$

In Problems 6 and 7, remove the grouping symbols.

6. $5 - (x - y)$ $5 - x + y$ **7.** $-2[-4(3c - d) + a] - b$
                                                $-12c + 4d + a$
                                         $24c - 8d - 2a - b$

In Problems 8 and 9, combine like terms.

8. $4x - 3x + 5x$ $-4x$ **9.** $2a - 5b - 7 - 3b + 4 - 5a$
                                       $-3a - 8b - 3$

In Problems 10–25, simplify each expression.

10. $x^3 \cdot x^4$ x^{12} **11.** $(x^2)^3$ x^6 **12.** $\dfrac{x^5}{x^2}$ x^3 **13.** x^{-4} $\dfrac{1}{x^4}$

14. x^2y^{-3} $\dfrac{x^2}{y^3}$ **15.** $\dfrac{a^{-3}}{b}$ $\dfrac{1}{a^3b}$ **16.** $\dfrac{x^{5a}}{x^{3a}}$ **17.** $(4^{3x})^0$

18. $(x^2y^4)^3$ x^6y^{12} **19.** $(a^{-3}b)^2$ $\dfrac{b^2}{a^6}$ **20.** $\left(\dfrac{p^3}{q^2}\right)^2$ **21.** $\left(\dfrac{x}{y^2}\right)^{-3}$ $\dfrac{y^6}{x^3}$

22. $\left(\dfrac{4x^{-2}}{2x^{-3}}\right)^{-1}$ $\dfrac{2x^3}{4x^2}$ $\dfrac{4x^2}{2x^3}$ **23.** $5x - 3(y - x)$ $5x$
                                                 $5x - 3y$

24. $3h(2k^2 - 5h) - h(2h - 3k^2)$

25. $x(x^2 + 2x + 4) - 2(x^2 + 2x + 4)$

In Problems 26 and 27, write the expression without fractions, using negative exponents when necessary.

26. $\dfrac{a^3}{b}$ a^3b^{-1} **27.** $\dfrac{h^{-2}}{k^{-3}h^{-4}}$ $\dfrac{k^3h^2}{h^2}$ k^3h^2

In Problems 28–35, evaluate each expression.

28. $2^3 \cdot 2^2$ 2^5 **29.** $10^{-4} \cdot 10^2$ $\dfrac{1}{100}$ **30.** 5^{-2} $\dfrac{1}{25}$ **31.** $(2^{-3})^2$ $\dfrac{1}{26}$

32. $(4^{-2})^{-1}$ $\dfrac{1}{4^2}$ 16 **33.** $\dfrac{10^{-3}}{10^{-4}}$ 10 **34.** $\dfrac{-3^2}{(-3)^2}$ -1 **35.** $(5^0)^2$ 1

36. Write each number in scientific notation:

   a. 1.326 b. 0.527

Cumulative Review Exercises: Chapters 1– 3

In Exercises 1–6, write "true" if the statement is always true; otherwise, write "false."

1. The additive identity is 1.

2. The additive inverse of $\frac{8}{5}$ is $-\frac{8}{5}$.

3. Subtraction is commutative.

4. Division is associative.

5. $3,870,000 = 3.87 \times 10^{-6}$ in scientific notation.

6. $3x$ is a factor of the expression $3x + 7$.

In Exercises 7–13, evaluate each expression.

7. $\dfrac{14 - 23}{-8 + 11}$

8. $\sqrt{13^2 - 5^2}$

9. $-4^2 + (-4)^2$

10. $(27 - 3^3)(8^2 + 5^2)$

11. $6(-3) - 4\sqrt{36}$

12. $\{-12 - [7 + (3 - 9)]\} - 15$

13. $\dfrac{0}{-11}$

In Exercises 14–16, evaluate each formula using the values of variables given with the formula.

14. $C = \dfrac{5}{9}(F - 32)$ $\qquad F = -13$

15. $A = P(1 + rt)$ $\qquad P = 1,200, r = 0.15, t = 4$

16. $S = 4\pi R^2$ $\qquad \pi \doteq 3.14, R = 10$

In Exercises 17–20, simplify the given expression.

17. $(-5p)(4p^3)$

18. $(-4h^1j^5)(-j^3k^4)(-5hk)(-10j^2h^7)$

19. $7 - 4(3xy - 3)$

20. $3xy^2(8x - 2xy) - 5xy(10xy - 3xy^2)$

4 Equations and Inequalities

The main reason for studying algebra is to equip oneself with the tools necessary for solving problems. Most problems are solved by the use of equations. In this chapter, we show how to solve simple equations. Methods for solving more difficult equations will be given in later chapters.

4.1 Conditional Equations, Solutions and Solution Sets, and Equivalent Equations

Recall from Section 2.4 that an equation is made up of three parts:

1. The equal sign $(=)$

2. The expression to the left of the equal sign, called the left side (or left member) of the equation

3. The expression to the right of the equal sign, called the right side (or right member) of the equation

Conditional Equations

An equation may be a *true* statement, such as $7 = 7$; a *false* statement, such as $8 = 3$; or a statement that is sometimes true and sometimes false, such as $x = -2$. An equation that is true for some values of the variable and false for other values is called a **conditional equation**. The equation $x = -2$ is a conditional equation, because it is a true statement if the value of x is -2 and a false statement otherwise.

A Solution of an Equation

A **solution** of an equation is any value of the variable that, when substituted for the variable, makes the two sides of the equation equal.

Example 1   Determine whether -2 is a solution of the equation $x = -2$.

Substituting -2 for x, we have $-2 = -2$, which is a *true* statement. Therefore, -2 is a solution of the equation $x = -2$. ∎

Example 2   Determine whether 5 is a solution of the equation $x = -2$.

Substituting 5 for x, we have $5 = -2$, which is a *false* statement. Therefore, 5 is not a solution of the equation $x = -2$. ∎

The Solution Set of an Equation

The **solution set** of an equation is the set of all the solutions of the equation. (Recall from Section 1.1 that we enclose the elements of a set within braces.) Thus, the solution set of the equation $x = -2$ is $\{-2\}$, because -2 is the only value of x that makes the statement $x = -2$ true.

Equivalent Equations

Equations that have the same solution set are called **equivalent equations**.

Example 3   Examples of equivalent equations:

a. The solution set of the equation $x + 2 = 0$ is $\{-2\}$, because $(-2) + 2 = 0$ is a true statement, and -2 is the only value of x that makes the statement $x + 2 = 0$ true. Since $\{-2\}$ was also the solution set for the equation $x = -2$, $x + 2 = 0$ and $x = -2$ are equivalent equations.

b. The solution set of the equation $x = 4$ is $\{4\}$, because $4 = 4$ is a true statement, and 4 is the only value of x that makes the statement $x = 4$ true.

The solution set of the equation $x - 4 = 0$ is $\{4\}$, because $4 - 4 = 0$ is a true statement, and 4 is the only value of x that makes the statement $x - 4 = 0$ true.

Therefore, $x = 4$ and $x - 4 = 0$ are equivalent equations. ■

EXERCISES 4.1

Set I
1. Is -4 a solution of the equation $x + 2 = 3$?

2. Is 3 a solution of the equation $x - 5 = 2$?

3. Is 4 a solution of the equation $x + 1 = 5$?

4. Is -4 a solution of the equation $x + 4 = 0$?

5. Is -3 in the solution set for $2 + x = -1$?

6. Is -5 in the solution set for $2 + x = -3$?

7. Is 3 in the solution set for $5 + x = 1$?

8. Is 4 in the solution set for $-1 + x = 1$?

Set II
1. Is 5 a solution of the equation $x + 3 = 4$?

2. Is -3 a solution of the equation $x + 5 = 2$?

3. Is -4 a solution of the equation $x + 1 = -3$?

4. Is 4 a solution of the equation $x + 3 = 7$?

5. Is 3 in the solution set for $2 + x = 5$?

6. Is 5 in the solution set for $2 - x = -3$?

7. Is -1 in the solution set for $5 + x = 1$?

8. Is 2 in the solution set for $-1 + x = 1$?

4.2 Solving Equations by Adding the Same Signed Number to Both Sides

In this chapter, we will be solving equations and inequalities that contain only one variable and in which the variable is not raised to any power. When we solve such an equation, our answer should be an equivalent equation of the form $\boxed{x = a}$, where a is some number. That is, an equation is solved when we succeed in getting the variable by itself on one side of the equal sign and a single number on the other side.

In this section, we solve equations that have the variable on one side of the equation only and in which the coefficient of the variable is an understood 1. The equations will have some number being added to or subtracted from the variable. We solve such equations by using the addition property of equality.

ADDITION PROPERTY OF EQUALITY

If the same number is added to both sides of an equation, the new equation is equivalent to the original equation.

When you use this property to solve an equation, always write the new equation *under* the previous equation.

Example 1 Solve the equation $2 + m = 7$ and graph the solution on the number line.
Solution Since 2 is being *added* to m, we add the *negative* of 2 to both sides.

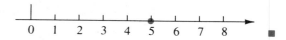

$$\begin{array}{rcl} 2 + m &=& 7 \\ -2 & & -2 \\ \hline m &=& 5 \end{array}$$

Adding -2 to both sides gets m by itself on the left side

Solution

Do not omit the equal sign

Note that the equation $m = 5$ was written *under* the equation $2 + m = 7$.
Graph We must graph the number 5:

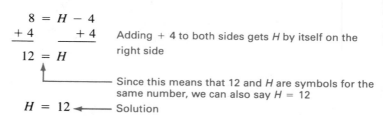

A WORD OF CAUTION   It is incorrect to write

$$2 + m = 7 = m = 5$$

An equal sign here implies that $m = 7$, which is *not* true, and implies that $7 = 5$, which is false

Example 2 Solve the equation $8 = H - 4$ and graph the solution on the number line.
Solution Since -4 is being *added* to H, we add the *negative* of -4 to both sides.

$$\begin{array}{rcl} 8 &=& H - 4 \\ +4 & & +4 \\ \hline 12 &=& H \end{array}$$

Adding $+4$ to both sides gets H by itself on the right side

Since this means that 12 and H are symbols for the same number, we can also say $H = 12$

$H = 12$ ◄——— Solution

Graph   We graph $H = 12$.

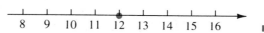

Checking the Solution of an Equation

TO CHECK THE SOLUTION OF AN EQUATION

1. Replace the variable in the given equation by the number obtained as a solution.

2. Perform the indicated operations on both sides of the equal sign.

3. If the resulting number on each side of the equal sign is the same, the solution is correct.

Example 3 Solve and check $x - 5 = 3$ and graph the solution on the number line.
Solution

$$x - 5 = 3$$
$$\underline{+5 \qquad +5}$$
$$x = 8$$

Check

$$x - 5 = 3$$
$$(8) - 5 \overset{?}{=} 3 \qquad x \text{ replaced by the solution 8}$$
$$3 = 3 \qquad \text{The solution is correct}$$

Graph We graph $x = 8$.

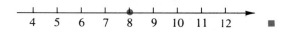

Example 4 Solve and check $\frac{1}{4} + x = 3\frac{1}{4}$ and graph the solution on the number line.
Solution

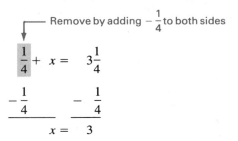

$$\frac{1}{4} + x = 3\frac{1}{4}$$
$$\underline{-\frac{1}{4} \qquad -\frac{1}{4}}$$
$$x = 3$$

Check

$$\frac{1}{4} + x = 3\frac{1}{4}$$
$$\frac{1}{4} + (3) \overset{?}{=} 3\frac{1}{4} \qquad x \text{ replaced by the solution 3}$$
$$3\frac{1}{4} = 3\frac{1}{4}$$

Graph We graph $x = 3$.

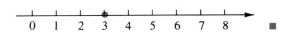

Example 5 Solve $9.08 = x - 5.47$ and graph the solution on the number line.
Solution

$$9.08 = x - 5.47$$
$$\underline{+5.47 \qquad +5.47}$$
$$14.55 = x$$

or

$$x = 14.55$$

Graph

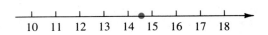

Notice that in all the examples, a number was being added to or subtracted from the variable in the original equation.

EXERCISES 4.2

Set I   Solve and check the following equations and graph each solution on the number line.

1. $x + 5 = 8$ **2.** $x + 4 = 9$ **3.** $x - 3 = 4$

4. $x - 7 = 2$ **5.** $3 + x = -4$ **6.** $2 + x = -5$

7. $x + 4 = 21$ **8.** $x + 15 = 24$ **9.** $x - 35 = 7$

10. $x - 42 = 9$ **11.** $9 = x + 5$ **12.** $11 = x + 8$

13. $12 = x - 11$ **14.** $14 = x - 15$ **15.** $-17 + x = 28$

16. $-14 + x = 33$ **17.** $-28 = -15 + x$ **18.** $-47 = -18 + x$

19. $x + \dfrac{1}{2} = 2\dfrac{1}{2}$ **20.** $x + \dfrac{3}{4} = 5\dfrac{3}{4}$ **21.** $5.6 + x = 2.8$

22. $3.04 + x = 2.96$ **23.** $7.84 = x - 3.98$ **24.** $4.99 = x - 2.08$

Set II   Solve and check the following equations and graph each solution on the number line.

1. $x + 7 = 12$ **2.** $x - 4 = -3$ **3.** $x - 5 = 8$

4. $3 + x = 1$ **5.** $6 + x = -9$ **6.** $-2 + x = 5$

7. $x + 11 = 25$ **8.** $-5 + x = 0$ **9.** $x - 18 = 13$

10. $5 + x = 5$ **11.** $14 = x + 6$ **12.** $3 = x - 1$

13. $17 = x - 11$ **14.** $x = 5 - 8$ **15.** $-21 + x = -41$

16. $-3 + x = 4$ **17.** $-51 = -37 + x$ **18.** $-3 + x = -4$

19. $x + 2\dfrac{1}{4} = 8\dfrac{1}{4}$ **20.** $x - 3\dfrac{1}{2} = 5\dfrac{1}{2}$ **21.** $8.4 + x = 6.2$

22. $3.5 + x = 1.07$ **23.** $5.36 = x - 4.82$ **24.** $x - 1.35 = -4.2$

4.3 Solving Equations by Dividing Both Sides by the Same Signed Number

In this section, the equations we need to solve will still have the variable on one side of the equation only, but the coefficient of the variable will no longer be a 1. We solve such equations by using the division property of equality.

DIVISION PROPERTY OF EQUALITY

If both sides of an equation are divided by the same nonzero number, the new equation is equivalent to the original one.

Example 1   Solve the equation $2x = 10$ and graph the solution on the number line.

Solution Since 2 is being *multiplied* by x, we can get x all by itself on the left side if we *divide* both sides of the equation by 2.

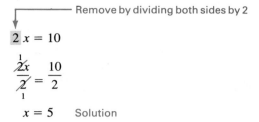

Remove by dividing both sides by 2

$$2\,x = 10$$

$$\frac{\overset{1}{\cancel{2}}x}{\underset{1}{\cancel{2}}} = \frac{10}{2}$$

$$x = 5 \qquad \text{Solution}$$

Graph

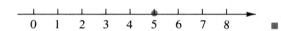

A WORD OF CAUTION   Notice the difference between the equations (a) $2x = 10$ and (b) $2 + x = 10$. In (a), 2 is *multiplied* by x; we get x by itself by *dividing* both sides by 2. In (b), 2 is *added* to x; we get x by itself by *adding* -2 to both sides.   ☑

Example 2   Solve the equation $9x = -27$ and graph the solution on the number line.

Solution

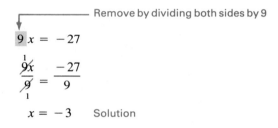

Remove by dividing both sides by 9

$$9\,x = -27$$

$$\frac{\overset{1}{\cancel{9}}x}{\underset{1}{\cancel{9}}} = \frac{-27}{9}$$

$$x = -3 \qquad \text{Solution}$$

Graph

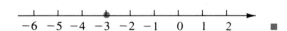

Example 3   Solve the equation $12x = 8$ and graph the solution on the number line.

Solution

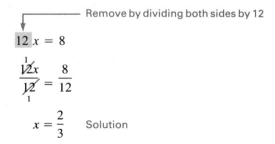

Remove by dividing both sides by 12

$$12\,x = 8$$

$$\frac{\overset{1}{\cancel{12}}x}{\underset{1}{\cancel{12}}} = \frac{8}{12}$$

$$x = \frac{2}{3} \qquad \text{Solution}$$

Graph

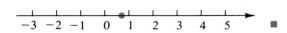

Example 4   Solve the equation $16 = 2x$ and graph the solution on the number line.
Solution

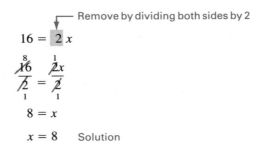

$16 = 2\,x$ — Remove by dividing both sides by 2

$$\dfrac{\overset{8}{\cancel{16}}}{\underset{1}{\cancel{2}}} = \dfrac{\overset{1}{\cancel{2}x}}{\underset{1}{\cancel{2}}}$$

$8 = x$

$x = 8$ Solution

Graph

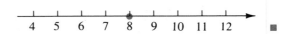

In Examples 5–8, we combine the methods learned in this section with the methods learned in Section 4.2; the variable is multiplied by some number *and* has some number added to or subtracted from it. In such problems, use the following procedure:

TO SOLVE AN EQUATION USING ADDITION AND DIVISION

All numbers on the same side as the variable must be removed.

1. First, remove those numbers being added or subtracted using the method of Section 4.2.

2. Second, remove the coefficient of the variable by dividing both sides by that coefficient.

Example 5   Solve and check the equation $2x + 3 = 11$ and graph the solution on the number line.
Solution The numbers 2 and 3 must be removed from the side with the x.

1. Since the 3 is added, it is removed first.

$$
\begin{array}{rcr}
2x + 3 & = & 11 \\
-\,3 & & -\,3 \\
\hline
2x & = & 8
\end{array}
$$

Adding -3 to both sides

2. Since 2 is the coefficient of x, it is removed by dividing both sides by 2.

$2x = 8$

$$\dfrac{\overset{1}{\cancel{2}x}}{\underset{1}{\cancel{2}}} = \dfrac{8}{2}$$

Dividing both sides by 2

$x = 4$

Check

$$2x + 3 = 11$$

$$2(4) + 3 \stackrel{?}{=} 11$$

$$8 + 3 \stackrel{?}{=} 11$$

$$11 = 11$$

Graph   We graph $x = 4$.

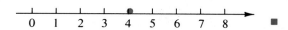

Example 6   Solve the equation $3x - 2 = 10$ and graph the solution on the number line.
Solution

$$
\begin{aligned}
3x - 2 &= \quad 10 \\
+ 2 \quad &\quad + 2 \qquad \text{Adding 2 to both sides} \\
\overline{\quad\quad\quad} \\
3x &= \quad 12
\end{aligned}
$$

$$\frac{\overset{1}{\cancel{3}}x}{\underset{1}{\cancel{3}}} = \frac{12}{3} \qquad \text{Dividing both sides by 3}$$

$$x = 4 \qquad \text{Solution}$$

Graph

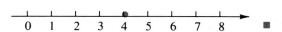

Example 7   Solve and check the equation $-12 = 3x + 15$ and graph its solution on the number line.
Solution 15 and 3 must be removed from the side with the x.

$$
\begin{aligned}
-12 &= 3x + 15 \\
-15 & \qquad\quad -15 \qquad \text{Adding } -15 \text{ to both sides} \\
\overline{\quad\quad\quad} \\
-27 &= 3x
\end{aligned}
$$

$$\frac{-27}{3} = \frac{\overset{1}{\cancel{3}}x}{\underset{1}{\cancel{3}}} \qquad \text{Dividing both sides by 3}$$

$$-9 = x$$

$$x = -9 \qquad \text{Solution}$$

Check

$$-12 = 3x + 15$$

$$-12 \stackrel{?}{=} 3(-9) + 15$$

$$-12 \stackrel{?}{=} -27 + 15$$

$$-12 = -12$$

Graph   We graph $x = -9$.

In Example 8, the variable has a negative coefficient.

Example 8   Solve and check $3 - 2x = 9$ and graph the solution on the number line.

Solution

$$
\begin{aligned}
3 - 2x &= 9 \\
-3 \qquad &\quad -3 \\
\hline
-2x &= 6
\end{aligned}
$$

$$\frac{(\cancel{-2})x}{(\cancel{-2})} = \frac{6}{-2} \qquad \text{Since the coefficient of } x \text{ is } -2, \text{ we divide both sides by } -2$$

$$x = -3 \qquad \text{Solution}$$

Check

$$3 - 2x = 9$$
$$3 - 2(-3) \overset{?}{=} 9$$
$$3 + 6 \overset{?}{=} 9$$
$$9 = 9$$

Graph   We graph $x = -3$.

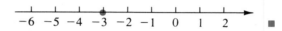

EXERCISES 4.3

Set I   Solve and check the following equations. Graph each solution on the number line.

1. $2x = 8$	**2.** $3x = 15$	**3.** $21 = 7x$
4. $42 = 6x$	**5.** $11x = 33$	**6.** $12x = 48$
7. $4x + 1 = 9$	**8.** $5x + 2 = 12$	**9.** $6x - 2 = 10$
10. $7x - 3 = 4$	**11.** $2x - 15 = 11$	**12.** $3x - 4 = 14$
13. $4x + 2 = -14$	**14.** $5x + 5 = -10$	**15.** $14 = 9x - 13$
16. $25 = 8x - 15$	**17.** $12x + 17 = 65$	**18.** $11x + 19 = 41$
19. $8x - 23 = 31$	**20.** $6x - 33 = 29$	**21.** $14 - 4x = -28$
22. $18 - 6x = -44$	**23.** $8 = 25 - 3x$	**24.** $10 = 27 - 2x$
25. $-73 = 24x + 31$	**26.** $-48 = 36x + 42$	
27. $18x - 4.8 = 6$	**28.** $15x - 7.5 = 8$	
29. $2.5x - 3.8 = -7.9$	**30.** $3.75x + 0.125 = -0.125$	

Set II   Solve and check the following equations. Graph each solution on the number line.

1. $5x = 35$	**2.** $8x = 32$	**3.** $24 = 6x$
4. $56 = 7x$	**5.** $12x = 36$	**6.** $8 = 24x$

7. $3x + 2 = 14$

8. $5x - 3 = -13$

9. $7x - 4 = 10$

10. $8x + 4 = -20$

11. $8x - 12 = 12$

12. $6x - 18 = -18$

13. $4x + 11 = -13$

14. $6x + 13 = -5$

15. $16 = 5x - 14$

16. $14 = 7x - 28$

17. $11x + 19 = 63$

18. $8x - 4 = 24$

19. $12x - 17 = 13$

20. $11x + 5 = 16$

21. $19 - 10x = -26$

22. $8 - 3x = -1$

23. $27 = 32 - 15x$

24. $17 = 14 - 3x$

25. $-67 = 18x + 29$

26. $-12 = 5x + 3$

27. $5x - 3.6 = 5$

28. $4x - 4.26 = 8$

29. $4.1x - 7.4 = -11.5$

30. $0.3x + 5.09 = -8.2$

4.4 Solving Equations by Multiplying Both Sides by the Same Signed Number

In this section, the equations will have the variable on one side of the equation only, but now we may have one fraction in the equation. We will be able to solve such equations by using the multiplication property of equality.

MULTIPLICATION PROPERTY OF EQUALITY

If both sides of an equation are multiplied by the same nonzero number, the new equation is equivalent to the original equation.

Examples 1–3 show how to solve simple equations containing fractions. Multiplying both sides of such equations by the denominator is called "clearing fractions."

Example 1 Solve and check $\dfrac{x}{3} = 5$ and graph the solution on the number line.

Solution Because x is *divided* by 3 in this equation, we will get x all by itself on the left side if we *multiply* both sides of the equation by 3.

$$\frac{x}{3} = 5$$

$$3\left(\frac{x}{3}\right) = 3(5) \qquad \text{Multiplying both sides by 3}$$

$$\overset{1}{\cancel{3}}\left(\frac{x}{\underset{1}{\cancel{3}}}\right) = 3(5)$$

$$x = 15 \qquad \text{Solution}$$

Check

$$\frac{x}{3} = 5$$

$$\frac{15}{3} \overset{?}{=} 5$$

$$5 = 5$$

Graph　　We graph $x = 15$.

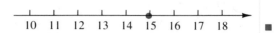

Example 2　　Solve and check $-8 = \dfrac{x}{6}$ and graph the solution on the number line.

Solution

$$-8 = \frac{x}{\boxed{6}}$$

$$6(-8) = \overset{1}{\cancel{6}}\left(\frac{x}{\cancel{6}}\right) \qquad \text{Multiplying both sides by 6}$$

$$-48 = x$$

$$x = -48 \qquad \text{Solution}$$

Check

$$-8 = \frac{x}{6}$$

$$-8 \overset{?}{=} \frac{-48}{6}$$

$$-8 = -8$$

Graph　　We graph $x = -48$.

Example 3　　Solve and check $\dfrac{2x}{3} = 4$ and graph the solution on the number line.

Solution

$$\frac{2x}{\boxed{3}} = 4$$

$$\overset{1}{\cancel{3}}\left(\frac{2x}{\cancel{3}}\right) = 3(4) \qquad \text{Multiplying both sides by 3}$$

$$\boxed{2}\,x = 12$$

$$\frac{\overset{1}{\cancel{2}}x}{\cancel{2}} = \frac{12}{2} \qquad \text{Dividing both sides by 2}$$

$$x = 6 \qquad \text{Solution}$$

Check

$$\frac{2x}{3} = 4$$

$$\frac{2(6)}{3} \overset{?}{=} 4$$

$$\frac{12}{3} \overset{?}{=} 4$$

$$4 = 4$$

Graph   We graph $x = 6$.

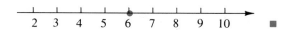

In Examples 4 and 5, we combine all the methods of solving equations learned so far. When more than one number must be removed from the same side as the variable, use the following procedure:

TO SOLVE AN EQUATION USING ADDITION, DIVISION, AND MULTIPLICATION

All numbers on the same side as the variable must be removed.

1. First, remove those numbers being added or subtracted.

2. Multiply both sides by the signed number the variable is divided by.

3. Divide both sides by the coefficient of the variable.

Example 4   Solve and check $4 = \dfrac{2x}{5} - 6$ and graph the solution on the number line.

Solution In this example, three numbers (-6, 5, and 2) must be removed from the right side of the equation.

Step 1.   Since the -6 is *added*, remove it first.

$$4 = \frac{2x}{5} - 6$$
$$\underline{+\,6 \qquad\quad +\,6} \qquad \text{Add } +6 \text{ to both sides}$$
$$10 = \frac{2x}{5}$$

Step 2.   Since the variable is *divided* by 5, remove the 5 next.

$$5(10) = \overset{1}{\cancel{5}}\left(\frac{2x}{\cancel{5}}\right) \qquad \text{Multiply both sides by 5}$$
$$ {}_{1}$$

$$50 = 2x$$

Step 3. Since 2 is the coefficient of x, remove the 2 last.

$$\frac{50}{2} = \frac{\overset{1}{\cancel{2}}x}{\underset{1}{\cancel{2}}} \qquad \text{Divide both sides by 2}$$

$$25 = x$$

$$x = 25 \qquad \text{Solution}$$

Check

$$4 = \frac{2x}{5} - 6$$

$$4 \overset{?}{=} \frac{2(25)}{5} - 6$$

$$4 \overset{?}{=} \frac{50}{5} - 6$$

$$4 \overset{?}{=} 10 - 6$$

$$4 = 4$$

Graph   We graph $x = 25$.

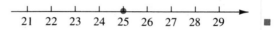

Example 5   Solve and check $\dfrac{3x}{4} + 2 = 11$ and graph the solution on the number line.

Solution The numbers 2, 4, and 3 must be removed from the left side of the equation.

Step 1. Since the 2 is added, remove it first.

$$\begin{array}{rcl} \dfrac{3x}{4} + 2 &=& 11 \\[2mm] \underline{\phantom{\dfrac{3x}{4}} - 2} & & \underline{-\ 2} \\[2mm] \dfrac{3x}{4} &=& 9 \end{array}$$

Step 2. Since the variable is divided by 4, multiply both sides by 4.

$$\frac{3x}{4} = 9$$

$$\overset{1}{\cancel{4}}\left(\frac{3x}{\underset{1}{\cancel{4}}}\right) = 4(9)$$

$$3x = 36$$

Step 3. Since 3 is the coefficient of x, divide both sides by 3.

$$\frac{\overset{1}{\cancel{3}}x}{\underset{1}{\cancel{3}}} = \frac{36}{3}$$

$$x = 12 \qquad \text{Solution}$$

Check

$$\frac{3x}{4} + 2 = 11$$

$$\frac{3(\overset{3}{\cancel{12}})}{\underset{1}{\cancel{4}}} + 2 \overset{?}{=} 11$$

$$9 + 2 \overset{?}{=} 11$$

$$11 = 11$$

Graph   We graph $x = 12$.

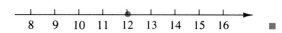

Equations with fractions in them can also be solved by a method given in Section 8.9.

Changing Signs in an Equation   Multiplying both sides of an equation by -1 is equivalent to changing the sign of every term in the equation. For example, consider the equation $5 - x = 7$:

Changing every sign	*Multiplying both sides by* -1
$5 - x = 7$	$5 - x = 7$
$-5 + x = -7$	$(-1)(5 - x) = (-1)7$
└ Same equation ──────→	$-5 + x = -7$

Example 6   Changing the sign of every term in an equation does not change the solution.

$$
\begin{array}{cc}
5 - x = & 7 \\
-5 & -5 \\
\hline
- x = & 2
\end{array}
\qquad
\begin{array}{cc}
5 - x = & 7 \\
-5 + x = & -7 \quad \text{Changed sign of every term} \\
+5 & +5 \\
\hline
\boxed{x = -2}
\end{array}
$$

$$\frac{-x}{-1} = \frac{2}{-1}$$

$$\boxed{x = -2} \longleftarrow \text{Same solution} \quad ■$$

In solving an equation, changing the sign of every term will not change the solution.

Example 7   Examples of changing the sign of every term in an equation:

a. If $-x = 3$

then $x = -3$ The sign of every term was changed

b. If $-x - 8 = -4$

then $x + 8 = 4$ ■

The following property can be included with the other properties of equality, though it is not essential.

> ### SUBTRACTION PROPERTY OF EQUALITY
>
> If the same number is subtracted from both sides of an equation, the new equation is equivalent to the original equation.

EXERCISES 4.4

Set I Solve and check the following equations and graph each solution on the number line.

1. $\dfrac{x}{3} = 4$ **2.** $\dfrac{x}{5} = 3$ **3.** $\dfrac{x}{5} = -2$

4. $\dfrac{x}{6} = -4$ **5.** $4 - \dfrac{x}{7} = 0$ **6.** $3 - \dfrac{x}{8} = 0$

7. $-13 = \dfrac{x}{9}$ **8.** $-15 = \dfrac{x}{8}$ **9.** $\dfrac{x}{10} = 3.14$

10. $\dfrac{x}{5} = 7.8$ **11.** $\dfrac{x}{4} + 6 = 9$ **12.** $\dfrac{x}{5} + 3 = 8$

13. $\dfrac{x}{10} - 5 = 13$ **14.** $\dfrac{x}{20} - 4 = 12$ **15.** $-14 = \dfrac{x}{6} - 7$

16. $-22 = \dfrac{x}{8} - 11$ **17.** $7 = \dfrac{2x}{5} + 3$ **18.** $9 = \dfrac{3x}{4} + 6$

19. $4 - \dfrac{7x}{5} = 11$ **20.** $3 - \dfrac{2x}{9} = 13$ **21.** $-24 + \dfrac{5x}{8} = 41$

22. $-16 + \dfrac{9x}{4} = 29$ **23.** $41 = 25 - \dfrac{4x}{5}$ **24.** $54 = 14 - \dfrac{8x}{7}$

Set II Solve and check the following equations and graph each solution on the number line.

1. $\dfrac{x}{7} = 3$ **2.** $\dfrac{x}{2} = 13$ **3.** $\dfrac{x}{4} = -5$

4. $-5 = \dfrac{x}{6}$ **5.** $6 - \dfrac{x}{3} = 0$ **6.** $8 + \dfrac{x}{2} = 4$

7. $-12 = \dfrac{x}{6}$ **8.** $16 = \dfrac{x}{3}$ **9.** $\dfrac{x}{8} = 12.5$

10. $\dfrac{x}{7} = -1$ **11.** $\dfrac{x}{3} + 5 = 7$ **12.** $\dfrac{x}{2} - 4 = -1$

13. $\dfrac{x}{10} - 4 = 11$ **14.** $\dfrac{x}{9} + 3 = 2$ **15.** $-16 = \dfrac{x}{7} - 9$

16. $5 = 3 + \dfrac{x}{2}$ **17.** $8 = \dfrac{3x}{5} + 4$ **18.** $\dfrac{2x}{4} - 1 = -7$

19. $6 - \dfrac{4x}{5} = 12$ **20.** $8 - \dfrac{2x}{3} = 4$ **21.** $-19 + \dfrac{7x}{4} = 23$

22. $-5 - \dfrac{3x}{4} = 4$ **23.** $38 = 14 - \dfrac{6x}{11}$ **24.** $5 = 2 - \dfrac{7x}{3}$

4.5 Solving Equations in Which Simplification of Algebraic Expressions Is Necessary

4.5A Equations in Which the Variable Appears on Both Sides

All the equations discussed in Chapter 4 so far have had the variable on only one side of the equation.

TO SOLVE AN EQUATION IN WHICH
THE VARIABLE APPEARS ON BOTH SIDES

1. First combine like terms on each side of the equation (if there are any).

2. Remove from one side the term containing the variable by adding the negative of that term to both sides.

3. Solve the resulting equation by the methods given in Sections 4.2, 4.3, and 4.4.

Example 1   Solve and check $6x - 15 = -23 + 2x$ and graph the solution on the number line.
Solution We first remove the entire term $2x$ from the right side of the equation.

$$\begin{array}{rcl} 6x - 15 &=& -23 + 2x \\ -2x & & \quad\ -2x \\ \hline 4x - 15 &=& -23 \end{array}$$ Adding $-2x$ to both sides

The numbers 4 and -15 must be removed from the side containing the x (-15 first, then 4).

$$\begin{array}{rcl} 4x - 15 &=& -23 \\ +15 & & +15 \\ \hline 4x &=& -8 \end{array}$$ Adding 15 to both sides

$$\frac{\overset{1}{\cancel{4}}x}{\underset{1}{\cancel{4}}} = \frac{-8}{4}$$ Dividing both sides by 4

$$x = -2$$

Check

$$6x - 15 = -23 + 2x$$
$$6(-2) - 15 \overset{?}{=} -23 + 2(-2)$$
$$-12 - 15 \overset{?}{=} -23 - 4$$
$$-27 = -27$$

The same answer is obtained whether the x-term is removed from the left side or the right side. We now solve the same problem by removing the x-term from the *left* side.

Alternate solution

$$
\begin{array}{rcl}
6x - 15 &=& -23 + 2x \\
-6x & & \quad -6x \\
\hline
-15 &=& -23 - 4x \\
+23 & & +23 \\
\hline
8 &=& \quad -4x
\end{array}
$$

$$
\frac{8}{-4} = \frac{-4x}{-4} \qquad \text{Dividing both sides by } -4
$$

$$
-2 = x \qquad \text{Same solution}
$$

Graph We graph $x = -2$.

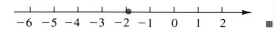

Example 2 Solve and check $4x - 5 - x = 13 - 2x - 3$ and graph the solution on the number line.

Solution

$$
4x - 5 - x = 13 - 2x - 3 \qquad \text{First combine like terms on each side}
$$

$$
\begin{array}{rcl}
3x - 5 &=& 10 - 2x \\
+2x & & \quad +2x \\
\hline
5x - 5 &=& 10 \\
+5 & & 5 \\
\hline
5x &=& 15
\end{array}
$$

Add $2x$ to both sides so an x-term remains on only one side

Add 5 to both sides

$$
\frac{5x}{5} = \frac{15}{5} \qquad \text{Divide both sides by 5}
$$

$$
x = 3 \qquad \text{Solution}
$$

Check

$$
4x - 5 - x = 13 - 2x - 3
$$
$$
4(3) - 5 - (3) \overset{?}{=} 13 - 2(3) - 3
$$
$$
12 - 5 - 3 \overset{?}{=} 13 - 6 - 3
$$
$$
4 = 4
$$

Graph We graph $x = 3$.

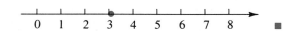

NOTE In the following exercises, if you're expressing answers in decimal form and the decimal repeats, round off the answer to three decimal places. When checking such solutions, the two sides usually will not be exactly equal to each other. ☑

EXERCISES 4.5A

Set I   Solve and check each of the following equations and graph each solution on the number line.

1. $3x + 11 = 14x$ **2.** $5x + 14 = 19x$

3. $9x - 7 = 2x$ **4.** $16x - 3 = 13x$

5. $2x - 7 = x$ **6.** $5x - 8 = x$

7. $5x = 3x - 4$ **8.** $7x = 4x - 9$

9. $9 - 2x = x$ **10.** $8 - 5x = 3x$

11. $3x - 4 = 2x + 5$ **12.** $5x - 6 = 3x + 6$

13. $6x + 7 = 3 + 8x$ **14.** $4x + 28 = 7 + x$

15. $7x - 8 = 8 - 9x$ **16.** $5x - 7 = 7 - 9x$

17. $3x - 7 - x = 15 - 2x - 6$

18. $5x - 2 - x = 4 - 3x - 27$

19. $8x - 13 + 3x = 12 + 5x - 7$

20. $9x - 16 + 6x = 11 + 4x - 5$

21. $7 - 9x - 12 = 3x + 5 - 8x$

22. $13 - 11x - 17 = 5x + 4 - 10x$

23. $7.84 - 1.15x = 2.45$

24. $6.09 - 3.75x = 5.45x$

Set II   Solve and check each of the following equations and graph each solution on the number line.

1. $4x + 15 = 19x$ **2.** $8x - 3 = 5x$

3. $9x - 42 = 7x$ **4.** $8 + x = 17x$

5. $3x - 10 = x$ **6.** $4x + 7 = x$

7. $6x = 2x - 8$ **8.** $9x = 3x + 18$

9. $12 - 5x = x$ **10.** $21 - 6x = x$

11. $5x - 7 = 4x + 6$ **12.** $7x + 2 = 8x - 2$

13. $8x + 5 = 14 + 11x$ **14.** $6x - 2 = 4x - 2$

15. $9x - 13 = 13 - 4x$ **16.** $-4x + 3 = 3x + 3$

17. $6x - 2 - x = 21 - 3x - 7$

18. $7x + 3 - 2x = 5 - 3x$

19. $4x + 14 + 2x = 12 - 3x - 8$

20. $8x - 3 - 5x = 5 - 2x - 9$

21. $16 - 7x - 4 = 5x + 6 - 4x$

22. $8 + 4x - 3 = 2x + 5 - 7x$

23. $8.42 - 2.35x = 1.25x$

24. $2.67x + 3.4 = -5.33x$

4.5B Equations Containing Grouping Symbols

When grouping symbols appear in an equation, first remove them and then solve the resulting equation by the methods discussed in the previous sections.

The complete procedure for solving an equation in one variable that has no exponents is as follows:

TO SOLVE AN EQUATION IN ONE VARIABLE

1. Remove grouping symbols.

2. Combine like terms on each side of the equation.

3. If the variable appears on both sides of the equation, remove from one side the term containing the variable by adding the negative of that term to both sides.

4. Remove all numbers that appear on the same side as the variable.

First, remove those numbers being added or subtracted.

Second, multiply both sides by the signed number the variable is divided by.

Third, divide both sides by the coefficient of the variable.

5. Check the solution in the original equation.

Example 3 Solve and check $10x - 2(3 + 4x) = 7 - (x - 2)$ and graph the solution on the number line.

Solution

$$10x - 2(3 + 4x) = 7 - (x - 2)$$

$$10x - 6 - 8x = 7 - x + 2 \qquad \text{Remove grouping symbols}$$

$$2x - 6 = 9 - x \qquad \text{Combine like terms on each side}$$
$$\underline{+ x \qquad\qquad + x} \qquad \text{Get the } x\text{-term on only one side}$$
$$3x - 6 = 9$$
$$\underline{+ 6 \quad +6}$$
$$3x \quad = 15$$

$$\frac{3x}{3} = \frac{15}{3} \qquad \text{Divide both sides by 3}$$

$$x = 5 \qquad \text{Solution}$$

Check

$$10x - 2(3 + 4x) = 7 - (x - 2)$$

$$10(5) - 2(3 + 4 \cdot 5) \stackrel{?}{=} 7 - (5 - 2)$$

$$10(5) - 2(3 + 20) \stackrel{?}{=} 7 - (3)$$

$$10(5) - 2(23) \stackrel{?}{=} 7 - 3$$

$$50 - 46 \stackrel{?}{=} 4$$

$$4 = 4$$

Graph We graph $x = 5$.

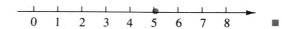

Example 4 Solve $7y - 3(2y - 5) = 6(2 + 3y) - 31$.
Solution

$$7y - 3(2y - 5) = 6(2 + 3y) - 31$$

$7y - 6y + 15 = 12 + 18y - 31$ Remove grouping symbols

$y + 15 = 18y - 19$ Combine like terms on each side

$\underline{ -y   -y }$ Get the y-term on only one side

$15 = 17y - 19$

$\underline{+ 19 + 19 }$

$34 = 17y$

$\dfrac{34}{17} = \dfrac{17y}{17}$ Divide both sides by 17

$y = 2$ ∎

Example 5 Solve $5(2 - 3x) - 4 = 5x + [-(2x - 10) + 8]$.
Solution

$5(2 - 3x) - 4 = \quad 5x + [-(2x - 10) + 8]$

$10 - 15x - 4 = \quad 5x + [-2x + 10 + 8]$

$10 - 15x - 4 = \quad 5x + [-2x + 18]$ Remove grouping symbols

$10 - 15x - 4 = \quad 5x - 2x + 18$

$-15x + 6 = \quad 3x + 18$ Combine like terms on both sides

$\underline{+ 15x + 15x }$ Get the x-term on only one side

$6 = \quad 18x + 18$

$\underline{- 18 - 18}$

$-12 = \quad 18x$

$\dfrac{-12}{18} = \dfrac{\cancel{18}^{1}x}{\cancel{18}_{1}}$

$-\dfrac{2}{3} = \quad x$

$x = -\dfrac{2}{3}$ ∎

EXERCISES 4.5B

Set I Solve and check the following equations and graph each solution on the number line.

1. $5x - 3(2 + 3x) = 6$

2. $7x - 2(5 + 4x) = 8$

3. $6x + 2(3 - 8x) = -14$

4. $4x + 5(4 - 5x) = -22$

5. $7x + 5 = 3(3x + 5)$

6. $8x + 6 = 2(7x + 9)$

7. $9 - 4x = 5(9 - 8x)$

8. $10 - 7x = 4(11 - 6x)$

9. $3y - 2(2y - 7) = 2(3 + y) - 4$

10. $4a - 3(5a - 14) = 5(7 + a) - 9$

11. $6(3 - 4x) + 12 = 10x - 2(5 - 3x)$

12. $7(2 - 5x) + 27 = 18x - 3(8 - 4x)$

13. $2(3x - 6) - 3(5x + 4) = 5(7x - 8)$

14. $4(7z - 9) - 7(4z + 3) = 6(9z - 10)$

15. $6(5 - 4h) = 3(4h - 2) - 7(6 + 8h)$

16. $5(3 - 2k) = 8(3k - 4) - 4(1 + 7k)$

17. $2[3 - 5(x - 4)] = 10 - 5x$

18. $3[2 - 4(x - 7)] = 26 - 8x$

19. $3[2h - 6] = 2\{2(3 - h) - 5\}$

20. $6(3h - 5) = 3\{4(1 - h) - 7\}$

21. $5(3 - 2x) - 10 = 4x + [-(2x - 5) + 15]$

22. $4(2 - 6x) - 6 = 8x + [-(3x - 11) + 20]$

23. $9 - 3(2x - 7) - 9x = 5x - 2[6x - (4 - x) - 20]$

24. $14 - 2(7 - 4x) - 4x = 8x - 3[2x - (5 - x) - 30]$

25. $-2\{5 - [6 - 3(4 - x)] - 2x\} = 13 - [-(2x - 1)]$

26. $-3\{10 - [7 - 5(4 - x) - 8]\} = 11 - [-(5x - 4)]$

In Exercises 27–30, round off answers to three decimal places.

27. $5.073x - 2.937(8.622 + 7.153x) = 6.208$

28. $21.35 - 27.06x = 34.19(19.22 - 37.81x)$

29. $8.23x - 4.07(6.75x - 5.59) = 3.84(9.18 - x) - 2.67$

30. $11.28(15.93x - 24.66) - 35.42(29.05 - 41.84x) = 22.41(32.56x - 16.29)$

Set II   Solve and check the following equations and graph each solution on the number line.

1. $4x - 5(3 + 2x) = 3$ **2.** $5x + 3(2 - x) = 8$

3. $8x + 3(4 - 5x) = -16$ **4.** $9x - 4(x + 3) = 3$

5. $9x + 12 = 2(4x + 5)$ **6.** $8x + 6 = 5(2x - 4)$

7. $10 - 6x = 4(8 - 7x)$ **8.** $7 - 3x = 5(7 - 2x)$

9. $2y - 3(5y - 8) = 2(5 + y) - 10$

10. $3x - 5(2x - 3) = 4(2 - x) + 7$

11. $5(6 - 3a) + 18 = -9a - 3(4 - 2a)$

12. $3(3 - 2x) + 5 = 9x - 4(3x - 1)$

13. $3(2z - 6) - 2(6z + 4) = 5(z + 8)$

14. $2(5 - 3x) + 6 = 8x - 5(4 - 2x)$

15. $7(3 - 5h) = 4(3h - 2) - 6(7 + 9h)$

16. $4(2 - x) = 5(3x - 2) - 4(2 + 5x)$

17. $3[2 - 4(k - 6)] = 12 - 6k$

18. $2[2 - 3(x - x)] = 3 - 8x$

19. $4[2x - 5] = 3\{6(7 - x) - 12\}$

20. $5(2x - 3) = 3\{6(1 - x) - 5\} + 10$

21. $5(1 - 2x) - 3 = 4x + [-(2x - 8) + 6]$

22. $20 = 18 - \{-2[3z - 2(z - 1)]\}$

23. $10 - 3(7 - 3x) - 2x = 6x - 5[3x - (4 - x) - 5]$

24. $12 = -\{-3[4z - 2(z - 2)]\}$

25. $-2\{3 - [2 - 4(5 - x) - 7]\} = 12 - [-(3x - 2)]$

26. $6(2 - 3y) - 5 = 5y + [-(2y - 7) + 14]$

In Exercises 27–30, round off answers to three decimal places.

27. $61.25 - 23.04x = 16.19(18.32 - 1.06x)$

28. $7.209x - 4.395(6.281 + 9.154x) = 8.013$

29. $21.82(39.51x - 62.46) - 24.53(50.29 - 48.14x) = 14.28(65.23x - 92.61)$

30. $5.06(18.13x - 4.021) - 6.12(3.062 - 4.31x) = 42.12(31.16x - 10.04)$

4.6 Conditional Equations, Identities, and Equations with No Solution

There are different kinds of equations. In this section, we discuss three types: conditional equations, identities, and equations with no solution.

Conditional Equations As mentioned in Section 4.1, a **conditional equation** is an equation whose two sides are equal only when certain numbers are substituted for the variable. All equations given in this chapter so far have been conditional equations.

Identities If the two sides of an equation are equal when *any* permissible number is substituted for the variable, the equation is called an **identity**. Therefore, an identity has an endless number of solutions.

Example 1   Verify that $0, -5, 0.5,$ and 7 are solutions of the identity $2(5x - 7) = 10x - 14$.
Check for 0

$$2(5[0] - 7) \stackrel{?}{=} 10[0] - 14$$
$$2(0 - 7) \stackrel{?}{=} 0 - 14$$
$$2(-7) \stackrel{?}{=} -14$$
$$-14 = -14$$

Check for -5

$$2(5[-5] - 7) \stackrel{?}{=} 10[-5] - 14$$
$$2(-25 - 7) \stackrel{?}{=} -50 - 14$$
$$2(-32) \stackrel{?}{=} -64$$
$$-64 = -64$$

Check for 0.5

$$2(5[0.5] - 7) \stackrel{?}{=} 10[0.5] - 14$$
$$2(2.5 - 7) \stackrel{?}{=} 5 - 14$$
$$2(-4.5) \stackrel{?}{=} -9$$
$$-9 = -9$$

Check for 7

$$2(5[7] - 7) \stackrel{?}{=} 10[7] - 14$$

$$2(35 - 7) \stackrel{?}{=} 70 - 14$$

$$2(28) \stackrel{?}{=} 56$$

$$56 = 56$$

NOTE   The two sides are equal if *any* real number is substituted for x.   ∎

Equations with No Solution   If *no* number will make the two sides of an equation equal, we say that the equation is an **equation with no solution**.

Example 2   Consider the equation $x + 1 = x + 2$.

Try 0 as a solution: $0 + 1 \neq 0 + 2$.

Try 1 as a solution: $1 + 1 \neq 1 + 2$.

Try 4 as a solution: $4 + 1 \neq 4 + 2$.

Try -6 as a solution: $-6 + 1 \neq -6 + 2$.

Will *any* number work? No. *Unequal* numbers have been added to the same number, x; therefore, the sums cannot be equal.   ∎

Usually, we cannot determine whether an equation is a conditional equation, an identity, or an equation with no solution simply by looking at it. Instead, we try to solve the equation by using the methods shown in the preceding sections. In those sections, the equations always reduced to the form $x = a$. Now, however, three outcomes are possible:

1. If the equation *can* be reduced to the form $x = a$, where a is some real number, the equation is a *conditional equation*.

2. If the variable drops out and the two sides of the equation reduce to the same constant so we obtain a *true* statement (for example, $0 = 0$), the equation is an *identity*.

3. If the variable drops out and the two sides of the equation reduce to unequal constants so we obtain a *false* statement (for example, $0 = 3$), the equation is an *equation with no solution*.

Example 3   Solve $4x - 2(3 - x) = 12$, or identify the equation as an identity or an equation with no solution.

$$4x - 2(3 - x) = 12$$

$$4x - 6 + 2x = 12$$

$$6x = 18$$

$$x = 3 \qquad \text{Conditional equation (single solution)}$$

Because the equation reduced to the form $x = a$, we know it is a conditional equation. The solution is $x = 3$.   ∎

Example 4 Solve $2(5x - 7) = 10x - 14$, or identify the equation as an identity or an equation with no solution.

$$2(5x - 7) = 10x - 14$$

$$
\begin{array}{rcl}
10x - 14 &=& 10x - 14 \\
-10x & & -10x \\
\hline
-14 &=& -14
\end{array}
$$

Add $-10x$ to both sides

True

When we tried to get the x-terms on one side, all the x's dropped out. Because the two sides of the equation reduced to the same constant and we obtained a *true statement* ($-14 = -14$), the equation is an identity. (The solution set is the set of all real numbers.) ∎

Example 5 Solve $3(2x - 5) = 2x + 4(x - 1)$, or identify the equation as an identity or an equation with no solution.

$$
\begin{array}{rcl}
3(2x - 5) &=& 2x + 4(x - 1) \\
6x - 15 &=& 2x + 4x - 4 \\
6x - 15 &=& 6x - 4 \\
-6x & & -6x \\
\hline
-15 &=& -4
\end{array}
$$

Add $-6x$ to both sides

False

When we tried to get the x-terms on one side, all the x's dropped out. Because the two sides of the equation reduced to different constants and we obtained a *false statement* ($-15 = -4$), the equation is an equation with no solution. (The solution set is the empty set, { }.) ∎

EXERCISES 4.6

Set I Find the solution of each conditional equation. Identify any equation that is *not* a conditional equation as either an identity or an equation with no solution.

1. $x + 3 = 8$ **2.** $4 - x = 6$

3. $2x + 5 = 7 + 2x$ **4.** $10 - 5y = 8 - 5y$

5. $6 + 4x = 4x + 6$ **6.** $7x + 12 = 12 + 7x$

7. $5x - 2(4 - x) = 6$ **8.** $8x - 3(5 - x) = 7$

9. $6x - 3(5 + 2x) = -15$ **10.** $4x - 2(6 + 2x) = -12$

11. $4x - 2(6 + 2x) = -15$ **12.** $6x - 3(5 + 2x) = -12$

13. $7(2 - 5x) - 32 = 10x - 3(6 + 15x)$

14. $6(3 - 4x) + 10 = 8x - 3(2 - 3x)$

15. $2(2x - 5) - 3(4 - x) = 7x - 20$

16. $3(x - 4) - 5(6 - x) = 2(4x - 21)$

17. $2[3 - 4(5 - x)] = 2(3x - 11)$

18. $3[5 - 2(7 - x)] = 6(x - 7)$

19. $460.2x - 23.6(19.5x - 51.4) = 1,213.04$

20. $46.2x - 23.6(19.5x - 51.4) = 213.04$

Set II   Find the solution of each conditional equation. Identify any equation that is *not* a conditional equation as either an identity or an equation with no solution.

1. $7 - x = 11$

2. $6y - 8 = 3 + 6y$

3. $7x - 4 = 3 + 7x$

4. $3(2 - x) = 5(2x + 1)$

5. $8 - 6x = -6x + 8$

6. $6(2x - 1) = 3(4x + 2) - 12$

7. $4x - 3(2 - x) = 8$

8. $7h - 3(5 - h) = 10$

9. $9x - 3(7 + 3x) = -21$

10. $4(3 - 4x) - 5 = 8(1 - 2x) - 1$

11. $2(x - 4) - (3 + 2x) = 3$

12. $2(7k + 9) - 18 = 14k$

13. $9(3 + 4x) - 17 = 14x + 2(6 + 11x)$

14. $3(2y - 7) - 2(5 - y) = 8y - 31$

15. $8(3x - 2) - (5 + 16x) = 8x - 5$

16. $5(4x - 3) + 6 = 2(3 + 10x)$

17. $5[2 - 3(2 - x)] = 7(2x - 1)$

18. $4(5x - 9) = 3[2 - 4(6 - x)]$

19. $460.2x - 23.6(19.5x - 51.4) = 213.04$

20. $3.76x - 1.02(5.21x - 10.7) = 21.45$

4.7 Graphing and Solving Inequalities in One Variable

Basic Definitions

An *equation* is a statement that two expressions are *equal*. An *inequality* is a statement that two expressions are *not equal*.

An inequality has three parts:

$$3x + 4 > 2 - x$$

Left side ⟶

Right side

Inequality symbol; other inequality symbols can be used here

In this text, we will discuss only inequalities that have the symbols $\neq$, $>$, $<$, $\leq$, or $\geq$.

The following three inequality symbols were introduced in Section 1.1:

"Unequal to" symbol ($\neq$) $a \neq b$ is read "a is unequal to b."

"Greater than" symbol ($>$) $a > b$ is read "a is greater than b."

"Less than" symbol ($<$) $a < b$ is read "a is less than b."

Example 1   Examples of reading the inequalities $\neq$, $>$, and $<$:

a. $5 \neq 7$ is read "5 is unequal to 7."

b. $5 \neq x - 3$ is read "5 is unequal to x minus 3."

c. $11 > 2$ is read "11 is greater than 2."

d. $3x - 4 > 7$ is read "$3x$ minus 4 is greater than 7."

e. $3 < 6$ is read "3 is less than 6."

f. $2x < 5 - x$ is read "2x is less than 5 minus x." ■

"Greater Than or Equal To" Symbol ($\geq$) The inequality $a \geq b$ is read "a is greater than *or* equal to b." This means if $\begin{cases} \text{either } a > b \\ \text{or} \quad\;\; a = b \end{cases}$ is true, then $a \geq b$ is true.

Example 2 The meaning of $\geq$:

a. $5 \geq 1$ is true because $5 > 1$ is true.

b. $1 \geq 1$ is true because $1 = 1$ is true. ■

Example 3 Examples of reading $\geq$:

a. $-9 \geq -16$ is read "negative 9 is greater than or equal to negative 16."

b. $x + 6 \geq 10$ is read "x plus 6 is greater than or equal to 10." ■

"Less Than or Equal To" Symbol ($\leq$) The inequality $a \leq b$ is read "a is less than *or* equal to b." This means if $\begin{cases} \text{either } a < b \\ \text{or} \quad\;\; a = b \end{cases}$ is true, then $a \leq b$ is true.

Example 4 The meaning of $\leq$:

a. $2 \leq 3$ is true because $2 < 3$ is true.

b. $3 \leq 3$ is true because $3 = 3$ is true. ■

Example 5 Examples of reading $\leq$:

a. $-7 \leq 0$ is read "negative 7 is less than or equal to 0."

b. $7 \leq 5x - 2$ is read "7 is less than or equal to 5x minus 2." ■

The Graph of an Inequality
An inequality in one variable can be graphed on the number line.

Example 6 Graph each of the following inequalities:

a. $x \geq 3$

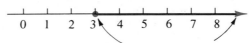

The solid circle and the arrow together indicate that the 3 and all numbers to the *right* of the 3 are solutions

b. $x < 1$

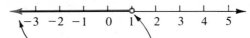

The arrow indicates that all numbers to the *left* of 1 are solutions

Hollow circle because 1 itself is *not* a solution

c. $x \neq 2$

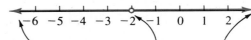

This arrow indicates that all numbers to the *left* of −2 are solutions

Hollow circle because −2 is *not* a solution

This arrow indicates that all numbers to the *right* of −2 are solutions ■

The Sense of an Inequality

The **sense** of an inequality symbol refers to the direction in which the symbol points.

$a > b$ | Same sense (both are >)

$c > d$

$a < b$ | Opposite sense (one is <, one is >)

$c > d$

$a \leq b$ | Opposite sense (one is ≤, one is ≥)

$c \geq d$

$a \leq b$ | Same sense (both are ≤)

$c \leq d$

Earlier in this chapter, we solved equations by adding the same number to both sides of the equation, multiplying both sides of the equation by the same number, and so on. We now examine how these procedures affect the *sense* of an *inequality*.

Example 7 Determine whether the sense is changed in each of the following:

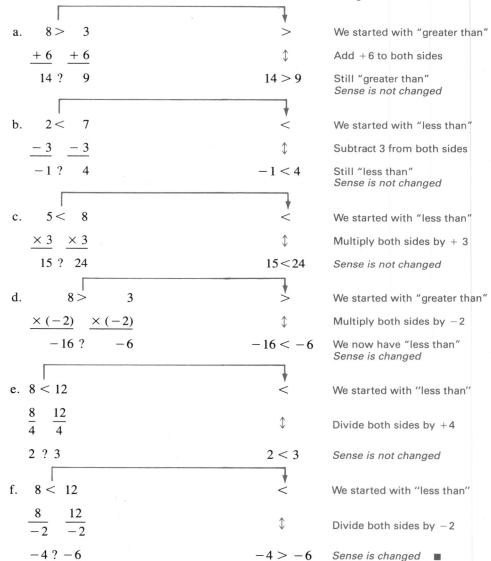

a. $8 > 3$

$+6 \quad +6$

$14\ ?\ 9$ $14 > 9$

> We started with "greater than"
> Add +6 to both sides
> Still "greater than" *Sense is not changed*

b. $2 < 7$

$-3 \quad -3$

$-1\ ?\ 4$ $-1 < 4$

> We started with "less than"
> Subtract 3 from both sides
> Still "less than" *Sense is not changed*

c. $5 < 8$

$\times 3 \quad \times 3$

$15\ ?\ 24$ $15 < 24$

> We started with "less than"
> Multiply both sides by + 3
> *Sense is not changed*

d. $8 > 3$

$\times (-2) \quad \times (-2)$

$-16\ ?\quad -6$ $-16 < -6$

> We started with "greater than"
> Multiply both sides by −2
> We now have "less than" *Sense is changed*

e. $8 < 12$

$\dfrac{8}{4} \quad \dfrac{12}{4}$

$2\ ?\ 3$ $2 < 3$

> We started with "less than"
> Divide both sides by +4
> *Sense is not changed*

f. $8 < 12$

$\dfrac{8}{-2} \quad \dfrac{12}{-2}$

$-4\ ?\ -6$ $-4 > -6$

> We started with "less than"
> Divide both sides by −2
> *Sense is changed* ∎

We can see that when we multiply or divide both sides of an inequality that contains a $<$, $\leq$, $>$, or $\geq$ by a *negative number*, the *sense* of the inequality changes.

The Solution and Solution Set of an Inequality

A conditional inequality is an inequality that is true for some values of the variable and false for others. (Examples 1b, 1d, 1f, 3b, and 5b are examples of conditional inequalities.)

A *solution* of a conditional inequality is any number that, when substituted for the variable, makes the inequality a true statement. The *solution set* of an inequality is the set of all numbers that are solutions of the inequality. While the equations we have solved in this chapter have had just one solution (except for identities), an inequality usually has many solutions.

When we solve an inequality, we must find all the values of the variable that satisfy the inequality. Therefore, in the problems in this book, we want our answer to be in the form $x < a$, $x > a$, $x \leq a$, $x \geq a$, or $x \neq a$, where a is some number. To accomplish this, we may need to add the same quantity to both sides of the inequality, multiply both sides by the same number, and so forth.

The basic rules used to solve inequalities that contain $<$, $\leq$, $>$, or $\geq$ are the same as those used to solve equations, with the exception that *the sense must be changed when multiplying or dividing both sides of an inequality by a negative number.* The following summary shows how the methods are alike or different.

In solving equations	*In solving inequalities containing* $>$, $\geq$, $<$, *or* $\leq$
<u>Addition Rule:</u> The same number may be added to both sides.	<u>Addition Rule:</u> The same number may be added to both sides.
<u>Subtraction Rule:</u> The same number may be subtracted from both sides.	<u>Subtraction Rule:</u> The same number may be subtracted from both sides.
<u>Multiplication Rule:</u> Both sides may be multiplied by the same number.	<u>Multiplication Rule:</u> **1.** Both sides may be multiplied by the same *positive* number. **2.** When both sides are multiplied by the same *negative* number, *the sense must be changed*.
<u>Division Rule:</u> Both sides may be divided by the same nonzero number.	<u>Division Rule:</u> **1.** Both sides may be divided by the same *positive* number. **2.** When both sides are divided by the same *negative* number, *the sense must be changed*.

To solve an inequality that contains the $\neq$ sign, we use the same method as for solving an equation, except that we write the $\neq$ sign in each step rather than an equal sign (see Example 12).

The methods used to solve inequalities may be summarized as follows:

TO SOLVE AN INEQUALITY THAT CONTAINS $\neq$

Proceed in exactly the same way used to solve equations.

TO SOLVE AN INEQUALITY THAT CONTAINS $<$, $\leq$, $>$, OR $\geq$

Proceed in the same way used to solve equations, with the exception that the sense must be changed when multiplying or dividing both sides by a negative number.

Example 8   Solve $x + 3 < 7$ and graph the solution on the number line.
Solution

$$\begin{array}{rl}
x + 3 < & 7 \\
\underline{-\,3} \quad & \underline{-\,3} \qquad \text{Add } -3 \text{ to both sides} \\
x \quad < & 4 \qquad \text{Solution}
\end{array}$$

This means that if we replace x by any number less than 4 in the given inequality, we get a true statement. For example, if we replace x by 3 (which is less than 4),

$$x + 3 < 7$$
$$\overset{?}{3 + 3 < 7}$$
$$6 < 7 \qquad \text{True}$$

Graph   We graph $x < 4$.

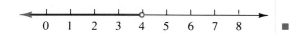

Example 9   Solve $2x - 5 > 1$ and graph the solution on the number line.
Solution

$$\begin{array}{rl}
2x - 5 > & 1 \\
\underline{+\,5} \quad & \underline{+\,5} \qquad \text{Add 5 to both sides} \\
2x \quad > & 6
\end{array}$$

$$\frac{2x}{2} > \frac{6}{2} \qquad \begin{array}{l}\text{Divide both sides by 2}\\ \text{Sense is not changed}\end{array}$$

$$x > 3 \qquad \text{Solution}$$

Graph

Example 10   Solve $3x - 2(2x - 7) \le 2(3 + x) - 4$ and graph the solution on the number line.
Solution

$$3x - 2(2x - 7) \le 2(3 + x) - 4$$
$$3x - 4x + 14 \le 6 + 2x - 4$$
$$\begin{array}{rl}
-x + 14 \le & 2 + 2x \\
\underline{-\,2x - 14} \quad & \underline{-\,14 - 2x} \qquad \text{Add } -2x - 14 \text{ to both sides.} \\
-3x \quad \le & -12
\end{array}$$

$$\frac{-3x}{-3} \ge \frac{-12}{-3} \qquad \begin{array}{l}\text{Divide both sides by } -3\\ \text{Sense is changed}\end{array}$$

$$x \ge 4 \qquad \text{Solution}$$

Alternate solution **The problem can also be done as follows:**

$$3x - 2(2x - 7) \le 2(3 + x) - 4$$

$$3x - 4x + 14 \le 6 + 2x - 4$$

$$
\begin{array}{ll}
-x + 14 \le & 2 + 2x \\
\underline{+x - \ 2} & \underline{-2 + \ x} \qquad \text{Add } +x - 2 \text{ to both sides} \\
12 \le & 3x
\end{array}
$$

$$\frac{12}{3} \le \frac{3x}{3} \qquad \begin{array}{l}\text{Divide both sides by 3}\\ \text{Sense is not changed}\end{array}$$

$$4 \le x$$

$$x \ge 4 \qquad x \ge 4 \text{ has the same meaning as } 4 \le x$$

Graph

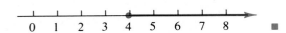

Example 11   Solve $4(y - 2) + 3 \ge 3(y + 3)$ and graph the solution on the number line.
Solution

$$4(y - 2) + 3 \ge 3(y + 3)$$

$$4y - 8 + 3 \ge 3y + 9 \qquad \text{Remove parentheses}$$

$$
\begin{array}{ll}
4y - 5 \ge & 3y + 9 \\
\underline{-3y + 5} & \underline{-3y + 5} \qquad \text{Add } -3y + 5 \text{ to both sides} \\
y \ge 14
\end{array}
$$

Graph

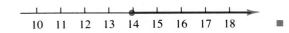

In Example 12, remember that when the inequality contains the $\ne$ sign, we proceed in exactly the same manner as for solving equations.

Example 12   Solve $4(x - 3) - 5 \ne 2x - 7$ and graph the solution on the number line.
Solution

$$4(x - 3) - 5 \ne 2x - 7$$

$$4x - 12 - 5 \ne 2x - 7 \qquad \text{Remove parentheses}$$

$$
\begin{array}{ll}
4x - 17 \ne & 2x - 7 \\
\underline{-2x} & \underline{-2x} \qquad \text{Get the } x\text{-term on only one side} \\
2x - 17 \ne & -\ 7 \\
\underline{+\ 17} & \underline{+\ 17} \\
2x \quad \ne & 10
\end{array}
$$

$$\frac{2x}{2} \ne \frac{10}{2}$$

$$x \ne 5 \qquad \text{Solution}$$

Graph

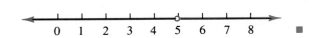

163

EXERCISES 4.7

Set I Solve each of the following inequalities and graph each solution set on the number line.

1. $x - 5 < 2$ **2.** $x - 4 < 7$

3. $5x + 4 \leq 19$ **4.** $3x + 5 \leq 14$

5. $6x + 7 > 3 + 8x$ **6.** $4x + 28 > 7 + x$

7. $2x - 9 > 3(x - 2)$ **8.** $3x - 11 > 5(x - 1)$

9. $6(3 - 4x) + 12 \geq 10x - 2(5 - 3x)$

10. $7(2 - 5x) + 27 \geq 18x - 3(8 - 4x)$

11. $4(6 - 2x) \neq 5x - 2$

12. $6(2x - 5) + 29 \neq 3x - 7(11 - 4x)$

13. $2[3 - 5(x - 4)] < 10 - 5x$

14. $3[2 - 4(x - 7)] < 26 - 8x$

15. $7(x - 5) - 4x > x - 8$

16. $8(x + 2) \neq 24 - 2(x - 1)$

17. $3x - 5(x + 2) \leq 4x + 8$

18. $5[6 - 2(3 - x)] - 3 < 3x + 4$

19. $12.85x - 15.49 \geq 22.06(9.66x - 12.74)$

20. $7.12(3.65x - 8.09) + 5.76 < 5.18x - 6.92(4.27 - 3.39x)$

Set II Solve each of the following inequalities and graph each solution set on the number line.

1. $x + 3 \geq -4$ **2.** $4 + x > -5$

3. $x + 4 < 7$ **4.** $5x + 3 < 18$

5. $5x + 7 > 13 + 11x$ **6.** $7x - 5 \leq 2x - 25$

7. $5x - 6 \leq 3(2 + 3x)$ **8.** $3x - 2 \geq -2(11 - x)$

9. $5(3 - 2x) + 25 \geq 4x - 6(10 - 3x)$

10. $3(x + 3) - 4 \neq 7x - 3(2 - x)$

11. $2(5 - 3x) \neq 7 - 4x$

12. $6(3 - 2x) - 3(x + 1) < 0$

13. $4[2 - 3(x - 5)] < 2 - 6x$

14. $2(3x - 4) + (x - 1) > 5$

15. $8(x - 3) - (x + 4) < 2x + 2$

16. $3[5 + 3(4 + x)] \leq 4(2x - 3)$

17. $5x - 3(3x - 4) \geq 2x + 7$

18. $5 - (3 - x) \neq 3(2 + x)$

19. $821.4x - 395.2 \geq 604.1(542.8x - 193.7)$

20. $4.01x + 62.1 \leq 3.04(5.143 - 6.21x)$

4.8 Review: 4.1–4.7

Parts of an Equation
4.1

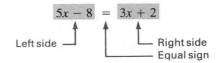

Left side ⌐ ⌐ Right side
Equal sign

Conditional Equations
4.1

A *conditional equation* is an equation whose two sides are equal only when certain numbers (called *solutions*) are substituted for the variable.

Solution of an Equation
4.1

A *solution* of an equation is a number that, when substituted for the variable, makes the two sides of the equation equal.

Equivalent Equations
4.1

Equations that have the same solution set are *equivalent equations*.

The Properties of Equality
4.2, 4.3, 4.4

Addition: If the same number is added to both sides of an equation, the new equation is equivalent to the original equation.

Division: If both sides of an equation are divided by the same nonzero number, the new equation is equivalent to the original one.

Multiplication: If both sides of an equation are multiplied by the same nonzero number, the new equation is equivalent to the original equation.

Subtraction: If the same number is subtracted from both sides of an equation, the new equation is equivalent to the original equation.

To Solve an Equation
4.5

1. Remove grouping symbols.

2. Combine like terms on each side of the equation.

3. If the variable appears on both sides of the equation, remove the term containing the variable from one side by adding the negative of that term to both sides.

4. Remove all numbers that appear on the same side as the variable.

   First, remove those numbers being added or subtracted.

   Second, multiply both sides by the number the variable is divided by.

   Third, divide both sides by the coefficient of the variable.

4.6
5. Three outcomes are possible:

   If the equation reduces to the form $x = a$, it is a *conditional equation*.

   If the variable drops out and the two sides reduce to the same constant and we're left with a *true* statement, the equation is an *identity*.

   If the variable drops out and the two sides reduce to unequal constants and we're left with a *false* statement, the equation has *no solution*.

6. If the equation was a conditional equation, check the solution in the *original* equation.

Inequalities
4.7

An *inequality* is a statement that two expressions are not equal.

Graphing Inequalities on the Number Line

To graph the inequality $x > c$:

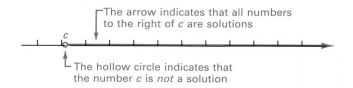

The arrow indicates that all numbers to the right of c are solutions

The hollow circle indicates that the number c is *not* a solution

To graph the inequality $x \leq b$:

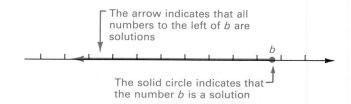

The arrow indicates that all numbers to the left of b are solutions

The solid circle indicates that the number b is a solution

Other types of inequalities, for example, $x < c$ and $x \geq c$, are graphed using the same procedures.

To Solve a Simple Inequality with Only One Variable
4.7

If it contains $\neq$, proceed in exactly the same way used to solve equations.

If it contains $<$, $\leq$, $>$, or $\geq$, proceed in the same way used to solve equations, with the exception that the sense must be changed when multiplying or dividing both sides by a negative number.

Review Exercises 4.8 Set I

In Exercises 1–20, find the solution of each conditional equation and check your answers. Identify any equation that is not a conditional equation as either an identity or an equation with no solution. Graph the solution of each conditional equation on the number line.

1. $3x - 5 = 4$

2. $22 - 8x = 6$

3. $2 = 20 - 9x$

4. $5x - 3 = 5x + 4$

5. $7.5 = \dfrac{A}{10}$

6. $\dfrac{D}{9} - 12 = 8$

7. $7 - 2(M - 4) = 5$

8. $20 - 3(4 - 5x) = 8 + 15x$

9. $6R - 8 = 6(2 - 3R)$

10. $7P - 15 = 7(3 - 2P)$

11. $56T - 18 = 7(8T - 4)$

12. $65 - 77S = 11(5 - 7S)$

13. $15(4 - 5V) = 16(4 - 6V) + 10$

14. $10 - 5(2x - 3) = 5(5 - 2x)$

15. $5x - 7(4 - 2x) + 8 = 10 - 9(11 - x)$

16. $18 - 6(5x - 4) - 13x = 11(12 - 3x) - 7$

17. $2[-7y - 3(5 - 4y) + 10] = 10y - 12$

18. $5[-13z - 8(4 - 2z) + 20] = 15z - 17$

19. $4[-24 - 6(3x - 5) + 22x] = 0$

20. $3[-53 - 7(4x - 9) + 18x] = 0$

In Exercises 21–26, solve each inequality and graph the solution on the number line.

21. $x + 7 > 2$ **22.** $x + 5 > 1$

23. $x - 3 \leq -8$ **24.** $x - 6 \leq -10$

25. $2(x - 4) - 5 \geq 7 + 3(2x - 1)$

26. $10 - 3(x + 2) \geq 9 - 2(4 - 3x)$

Review Exercises 4.8 Set II

NAME _____

In Exercises 1–20, find the solution of each conditional equation and check your answers. Identify any question that is not a conditional equation as either an identity or an equation with no solution. Graph the solution of each of the first six conditional equations on the number line. (Use the number lines at the bottom of this page. *You* insert the problem number and label the points on the number line.)

1. $7x - 8 = 6$

2. $12 - 6x = -12$

3. $3 = 18 - 5x$

4. $5 - 3x = 9 - 3x$

5. $\dfrac{x}{4} = 8.6$

6. $-27 = \dfrac{z}{6} - 18$

7. $-9 - 3(N - 8) = 30$

8. $19 - 5(3 - 2x) = 4 + 10x$

9. $12R - 9 = 8(3 - 4R)$

10. $42H + 12 = 6(7H - 2)$

11. $4(3x - 2) + 3x = 5(2 + 3x)$

12. $2(9x - 5) - 2 = 3(6x - 5)$

13. $11(3 - 4V) = 8(5 - 6V) + 17$

14. $7x - (3 - x) = 2(4x + 1) - 5$

ANSWERS

1. _____

2. _____

3. _____

4. _____

5. _____

6. _____

7. _____

8. _____

9. _____

10. _____

11. _____

12. _____

13. _____

14. _____

15. $4 - 5(x - 7) = 10(4 - x)$

16. $2 + 3(4 - x) = 6x - 13$

17. $25 - 2[4(x - 2) - 12x + 4] = 16x + 33$

18. $2\{3[10 - 4(3 - x) + x] - 5\} = 0$

19. $15 - 5[2(8x - 4) - 14x + 8] = 25$

20. $2\{3[2(5 - V) + 4V] - 20\} = 0$

In Exercises 21–26, solve each inequality and graph the solution on the number line. (*You label the points on the number line.*)

21. $x + 5 \geq -6$

22. $x - 7 < 2$

23. $3x - 2 > 6 - x$

24. $7x - 2(5 + 4x) \leq 8$

25. $5(6 - 2x) + 3 \geq 2(x + 1) + 11$

26. $3[2 - (4 - x)] + 7 < 4x$

15. _____

16. _____

17. _____

18. _____

19. _____

20. _____

21. _____

22. _____

23. _____

24. _____

25. _____

26. _____

Chapter 4 Diagnostic Test

The purpose of this test is to see how well you understand solving simple equations and inequalities. We recommend that you work this diagnostic test *before* your instructor tests you on this chapter. Allow yourself about 50 minutes.

Complete solutions for all the problems on this test, together with section references, are given in the answer section in the back of this book. For the problems you do incorrectly, study the sections referred to.

In Problems 1–13, find the solution of each conditional equation and check your answers. Identify any equation that is not a conditional equation as either an identity or an equation with no solution.

1. $x - 5 = 3$ (Graph the solution on the number line.)

2. $5y + 7 = 22$ (Show your check on this problem.)

3. $17 - 3z = -1$

4. $\dfrac{x}{6} = 5.1$

5. $-6 = \dfrac{w}{7}$

6. $\dfrac{x}{4} - 5 = 3$

7. $6x + 1 = 17 - 2x$

8. $3z - 21 + 5z = 4 - 6z + 17$ (Graph the solution on the number line.)

9. $5k - 9(7 - 2k) = 6$ (Show your check on this problem.)

10. $10x - 2(5x - 7) = 14$

11. $2y - 4(3y - 2) = 5(6 + y) - 7$

12. $7(3z + 4) = 14 + 3(7z - 1)$

13. $3[7 - 6(x - 2)] = -3 + 2x$ (Graph the solution on the number line.)

14. Solve the inequality $4x + 5 > -3$ and graph the solution on the number line.

15. Solve the inequality $5x - 2 \le 10 - x$ and graph the solution on the number line.

Cumulative Review Exercises
Chapters 1– 4

In Exercises 1–4, evaluate each expression.

1. $\dfrac{-7}{0}$

2. $\sqrt{8^2 + 6^2}$

3. $7\sqrt{16} - 5(-4)$

4. $25 - \{-16 - [(11 - 7) - 8]\}$

In Exercises 5 and 6, evaluate each formula using the values of the variables given with the formula.

5. $V = \dfrac{4}{3}\pi R^3$ $\pi \doteq 3.14, R = 9$

6. $S = P(1 + i)^n$ $P = 3{,}000, i = 0.10, n = 2$

In Exercises 7–10, simplify each algebraic expression.

7. $15x - [9y - (7x - 10y)]$

8. $5(2a - 3b) - 6(4a + 7b)$

9. $x(2x^2 - 5x + 12) - 6(3x^2 + 2x - 5)$

10. $4hk(3hk - hk^2) - hk^2(2h - hk)$

In Exercises 11–17, solve each conditional equation. Identify any equation that is not a conditional equation as either an identity or an equation with no solution.

11. $2x + 1 = 2x + 7$

12. $3 = 33 - 10x$

13. $10 - 4(2 - 3x) = 2 + 12x$

14. $6 - 4(N - 3) = 2$

15. $\dfrac{C}{7} - 15 = 13$

16. $12(4W - 5) = 9(7W - 8) - 13$

17. $9 - 3(x - 2) = 3(5 - x)$

In Exercises 18–20, solve each inequality.

18. $2x - 5 < 3$

19. $3 - x \geq 4$

20. $5x - 2 \leq 8x + 4$

Critical Thinking

Each of the following problems has an error. Can you find it?

1. Evaluate $2^2 \cdot 2^3$.

$$2^2 \cdot 2^3 = 4^5 = 1{,}024$$

2. Evaluate $\sqrt{5^2 + 12^2}$.

$$\sqrt{5^2 + 12^2}$$
$$= \sqrt{25 + 144}$$
$$= \sqrt{25} + \sqrt{144}$$
$$= 5 + 12 = 17$$

3. Evaluate $2^3 - 3 \cdot 4 \div 2$.

$$2^3 - 3 \cdot 4 \div 2$$
$$= 8 - 3 \cdot 4 \div 2$$
$$= 5 \cdot 4 \div 2$$
$$= 20 \div 2 = 10$$

4. Simplify $2x(3x^2 - 4x) + 5x^2(x - 2)$.

$$2x(3x^2 - 4x) + 5x^2(x - 2)$$
$$= 6x^3 - 8x^2 + 5x^3 - 10x^2$$
$$= 11x^6 - 18x^4$$

5. Simplify $x - 2[x - (4 - x) - 8]$.

$$x - 2[x - (4 - x) - 8]$$
$$= x - 2[x - 4 + x - 8]$$
$$= x - 2[2x - 12]$$
$$= x - 4x - 24$$
$$= -3x - 24$$

6. Solve $7 - 4x = -9$.

$$7 - 4x = -9$$
$$\underline{-7 \qquad\quad -7}$$
$$4x = -16$$
$$\frac{4x}{4} = \frac{-16}{4}$$
$$x = -4$$

5 Word Problems

As we stated in Chapter 4, the main reason for studying algebra is to equip oneself with the tools necessary to solve problems. Most problems are expressed in words. In this chapter, we show how to change the words of a written problem into an equation. The resulting equation can then be solved by the methods learned in Chapter 4.

5.1 Changing Word Expressions into Algebraic Expressions

5.1A Key Word Expressions and Their Corresponding Algebraic Operations

In solving word problems, it is helpful to break them up into smaller expressions. In this section, we show how you can change these small *word* expressions into *algebraic* expressions. Below is a list of key word expressions and their corresponding algebraic operations.

+	−	×	÷
add	subtract	multiply	divided by
sum	difference	times	quotient
plus	minus	product	
increased by	decreased by		
more than	diminished by		
	less than		
	subtracted from		

NOTE   Because subtraction is not commutative, care must be taken to put the numbers in a subtraction word problem in the correct order. For example, while the statements "m minus n," "m decreased by n," and "the difference of m and n" are translated as $m - n$, the expressions "m subtracted from n" and "m less than n" are translated as $n - m$.   ☑

Example 1   Change each of the following word expressions into an algebraic expression.

a. "The sum of A and B"
   Solution $A + B$

b. "The product of L and W"
   Solution LW

c. "Two decreased by C"
   Solution $2 - C$

d. "Two less than C"
   Solution $C - 2$

e. "Three times the square of x, plus ten"
   Solution $3x^2 + 10$

f. "Five subtracted from the quotient of S by T"
   Solution $\dfrac{S}{T} - 5$ ■

EXERCISES 5.1A

Set I   Change each of the following word expressions into an algebraic expression.

1. The sum of x and 10

2. A added to B

3. Five less than A

4. B diminished by C

5. The product of 6 and z

6. A multiplied by B

7. x decreased by 7

8. Nine increased by A

9. Four less than 3 times x

10. Six more than twice x

11. Subtract the product of u and v from x

12. Subtract x from the product of P and Q

13. The product of 5 and the square of x

14. The product of 10 and the cube of x

15. The square of the sum of A and B

16. The square of the quotient of A divided by B

17. The sum of x and 7, divided by y

18. T divided by the sum of x and 9

19. The product of x and the difference, 6 less than y

20. The product of A and the sum, 3 plus B

Set II   Change each of the following word expressions into an algebraic expression.

1. Ten added to x

2. The product of s and t

3. Three less than w

4. x diminished by 4

5. The sum of u and v

6. Five increased by x

7. Five decreased by y

8. Ten plus x

9. Seven more than z

10. Five times b

11. Twice F subtracted from 15

12. The quotient of A by the sum of C and 10

13. The sum of x and the square of y

14. The quotient of the sum of A and C by B

15. The sum of the squares of A and B

16. The square of the sum of x and 4

17. The sum of x and y divided by z

18. Twice the sum of x and y

19. The product of 7 and the sum of x and y

20. Three less than 5 times y

5.1B Translating Word Expressions into Algebraic Expressions

A word expression often contains unknown numbers. In algebra, we must translate such an expression into an algebraic expression, and we must represent the unknown numbers by *variables*. In this chapter, when there are two or more unknowns, we must express each unknown in terms of the same variable.

TO CHANGE A WORD EXPRESSION INTO AN ALGEBRAIC EXPRESSION

1. Identify which number or numbers are unknowns.

2. Represent *one* of the unknown numbers by a variable. Express any other unknown number in terms of the same variable.

3. Change the word expression into an algebraic expression, using the variable(s) in place of the unknown number(s).

Example 2   Change the word expression "twice Albert's salary" into an algebraic expression.

Step 1. Albert's salary is the unknown number.

Step 2. Let S represent Albert's salary.

Step 3. Then $2S$ is the algebraic expression for "twice Albert's salary." ■

Example 3   Change the word expression "the cost of five stamps" into an algebraic expression.

Step 1. The cost of one stamp is the unknown number.

Step 2. Let c represent the cost of one stamp.

Step 3. Then $5c$ is the algebraic expression for "the cost of five stamps." Since one stamp costs c cents, then 5 stamps will cost 5 times c cents $= 5c$. ■

Example 4   In the word expression "Mary is 10 years older than Nancy," represent both unknown numbers in terms of the same variable.
First Solution

Step 1. There are two unknown numbers in this expression: Mary's age and Nancy's age.

Step 2. Let N represent Nancy's age.

Then $N + 10$ represents Mary's age, because Mary is 10 years older than Nancy.

Second Solution

Step 1. There are two unknown numbers in this expression: Mary's age and Nancy's age.

Step 2. Let M represent Mary's age.

Then $M - 10$ represents Nancy's age, because Nancy is 10 years younger than Mary. ■

Example 5   In the word expression "The sum of two numbers is 10," represent both unknown numbers in terms of the same variable.

Step 1. There are two unknown numbers.

Step 2. Let $x =$ one of the unknown numbers
Then $10 - x =$ the other unknown number

Notice that $x + (10 - x) = 10$; the sum of the two numbers is 10. ■

Recall from Section 1.2 that integers that follow one another in sequence (without interruption) are called *consecutive integers*. Consecutive integers can be represented by variables.

If the problem deals with two or more consecutive integers, let $x =$ the first integer, $x + 1 =$ the second one, $x + 2 =$ the third one, and so on.

Adding 2 to any even integer gives the next even integer; for example, $8 + 2 = 10$, $-16 + 2 = -14$, and so on. Therefore, if a problem deals with consecutive *even* integers, let $x =$ the first integer, $x + 2 =$ the second one, $x + 4 =$ the third one, and so on.

Adding 2 to any odd integer gives the next odd integer; for example, $9 + 2 = 11$, $-5 + 2 = -3$, and so forth. Therefore, if a problem deals with consecutive *odd* integers, let $x =$ the first integer, $x + 2 =$ the second one, $x + 4 =$ the third one, and so on.

Example 6 In the word expression "The sum of three consecutive integers," represent the sum of the three integers in terms of the same variable.

Step 1. There are three unknown integers.

Step 2. Let $x =$ the first integer
$x + 1 =$ the second integer
$x + 2 =$ the third integer

Step 3. Then the sum of the three integers is $x + (x + 1) + (x + 2)$. ∎

Example 7 In the word expression "The sum of three consecutive odd integers," represent the sum of the three integers in terms of the same variable.

Step 1. There are three unknown *odd* integers.

Step 2. Let $x =$ the first odd integer
$x + 2 =$ the second odd integer
$x + 4 =$ the third odd integer

Step 3. Then the sum of the three integers is $x + (x + 2) + (x + 4)$. ∎

Word problems sometimes deal with geometric figures. Several formulas relating to geometric figures are given below. The *perimeter* of a geometric figure is the sum of the lengths of all its sides.

Rectangle

Area: $A = LW$

Perimeter: $P = 2L + 2W$

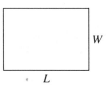

Square

Area: $A = s^2$

Perimeter: $P = 4s$

Triangle

Area: $A = \frac{1}{2}bh$

where b is the base and h is the altitude

Perimeter: $P = a + b + c$

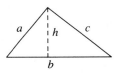

179

Circle

Area $A = \pi r^2$

Circumference $C = 2\pi r$

where $\pi \doteq 3.14$ and r is the radius

Rectangular Solid

Volume: $V = LWH$

Surface area: $S = 2(LW + LH + WH)$

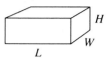

Cube

Volume: $V = s^3$

Surface area: $S = 6s^2$

Sphere

Volume: $V = \frac{4}{3}\pi r^3$

Surface area: $S = 4\pi r^2$

where $\pi \doteq 3.14$ and r is the radius

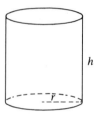

Cylinder

Volume: $V = \pi r^2 h$

where $\pi \doteq 3.14$, r is the radius,

and h is the height

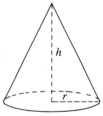

Cone

Volume: $V = \frac{1}{3}\pi r^2 h$

where $\pi \doteq 3.14$, r is the radius,

and h is the height

Example 8   In the word expression "The height of a cone is 2 ft more than the radius," represent the height and the radius in terms of the same variable.
First Solution

Step 1. There are two unknowns: the height and
the radius.

Step 2. Let r = the radius.

Then $r + 2$ = the height, because the
height is 2 ft more than the radius.

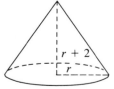

Second Solution

Step 1. There are two unknowns: the height and the radius.

Step 2. Let h = the height.

Then $h - 2$ = the radius, because the radius is 2 ft less than the height. ■

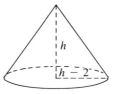

Example 9   If the length of a rectangular solid is 5 yd more than the height and the width is 2 yd less than the height, express the *volume* in terms of one variable.

Step 1. There are four unknowns: the height, the length, the width, and the volume.

Step 2. Let H = the height
$H + 5$ = the length
$H - 2$ = the width

Step 3. Then the volume is $H(H + 5)(H - 2)$.

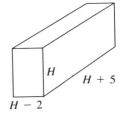

NOTE   It is also possible to let L = the length and then to express the volume in terms of L, or to let W = the width and then to express the volume in terms of W. ■

EXERCISES 5.1B

Set I   In Exercises 1–36, if there is only one unknown number, represent it by a variable and then change the word expression into an algebraic expression; if there is more than one unknown number, represent all unknowns in terms of the *same* variable, and then change the word expression into an algebraic expression.

1. Fred's salary plus seventy-five dollars

2. Jaime's salary minus forty-two dollars

3. Two less than the number of children in Mr. Moore's family

4. The number of players on Jerry's team plus two more players

5. Four times Joyce's age

6. One-fourth of Rene's age

7. Twenty times the cost of a record increased by eighty-nine cents

8. Seventeen cents less than 5 times the cost of a ballpoint pen

9. One-fifth the cost of a hamburger

10. The length of the building divided by 8

11. Five times the speed of the car plus 100 miles per hour

12. Twice the speed of a car diminished by 40 miles per hour

13. Ten less than 5 times the square of an unknown number

14. Eight more than 4 times the cube of an unknown number

15. The length of a rectangle is 12 centimeters more than its width.

16. The altitude of a triangle is 7 centimeters less than its base.

17. Take the quotient of 7 by an unknown number away from 50.

18. Add the quotient of an unknown number by 9 to 15.

19. Eleven feet more than twice the length of a rectangle

20. One inch less than the diameter of a circle

21. The combined weight of Walter and Carlos is 320 lb.

22. The combined weight of Teresa and Lucy is 224 lb.

23. 2.2 times the weight in kilograms

24. 0.62 times the distance in kilometers

25. The sum of 8 and an unknown number is divided by the square of that unknown number.

26. The sum of the square of an unknown number and 11 is divided by the unknown number.

27. The sum of 32 and nine-fifths the Celsius temperature

28. Five-ninths times the result of subtracting 32 from the Fahrenheit temperature

29. The sum of three consecutive integers

30. The sum of two consecutive integers

31. The sum of four consecutive odd integers

32. The sum of three consecutive even integers

33. Six plus the radius of a sphere

34. Five plus the radius of a circle

35. The height of a cylinder is 2 times the radius.

36. The radius of a cone is 4 times the height.

37. If the radius of a cylinder is 4 cm less than its height, express the volume of the cylinder in terms of one variable.

38. If the base of a triangle is 5 m less than the altitude, express the area of the triangle in terms of one variable.

39. If the height of a cone is 4 times the radius, express the volume of the cone in terms of one variable.

40. If the radius of a cylinder is 6 times the height, express the volume of the cylinder in terms of one variable.

Set II   In Exercises 1–36, if there is only one unknown number, represent it by a variable and then change the word expression into an algebraic expression; if there is more than one unknown number, represent all unknowns in terms of the *same* variable, and then change the word expression into an algebraic expression.

1. The cost of a television set plus thirty-eight dollars

2. The sum of two numbers is 60.

3. Eight less than the length of a certain bridge

4. One-third of Carol's age

5. Eight times the width of a certain window

6. The cost of seven stamps

7. Three times the cost of a video tape increased by forty-five cents

8. Henry is five years younger than his brother Brian.

9. Two-thirds the cost of a compact disc

10. Nine less than one-fourth of Jo's height

11. Ten times the height of a building minus 32 meters

12. Mrs. Lopez is twenty-one years older than her daughter Flora.

13. Five less than the product of 4 and x

14. One-half the sum of 7 and an unknown number

15. The altitude of a triangle is 8 centimeters less than its base.

16. Eight added to the sum of 10 and an unknown number

17. Subtract 51 from the sum of x and y.

18. The sum of two numbers is -22.

19. Three meters less than 3 times the height of the building

20. The speed of the train divided by 6

21. The combined age of Esther and Marge is 43.

22. Twice the result of subtracting an unknown number from 5

23. 8.6 times the length in kilometers

24. Pete has fifty-three dollars less than Ann.

25. The square of the result of subtracting 12 from an unknown number

26. The sum of the squares of 3 and an unknown number

27. The square of the sum of 3 and an unknown number

28. The product of 4 and the square of the length of a side of a square

29. The sum of four consecutive integers

30. The product of two consecutive integers

31. The sum of four consecutive even integers

32. The product of two consecutive odd integers

33. Fourteen plus the radius of a circle

34. Five times the radius of a sphere

35. The radius of a cylinder is 2 times the height.

36. The length of a rectangle is 3 times the width.

37. If the height of a cone is 6 cm more than its radius, express the volume of the cone in terms of one variable.

38. If the length of a rectangle is 9 yd more than its width, express the perimeter of the rectangle in terms of one variable.

39. If the altitude of a triangle is 3 times the base, express the area of the triangle in terms of one variable.

40. If the height of a cylinder is 5 ft less than the radius, express the volume of the cylinder in terms of one variable.

5.2 Changing Word Problems into Equations

In this section, we show how to translate an English sentence into an algebraic equation. (We will not yet be *solving* word problems.)

TO CHANGE A WORD PROBLEM INTO AN EQUATION

1. Represent the unknown number by x (Section 5.1).

2. Break the word problem into small pieces.

3. Represent each piece by an algebraic expression (Section 5.1).

4. Arrange the algebraic expressions into an equation.

In Examples 1–4, translate each sentence into an equation.

Example 1   Fifteen plus twice an unknown number is 37.
Solution Let $x =$ the unknown number ← *Do not omit this step*

Fifteen	plus	twice an unknown number	is	37
15	+	$2x$	=	37

Equation $15 + 2x = 37$ ∎

Example 2   Three times an unknown number is equal to 12 increased by the unknown number.
Solution Let $x =$ the unknown number.

Three	times	an unknown number	is equal to	12	increased by	the unknown number
3	·	x	=	12	+	x

$$3x = 12 + x \quad \blacksquare$$

Example 3   One-third of an unknown number is 7.
Solution Let $x =$ the unknown number.

One-third	of	an unknown number	is	7
$\dfrac{1}{3}$	·	x	=	7

$$\frac{1}{3}x = 7 \quad \blacksquare$$

Example 4   Twice the sum of 6 and an unknown number is equal to 20.
Solution Let $x =$ the unknown number.

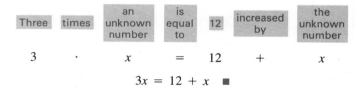

Twice	the sum of 6 and an unknown number	is	20
2 ·	$(6 + x)$	=	20

$$2(6 + x) = 20 \quad \blacksquare$$

Example 5   Write the equation that represents these facts: A piece of wire 63 cm long is to be cut into two pieces. One piece is to be 15 cm longer than the other piece.
Solution (There are *two* unknowns.)

Let x = the length of the shorter piece (in centimeters)

$x + 15$ = the length of the longer piece

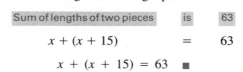

Sum of lengths of two pieces	is	63
$x + (x + 15)$	=	63

$$x + (x + 15) = 63 \quad \blacksquare$$

Example 6   Write an equation that represents these facts: The sum of three consecutive odd integers is 75.
Solution

Let x = the first odd integer

$x + 2$ = the second odd integer

$x + 4$ = the third odd integer

Sum of three consecutive odd integers	is	75
$x + (x + 2) + (x + 4)$	=	75 $\blacksquare$

Example 7   Write an equation that represents these facts: The length of a rectangle is 5 ft more than the width, and the perimeter is 44 ft.
Solution (The formula for the perimeter of a rectangle is $P = 2L + 2W$.)

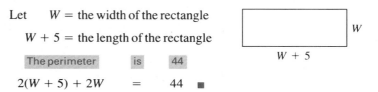

Let W = the width of the rectangle

$W + 5$ = the length of the rectangle

The perimeter	is	44
$2(W + 5) + 2W$	=	44 $\blacksquare$

EXERCISES 5.2

Set I   Write the equation that represents the given facts. Do not solve the equations. Be sure to state what your variables represent.

1. Thirteen more than twice an unknown number is 25.

2. Twenty-five more than 3 times an unknown number is 34.

3. Five times an unknown number, decreased by 8, is 22.

4. Four times an unknown number, decreased by 5, is 15.

5. Seven minus an unknown number is equal to the unknown number plus 1.

6. Six plus an unknown number is equal to 12 decreased by the unknown number.

7. One-fifth of an unknown number is 4.

8. An unknown number divided by 12 equals 6.

9. When 4 is subtracted from one-half of an unknown number, the result is 6.

10. When 5 is subtracted from one-third of an unknown number, the result is 4.

11. Twice the sum of 5 and an unknown number is equal to 26.

12. Four times the sum of 9 and an unknown number is equal to 18.

13. When the sum of an unknown number and itself is multiplied by 3, the result is 24.

14. Five times the sum of an unknown number and itself is 40.

15. A 75-m rope is to be cut into two pieces. One piece is to be 13 m longer than the other piece.

16. A 46-m wire is to be cut into two pieces. One piece is to be 8 m shorter than the other piece.

17. A 72-cm piece of wire is to be cut into two pieces. One piece is to be twice as long as the other piece.

18. A 72-cm piece of wire is to be cut into two pieces. One piece is to be 3 times as long as the other piece.

19. Ingrid buys eight more cans of peaches than cans of pears. Altogether, she buys forty-two cans of these two fruits.

20. Jane buys eleven more cans of peas than cans of corn. Altogether, she buys forty-nine cans of these two vegetables.

21. The sum of four consecutive integers is 106.

22. The sum of three consecutive integers is -72.

23. The length of a rectangle is 4 cm more than its width, and the perimeter is 36 cm.

24. The width of a rectangle is 6 ft less than its length, and the perimeter is 64 ft.

Set II Write the equation that represents the given facts. Do not solve the equations. Be sure to state what your variables represent.

1. Fifteen more than twice an unknown number is 27.

2. Twice the sum of an unknown number and 9 is 46.

3. Four times an unknown number, decreased by 9, is 19.

4. When 5 is subtracted from one-half of an unknown number, the result is 19.

5. Eighteen minus an unknown number is equal to 4 plus the unknown number.

6. When 7 is added to an unknown number, the result is twice that unknown number.

7. Three-eighths of an unknown number is 27.

8. When a number is decreased by 5, the difference is half of the number.

9. When 8 is subtracted from two-thirds of an unknown number, the result is 16.

10. When 3 times an unknown number is subtracted from 20, the result is the unknown number.

11. Three times the result of subtracting an unknown number from 8 is 12.

12. Four times the result of adding an unknown number to itself is 96.

13. When the sum of an unknown number and itself is multiplied by 4, the result is 56.

14. When 6 is subtracted from 5 times an unknown number, the result is the same as when 4 is added to 3 times the unknown number.

15. A 7-yd piece of fabric is to be cut into two pieces. One piece is to be 3 yd longer than the other piece.

16. Terri buys 3 times as many bottles of catsup as jars of mustard. Altogether, she buys twelve containers of these two products.

17. A 42-m rope is to be cut into two pieces. One piece is to be 6 times as long as the other piece.

18. Alice buys 5 times as many skeins (balls) of pink yarn as skeins of white yarn. Altogether, she buys eighteen skeins of these two colors.

19. Todd buys four more cans of car wax than of rubbing compound. Altogether, he buys eighteen cans of these two products.

20. A 30-m cable is to be cut into two pieces. One piece is to be 8 m longer than the other.

21. The sum of four consecutive integers is 2.

22. The sum of three consecutive even integers is 78.

23. The length of a rectangle is 19 yd more than its width, and the perimeter is 62 yd.

24. The lengths of all three sides of a triangle are equal, and the perimeter is 42 cm.

5.3 Solving Word Problems

In Section 5.2, we showed how to change the words of a written problem into an equation. In this section, we show how to solve word problems, some of which lead to inequalities rather than to equations.

The method is summarized below.

METHOD FOR SOLVING WORD PROBLEMS

1. To solve a word problem, first read it very carefully and determine what *type* of problem it is, if possible.* *Be sure you understand the problem.* Don't try to solve it at this point.

2. Determine what is unknown. What is being asked for is often found in the last sentence of the problem, which may begin with "What is the . . ." or "Find the . . . ."

3. Represent one unknown number by a variable, and declare its meaning in a sentence of the form "Let $x = $ . . . ." Then reread the problem to see how you can represent any other unknown numbers in terms of the same variable.

4. Reread the entire word problem, translating each English phrase into an algebraic expression.

5. Fit the algebraic expressions together into an equation or inequality.

6. Using the methods described in Chapter 4, solve the equation or inequality for the variable.

7. Answer *all* the questions asked in the problem.

8. Check the solution(s) in the word statement.

---

*We discuss several general types of word problems in this chapter, for example, money problems, mixture problems, and so on.

Example 1 Seven increased by 3 times an unknown number is 13. What is the unknown number?

Solution Let $x =$ the unknown number.

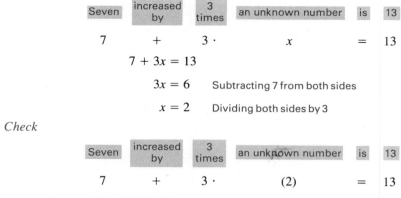

$$7 + 3x = 13$$

$$3x = 6 \qquad \text{Subtracting 7 from both sides}$$

$$x = 2 \qquad \text{Dividing both sides by 3}$$

Check

Seven	increased by	3 times	an unknown number	is	13
7	+	3 ·	(2)	=	13

$$7 + 3(2) \stackrel{?}{=} 13 \qquad \text{The unknown number was replaced by 2}$$

$$7 + 6 \stackrel{?}{=} 13$$

$$13 = 13 \quad \blacksquare$$

NOTE To check a word problem, you must check the solution in the *word statement*. Any error that may have been made in writing the equation will not be discovered if you simply substitute the solution into the equation. ☑

Example 2 Four times an unknown number is equal to twice the sum of 5 and that unknown number. Find the unknown number.

Solution Let $x =$ the unknown number.

Four times	an unknown number	is equal to	twice the sum of 5 and that unknown number
4 ·	x	=	$2 \cdot (5 + x)$

$$4x = 2(5 + x)$$

$$4x = 10 + 2x \qquad \text{Using the distributive rule}$$

$$2x = 10 \qquad \text{Subtracting } 2x \text{ from both sides}$$

$$x = 5 \qquad \text{Dividing both sides by 2}$$

Check Four times 5 is 20. The sum of 5 and 5 is 10, and twice 10 is 20. ■

In Examples 3–5, the checking is left to the student.

Example 3 When 7 is subtracted from one-half of an unknown number, the result is 11. What is the unknown number?

Solution Let $x =$ the unknown number.

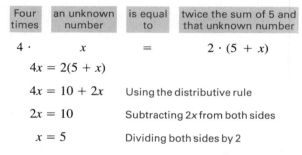

$$\frac{1}{2}x - 7 = 11$$

$$\frac{1}{2}x = 18 \qquad \text{Adding 7 to both sides}$$

$$x = 36 \qquad \text{Multiplying both sides by 2} \quad \blacksquare$$

It is shown in geometry that the sum of the measures of the angles of a triangle is always 180°. Example 4 uses this property.

Example 4   If the measure of one angle of a triangle is 73° and the measure of another angle is 54°, what is the measure of the third angle?
Solution Let x = the number of degrees of the third angle.

The sum of all the angles	equals	180
73 + 54 + x	=	180

$$127 + x = 180$$
$$-127 \qquad\qquad -127 \qquad \text{Adding } -127 \text{ to both sides}$$
$$x = 53$$

Therefore, the measure of the third angle is 53°. ■

Example 5   The length of a rectangle is 25 in. and the perimeter is 84 in. Find the width of the rectangle.
Solution The formula for the perimeter of a rectangle is $P = 2L + 2W$.
Let W = the width of the rectangle.

$$P = 2L \quad + 2W$$
$$84 = 2(25) + 2W$$
$$84 = \quad 50 + 2W$$
$$-50 \quad -50$$
$$\text{Adding } -50 \text{ to both sides}$$
$$34 = \qquad 2W$$
$$W = 17 \qquad \text{Dividing both sides by 2}$$

Therefore, the width of the rectangle is 17 in. ■

Example 6 illustrates the solution of a word problem that leads to an inequality.

Example 6   In an English class, any student needs at least 730 points in order to earn an A, and the final exam is worth 200 points. If Shirley has 545 points just before the final exam, what range of scores will give her an A for the course?
Solution Let x = her score on the final.

Score going into final	plus	score on final	is greater than or equal to	730
545	+	x	$\geq$	730

$$545 + x \geq 730$$
$$-545 \qquad -545$$
$$x \geq 185$$

Therefore, the range of scores that will give Shirley an A for the course is any score greater than or equal to 185 (and less than or equal to 200, since the final exam is worth 200 points).

Check We will show the check for three different numbers:

Score of 185: 185 + 545 = 730, and 730 ≥ 730

Score of 192: 192 + 545 = 737, and 737 ≥ 730

Score of 200: 200 + 545 = 745, and 745 ≥ 730

Any other numbers between 185 and 200 would also have checked. ■

EXERCISES 5.3

Set I   In all exercises, set up the problem algebraically and solve. Be sure to state what your variables represent.

In Exercises 1–10, find the unknown number and check your solution.

1. When twice an unknown number is added to 13, the sum is 25.

2. When 25 is added to 3 times an unknown number, the sum is 34.

3. Five times an unknown number, decreased by 8, is 22.

4. Four times an unknown number, decreased by 5, is 15.

5. Seven minus an unknown number is equal to the unknown number plus 1.

6. Six plus an unknown number is equal to 12 decreased by the unknown number.

7. When 4 is subtracted from one-half of an unknown number, the result is 6.

8. When 5 is subtracted from one-third of an unknown number, the result is 4.

9. Twice the sum of 5 and an unknown number is equal to 26.

10. Four times the sum of 9 and an unknown number is equal to 20.

In Exercises 11–28, solve for the unknown and check your solution.

11. A 36-cm piece of wire is to be cut into two pieces. One of the pieces is to be 10 cm longer than the other piece. Find the length of each piece.

12. A 10-yd piece of fabric is to be cut into two pieces. One of the pieces is to be 2 yd longer than the other piece. Find the length of each piece.

13. Rebecca buys seven more packages of dried apples than packages of dried apricots. If she buys fifteen packages of these two fruits altogether, how many of each does she buy?

14. Susan buys three more skeins of blue yarn than skeins of green yarn. If she buys thirteen skeins of these two colors altogether, how many of each color does she buy?

15. Find the area of this triangle:

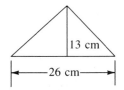

16. Find the volume and the surface area of this rectangular solid:

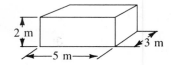

17. Find the approximate volume and surface area of this sphere:

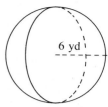

 18. Find the approximate area and circumference of this circle:

19. The sum of three consecutive odd integers is 177. Find the integers.

20. The sum of three consecutive even integers is −144. Find the integers.

21. Four times the sum of the first and third of three consecutive integers is 140 more than the second integer. Find the integers.

22. Three times the sum of the first and third of three consecutive integers is seventy-five more than the second integer. Find the integers.

23. If the measure of one angle of a triangle is 62° and the measure of another angle is 47°, what is the measure of the third angle?

24. If the measure of one angle of a triangle is 18° and the measure of another angle is 37°, what is the measure of the third angle?

25. The length of a rectangle is twice the width, and the perimeter is 102 ft. Find the length and the width.

26. The length of a rectangle is 4 times the width, and the perimeter is 160 cm. Find the length and the width.

27. The width of a rectangle is 5 ft and the perimeter is 44 ft. Find the length of the rectangle.

28. The length of a rectangle is 7 yd and the perimeter is 38 yd. Find the width of the rectangle.

In Exercises 29−36, find the unknown number and check your solution.

29. When twice the sum of 4 and an unknown number is added to the unknown number, the result is the same as when 10 is added to the unknown number.

30. When 6 is subtracted from 5 times an unknown number, the result is the same as when 4 is added to 3 times the unknown number.

31. Three times the sum of 8 and twice an unknown number is equal to 4 times the sum of 3 times the unknown number and 8.

32. Five times the sum of 4 and 6 times an unknown number is equal to 4 times the sum of twice the unknown number and 10.

33. When 3 times the sum of 4 and an unknown number is subtracted from 10 times the unknown number, the result is equal to 5 times the sum of 9 and twice the unknown number.

34. When twice the sum of 5 and an unknown number is subtracted from 5 times the sum of 6 and twice the unknown number, the result is equal to zero.

35. When 5.75 times the sum of 6.94 and an unknown number is subtracted from 8.66 times the unknown number, the result is equal to 4.69 times the sum of 8.55 and 3.48 times the unknown number.

36. When 8.23 is subtracted from 4.85 times an unknown number, the result is the same as when 12.62 is added to 5.49 times the unknown number.

In Exercises 37–40, find all possible solutions for each problem.

37. The sum of an unknown number and 18 is to be at least 5. What is the range of values that the unknown number can have?

38. The sum of an unknown number and 12 is to be at least 47. What is the range of values that the unknown number can have?

39. A rope less than 80 m long is to be cut into two pieces. One piece must be 37 m long. What will the length of the other piece be?

40. A chain more than 25 m long is to be cut into two pieces. One piece must be 8 m long. What will the length of the other piece be?

Set II In all exercises, set up the problem algebraically and solve. Be sure to state what your variables represent.

In Exercises 1–10, find the unknown number and check your solution.

1. When 4 times an unknown number is added to 21, the sum is 105.

2. When 7 is added to an unknown number, the result is twice that unknown number.

3. Three times an unknown number, decreased by 12, is 6.

4. When 3 times an unknown number is subtracted from 20, the result is the unknown number.

5. Eighteen minus an unknown number is equal to the unknown number minus 16.

6. One-fifth of an unknown number is 4.

7. When 3 is subtracted from one-third of an unknown number, the result is 7.

8. An unknown number divided by 12 equals 6.

9. Five times the sum of an unknown number and itself is 40.

10. When an unknown number is subtracted from 12, the difference is one-third of the number.

In Exercises 11–28, solve for the unknown and check your solution.

11. A 12-yd piece of fabric to be cut into two pieces. One of the pieces must be 3 yd longer than the other piece. Find the length of each piece.

12. Tom buys 5 times as many cassette tapes as compact discs. If he purchases twelve of these items altogether, how many of each kind does he buy?

13. Irene buys four more packages of unsalted crackers than packages of salted crackers. If she buys sixteen packages of crackers altogether, how many of each kind does she buy?

14. A 42-m piece of cord is to be cut into two pieces. If one piece must be 15 m long, what will the length of the other piece be?

15. Find the area and perimeter of this rectangle:

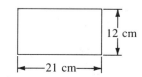

12 cm

21 cm

16. Find the area and perimeter of this square:

├── 14 yd ──┤

17. Find the approximate volume of this cone:

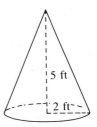

5 ft

2 ft

18. Find the approximate volume of this cylinder:

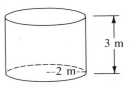

3 m

2 m

19. The sum of three consecutive even integers is − 102. Find the integers.

20. The sum of three consecutive odd integers is twenty-seven. Find the integers.

21. Twice the sum of the first and third of three consecutive integers is 102 more than the second integer. Find the integers.

22. Three times the sum of the first and third of three consecutive odd integers is fifty-five more than the second integer. Find the integers.

23. If the measure of one angle of a triangle is 90° and the measure of another angle is 37°, what is the measure of the third angle?

24. If the measures of the three angles of a triangle are equal, find the measure of the angles.

25. The length of a rectangle is 5 times the width, and the perimeter is 204 ft. Find the length and the width.

26. The length of a rectangle is 4 m and the area is 20 sq. m. Find the width of the rectangle.

27. The length of a rectangle is 7 cm and the perimeter is 24 cm. Find the width of the rectangle.

28. The base of a triangle is 4 yd and the area is 38 sq. yd. Find the altitude of the triangle.

In Exercises 29 – 36, find the unknown number and check your solution.

29. When 4 times the sum of 5 and an unknown number is added to the unknown number, the result is the same as when 32 is added to the unknown number.

30. Twice the result of subtracting an unknown number from 5 is 8.

31. Twice the sum of 3 and twice an unknown number is equal to 5 times the sum of the unknown number and 1.

32. When an unknown number is subtracted from 11, the result is the same as when the unknown number is added to 3.

33. If the sum of an unknown number and 12 is subtracted from 4 times the unknown number, the result is the unknown number less 4.

34. When 4 times an unknown number is subtracted from 16, the result is twice the sum of 12 and twice the unknown number.

35. When 3.48 times the sum of 9.06 and an unknown number is subtracted from 5.37 times the unknown number, the result is equal to 4.65 times the sum of 2.83 and 8.34 times the unknown number.

36. Five times the sum of 18 and an unknown number is 2 less than the unknown number.

In Exercises 37–40, find all possible solutions for each problem.

37. The sum of an unknown number and 7 is to be at least 3. What is the range of values that the unknown number can have?

38. Five times an unknown number plus 7 is greater than 42. What is the range of values that the unknown number can have?

39. A piece of wire less than 35 cm long is to be cut into two pieces. One piece must be 13 cm long. What is the range of values for the length of the other piece?

40. A rope less than 45 m long is to be cut into two pieces. One piece must be 4 times as long as the other piece. What is the range of values for the *shorter* piece?

5.4 Money Problems

In this section, we discuss a type of word problem commonly referred to as a "coin problem." Not all problems in this section deal with coins, but the method of solving the problems is essentially the same.

5.4A Getting Ready to Solve Money Problems

The main idea used in solving problems of this type is as follows:

$$\begin{pmatrix} \text{The value} \\ \text{of one kind} \\ \text{of item} \end{pmatrix} \cdot \begin{pmatrix} \text{The number} \\ \text{of} \\ \text{those items} \end{pmatrix} = \begin{pmatrix} \text{The total value} \\ \text{of that kind} \\ \text{of item} \end{pmatrix}$$

Example 1   If you have eight nickels, the total number of cents is

$$5(8) = 40$$

     └── Number of nickels
     └── Value of one nickel ∎

Example 2   If you have seven quarters, the total number of cents is

$$25(7) = 175$$

     └── Number of quarters
     └── Value of one quarter ∎

Example 3 If you have x dimes, the total number of cents is

$$10(x) = 10x$$

└─ Number of dimes
└─ Value of one dime ■

Example 4 If you have y 35¢ candy bars, the total value in cents is

$$35(y) = 35y$$

└─ Number of candy bars
└─ Value of one candy bar ■

Example 5 If you have two adults' movie tickets costing $5.00 each and four children's tickets costing $2.50 each, the total value in dollars is

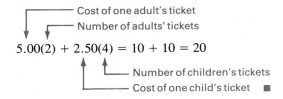

┌─ Cost of one adult's ticket
┌─ Number of adults' tickets

$$5.00(2) + 2.50(4) = 10 + 10 = 20$$

└─ Number of children's tickets
└─ Cost of one child's ticket ■

Example 6 If you have x 10¢ stamps and y 12¢ stamps, the total value in cents is

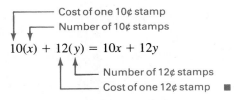

┌─ Cost of one 10¢ stamp
┌─ Number of 10¢ stamps

$$10(x) + 12(y) = 10x + 12y$$

└─ Number of 12¢ stamps
└─ Cost of one 12¢ stamp ■

EXERCISES 5.4A

Set I   In each exercise, express the total amount either as a number or as an algebraic expression.

1. If you have seven nickels, what amount of money is this?

2. If you have eleven quarters, what amount of money is this?

3. If you have nine 50¢ pieces, what amount of money is this?

4. If you have twelve dimes, what amount of money is this?

5. If you have x quarters, what amount of money is this?

6. If you have y nickels, what amount of money is this?

7. Find the total cost of seven adults' tickets and five children's tickets if adults' are $1.50 each and children's are 75¢ each.

8. If adults' tickets are $2.50 each and children's tickets are $1.25 each, what will four adults' and nine children's tickets cost?

9. If adults' tickets are $3.50 each and children's tickets are $1.90 each, what will x adults' and y children's tickets cost?

10. Find the total cost of x adults' and y children's tickets if adults' tickets cost $2.75 each and children's tickets cost $1.50 each.

11. Find the total cost of x 25¢ stamps and y 6¢ stamps.

12. Find the total cost of x 4¢ stamps and y 2¢ stamps.

Set II   In each exercise, express the total amount either as a number or as an algebraic expression.

1. If you have nine nickels, what amount of money is this?

2. If you have eleven dimes, what amount of money is this?

3. If you have fifteen dimes, what amount of money is this?

4. If you have twenty-three quarters, what amount of money is this?

5. If you have x 50¢ coins, what amount of money is this?

6. If you have x dimes and $(x + 4)$ nickels, what amount of money is this?

7. If adults' tickets cost $4.75 each and children's tickets cost $2.75 each, what will five adults' and three children's tickets cost?

8. If adults' tickets cost $5.25 each and children's tickets cost $3.00 each, what will x adults' tickets and $4x$ children's tickets cost?

9. Find the total cost of x adults' and y children's tickets if adults' tickets cost $4.50 each and children's tickets cost $2.25 each.

10. Find the total cost of x adults' and $(x + 6)$ children's tickets if adults' tickets cost $5.00 each and children's tickets cost $2.50 each.

11. Find the total cost of x 13¢ stamps and y 15¢ stamps.

12. Find the total cost of x 22¢ stamps and $5x$ 18¢ stamps.

5.4B Solving Money Problems

Example 7   A piggy bank contains seventeen coins with a total value of $1.15. If the coins are all nickels and dimes, how many of each kind of coin are there?

Solution There are two unknown numbers: the number of nickels and the number of dimes.

$$\text{Let} \qquad D = \text{number of dimes}$$

$$17 - D = \text{number of nickels}$$

Each dime is worth 10¢; therefore, D dimes are worth $10D$¢. Each nickel is worth 5¢; therefore, $(17 - D)$ nickels are worth $5(17 - D)$¢.

$$\underbrace{\text{Value of dimes}}_{10D} + \underbrace{\text{value of nickels}}_{5(17 - D)} = \underbrace{115¢}_{115}$$

$$10D + 5(17 - D) = 115$$

$$10D + 85 - 5D = 115$$

$$5D = 30$$

$$D = 6 \text{ dimes}$$

$$17 - D = 17 - 6 = 11 \text{ nickels}$$

Check

$$6 \text{ dimes} = 6(10¢) = 60¢$$

$$11 \text{ nickels} = 11(5¢) = \underline{55¢}$$

$$115¢ = \$1.15 \quad \blacksquare$$

Example 8   Dianne has $3.20 in nickels, dimes, and quarters. If she has seven more dimes than quarters and 3 times as many nickels as quarters, how many of each kind of coin does she have?

Solution

Let Q = number of quarters

$Q + 7$ = number of dimes   Because there are seven more dimes than quarters

$3Q$ = number of nickels   Because there are 3 times as many nickels as quarters

Q quarters are worth $25Q¢$. $(Q + 7)$ dimes are worth $10(Q + 7)¢$. $3Q$ nickels are worth $5(3Q)¢$.

Value of quarters	+	value of dimes	+	value of nickels	=	320¢
$25Q$	+	$10(Q + 7)$	+	$5(3Q)$	=	320

$$25Q + 10(Q + 7) + 5(3Q) = 320$$
$$25Q + 10Q + 70 + 15Q = 320$$
$$50Q = 250$$
$$Q = 5 \text{ quarters}$$
$$Q + 7 = 5 + 7 = 12 \text{ dimes}$$
$$3Q = 3(5) = 15 \text{ nickels}$$

Check

$$5 \text{ quarters} = 5(25¢) = 125¢$$
$$12 \text{ dimes} = 12(10¢) = 120¢$$
$$15 \text{ nickels} = 15(5¢) = \underline{75¢}$$
$$320¢ = \$3.20 \quad \blacksquare$$

EXERCISES 5.4B

Set I   Set up each problem algebraically, solve, and check. Be sure to state what your variables represent.

1. Bill has thirteen coins in his pocket that have a total value of 95¢. If these coins consist of nickels and dimes, how many of each kind are there?

2. Miko has eleven coins that have a total value of 85¢. If the coins are only nickels and dimes, how many of each kind are there?

3. Jennifer has twelve coins that have a total value of $2.20. The coins are nickels and quarters. How many of each kind of coin are there?

4. Brian has eighteen coins consisting of nickels and quarters. If the total value of the coins is $2.50, how many of each kind of coin does he have?

5. Derek has $4.00 in nickels, dimes, and quarters. If he has four more quarters than nickels and 3 times as many dimes as nickels, how many of each kind of coin does he have?

6. Staci has $5.50 in nickels, dimes, and quarters. If she has seven more dimes than nickels and twice as many quarters as dimes, how many of each kind of coin does she have?

7. Michael has $2.25 in nickels, dimes, and quarters. If he has three fewer dimes than quarters and as many nickels as the sum of the dimes and quarters, how many of each kind of coin does he have?

8. Muriel has $2.57 in dimes, nickels, and pennies. If she has five fewer pennies than nickels and as many dimes as the sum of the nickels and pennies, how many of each kind of coin does she have?

9. The total receipts for a concert were $6,600 for the 1,080 tickets sold. They sold orchestra seats for $7 each, box seats for $10 each, and balcony seats for $4 each. If there were 5 times as many balcony seats as box seats sold, how many of each kind were sold?

10. The total receipts for a football game were $360,800. General admission tickets cost $5 each, reserved seat tickets were $8 each, and box seat tickets $10 each. If there were twice as many general admission as reserved seat tickets sold and 4 times as many reserved as box seat tickets sold, how many of each kind were sold?

11. Christy spent $3.80 for sixty stamps. She bought only 2¢, 10¢, and 12¢ stamps. If she bought twice as many 10¢ stamps as 12¢ stamps, how many of each kind did she buy?

12. Mark spent $9.80 for 100 stamps. He bought only 6¢, 8¢, and 12¢ stamps. If there were 3 times as many 6¢ stamps as 8¢ stamps, how many of each kind did he buy?

Set II Set up each problem algebraically, solve, and check. Be sure to state what your variables represent.

1. Don spent $2.36 for twenty-two stamps. If he bought only 10¢ stamps and 12¢ stamps, how many of each kind did he buy?

2. Karla has fifteen coins with a total value of $2.55. If the coins are only dimes and quarters, how many of each kind of coin does she have?

3. Jill has fifteen coins with a total value of $2.70. If the coins are only dimes and quarters, how many of each kind of coin does she have?

4. Several families went to a movie together. They spent $24.75 for eight tickets. If adults' tickets cost $4.50 each and children's tickets cost $2.25 each, how many of each kind of ticket were bought?

5. Rachelle has $4.75 in nickels, dimes, and quarters. If she has four more nickels than dimes and twice as many quarters as dimes, how many of each kind of coin does she have?

6. Jason has $4.80 in nickels, dimes, and quarters. If he has three more dimes than quarters and 3 times as many nickels as quarters, how many of each kind of coin does he have?

7. Kevin has $4.75 in nickels, dimes, and quarters. If he has two more dimes than nickels and twice as many quarters as nickels, how many of each kind of coin does he have?

8. Anne spent $2.88 for fifteen stamps. She bought only 22¢ and 15¢ stamps. How many of each kind did she buy?

9. A class received $233 for selling 200 tickets to the school play. If students' tickets cost $1 each and nonstudents' tickets cost $2 each, how many nonstudent tickets were sold?

10. Heather spent $3.02 for stamps. She bought only 22¢ and 15¢ stamps, and she bought seven more 22¢ stamps than 15¢ stamps. How many of each kind did she buy?

11. Tricia spent $5.72 for twenty-six stamps. She bought only 22¢, 18¢, and 25¢ stamps. If she bought twice as many 22¢ stamps as 18¢ stamps and two more 25¢ stamps than 18¢ stamps, how many of each kind did she buy?

12. The total receipts for a concert were $13,000. Some tickets cost $10 each, some cost $14 each, and the rest cost $18 each. If there were eighty more $14 tickets sold than $18 tickets, and 10 times as many $10 as $18 tickets, how many of each kind were sold?

5.5 Ratio and Rate Problems

Ratio

A **ratio** is the quotient of one quantity divided by another quantity of the same kind. "The ratio of a to b" is written as $\frac{a}{b}$, and a and b are called the **terms** of the ratio. The terms of a ratio can be any kind of number, the only restriction being that the denominator cannot be zero. The ratio of a to b is also sometimes written as $a : b$, but this hides the fact that a ratio is a fraction or a rational number. (Notice the word *ratio* in the word *ratio nal*.) We will emphasize the fraction meaning of ratio.

A WORD OF CAUTION   "The ratio of a to b" is *not* $\frac{b}{a}$. It is $\frac{a}{b}$. The number that is *before* the word *to* is always in the numerator. The number that is *after* the word *to* is always in the denominator. ☑

We can give three different meanings to an expression such as $\frac{3}{4}$:

1. 3 of the 4 equal parts a unit has been divided into (fraction meaning)

2. $3 \div 4$ (division meaning)

3. The ratio of 3 to 4 (ratio meaning)

The meaning chosen depends on how the expression is used. The ratio meaning is used when we *compare* two numbers by division.

Example 1   An algebra class has thirteen men and seventeen women.

a. The ratio of men to women is $\frac{13}{17}$.

b. The ratio of women to men is $\frac{17}{13}$.   ■

The key to solving ratio problems is to use the given ratio to help represent the unknown numbers.

TO REPRESENT THE UNKNOWNS IN A RATIO PROBLEM

1. Multiply each term of the ratio by x.

2. Let the resulting products represent the unknowns.

A WORD OF CAUTION   In ratio problems, you are not finished when you have found x. You must multiply the value of x by the terms of the ratio in order to find the unknown numbers. ☑

Example 2   Two numbers are in the ratio of 3 to 5. Their sum is 32. Find the numbers. (The same problem could have been worded as follows: "Divide 32 into two parts whose ratio is 3 : 5.")

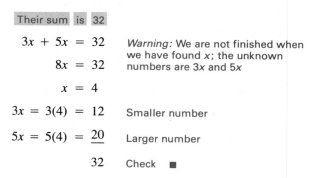

3 : 5   The ratio

$3x : 5x$ Multiply each term of the ratio by x
Let the resulting products represent the unknowns

Let $5x$ = one number

Let $3x$ = the other number

Their sum   is   32

$3x + 5x = 32$ *Warning:* We are not finished when we have found x; the unknown

$8x = 32$ numbers are $3x$ and $5x$

$x = 4$

$3x = 3(4) = 12$ Smaller number

$5x = 5(4) = \underline{20}$ Larger number

32   Check   ∎

Example 3   The three sides of a triangle are in the ratio 2 : 3 : 4. The perimeter is 63. Find the three sides.
Solution

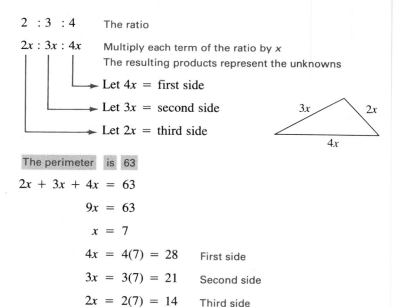

2 : 3 : 4   The ratio

$2x : 3x : 4x$ Multiply each term of the ratio by x
The resulting products represent the unknowns

Let $4x$ = first side

Let $3x$ = second side

Let $2x$ = third side

The perimeter   is   63

$2x + 3x + 4x = 63$

$9x = 63$

$x = 7$

$4x = 4(7) = 28$ First side

$3x = 3(7) = 21$ Second side

$2x = 2(7) = 14$ Third side

Check

28

21

$+ \, 14$

63   Perimeter   ∎

Example 4   Sixty-six hours of a student's week are spent in study, in class, and in work. The times spent in these activities are in the ratio 4 : 2 : 5. How many hours are spent in each activity?

Solution

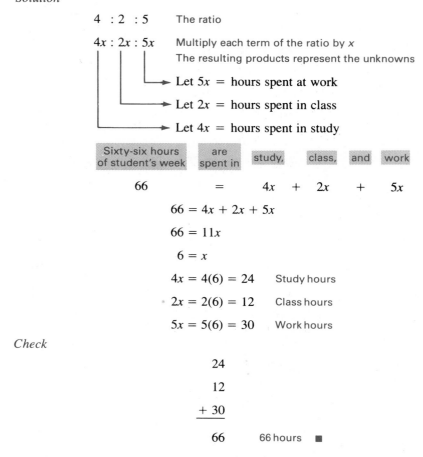

$4 : 2 : 5$ The ratio

$4x : 2x : 5x$ Multiply each term of the ratio by x
The resulting products represent the unknowns

→ Let $5x$ = hours spent at work

→ Let $2x$ = hours spent in class

→ Let $4x$ = hours spent in study

Sixty-six hours of student's week	are spent in	study,	class,	and	work
66	=	$4x$ +	$2x$	+	$5x$

$66 = 4x + 2x + 5x$

$66 = 11x$

$6 = x$

$4x = 4(6) = 24$ Study hours

$2x = 2(6) = 12$ Class hours

$5x = 5(6) = 30$ Work hours

Check

$$24$$
$$12$$
$$\underline{+\ 30}$$
$$66 \qquad 66 \text{ hours} \quad \blacksquare$$

Example 5   The length and width of a rectangle are in the ratio of 7 to 5. The perimeter is to be greater than 72. What is the range of values for the *length*?

Solution

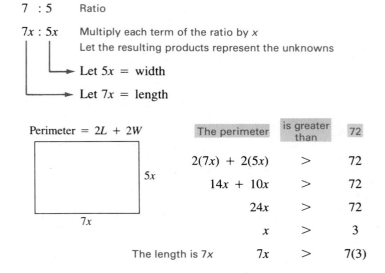

$7 : 5$ Ratio

$7x : 5x$ Multiply each term of the ratio by x
Let the resulting products represent the unknowns

→ Let $5x$ = width

→ Let $7x$ = length

Perimeter = $2L + 2W$

The perimeter	is greater than	72
$2(7x) + 2(5x)$	>	72
$14x + 10x$	>	72
$24x$	>	72
x	>	3
The length is $7x$ $7x$	>	$7(3)$

Therefore, the length must be greater than 21.

Check We will show the check for two values of the length:

Length = 28: If the length is 28, $7x = 28$, so $x = 4$ and the width must be 5(4), or 20. The perimeter, then, is

$$2(28) + 2(20) = 56 + 40 = 96$$

which is greater than 72.

Length = 47: If the length is 47, $7x = 47$, so $x = \frac{47}{7}$ and the width must be $5(\frac{47}{7})$, or $\frac{235}{7}$. The perimeter is

$$2(47) + 2\left(\frac{235}{7}\right) = 94 + \frac{470}{7} = \frac{1128}{7} \doteq 161.14$$

which is greater than 72. ■

We will consider ratios again in Section 8.9.

Rate

We often find it necessary or desirable to express in fraction form the quotient of one quantity divided by a quantity of a *different* kind. These fractions are technically called *rates*, although many authors do not distinguish between a ratio and a rate. You are undoubtedly familiar with rates such as $\frac{\text{miles}}{\text{hour}}$ (miles per hour), $\frac{\text{miles}}{\text{gallon}}$ (miles per gallon), $\frac{\text{dollars}}{\text{hour}}$ (dollars per hour), and so on.

Example 6 A man earns $103 for every 8 hours he works. Express the rate of dollars to hours as a fraction.

Solution The rate is $\frac{103\,\text{dollars}}{8\,\text{hours}}$. ■

Example 7 Gloria can drive about 533 miles on 14 gallons of gasoline. Express the rate of miles to gallons as a fraction.

Solution The rate of miles to gallons is $\frac{533\,\text{miles}}{14\,\text{gallons}}$. ■

When you solve problems involving rates, it is helpful to cancel the units whenever possible (see Example 8).

Example 8 Mrs. Norton's car uses gasoline at the rate of $16\frac{\text{miles}}{\text{gallon}}$. How many gallons of gasoline will she need in order to drive 464 miles?

Solution Let $x =$ the number of gallons of gasoline.

$$\left(16\frac{\text{miles}}{\text{gallon}}\right)(x\,\text{gallons}) = 464\,\text{miles} \qquad \text{We cancel ``gallons''}$$

We now drop the word "miles"

$$16x\,\text{miles} = 464\,\text{miles}$$

$$16x = 464$$

$$x = 29$$

Therefore, she will need 29 gallons of gasoline. (The check will not be shown.) ■

We will consider rates again in Sections 5.7, 8.9, and 8.11.

EXERCISES 5.5

Set I In Exercises 1–14, set up the problem algebraically, solve, and check. Be sure to state what your variables represent.

1. Two numbers are in the ratio of 3 to 4. Their sum is 35. Find the numbers.

2. Two numbers are in the ratio of 7 to 3. Their sum is 130. Find the numbers.

3. The length and width of a rectangle are in the ratio of 9 to 4. The perimeter is 78. Find the length and width.

4. The length and width of a rectangle are in the ratio 7 : 2. The perimeter is 72. Find the length and width.

5. The amounts Beverly spends for food, rent, and clothing are in the ratio 4 : 5 : 1. She spends an average of $460 a month for these three items. What can she expect to spend for each of these items in a month?

6. Frank spends $6,075 for tuition, housing, and food each school year. The amounts spent for these three items are in the ratio 4 : 4 : 1. How much does he spend for each item?

7. Divide 88 into two parts whose ratio is 6 to 5.

8. Divide 99 into two parts whose ratio is 2 to 7.

9. The three sides of a triangle are in the ratio 4 : 5 : 6. The perimeter is 90 ft. Find the three sides.

10. The three sides of a triangle are in the ratio 5 : 6 : 7. The perimeter is 72 yd. Find the three sides.

11. An uncle divided $27,000 among his three nephews in the ratio 2 : 3 : 4. How much did each receive?

12. A pension fund of $150,000 is invested in stocks, bonds, and real estate in the ratio 5 : 1 : 4. How much is invested in each?

13. When mixing the concrete for his patio floor, Mr. Mora used a cement-sand-gravel ratio of $1 : 2\frac{1}{2} : 4$. If he used 5 cubic yards of gravel, how much sand and cement did he use?

14. In one week's time, a restaurant served 865 dinners. For every five dinners served, three were seafood dinners. How many seafood dinners were served during the week?

In Exercises 15–18, set up the problem algebraically and solve.

15. The length and width of a rectangle are in the ratio of 7 to 4. The perimeter is less than 88. What is the range of values for the width?

16. The length and width of a rectangle are in the ratio of 8 to 5. The perimeter is greater than 78. What is the range of values for the length?

17. The three sides of a triangle are in the ratio 2 : 4 : 7. The perimeter is greater than 117. What is the range of values for the longest side?

18. The three sides of a triangle are in the ratio 3 : 5 : 6. The perimeter is less than 112. What is the range of values for the shortest side?

19. Ruby drives 360 miles in 7 hours. Express the rate of miles to hours as a fraction.

20. Justin bicycles 35 miles in 4 hours. Express the rate of miles to hours as a fraction.

21. Susan can crochet three afghans in 25 days. Express the rate of afghans to days as a fraction.

22. Cathy types 500 words in 11 minutes. Express the rate of words to minutes as a fraction.

In Exercises 23–26, set up the problem algebraically and solve.

23. If $805 was spent for carpeting that cost $23 \dfrac{\text{dollars}}{\text{sq. yd}}$, how many square yards were purchased?

24. If Florence earned $432 one week and if her pay was $12 \dfrac{\text{dollars}}{\text{hour}}$, how many hours did she work?

25. Mr. Lee's car uses gasoline at the rate of $23 \dfrac{\text{miles}}{\text{gallon}}$. How much gasoline will he use in driving 368 miles?

26. If Jennifer uses wallpaper that has a rate of coverage of $56 \dfrac{\text{sq. ft}}{\text{roll}}$, how many rolls will she need to cover 672 sq. ft?

Set II In Exercises 1–14, set up the problem algebraically and solve. Be sure to state what your variables represent.

1. Two numbers are in the ratio of 3 to 5. Their sum is 40. Find the numbers.

2. The length and width of a rectangle are in the ratio of 7 to 4. The perimeter is 88. Find the length and width.

3. The length and width of a rectangle are in the ratio of 9 to 5. The perimeter is 140. Find the length and width.

4. A wire 135 cm long was cut into two pieces. The ratio of the lengths of the pieces is 4 to 5. Find the length of each piece.

5. The amounts Diane spends for food, rent, and clothing are in the ratio 4 : 7 : 2. She spends an average of $715 a month for these three items. What can she expect to spend for each of these items in a *year*?

6. A farmer plants 162 acres of corn and soy beans in the ratio of 5 to 4. How many acres of each does he plant?

7. Divide 126 into two parts whose ratio is 5 to 13.

8. Separate 231 into two parts whose ratio is 17 to 4.

9. The three sides of a triangle are in the ratio 4 : 5 : 7. The perimeter is 128. Find the three sides.

10. A man cuts a 65-in. board into two parts whose ratio is 9 : 4. Find the length of the pieces.

11. The gold solder used in a crafts class has gold, silver, and copper in the ratio 5 : 3 : 2. How much gold is there in 30 g of this solder?

12. A civil service office tries to employ people with no discrimination because of age. If the ratio of people over 40 to those under 40 is 3 : 4, how many out of 259 people hired would have to be over 40?

13. Fast Eddie's Pro Shop sells four types of bowling balls: type A, B, C, and D. Sales of the balls totaled $3,600 for the year. If the balls sold in the ratio 3 : 2 : 1 : 4, how much of the sales, in dollars, can be attributed to each type of ball?

14. Alice is making an afghan that requires thirty-two skeins of yarn. The colors brown, beige, and rust are to be used in the ratio 5 : 8 : 3. How many skeins of each color should she buy?

In Exercises 15–18, set up the problem algebraically and solve.

15. The length and width of a rectangle are in the ratio of 9 to 4. The perimeter is greater than 156. What is the range of values for the width?

16. The ratio of Diane's age to Michelle's age is 3 to 2. The sum of their ages is less than 90 (years). What is the range of values for Diane's age?

17. The three sides of a triangle are in the ratio 8 : 9 : 12. The perimeter is less than 174. What is the range of values for the shortest side?

18. The amounts Carol spends for clothing, entertainment, and food are in the ratio 3 : 2 : 6. If she can spend less than $198 a month for these three items altogether, what range of values should she allow for entertainment?

19. George drives 323 miles in 6 hours. Express the rate of miles to hours as a fraction.

20. A machine produces sixty-three parts in 2 hours. Express the rate of parts to hours as a fraction.

21. Alan paid $682 for 23 sq. yd of carpeting. Express the rate of dollars to square yards as a fraction.

22. A plane flies 361 miles in 3 hours. Express the rate of miles to hours as a fraction.

In Exercises 23–26, set up the problem algebraically and solve.

23. If Ralph earned $561 one week and if his pay was $17 \dfrac{\text{dollars}}{\text{hour}}$, how many hours did he work?

24. If Elizabeth's car uses gasoline at the rate of $29 \dfrac{\text{miles}}{\text{gallon}}$, how much gasoline will she use in driving 493 miles?

25. If Dean uses paint that has a rate of coverage of $325 \dfrac{\text{sq. ft}}{\text{gallon}}$, how many gallons will he use in painting 2,275 sq. ft?

26. If $212 was spent for fencing that cost $4 \dfrac{\text{dollars}}{\text{ft}}$, how many feet of fencing were purchased?

5.6 Percent Problems

The Meaning of Percent

Percent means "per hundred" (*cent um* means 100 in Latin). For example, $5\% = \frac{5}{100}$, or $5\% = 0.05$. (If necessary, refer to Appendix B for the method for changing a number from a percent to a decimal, from a common fraction to a percent, and so forth.)

Example 1   A 5-gal paint can contains 2 gal of paint. Therefore, the paint can is $\frac{2}{5}$ full. Express this fraction as a percent.

Solution   Method 1: $\dfrac{2}{5} = \dfrac{2 \cdot 20}{5 \cdot 20} = \dfrac{40}{100} = 40\%$

⌐— Multiplying 5 by 20 makes the denominator 100

Method 2: $\dfrac{2}{5} = 0.4 = 0.40 = 40\%$

We can say "The can is $\frac{2}{5}$ full," or "The can is 40% full," or "The can is 0.4 full." ■

In algebra, word problems involving percents can be done by letting some variable represent the unknown number and using the fact that the word *of* in such problems indicates *multiplication*. The percent must be changed to its decimal form or to its common fraction form before the problem can be solved.

Checks will not be shown in this section.

Example 2   What number is 8% of 40?
Solution Let x = the unknown number.

What number | is | 8% | of 40
x = (0.08) · (40)

⌐— 8% changed to its decimal form

$x = 3.2$

Therefore, 3.2 is 8% of 40. ■

Example 3   16 is 40% of what number?
Solution Let x = the unknown number.

16 | is | 40% | of | what number
16 = (0.40) · x

⌐— 40% changed to its decimal form

$\dfrac{16}{0.4} = x$

$x = 40$

Therefore, 16 is 40% of 40. ■

Example 4   15 is what percent of 60?

Solution Let x = the fractional part (x must later be converted to its percent form).

$$\boxed{15} \; \boxed{\text{is}} \; \boxed{\text{what percent}} \; \boxed{\text{of}} \; \boxed{60}$$

$$15 \;=\; x \;\cdot\; 60$$

$$\frac{15}{60} \;=\; x$$

$$x \;=\; \frac{15}{60} = \frac{1}{4} = \frac{25}{100} = 0.25 = 25\%$$

Therefore, 15 is 25% of 60. ■

Example 5   What number is 175% of 80?

Solution Let x = the unknown number.

$$\boxed{\text{What number}} \; \boxed{\text{is}} \; \boxed{175\%} \; \boxed{\text{of}} \; \boxed{80}$$

$$x \;=\; (1.75) \;\cdot\; (80)$$

↑— 175% changed to its decimal form

$$x \;=\; 140$$

Therefore, 140 is 175% of 80. ■

Example 6   25 is what percent of 5?

Solution Let x = the fractional part (x must later be converted to its percent form).

$$\boxed{25} \; \boxed{\text{is}} \; \boxed{\text{what percent}} \; \boxed{\text{of}} \; \boxed{5}$$

$$25 \;=\; x \;\cdot\; 5$$

$$\frac{25}{5} \;=\; x$$

$$x \;=\; 5 = \frac{5}{1} = \frac{500}{100} = 500\%$$

Therefore, 25 is 500% of 5. ■

Example 7   25.2 is 140% of what number?

Solution Let x = the unknown number.

$$\boxed{25.2} \; \boxed{\text{is}} \; \boxed{140\%} \; \boxed{\text{of}} \; \boxed{\text{what number}}$$

$$25.2 \;=\; (1.40) \;\cdot\; x$$

↑— 140% changed to its decimal form

$$\frac{25.2}{1.40} \;=\; x$$

$$x \;=\; 18$$

Therefore, 25.2 is 140% of 18. ■

Example 8   In an examination, a student worked fifteen problems correctly. This was 75% of the problems. Find the total number of problems on the examination.
Solution In this problem, we are saying "15 is 75% of what number?"

Let $x =$ the unknown number.

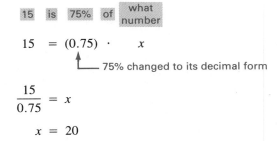

$$15 = (0.75) \cdot x$$

75% changed to its decimal form

$$\frac{15}{0.75} = x$$

$$x = 20$$

Therefore, twenty problems were on the examination. ■

Example 9   Mr. Delgado, a salesman, makes a 6% commission on all items he sells. One week he made $390. What were his gross sales for the week?
Solution 390 is 6% of what number?

Let $x =$ the unknown number.

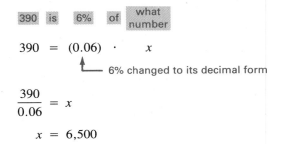

$$390 = (0.06) \cdot x$$

6% changed to its decimal form

$$\frac{390}{0.06} = x$$

$$x = 6,500$$

Therefore, his gross sales for the week were $6,500. ■

Markup

In a business that buys and sells merchandise, the merchandise must be sold at a price high enough to return to the merchant (1) the price paid for the goods; (2) the expenses, salaries, rents, taxes, and so on; and (3) a reasonable profit. To accomplish this, the cost of each item must be *marked up* before it is sold. We use markup based on cost. Selling price, cost, and markup are related by the following formula:

$$\text{Selling price} = \text{cost} + \text{markup}$$
$$S = C + M$$

Example 10   If a business pays $75 for an item, what is the selling price of the item if it is marked up 40%?
Solution Markup = 40% of $75 (cost)

Let $M =$ the markup. (*Selling* price will be $75 + M$.)

Then, $M = 40\%$ of 75

$$M = 0.40(75) = 30$$

Selling price = cost + markup

$$= \$75 + \$30 = \$105 \quad ■$$

EXERCISES 5.6

Set I Set up each of the following problems algebraically and solve, even if you could do the problem using methods learned in an arithmetic class. Be sure to state what your variables represent.

1. 15 is 30% of what number?

2. 16 is 20% of what number?

3. 115 is what percent of 250?

4. 330 is what percent of 225?

5. What is 25% of 40?

6. What is 45% of 65?

7. 15% of what number is 127.5?

8. 32% of what number is 256?

9. What percent of 8 is 17?

10. What percent of 6 is 12?

11. 63% of 48 is what number?

12. 87% of 49 is what number?

13. 750 is 125% of what number?

14. 325 is 130% of what number?

15. 23 is what percent of 16?

16. 57 is what percent of 23?

17. What is 200% of 12?

18. What is 300% of 9?

19. 15% of a number is 37.5. What is the number?

20. What is 27% of $135?

21. 42 is $66\frac{2}{3}$% of what number?

22. 36 is $16\frac{2}{3}$% of what number?

23. A team wins 80% of its games. If it wins sixty-eight games, how many games has it played?

24. Seventy percent of the 46,000 burglaries in a city were committed by persons who had previously been convicted at least three times for the same crime. If all of these criminals had been kept in jail, how many burglaries could have been prevented?

25. In a class of forty-two students, seven students received a grade of B. What percent of the class received a grade of B?

26. John's weekly gross pay is $110, but 23% of his check is withheld. How much is withheld?

27. Fifty-four out of 210 civil service applicants pass their exams. What percent of the applicants pass?

28. A 4,200-lb automobile contains 462 lb of rubber. What percent of the car's total weight is rubber?

29. A merchant pays $125 for an item. What is the selling price of this item if it is marked up 35% of the cost?

30. A suit costing a merchant $72 is marked up 20%. What is its selling price?

31. A truckload of fifteen steers and eighteen heifers was hauled to market. The average weights of the steers and heifers were 1,027 lb and 956 lb, respectively. Due to weight loss during shipment, a 3% deduction is made from the total live weight, and the seller is paid on this reduced weight. If the producer received 84¢ per pound for the steers and 78¢ per pound for the heifers, how much was his check from the buyer?

32. A timber company paid $632,000 for 1.27 million board feet of timber in the Mt. Hood National Forest. The timber had been appraised at $456,000. What percent of the appraised value was the amount that the company paid above the appraised value?

Set II Set up each of the following problems algebraically and solve, even if you could do the problem using methods learned in an arithmetic class. Be sure to state what your variables represent.

1. 24 is 40% of what number?

2. 650 is what percent of 325?

3. 34 is 68% of what number?

4. 335 is 134% of what number?

5. What is 55% of 82?

6. 102 is what percent of 17?

7. 23% of what number is 13.11?

8. 156% of what number is 195?

9. 39 is what percent of 27? (Round off your answer to one decimal place.)

10. 86 is what percent of 344?

11. 91% of 64 is what number?

12. 63% of what number is 10.71?

13. 930 is 124% of what number?

14. What is 134% of 335?

15. 67 is what percent of 32?

16. 89 is what percent of 20?

17. What is 400% of 19?

18. What is $16\frac{2}{3}\%$ of 78?

19. 12% of a number is 9.36. What is the number?

20. 136% of what number is 10.88?

21. 28 is $33\frac{1}{3}\%$ of what number?

22. 29 is $12\frac{1}{2}\%$ of what number?

23. A team wins 105 games. This is 70% of the games played. How many games were played?

24. 85% of the members of the golf club voted for Brian for president. He received 238 votes. How many members are in the club?

25. Michelle is on a 1,200-calorie-a-day diet. She consumed 360 calories at breakfast. What percent of her entire daily calorie allowance was this?

26. Heather sleeps 15 hr a day (including naps). What percent of the day does she sleep?

27. Rosie's weekly salary is $225. If her deductions amount to $63, what percent of her salary is take-home pay?

28. Trisha found a television set on sale. The original price was $379, but it was on sale for 30% off. What was the sale price?

29. A camera costing a merchant $244 is marked up 30%. What is its selling price?

30. A saleswoman makes a 7% commission on all the items she sells. One week she made $504. What were her gross sales for the week?

31. The largest meteorite found in the United States was discovered in a forest near Willamette, Oregon, in 1902. This 13.5-ton meteorite is about 91% iron and 8.3% nickel. How much iron and how much nickel does this meteorite contain?

32. The largest meteorite found in the world was discovered in southwest Africa. This 60.2-ton meteorite is about 81% iron and 17.5% nickel. How much iron and how much nickel does this meteorite contain?

5.7 Distance-Rate-Time Problems

Distance-rate-time problems are used in any field involving motion. A physical law relating *distance* traveled d, *rate* of travel r, and *time* of travel t is

$$d = r \cdot t$$

For example, you know that if you are driving your car at an average speed of 50 mph, then

you travel a distance of 100 miles in 2 hours: $100 \, \text{mi} = 50\dfrac{\text{mi}}{\text{hr}}(2\,\text{hr})$

you travel a distance of 150 miles in 3 hours: $150 \, \text{mi} = 50\dfrac{\text{mi}}{\text{hr}}(3\,\text{hr})$

and so on.

Example 1   If Mrs. Petersen drives at an average rate of 47 mph (miles per hour), how long will it take her to drive 329 miles?
Solution Let x = the number of hours.

$$r \cdot t = d$$

$$\left(47\frac{\text{miles}}{\text{hour}}\right)(x\,\text{hours}) = 329 \text{ miles} \qquad \text{We cancel "hours"}$$

We now drop the word "miles"

$$47x \text{ miles} = 329 \text{ miles}$$

$$47x = 329$$

$$x = 7$$

Therefore, it will take her 7 hours to drive 329 miles. (The check will not be shown.) ∎

Many distance-rate-time problems involve two different rates. Use the method in the following box for such problems.

METHOD FOR SOLVING DISTANCE-RATE-TIME PROBLEMS

1. Draw the blank chart:

	d	$=$	r	$\cdot$	t

2. Fill in as many of the boxes as possible, using the information given.

3. Let x represent an unknown. Express other unknowns in terms of x, if possible.

4. Fill in the remaining boxes, using x and the formula $d = r \cdot t$.

5. Write an equation, using the information in the chart along with any unused facts given in the problem.

6. Solve the resulting equation.

7. Answer all the questions asked in the problem.

8. Check your solution(s) in the original word problem.

Example 2 Mr. Maxwell takes 1 hr to drive to work in the morning, but he takes $1\frac{1}{2}$ hr to return home over the same route during the evening rush hour. If his average morning speed is 10 mph faster than his average evening speed, how far is it from his home to his work?

Solution Although we are asked to find a *distance*, the problem is solved more easily by letting x equal a rate of speed.

Let $\qquad x = $ speed returning from work (the slower speed)

$\qquad x + 10 = $ speed going to work (the faster speed)

	d	$=$	r	$\cdot$	t
Going to work			$x + 10$		1 hr
Returning from work			x		$1\frac{1}{2}$ hr

Use the given information to fill in these boxes

Use the formula $d = r \cdot t$ to find what goes here

	d	$=$	r	$\cdot$	t
Going to work	$1(x + 10)$		$x + 10$		1 hr
Returning from work	$\frac{3}{2}(x)$		x		$1\frac{1}{2}$ hr

We now make use of the "unused fact" that the two distances are equal:

$$\underbrace{\text{Distance going to work}} = \underbrace{\text{Distance returning from work}}$$

$$1(x + 10) = \frac{3}{2}(x)$$

$$x + 10 = \frac{3}{2}x$$

$$\underline{-x \qquad\quad = -x} \qquad\qquad \text{Add } -x \text{ to both sides}$$

$$10 = \frac{1}{2}x$$

$$x = 20 \text{ mph} \qquad \text{Speed returning from work}$$

$$x + 10 = 30 \text{ mph} \qquad \text{Speed going to work}$$

However, we were asked to find the *distance* from his home to his work.

$$\text{Distance} = (\text{rate})(\text{time}) = \left(20\,\frac{\text{mi}}{\text{hr}}\right)\left(\frac{3}{2}\,\text{hr}\right) = 30\,\text{mi}$$

Check The distance *going* $= \left(30\,\dfrac{\text{mi}}{\text{hr}}\right)(1\,\text{hr}) = 30$ mi. The distances are equal. ∎

Example 3 A boat cruises downstream for 4 hr before heading back. After traveling upstream for 5 hr, it is still 16 mi short of the starting point. If the speed of the stream is 4 mph, find the speed of the boat in still water.
Solution

$$\text{Let} \quad x = \text{speed of boat in still water}$$

$$x + 4 = \text{speed of boat downstream}$$

$$x - 4 = \text{speed of boat upstream}$$

	d	$=$	r	$\cdot$	t
Downstream			$x + 4$		4 hr
Upstream			$x - 4$		5 hr

Use the formula $d = r \cdot t$ to fill in the two empty boxes

	d	$=$	r	$\cdot$	t
Downstream	$(x + 4)(4)$		$x + 4$		4
Upstream	$(x - 4)(5)$		$x - 4$		5

$d = r \cdot t$
$d = (x + 4)(4)$
$d = r \cdot t$
$d = (x - 4)(5)$

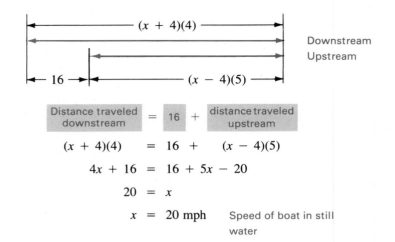

$$
\begin{array}{cccc}
\text{Distance traveled downstream} & = & 16 & + & \text{distance traveled upstream} \\
(x + 4)(4) & = & 16 & + & (x - 4)(5) \\
4x + 16 & = & 16 + 5x - 20 \\
20 & = & x \\
x & = & 20 \text{ mph} & & \text{Speed of boat in still water}
\end{array}
$$

Check

$$\text{Speed of boat downstream} = (20 + 4)\text{ mph} = 24 \text{ mph}$$

$$\text{Speed of boat upstream} = (20 - 4)\text{ mph} = 16 \text{ mph}$$

$$\text{Distance downstream} = \left(24\,\frac{\text{mi}}{\text{hr}}\right)(4\,\text{hr}) = 96 \text{ mi}$$

$$\text{Distance upstream} = \left(16\,\frac{\text{mi}}{\text{hr}}\right)(5\,\text{hr}) = 80 \text{ mi}$$

$$
\begin{array}{ccccc}
\text{Distance downstream} & = & 16 \text{ mi} & \text{plus} & \text{distance upstream} \\
96 \text{ mi} & \overset{?}{=} & 16 \text{ mi} & + & 80 \text{ mi} \\
96 \text{ mi} & = & 96 \text{ mi} & & \blacksquare
\end{array}
$$

Example 4 Terri drove from her home to San Diego to pick up her husband, Nick. She encountered heavy traffic all the way and averaged 45 mph. Nick encountered no traffic and averaged 63 mph on the trip home. If the total driving time for the round trip was 12 hr, how far is their home from San Diego?

Solution

$$\text{Let} \quad x = \text{the number of hours Terri drove}$$

$$12 - x = \text{the number of hours Nick drove}$$

	d	$=$	r	$\cdot$	t
Terri			45 mph		x hr
Nick			63 mph		$(12 - x)$ hr

Use the formula $d = r \cdot t$ to fill in these boxes

	d	$=$	r	$\cdot$	t
Terri	$(45)(x)$		45 mph		x hr
Nick	$(63)(12 - x)$		63 mph	$\cdot$	$(12 - x)$ hr

We now use the unused fact that the two distances are equal:

$$45x = 63(12 - x)$$

$$45x = 756 - 63x$$

$$\underline{63x \qquad\qquad 63x} \qquad \text{Add } 63x \text{ to both sides}$$

$$108x = 756$$

$$\frac{108x}{108} = \frac{756}{108} \qquad \text{Divide both sides by 108}$$

$$x = 7 \qquad \text{Number of hours Terri drove}$$

$$12 - x = 5 \qquad \text{Number of hours Nick drove}$$

Find the distance:

$$d = \left(45\frac{\text{mi}}{\text{hr}}\right)(7 \text{ hr}) = 315 \text{ mi} \qquad \text{Distance Terri drove}$$

Check Distance Nick drove: $d = \left(63\frac{\text{mi}}{\text{hr}}\right)(5 \text{ hr}) = 315$ mi. The distances are equal, and the sum of the driving times is 12 hr. ∎

EXERCISES 5.7

Set I   Set up each problem algebraically, solve, and check. Be sure to state what your variables represent.

1. If Robbie drives at an average rate of 51 mph, how long will it take him to drive 408 mi?

2. If Janine bicycles at an average rate of 8 mph, how long will it take her to go 72 mi?

3. Jennie drives to her son's house in 5 hr when there is heavy traffic. When there is no traffic, the same trip takes 4 hr. She can drive an average of 10 mph faster when there is no traffic than when there is heavy traffic.

   a. What is her average speed when there is no traffic?

   b. What is the distance from her house to her son's house?

4. Bud occasionally has to drive to a company that is some distance from his home. If there is no traffic, the trip takes 5 hr, but in heavy traffic it takes 6 hr. He can drive an average of 10 mph faster when there is no traffic than when there is heavy traffic.

   a. What is his average speed when there is no traffic?

   b. What is the distance from his home to the company?

5. Mr. Robinson left San Diego at 7 A.M., heading toward Los Angeles. His neighbor Mr. Reid left at 8 A.M. on the same highway, also heading toward Los Angeles. By driving 9 mph faster, Mr. Reid overtook Mr. Robinson at noon.

   a. Find Mr. Robinson's average speed.

   b. Find Mr. Reid's average speed.

   c. Find the distance each traveled before they met.

6. Mr. Curtis left Fillmore at 6 A.M., heading toward San Francisco. His neighbor Mr. Castillo left Fillmore at 7 A.M., also heading toward San Francisco. By driving 9 mph faster, Mr. Castillo overtook Mr. Curtis at noon.

   a. Find Mr. Curtis's average speed.

   b. Find Mr. Castillo's average speed.

   c. Find the distance each traveled before they met.

7. A boat cruises downstream for 3 hr before heading back. It takes 4 hr going upstream for the boat to get back to its starting point. If the speed of the stream is 3 mph, find the speed of the boat in still water.

8. A boat cruises downstream for 5 hr before stopping for the night. The next day, it took 7 hr for the boat to get back to its starting point. If the speed of the stream both days was 4 mph, find the speed of the boat in still water.

9. A boat cruised downstream for 3 hr before heading back. After traveling 4 hr back upstream, the boat was still 6 mi short of the starting point. If the speed of the stream was 4 mph, find the speed of the boat in still water.

10. A boat cruised downstream for 4 hr before heading back. After traveling 5 hr back upstream, the boat was still 3 mi short of the starting point. If the speed of the stream was 3 mph, find the speed of the boat in still water.

11. Matthew drove from his home to a friend's house on a Saturday. Because traffic was very light, his average speed was 54 mph. He returned home the next day, and his average speed was 48 mph. His total driving time was 17 hr.

   a. How long did it take him to get to his friend's house?

   b. How far was it from his home to his friend's home?

12. Brian bicycled from his home to a picnic area at Lake Perris. Because much of the path was uphill, he averaged only 16 mph. On the return trip, he averaged 24 mph. His total bicycling time was 5 hr.

   a. How long did it take him to get to the picnic area?

   b. How far was it from his home to the picnic area?

Set II Set up each problem algebraically, solve, and check. Be sure to state what your variables represent.

1. If Mrs. Gragson drives at an average rate of 49 mph, how long will it take her to drive 441 miles?

2. If Melanie drives 432 miles in 9 hours, what is her average rate of speed?

3. If driving conditions are poor, Sherma can drive to her sister's house in Arizona in 12 hr. However, if conditions are good, her average speed is 9 mph faster, and she can make the trip in 10 hr.

   a. What is her average speed when conditions are poor?

   b. What is her average speed when conditions are good?

   c. What is the distance from her house to her sister's house?

4. Tom drives to his favorite ski area in 5 hr when there is no traffic. When there is heavy traffic, his average speed is 10 mph slower than when there is no traffic, and the same trip takes him 6 hr.

   a. What is his average speed when there is heavy traffic?

   b. What is the distance from his house to the ski area?

5. Mark leaves his house at 8 A.M., heading toward Steve's house. Mark's brother, Brian, leaves their house at 9 A.M., also heading toward Steve's house. By bicycling 4 mph faster, Brian overtakes Mark at 3 P.M.

a. What is Mark's average speed?

b. What is Brian's average speed?

c. Find the distance traveled before Brian overtakes Mark.

6. Lori hikes from her campsite to Green Lake. She hiked at a rate of 2 mph going to the lake and 3 mph coming back. The trip to the lake took 3 hr longer than the trip back.

a. How long did it take her to hike to the lake?

b. What was the distance between her campsite and the lake?

7. An airplane flies from point A to point B in 2 hr, flying with the wind. It takes the plane 3 hr to make the return trip. If the speed of the wind is 30 mph, find the speed of the plane in still air.

8. Todd and Jo live 102 mi apart. Both leave their homes by bicycle, riding toward one another. They meet 2 hr later. Todd bicycles 5 mph faster than Jo.

a. What is Jo's average speed?

b. What is Todd's average speed?

9. Nick drove his boat upstream for 3 hr before heading back. After cruising back downstream for 2 hr, he was only 3 mi short of his starting point. If the speed of the stream is 4 mph, find the speed of his boat in still water.

10. Danny and Cathy live 60 mi apart. Both leave their homes at 10 A.M. by bicycle, riding toward one another. They meet at 2 P.M. If Cathy's average speed is two-thirds of Danny's, how fast does each cycle?

11. Leon hiked from his camp to a lookout point at the rate of 3 mph. On the trip back to his camp, he hiked at the rate of 4 mph. He hiked a total of 7 hr.

a. How long did it take him to hike to the lookout point?

b. How far was it from his camp to the lookout point?

12. The Wright family sails their houseboat upstream for 4 hr. After lunch, they motor downstream for 2 hr. At that time, they are still 12 mi away from the marina where they began. If the speed of the houseboat in still water is 15 mph, what is the speed of the stream? How far upstream did the Wrights travel?

5.8 Mixture Problems

Mixture problems usually involve mixing two or more dry ingredients. Because the cost of each item is often important, mixture problems are actually *money problems*.

TWO IMPORTANT RELATIONSHIPS NECESSARY TO SOLVE
MIXTURE PROBLEMS

$$\begin{pmatrix} \text{Amount of} \\ \text{ingredient A} \end{pmatrix} + \begin{pmatrix} \text{Amount of} \\ \text{ingredient B} \end{pmatrix} = \begin{pmatrix} \text{Amount of} \\ \text{mixture} \end{pmatrix}$$

$$\begin{pmatrix} \text{Cost of} \\ \text{ingredient A} \end{pmatrix} + \begin{pmatrix} \text{Cost of} \\ \text{ingredient B} \end{pmatrix} = \begin{pmatrix} \text{Cost of} \\ \text{mixture} \end{pmatrix}$$

Students often find a chart helpful in solving mixture problems.

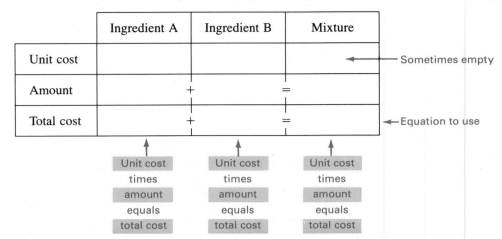

	Ingredient A	Ingredient B	Mixture
Unit cost				←— Sometimes empty
Amount		+	=
Total cost		+	=	←— Equation to use

In each column (that is, in each vertical line), the total cost is found by multiplying the unit cost by the amount. The *bottom row* of the chart gives the equation.

Example 1 A wholesaler makes up a 50-lb mixture of two kinds of coffee. Brand A costs $3.30 per pound and Brand B costs $3.70 per pound. How many pounds of each kind of coffee must be used if the mixture is to cost $3.58 per pound?

Solution

Let x = number of pounds of Brand A

$50 - x$ = number of pounds of Brand B

	Brand A	Brand B	Mixture
Unit cost	3.30	3.70	3.58
Amount	x +	$(50 - x)$ =	50
Total cost	$(3.30)x$ +	$3.70(50 - x)$ =	$3.58(50)$

Notice that

$$\text{Amount of Brand A} + \text{Amount of Brand B} = \text{Amount of mixture}$$
$$x + (50 - x) = 50$$

and that

$$\text{Total cost of Brand A} + \text{Total cost of Brand B} = \text{Total cost of mixture}$$
$$3.30x + 3.70(50 - x) = 3.58(50)$$

$$3.30x + 3.70(50 - x) = 3.58(50)$$
$$330x + 370(50 - x) = 358(50) \qquad \text{Multiply both sides by 100}$$
$$330x + 18{,}500 - 370x = 17{,}900 \qquad \text{Remove parentheses}$$
$$18{,}500 - 40x = 17{,}900 \qquad \text{Combine like terms}$$
$$\underline{-17{,}900 + 40x \qquad -17{,}900 + 40x}$$
$$600 = 40x$$

$$\frac{600}{40} = \frac{40x}{40} \qquad \text{Divide both sides by 40}$$

$$x = 15 \qquad \text{Number of pounds of Brand A}$$

$$50 - x = 35 \qquad \text{Number of pounds of Brand B}$$

Check

$$15 \text{ lb @ } \$3.30 = \$\ 49.50$$
$$35 \text{ lb @ } \$3.70 = \underline{\$129.50} \qquad 50 \text{ lb @ } \$3.58 = \$179.00$$

$$\text{Total} = \underline{\$179.00} \longleftarrow \text{Total cost} \longrightarrow \blacksquare$$

Example 2   A 10-lb mixture of walnuts and almonds costs $35.20. If the walnuts cost $4.50 per pound and the almonds cost $3.10 per pound, how many pounds of each kind are in the mixture?
Solution Notice that in this problem, we were given the *total* cost of the mixture, rather than the unit cost of the mixture.

$$\text{Let } x = \text{ number of pounds of walnuts}$$

$$10 - x = \text{ number of pounds of almonds}$$

	Walnuts	Almonds	Mixture
Unit cost	4.50	3.10	/////////
Amount	x $+$	$(10 - x)$ $=$	10
Total cost	$(4.50)x$ $+$	$3.10(10 - x)$ $=$	35.20

$$4.50x + 3.10(10 - x) = 35.20$$

$$450x + 310(10 - x) = 3{,}520 \qquad \text{Multiply both sides by 100}$$

$$45x + 31(10 - x) = 352 \qquad \text{Divide both sides by 10}$$

$$45x + 310 - 31x = 352$$

$$\begin{array}{r} 14x + 310 = \quad 352 \\ -310 \quad\ -310 \\ \hline 14x = \quad 42 \end{array}$$

$$\frac{14x}{14} = \frac{42}{14}$$

$$x = 3 \qquad \text{Number of pounds of walnuts}$$

$$10 - x = 7 \qquad \text{Number of pounds of almonds}$$

Check

$$3 \text{ lb @ } \$4.50 = \$13.50$$
$$7 \text{ lb @ } \$3.10 = \underline{\$21.70}$$

$$\text{Total} = \$35.20 \qquad \text{Total cost of mixture} \quad \blacksquare$$

Example 3   Mrs. Reid needs to mix 20 lb of macadamia nuts that cost $8.10 per pound with pecans that cost $5.40 per pound. How many pounds of pecans should she use if the mixture is to cost $6.48 per pound?

Solution Let x = number of pounds of pecans.

	Pecans	Macadamia nuts	Mixture
Unit cost	5.40	8.10	6.48
Amount	x +	20 =	$x + 20$
Total cost	$(5.40)x$ +	$8.10(20)$ =	$6.48(x + 20)$

$$5.40x + 8.10(20) = 6.48(x + 20)$$

$$540x + 810(20) = 648(x + 20)$$

$$\begin{array}{rl} 540x + 16{,}200 = & 648x + 12{,}960 \\ -540x - 12{,}960 & -540x - 12{,}960 \end{array}$$

$$3{,}240 = 108x$$

$$\frac{3{,}240}{108} = \frac{108x}{108}$$

$$x = 30 \quad \text{Number of pounds of pecans}$$

Check

20 lb @ $8.10 = $162   Value of macadamia nuts

30 lb @ $5.40 = $162   Value of pecans

50 lb ◄——— Totals ——► $324

$$\text{Price per pound of mixture} = \frac{324 \text{ dollars}}{50 \text{ pounds}} = \$6.48 \text{ per pound} \quad \blacksquare$$

NOTE   All money problems can be done by using a chart, as shown in Examples 1, 2, and 3.   ☑

EXERCISES 5.8

Set I   Set up each problem algebraically, solve, and check. Be sure to state what your variables represent.

1. A grocer makes a 10-lb mixture of granola and dried apple chunks, and he wants the mixture to cost $3.00 per pound. If the granola costs $2.20 per pound and the apple chunks cost $4.20 per pound, how many pounds of each kind should he use?

2. Jim wants to make up a 50-lb mixture of macadamia nuts at $8.50 per pound and peanuts at $3.50 per pound. How many pounds of each kind should he use if the mixture is to cost $5.80 per pound?

3. Alice wants to make up a 30-lb mixture of apples that will be worth 78¢ per pound. One kind of apple costs 95¢ per pound, and the other kind costs 65¢ per pound. How many pounds of each kind should she use?

4. A grocer must make a 60-lb mixture of peanut brittle and English toffee that will

be worth $5.60 per pound. If peanut brittle costs $4.20 per pound and English toffee costs $6.60 per pound, how many pounds of each kind should he use?

5. Margie makes a 27-lb mixture of gumdrops and caramels that is to be worth $66.00. If the gumdrops cost $2.80 per pound and the caramels cost $2.20 per pound, how many pounds of each kind should she use?

6. A grocer makes a 40-lb mixture of two kinds of coffee that is worth $151. If Brand C costs $4.20 per pound and Brand D costs $3.20 per pound, how many pounds of each kind does he use?

7. Dorothy needs a mixture of candy and nuts. How many pounds of candy at $3.00 per pound should she mix with 50 lb of nuts at $2.40 per pound to obtain a mixture worth $2.50 per pound?

8. How many pounds of cashews at $7.60 per pound should be mixed with 30 lb of macadamia nuts at $8.20 per pound in order to obtain a mixture worth $7.96 per pound?

Set II   Set up each problem algebraically, solve, and check. Be sure to state what your variables represent.

1. A 40-lb mixture of peanuts and cashews is to be made; it is to be worth $5.55 per pound. If peanuts cost $2.25 per pound and cashews cost $7.53 per pound, how many pounds of each should be used?

2. A grocer needs to make a 25-lb mixture of coffee worth $3.52 per pound. If Brand A costs $3.80 per pound and Brand B costs $3.30 per pound, how many pounds of each kind should he use?

3. Walt needs a mixture of walnuts and almonds to be worth $3.92 per pound. If walnuts cost $4.50 per pound and almonds cost $3.50 per pound, how many pounds of each kind should he use to obtain a 50-lb mixture?

4. Herb makes a 25-lb mixture of peanuts and raisins; it is worth $2.88 per pound. If peanuts cost $3.69 per pound and raisins cost $1.44 per pound, how many pounds of each kind did he use?

5. A 35-lb mixture of two kinds of coffee is worth $118.00. If Brand A costs $3.70 per pound and Brand B costs $3.20 per pound, how many pounds of each kind of coffee was used?

6. Sherma made a 32-lb mixture of nougats and peppermint candies that was worth $70.00. The nougats cost $2.50 per pound and the peppermints cost $2.10 per pound. How many pounds of each did she use?

7. How much Brand A coffee costing $3.95 per pound should be mixed with 25 lb of Brand B coffee costing $3.60 per pound in order to obtain a mixture worth $3.70 per pound?

8. How many pounds of Delicious apples at 95¢ per pound should be mixed with 30 lb of Spartan apples at 90¢ per pound in order to obtain a mixture worth 92¢ per pound?

5.9 Solution Problems

Another type of mixture problem involves the mixing of liquids. Such problems are often called *solution problems* because a mixture of two or more liquids is, under certain conditions, a *solution*.*

---

*A solution is a homogeneous mixture of two or more substances.

Before we discuss solution mixture problems, we will discuss chemical solutions in general. A "30% solution of alcohol" is a mixture that is 30% pure alcohol and 70% (that is, 100% − 30%) water.

Example 1 How many liters of pure alcohol are there in 2 ℓ of a 30% solution of alcohol?
Solution We must find 30% of 2.

Let x = the amount of pure alcohol.

$$x = (0.30)(2) = 0.6 \qquad \text{Number of liters of pure alcohol present} \quad \blacksquare$$

Example 2 In a 5% glycerin solution, there are 3 ml of pure glycerin present. How many milliliters of solution are there?
Solution We must answer this question: 3 is 5% of what number?

Let x = the unknown number.

$$3 = (0.05) \cdot x$$

$$\frac{3}{0.05} = \frac{0.05x}{0.05}$$

$$x = 60 \qquad \text{Number of ml of solution} \quad \blacksquare$$

Example 3 Fifty milliliters of an alcohol solution contain 6 ml of pure alcohol. What percent of the solution is alcohol?
Solution We must find what percent 6 is of 50.

Let x = the fractional part.

$$6 = x \cdot (50)$$

$$\frac{6}{50} = \frac{50x}{50}$$

$$x = \frac{6}{50} = 0.12 = 12\% \quad \blacksquare$$

Let us now discuss mixtures that involve the mixing of solutions. Solution problems can be solved by using a method similar to that used for solving other mixture problems. The chart is set up as follows:

	Ingredient A	Ingredient B	Mixture	
Percent				← In decimal form
Amount	+	=		
Total amount of pure substance	+	=		

Percent	Percent	Percent
times	times	times
amount	amount	amount
equals	equals	equals
total amount	total amount	total amount

In each column (vertical line), the total amount of the substance is found by multiplying the percent (changed to a decimal or common fraction) by the amount.

TWO RELATIONSHIPS NECESSARY TO SOLVE SOLUTION PROBLEMS

$$\begin{pmatrix} \text{Amount of} \\ \text{ingredient A} \end{pmatrix} + \begin{pmatrix} \text{Amount of} \\ \text{ingredient B} \end{pmatrix} = \begin{pmatrix} \text{Amount of} \\ \text{mixture} \end{pmatrix}$$

$$\begin{pmatrix} \text{Total amount of} \\ \text{pure substance} \\ \text{from ingredient A} \end{pmatrix} + \begin{pmatrix} \text{Total amount of} \\ \text{pure substance} \\ \text{from ingredient B} \end{pmatrix} = \begin{pmatrix} \text{Total amount of} \\ \text{pure substance} \\ \text{in mixture} \end{pmatrix}$$

Example 4   How many liters of a 20% alcohol solution must be added to 3 ℓ of a 90% alcohol solution to make an 80% solution?

Solution

Let x = number of liters of 20% solution

	20% solution	90% solution	Mixture
Percent	0.20	0.90	0.80
Amount	x +	3 =	$x + 3$
Total amount of pure substance	$0.20x$ +	$0.90(3)$ =	$0.80(x + 3)$

Amount of alcohol in 20% solution		Amount of alcohol in 90% solution		Amount of alcohol in 80% solution

$$\begin{array}{lcccll} 0.20x & + & 0.90(3) & = & 0.80(x + 3) & \\ 2x & + & 9(3) & = & 8(x + 3) & \text{Multiply by 10} \\ 2x & + & 27 & = & 8x + 24 & \end{array}$$

$$3 = 6x$$

$$x = \frac{1}{2} \text{ liter of 20\% alcohol}$$

Check

$$\left(\frac{1}{2}\right)(0.20) = 0.10\,\ell \qquad \text{Amount of pure alcohol in 20\% solution}$$

$$(3)(0.90) = \underline{2.70\,\ell} \qquad \text{Amount of pure alcohol in 90\% solution}$$

$$2.80\,\ell \qquad \text{Total number of liters of pure alcohol}$$

$$3 + \frac{1}{2} = 3.5\,\ell \qquad \text{Total number of liters in mixture}$$

$$(3.5)(0.80) = 2.80\,\ell \qquad \text{Amount of pure alcohol in mixture} \quad \blacksquare$$

NOTE   If we add *pure* alcohol to an alcohol solution to get a stronger solution, we are adding a *100% solution* of alcohol. If we add *water* to an alcohol solution to get a weaker solution, we are adding a *0% solution*.   ☑

Example 5   How much water should be added to a 10% solution of alcohol to obtain 20 ml of an 8% solution?
Solution

Let x = number of milliliters of water

$20 - x$ = number of milliliters of 10% solution

	10% solution	Water	Mixture
Percent	0.10	0.00	0.08
Amount	$20 - x$ +	x =	20
Total amount of pure substance	$0.10(20 - x)$ +	0 =	$0.08(20)$

$$0.10(20 - x) + 0 = 0.08(20)$$

$$10(20 - x) = 8(20)$$

$$200 - 10x = 160$$

$$\underline{-160 + 10x \quad -160 + 10x}$$

$$40 = 10x$$

$$\frac{40}{10} = \frac{10x}{10}$$

$$x = 4 \qquad \text{Number of milliliters of water}$$

$$20 - x = 16 \qquad \text{Number of millilters of 10\% solution}$$

Check In 16 ml of a 10% solution, there are 1.6 ml of pure alcohol.

In 20 ml of an 8% solution, there are 1.6 ml of pure alcohol.   ∎

EXERCISES 5.9

Set I   Set up each problem algebraically and solve. Be sure to state what your variables represent.

1. How many milliliters of pure alcohol are there in 16 ml of a 20% solution of alcohol?

2. How many liters of pure sulfuric acid are there in 2 ℓ of a 20% solution?

3. There are 43 g of sulfuric acid in 500 g of solution. Find the percent of acid in the solution.

4. Five hundred grams of a solution contain 27 g of a drug. Find the percent of drug strength.

5. A 40% solution of hydrochloric acid contains 24 ml of pure acid. How many milliliters of solution are there?

6. A 30% solution of potassium chloride contains 15 ml of pure potassium chloride. How many milliliters of solution are there?

7. How many cubic centimeters (cc) of a 20% solution of sulfuric acid must be mixed with 100 cc of a 50% solution to make a 25% solution of sulfuric acid?

8. How many pints of a 2% solution of disinfectant must be mixed with 5 pt of a 12% solution to make a 4% solution of disinfectant?

9. How many milliliters of water must be added to 500 ml of a 40% solution of sodium bromide to reduce it to a 25% solution?

10. How many milliliters of water must be added to 600 ml of a 30% solution of antifreeze to reduce it to a 25% solution?

11. How many liters of pure alcohol must be added to 10 ℓ of a 20% solution of alcohol to make a 50% solution?

12. How many liters of pure alcohol must be mixed with 20 ℓ of a 30% alcohol solution in order to obtain a 44% solution?

Set II   Set up each problem algebraically, solve, and check. Be sure to state what your variables represent.

1. How many liters of pure acetic acid are there in 3 ℓ of a 15% solution of acetic acid?

2. There are 90 ml of pure sulfuric acid in a 15% sulfuric acid solution. How many milliliters of solution are there?

3. There are 66 g of hydrochloric acid in 120 g of solution. What is the percent of hydrochloric acid in the solution?

4. How many milliliters of pure antifreeze are there in 1,600 ml of a 20% solution of antifreeze?

5. A 35% glycerin solution contains 28 ℓ of pure glycerin. How many liters of the glycerin solution are there?

6. There are 48 ml of pure acetic acid in 64 ml of an acetic acid solution. What is the percent of acetic acid in the solution?

7. If 100 gal of 75% glycerin solution is made up by combining a 30% glycerin solution with a 90% glycerin solution, how much of each solution must be used?

8. How many milliliters of a 25% solution of potassium chloride must be added to 8 ml of pure potassium chloride to obtain a 35% solution?

9. How much water must be added to 60 ml of a 40% alcohol solution to obtain a 24% solution?

10. If 1,600 cc of 10% dextrose solution is made up by combining a 20% dextrose solution with a 4% dextrose solution, how much of each solution must be used?

11. How many liters of pure alcohol must be added to 15 ℓ of a 60% solution in order to obtain a 70% solution?

12. How many liters of water should be added to 20 ℓ of a 40% solution of hydrochloric acid to reduce it to a 25% solution?

5.10 Review: 5.1–5.9

Method for Solving Word Problems 5.1–5.9

1. To solve a word problem, first read it very carefully. Determine what *type* of problem it is, if possible. *Be sure you understand the problem.* Don't try to solve it at this time.

2. Determine what is unknown.

3. Represent one unknown number by a variable, and declare its meaning in a sentence of the form "Let x = ..." Then reread the problem to see how you can represent any other unknown numbers in terms of the same variable.

4. Reread the entire word problem, translating each English phrase into an algebraic expression.

5. Fit the algebraic expressions together into an equation or inequality.

6. Solve the equation or inequality for the variable.

7. Answer *all* the questions asked in the problem.

8. Check the solution(s) in the word statement.

Review Exercises 5.10 Set I

Set up each problem algebraically, solve, and check. Be sure to state what your variables represent.

1. Two numbers are in the ratio of 6 to 7. Their sum is 52. Find the numbers.

2. What is 245% of $450?

3. 77.5 is 31% of what number?

4. Six is what percent of 16?

5. The rent on a $380-a-month apartment was raised 5%. Find the new rent.

6. Solder used for soldering zinc is composed of lead and tin in the ratio 3 : 5. How many ounces of each are present in 24 oz of solder?

7. The three sides of a triangle are in the ratio of 3 : 4 : 5. The perimeter is 108. Find the three sides of the triangle.

8. Raul has twenty-two coins with a total value of $5.00. If these coins are dimes and 50¢ pieces, how many of each kind are there?

9. Roger bought $2.10 worth of stamps. He bought only 2¢, 10¢, and 12¢ stamps. If there were twice as many 12¢ stamps as 10¢ stamps and twice as many 2¢ stamps as 12¢ stamps, how many of each did he buy?

10. The total receipts for a football game were $164,000. General admission tickets cost $5 each, reserved seat tickets $8, and box seat tickets $10. If they sold 4,500 more reserved seats than box seats and 4 times as many general admissions as reserved seats, how many of each did they sell?

11. An affirmative action committee demands that 24% of the 1,800 entering freshmen be minority students. How many more than the actual 256 minority freshmen entering would satisfy the committee?

12. After sailing downstream for 2 hr, it takes a boat 7 hr to return to its starting point. If the speed of the boat in still water is 9 mph, what is the speed of the stream?

13. Mrs. Koontz left Downey at 5 A.M., heading toward Washington. Her neighbor Mrs. Fowler left at 6 A.M., also heading toward Washington. By driving 5 mph faster, Mrs. Fowler overtook Mrs. Koontz at 3 P.M. (the same day).

   a. What was Mrs. Koontz's average speed?

   b. What was Mrs. Fowler's average speed?

   c. How far had they driven before Mrs. Fowler overtook Mrs. Koontz?

14. A dealer makes up a 30-lb mixture of nuts costing 85¢ and 95¢ a pound. How many pounds of each must be used in order for the mixture to cost 91¢ a pound?

15. How many cubic centimeters of water must be added to 500 cc of a 25% solution of potassium chloride to reduce it to a 5% solution?

Review Exercises 5.10 Set II

NAME _____

Set up each problem algebraically, solve, and check. Be sure to state what your variables represent.

ANSWERS

1. A man cuts a 5-ft board into two parts whose ratio is 1 : 3. Find the lengths of the pieces (in inches).

2. What is 175% of $350?

3. Ninety-three is what percent of 124?

4. A 26% solution of hydrochloric acid contains 62.4 ml of pure acid. How many milliliters of solution are there?

5. There are 45 ml of sulfuric acid in 300 ml of solution. Find the percent of acid in the solution.

6. Mr. Edmonson's new contract calls for an 8% raise in salary. His present salary is $32,000. What will his new salary be?

7. A disc jockey plays folk, rock, country-western, and Latino records in the ratio 2 : 5 : 3 : 4. If he played 126 records in a week, how many of each type were included?

1. _____

   _____

2. _____

3. _____

4. _____

5. _____

6. _____

7. _____

   _____

   _____

   _____

8. Alice has sixty coins with a total value of $13.35. The coins are dimes, quarters, and 50¢ pieces. If there are 4 times as many dimes as quarters, how many coins of each kind does she have?

9. A 10-lb mixture of nuts and raisins costs $25. If raisins cost $1.90 per pound and nuts $3.40 per pound, how many pounds of each are used?

10. Randy wants to mix dried figs with dried apricots to make an 8-lb mixture costing $2.70 per pound. If dried figs cost $1.80 per pound and dried apricots cost $4.20 per pound, how many pounds of each are used?

11. Ms. Rennie invested part of $18,000 at 16% and the remainder at 20%. Her total yearly income from these investments is $3,120. How much is invested at each rate?

12. Todd paddles his kayak downstream for 4 hr. After a lunch break, he paddles back upstream, but it takes him 7 hr to get back to his starting point. The speed of the stream is 3 mph.

a. How fast does he paddle in still water?

b. How far downstream did he travel?

8. _____

_____

_____

9. _____

_____

10. _____

_____

11. _____

_____

12a. _____

b. _____

13. A 100-lb mixture of two different kinds of apples costs $72.50. If one kind of apple costs 90¢ a pound and the other costs 65¢ a pound, how many pounds of each kind were used?

14. How many pounds of Spanish peanuts costing $2.50 per pound should be mixed with 35 lb of other peanuts costing $2.60 per pound in order for the mixture to be worth $2.57 per pound?

15. How many cubic centimeters of a 50% phenol solution must be added to 400 cc of a 5% solution to make it a 10% solution?

Chapter 5 Diagnostic Test

The purpose of this test is to see how well you understand solving word problems. We recommend that you work this diagnostic test *before* your instructor tests you on this chapter. Allow yourself about 50 minutes.

Complete solutions for all the problems on this test, together with section references, are given in the answer section at the back of this book. For the problems you do incorrectly, study the sections referred to.

Set up each problem algebraically, solve, and check. Be sure to state what your variables represent.

1. A solution contains 28 ml of antifreeze. If it is a 40% solution, how many milliliters of the solution are there?

2. A 20-ml solution of alcohol contains 17 ml of pure alcohol. What is the percent of alcohol in the solution?

3. When 16 is added to 3 times an unknown number, the sum is 37. Find the unknown number.

4. Cheryl has twenty-five coins in her purse with a total value of $1.65. If these coins consist of nickels and dimes, how many of each kind are there?

5. A 50-lb mixture of granola and dried apricots is to be worth $2.34 per pound. If the granola costs $2.20 per pound and the apricots cost $2.70 per pound, how many pounds of each should be used?

6. Kevin leaves his home at 7 A.M., driving toward St. George. His brother Jason leaves their home one hour later, also driving toward St. George. By driving 9 mph faster than Kevin, Jason overtakes Kevin at 1 P.M. the same day.

    a. What is Kevin's average speed?

    b. What is Jason's average speed?

    c. How far do they drive before Jason overtakes Kevin?

7. How many pounds of nuts costing $3.50 per pound should be mixed with 30 lb of macadamia nuts costing $8.00 per pound in order to obtain a mixture worth $4.85 per pound?

8. Two hundred milliliters of a 45% alcohol solution are to be made by combining a 30% solution with an 80% solution.

    a. How many milliliters of the 30% solution should be used?

    b. How many milliliters of the 80% solution should be used?

9. The three sides of a triangle are in the ratio 7 : 9 : 11. The perimeter is 135 m. Find the lengths of three sides.

10. An item costing a merchant $240 is marked up 30%. What is its selling price?

Cumulative Review Exercises: Chapters 1–5

In Exercises 1–3, evaluate each expression if possible.

1. $\dfrac{0}{0}$

2. $3\sqrt{36} - 4^2(-5)$

3. $46 - 2\{4 - [3(5 - 8) - 10]\}$

In Exercises 4 and 5, evaluate each formula using the values of the variables given with the formula.

4. $A = \dfrac{h}{2}(b + B)$ $\quad h = 5, b = 7, B = 11$

5. $S = 4\pi R^2$ $\quad \pi \doteq 3.14, R = 5$

In Exercises 6–10, simplify each expression and write your answers using only positive exponents.

6. $\dfrac{10^2 \cdot 10^0}{10^{-3}}$

7. $\dfrac{x^{3c}}{x^c}$

8. $(2a^2b^{-1})^3$

9. $\left(\dfrac{6y^{-1}}{3y^3}\right)^2$

10. $\left(\dfrac{12x^3}{4x^5}\right)^{-2}$

In Exercises 11–15, solve each equation.

11. $32 - 4x = 15$

12. $-11 = \dfrac{x}{3}$

13. $5w - 12 + 7w = 6 - 8w - 3$

14. $4(2y - 5) = 16 + 3(6y - 2)$

15. $6z - 22 = 2[8 - 4(5z - 1)]$

16. Is addition associative?

17. What is the multiplicative identity?

In Exercises 18–20, set up each problem algebraically, solve, and check. Be sure to state what your variables represent.

18. Two numbers are in the ratio of 11 to 9. Their sum is 160. Find the numbers.

19. Leona has fifteen coins with a total value of $1.75. If these coins are all nickels and quarters, how many of each kind of coin are there?

20. Jodi left her home in Lafayette at 6 A.M., heading up the coast. Her husband, Bud, left at 7 A.M., also heading up the coast. By driving 10 mph faster, Bud overtook Jodi at noon. How fast was Bud driving?

6 Polynomials

In this chapter, we look in detail at a particular type of algebraic expression called a *polynomial*. Polynomials have the same importance in algebra that whole numbers have in arithmetic. Just as much of the work in arithmetic involves operations with whole numbers, much of the work in algebra involves operations with polynomials.

Because a polynomial is a special kind of algebraic expression, we have already discussed some of the work with polynomials. In this chapter, we review and extend these concepts.

6.1 Basic Definitions

Because of its importance, we repeat here the definition of a *term* of an algebraic expression. The plus and minus signs in an algebraic expression break it into smaller pieces called *terms*. Each plus and minus sign is part of the term that follows it. *Exception:* An expression within grouping symbols is considered as a single piece, even though it may contain plus and minus signs.

Polynomials

A **polynomial in one variable** is an algebraic expression that has only terms of the form ax^n, where a stands for any real number, n stands for any whole number, and x stands for any variable. For example, $8x^4 + 3x^2 - 3x$ is a polynomial in x, because each of its terms is of the form ax^n.

A polynomial with only one term is called a **monomial**, a polynomial with two unlike terms is called a **binomial**, and a polynomial with three unlike terms is called a **trinomial**. We will not use any special names for polynomials with four or more terms.

Example 1   The following algebraic expressions are all polynomials in one variable:

a. $-2x$ This polynomial is a monomial in x.

b. $7y + 3y^3$ This polynomial is a binomial in y.

c. 8 This polynomial is a monomial; it is of the form $8x^0$, because $8x^0 = 8 \cdot 1 = 8$.

d. $z^7 - 2z^2 + 3$ This polynomial is a trinomial in z.

e. $x^3 + 3x^2 + 3x + 1$ This is a polynomial in x. ■

If an algebraic expression contains terms with negative (or fractional*) exponents *on the variables*, or if it contains terms with variables in a denominator or under a radical sign, then the algebraic expression is *not* a polynomial.

Example 2   The following algebraic expressions are *not* polynomials:

a. $3x^{-5}$ This is *not* a polynomial because it has a negative exponent on a variable.

b. $\dfrac{1}{x + 2}$ This is *not* a polynomial because it has a variable in the denominator.

c. $\sqrt{3 + x}$ This is *not* a polynomial because the variable is under a radical sign. ■

An algebraic expression with two variables is a **polynomial in two variables** if (1) it contains no negative or fractional exponents on the variables, (2) no variables are in denominators, and (3) no variables are under radical signs.

---

*Fractional exponents will not be discussed in this book.

Example 3   The following algebraic expressions are polynomials in two variables:

a. $\sqrt{7}xy^3$ This polynomial is a monomial; note that *constants* can be under radical signs.

b. $-4xy + \frac{1}{2}x^2y$ This polynomial is a binomial; note that *constants* can be in denominators.

c. $(x + y)^2 - 3(x + y) + 4$ This polynomial is a trinomial in x and y.

d. x^3y^2 $- 2x$ $+ 3y^2$ $- 1$ This is a polynomial in x and y.

$$-1 = -1x^0y^0$$
$$3y^2 = 3x^0y^2$$
$$-2x = -2x^1y^0$$

e. $7uv^4 - 5u^2v + 2u$ Polynomials can be in any variables; this is a polynomial in u and v. ∎

Degree of a Term of a Polynomial   If a polynomial has only one variable, then the **degree of any term** of that polynomial is the exponent on the variable in that term. If a polynomial has more than one variable, the degree of any term of that polynomial is the *sum* of the exponents on the variables in that term.

Example 4   Find the degree of each term:

a. $5x^3$ 3rd degree

b. $6x^2y$ 3rd degree because $6x^2y = 6x^2y^1$ $2 + 1 = 3$

c. 14 0 degree because $14 = 14x^0$

d. $-2u^3vw^2$ 6th degree because $-2u^3vw^2 = -2u^3v^1w^2$

$3 + 1 + 2 = 6$ ∎

Degree of a Polynomial   The **degree of a polynomial** is defined to be the degree of its highest-degree term. Therefore, to find the degree of a polynomial, first find the degree of each of its terms. The *largest* of these numbers will be the degree of the polynomial.*

Example 5   Find the degree of each polynomial:

a. $9x^3 - 7x + 5$

0-degree term
1st-degree term
3rd-degree term   Highest-degree term

Therefore, $9x^3 - 7x + 5$ is a third-degree polynomial.

4th-degree term
6th-degree term   Highest-degree term
0-degree term

b. $14xy^3 - 11x^5y + 8$

Therefore, $14xy^3 - 11x^5y + 8$ is a sixth-degree polynomial.

c. $6a^2bc^3 + 12ab^6c^2$

This is a ninth-degree polynomial, because the term with the highest degree ($12ab^6c^2$) is of degree 9. ∎

---

*Mathematicians define the zero polynomial, 0, as having no degree.

Leading Coefficient   The **leading coefficient** of a polynomial is defined to be the numerical coefficient of its highest-degree term.

Example 6   Name the leading coefficient for each polynomial from Example 5:

   a. The leading coefficient of $9x^3 - 7x + 5$ is 9.

   b. The leading coefficient of $14xy^3 - 11x^5y + 8$ is -11.

   c. The leading coefficient of $6a^2bc^3 + 12ab^6c^2$ is 12.   ∎

Descending and Ascending Powers   If the exponents on one variable get smaller as we read from left to right, the polynomial is arranged in **descending powers** of that variable. If the exponents on one variable get larger as we read from left to right, the polynomial is arranged in **ascending powers** of that variable. For example,

$$8x^3 - 3x^2 + 5x^1 + 7$$

Exponents get smaller from left to right

$7 = 7x^0$

Therefore, $8x^3 - 3x^2 + 5x + 7$ is arranged in descending powers of x.

Example 7   Arrange $5 - 2x^2 + 4x$ in descending powers of x.
Solution $-2x^2 + 4x + 5$ ∎

A polynomial with more than one variable can be arranged in descending or ascending powers of any one of its variables.

Example 8   Arrange $3x^3y - 5xy + 2x^2y^2 - 10$ (a) in descending powers of x, then (b) in descending powers of y.

   a. $3x^3y + 2x^2y^2 - 5xy - 10$ Arranged in descending powers of x

   b. $2x^2y^2 + 3x^3y - 5xy - 10$ Arranged in descending powers of y

Since y is the same power in both terms, the higher-degree *term* is written first   ∎

EXERCISES 6.1

Set I   In Exercises 1–16, if the expression is a polynomial, (a) find the degree of the first term and (b) find the degree of the polynomial. If it is *not* a polynomial, write "Not a polynomial."

   1. $3x + 2x^2$

   2. $4y + 5y^3$

   3. $\dfrac{1}{x} - 3x^2 + 3$

   4. $8x^4 - \dfrac{3}{2x} - 2x$

   5. $3xy^2 + x^3y^3 - 3x^2y - y^3$

   6. $6mn^2 + 8m^3 - 12m^2n - n^3$

   7. $xy + 3x^4 + 5$

   8. $xy + 5x^3 + 2$

   9. $x^{-2} + 5x^{-1} + 4$

   10. $y^{-3} + y^{-2} + 6$

   11. $\sqrt{x-3} + x^3$

   12. $y^2 + \sqrt{2+x}$

   13. $\sqrt{5}x^2 - xy + 3$

   14. $\sqrt{6x^3} - xy + 2$

   15. $\dfrac{3}{2x^2 - 3x + 1}$

   16. $\dfrac{8}{3x^3 - 2x - 5}$

In Exercises 17–20, write each polynomial in descending powers of the indicated variable and find the leading coefficient.

17. $7x^3 - 4x - 5 + 8x^5$ Powers of x

18. $10 - 3y^5 + 4y^2 - 2y^3$ Powers of y

19. $8xy^2 + xy^3 - 4x^2y$ Powers of y

20. $3x^3y + x^4y^2 - 3xy^3$ Powers of x

Set II   In Exercises 1–16, if the expression is a polynomial, (a) find the degree of the first term and (b) find the degree of the polynomial. If it is *not* a polynomial, write "Not a polynomial."

1. $8x + 5x^3$

2. $4y^3 + \dfrac{5}{4x}$

3. $7x^3 + \dfrac{9}{2x} - 3x^2$

4. $5x^{-3} - 2x + 4$

5. $x^3y - 8x^4y^3 + xy - y^5$

6. $8xyz^2 + 3x^2y - z^3$

7. $xy + 8x^3 + 5$

8. $7uv + 8$

9. $3x^{-4} + 2x^{-2} + 6$

10. $\dfrac{1}{2x^2 - 5x}$

11. $x^2 - \sqrt{2x - 5}$

12. $\sqrt{7}x^2 - 3 + 4x$

13. $5 - xy + \sqrt{2}x^2$

14. $\dfrac{5}{2x^4 + 3x^2 - 3}$

15. $\dfrac{2}{6x^2 - 5x + 8}$

16. $\sqrt{9x + 5} - x^2$

In Exercises 17–20, write each polynomial in descending powers of the indicated variable and find the leading coefficient.

17. $17a - 15a^3 + a^{10} - 4a^5$ Powers of a

18. $3x^2y + 8x^3 + y^3 - xy^5$ Powers of y

19. $5st - 9rs^2t - rt^2 - 3rs$ Powers of r

20. $9x^2y^4 - 8x^5y^2 + 3x^4 - 3$ Powers of x

6.2 Addition and Subtraction of Polynomials

Addition of Polynomials

Polynomials can be added horizontally by removing the grouping symbols and combining like terms, as was done in Section 3.6. It is often helpful to underline all like terms with the same kind of line before you add.

Example 1   Add the polynomials; write the answers in descending powers of x:

a. $(3x^2 + 5x - 4) + (2x + 5) + (x^3 - 4x^2 + x)$

$= \underline{3x^2} + \underline{\underline{5x}} - 4 + \underline{\underline{2x}} + 5 + x^3 - \underline{4x^2} + \underline{\underline{x}}$

$= x^3 - x^2 + 8x + 1$

b.　$(5x^3y^2 - 3x^2y^2 + 4xy^3) + (4x^2y^2 - 2xy^2) + (-7x^3y^2 + 6xy^2 - 3xy^3)$

$= \underline{5x^3y^2} - \underline{\underline{3x^2y^2}} + \underset{\sim}{4xy^3} + \underline{\underline{4x^2y^2}} - \underline{\underline{2xy^2}} - \underline{7x^3y^2} + \underline{\underline{6xy^2}} - \underset{\sim}{3xy^3}$

$= -2x^3y^2 + x^2y^2 + xy^3 + 4xy^2$ ∎

Vertical addition is sometimes desirable or necessary. In this case, it is important to have all like terms lined up vertically.

TO ADD POLYNOMIALS VERTICALLY

1. Arrange them under one another so that like terms are in the same vertical line.

2. Find the sum of the terms in each vertical line by adding their numerical coefficients.

Example 2　Add $(3x^2 + 2x - 1)$, $(2x + 5)$, and $(4x^3 + 7x^2 - 6)$ vertically.
Solution

$$
\begin{array}{r}
3x^2 + 2x - 1 \\
2x + 5 \\
\underline{4x^3 + 7x^2 \qquad - 6} \\
4x^3 + 10x^2 + 4x - 2
\end{array}
$$
∎

Example 3　Add $(8x^2y - 3xy^2 + xy - 2)$, $(4xy^2 - 7xy)$, and $(-5x^2y + 9)$ vertically.
Solution

$$
\begin{array}{r}
8x^2y - 3xy^2 + xy - 2 \\
4xy^2 - 7xy \\
\underline{-5x^2y \qquad\qquad + 9} \\
3x^2y + xy^2 - 6xy + 7
\end{array}
$$
∎

Subtraction of Polynomials

We subtract polynomials the same way we subtract signed numbers: Change the signs in the polynomial being subtracted, then add the resulting polynomials.

TO SUBTRACT POLYNOMIALS HORIZONTALLY

1. Change the sign of each term in the polynomial being subtracted.

2. Add the resulting terms horizontally.

Example 4   Examples of subtracting polynomials:

This is the polynomial being subtracted

a. $(-4x^3 + 8x^2 - 2x - 3) - (4x^3 - 7x^2 + 6x + 5)$
   Solution

$$(-4x^3 + 8x^2 - 2x - 3) - (4x^3 - 7x^2 + 6x + 5)$$
$$= -4x^3 + 8x^2 - 2x - 3 - 4x^3 + 7x^2 - 6x - 5$$
$$= -8x^3 + 15x^2 - 8x - 8$$

b. Subtract $(-4x^2y + 10xy^2 + 9xy - 7)$ from $(11x^2y - 8xy^2 + 7xy + 2)$.

This is the polynomial being subtracted

   Solution

$$(11x^2y - 8xy^2 + 7xy + 2) - (-4x^2y + 10xy^2 + 9xy - 7)$$
$$= 11x^2y - 8xy^2 + 7xy + 2 + 4x^2y - 10xy^2 - 9xy + 7$$
$$= 15x^2y - 18xy^2 - 2xy + 9$$

c. Subtract $(2x^2 - 5x + 3)$ from the sum of $(8x^2 - 6x - 1)$ and $(4x^2 + 7x - 9)$.
   Solution

$$(8x^2 - 6x - 1) + (4x^2 + 7x - 9) - (2x^2 - 5x + 3)$$
$$= 8x^2 - 6x - 1 + 4x^2 + 7x - 9 - 2x^2 + 5x - 3$$
$$= 10x^2 + 6x - 13$$

d. $(x^2 + 5) - [(x^2 - 3) + (2x^2 - 1)]$
   Solution

$$(x^2 + 5) - [(x^2 - 3) + (2x^2 - 1)]$$
$$= (x^2 + 5) - [x^2 - 3 + 2x^2 - 1] \qquad \text{Removing innermost ( )}$$
$$= (x^2 + 5) - [3x^2 - 4] \qquad \text{Combining like terms inside [ ]}$$
$$= x^2 + 5 - 3x^2 + 4 \qquad \text{Removing grouping symbols}$$
$$= -2x^2 + 9 \qquad \text{Combining like terms} \quad \blacksquare$$

Because subtraction in long division problems is always done vertically, it is essential that you know how to subtract vertically.

TO SUBTRACT POLYNOMIALS VERTICALLY

1. Write the polynomial being subtracted under the polynomial it is being subtracted from. Write like terms in the same vertical line.

2. Change the sign of each term in the polynomial being subtracted.

3. Find the sum of the resulting terms in each vertical line by adding their numerical coefficients.

Example 5   Subtract the lower polynomial from the upper one: $5x^2 + 3x - 6$

$$3x^2 - 5x + 2$$

Solution

$$
\begin{array}{ll}
5x^2 + 3x - 6 & \quad 5x^2 + 3x - 6 \\
\underline{-\,(3x^2 - 5x + 2)} \rightarrow & \underline{-\,3x^2 + 5x - 2} \\
 & \quad 2x^2 + 8x - 8
\end{array}
$$

┌ Change the sign of *each* term
  in the polynomial being
  subtracted, then *add* the
  resulting terms   ∎

Example 6   Subtract $(3x^2 - 2x - 7)$ from $(x^3 - 2x - 5)$ vertically.
Solution

$$
\begin{array}{ll}
x^3 \qquad - 2x - 5 & \quad x^3 \qquad - 2x - 5 \\
\underline{-\,(3x^2 - 2x - 7)} \rightarrow & \underline{-\,3x^2 + 2x + 7} \\
 & \quad x^3 - 3x^2 \qquad + 2
\end{array}
$$

← Change signs and *add*

∎

NOTE   Your instructor may require you to change the signs *mentally* and may not allow you to show the sign changes.   ☑

EXERCISES 6.2

Set I   Perform the indicated operations.

1. $(2m^2 - m + 4) + (3m^2 + m - 5)$

2. $(5n^2 + 8n - 7) + (6n^2 - 6n + 10)$

3. $(2x^3 - 4) + (4x^2 + 8x) + (-9x + 7)$

4. $(5 + 8z^2) + (4 - 7z) + (z^2 + 7z)$

5. $(3x^2 + 4x - 10) - (5x^2 - 3x + 7)$

6. $(2a^2 - 3a + 9) - (3a^2 + 4a - 5)$

7. Subtract $(-5b^2 + 4b + 8)$ from $(8b^2 + 2b - 14)$.

8. Subtract $(-8c^2 - 9c + 6)$ from $(11c^2 - 4c + 7)$.

9. $(6a - 5a^2 + 6) + (4a^2 + 6 - 3a)$

10. $(2b + 7b^2 - 5) + (4b^2 - 2b + 8)$

11. Subtract $(5a + 3a^2 - 4)$ from $(4a^2 + 6 - 3a)$.

12. Subtract $(2b + b^2 - 7)$ from $(8 + 3b^2 - 7b)$.

13. Add: $17a^3 \qquad + 4a - 9$

$$\underline{\qquad 8a^2 - 6a + 9}$$

14. Add: $\qquad -\ b^3 + 5b^2 - 8$

$$\underline{-20b^4 + 2b^3 \qquad + 7}$$

15. Add: $14x^2y^3 - 11xy^2 + 8xy$

$-9x^2y^3 + \ 6xy^2 - 3xy$

$$\underline{\ 7x^2y^3 - \ 4xy^2 - 5xy}$$

16. Add: $12a^2b - 8ab^2 + 6ab$

$-7a^2b + 11ab^2 - 3ab$

$4a^2b - ab^2 - 13ab$

In Exercises 17–20, subtract the lower polynomial from the upper one.

17. Subtract: $15x^3 - 4x^2 + 12$

$8x^3 + 9x - 5$

18. Subtract: $-14y^2 + 6y - 24$

$7y^3 + 14y^2 - 13y$

19. Subtract: $10a^2b - 6ab + 5ab^2$

$3a^2b + 6ab - 7ab^2$

20. Subtract: $14m^3n^2 - 9m^2n^2 - 6mn$

$-8m^3n^2 - 5m^2n^2 + 3mn$

21. $(7m^8 - 4m^4) + (4m^4 + m^5) + (8m^8 - m^5)$

22. $(8h - 4h^6) + (5h^7 + 3h^6) + (9h - 5h^7)$

23. $(6r^3t + 14r^2t - 11) + (19 - 8r^2t + r^3t) + (8 - 6r^2t)$

24. $(13m^2n^2 + 4mn + 23) + (17 + 4mn - 9m^2n^2) + (-29 - 8mn)$

25. $(7x^2y^2 - 3x^2y + xy + 7) - (3x^2y^2 - 5xy + 4 + 7x^2y)$

26. $(4x^2y^2 + x^2y - 5xy - 4) - (9 - 5x^2y^2 - xy + 3x^2y)$

27. $(x^2 + 4) - [(x^2 - 5) - (3x^2 + 1)]$

28. $(3x^2 - 2) - [(4 - x^2) - (2x^2 - 1)]$

29. Subtract $(2x^2 - 4x + 3)$ from the sum of $(5x^2 - 2x + 1)$ and $(-4x^2 + 6x - 8)$.

30. Subtract $(6y^2 + 3y - 4)$ from the sum of $(-2y^2 + y - 9)$ and $(8y^2 - 2y + 5)$.

31. Subtract the sum of $(x^3y + 3xy^2 - 4)$ and $(2x^3y - xy^2 + 5)$ from the sum of $(5 + xy^2 + x^3y)$ and $(-6 - 3xy^2 + 4x^3y)$.

32. Subtract the sum of $(2m^2n - 4mn^2 + 6)$ and $(-3m^2n + 5mn^2 - 4)$ from the sum of $(5 + m^2n - mn^2)$ and $(3 + 4m^2n + 2mn^2)$.

33. $(7.239x^2 - 4.028x + 6.205) + (-2.846x^2 + 8.096x + 5.307)$

34. $(29.62x^2 + 35.78x - 19.80) + (7.908x^2 - 29.63x - 32.84)$

Set II   Perform the indicated operations.

1. $(2x^2 - 3x + 1) + (4x^2 + 5x - 3)$

2. $(5z + 7) - (7z^2 - 8)$

3. $(8x^3 - 4x^2) + (x^2 - 3x) + (8 - 2x^2)$

4. $(3x^2 + 5x) + (2x^3 - 6x^2) - (3 - 5x)$

5. $(9x^2 - 2x + 3) - (12x^2 - 8x - 9)$

6. $(3 - x^3 + 5x) - (-8x - x^2 + 8x)$

7. Subtract $(8y^3 - 3y)$ from $(y^2 - 3y + 12)$.

8. Subtract $(-3x^3 + 2x^2 - 3x + 2)$ from $(1 - x - x^2 - x^3)$.

9. $(9c + 2c^2 - 8) + (3 - 17c - 8c^2)$

10. $(8x^3 - x + 2) - (x^2 - x + 2)$

11. Subtract $(9 - z + z^2)$ from $(-3z^2 - z + 9)$.

12. Subtract $(2x - 7x^2 + 3x^3)$ from $(-x + 9x^2 - 1 + x^3)$.

13. Add: $\quad 8x^3 \qquad\quad + 3x - 7$
$$\underline{\qquad\quad 5x^2 - 5x + 7}$$

14. Add: $\quad 4x^3 + 7x^2 - 5x + 4$
$$\underline{\quad 2x^3 - 5x^2 + 5x - 6}$$

15. Add: $\qquad 3y^4 - 2y^3 + 4y + 10$
$$-5y^4 + 2y^3 + 4y - \ 6$$
$$\underline{\quad 7y^4 \qquad\quad - 6y - \ 8}$$

16. Add: $\qquad 18x^2y - 3xy^2 + 4xy$
$$-6x^2y + 8xy^2 - 9xy$$
$$\underline{\quad -4x^2y - 9xy^2 + \ xy}$$

In Exercises 17–20, subtract the lower polynomial from the upper one.

17. Subtract: $\quad 10x^3 - 5x^2 + 6x - 1$
$$\underline{\ -2x^3 + 3x^2 + 9x - 5}$$

18. Subtract: $\quad 5z^3 \qquad\ - 7z + 8$
$$\underline{\quad 8z^3 - 10z^2 + 7z}$$

19. Subtract: $\quad 7x^2y^2 - 8xy^2 + \ xy - 6$
$$\underline{\quad 2x^2y^2 - 6xy^2 - 5xy + 9}$$

20. Subtract: $\quad -3x^2 + 3x$
$$\underline{\qquad 5x^2 - 2x + 6}$$

21. $(8x^6 - 3x^4) + (x - 7x^4) + (3x^4 - x)$

22. $(5x^2 - 3x + 1) - (5x^2 - 3x + 1)$

23. $(7x^2y + 4xy^2 - 5) + (8xy^2 - 7x^2y + xy) + (-2 - xy^2)$

24. $(5x^2y - 4x + 3y^2) + (-yx^2 + 6x - y^2)$

25. $(9x^4y - x^2y^2 - 5 + 8xy^3) - (3x^2y^2 - 4xy^3 + 7 - 6x^3)$

26. $(3y^2z - 4x^2y^2 + 5) - (5x^2y^2 - 4y^2z)$

27. $(3x^2 + 5) - [(x^2 - 3) - (2x^2 + 8)]$

28. $(x^2 - 3xy + 4) - [x^2 - 3xy - (6 + x^2)]$

29. Subtract $(3x^2 - 7x - 1)$ from the sum of $(x^2 - x + 2)$ and $(x^2 - 3x - 6)$.

30. Subtract $(2y^2 + 3y - 9)$ from the sum of $(y - y^2 + 2)$ and $(5y - y^2 + 7)$.

31. Subtract the sum of $(xy^2 - 2x^2y - 2)$ and $(3x^2y - 4xy^2 - 6)$ from the sum of $(3 - x^2y)$ and $(2xy^2 - 5x^2y - 8)$.

32. Subtract $(-3m^2n^2 + 2mn - 7)$ from the sum of $(6m^2n^2 - 8mn + 9)$ and $(-10m^2n^2 + 18mn - 11)$.

33. $(5.416x - 34.54x^2 + 7.806) + (51.75x^2 - 1.644x - 9.444)$

34. $(5.886x^2 - 3.009x + 7.966) - [4.961x^2 - 54.51x - (7.864 - 1.394x^2)]$

6.3 Simplifying and Multiplying Polynomials

6.3A Simplifying Polynomials

We simplify a polynomial the same way we simplify an algebraic expression: We remove all grouping symbols, simplify each term, and combine all like terms. In Section 6.2, we discussed removing grouping symbols when addition and subtraction were indicated. In this section, we discuss removing grouping symbols when multiplication is indicated.

When we multiply a monomial by a monomial, we use the methods learned in Section 3.3 (see Example 1).

Example 1 Multiply $(3x^3y)(2xy)(-5xy^2)$.
Solution We have a *monomial* inside each set of parentheses.

$$(3x^3y)(2xy)(-5xy^2) = -30x^5y^4 \quad \blacksquare$$

When we multiply a polynomial with more than one term by a monomial, we use the distributive property (see Examples 2 and 3).

Example 2 Multiply $(3x^3y)(2xy - 5xy^2)$.
Solution We have a *binomial* inside the second set of parentheses.

$$(3x^3y)(2xy - 5xy^2) = 6x^4y^2 - 15x^4y^3 \quad \blacksquare$$

Example 3 Simplify the following polynomials:

a. $a - 2(a + b)$

$= a - 2a - 2b$ Using the distributive property

$= -a - 2b$ Combining like terms

It often helps to underline like terms when combining them.

b. $3(5x - 2y) - 4(3x - 6y)$

$= 15x - 6y - 12x + 24y$ Using the distributive property

$= 15x - 12x - 6y + 24y$ Collecting like terms

$= 3x + 18y$ Combining like terms

c. $x(x^2 + xy + y^2) - y(x^2 + xy + y^2)$

$= x^3 + x^2y + xy^2 - x^2y - xy^2 - y^3$

$= x^3 + x^2y - x^2y + xy^2 - xy^2 - y^3$

$= x^3 + 0 + 0 - y^3$

$= x^3 - y^3$

d. $2x^2y(4xy - 3xy^2) - 5xy(3x^2y^2 - 2x^2y)$

$= 8x^3y^2 - 6x^3y^3 - 15x^3y^3 + 10x^3y^2$

$= 8x^3y^2 + 10x^3y^2 - 6x^3y^3 - 15x^3y^3$

$= 18x^3y^2 - 21x^3y^3$

Sometimes it is helpful to combine like terms within a pair of grouping symbols *before* removing that pair of grouping symbols, as is done in Examples 3e and 3f.

e. $-8[-5(3x - 2) + 13] - 11x$

  $= -8[-15x + \underline{10} + \underline{13}] - 11x$ Remove innermost grouping symbol first

  $= -8[-15x + 23] \qquad - 11x$

  $= +120x \qquad - 184 \qquad - 11x$

  $= \qquad 109x \qquad - 184$

f. $(8x + 10y) - 2\{[4x - 5(8 - y)] - 15\}$

  $= (8x + 10y) - 2\{[4x - 40 + 5y] - 15\}$

  $= (8x + 10y) - 2\{4x - \underline{40} + 5y \ - \underline{15}\}$

  $= (8x + 10y) - 2\{4x \qquad + 5y \ - 55\}$

  $= \underline{8x} + 10y - \underline{8x} \qquad - 10y + 110$

  $= 110$ ■

EXERCISES 6.3A

Set I   Simplify.

1. $x - 3(x + y)$ **2.** $y - 2(y + z)$

3. $2a - 4(a - b)$ **4.** $3c - 2(c - d)$

5. $u(u^2 + 2u + 4) - 2(u^2 + 2u + 4)$

6. $x(x^2 - 3x + 9) + 3(x^2 - 3x + 9)$

7. $x^2(x^2 + y^2) - y^2(x^2 + y^2)$

8. $w^2(w^2 - 4) + 4(w^2 - 4)$

9. $2x(3x^2 - 5x + 1) - 4x(2x^2 - 3x - 5)$

10. $3x(4x^2 - 2x - 3) - 2x(3x^2 - x + 1)$

11. $-3(a - 2b) + 2(a - 3b)$ **12.** $-2(m - 3n) + 4(m - 2n)$

13. $-5(2x - 3y) - 10(x + 5y)$ **14.** $-4(3s - 7t) - 8(3s - 4t)$

15. $3xy(2x)(-5y)$ **16.** $2ac(5b)(-3c)$

17. $3xy(2x - 5y)$ **18.** $2ac(5b - 3c)$

19. $(3xy + 2x)(-5y)$ **20.** $(2ac + 5b)(-3c)$

21. $(3xy + 2x) - 5y$ **22.** $(2ac + 5b) - 3c$

23. $x^2y(3xy^2 - y) - 2xy^2(4x - x^2y)$

24. $ab^2(2a - ab) - 3ab(2ab - ab^2)$

25. $2h(3h^2 - k) - k(h - 3k^3)$

26. $4x(2y^2 - 3x) - x(2x - 3y^2)$

27. $3f(2f^2 - 4g) - g(2f - g^2)$

28. $4a(b^2 - 2b) - b(ab - a)$

29. $2x - [3a + (4x - 5a)]$ **30.** $2y - [5c + (3y - 4c)]$

31. $5x + [-(2x - 10) + 7]$ **32.** $4x + [-(3x - 5) + 4]$

33. $25 - 2[3g - 5(2g - 7)]$ **34.** $40 - 3[2h - 8(3h - 10)]$

35. $-2\{-3[-5(-4 - 3z) - 2z] + 30z\}$

36. $-3\{-2[-3(-5 - 2z) - 3z] - 40z\}$

37. $(4x)(3x)^2$ **38.** $(5y)^2(2y)$ **39.** $(9x - 2y)(-5x)$

40. $(3c - 7d)(-9d)$ **41.** $(2xy^2)^2(3y)$ **42.** $(3a^2b)^2(5b)$

43. $(9x - 2y) - 5x$ **44.** $(3c - 7d) - 9d$

Set II   Simplify.

1. $a - 4(a + b)$ **2.** $2x - 3(x - y)$

3. $5x - 2(4x - 3y)$ **4.** $2(a - 3b) - 3(4a - b)$

5. $2h(4h^2 - k) - 3k(h - 2k^2)$

6. $x(x^2 + 2x + 4) - 2(x^2 + 2x + 4)$

7. $a^2(a^2 - b^2) + b^2(a^2 - b^2)$

8. $2x^2y(3xy^2 - 2y) - 3xy^2(2x^2y - 5x)$

9. $3xy^2(2x - xy + 4) + x(3xy^3 - 12y^2)$

10. $5r^2s(3r - 5 - 2s) + 5r^2s(5 + 2s)$

11. $-5(x - 3y + z) - 2(3y - x - z)$

12. $(a + b + c) - 4$

13. $-8(3x - 2y) + (3x - 4y)(-3)$

14. $-8(3x - 2y) + (3x - 4y) - 3$

15. $8xy(3x)(-2y)$ **16.** $4x(3x)^2$

17. $8xy(3x - 2y)$ **18.** $4 + 3(2x - 7)$

19. $(8xy + 3x)(-2y)$ **20.** $8 + 3x(9x - 2)$

21. $(8xy + 3x) - 2y$ **22.** $(5st - 3s) - 9s$

23. $a^2b(3ab - b^2 + a^2b) - (5a^3b^2 + a^2b^3)$

24. $3 + 2(4x^4 - 3x^3 + 2x^2) - (4x^3 - 2x^2)$

25. $5x(4x^2 - 3x + 1) - 3(2x^2 + x - 1)$

26. $4x^0(3x^3y - 4xy^2 + 3y^4) + 2(xy^2 - 12x^3y)$

27. $5x(2x^2 + x - 1) + (3x^2 - 4)$

28. $(3x - 1)^0 - (5x^0 + 5y - 2)$

29. $8x - [3x - (5x - 2)]$ **30.** $9 - 3[5x - (8x - 2\{x + 3\})]$

31. $6x + [-2(5 - x) + 3]$ **32.** $7^0 - \{3x[2 - (x^2 - 1) + 5]\}$

33. $15 + 3[2 - (5 + x^2)]$ **34.** $21 - 5[3 + 2(x - 5) - x]$

35. $-3\{-2[-1(3 - x) + 5x]\}$ **36.** $-5\{3[8 + 2(x - 6) - 5x] + 1\}$

37. $(9t)(3t)^2$

38. $(7a)^2(2a)$

39. $(8x - 3y)(-2x)$

40. $(c - 8d)(-6d)$

41. $(6xy^2)^2(2y)$

42. $(2a^2b)^3(1b)$

43. $(6a - b) - 4b$

44. $(7 - x) - 9x$

6.3B Products of Two Binomials

Since we often need to find the product of two binomials, it is helpful to be able to find their product by inspection (that is, without writing anything down except the answer). First, however, we show the step-by-step procedure for multiplying two binomials; it is necessary to use the distributive property (or rule) *more than once*.

Example 4

Multiply $(3x + 2y)(4x + 5y)$.

Solution We first treat $3x + 2y$ as if it were a single number.

Step 1: $(3x + 2y)(4x + 5y) = (3x + 2y)(4x) + (3x + 2y)(5y)$

Step 2: We now use the distributive rule again on $(3x + 2y)(4x)$ and on $(3x + 2y)(5y)$:

This step need not be shown

$$= (3x)(4x) + (2y)(4x) + (3x)(5y) + (2y)(5y)$$

Step 3: $= 12x^2 + 8xy + 15xy + 10y^2$

Step 4: We combine like terms: $= 12x^2 + 23xy + 10y^2$ ∎

Let us agree on some terminology so we can more easily discuss finding products of two binomials by inspection.

In Step 3 of finding the product $(3x + 2y)(4x + 5y)$, the two *middle terms* are $8xy$ and $15xy$. Since $8xy$ is the product of the two "inside" terms, we call it the *inner product*, and since $15xy$ is the product of the two "outside" terms, we call it the *outer product*.

Outer product $= 15xy$

$(3x + 2y) \cdot (4x + 5y)$

Inner product $= 8xy$

Product of two *first* terms is $(3x)(4x) = 12x^2$
Product of two *last* terms is $(2y)(5y) = 10y^2$
Sum of *outer* and *inner* products is $23xy$

When a multiplication problem is in the form $(ax + by)(cx + dy)$, as is Example 4, we can quickly find the product by using the following rules:

TO MULTIPLY $(ax + by)(cx + dy)$

1. The *first term* of the product is the product of the first terms of the binomials.

2. The *middle term* of the product is the sum of the inner and outer products.

3. The *last term* of the product is the product of the last terms of the binomials.

NOTE Plus (or minus) signs between the *terms* of the product are essential. ☑

When we use this method of multiplying binomials, we find the product of the two *F*irst terms, the *O*uter product, the *I*nner product, and the product of the two *L*ast terms. For this reason, this procedure is often called the *FOIL* method.

Example 5   Multiply $(x + 2)(x - 5)$.
Solution **Method 1:**

$$(x + 2)(x - 5)$$

$$= (x + 2)(x) + (x + 2)(-5) \qquad \text{Using the distributive rule}$$

$$= (x)(x) + (2)(x) + (x)(-5) + (2)(-5) \qquad \text{Using the distributive rule again}$$

$$= x^2 + \underline{2x} - \underline{5x} - 10$$

$$= x^2 - 3x - 10 \qquad \text{Combining like terms}$$

Method 2: Because the problem is in the form $(ax + by)(cx + dy)$, we can find the product by using the FOIL method.

First	Outer	Inner	Last
(x + 2)(x − 5)	(x + 2)(x − 5)	(x + 2)(x − 5)	(x + 2)(x − 5)
x^2	− 5x	2x	− 10
x^2		− 3x	− 10 ∎

When Method 2 (the FOIL method) is used, we can add the inner and outer products mentally. No intermediate steps need to be shown. (In the next few examples, we will continue to show steps that don't have to be shown.)

Example 6   Multiply $(x - 3)(x - 4)$.
Solution **Method 1:**

$$(x - 3)(x - 4) = (x - 3)(x) + (x - 3)(-4)$$

$$= x^2 - 3x + (-4x) + 12$$

$$= x^2 - 7x + 12$$

Method 2:

First	Outer	Inner	Last
(x − 3)(x − 4)	(x − 3)(x − 4)	(x − 3)(x − 4)	(x − 3)(x − 4)
x^2	−4x	−3x	+ 12
x^2		− 7x	+ 12

Therefore, $(x - 3)(x - 4) = x^2 - 7x + 12$. ∎

Example 7   Multiply $(3x + 2)(4x - 5)$.
Solution **Method 1:**

$$(3x + 2)(4x - 5) = (3x + 2)(4x) + (3x + 2)(-5)$$

$$= 12x^2 + 8x - 15x - 10$$

$$= 12x^2 - 7x - 10$$

Method 2:

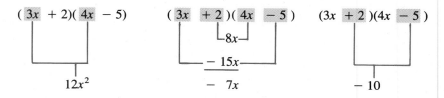

Therefore, $(3x + 2)(4x - 5) = 12x^2 - 7x - 10$. ■

In the remaining examples, we show Method 2 only.

Example 8 Multiply $(5x - 4y)(6x + 7y)$.

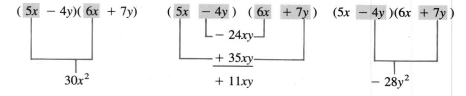

Therefore, $(5x - 4y)(6x + 7y) = 30x^2 + 11xy - 28y^2$. ■

Example 9 Simplify $(3x - 8y)^2$.
Solution Because raising to a power is repeated multiplication, we have

$$(3x - 8y)^2 = (3x - 8y)(3x - 8y)$$

$$- 24xy$$

$$- 24xy$$

$$9x^2 - 48xy + 64y^2$$

Therefore, $(3x - 8y)^2 = 9x^2 - 48xy + 64y^2$. ■

We can use the FOIL method to find the product of two binomials even when the problem is *not* in the form $(ax + by)(cx + dy)$. In such problems, however, the inner and outer products are not necessarily like terms (see Examples 11 and 12).

Example 10 Multiply $(2x^2 - 5x)(3x + 7)$.
Solution

The inner and outer products are
like terms and must be combined

$$(2x^2 - 5x)(3x + 7) = 6x^3 + 14x^2 - 15x^2 - 35x$$

$$= 6x^3 - x^2 - 35x$$ ■

Example 11 Multiply $(3x + 7)(4y + 2)$.
Solution

The inner and outer products are
not like terms; they cannot be combined

$$(3x + 7)(4y + 2) = 12xy + 6x + 28y + 14$$ ■

Example 12   Multiply $(2x^2 + 7)(4x + 2)$.
Solution

The inner and outer products are
not like terms; they cannot be combined

$$(2x^2 + 7)(4x + 2) = 8x^3 + 4x^2 + 28x + 14 \quad \blacksquare$$

EXERCISES 6.3B

Set I   Simplify. Be careful! Some problems are products of monomials, some are products of a monomial and a polynomial, some are products of two binomials, and some are raising polynomials to a power.

1. $(7x^3y)(-2y^2)(4x)$ **2.** $(-4ab^2)(3a^4)(2b^3)$

3. $(x + 3)(x - 2)$ **4.** $(a - 4)(a + 3)$

5. $2x(3x^2 + 7)$ **6.** $3y(4y^3 - 2)$

7. $(y + 8)(y - 9)$ **8.** $(z - 3)(z + 10)$

9. $(5x^2 - 2y)(3x + y)$ **10.** $(2s + 3t)(s^2 - 4t)$

11. $(5a^2 - b^2)(2a)$ **12.** $(7c^2 - d^2)(3c)$

13. $(x + 1)(x + 4)$ **14.** $(x + 3)(x + 1)$

15. $(a + 5)(a + 2)$ **16.** $(a + 7)(a + 1)$

17. $(m - 4)(m + 2)$ **18.** $(n - 3)(n + 7)$

19. $(y + 9)(y - 12)$ **20.** $(z - 5)(z - 11)$

21. $(x + y)(-ab)$ **22.** $(x - y)(ab)$

23. $(x + y)(a - b)$ **24.** $(x - y)(a + b)$

25. $(4x)^2$ **26.** $(3x)^2$

27. $(4 + x)^2$ **28.** $(3 + x)^2$

29. $(b - 4)^2$ **30.** $(b - 6)^2$

31. $(3x + 1)(x + 2)$ **32.** $(2x + 3)(x + 2)$

33. $(2x + 4)(4x - 3)$ **34.** $(3x + 5)(2x - 1)$

35. $(4x - 6)(5x - 2)$ **36.** $(5x - 4)(3x - 4)$

37. $(2x + 5)^2$ **38.** $(3x + 4)^2$

39. $(3x^2 - 2)(2x + 1)$ **40.** $(8a^2 + 1)(3a - 2)$

41. $(4x - y)(2x + 7y)$ **42.** $(3x - 2y)(4x + 5y)$

43. $(7x - 10y)(7x - 10y)$ **44.** $(4u - 9v)(4u - 9v)$

45. $(3a + 2b)^2$ **46.** $(2x - 6y)^2$

47. $(4c - 3d)(4c + 3d)$ **48.** $(5e + 2f)(5e - 2f)$

Set II   Simplify. Be careful! Some problems are products of monomials, some are products of a monomial and a polynomial, some are products of two binomials, and some are raising polynomials to a power.

1. $(8x)(2y^2)(-3x)$ **2.** $(8x)(2y^2 - 3x)$

3. $(a + 2)(a - 5)$ **4.** $(a + 2)(a)(-5)$

5. $3m(4m^3 - 5)$ **6.** $3m(4m^3)(-5)$

7. $(x - 4)(x + 7)$ **8.** $(x^2 - 4)(x + 7)$

9. $(3m^2 + 2n)(m - n)$ **10.** $(3m + 2n)(m - n)$

11. $(2x^2 - y^2)(12x)$ **12.** $(2x - y)^2$

13. $(x + 2)(x + 3)$ **14.** $(y - 1)(y - 1)$

15. $(m + 3)(m + 5)$ **16.** $(m - 3)(m + 5)$

17. $(h - 6)(h + 3)$ **18.** $(h + 6)(h - 3)$

19. $(w + 7)(w - 8)$ **20.** $(w - 7)(w + 8)$

21. $(a + b)(-cd)$ **22.** $(a + b)(a - b)$

23. $(a + b)(c - d)$ **24.** $(a + b)^2$

25. $(7x)^2$ **26.** $(2 + y)^2$

27. $(7 + x)^2$ **28.** $(2y)^2$

29. $(y - 5)^2$ **30.** $(-5y)^2$

31. $(5x + 1)(x + 3)$ **32.** $(2a + 5b)(a + b)$

33. $(3c + 2)(c + 1)$ **34.** $(4x^2 + 1)(x + 2)$

35. $(y - 5)(y - 5)$ **36.** $(4x - 7)^2$

37. $(a + 4)^2$ **38.** $(5x - 3)^2$

39. $(5x^2 - 3)(2x + 3)$ **40.** $(x - 1)(x^2 + 2)$

41. $(11x + 10y)(3x - 4y)$ **42.** $(10x - 7y)(8x + 9y)$

43. $(2x - 3y)(2x + 3y)$ **44.** $(2x - 3y)^2$

45. $(5m + 2n)^2$ **46.** $(2y + 5z)^2$

47. $(2y + 5z)(2y - 5z)$ **48.** $(3x + 4)(2x - 5)$

6.3C Multiplying a Polynomial by a Polynomial

When we're finding the product of two polynomials and both polynomials have two or more terms, it is necessary to use the distributive property more than once.

Example 13   Multiply $(3x^2 - 4x + 6)(5x - 2)$.
Solution

$$(3x^2 - 4x + 6)(5x - 2) = (3x^2 - 4x + 6)(5x) + (3x^2 - 4x + 6)(-2)$$
$$= 15x^3 - 20x^2 + 30x - 6x^2 + 8x - 12$$
$$= 15x^3 - 26x^2 + 38x - 12 \quad \blacksquare$$

The multiplication can, however, be conveniently arranged vertically if we're careful to line up like terms in the same vertical column as we perform the multiplications.

It helps to compare this procedure with the one used in arithmetic for multiplying whole numbers.

$$
\begin{array}{r}
56 \\
23 \\
\hline
168 \\
112 \\
\hline
1288
\end{array}
$$

168   Product of 56 and 3

112   Product of 56 and 2

1288   Notice that the second line in this arithmetic example is positioned so that the tens digits are lined up and the hundreds digits are lined up

Example 14   Multiply $(3x^2 - 4x + 6)(5x - 2)$, setting up the problem vertically.

Solution We usually place the polynomial that has more terms above the other polynomial.

$$
\begin{array}{r}
3x^2 - 4x + 6 \\
5x - 2 \\
\hline
-6x^2 + 8x - 12 \\
15x^3 - 20x^2 + 30x \\
\hline
15x^3 - 26x^2 + 38x - 12
\end{array}
$$

$-6x^2 + 8x - 12$ This is $(3x^2 - 4x + 6)(-2)$

$15x^3 - 20x^2 + 30x$ This is $(3x^2 - 4x + 6)(5x)$

$15x^3 - 26x^2 + 38x - 12$ Notice that we have like terms in the same vertical column ∎

The vertical method can be used to multiply two binomials (see Example 15).

Example 15   Multiply $(2x^2 - 5x)(3x + 7)$, setting up the problem vertically. (This is the same problem as in Example 10.)

$$
\begin{array}{r}
2x^2 - 5x \\
3x + 7 \\
\hline
14x^2 - 35x \\
6x^3 - 15x^2 \\
\hline
6x^3 - x^2 - 35x
\end{array}
$$

$14x^2 - 35x$ This is $(2x^2 - 5x)(7)$

$6x^3 - 15x^2$ This is $(2x^2 - 5x)(3x)$

$6x^3 - x^2 - 35x$ ∎

Example 16   Multiply $(2m + 3m^2 - 5)(2 + m^2 - 3m)$.

$$
\begin{array}{r}
3m^2 + 2m - 5 \\
m^2 - 3m + 2 \\
\hline
6m^2 + 4m - 10 \\
-9m^3 - 6m^2 + 15m \\
3m^4 + 2m^3 - 5m^2 \\
\hline
3m^4 - 7m^3 - 5m^2 + 19m - 10
\end{array}
$$

Multiplication is simplified by first arranging the polynomials in descending powers of m

$3m^4 - 7m^3 - 5m^2 + 19m - 10$ ∎

Example 17   Multiply $(a^2 + 3a + 9)(a - 3)$.

$$
\begin{array}{r}
a^2 + 3a + 9 \\
a - 3 \\
\hline
-3a^2 - 9a - 27 \\
a^3 + 3a^2 + 9a \\
\hline
a^3 \qquad\qquad - 27
\end{array}
$$

∎

Example 18   Multiply $(x^3 - 1 + x)(2x^2 + 2 - x)$.

Note that $0x^2$ was written in to save a place for the x^2 terms that arise in the multiplication

$$
\begin{array}{r}
x^3 + 0x^2 + \ x - 1 \\
2x^2 - \ x + 2 \\
\hline
2x^3 + 0x^2 + 2x - 2 \\
- \ x^4 + 0x^3 - \ x^2 + \ x \\
2x^5 + 0x^4 + 2x^3 - 2x^2 \\
\hline
2x^5 - \ x^4 + 4x^3 - 3x^2 + 3x - 2
\end{array}
$$ ∎

Powers of Polynomials

We can raise any polynomial to any power by using repeated multiplication, just as we found $(3x - 8y)^2$ in Example 9. (In Section 6.4B, we will show an alternate method for squaring a binomial.)

Example 19   Simplify $(a - b)^3$.

Solution To simplify $(a - b)^3$, we must remove the grouping symbols; therefore, we must multiply $a - b$ by itself three times.

$$(a - b)^3 = \underbrace{(a - b)(a - b)}_{\substack{\text{First find} \\ (a - b)^2}}\underbrace{(a - b)}_{\substack{\text{Then multiply} \\ \text{by } (a - b)}}$$

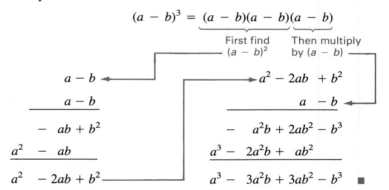

A WORD OF CAUTION

$$(3x)^2 = 3^2 x^2 = 9x^2 \qquad \text{By the rules for exponents}$$

3 and x are *factors*

$$(x + 3)^2 \neq x^2 + 3^2$$

x and 3 are *terms*

For example, suppose x is 4. Then $(x + 3)^2 = (4 + 3)^2 = 7^2 = 49$, but $x^2 + 3^2 = 4^2 + 3^2 = 16 + 9 = 25$, and $49 \neq 25$.

Correct method

$$
\begin{aligned}
(x + 3)^2 &= (x + 3)(x + 3) \\
&= (x + 3)x + (x + 3)(3) &&\text{Distributive rule} \\
&= x^2 + 3x + 3x + 9 &&\text{Distributive rule} \\
&= x^2 + \boxed{6x} + 9 &&\text{Combining like terms}
\end{aligned}
$$

This is the term that students sometimes leave out

Again, if x is 4, $x^2 + 6x + 9 = 4^2 + 6(4) + 9 = 16 + 24 + 9 = 49$.   ☑

EXERCISES 6.3C

Set I   Simplify.

1. $(x - 3)(2x^2 + x - 1)$ **2.** $(x - 2)(3x^2 + x - 1)$

3. $(-3x^2y + xy^2 - 4y^3)(-2xy)$ **4.** $(-4xy^2 - x^2y + 3x^3)(-3xy)$

5. $(x^2 + x + 1)(x^2 + x + 1)$ **6.** $(x^2 - x - 1)(x^2 - x - 1)$

7. $(4z)(z^2 - 4z + 16)$ **8.** $(-5a)(a^2 + 5a + 25)$

9. $(z + 4)(z^2 - 4z + 16)$ **10.** $(a - 5)(a^2 + 5a + 25)$

11. $(4 - 3z^3 + z^2 - 5z)(4 - z)$

12. $(3 + 2v^2 - v^3 + 4v)(2 - v)$

13. $(2x^2 - x + 4)(x^2 - 5x - 3)$

14. $(6a^2 + 2a - 3)(2a^2 - 3a + 5)$

15. $(7 - 2y + 3y^2)(8y + 4y^2 - 3)$

16. $(x - 4 + 6x^2)(9 - 3x + x^2)$

17. $(5x - 2)^2$ **18.** $(2x - 5)^2$

19. $(x + y)^2(x - y)^2$ **20.** $(x - 2)^2(x + 2)^2$

21. $(x + 2)^3$ **22.** $(x + 3)^3$

23. $(x^2 + 2x - 3)^2$ **24.** $(y^2 - 4y - 5)^2$

25. $(x + 3)^4$ **26.** $(x + 2)^4$

Set II   Simplify.

1. $(x - 1)(5x^2 + x - 1)$ **2.** $(x + 1)(2x^2 + x + 1)$

3. $(3xy^2 - 5x^2y + 4)(-2x^2y)$ **4.** $(a^3 - 3a^2b + 3ab^2 - b^3)(-5a^2b)$

5. $(x^2 + 2x + 1)(x^2 + 2x + 1)$ **6.** $(x^2 - 2x + 1)(x^2 - 2x - 1)$

7. $(3x)(x^2 - 3x + 9)$ **8.** $(-2x)(4 + 2x + x^2)$

9. $(3 + x)(x^2 - 3x + 9)$ **10.** $(2 - x)(4 + 2x + x^2)$

11. $(2 - 2x^3 + x^2 - 3x)(3 - x)$

12. $(x + 3x^3 + 4)(5 - x)$

13. $(7 + 2x^3 + 3x)(4 - x)$

14. $(4 + a^4 + 3a^2 - 2a)(a + 3)$

15. $(5y^2 - 2y - 6)(2y^2 - 4y + 3)$

16. $(2 + 3x^2 - 4x)(6x - 7 + 2x^2)$

17. $(3x + 4)^2$ **18.** $(x^2 - 2x + 1)(x^2 - 2x + 1)$

19. $(x - 1)^2(x + 1)^2$ **20.** $(y + 2)^2(y - 2)^2$

21. $(x + 1)^3$ **22.** $(2w - 3)^3$

23. $(x^2 + 4x + 4)^2$ **24.** $(z^2 - 3z - 4)^2$

25. $(x + 1)^4$ **26.** $(x - 1)^4$

6.4 Special Products

Some products are especially important and have special formulas. You *must* learn these formulas and how to use them so you will be able to do the factoring problems that are so important in all of higher mathematics.

6.4A The Product of the Sum and Difference of Two Terms

Because $a + b$ is the *sum of two terms* and $a - b$ is the *difference of two terms*, we call the product $(a + b)(a - b)$ the *product of the sum and difference of two terms*.

RULE 6.1

$$(a + b)(a - b) = a^2 - b^2$$

Proof:

$$(a + b)(a - b) = (a + b)a + (a + b)(-b) \qquad \text{Distributive rule}$$
$$= a^2 + ba - ab - b^2$$
$$= a^2 - b^2$$

Rule 6.1 can be stated in words as follows:

The product of the sum and difference of two terms is equal to the square of the first term minus the square of the second term.

When we use Rule 6.1 to find a product, we say we are finding the product by inspection.

Example 1　Find the products:

a. $(x + 2)(x - 2) = (x)^2 - (2)^2 = x^2 - 4$

b. $(2x + 3y)(2x - 3y) = (2x)^2 - (3y)^2 = 4x^2 - 9y^2$

c. $(10x^2 - 7y^3)(10x^2 + 7y^3) = (10x^2)^2 - (7y^3)^2 = 100x^4 - 49y^6$

d. $(5a^3b^2 + 6cd^4)(5a^3b^2 - 6cd^4) = (5a^3b^2)^2 - (6cd^4)^2 = 25a^6b^4 - 36c^2d^8$　∎

EXERCISES 6.4A

Set I　Find the products by inspection.

1. $(x + 3)(x - 3)$　　　　　　　**2.** $(z + 4)(z - 4)$

3. $(w - 6)(w + 6)$　　　　　　　**4.** $(y - 5)(y + 5)$

5. $(5a + 4)(5a - 4)$　　　　　　**6.** $(6a - 5)(6a + 5)$

7. $(2u + 5v)(2u - 5v)$　　　　　**8.** $(3m - 7n)(3m + 7n)$

9. $(4b - 9c)(4b + 9c)$　　　　　**10.** $(7a - 8b)(7a + 8b)$

11. $(2x^2 - 9)(2x^2 + 9)$ **12.** $(10y^2 - 3)(10y^2 + 3)$

13. $(1 + 8z^3)(1 - 8z^3)$ **14.** $(9v^4 - 1)(9v^4 + 1)$

15. $(5xy + z)(5xy - z)$ **16.** $(10ab + c)(10ab - c)$

17. $(7mn + 2rs)(7mn - 2rs)$ **18.** $(8hk + 5ef)(8hk - 5ef)$

Set II   Find the products by inspection.

1. $(h - 7)(h + 7)$ **2.** $(u + 9)(u - 9)$

3. $(3m + 5)(3m - 5)$ **4.** $(2x - 3y)(2x + 3y)$

5. $(6a - 7b)(6a + 7b)$ **6.** $(3x + 7)(3x - 7)$

7. $(8x - 2y)(8x + 2y)$ **8.** $(5x - 12y)(5x + 12y)$

9. $(7a - 9b)(7a + 9b)$ **10.** $(13c + d)(13c - d)$

11. $(4h^2 - 5)(4h^2 + 5)$ **12.** $(8x^2 + 1)(8x^2 - 1)$

13. $(1 + 9k^3)(1 - 9k^3)$ **14.** $(3x^2y - 8xy^2)(3x^2y + 8xy^2)$

15. $(4w + 3xy)(4w - 3xy)$ **16.** $(9ab - 2c)(9ab + 2c)$

17. $(5uv - 8ef)(5uv + 8ef)$ **18.** $(9st + 4z^2)(9st - 4z^2)$

6.4B   The Square of a Binomial

Rules 6.2 and 6.3 help us quickly square a binomial.

RULE 6.2

$$(a + b)^2 = a^2 + 2ab + b^2$$

RULE 6.3

$$(a - b) = a^2 - 2ab + b^2$$

Proof of Rule 6.2:

$$(a + b)^2 = (a + b)\ (a + b) = a^2 + 2ab + b^2$$

$$a^2 + 2ab + b^2$$

Proof of Rule 6.3:

$$(a - b)^2 = (a - b)(a - b) = a^2 - 2ab + b^2$$

$$a^2 - 2ab + b^2$$

Rules 6.2 and 6.3 can be stated in words as follows:

TO SQUARE A BINOMIAL

1. The *first term* of the product is the square of the first term of the binomial.

2. The *middle term* of the product is twice the product of the two terms of the binomial.

3. The *last term* of the product is the square of the last term of the binomial.

While the square of a binomial *can* be found by using the methods learned in Section 6.3, you are strongly urged to use Rule 6.2 or 6.3 in solving such problems.

Example 2 Simplify each of the following:

a. $(m + n)^2 = (m)^2 + 2(m)(n) + (n)^2 = m^2 + 2mn + n^2$

b. $(a - 3)^2 = (a)^2 - 2(a)(3) + (3)^2 = a^2 - 6a + 9$

c. $(2x - 5)^2 = (2x)^2 - 2(2x)(5) + (5)^2 = 4x^2 - 20x + 25$ ■

A WORD OF CAUTION Using the rules for exponents, students remember that
$(ab)^2 = a^2b^2$

——————————————————— Here, *a* and *b* are *factors*

They try to apply this rule of exponents to the expression $(a + b)^2$. However, $(a + b)^2$ cannot be found simply by squaring *a* and *b*.

Here, *a* and *b* are *terms* ———————

Correct method	*Incorrect method*
$(a + b)^2 = (a + b)(a + b)$	$(a + b)^2 = a^2 + b^2$
$= a^2 + \boxed{2ab} + b^2$	

—— When squaring a binomial, do not forget this middle term ■

EXERCISES 6.4B

Set I Simplify.

1. $(x - 1)^2$ **2.** $(x - 5)^2$

3. $(x + 3)^2$ **4.** $(x + 4)^2$

5. $(4x - 1)^2$ **6.** $(7x - 1)^2$

7. $(5x - 3y)^2$ **8.** $(4x - 7y)^2$

9. $(3x + 2z)^2$ **10.** $(2x + 7s)^2$

Set II Simplify.

1. $(x + 1)^2$ **2.** $(x - 12)^2$

3. $(x + 15)^2$ **4.** $(x - 9)^2$

5. $(9x - 1)^2$ **6.** $(8x + 1)^2$

7. $(10x - 3t)^2$

8. $(12x + 5y)^2$

9. $(11x + 2y)^2$

10. $(6x - 5a)^2$

6.5 Division of Polynomials

6.5A Division of a Polynomial by a Monomial

The rule for dividing a polynomial by a monomial is based on a property of fractions you should recall from arithmetic, $\dfrac{a}{b} = a \cdot \dfrac{1}{b}$, and on the distributive property. Consider this example:

$$\frac{4x^3 - 6x^2}{2x} = \frac{1}{2x} \cdot \frac{4x^3 - 6x^2}{1} = \frac{1}{2x}(4x^3 - 6x^2)$$

$$= \left(\frac{1}{2x}\right)(4x^3) + \left(\frac{1}{2x}\right)(-6x^2) \quad \text{By the distributive rule}$$

$$= \frac{4x^3}{2x} + \frac{-6x^2}{2x} \qquad \text{Dividing each term of the polynomial by the monomial}$$

$$= 2x^2 - 3x$$

The rule for dividing a polynomial by a monomial is as follows:

TO DIVIDE A POLYNOMIAL BY A MONOMIAL

Divide *each* term in the polynomial by the monomial, simplify each term, and then add the resulting quotients.

Example 1 Divide:

a. $\dfrac{6 + 8}{2} = \dfrac{6}{2} + \dfrac{8}{2} = 3 + 4 = 7$

b. $\dfrac{4x + 2}{2} = \dfrac{4x}{2} + \dfrac{2}{2} = 2x + 1$

c. $\dfrac{9x^3 - 6x^2 + 12x}{3x} = \dfrac{9x^3}{3x} + \dfrac{-6x^2}{3x} + \dfrac{12x}{3x} = 3x^2 - 2x + 4$

d. $\dfrac{4x^4 - 8x^3 + 16x}{-4x} = \dfrac{4x^4}{-4x} + \dfrac{-8x^3}{-4x} + \dfrac{16x}{-4x} = -x^3 + 2x^2 - 4$

e. $\dfrac{15x^4y^2z + 20xy^3z - 10xyz^2}{5xyz} = \dfrac{15x^4y^2z}{5xyz} + \dfrac{20xy^3z}{5xyz} + \dfrac{-10xyz^2}{5xyz}$

$$= 3x^3y + 4y^2 - 2z$$

f. $\dfrac{4a^2bc^2 - 6ab^2c^2 + 6abc}{-6abc} = \dfrac{4a^2bc^2}{-6abc} + \dfrac{-6ab^2c^2}{-6abc} + \dfrac{6abc}{-6abc}$

$$= -\frac{2}{3}ac + bc - 1 \quad \blacksquare$$

EXERCISES 6.5A

Set I Perform the indicated divisions.

1. $\dfrac{3x + 6}{3}$ **2.** $\dfrac{10x + 15}{5}$

3. $\dfrac{4 + 8x}{4}$ **4.** $\dfrac{5 - 10x}{5}$

5. $\dfrac{6x - 8y}{2}$ **6.** $\dfrac{5x - 10y}{5}$

7. $\dfrac{2x^2 + 3x}{x}$ **8.** $\dfrac{4y^2 - 3y}{y}$

9. $\dfrac{15x^3 - 5x^2}{5x^2}$ **10.** $\dfrac{12y^4 - 6y^2}{6y^2}$

11. $\dfrac{3a^2b - ab}{ab}$ **12.** $\dfrac{5mn^2 - mn}{mn}$

13. $\dfrac{5x^5 - 4x^3 + 10x^2}{-5x^2}$ **14.** $\dfrac{7y^4 - 5y^3 + 14y^2}{-7y^2}$

15. $\dfrac{-15x^2y^2z^2 - 30xyz}{-5xyz}$ **16.** $\dfrac{-24a^2b^2c^2 - 16abc}{-8abc}$

17. $\dfrac{13x^4y^2 - 26x^2y^3 + 39x^2y^2}{13x^2y^2}$ **18.** $\dfrac{21m^2n^5 - 35m^3n^2 - 14m^2n^2}{7m^2n^2}$

Set II Perform the indicated divisions.

1. $\dfrac{9x + 12}{3}$ **2.** $\dfrac{15y + 20}{5}$

3. $\dfrac{10 + 20x}{10}$ **4.** $\dfrac{8 - 10y}{8}$

5. $\dfrac{5x - 15y}{5}$ **6.** $\dfrac{4x - 6}{2}$

7. $\dfrac{3x^2 - 6x}{3x}$ **8.** $\dfrac{8x^3 - 10x}{4x}$

9. $\dfrac{8x^3 - 4x^2}{4x^2}$ **10.** $\dfrac{15x^3 - 30x}{15x}$

11. $\dfrac{6ab^2 - ab}{ab}$ **12.** $\dfrac{12xy^2 - 6x^2y}{9xy}$

13. $\dfrac{8x^4 - 4x^2 + 12x^3}{-8x^2}$ **14.** $\dfrac{9y^2 - 3y^3 + 18y^5}{-9y^2}$

15. $\dfrac{-16a^3b^2c^3 - 32abc}{-16abc}$ **16.** $\dfrac{-18s^2t^3u^2 - 9stu^2}{-9stu}$

17. $\dfrac{12a^5b^3 - 24a^3b^3 + 48a^2b^2}{12a^2b^2}$ **18.** $\dfrac{15x^5y^2 - 25x^2y^3 - 10x^2y^2}{15x^2y^2}$

6.5B   Division of a Polynomial by a Polynomial

The method used to divide one polynomial (the dividend) by a polynomial (the divisor) with two or more terms is similar to the method used to divide one whole number by another (using long division) in arithmetic. The long-division procedure for polynomials will be demonstrated step-by-step; it can be summarized as follows:

(*Hint:* Read the steps in the summary, and *while you are doing so* follow the steps in Example 2 to see how the step-by-step method is followed.)

TO DIVIDE ONE POLYNOMIAL BY ANOTHER

1. Arrange the divisor and the dividend in descending powers of one variable. In the *dividend*, leave spaces for any missing terms.

2. Find the first term of the quotient by dividing the first term of the dividend by the first term of the divisor.

3. Multiply the *entire* divisor by the first term of the quotient. Place the product under the dividend, lining up like terms.

4. Subtract the product found in step 3 from the dividend, bringing down at least one term. This difference is the remainder. If the degree of the remainder is not less than the degree of the divisor, continue with steps 5–8.

5. Find the next term of the quotient by dividing the first term of the remainder by the first term of the divisor.

6. Multiply the entire divisor by the term found in step 5.

7. Subtract the product found in step 6 from the polynomial above it, bringing down at least one more term.

8. Repeat steps 5 through 7 until the remainder is 0 *or* until the degree of the remainder is less than the degree of the divisor.

9. Check your answer. (Divisor × quotient + remainder = dividend)

Example 2   Divide $(5x^2 + 6x^3 - 5 - 4x)$ by $(3 + 2x)$.

Solution

Step 1: First arrange the divisor and the dividend in descending powers of x.

$$2x + 3 \overline{)6x^3 + 5x^2 - 4x - 5}$$

First term in quotient is $\dfrac{6x^3}{2x} = 3x^2$

$$3x^2$$

Step 2: $2x + 3 \overline{)\,6x^3 + 5x^2 - 4x - 5}$

Multiply

$$3x^2$$

Step 3: $2x + 3 \overline{)6x^3 + 5x^2 - 4x - 5}$
$$\underline{6x^3 + 9x^2} \qquad \text{This is } 3x^2(2x + 3)$$

Step 4:

$$3x^2$$
$$2x + 3\overline{)6x^3 + 5x^2 - 4x - 5}$$
$$\ominus 6x^3 \ominus 9x^2$$

Subtract (change signs and add) and bring down a term

Remainder → $-4x^2 - 4x$

Degree is 2

The division must be continued, because the degree of the remainder is not less than the degree of the divisor.

Second term in quotient is $\dfrac{-4x^2}{2x} = -2x$

Step 5:

$$3x^2 - 2x$$
$$2x + 3\overline{)6x^3 + 5x^2 - 4x - 5}$$
$$\ominus 6x^3 \ominus 9x^2$$
$$-4x^2 - 4x$$

Multiply

Step 6:

$$3x^2 - 2x$$
$$2x + 3\overline{)6x^3 + 5x^2 - 4x - 5}$$
$$\ominus 6x^3 \ominus 9x^2$$
$$-4x^2 - 4x$$
$$-4x^2 - 6x$$ ← This is $(-2x)(2x + 3)$

Step 7:

$$3x^2 - 2x$$
$$2x + 3\overline{)6x^3 + 5x^2 - 4x - 5}$$
$$\ominus 6x^3 \ominus 9x^2$$
$$-4x^2 - 4x$$
$$\oplus 4x^2 \oplus 6x$$ ← Subtract and bring down a term

Remainder → $2x - 5$

Degree is 1

Step 8: The degree of the remainder is still not less than the degree of the divisor; there-fore, we must repeat steps 5, 6, and 7.

Third term in quotient is $\dfrac{2x}{2x} = 1$

$$3x^2 - 2x + 1$$
$$2x + 3\overline{)6x^3 + 5x^2 - 4x - 5}$$
$$\ominus 6x^3 \ominus 9x^2$$
$$-4x^2 - 4x$$
$$\oplus 4x^2 \oplus 6x$$
$$2x - 5$$
$$\ominus 2x \ominus 3$$ ← This is $1(2x + 3)$

Remainder → -8 ← Degree is 0

The division is now finished, because the degree of the remainder is less than the degree of the divisor.

Check

$$(2x + 3)(3x^2 - 2x + 1) + (-8) \overset{?}{=} 6x^3 + 5x^2 - 4x - 5$$

$$(6x^3 + 5x^2 - 4x + 3) + (-8) \overset{?}{=} 6x^3 + 5x^2 - 4x - 5$$

$$6x^3 + 5x^2 - 4x - 5 = 6x^3 + 5x^2 - 4x - 5$$

Answer: $\quad 3x^2 - 2x + 1 - \dfrac{8}{2x + 3}$

or $\quad 3x^2 - 2x + 1 + \dfrac{-8}{2x + 3}$ ∎

Example 3 Divide $(-3x + x^2 - 10)$ by $(2 + x)$.

Solution

Step 1: First arrange the divisor and the dividend in descending powers of x.

$$x + 2 \overline{)x^2 - 3x - 10}$$

First term in quotient $= \dfrac{x^2}{x} = x$

Step 2: $x + 2 \overline{)x^2 - 3x - 10}$

Multiply

Step 3: $x + 2 \overline{)x^2 - 3x - 10}$

$\underline{x^2 + 2x}\quad$ This is $x(x + 2)$

Step 4: $x + 2 \overline{)x^2 - 3x - 10}$

$\underline{\ominus x^2 \ominus 2x}$ Change all signs and add

$- 5x - 10\quad$ Bring down next term

Second term in quotient $= \dfrac{-5x}{x} = -5$

Step 5: $x + 2 \overline{)x^2 - 3x - 10}$

$\underline{\ominus x^2 \ominus 2x}$

$- 5x - 10$

Multiply

Step 6: $x + 2 \overline{)x^2 - 3x - 10}$

$\underline{\ominus x^2 \ominus 2x}$

$- 5x - 10$

$\underline{- 5x - 10}\quad$ This is $-5(x + 2) = -5x - 10$

Step 7:

$$x + 2{\overline{)x^2 - 3x - 10}}$$

with quotient $x - 5$

$$\ominus x^2 \mp 2x$$

$$-5x - 10$$

$$\oplus 5x \oplus 10 \qquad \text{Change signs and add}$$

$$0$$

Check

$$(x + 2)(x - 5) + 0 \overset{?}{=} x^2 - 3x - 10$$

$$x^2 - 3x - 10 = x^2 - 3x - 10 \quad \blacksquare$$

NOTE When the final remainder is zero, we can say that the divisor is a *factor* of the dividend and that the quotient is a *factor* of the dividend. In Example 3, $(x + 2)$ and $(x - 5)$ are *factors* of $x^2 - 3x - 10$. ☑

Example 4 $(6x^2 + x - 10) \div (2x + 3)$

Solution

$$\begin{array}{r} 3x - 4 \text{ R 2} \ or \ 3x - 4 + \dfrac{2}{2x + 3} \\ 2x + 3{\overline{)6x^2 + x - 10}} \\ \ominus 6x^2 \ominus 9x \\ \hline -8x - 10 \\ \oplus 8x \oplus 12 \\ \hline 2 \quad \text{Remainder} \end{array}$$

Check

$$(2x + 3)(3x - 4) + 2 \overset{?}{=} 6x^2 + x - 10$$

$$(6x^2 + x - 12) + 2 \overset{?}{=} 6x^2 + x - 10$$

$$6x^2 + x - 10 = 6x^2 + x - 10 \quad \blacksquare$$

Example 5 $(27x - 19x^2 + 6x^3 + 10) \div (5 - 3x)$

Solution

$$\begin{array}{r} -2x^2 + 3x - 4 \text{ R 30} \ or \ -2x^2 + 3x - 4 + \dfrac{30}{-3x + 5} \\ -3x + 5{\overline{)6x^3 - 19x^2 + 27x + 10}} \\ \ominus 6x^3 \oplus 10x^2 \\ \hline -9x^2 + 27x \\ \oplus 9x^2 \ominus 15x \\ \hline 12x + 10 \\ \ominus 12x \oplus 20 \\ \hline 30 \end{array}$$

Check

$$(-3x + 5)(-2x^2 + 3x - 4) + 30 \stackrel{?}{=} 6x^3 - 19x^2 + 27x + 10$$

$$(6x^3 - 19x^2 + 27x - 20) + 30 \stackrel{?}{=} 6x^3 - 19x^2 + 27x + 10$$

$$6x^3 - 19x^2 + 27x + 10 = 6x^3 - 19x^2 + 27x + 10 \quad \blacksquare$$

In Example 6, there are two variables. Because the coefficient of a^3 is positive and the coefficient of b^3 is negative, we will arrange the divisor and dividend in descending powers of a before beginning the division.

Example 6 $(17ab^2 + 12a^3 - 10b^3 - 11a^2b) \div (3a - 2b)$

$$
\begin{array}{r}
4a^2 - ab + 5b^2 \\
3a - 2b \overline{)\,12a^3 - 11a^2b + 17ab^2 - 10b^3} \\
\ominus \quad 12a^3 \stackrel{\oplus}{-} 8a^2b \\
\hline
-3a^2b + 17ab^2 \\
\stackrel{\oplus}{-} 3a^2b \stackrel{\ominus}{\mp} 2ab^2 \\
\hline
15ab^2 - 10b^3 \\
\ominus \, 15ab^2 \stackrel{\oplus}{-} 10b^3 \\
\hline
0
\end{array}
$$

The checking is left to the student. $\blacksquare$

In Example 7, the dividend has no x^2-term and no x-term. Therefore, we write $0x^2$ and $0x$ as placeholders when we prepare for the division.

Example 7 $(x^3 - 1) \div (x - 1)$

$$
\begin{array}{r}
x^2 + x + 1 \\
x - 1 \overline{)\,x^3 + 0x^2 + 0x - 1} \\
\ominus \;\; x^3 \stackrel{\oplus}{-} x^2 \\
\hline
x^2 + 0x \\
\ominus \;\; x^2 \stackrel{\oplus}{-} x \\
\hline
x - 1 \\
\ominus \;\; x \stackrel{\oplus}{-} 1 \\
\hline
0
\end{array}
$$

It is helpful to leave space for missing powers by using zeros in this way

The checking is left to the student. $\blacksquare$

In Example 8, the divisor has more than two terms; we still use the procedure outlined in the summary for long division.

Example 8 $(2x^4 + x^3 - 8x^2 - 5x - 2) \div (x^2 - x - 2)$

$$
\begin{array}{r}
2x^2 + 3x - 1 \quad R -4 \\
x^2 - x - 2\overline{)2x^4 + x^3 - 8x^2 - 5x - 2} \\
\ominus 2x^4 \oplus 2x^3 \ominus 4x^2 \\
\hline
3x^3 - 4x^2 - 5x \\
\ominus 3x^3 \ominus 3x^2 \oplus 6x \\
\hline
-x^2 + x - 2 \\
\oplus x^2 \ominus x \mp 2 \\
\hline
-4
\end{array}
$$

The checking is left to the student. ∎

EXERCISES 6.5B

Set I   Perform the indicated divisions.

1. $(x^2 + 5x + 6) \div (x + 2)$ **2.** $(x^2 + 5x + 6) \div (x + 3)$

3. $(x^2 - x - 12) \div (x - 4)$ **4.** $(x^2 - x - 12) \div (x + 3)$

5. $(6x^2 + 5x - 6) \div (3x - 2)$ **6.** $(20x^2 + 13x - 15) \div (5x - 3)$

7. $(15v^2 + 19v + 10) \div (5v - 7)$ **8.** $(15v^2 + 19v - 4) \div (3v + 8)$

9. $(x + 6x^2 - 15) \div (2x - 3)$ **10.** $(10 - 26x + 12x^2) \div (3x - 5)$

11. $(8x - 4x^3 + 10) \div (2 - x)$ **12.** $(12x - 15 - x^3) \div (3 - x)$

13. $(6a^2 + 5ab + b^2) \div (2a + 3b)$ **14.** $(6a^2 + 5ab - b^2) \div (3a - 2b)$

15. $(a^3 - 8) \div (a - 2)$ **16.** $(c^3 - 27) \div (c - 3)$

17. $(x^3 - 8x - 15) \div (x - 4)$ **18.** $(2x^3 + x^2 + 9) \div (x + 2)$

19. $(x^4 + 2x^3 - x^2 - 2x + 4) \div (x^2 + x - 1)$

20. $(x^4 - 2x^3 + 3x^2 - 2x + 7) \div (x^2 - x + 1)$

21. $(x^4 + 3x^3 + 6x^2 + 5x - 5) \div (x^2 + 2x + 3)$

22. $(x^4 + 3x^3 + 6x^2 + 5x + 5) \div (x^2 + x + 1)$

Set II   Perform the indicated divisions.

1. $(x^2 + 9x + 14) \div (x + 2)$ **2.** $(x^2 - 11x + 24) \div (x - 3)$

3. $(x^2 + x - 12) \div (x - 3)$ **4.** $(x^2 + x - 12) \div (x + 4)$

5. $(6m^2 - m - 30) \div (2m - 5)$ **6.** $(2x^2 - 14x - 16) \div (x - 8)$

7. $(6m^2 - m + 30) \div (3m + 7)$ **8.** $(6x^2 + 13x - 8) \div (2x + 5)$

9. $(13x + 3x^2 - 10) \div (3x - 2)$ **10.** $(7x + x^2 - 8) \div (x + 8)$

11. $(7x - 3x^3 + 5) \div (2 - x)$ **12.** $(10x^2 - 3x^3 + 100) \div (5 - x)$

13. $(8a^2 - 2ab - b^2) \div (2a - b)$ **14.** $(8x - 4x^3 - 3) \div (2 - x)$

15. $(x^3 + 1) \div (x + 1)$ **16.** $(x^3 + 8) \div (x + 2)$

17. $(x^3 - 5x + 1) \div (x - 3)$ **18.** $(x^3 + x^2 + 3) \div (x + 2)$

19. $(x^4 - x^3 - 5x^2 + 3x + 6) \div (x^2 - 2x - 1)$

20. $(x^4 + 2x^3 + 5x^2 + 4x + 4) \div (x^2 + x + 2)$

21. $(x^4 + 2x^3 + 5x^2 + 4x + 1) \div (x^2 + x + 2)$

22. $(x^4 + 2x^3 + x^2 - 4) \div (x^2 + x - 2)$

6.6 Review: 6.1–6.5

Polynomials
6.1

A *polynomial in x* is an algebraic expression having only terms of the form ax^n, where a is any real number and n is a whole number. A polynomial cannot have negative exponents on the variables and cannot have variables in a denominator or under a radical sign.

A *monomial* is a polynomial with one term.

A *binomial* is a polynomial with two unlike terms.

A *trinomial* is a polynomial with three unlike terms.

Degree of Polynomials
6.1

The *degree of a term* in a polynomial is the sum of the exponents of its variables. The *degree of a polynomial* is the same as that of its highest-degree term.

Operations on
Polynomials
6.2

To add polynomials, add *like* terms.

To subtract polynomials, change the sign of each term in the polynomial being subtracted; then add the resulting *like* terms.

6.3

To multiply a polynomial by a monomial, multiply *each* term in the polynomial by the monomial; then add the results.

6.3A

To multiply two binomials, the FOIL method may be used.

6.3

To multiply a polynomial by a polynomial, multiply the first polynomial by *each term* of the second polynomial, then add the results.

Special products:

6.4A $(a + b)(a - b) = a^2 - b^2$ Sum and difference of two terms

6.4B $(a + b)^2 = a^2 + 2ab + b^2$
$(a - b)^2 = a^2 - 2ab + b^2$ Square of a binomial

6.5A   *To divide a polynomial by a monomial*, divide *each* term in the polynomial by the monomial; then add the results.

6.5B   *To divide a polynomial by a polynomial*, see Section 6.5B.

Review Exercises 6.6   Set I

In Exercises 1–4, if the expression is a polynomial, (a) find the degree of the polynomial and (b) find the degree of the first term. If the expression is not a polynomial, write "Not a polynomial."

1. $5xy^2 + 3x$

2. $5u^2 + \dfrac{3}{u + v}$

3. $\sqrt{16 + x^2}$

4. $4u - 2u^2v^2$

In Exercises 5 and 6, (a) write each polynomial in descending powers of the indicated variable and (b) find the leading coefficient.

5. $3x^4 - 6 + 7x^2 + x$ Powers of x

6. $x^3 + 3x^2y + y^4 + 3xy^2$ Powers of y

In Exercises 7–28, perform the indicated operations and simplify the results.

7. $(5x^2y + 3xy^2 - 4y^3) + (2xy^2 + 4y^3 + 3x^2y)$

8. $(9xy^2)(x^2y)(-2xy)$

9. $(9xy^2)(x^2y - 2xy)$

10. $(8a + 5a^2b - 3 + 6ab^2) - (5 + 4a - 3ab^2 + 2b^2)$

11. $\dfrac{8x + 2}{2}$

12. $(5x^2y - 3y)(-4xy^2)$

13. $(x - 5)(x + 7)$

14. $(x - 2)(x^2 + 2x + 4)$

15. $(6x^2 - 9x + 10) \div (2x - 3)$

16. $(x - 8)^2$

17. Subtract $(4x^3 - x + 4x^2 - 1)$ from $(x^2 + 3 - 5x)$.

18. $(2a^4 - a^3 + a^2 + a - 3) \div (2a^2 - a + 3)$

19. $(2m^2 - 5) - [(7 - m^2) - (4m^2 - 3)]$

20. Subtract $(5x^2 - 9x + 6)$ from the sum of $(2x - 8 - 7x^2)$ and $(12 - 2x^2 + 11x)$.

21. $3xy^2(4x^2y^2 - 5xy - 10)$

22. $(x - 2)^3$

23. $(3a - 5)(3a + 5)$

24. $(3 + x + y)^2$

25. $(4 - 9y^2) \div (3y + 2)$

26. $\dfrac{-15a^2b^3 + 4ab^2 - 10ab}{-5ab}$

27. $(3xy + z)(-z)$

28. $(3xy + z) - z$

In Exercises 29 and 30, subtract the lower polynomial from the upper one.

29. $7a^2 - 3ab + 5 - b^2$
$\underline{9a^2 + 2ab - 8 + b^2}$

30. $3x^3 - 5x^2 + 2$
$\underline{-3x^3 + 2x^2 - x + 6}$

Review Exercises 6.6   Set II

In Exercises 1–4, if the expression is a polynomial, (a) find the degree of the polynomial and (b) find the degree of the first term. If the expression is not a polynomial, write "Not a polynomial."

1. $\sqrt{x^2 + 9}$

2. $8x^3y^2 - 5x^4$

3. $2 + 3x^4y$

4. $\dfrac{1}{x^2 + 2}$

In Exercises 5 and 6, (a) write each polynomial in descending powers of the indicated variable and (b) find the leading coefficient.

5. $5 - 3y^2 + y$ Powers of y

6. $5 - a^2b^3 + a^4b^2 + ab$ Powers of a

In Exercises 7–28, perform the indicated operations and simplify the results.

7. $(13x - 6x^3 + 14 - 15x^2) + (-17 - 23x^2 + 4x^3 + 11x)$

8. $(6x^3y - 4x^2y^2 + xy + 5) - (-4x^2y^2 + 3xy + 7 + 6x^3y)$

9. $5x^2y(3xy^3 + 4x - 2z)$

10. $(z + 3)(z^2 - 3z + 9)$

11. $(2xy^3 - 3x^2y - 4) - 5x^2y$

12. $(2xy^3 - 3x^2y - 4)(-5x^2y)$

ANSWERS

1a. _____

b. _____

2a. _____

b. _____

3a. _____

b. _____

4a. _____

b. _____

5a. _____

b. _____

6a. _____

b. _____

7. _____

8. _____

9. _____

10. _____

11. _____

12. _____

13. $(c - 2)(5c + 2)$

14. $(20a^2 - 7a + 5) \div (4a - 3)$

15. $\dfrac{5mn^2 - 10m^2n^3}{5mn^2}$

16. $(2 - x + y)^2$

17. Subtract $(-2x + 7x^3 - x^2 + 4)$ from $(x^3 + 5x - 2 + 4x^2)$.

18. Subtract $(3x^2 - 4x + 8)$ from the sum of $(3x + 5x^2 - 2)$ and $(7 - 3x - 4x^2)$.

19. $(3x^4 - 2x^3 + 2x^2 + 2x - 5) \div (3x^2 - 2x + 5)$

20. $(6b + 5)(6b - 5)$

21. $(6b + 5)^2$

22. $\dfrac{-12x^2y^2 + 4xy^3 - 3xy^2}{-3xy^2}$

13. _____

14. _____

15. _____

16. _____

17. _____

18. _____

19. _____

20. _____

21. _____

22. _____

23. $(3x^2 - 4) - [(2 - x^2) - (5x^2 + 1)]$

24. $(4x^2 - 1) \div (2x + 1)$

25. $(7xy^2)(3x)(-2y)$

26. $(7xy^2)(3x - 2y)$

27. $(15x^2 - 29xy - 14y^2) \div (3x - 7y)$

28. $(8x + y)(3x + 2)$

In Exercises 29 and 30, subtract the lower polynomial from the upper one.

29. $2x^3 - 4x^2 + x - 3$

$\underline{5x^3 + 2x^2 + x - 5}$

30. $\quad 4xy^2 + 2x^2y - 3x + 4$

$\underline{-2xy^2 - 2x^2y - 5x + 9}$

23. _____

24. _____

25. _____

26. _____

27. _____

28. _____

29. _____

30. _____

Chapter 6 Diagnostic Test

The purpose of this test is to see how well you understand operations with polynomials. We recommend that you work this diagnostic test *before* your instructor tests you on this chapter. Allow yourself about 50 minutes.

Complete solutions for all the problems on this test, together with section references, are given in the answer section in the back of this book. For the problems you do incorrectly, study the sections referred to.

1. In the polynomial $x^2 - 4xy^2 + 5$, find:

   a. the degree of the first term

   b. the degree of the polynomial

   c. the numerical coefficient of the second term

   d. the leading coefficient

2. Add: $\quad -7x^3 + 4x^2 \qquad\quad + 3$
$$3x^3 \qquad\quad + 6x - 5$$
$$\underline{\qquad\quad 7x^2 - 4x + 8}$$

3. Add: $(6xy^2 - 5xy) + (17xy - 7x^2y) + (3xy^2 - y^3)$

4. Subtract $(8 - 2x + 5x^2)$ from $(-3x^2 - 6x + 9)$.

5. Subtract the lower polynomial from the upper one:
$$-6a^3 + 5a^2 \qquad\quad + 4$$
$$\underline{4a^3 \qquad\quad + 6a - 7}$$

6. Subtract: $(8x^2y^2 - 5xy + 3) - (8xy - 6xy^2 + 4)$

In Problems 7–14, simplify.

7. $(8x - 3) - (10x - 5) + (9 - 5x)$

8. $-6xy(3x^2 - 5xy^2 + 8y)$

9. $(9x - 7)(8x + 9)$

10. $(3x - 8)^2$

11. $(x + 3)(x^2 - 3x + 9)$

12. $(x^2 + 3x - 5)(2x^2 - x - 4)$

13. $(4abc^2)(3b)(-2a^2c)$

14. $(x - 1)^4$

In Problems 15–18, perform the divisions.

15. $\dfrac{8x^4 - 4x^3 + 12x^2}{4x^2}$

16. $(15x^2 + x + 1) \div (3x - 1)$

17. $(12 - 6x^2 + x^3) \div (x - 4)$

18. $(3x^3 - 14x^2 + 5x + 12) \div (x - 4)$

Cumulative Review Exercises: Chapters 1–6

1. In the formula $F = \frac{9}{5}C + 32$, find the value of F when $C = -20$.

In Exercises 2 and 3, evaluate each expression.

2. $16 \div 2 \cdot 4 - 9\sqrt{25}$

3. $5[11 - 2(6 - 9)] - 4(13 - 5)$

4. Simplify: $2x(3x^2 + 6x + 8) - 4(3x^2 + 5x - 6)$

5. Evaluate the expression $-5^2 \cdot 4 - 15 \div 3\sqrt{25}$.

6. Solve the equation $5(x - 6) + 2(3 - 4x) = 6 - (2x + 8)$.

7. Find the prime factorization of 294.

In Exercises 8–11, perform the indicated operations.

8. $(3x^2 - 2x + 3)(2x - 5)$

9. $(9x - 2)(9x + 2)$

10. $(5x - 3)^2$

11. $(5x^3 - 2x^2 + 4x - 1) \div (x - 1)$

In Exercises 12–17, set up the problem algebraically and solve. Be sure to state what your variables represent.

12. The three sides of a triangle are in the ratio 3 : 5 : 7. The perimeter is 75. Find the three sides.

13. Susan worked twenty-two problems correctly on a math test having twenty-five problems. Find her percent score.

14. A business pays $125 for an item. What is the selling price of the item if it is marked up 40%?

15. Several families went to a movie together. They spent $16.25 for nine tickets. If adults' tickets cost $2.50 and children's tickets cost $1.25, how many of each kind of ticket was bought?

16. When twice the sum of 11 and an unknown number is subtracted from 6 times the sum of 8 and twice the unknown number, the result is 6. What is the unknown number?

17. A 10-lb mixture of walnuts and almonds is to be worth $3.52 per pound. If walnuts cost $4.50 per pound and almonds cost $3.10 per pound, how many pounds of each kind should be used?

Critical Thinking

Each of the following problems has an error. Can you find it?

1. Simplify $2^3 \cdot 3^2$.

$$2^3 \cdot 3^2 = 6^5$$

2. Simplify -5^{-2}.

$$-5^{-2} = \frac{1}{25}$$

3. Simplify $(2x^2y^4)^3$.

$$(2x^2y^4)^3 = 2x^6y^{12}$$

4. Simplify $(x + 4)^2$.

$$(x + 4)^2 = x^2 + 16$$

5. Divide, using long division: $(x^3 - 7x + 12) \div (x - 3)$

$$
\begin{array}{r}
x^2 - 4 \\
x - 3 \overline{) x^3 - 7x + 12} \\
\ominus x^3 \oplus 3x \\
\hline
-4x + 12 \\
\oplus 4x \ominus 12 \\
\hline
0
\end{array}
$$

6. The sum of 4 times a number and 12 is 6 less than twice the number. Find the number.

Let $x = $ the number.

$$4x + 12 = 6 - 2x$$
$$6x + 12 = 6$$
$$6x = -6$$
$$x = -1$$

7 Factoring

This chapter deals with factoring polynomials, solving equations that can be solved by factoring, and solving word problems that lead to such equations.

7.1 Greatest Common Factor (GCF)

Factoring a sum of terms means rewriting it, if possible, as a single term that is a *product of prime factors*. It is essential that the techniques of factoring be mastered, because factoring is used a great deal in the remainder of this course and in higher-level mathematics courses.

7.1A Factoring Expressions with a Common Monomial Factor

Finding the Greatest Common Factor (GCF)
In Section 1.11, we discussed factoring an integer. The **greatest common factor (GCF)** of two integers is the largest integer that is a factor of *both* integers.

Example 1   Find the GCF of 12 and 16.

$$\underbrace{2 \cdot 2 \cdot 3}_{12} \qquad \underbrace{2 \cdot 2 \cdot 2 \cdot 2}_{16}$$

The greatest factor common to both 12 and 16 is $2 \cdot 2 = 4$

Therefore, the GCF of 12 and 16 is 4.   ∎

The *greatest common factor (GCF) of a polynomial* is the largest term that is a factor of all the terms in the polynomial. We find the GCF as follows:

TO FIND THE GREATEST COMMON FACTOR (GCF)

1. Write each numerical coefficient in prime factored form. Repeated factors must be expressed in exponential form.

2. Write down each different base, numerical or literal, that is common to all terms.

3. Raise each of the bases in step 2 to the *lowest* power to which it occurs in any of the terms.

4. The *greatest common factor* (GCF) is the *product* of all the factors found in step 3. It may be positive or negative.

Example 2   Find the GCF for $15x^3 + 9x$.
Solution

$$15x^3 + 9x = 3^1 \cdot 5^1 \cdot x^3 + 3^2 \cdot x^1 \quad \longleftarrow \text{Each numerical coefficient is in prime factored form}$$

The bases common to both terms are 3 and x. The lowest power on the 3 is a 1, and the lowest power on the x is a 1. Therefore, the GCF is $3x$ or $-3x$.   ∎

Factoring Polynomials That Have a Common Factor

In Section 3.4, we used the distributive rule to rewrite a product of factors as a sum of terms; that is, $a(b + c) = ab + ac$. In this section, we use the distributive rule to rewrite a sum of terms as a product of factors (that is, as a single term) whenever possible. It is not always possible to do this, since a sum of terms is not always factorable. When we use the distributive rule to get

$$ab + ac = a(b + c)$$

we say we are factoring out the greatest common factor. Notice that the right side of the equation has only one term.

We now consider how to factor a polynomial that has some term that is a factor of all the terms of the polynomial.

TO FACTOR OUT THE GCF

1. Combine like terms, if there are any.

2. Find the GCF for all the terms. It will often, but not always, be a monomial.

3. Find the *polynomial factor** by dividing each term of the polynomial being factored by the GCF. The polynomial factor will always have as many terms as the expression in step 1. It should have only *integer* coefficients.

4. Rewrite the expression as the product of the factors found in steps 2 and 3.

5. Check the result by using the distributive rule to remove the parentheses; you should get back the polynomial you started with.

Example 3   Factor $15x^3 + 9x$.

Solution We found in Example 2 that the GCF is $3x$. To find the polynomial factor, divide $15x^3$ by $3x$ and then divide $9x$ by $3x$.

$$15x^3 + 9x$$

$$= 3x(\ \boxed{}\ +\ \boxed{}\ ) \qquad \text{The polynomial factor has as many terms as the original expression}$$

$$= 3x(5x^2 + 3)$$

This term is $\dfrac{9x}{3x} = 3$

This term is $\dfrac{15x^3}{3x} = 5x^2$

Therefore, the factors of $15x^3 + 9x$ are $3x$ and $(5x^2 + 3)$.

$$15x^3 + 9x = \boxed{3x}\ \boxed{(5x^2 + 3)}$$

GCF ⟶   ⟵ Polynomial factor

The answer, $3x(5x^2 + 3)$, has only one term, and the expression within the parentheses has no common factor left.

Check $3x(5x^2 + 3) = (3x)(5x^2) + (3x)(3) = 15x^3 + 9x$ ∎

---

*We will call this factor the *polynomial factor* because it will be a polynomial and will always have more than one term.

A WORD OF CAUTION   In Example 3, neither $3x \cdot 5x^2 + 3$ nor $(3x) \cdot 5x^2 + 3$ is the correct factored form for $15x^3 + 9x$. There *must* be parentheses around $5x^2 + 3$. Note that

$$3x \cdot 5x^2 + 3 = 15x^3 + 3 \neq 15x^3 + 9x$$

$$(3x) \cdot 5x^2 + 3 = 15x^3 + 3 \neq 15x^3 + 9x$$
☑

Example 4   Factor $6x + 4$.

Solution

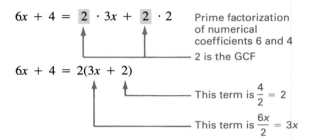

Therefore, the factors of $6x + 4$ are 2 and $(3x + 2)$.

$$6x + 4 = \underset{\text{GCF}}{2} (\underset{\text{Polynomial factor}}{3x + 2})$$

The answer, $2(3x + 2)$ has only one term, and the expression within the parentheses has no common factor left.

Check $2(3x + 2) = 6x + 4$ ∎

Prime (or Irreducible) Polynomials

It is important to realize that most polynomials cannot be factored. A polynomial is said to be **prime** (or **irreducible**) over the integers if it *cannot* be expressed as a product of polynomials of lower degree such that all the constants in the new polynomials are integers. In this book, when we say to factor a polynomial, we mean to factor over the integers.

Example 5   Factor $5x + 2y$.

Solution Although 5 and x are factors of the *first term* of $5x + 2y$, neither one is a factor of the *second term*. In fact, $5x$ and $2y$ have no common integral factors. While it is true that

$$5x + 2y = 5\left(x + \frac{2}{5}y\right)$$

and that

$$5x + 2y = 2\left(\frac{5}{2}x + y\right)$$

the parentheses in both cases contain constants that are *not integers*. Therefore, $5x + 2y$ is not factorable over the integers. It is a prime polynomial.   ∎

The Uniqueness of Factorization

We mentioned in Section 1.11 that a composite number can be expressed in prime factored form in one and only one way, except for the order in which the factors are written. A similar property holds for the factorization of polynomials.

> Factorization of polynomials over the integers is *unique*; that is, a polynomial can be completely factored over the integers in one and only one way (except for the order in which the factors are written and the sign of the product).

Example 6 Factor $12x^3 - 16x^5$.
Solution

Coefficients in prime factored form

$$12x^3 - 16x^5 = 2^2 \cdot 3x^3 - 2^4x^5$$

The bases common to both terms are 2 and x. The lowest power that occurs on the 2 is a 2, and the lowest power that occurs on the x is a 3. Therefore, the GCF is 2^2x^3 or $4x^3$.

$$\text{The polynomial factor} = \frac{12x^3}{4x^3} - \frac{16x^5}{4x^3} = 3 - 4x^2$$

Therefore,

GCF ——   —— Polynomial factor

$$12x^3 - 16x^5 = 4x^3(3 - 4x^2)$$

The answer, $4x^3(3 - 4x^2)$, has only one term, and $(3 - 4x^2)$ has no common factor left.
Check

$$4x^3(3 - 4x^2) = (4x^3)(3) - (4x^3)(4x^2) = 12x^3 - 16x^5$$

The factoring could also be done using $-4x^3$ as the GCF:

$$12x^3 - 16x^5 = -4x^3(-3 + 4x^2) = -4x^3(4x^2 - 3)$$

Check

$$-4x^3(-3 + 4x^2) = (-4x^3)(-3) + (-4x^3)(4x^2) = 12x^3 - 16x^5$$

These answers are all acceptable:
$4x^3(3 - 4x^2)$, $(3 - 4x^2)(4x^3)$, $-4x^3(-3 + 4x^2)$, $-4x^3(4x^2 - 3)$, $(-3 + 4x^2)(-4x^3)$, $(4x^2 - 3)(-4x^3)$ ∎

A WORD OF CAUTION $(-3 + 4x^2) - 4x^3$ and $(4x^2 - 3) - 4x^3$ are *not* acceptable answers for Example 6. Both equal $-3 + 4x^2 - 4x^3$ and *not* $12x^3 - 16x^5$. ☑

If an expression has been factored, it has only one term. However, even if an expression has only one term, it still may not be factored completely. Any expression inside parentheses should always be examined carefully to see whether it can be factored further.

Example 7 Factor $2x^4 + 4x^3 - 8x^2$.
Solution

$$2x^4 + 4x^3 - 8x^2$$

$$= 2^1x^4 + 2^2x^3 - 2^3x^2 \longleftarrow \text{Numerical coefficients written in prime factored form}$$

2^1 is the lowest power of 2 that occurs in any term. x^2 is the lowest power of x that occurs in any term. Therefore, GCF $= 2^1 x^2 = 2x^2$.

$$\text{The polynomial factor} = \frac{2x^4}{2x^2} + \frac{4x^3}{2x^2} - \frac{8x^2}{2x^2} = x^2 + 2x - 4$$

Therefore,

GCF — Polynomial factor

$$2x^4 + 4x^3 - 8x^2 = 2x^2(x^2 + 2x - 4)$$

Check $2x^2(x^2 + 2x - 4) = 2x^4 + 4x^3 - 8x^2$ ∎

A WORD OF CAUTION Suppose we had factored the trinomial in Example 7 as follows:

$$2x^4 + 4x^3 - 8x^2 = 2(x^4 + 2x^3 - 4x^2)$$

This *is* in factored form, since it has only one term, but it is not factored *completely*, because $x^4 + 2x^3 - 4x^2$ can still be factored. Similarly, if the trinomial is factored as follows:

$$2x^4 + 4x^3 - 8x^2 = x^2(2x^2 + 4x - 8)$$

the factoring is not complete, since $2x^2 + 4x - 8$ is not a prime polynomial. ☑

Example 8 Factor $6a^3b^3 - 8a^2b^2 + 10a^3b$.
Solution
$$6a^3b^3 - 8a^2b^2 + 10a^3b = 2 \cdot 3a^3b^3 - 2^3 a^2 b^2 + 2 \cdot 5a^3 b^1$$
$$\text{GCF} = 2^1 \cdot a^2 \cdot b^1 = 2a^2 b$$
$$\text{The polynomial factor} = \frac{6a^3b^3}{2a^2b} - \frac{8a^2b^2}{2a^2b} + \frac{10a^3b}{2a^2b}$$
$$= 3ab^2 - 4b + 5a$$

Therefore, $6a^3b^3 - 8a^2b^2 + 10a^3b = 2a^2b(3ab^2 - 4b + 5a)$.

Check $2a^2b(3ab^2 - 4b + 5a) = 6a^3b^3 - 8a^2b^2 + 10a^3b$. ∎

Example 9 Factor $3xy - 6y^2 - 3y$.
Solution
$$3xy - 6y^2 - 3y = 3xy - 2 \cdot 3y^2 - 3y$$

$$\text{GCF} = 3 \cdot y = 3y$$
$$\text{The polynomial factor} = \frac{3xy}{3y} - \frac{6y^2}{3y} - \frac{3y}{3y} = x - 2y - 1$$

Therefore, $3xy - 6y^2 - 3y = 3y(x - 2y - 1)$.

Check $3y(x - 2y - 1) = 3xy - 6y^2 - 3y$ ∎

Example 10   Factor $-15x^2y^3 - 20xy^4 + 25x^3y^2$.
Solution

$$-15x^2y^3 - 20xy^4 + 25x^3y^2 = -3 \cdot 5x^2y^3 - 2^2 \cdot 5xy^4 + 5^2x^3y^2$$

$$GCF = 5xy^2 \; or \; -5xy^2$$

First solution, using $5xy^2$ as the GCF:

$$\text{The polynomial factor} = \frac{-15x^2y^3}{5xy^2} + \frac{-20xy^4}{5xy^2} + \frac{25x^3y^2}{5xy^2}$$

$$= -3xy - 4y^2 + 5x^2$$

Therefore, $-15x^2y^3 - 20xy^4 + 25x^3y^2 = 5xy^2(-3xy - 4y^2 + 5x^2)$.
Second solution, using $-5xy^2$ as the GCF:

$$\text{The polynomial factor} = \frac{-15x^2y^3}{-5xy^2} + \frac{-20xy^4}{-5xy^2} + \frac{25x^3y^2}{-5xy^2}$$

$$= 3xy + 4y^2 - 5x^2$$

Therefore, $-15x^2y^3 - 20xy^4 + 25x^3y^2 = -5xy^2(3xy + 4y^2 - 5x^2)$. Answers from the first solution and from the second are correct.
Checks

$$5xy^2(-3xy - 4y^2 + 5x^2) = -15x^2y^3 - 20xy^4 + 25x^3y^2$$

$$-5xy^2(3xy + 4y^2 - 5x^2) = -15x^2y^3 - 20xy^4 + 25x^3y^2 \quad \blacksquare$$

EXERCISES 7.1A

Set I   Factor each expression completely, or write "Not factorable."

1. $12x + 8$ **2.** $6x + 9$

3. $5a - 8$ **4.** $7b - 2$

5. $2x + 8$ **6.** $3x + 9$ **7.** $5a - 10$

8. $7b - 14$ **9.** $6y - 3$ **10.** $15z - 5$

11. $9x^2 + 3x$ **12.** $8y^2 - 4y$ **13.** $10a^3 - 25a^2$

14. $27b^2 - 18b^4$ **15.** $21w^2 - 20z^2$ **16.** $15x^3 - 16y^3$

17. $2a^2b + 4ab^2$ **18.** $3mn^2 + 6m^2n^2$

19. $12c^3d^2 - 18c^2d^3$ **20.** $15ab^3 - 45a^2b^4$

21. $4x^3 - 12x - 24x^2$ **22.** $18y - 6y^2 - 30y^3$

23. $4x^4 - 7x^2 + 1$ **24.** $3x^2 + 2x - 2$

25. $8x^3 - 6x^2 + 2x$ **26.** $9y^4 + 6y^3 - 3y^2$

27. $24a^4 + 8a^2 - 40$ **28.** $45b^3 - 15b^4 - 30$

29. $-14x^8y^9 + 42x^5y^4 - 28xy^3$

30. $-21u^7v^8 - 63uv^5 + 35u^2v^5$

31. $15h^2k - 8hk^2 + 9st$

32. $10uv^3 + 5u^2v - 4wz$

33. $-44a^{14}b^7 - 33a^{10}b^5 + 22a^{11}b^4$

34. $-26e^8f^6 + 13e^{10}f^8 - 39e^{12}f^5$

35. $18u^{10}v^5 + 24 - 14u^{10}v^6$

36. $30a^3b^4 - 15 + 45a^8b^7$

37. $18x^3y^4 - 12y^2z^3 - 48x^4y^3$

38. $32m^5n^7 - 24m^8p^9 - 40m^3n^6$

Set II Factor each expression completely, or write "Not factorable."

1. $6h + 9$ **2.** $12x - 5$

3. $2x - 7$ **4.** $8k - 12$

5. $6h + 18$ **6.** $8x + 9$ **7.** $8k - 16$

8. $6 - 12x$ **9.** $8k - 4$ **10.** $3x - 22$

11. $4a^2 + 2a$ **12.** $10x + 25$ **13.** $8x^3 - 12x^2$

14. $4y - 12y^2$ **15.** $9w^2 + 16z^2$ **16.** $14a^2 - 21a^3$

17. $5xy^2 + 10x^2y$ **18.** $x^2 + 4$

19. $15a^2b^3 - 12ab^2$ **20.** $12x^2y + 9xy^2$

21. $16z - 8z^3 - 12z^2$ **22.** $12x^3 - 5y + 3x$

23. $8x^4 - 3x^2 + 5$ **24.** $12x^3 - 28x^2 - 8x$

25. $12x^4 - 6x^3 + 3x^2$ **26.** $4y^2 - 8y^4 + 16y^5$

27. $42x^5 - 6x^4 + 12x^3$ **28.** $-7a^2 + 14a^3 - 35a^4$

29. $-12ab^2 + 9a^2b - 36ab$ **30.** $20z^5 - 30z^3 - 10z^2$

31. $12h^2k - 18hk^2 - 35mp$ **32.** $30e^3f + 18 - 12ef^2$

33. $10m^2n - 21mn^3 - 13mn$ **34.** $-16u^3v^2 + 24uv^3 - 40v^4w^2$

35. $25x^2y^3 + 10 - 15x^3y^2$ **36.** $-5xy^2 + 7x - 3y$

37. $15a^3y^4 - 18ay^3 + 12a^2y$ **38.** $x^4 + 16$

7.1B   Factoring Expressions with a Common Binomial Factor

Sometimes an expression has a GCF that is not a monomial. Such an expression can still be factored, using the same rules given in Section 7.1A. This type of factoring is used in *factoring by grouping* (see Section 7.5) and also in higher-level mathematics courses.

Example 11   Factor $a(x + y) + b(x + y)$.

Solution This expression has two terms, and therefore it is *not* in factored form. The common factor, $x + y$, is not a monomial; it is a binomial.

Common factor is $(x + y)$

$$a(\,x + y\,) + b(\,x + y\,) = (x + y)(\,a\,+\,b\,)$$

This term is $\dfrac{b(x + y)}{x + y} = b$

This term is $\dfrac{a(x + y)}{x + y} = a$

Therefore, $a(x + y) + b(x + y) = (x + y)(a + b)$.

Check $(x + y)(a + b) = (x + y)a + (x + y)b = a(x + y) + b(x + y)$

The answer, $(x + y)(a + b)$, has only one term, so it is in factored form. ■

Example 12 Factor $b(a - 1) + (a - 1)$.
Solution The expression has two terms. The common binomial factor is $a - 1$.

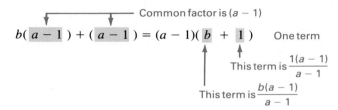

Therefore, $b(a - 1) + (a - 1) = (a - 1)(b + 1)$.

Check $(a - 1)(b + 1) = (a - 1)b + (a - 1)(1) = b(a - 1) + (a - 1)$ ■

EXERCISES 7.1B

Set I   Factor each expression completely, or write "Not factorable."

1. $c(s + t) + b(s + t)$

2. $a(b + c) + d(b + c)$

3. $x(a - b) + 5(a - b)$

4. $y(s - t) + 7(s - t)$

5. $x(u + v) - 3(u + v)$

6. $s(t + u) - 2(t + u)$

7. $8(x - y) - a(x - y)$

8. $7(a - b) - c(a - b)$

9. $4(s - t) + u(s - t) - v(s - t)$

10. $a(x - y) + 5(x - y) - b(x - y)$

11. $3x^2y^3(a + b) + 9xy^2(a + b)$

12. $10ab^3(s + t^2) + 15a^2b(s + t^2)$

Set II   Factor each expression completely, or write "Not factorable."

1. $x(f + g) + s(f + g)$

2. $y(u - v) - z(u - v)$

3. $u(x - y) + 4(x - y)$

4. $d(a - b) - 3(a - b)$

5. $a(b + c) - 9(b + c)$

6. $x(y + z) - w(y + z)$

7. $9(a - b) - c(a - b)$

8. $y(x - z) - 9(x - z)$

9. $8(x - y) + z(x - y) - w(x - y)$

10. $f(a + b) - g(a + b) + (a + b)$

11. $8x^3y^2(c - d) + 4x^2y^3(c - d)$

12. $14a^3b^3(x - 2) + 7ab^2(x - 2)$

7.2 Factoring the Difference of Two Squares

Any polynomial that can be expressed in the form $a^2 - b^2$ is called a **difference of two squares.**

7.2A Principal Square Root of a Term

We discussed principal square roots of integers in Section 1.10. In this section, we discuss finding the principal square root of an algebraic term with even exponents on the variables. We assume that all variables represent positive integers.

Example 1   Find the principal square root of each term:

a. $\sqrt{16} = 4$, because $(4)^2 = 16$.

b. $\sqrt{25x^2} = 5x$, because $(5x)^2 = 25x^2$.

c. $\sqrt{100a^4b^6} = 10a^2b^3$, because $(10a^2b^3)^2 = 100a^4b^6$. ■

TO FIND THE PRINCIPAL SQUARE ROOT OF A TERM

1. The square root of the numerical coefficient is found by inspection.

2. The square root of each literal factor is found by dividing its exponent by 2.

Example 2   Use the information given in the preceding box to find the principal square root for each of the following terms:

a. $\sqrt{36e^8f^4} = 6e^{8/2}f^{4/2} = 6e^4f^2$

b. $\sqrt{9x^{10}y^6} = 3x^{10/2}y^{6/2} = 3x^5y^3$ ■

In Chapter 11, we will discuss simplifying square roots when the exponents on the variables are not exactly divisible by 2.

EXERCISES 7.2A

Set I   Find the principal square roots by inspection.

1. $\sqrt{64}$ **2.** $\sqrt{81}$ **3.** $\sqrt{4x^2}$

4. $\sqrt{9y^2}$ **5.** $\sqrt{100a^8}$ **6.** $\sqrt{49b^6}$

7. $\sqrt{m^4n^2}$ **8.** $\sqrt{u^{10}v^6}$ **9.** $\sqrt{x^{10}y^4}$

10. $\sqrt{x^{12}y^8}$ **11.** $\sqrt{25a^4b^2}$ **12.** $\sqrt{100b^4c^2}$

13. $\sqrt{36e^8f^2}$ **14.** $\sqrt{81h^{12}k^{14}}$ **15.** $\sqrt{100a^{10}y^2}$

16. $\sqrt{121a^{24}b^4}$ **17.** $\sqrt{9a^4b^2c^6}$ **18.** $\sqrt{144x^8y^2z^6}$

Set II   Find the principal square roots by inspection.

1. $\sqrt{49}$ **2.** $\sqrt{144}$ **3.** $\sqrt{16z^2}$

4. $\sqrt{36a^8}$ **5.** $\sqrt{64a^6}$ **6.** $\sqrt{169x^4y^8}$

7. $\sqrt{x^2y^6}$ **8.** $\sqrt{400x^4}$ **9.** $\sqrt{h^8k^4}$

10. $\sqrt{4x^6y^{12}}$ **11.** $\sqrt{81m^6n^8}$ **12.** $\sqrt{25a^8b^8c^2}$

13. $\sqrt{144u^{12}v^8}$ **14.** $\sqrt{x^{14}y^{16}}$ **15.** $\sqrt{121c^{10}d^6}$

16. $\sqrt{4a^2b^4c^6}$ **17.** $\sqrt{25r^6s^8t^4}$ **18.** $\sqrt{121x^{20}y^{30}}$

7.2B Factoring the Difference of Two Squares

We now consider factoring a difference of two squares, that is, factoring $a^2 - b^2$. Factoring $a^2 - b^2$ depends on the product $(a + b)(a - b) = a^2 - b^2$.

Finding the product

$$(a + b)(a - b) = a^2 - b^2$$

Finding factors

Therefore, $a^2 - b^2$ *factors into* $(a + b)(a - b)$.

TO FACTOR THE DIFFERENCE OF TWO SQUARES

1. Make the following blank form for the factors:

$$(\blacksquare + \blacksquare)(\blacksquare - \blacksquare)$$

One factor has $+$, the other has $-$

2. Put the principal square root of the *first* term here.

Put the principal square root of the *second* term here.

$$(\blacksquare + \blacksquare)(\blacksquare - \blacksquare)$$

The product found in step 2 is the factored form of the difference of two squares. In symbols,

$$a^2 - b^2 = (a + b)(a - b)$$

Example 3 Factor $x^2 - 4$.
Solution

Difference of

$$x^2 - 4 = x^2 - 2^2$$

two squares

Step 1: $x^2 - 4 = (\blacksquare + \blacksquare)(\blacksquare - \blacksquare)$

$\sqrt{x^2}$

Step 2: $x^2 - 4 = (\,x\, + \,2\,)(\,x\, - \,2\,)$

$\sqrt{4}$

Therefore, $x^2 - 4 = (x + 2)(x - 2)$.
Check $(x + 2)(x - 2) = x^2 - 4$ ∎

Example 4 Factor $25y^4 - 9z^2$.
Solution $25y^4 - 9z^2 = (5y^2)^2 - (3z)^2$, which is in the form $a^2 - b^2$.

Step 1: $25y^4 - 9z^2 = (\blacksquare + \blacksquare)(\blacksquare - \blacksquare)$

$\sqrt{9z^2}$

Step 2: $25y^4 - 9z^2 = (\,5y^2\, + \,3z\,)(\,5y^2\, - \,3z\,)$

$\sqrt{25y^4}$

Therefore, $25y^4 - 9z^2 = (5y^2 + 3z)(5y^2 - 3z)$.
Check $(5y^2 + 3z)(5y^2 - 3z) = 25y^4 - 9z^2$ ∎

Example 5   Factor $49a^6b^2 - 81c^4d^8$.
Solution

Step 1: $49a^6b^2 - 81c^4d^8 = (\blacksquare + \blacksquare)(\blacksquare - \blacksquare)$

Step 2: $49a^6b^2 - 81c^4d^8 = (7a^3b + 9c^2d^4)(7a^3b - 9c^2d^4)$ $\sqrt{81c^4d^8}$

$\sqrt{49a^6b^2}$

Therefore, $49a^6b^2 - 81c^4d^8 = (7a^3b + 9c^2d^4)(7a^3b - 9c^2d^4)$.
Check $(7a^3b + 9c^2d^4)(7a^3b - 9c^2d^4) = 49a^6b^2 - 81c^4d^8$ ∎

A WORD OF CAUTION   A *sum* of two squares (that is, a polynomial that can be expressed in the form $a^2 + b^2$) is *not factorable* over the integers. (Exception: If the exponents on the variables are even numbers greater than 2, the polynomial may be factorable; however, we will not discuss the methods of such factoring in this book.) ☑

Example 6   Factor $x^2 + 4$.
Solution There is no common monomial factor. Then, because $x^2 + 4$ is a sum of two squares, we conclude that it is not factorable. ∎

You should begin every factoring problem by looking for and factoring out any factor common to every term. Then try to factor the polynomial in the parentheses (see Example 7).

Example 7   Factor $16x^4 - 4x^2$.
Solution There *is* a common monomial factor: $4x^2$. Therefore,

$16x^4 - 4x^2 = 4x^2(4x^2 - 1)$ ← This has been factored, but not completely

Difference of two squares

$= 4x^2(2x + 1)(2x - 1)$

Check $4x^2(2x + 1)(2x - 1) = 4x^2(4x^2 - 1) = 16x^4 - 4x^2$ ∎

EXERCISES 7.2B

Set I   Factor each polynomial completely, or write "Not factorable."

1. $m^2 - n^2$ **2.** $u^2 - v^2$ **3.** $x^2 - 9$

4. $x^2 - 25$ **5.** $a^2 - 1$ **6.** $1 - b^2$

7. $4c^2 - 1$ **8.** $16d^2 - 1$ **9.** $16x^2 - 9y^2$

10. $25a^2 - 4b^2$ **11.** $9h^2 - 10k^2$ **12.** $16e^2 - 15f^2$

13. $4x^4 - 2x$ **14.** $9a^4 - 3a$ **15.** $16x^2 + 1$

16. $25y^2 + 1$ **17.** $49u^4 - 36v^4$ **18.** $81m^6 - 100n^4$

19. $x^6 - a^4$ **20.** $b^2 - y^6$ **21.** $2x^2 - 18$

22. $3x^2 - 12$ **23.** $2x^2 + 9$ **24.** $5y^2 + 2$

25. $a^2b^2 - c^2d^2$ **26.** $m^2n^2 - r^2s^2$ **27.** $49 - 25w^2z^2$

28. $36 - 25u^2v^2$ **29.** $4h^4k^4 - 1$ **30.** $9x^4y^4 - 1$

31. $81a^4b^6 - 16m^2n^8$ **32.** $49c^8d^4 - 100e^6f^2$

Set II Factor each polynomial completely, or write "Not factorable."

1. $h^2 - k^2$

2. $36 - m^2$

3. $x^2 - 1$

4. $1 - 25a^2$

5. $9w^2 - 49z^2$

6. $16x^2 + 64x$

7. $25x^2 - 1$

8. $9x^2 - 5$

9. $100x^2 - y^2$

10. $2y^3 - 2y$

11. $16x^2 - 11y^2$

12. $a^2 + 16b^2$

13. $9x^4 + 3x$

14. $3x^2 - 48x$

15. $64x^2 + 1$

16. $4x^2 - 5$

17. $49m^4 - 64n^2$

18. $1 - 81y^2$

19. $e^6 - 4$

20. $f^2 + 4$

21. $7a^2 - 63$

22. $3x^3 - 48x$

23. $5y^2 + 16$

24. $3x^4 - 12x^2$

25. $r^2s^2 - t^2u^2$

26. $r^2s^2 + t^2u^2$

27. $81 - 16u^2v^2$

28. $81 + 9u^2v^2$

29. $25x^4y^4 - 1$

30. $9 - 16a^4b^4$

31. $144w^4x^6 - 121y^8z^2$

32. $20y^4 - 5y^2$

7.3 Factoring Trinomials

7.3A Factoring a Trinomial with a Leading Coefficient of 1

While many trinomials are not factorable, many others will factor into the product of two binomials. Recall from Section 6.1 that the leading coefficient of a polynomial is the numerical coefficient of its highest-degree term. The easiest type of trinomial to factor is one with a leading coefficient of 1.

We will first be concerned with what the *signs* of a trinomial tell us about the (possible) factors of that trinomial. Let us consider four similar products:

1. $(x + 2)(x + 5)$

2. $(x - 2)(x - 5)$

3. $(x + 2)(x - 5)$

4. $(x - 2)(x + 5)$

In 1,

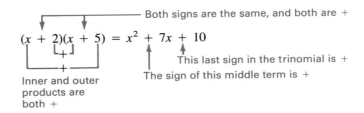

In 2,

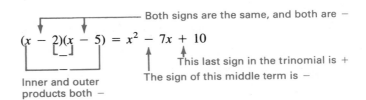

In 3,

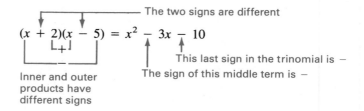

In 4,

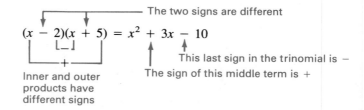

A careful examination of the four products above shows that the Rule of Signs for Factoring Trinomials that follows is reasonable.

RULE OF SIGNS FOR FACTORING TRINOMIALS

(It is assumed that the trinomial has been arranged in descending powers of a variable.)

1. If the last sign in the trinomial is $+$, the signs in the binomial factors are the same.
   a. If the middle term of the trinomial is $+$, the signs in both binomials are $+$.
   b. If the middle term of the trinomial is $-$, the signs in both binomials are $-$.

2. If the last sign in the trinomial is $-$, the signs in the binomials are different, and we must put the signs in last.

Let's discuss the product $(x + 2)(x + 5)$ a little more.

$$(x + 2)(x + 5) = x^2 + 7x + 10$$

Therefore, $x^2 + 7x + 10$ factors into $(x + 2)(x + 5)$.
$$x^2 = \sqrt{x^2} \cdot \sqrt{x^2} = x \cdot x$$

When we try to factor a trinomial, we must *assume* that it will factor into a product of two binomials (we may decide later that it does *not*) and first make a blank outline:
$(\quad)(\quad)$.

When the leading coefficient is 1, the first term in each binomial factor is the square root of the first term of the trinomial.

Example 1   a. $x^2 + 6x + 8 = (x\quad)(x\quad)$

b. $z^2 - z - 6 = (z\quad)(z\quad)$ } Partially filled-in outlines

c. $m^2 + 7m + 12 = (m\quad)(m\quad)$ ∎

We continue our discussion of the product $(x + 2)(x + 5)$.

$$(x + 2)(x + 5) = x^2 + 7x + 10$$

Therefore, $x^2 + 7x + \boxed{10}$ factors into $(x + \boxed{2})(x + \boxed{5})$.

$$10 = 2 \cdot 5$$

The last terms of the binomial factors must be factors of the last term of the trinomial.

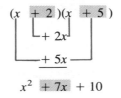

When we multiply the binomials, notice that the sum of the inner and outer products must equal the middle term of the trinomial.

The method of factoring a trinomial with a leading coefficient of 1 is summarized as follows:

TO FACTOR A TRINOMIAL WITH A LEADING COEFFICIENT OF 1

(It is assumed that like terms have been combined.)

1. Factor out the GCF.

2. Arrange the trinomial in descending powers of the variable.

3. Make a blank outline. If the Rule of Signs indicates that the signs in the binomials will be the same, put them in the outline.

4. The first term inside each set of parentheses will be the square root of the first term of the trinomial. If the third term of the trinomial has a variable, the square root of that variable must be a factor of the second term in each binomial.

5. To find the last term of each binomial:
  a. List (mentally, at least) *all* pairs of integral factors of the coefficient of the last term of the trinomial.
  b. Select the particular pair of factors that has a sum equal to the coefficient of the middle term of the trinomial. If no such pair of factors exists, the trinomial is not factorable.

6. Check your result by multiplying the binomials together. You should get back the trinomial from step 2.

Example 2 Factor $x^2 + 5x + 6$.

Solution There are no like terms and no common factor.

$$(\quad)(\quad)\,\text{←The blank outline}$$

The first term in each binomial is $\sqrt{x^2} = x$:

$$(x\quad)(x\quad)\,\text{← Partially filled-in outline}$$

The last sign in the trinomial is $+$; the rule of signs for factoring a trinomial tells us that the signs in the binomials will be the same. The middle term of the trinomial is $+$; therefore, both signs will be $+$.

$$(x +\quad)(x +\quad)\,\text{←Partially filled-in outline}$$

The product of the last terms of the binomials must be 6. Since $6 = 1 \cdot 6$ or $6 = 2 \cdot 3$, we could have either

$$(x + 1)(x + 6) \qquad or \qquad (x + 2)(x + 3)$$

The sum of the inner and outer products must equal the middle term of the trinomial, $5x$.

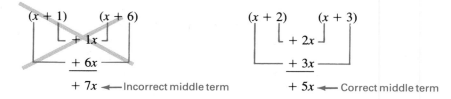

Therefore,

$$x^2 + 5x + 6 = (x + 2)(x + 3) \quad or \quad (x + 3)(x + 2)$$

The order of the factors is unimportant. Notice that $(x + 2)(x + 3)$ has only one term; it is in factored form.

Check $(x + 2)(x + 3) = x^2 + 5x + 6$ ∎

A WORD OF CAUTION   Both sets of parentheses are essential.

$$(x + 2)x + 3 = x^2 + 2x + 3 \neq x^2 + 5x + 6$$
$$x + 2(x + 3) = x + 2x + 6 = 3x + 6 \neq x^2 + 5x + 6$$
$$x + 2 \cdot x + 3 = x + 2x + 3 = 3x + 3 \neq x^2 + 5x + 6 \qquad ☑$$

Example 3 Factor $m^2 - 9m + 8$.

Solution

$$m^2 - 9m + 8$$

$+$ here tells us that the signs in the binomials will be the same
$-$ here tells us that both signs will be $-$

The first terms of the binomial factors must be m.

$$(m -\quad)(m -\quad)\,\text{← Partially filled-in outline}$$

The pairs of negative factors of 8 are $(-1)(-8)$ and $(-2)(-4)$. We must select the pair whose sum is -9. This is the pair -1 and -8. Therefore,

$$m^2 - 9m + 8 = (m - 1)(m - 8) = (m - 8)(m - 1)$$

Check $(m - 1)(m - 8) = m^2 - 9m + 8$ ■

A WORD OF CAUTION   While $(m - 2)(m - 4)$ gives us the correct *first* term and the correct *third* term, the *middle term* is incorrect:

$$(m - 2)(m - 4) = m^2 - 6m + 8 \neq m^2 - 9m + 8 \qquad \boxed{\checkmark}$$

Example 4   Factor $4a - 12 + a^2$.
Solution

$$a^2 + 4a - 12 \qquad \text{Arranged in descending powers of } a$$

— here tells us that the signs in the binomials will be different; we must put them in last

The first terms of the binomial factors must be a.

$$(a \quad)(a \quad) \longleftarrow \text{Partially filled-in outline}$$

The pairs of factors of -12 are $(1)(-12)$, $(-1)(12)$, $(2)(-6)$, $(-2)(6)$, $(3)(-4)$, and $(-3)(4)$. We must select the pair whose sum is $+4$; this is the pair -2 and 6. Therefore,

$$a^2 + 4a - 12 = (a - 2)(a + 6) = (a + 6)(a - 2)$$

Check $(a - 2)(a + 6) = a^2 + 4a - 12 = 4a - 12 + a^2$ ■

Example 5   Factor $a^2 - 4a - 12$.
Solution

$$a^2 - 4a - 12$$

— here tells us that the signs in the binomials will be different

See Example 4 for pairs of factors of -12. We must select the pair whose sum is -4; this is the pair 2 and -6. Therefore,

$$a^2 - 4a - 12 = (a + 2)(a - 6) = (a - 6)(a + 2)$$

Check $(a + 2)(a - 6) = a^2 - 4a - 12$ ■

Example 6   Factor $x^2 - 8x - 20$.
Solution

$$x^2 - 8x - 20$$

— here tells us that the signs in the binomials will be different

The pairs of factors of -20 are $(1)(-20)$, $(-1)(20)$, $(2)(-10)$, $(-2)(10)$, $(4)(-5)$, and $(-4)(5)$. The pair whose sum is -8 is the pair 2 and -10. Therefore,

$$x^2 - 8x - 20 = (x + 2)(x - 10)$$

Check $(x + 2)(x - 10) = x^2 - 8x - 20$ ■

Example 7 Factor $x^4 + 9x^3 + 20x^2$.

Solution There is a common factor. We factor out x^2 (the GCF) first.

$$x^4 + 9x^3 + 20x^2 = x^2(x^2 + 9x + 20)$$

The expression $x^2(x^2 + 9x + 20)$ has been factored; however, the polynomial factor is still factorable:

$$x^2 + 9x + 20$$

— $+$ here tells us that the signs in the binomials will be the same
— $+$ here tells us that both signs will be $+$

The pairs of positive factors of 20 are (1)(20), (2)(10), and (4)(5). The pair whose sum is 9 is the pair 4 and 5. Therefore,

$$x^4 + 9x^3 + 20x^2 = x^2(x^2 + 9x + 20) = x^2(x + 4)(x + 5)$$

Check $x^2(x + 4)(x + 5) = x^2(x^2 + 9x + 20) = x^4 + 9x^3 + 20x^2$ ∎

Example 8 Factor $x^2 - 11xy + 24y^2$.

Solution

$$x^2 - 11xy + 24y^2$$

— $+$ here tells us that both signs will be the same
— $-$ here tells us that they will both be $-$

Because the third term contains the variable y^2, we must be sure to have $\sqrt{y^2} = y$ as a *factor* of the second term in each binomial.

$$\sqrt{y^2} = y$$
$$(x - \blacksquare\, y)(x - \blacksquare\, y) \leftarrow \text{Partially filled-in outline}$$

The pairs of negative factors of 24 are $(-1)(-24)$, $(-2)(-12)$, $(-3)(-8)$, and $(-4)(-6)$. The pair whose sum is -11 is the pair -3 and -8. Therefore,

$$x^2 - 11xy + 24y^2 = (x - 3y)(x - 8y)$$

Check $(x - 3y)(x - 8y) = x^2 - 11xy + 24y^2$ ∎

Example 9 Factor $x^2 - 5x + 3$.

Solution

$$x^2 - 5x + 3$$

— $+$ here tells us that both signs will be the same
— $-$ here tells us that they will both be $-$

$$(x - \quad)(x - \quad) \leftarrow \text{Partially filled-in outline}$$

We must find two numbers whose product is 3 and whose sum is -5. There are no such numbers. Therefore, $x^2 - 5x + 3$ is *not factorable*. ∎

EXERCISES 7.3A

Set I   Factor each polynomial completely, or write "Not factorable."

1. $x^2 + 6x + 8$ **2.** $x^2 + 9x + 8$

3. $x^2 + 5x + 4$ **4.** $x^2 + 4x + 4$

5. $k^2 + 7k + 6$ **6.** $k^2 + 5k + 6$

7. $7u + u^2 + 10$ **8.** $11u + u^2 + 10$

9. $y^2 - 2y + 8$ **10.** $y^2 - 7y + 8$

11. $b^2 - 9b + 14$ **12.** $b^2 - 15b + 14$

13. $z^2 - 9z + 20$ **14.** $z^2 - 12z + 20$

15. $18 + x^2 - 11x$ **16.** $18 + x^2 - 9x$

17. $x^2 + 9x - 10$ **18.** $y^2 - 3y - 10$

19. $z^2 - z - 6$ **20.** $m^2 + 5m - 6$

21. $5x^2 + 10x$ **22.** $8y^3 + 4y^2$

23. $x^2 + 4x - 5$ **24.** $y^2 + 6y - 7$

25. $x^2 + 100$ **26.** $z^2 + 1$

27. $z^5 + 9z^4 - 10z^3$ **28.** $x^4 + 7x^3 - 8x^2$

29. $t^2 + 11t - 30$ **30.** $m^2 - 17m - 30$

31. $u^4 + 12u^2 - 64$ **32.** $v^4 - 30v^2 - 64$

33. $16 - 8v + v^2$ **34.** $16 - 10v + v^2$

35. $b^2 - 11bd - 60d^2$ **36.** $c^2 + 17cx - 60x^2$

37. $r^2 - 13rs - 48s^2$ **38.** $s^2 + 22st - 48t^2$

39. $x^4 + 2x^3 - 35x^2$ **40.** $x^4 + 2x^3 - 48x^2$

41. $14x^2 - 15x + x^3$ **42.** $8x^3 - 9x^2 + x^4$

43. $x^4 + 6x^3 + x^2$ **44.** $y^4 + 5y^3 + y^2$

45. $x^4 - 9$ **46.** $a^4 - 25$

Set II   Factor each polynomial completely, or write "Not factorable."

1. $m^2 + 7m + 12$ **2.** $x^2 + 9x + 20$

3. $h^2 + 11h + 18$ **4.** $x^2 + 13x + 12$

5. $a^2 + 10a + 16$ **6.** $x^2 + x + 1$

7. $10k + k^2 + 24$ **8.** $21y + y^2 + 20$

9. $x^2 - x + 12$ **10.** $x^2 - 7x + 12$

11. $n^2 - 9n + 18$ **12.** $x^2 - 5x + 4$

13. $z^2 - 8z + 16$ **14.** $y^2 - 2y + 9$

15. $15 + x^2 - 8x$ **16.** $13n + n^2 - 14$

17. $w^2 + 2w - 24$ **18.** $3z - 18 + z^2$

19. $r^2 - 9r - 20$ **20.** $y^2 - 8y - 20$

21. $3x^2 + 12x$ **22.** $c^2 + 121$

23. $x^2 - 4x - 45$ **24.** $a^4 + a^3 + a^2$

25. $a^2 + 25$ **26.** $5a^2 + 25$

27. $z^5 - 5z^4 - 6z^3$ **28.** $x^4 + 5x^3 + 5x^2$

29. $n^2 - 10n - 24$ **30.** $x^2 + 4x - 45$

31. $v^4 - 18v^2 + 45$ **32.** $u^4 - 4u^3 - 21u^2$

33. $36 - 15y + y^2$ **34.** $u^4 - 4u^2 - 21$

35. $w^2 - 8wz - 48z^2$ **36.** $x^2 - 15xy + 54y^2$

37. $p^2 - 9pt - 52t^2$ **38.** $a^2 + 10ab + 16b^2$

39. $s^4 + 3s^3 - 28s^2$ **40.** $s^2 - 3st + 28t^2$

41. $y^3 - 30y^2 + y^4$ **42.** $y^2 - 11yz + 30z^2$

43. $t^4 + 5t^3 + t^2$ **44.** $x^2 - 5x - 24$

45. $x^2 - 121$ **46.** $y^2 + 121$

7.3B   Factoring a Trinomial with a Leading Coefficient Unequal to 1

The method we now show for factoring a trinomial with a leading coefficient that is not 1 is often called the *trial method*. In deciding what the literal parts of each term in each binomial should be, we use the same method used in Section 7.3A; we continue to use the Rule for Signs for Factoring Trinomials given in that section.

    The rules that follow assume that like terms have been combined and that any common factors have been factored out. It is also assumed that the trinomial has been arranged in descending powers of the variable and that the leading coefficient is *positive*. If the leading coefficient is negative either arrange the trinomial in ascending powers or factor out -1 before proceeding.

TO FACTOR A TRINOMIAL WITH A LEADING COEFFICIENT UNEQUAL TO 1

1. Make a blank outline and fill in the *literal parts* of each term of each binomial. Fill in the signs in each binomial if they are both positive or both negative.

2. List (mentally, at least) all pairs of factors of the coefficient of the first term of the trinomial and of the last term of the trinomial.

3. By trial and error, select the pairs of factors (if they exist) from step 2 that make the sum of the inner and outer products of the binomials equal to the middle term of the trinomial. If no such pairs exist, the trinomial is *not factorable*.

4. Check the result.

Example 10   Factor $2x^2 + 7x + 5$.

    Solution There are no like terms and no common factor; the trinomial is already in descending powers of x. The signs for both binomial factors are $+$.

$$(\quad + \quad)(\quad + \quad)$$

Each first term of the binomial factors must contain an x in order to give the x^2 in the first term of the trinomial $2x^2$.

$$(x + \quad)(\; x + \quad)$$

The only factors of 2 are 1 and 2. We put these numbers in the outline.

$$(1x + \quad)(2x + \quad)$$

We have two ways to try the factors of 5:

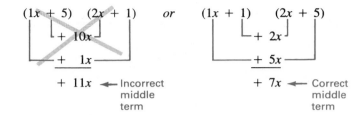

The sum of the inner and outer products must equal the middle term of the trinomial. We found the correct pair by trial. Therefore, $2x^2 + 7x + 5$ factors into $(x + 1)(2x + 5)$.
Check $(x + 1)(2x + 5) = 2x^2 + 7x + 5$ ∎

Example 11   Factor $13x + 5x^2 + 6$.
Solution There are no like terms and no common factor.

$$13x + 5x^2 + 6 = 5x^2 + 13x + 6 \;\longleftarrow\text{Arranged in descending powers of } x$$

$$(x + \quad)(\; x + \quad) \;\longleftarrow\text{Partially filled-in outline}$$

Signs are both $+$

Next, list all pairs of factors of the first coefficient and the last term of the trinomial.

Factors of the first coefficient	Factors of the last term
$5 = 1 \cdot 5$	$6 = 1 \cdot 6$
	$= 2 \cdot 3$

There are two pairs of factors to try for 6: $(1)(6)$ or $(2)(3)$. Therefore, we must try the following combinations:

$(5x + 1)(x + 6)$

$(5x + 6)(x + 1)$

$(5x + 2)(x + 3)$

$(5x + 3)(x + 2)$

In all these combinations, the product of the first two terms is $5x^2$ and the product of the two last terms is 6. However, only in the last combination is the sum of the inner and outer products equal to $13x$. Therefore,

$$5x^2 + 13x + 6 = (5x + 3)(x + 2)$$

Check $(5x + 3)(x + 2) = 5x^2 + 13x + 6 = 13x + 5x^2 + 6$ ∎

When the leading coefficient or the coefficient of the third term is a prime number, it is probably easiest to insert the factors of that number first (see Example 12).

Example 12 Factor $4z^2 - 3z - 7$.

Solution The polynomial is in descending powers of z; there are no like terms and no common factors. The signs will be different. Because 7 is a prime number, we will fill in the factors of 7 first.

$$(\blacksquare z \; \blacksquare \; 7)(\blacksquare z \; \blacksquare \; 1) \longleftarrow \text{Partially filled-in outline}$$

Now we must insert the factors of 4 (either 1 and 4 or 2 and 2) and the signs so the sum of the inner and outer products is $-3z$. We must consider the following combinations:

$$(2z + 7)(2z - 1) \qquad (2z - 7)(2z + 1)$$
$$(1z + 7)(4z - 1) \qquad (1z - 7)(4z + 1)$$
$$(4z + 7)(1z - 1) \qquad (4z - 7)(1z + 1) \longleftarrow \begin{array}{l}\text{This is the only combination in}\\ \text{which the sum of the inner and}\\ \text{outer products is } -3z\end{array}$$

Therefore, $4z^2 - 3z - 7 = (4z - 7)(z + 1)$.
Check $(4z - 7)(z + 1) = 4z^2 - 3z - 7$ ∎

Example 13 Factor $5x^2 - x + 2$.

Solution These are the only combinations we need to consider:

$$(5x - 1)(x - 2)$$
$$(5x - 2)(x - 1)$$

In both of these combinations, the product of the first two terms is $5x^2$ and the product of the two last terms is $+2$. However, in neither one is the sum of the inner and outer products $-x$. Therefore, $5x^2 - x + 2$ is *not factorable*. ∎

Example 14 Factor $19xy - 15y^2 - 6x^2$.

Solution Because the coefficients of x^2 and y^2 are both negative, we factor out -1 before proceeding.

$$19xy - 15y^2 - 6x^2 = -6x^2 + 19xy - 15y^2 = -1(6x^2 - 19xy + 15y^2)$$

We now factor $6x^2 - 19xy + 15y^2$.

$$(\blacksquare x - \blacksquare y)(\blacksquare x - \blacksquare y) \longleftarrow \text{Partially filled-in outline}$$

with $\sqrt{x^2}$ and $\sqrt{y^2}$ indicated.

These are the combinations we must consider:

$$(6x - 15y)(1x - 1y) \qquad (6x - 1y)(1x - 15y)$$
$$(6x - 3y)(1x - 5y) \qquad (6x - 5y)(1x - 3y)$$
$$(2x - 15y)(3x - 1y) \qquad (2x - 1y)(3x - 15y)$$
$$(2x - 5y)(3x - 3y) \qquad (2x - 3y)(3x - 5y)$$

The last combination, $(2x - 3y)(3x - 5y)$, is the only one in which the sum of the inner and outer products is $-19xy$. Therefore,

We factored out -1 earlier

$$19xy - 15y^2 - 6x^2 = -(2x - 3y)(3x - 5y)$$

Other acceptable answers include $(3y - 2x)(3x - 5y)$ and $(2x - 3y)(5y - 3x)$.

Check

$$-(2x - 3y)(3x - 5y) = -(6x^2 - 19xy + 15y^2)$$
$$= 19xy - 15y^2 - 6x^2 \quad \blacksquare$$

Example 15 Factor $10u^2 - 12u + 8u^3$.

Solution We first arrange the trinomial in descending powers of u and factor out the GCF.

$$10u^2 - 12u + 8u^3 = 8u^3 + 10u^2 - 12u = 2u(4u^2 + 5u - 6)$$

We now try to factor the polynomial in the parentheses.

$$(\quad u \quad)(\quad u \quad) \longleftarrow \text{Partially filled-in outline}$$

The combinations to consider:

$(2u + 6)(2u - 1)$ $(2u - 6)(2u + 1)$

$(2u + 2)(2u - 3)$ $(2u - 2)(2u + 3)$

$(4u + 1)(1u - 6)$ $(4u - 1)(1u + 6)$

$(4u + 6)(1u - 1)$ $(4u - 6)(1u + 1)$

$(4u + 2)(1u - 3)$ $(4u - 2)(1u + 3)$

$(4u + 3)(1u - 2)$ $(4u - 3)(1u + 2)$ $\longleftarrow$ This is the only combination in which the sum of the inner and outer products is $+5u$

Therefore,

$$4u^2 + 5u - 6 = (u + 2)(4u - 3)$$
$$10u^2 - 12u + 8u^3 = 2u(u + 2)(4u - 3)$$

Check

$$2u(u + 2)(4u - 3) = 2u(4u^2 + 5u - 6)$$
$$= 8u^3 + 10u^2 - 12u$$
$$= 10u^2 - 12u + 8u^3 \quad \blacksquare$$

NOTE It is not necessary to write down all possible combinations as was done in Examples 11–15. ☑

If the original trinomial does not have a common factor, then neither of the *binomial factors* can have a common factor. This fact can help you eliminate some combinations from consideration (see Example 16).

Example 16 Factor $12a^2 + 7ab - 10b^2$.
Solution The outline:

$$(\ \ a\ \ \ \ b)(\ \ a\ \ \ \ b)$$

If we first try 3 and 4 as the factors of 12, and 2 and 5 as the factors of 10, we have these combinations to consider:

★(3a ± 5b)(4a − 2b)   ★ (3a − 5b)(4a ± 2b)

(3a ∓ 2b)(4a − 5b)   (3a − 2b)(4a + 5b) ◂── In this combination, the sum of the inner and outer products is $+7ab$

Therefore, $12a^2 + 7ab - 10b^2 = (3a - 2b)(4a + 5b)$.
Check $(3a - 2b)(4a + 5b) = 12a^2 + 7ab - 10b^2$

The starred combinations both have a common factor of 2 in one of the binomials. Since the original trinomial had no common factor, these combinations need not be given any consideration. ■

Sometimes, the first term is a constant and the last term contains the variable. In this case, we proceed in almost the same way.

Example 17 Factor $8 + 2x - x^2$.
Solution If we arranged this trinomial in descending powers of x, the first term would be *negative*. Therefore, we will leave it in ascending powers of x. The outline:

$$(\ \ \ \ x)(\ \ \ \ x)$$

If we first try 2 and 4 as the factors of 8, we have this combination:

$$(2\ \ x)(4\ \ x)$$

If we put the signs as follows:

$$(2 - x)(4 + x)$$

we do *not* get the correct middle term. Let's try $(2 + x)(4 - x)$. The sum of the inner and outer products *is* $+2x$. Therefore,

$$8 + 2x - x^2 = (2 + x)(4 - x)$$

Check $(2 + x)(4 - x) = 8 + 2x - x^2$ ■

When the first and third terms of the trinomial are perfect squares, the trinomial *may* be the square of a binomial (see Section 6.4B).

Example 18 Factor $9x^2 - 12xy + 4y^2$.
Solution Because $9x^2 = (3x)^2$ and $4y^2 = (2y)^2$, it is possible that $9x^2 - 12xy + 4y^2$ is the square of a binomial. We'll try

$$(3x - 2y)^2$$

└──────── − because middle term of the trinomial was −

$$(3x - 2y)^2 = 9x^2 - 12xy + 4y^2$$

Therefore, $9x^2 - 12xy + 4y^2 = (3x - 2y)^2$ or $(3x - 2y)(3x - 2y)$.

($9x^2 - 12xy - 4y^2$ could also have been factored by using the trial method.)
Check $(3x - 2y)^2 = 9x^2 - 12xy + 4y^2$ ■

EXERCISES 7.3B

Set I   Factor each expression completely, or write "Not factorable."

1. $3x^2 + 7x + 2$ **2.** $3x^2 + 5x + 2$

3. $5x^2 + 7x + 2$ **4.** $5x^2 + 11x + 2$

5. $7x + 4x^2 + 3$ **6.** $13x + 4x^2 + 3$

7. $5x^2 + 20x + 4$ **8.** $5x^2 + 11x + 4$

9. $5a^2 - 16a + 3$ **10.** $5m^2 - 8m + 3$

11. $3b^2 - 22b + 7$ **12.** $3u^2 - 10u + 7$

13. $5z^2 - 36z + 7$ **14.** $5z^2 - 12z + 7$

15. $14n - 5 + 3n^2$ **16.** $2k - 7 + 5k^2$

17. $9x^2 - 49$ **18.** $16y^2 - 1$

19. $7x^2 + 23xy + 6y^2$ **20.** $7a^2 + 43ab + 6b^2$

21. $7h^2 - 11hk + 4k^2$ **22.** $7h^2 - 16hk + 4k^2$

23. $3t^2 + 19tz - 6z^2$ **24.** $3w^2 - 11wx - 6x^2$

25. $49x^2 - 42x + 9$ **26.** $25x^2 - 20x + 4$

27. $18u^3 + 39u^2 - 15u$ **28.** $12y^3 + 14y^2 - 10y$

29. $4x^2 + 4x + 4$ **30.** $6x^2 + 6x + 18$

31. $7x^2 - 49$ **32.** $5x^2 - 25$

33. $6 - 17v + 5v^2$ **34.** $6 - 11v + 5v^2$

35. $20x + 3x^2 + 12$ **36.** $13x + 3x^2 + 12$

37. $45x^2 - 120x + 80$ **38.** $64x^2 - 96x + 36$

39. $8x^2 - 2x - 15$ **40.** $8x^2 - 19x - 15$

41. $6y^2 - 19y + 10$ **42.** $6y^2 - 23y + 10$

43. $9a^2 + 24a - 20$ **44.** $9b^2 + 3b - 20$

45. $6e^4 - 7e^2 - 20$ **46.** $10f^4 - 29f^2 - 21$

47. $x^2 + 144$ **48.** $y^2 + 9$

Set II   Factor each expression completely, or write "Not factorable."

1. $3y^2 + 16y + 5$ **2.** $2z^2 + 9z + 7$

3. $4a^2 + 9a + 5$ **4.** $3b^2 + b + 4$

5. $13x + 6x^2 + 2$ **6.** $2 + 6x^2 + 7x$

7. $2x^2 + 6x + 3$ **8.** $6x^2 + 15x + 9$

9. $7a^2 - 22a + 3$ **10.** $7a^2 - 20a - 3$

11. $2x^2 - 15x + 7$ **12.** $7 + 2x^2 + 9x$

13. $3t^2 - 16t + 5$ **14.** $3t^2 - 16t - 5$

15. $13x - 7 + 2x^2$ **16.** $9x + 2x^2 - 7$

17. $25x^2 - 16$ **18.** $16x^2 + 25$

19. $5x^2 + 13x + 6$ **20.** $10a^3 + 14a^2 - 12a$

21. $5x^2 - 17x + 6$

22. $5x^2 - 13x - 6$

23. $5x^2 + 30x + 6$

24. $5x^2 + 11x + 6$

25. $25s^2 - 30s + 9$

26. $9x^2 - 37x + 4$

27. $14m^3 - 20m^2 + 6m$

28. $3e^2 - 20e - 7$

29. $5s^2 - 5s + 5$

30. $5h^2 - 8h + 3$

31. $6x^2 - 36$

32. $36x^2 - 1$

33. $12 - 17u - 5u^2$

34. $8ab + 5b^2 + 3a^2$

35. $15x + 2x^2 + 18$

36. $7x^2 + 2x - 5$

37. $70k^2 - 24k + 2$

38. $6s^2 - 17st - 5t^2$

39. $12x^2 + 7x - 10$

40. $8x^2 - 6x - 9$

41. $5x^2 - 16x + 12$

42. $6t^2 - 5t - 21$

43. $4m^2 - 16m + 7$

44. $14x^2 - 4x - 10$

45. $8v^4 - 14v^2 - 15$

46. $8x^3 - 8x^2 + 8x$

47. $z^2 + 4$

48. $2x^2 - 72$

7.4 Review: 7.1–7.3

Greatest Common Factor
7.1

The greatest common factor (GCF) of two integers is the greatest integer that is a factor of both integers.

Methods of Factoring

7.1   1. Greatest common factor (GCF)

7.2   2. Difference of two squares:

$$a^2 - b^2 = (a + b)(a - b)$$

7.3A
7.3B   3. Trinomial $\begin{cases} \text{Leading coefficient equal to 1} \\ \text{Leading coefficient greater than 1} \end{cases}$

Review Exercises 7.4   Set I

Factor each expression completely, or write "Not factorable."

1. $8x - 4$

2. $5 + 7a$

3. $m^2 - 4$

4. $25 + n^2$

5. $x^2 + 10x + 21$

6. $y^2 - 10y + 15$

7. $2u^2 + 4u$

8. $3b - 6b^2 + 12b^3$

9. $z^2 - 7z - 18$

10. $7x - 30 - x^2$

11. $4x^2 - 25x + 6$

12. $4x^2 + 20x + 25$

13. $9k^2 - 144$

14. $100 + 81p^2$

15. $8 - 2a^2$

16. $10c^2 - 44c + 16$

17. $15u^2v - 3uv$

18. $5ab^2 - 10a^2b - 5ab$

19. $10x^2 - xy - 24y^2$

20. $8x^5y^2 - 12x^2y^4 - 16x^2y^2$

Review Exercises 7.4   Set II

NAME

Factor each expression completely, or write "Not factorable."

1. $6 - 30u$ **2.** $36 - m^2$

3. $8x + 5y$ **4.** $x^2 + 9x + 18$

5. $12v^2 - 8v$ **6.** $z^2 + 2z - 35$

7. $4e^2 - 27e + 18$ **8.** $x^2 - 8x + 8$

9. $64 - 49p^2$ **10.** $48 - 3a^2$

11. $16x^2 - 8x + 1$ **12.** $3x^2 + 6xy + 9y^2$

13. $3a^2b - 15ab^2 - 3ab$ **14.** $24hk^2 - 6hk$

15. $40x^2 + 10xy - 50y^2$ **16.** $21u^2 + 13uv - 20v^2$

ANSWERS

1. _____
2. _____
3. _____
4. _____
5. _____
6. _____
7. _____
8. _____
9. _____
10. _____
11. _____
12. _____
13. _____
14. _____
15. _____
16. _____

17. $8m^3n^2 - 14mn^4 - 6mn^2$

18. $4x^2 + 8x + 8$

19. $28x^2 - 13xy - 6y^2$

20. $21ab^2 - 7ab$

17. _____

18. _____

19. _____

20. _____

7.5 Factoring by Grouping

If a polynomial has four terms, we can sometimes factor it by first separating its terms into two groups and factoring each group separately. Since we will still have more than one term at this point, the expression will not yet be factored. However, if we then see that each of the groups has a common factor, we will be able to factor the polynomial.

The rules that follow assume that any factor common to all *four* terms has already been factored out.

TO FACTOR AN EXPRESSION OF FOUR TERMS BY GROUPING TWO AND TWO

1. Arrange the four terms into two groups of two terms each so that each group of two terms is factorable.

2. Factor each *group*. You will now have two terms. *The expression will not yet be factored.*

3. Factor the two-term expression resulting from step 2 if the two terms now have a common binomial factor.

4. If the two terms resulting from step 2 do *not* have a GCF, try a different arrangement of the original four terms.

Example 1   Factor $ax + ay + bx + by$.
Solution

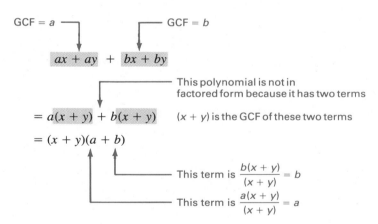

Therefore, $ax + ay + bx + by = (x + y)(a + b)$.
Check $(x + y)(a + b) = (x + y)a + (x + y)b = ax + ay + bx + by$ ■

It is often possible to group terms differently and still be able to factor the expression. *The same factors are obtained no matter what grouping is used,* because factorization over the integers is unique.

Example 2   Factor $ab - b + a - 1$.

One grouping	*A different grouping*
$ab - b$ + $a - 1$	$ab + a$ - $b - 1$
$= b(a - 1) + 1(a - 1)$	$= a(b + 1) - 1(b + 1)$
$= (a - 1)(b + 1)$	$= (b + 1)(a - 1)$

Same factors

Therefore, $ab - b + a - 1 = (a - 1)(b + 1) = (b + 1)(a - 1)$.
Check $(a - 1)(b + 1) = (a - 1)b + (a - 1)(1) = ab - b + a - 1$ ∎

Example 3   Factor $2x^2 - 6xy + 3x - 9y$.
Solution

GCF = 2x   GCF = 3

$2x^2 - 6xy$ + $3x - 9y$

$= 2x(x - 3y) + 3(x - 3y)$ ($x - 3y$) is the GCF

$= (x - 3y)(2x + 3)$

Therefore, $2x^2 - 6xy + 3x - 9y = (x - 3y)(2x + 3)$.
Check

$$(x - 3y)(2x + 3) = (x - 3y)(2x) + (x - 3y)(3)$$
$$= 2x^2 - 6xy + 3x - 9y$$ ∎

In Example 4, one group of two terms does not have a common factor. Instead, it is a difference of two squares.

Example 4   Factor $a^2 - b^2 + 3a - 3b$.
Solution

Difference of two squares
3 is a common factor

$a^2 - b^2 + 3a - 3b$

$a - b$ is the common binomial factor

$= (a + b)(a - b) + 3(a - b)$

$= (a - b)[(a + b) + 3]$

This term is $\dfrac{3(a - b)}{(a - b)}$

This term is $\dfrac{(a + b)(a - b)}{(a - b)}$

Therefore, $a^2 - b^2 + 3a - 3b = (a - b)(a + b + 3)$.
Check

$$(a - b)(a + b + 3) = (a - b)a + (a - b)b + (a - b)(3)$$
$$= a^2 - ab + ab - b^2 + 3a - 3b$$
$$= a^2 - b^2 + 3a - 3b$$ ∎

A WORD OF CAUTION   An expression is not factored until it has been written as a single term that is a product of factors. To illustrate this, consider Example 1 again.

$$ax + ay + bx + by$$

$$= \boxed{a(x + y)} + \boxed{b(x + y)}$$ This expression is not in factored form because it has two terms

First   Second
term   term

$$= \boxed{(x + y)(a + b)}$$ Factored form of $ax + ay + bx + by$

Single term

Expressions with more than four terms may also be factored by grouping; however, in this book we consider factoring only expressions of four terms by grouping.

EXERCISES 7.5

Set I   Factor each polynomial completely, or write "Not factorable."

1. $am + bm + an + bn$

2. $cu + cv + du + dv$

3. $mx - nx - my + ny$

4. $ah - ak - bh + bk$

5. $xy + x - y - 1$

6. $ad - d + a - 1$

7. $3a^2 - 6ab + 2a - 4b$

8. $2h^2 - 6hk + 5h - 15k$

9. $6e^2 - 2ef - 9e + 3f$

10. $8m^2 - 4mn - 6m + 3n$

11. $h^2 - k^2 + 2h + 2k$

12. $x^2 - y^2 + 4x + 4y$

13. $x^3 + 3x^2 - 4x + 12$

14. $a^3 + 5a^2 - 2a + 10$

15. $a^3 - 2a^2 - 4a + 8$

16. $x^3 - 3x^2 - 9x + 27$

17. $10xy - 15y + 8x - 12$

18. $35 - 42m - 18mn + 15n$

19. $a^2 - 4 + ab - 2b$

20. $x^2 - 25 - xy + 5y$

Set II   Factor each polynomial completely, or write "Not factorable."

1. $cz + dz + cw + dw$

2. $sx + sy + tx + ty$

3. $hw - kw - hz + kz$

4. $bx - by - cx + cy$

5. $ef + f - e - 1$

6. $ef + f - e + 1$

7. $2s^2 - 6st + 5s - 15t$

8. $6cd + 4c - 15de - 10e$

9. $6a^2 - 3ab - 14a + 7b$

10. $10x^2 - 5xy - 4x + 2y$

11. $x^2 - y^2 + 5x - 5y$

12. $s^2 - t^2 + 2s + 2t$

13. $y^3 - 2y^2 - 4y - 8$

14. $a^3 - 3a^2 - 5a + 15$

15. $u^3 - 3u^2 - 9u + 27$

16. $v^3 + 2v^2 - 25v - 50$

17. $12xy - 8y + 15x - 10$

18. $x^3 - 2x^2 + 9x - 18$

19. $x^2 - 9 + xy + 3y$

20. $y^2 - 16 - xy - 4x$

7.6 The Master Product Method for Factoring Trinomials (Optional)

The Master Product method, which makes use of factoring by grouping, is a method for factoring trinomials that eliminates some of the guesswork. It can also be used to determine whether a trinomial is factorable or not. The rules that follow assume that any common factors have been factored out.

TO FACTOR A TRINOMIAL BY THE MASTER PRODUCT METHOD

Arrange the trinomial in descending powers:

$$ax^2 + bx + c$$

1. Find the Master Product (MP) by multiplying the first and last coefficients of the trinomial being factored (MP $= a \cdot c$).

2. Write the pairs of factors of the Master Product (MP).

3. Choose the pair of factors whose sum is the coefficient of the middle term (b).

4. Rewrite the given trinomial, replacing the middle term by the sum of two terms whose coefficients are the pair of factors found in step 3.

5. Factor the expression in step 4 by grouping.

6. Check your factoring by multiplying the binomial factors to see if their product is the given trinomial.

Example 1   Factor $7x + 2x^2 + 5$ by the Master Product method.
Solution

$$7x + 2x^2 + 5 = 2x^2 + 7x + 5 \longleftarrow \text{Arranged in descending powers}$$

Master Product $= (2)(+5) = 10$; the middle coefficient is $+7$.

$$10 = (-1)(-10) = (1)(10)$$
$$= (-2)(-5) = \boxed{(2)(5)}$$

The pair whose sum is $+7$ is $(2)(5)$.

$$2x^2 + 7x + 5 = \underbrace{2x^2 + 2x}_{} + \underbrace{5x + 5}_{} \qquad \text{Factoring by grouping}$$
$$= 2x(x + 1) + 5(x + 1)$$
$$= (x + 1)(2x + 5)$$

Therefore, $2x^2 + 7x + 5 = (x + 1)(2x + 5)$.
Check $(x + 1)(2x + 5) = 2x^2 + 7x + 5 = 7x + 2x^2 + 5$ ∎

Example 2   Factor $4z^2 - 3z - 7$ by the Master Product method.
Solution

Master Product $= (4)(-7) = -28$; middle coefficient is -3.

$$-28 = (-1)(28) = (1)(-28)$$
$$= (-2)(14) = (2)(-14)$$
$$= (-4)(7) = \boxed{(4)(-7)} \longleftarrow \text{This is the pair whose sum is } -3$$

$$4z^2 - 3z - 7 = \underbrace{4z^2 + 4z}_{} - \underbrace{7z - 7}_{} \qquad \text{Factoring by grouping}$$

$$= 4z(z + 1) - 7(z + 1)$$

$$= (z + 1)(4z - 7)$$

Therefore, $4z^2 - 3z - 7 = (z + 1)(4z - 7)$.
Check $(z + 1)(4z - 7) = 4z^2 - 3z - 7$ ∎

Example 3   Factor $12a^2 + 5a - 10$.
Solution Master Product $= (12)(-10) = -120$; middle coefficient is $+5$.

$$-120 = (-1)(120) = (1)(-120)$$
$$= (-2)(60) = (2)(-60)$$
$$= (-3)(40) = (3)(-40)$$
$$= (-4)(30) = (4)(-30)$$
$$= (-5)(24) = (5)(-24)$$
$$= (-6)(20) = (6)(-20)$$
$$= (-8)(15) = (8)(-15)$$
$$= (-10)(12) = (10)(-12)$$

None of the sums of these pairs is $+5$. Therefore, the trinomial is *not factorable*. ∎

NOTE   The Master Product method can also be used with trinomials whose leading coefficient is 1; however, we think the method presented in Section 7.3A is shorter and simpler for trinomials of that type. ☑

EXERCISES 7.6   Factor each polynomial in Exercises 7.3B by the Master Product method.

7.7 How to Select the Method of Factoring

The following procedure can be used to select the correct method for factoring a polynomial:

First, check for a common factor, no matter how many terms the expression has. If there is a common factor, factor it out. (Section 7.1)

If the expression to be factored has two terms:

1. Is it a difference of two squares? (Section 7.2B)

2. Is it a sum of two squares? (If so, it is *not* factorable.)

If the expression to be factored has three terms:

1. Is the leading coefficient 1? (Section 7.3A)

2. Is the leading coefficient unequal to 1? (Section 7.3B or 7.6)

3. Is the trinomial the square of a binomial? (Section 7.3B)

If the expression to be factored has four terms, can it be factored by grouping? (Section 7.5)

Check to see if any factor can be factored again. When the expression is completely factored, the same factors are obtained no matter what method is used.

Check the result by multiplying the factors together.

Example 1   Factor $27x^2 - 12y^2$.
Solution

$$27x^2 - 12y^2 \qquad \text{3 is the GCF}$$

This factor can be factored again

$$= 3(\boxed{9x^2 - 4y^2}) \qquad (9x^2 - 4y^2) = (3x + 2y)(3x - 2y)$$

$$= 3(3x + 2y)(3x - 2y)$$

Check $3(3x + 2y)(3x - 2y) = 3(9x^2 - 4y^2) = 27x^2 - 12y^2$ ∎

Example 2   Factor $3x^3 - 27x + 5x^2 - 45$.
Solution

$$\underbrace{3x^3 - 27x} + \underbrace{5x^2 - 45} \qquad \text{Factor by grouping}$$

$$= 3x(x^2 - 9) + 5(x^2 - 9)$$

This factor can be factored again

$$= (\boxed{x^2 - 9})(3x + 5) \qquad x^2 - 9 = (x + 3)(x - 3)$$

$$= (x + 3)(x - 3)(3x + 5)$$

Check

$$(x + 3)(x - 3)(3x + 5) = (x^2 - 9)(3x + 5)$$
$$= (x^2 - 9)(3x) + (x^2 - 9)(5)$$
$$= 3x^3 - 27x + 5x^2 - 45 \quad ∎$$

Example 3   Factor $5x^2 + 10x - 40$.
Solution

$$5x^2 + 10x - 40 \qquad \text{5 is the GCF}$$

This factor can be factored again

$$= 5(\boxed{x^2 + 2x - 8}) \qquad (x^2 + 2x - 8) = (x - 2)(x + 4)$$

$$= 5(x - 2)(x + 4)$$

Check $5(x - 2)(x + 4) = 5(x^2 + 2x - 8) = 5x^2 + 10x - 40$ ∎

Example 4   Factor $a^4 - b^4$.
Solution

$$a^4 - b^4$$

This factor can be factored again

$$= (a^2 + b^2)(\,a^2 - b^2\,) \qquad (a^2 - b^2) = (a + b)(a - b)$$

$$= (a^2 + b^2)(a + b)(a - b)$$

Check $(a^2 + b^2)(a + b)(a - b) = (a^2 + b^2)(a^2 - b^2) = a^4 - b^4$ ∎

EXERCISES 7.7

Set I   Factor each polynomial completely, or write "Not factorable."

1. $2x^2 - 8y^2$
2. $3x^2 - 27y^2$
3. $5a^4 - 20b^2$
4. $6m^2 - 54n^4$
5. $x^4 - y^4$
6. $a^4 - 16$
7. $4v^2 + 14v - 8$
8. $6v^2 - 27v - 15$
9. $8z^2 - 12z - 8$
10. $18z^2 - 21z - 9$
11. $4x^2 - 100$
12. $9x^2 - 36$
13. $12x^2 + 10x - 8$
14. $45x^2 - 6x - 24$
15. $ab^2 - 2ab + a$
16. $au^2 - 2au + a$
17. $x^4 - 81$
18. $16y^8 - z^4$
19. $16x^2 + 16$
20. $25b^2 + 100$
21. $2u^3 + 2u^2v - 12uv^2$
22. $3m^3 - 3m^2n - 36mn^2$
23. $8h^3 - 20h^2k + 12hk^2$
24. $15h^2k - 35hk^2 + 10k^3$
25. $a^5b^2 - 4a^3b^4$
26. $x^2y^4 - 100x^4y^2$
27. $2ax^2 - 8a^3y^2$
28. $3b^2x^4 - 12b^2y^2$
29. $12 + 4x - 3x^2 - x^3$
30. $45 - 9z - 5z^2 + z^3$
31. $6my - 4nz + 15mz - 5zn$
32. $10xy + 5mn - 6xy - nm$
33. $x^4 - 8x^2 + 16$
34. $y^4 - 18y^2 + 81$
35. $x^8 - 1$
36. $a^8 - b^8$
37. $x^2 - a^2 - 4x + 4a$
38. $m^2 - 25 + mn - 5n$
39. $6ac - 6bd + 6bc - 6ad$
40. $10cy - 6cz + 5dy - 3dz$

Set II   Factor each polynomial completely, or write "Not factorable."

1. $3a^2 - 75b^2$
2. $7c^2 - 63b^2$
3. $4h^4 - 36b^2$
4. $9x^4 - 36y^2$
5. $x^4 - 16y^4$
6. $m^4 - 1$
7. $10x^2 + 25x - 15$
8. $6x^2 + 15x - 9$
9. $10y^2 + 14y - 12$
10. $6x^2 + 6x + 12$

11. $16x^2 - 36$ **12.** $25a^2 - 100$

13. $30w^2 + 27w - 21$ **14.** $8x^2 + 22x - 6$

15. $h^2k - 4hk + 4k$ **16.** $x^2y + 4xy + 4$

17. $81c^4 - 16$ **18.** $81c^2 + 16$

19. $5x^2 + 20$ **20.** $4m^3n^3 - mn^5$

21. $5wz^2 + 5w^2z - 10w^3$ **22.** $2t^2r^4 - 18t^4$

23. $12x^2y - 42xy^2 + 36y^3$ **24.** $xy + 3y - 4x - 12$

25. $6x^3y^2 - 12xy^4$ **26.** $5x^4y + 20x^2y^3$

27. $3bx^2 - 12b^3y^2$ **28.** $9x^2 + 36$

29. $45 + 9b - 5b^2 - b^3$ **30.** $12 + 4x - 3x^2 - x^3$

31. $3xy + 2xz - 8xw + 3xz$ **32.** $8wx + 5xy - 4yz - 11yz$

33. $x^4 - 2x^2 + 1$ **34.** $x^3 + 3x^2 - 25x - 75$

35. $y^8 - 1$ **36.** $x^2 + 1$

37. $a^2 - 4b^2 + 2a + 4b$ **38.** $10ac + 10ad - 5bc - 5bd$

39. $6ef + 3gf - 12eh - 9gh$ **40.** $x^2 - 9y^2 + 2x - 6y$

7.8 Solving Equations by Factoring

Factoring has many applications. In this section, we use factoring to solve equations.

A *polynomial equation* is an equation that has a polynomial on both sides of the equal sign; the polynomial on one side of the equal sign can be the zero polynomial, 0. The *degree* of the equation equals the degree of the highest-degree *term* in the equation.

Polynomial equations with a first-degree term as the highest-degree term are called *first-degree* or *linear equations*. (All the equations that were solved in Chapter 4 were first-degree equations.) Polynomial equations with a second-degree term as the highest-degree term are called *second-degree* or *quadratic equations*.

Example 1   Examples of polynomial equations:

a. $5x - 3 = 0$ Linear (or first-degree) equation in one variable

b. $2x^2 - 4x + 7 = 0$ Quadratic (or second-degree) equation in one variable   ■

We are now ready to solve quadratic and higher-degree equations. The method of solving such equations is based on Rule 7.1, which is stated without proof.

RULE 7.1

If the product of two factors is zero, then one or both of the factors must be zero.

$$\text{If } a \cdot b = 0, \text{ then } \begin{cases} a = 0 \\ \text{or } b = 0 \\ \text{or both } a \text{ and } b = 0 \end{cases}$$

Rule 7.1 can be extended to include more than two factors; that is, if a product of factors is zero, at least one of the factors must be zero. We use Rule 7.1 in solving higher-degree equations. The method is summarized below.

TO SOLVE AN EQUATION BY FACTORING

1. Write all nonzero terms on one side of the equation by adding the same expression to both sides. *Only zero must remain on the other side.* Then arrange the polynomial in descending powers.

2. Factor the polynomial completely.

3. Set each factor equal to zero.*

4. Solve each resulting first-degree equation.

5. Check apparent solutions in the original equation.

Example 2 Solve $(x - 1)(x - 2) = 0$.
Solution We already have 0 on one side of the equal sign, and the polynomial has already been factored. We proceed with step 3.

$$\text{Since } (x - 1)(x - 2) = 0,$$
$$\text{then } (x - 1) = 0 \; or \; (x - 2) = 0.$$

If	$x - 1 =$	0		If	$x - 2 =$	0
	$+ 1$	$+ 1$			$+ 2$	$+ 2$
then x		$= 1$		then x		$= 2$

Therefore, 1 and 2 are solutions for the equation $(x - 1)(x - 2) = 0$.

Check for x = 1	*Check for x = 2*
$(x - 1)(x - 2) = 0$	$(x - 1)(x - 2) = 0$
$(1 - 1)(1 - 2) \stackrel{?}{=} 0$	$(2 - 1)(2 - 2) \stackrel{?}{=} 0$
$(0)(- 1) \stackrel{?}{=} 0$	$(1)(0) \stackrel{?}{=} 0$
$0 = 0$	$0 = 0$ ∎

Example 3 Solve $3(x + 2)(x - 1) = 0$.
Solution $3(x + 2)(x - 1) = 0$

$3 \neq 0$	$x + 2 =$	0	$x - 1 =$	0
	$- 2$	$- 2$	$+ 1$	$+ 1$
	x	$= - 2$	x	$= 1$

We leave the checking of these solutions to the student. ∎

The same method can be used when a product of more than two factors is equal to zero.

---

*If any of the factors are not first-degree polynomials, we cannot solve the equation at this time.

Example 4 Solve $2x(x - 3)(x + 4) = 0$.
Solution $2x(x - 3)(x + 4) = 0$

$$
\begin{array}{c|c|c}
2x = 0 & x - 3 = 0 & x + 4 = 0 \\
& \underline{+3 \quad +3} & \underline{-4 \quad -4} \\
\dfrac{2x}{2} = \dfrac{0}{2} & & \\
& & \\
x = 0 & x \quad = 3 & x \quad = -4
\end{array}
$$

We leave the checking of these solutions to the student. ■

Example 5 Solve $x^2 - 9x = 0$.
Solution We have 0 on one side of the equal sign. We proceed with step 2.

$$x^2 - 9x = 0$$
$$x(x - 9) = 0$$

$$
\begin{array}{c|c}
x = 0 & x - 9 = 0 \\
& \underline{9 \quad 9} \\
& x \quad = 9
\end{array}
$$

Check for $x = 0$	*Check for $x = 9$*
$x^2 - 9x = 0$	$x^2 - 9x = 0$
$0^2 - 9(0) \overset{?}{=} 0$	$9^2 - 9(9) \overset{?}{=} 0$
$0 - 0 \overset{?}{=} 0$	$81 - 81 \overset{?}{=} 0$
$0 = 0$	$0 = 0$ ■

Example 6 Solve $6x^2 = 5 - 7x$.
Solution Adding $7x - 5$ to both sides and arranging the polynomial in descending powers of x, we have

$$6x^2 + 7x - 5 = 0$$

Factoring: $(2x - 1)(3x + 5) = 0$

$$
\begin{array}{c|c}
2x - 1 = 0 & 3x + 5 = 0 \\
\underline{+1 \quad +1} & \underline{-5 \quad -5} \\
2x \quad = 1 & 3x \quad = -5
\end{array}
$$

Dividing both sides by 2 $\quad x = \dfrac{1}{2}$ $\qquad$ $x = -\dfrac{5}{3}$ $\quad$ Dividing both sides by 3

Check for $x = \dfrac{1}{2}$	*Check for $x = -\dfrac{5}{3}$*
$6x^2 = 5 - 7x$	$6x^2 = 5 - 7x$
$6\left(\dfrac{1}{2}\right)^2 \overset{?}{=} 5 - 7\left(\dfrac{1}{2}\right)$	$6\left(-\dfrac{5}{3}\right)^2 \overset{?}{=} 5 - 7\left(-\dfrac{5}{3}\right)$
$6\left(\dfrac{1}{4}\right) \overset{?}{=} 5 - \dfrac{7}{2}$	$6\left(\dfrac{25}{9}\right) \overset{?}{=} 5 + \dfrac{35}{3}$
$\dfrac{6}{4} \overset{?}{=} \dfrac{10 - 7}{2}$	$\dfrac{2 \cdot 25}{3} \overset{?}{=} \dfrac{15 + 35}{3}$
$\dfrac{3}{2} = \dfrac{3}{2}$	$\dfrac{50}{3} = \dfrac{50}{3}$ ■

A WORD OF CAUTION The product must equal zero, or no conclusions can be drawn about the factors.

Suppose $(x - 1)(x - 3) = \boxed{8}$.

⬑ No conclusion can be drawn because the product $\neq 0$

Students sometimes think that

$$\text{if} \qquad (x - 1)(x - 3) = 8$$
$$\text{then } x - 1 = 8 \qquad x - 3 = 8$$
$$x = 9 \qquad x = 11$$

This "solution" is incorrect, because

if $x = 9$	*or*	if $x = 11$

$$\text{then } (x - 1)(x - 3) \qquad\qquad \text{then } (x - 1)(x - 3)$$
$$(9 - 1)(9 - 3) = \qquad\qquad (11 - 1)(11 - 3) =$$
$$8 \cdot 6 = 48 \neq 8 \qquad\qquad 10 \cdot 8 = 80 \neq 8$$

The correct solution is

$$(x - 1)(x - 3) = 8$$
$$x^2 - 4x + 3 = 8$$
$$x^2 - 4x - 5 = 0$$
$$(x - 5)(x + 1) = 0$$

$$
\begin{array}{c|c}
x - 5 = 0 & x + 1 = 0 \\
\underline{+5 \quad +5} & \underline{-1 \quad -1} \\
x = 5 & x = -1
\end{array}
$$

☑

Example 7 Solve $(x - 5)(x + 4) = -14$.

Solution

$$(x - 5)(x + 4) = -14$$
$$x^2 - x - 20 = -14 \qquad \text{Removing parentheses}$$
$$x^2 - x - 6 = 0 \qquad \text{Adding 14 to both sides}$$
$$(x + 2)(x - 3) = 0$$

$$
\begin{array}{c|c}
x + 2 = 0 & x - 3 = 0 \\
\underline{-2 \quad -2} & \underline{+3 \quad +3} \\
x = -2 & x = 3
\end{array}
$$

Check for $x = -2$	*Check for $x = 3$*

$$(x - 5)(x + 4) = -14 \qquad\qquad (x - 5)(x + 4) = -14$$
$$(-2 - 5)(-2 + 4) \stackrel{?}{=} -14 \qquad\qquad (3 - 5)(3 + 4) \stackrel{?}{=} -14$$
$$(-7)(2) \stackrel{?}{=} -14 \qquad\qquad (-2)(7) \stackrel{?}{=} -14$$
$$-14 = -14 \qquad\qquad\qquad -14 = -14 \quad ■$$

Example 8   Solve $3x^3 = 4x - x^2$.
Solution

$$3x^3 = 4x - x^2$$
$$3x^3 + x^2 - 4x = 0 \quad \text{Adding } x^2 - 4x \text{ to both sides}$$
$$x(3x^2 + x - 4) = 0 \quad \text{Factoring out the GCF}$$
$$x(x - 1)(3x + 4) = 0 \quad \text{Factoring the polynomial factor}$$

$$x = 0 \quad \bigg| \quad \begin{array}{c} x - 1 = 0 \\ \underline{+1 \quad +1} \\ x = 1 \end{array} \quad \bigg| \quad \begin{array}{c} 3x + 4 = 0 \\ \underline{-4 \quad -4} \\ 3x = -4 \\ x = -\dfrac{4}{3} \end{array}$$

Check for $x = 0$

$$3x^3 = 4x - x^2$$
$$3(0)^3 \overset{?}{=} 4(0) - 0^2$$
$$3(0) \overset{?}{=} 0 - 0$$
$$0 = 0$$

Check for $x = 1$

$$3x^3 = 4x - x^2$$
$$3(1)^3 \overset{?}{=} 4(1) - 1^2$$
$$3(1) \overset{?}{=} 4 - 1$$
$$3 = 3$$

Check for $x = -\dfrac{4}{3}$

$$3x^3 = 4x - x^2$$
$$3\left(-\dfrac{4}{3}\right)^3 \overset{?}{=} 4\left(-\dfrac{4}{3}\right) - \left(-\dfrac{4}{3}\right)^2$$
$$3\left(-\dfrac{64}{27}\right) \overset{?}{=} -\dfrac{16}{3} - \dfrac{16}{9}$$
$$-\dfrac{64}{9} \overset{?}{=} -\dfrac{48}{9} - \dfrac{16}{9}$$
$$-\dfrac{64}{9} = -\dfrac{64}{9} \quad \blacksquare$$

EXERCISES 7.8

Set I   Solve each of the following equations.

1. $(x - 5)(x + 4) = 0$ **2.** $(x + 7)(x - 2) = 0$
3. $3x(x - 4) = 0$ **4.** $5x(x + 6) = 0$
5. $(x + 10)(2x - 3) = 0$ **6.** $(x - 8)(3x + 2) = 0$
7. $x^2 + 9x + 8 = 0$ **8.** $x^2 + 6x + 8 = 0$
9. $x^2 - x - 12 = 0$ **10.** $x^2 + x - 12 = 0$
11. $x^2 = 64$ **12.** $x^2 = 144$
13. $6x^2 - 10x = 0$ **14.** $6y^2 - 21y = 0$
15. $24w = 4w^2$ **16.** $20m = 5m^2$
17. $5a^2 = 16a - 3$ **18.** $3z^2 = 22z - 7$
19. $3u^2 = 2u + 5$ **20.** $5k^2 = 34k + 7$
21. $(x - 2)(x - 3) = 2$ **22.** $(x - 3)(x - 5) = 3$
23. $x(x - 4) = 12$ **24.** $x(x - 2) = 15$
25. $4x(2x - 1)(3x + 7) = 0$ **26.** $5x(4x - 3)(7x - 6) = 0$

27. $2x^3 + x^2 = 3x$

28. $4x^3 = 10x - 18x^2$

29. $2a^3 - 10a^2 = 0$

30. $4b^3 - 24b^2 = 0$

Set II   Solve each of the following equations.

1. $(x + 3)(x - 5) = 0$

2. $(y - 8)(y + 9) = 0$

3. $2y(y - 7) = 0$

4. $7x(x + 4) = 0$

5. $(z - 6)(3z + 2) = 0$

6. $(5x - 10)(4x + 5) = 0$

7. $a^2 + 8a + 12 = 0$

8. $x^2 + 2x = 48$

9. $m^2 + 3m - 18 = 0$

10. $2x^2 + 11x + 15 = 0$

11. $w^2 - 24 = 5w$

12. $x^2 = 4x + 5$

13. $5h^2 - 20h = 0$

14. $k^3 = 5k^2$

15. $12t = 6t^2$

16. $x = 10 - 2x^2$

17. $3n^2 = 7n + 6$

18. $(x - 4)(x + 2) = -9$

19. $13x + 3 = -4x^2$

20. $y(3y - 2)(y + 5) = 0$

21. $(y - 3)(y - 6) = -2$

22. $10x^2 + 11x = 6$

23. $u(u - 9) = -14$

24. $(x - 2)(x + 4) = 7$

25. $2x(3x - 2)(5x + 9) = 0$

26. $8y(2y^2 + 5y - 3) = 0$

27. $21x^2 + 60x = 18x^3$

28. $12y^2 = 4y^3 + 5y$

29. $3a^2 + 18a = 0$

30. $(x - 5)(x + 6) = -10$

7.9 Word Problems Solved by Factoring

Many word problems leads to equations that can be solved by factoring. Several examples of such problems follow.

Example 1   The difference of two numbers is 3. Their product is 10. What are the two numbers?

$$\left. \begin{array}{l} \text{Let} \quad x = \text{smaller number} \\ \quad x + 3 = \text{larger number} \end{array} \right\} \text{Since their difference is 3}$$

$$\underbrace{\text{Their product}}\quad \underbrace{\text{is}}\quad \underbrace{10}$$

$$x(x + 3) \quad = \quad 10$$

$$x^2 + 3x = \quad 10 \qquad \text{Remember, it is incorrect to say that}$$
$$\qquad\qquad\qquad\qquad\qquad x = 10 \text{ or } x + 3 = 10$$

$$x^2 + 3x - 10 = \quad 0$$

$$(x - 2)(x + 5) = \quad 0$$

$x - 2 = 0$	$x + 5 = 0$
$x = 2$	$x = -5$ Smaller number
$x + 3 = 5$	$x + 3 = -2$ Larger number

Therefore the numbers 2 and 5 are a solution, and the numbers -5 and -2 are another solution.

Check for 2 and 5 The difference is $5 - 2 = 3$. The product is $5 \cdot 2 = 10$.

Check for -5 and -2 The difference is $-2 - (-5) = -2 + 5 = 3$. The product is $(-2)(-5) = 10$. ∎

Example 2 Find two consecutive integers whose product is 19 more than their sum.

Solution   Let $x =$ the smaller integer

$$x + 1 = \text{the larger integer}$$

Then $x(x + 1)$ represents their product, and $x + (x + 1)$ represents their sum.

Their product	is	19	more than	their sum

$$x(x + 1) = 19 + x + (x + 1)$$
$$x^2 + x = 19 + 2x + 1$$
$$x^2 - x - 20 = 0$$
$$(x + 4)(x - 5) = 0$$

$x + 4 = 0$	$x - 5 = 0$
$x = -4$	$x = 5$
$x + 1 = -3$	$x + 1 = 6$

There are two answers: the numbers are -4 and -3, or the numbers are 5 and 6.
Check for -4 and -3 Their sum is -7. Their product is 12. 12 is 19 more than -7.
Check for 5 and 6 Their sum is 11. Their product is 30. 30 is 19 more than 11. ∎

Example 3 Find three consecutive odd integers such that the product of the first two is 21 more than 6 times the third.

Solution   Let $x =$ the first odd integer

$$x + 2 = \text{the second odd integer}$$
$$x + 4 = \text{the third odd integer}$$

The product of the first two numbers	is	21	more than	6 times the third

$$x(x + 2) = 21 + 6(x + 4)$$
$$x^2 + 2x = 21 + 6x + 24$$
$$x^2 + 2x = 45 + 6x$$
$$x^2 - 4x - 45 = 0$$
$$(x + 5)(x - 9) = 0$$

$x + 5 = 0$		$x - 9 = 0$	
$x = -5$	The first integer	$x = 9$	The first integer
$x + 2 = -3$	The second	$x + 2 = 11$	The second
$x + 4 = -1$	The third	$x + 4 = 13$	The third

Check for $x = -5$	*Check for $x = 9$*
$(-5)(-3) \stackrel{?}{=} 21 + 6(-1)$	$(9)(11) \stackrel{?}{=} 21 + 6(13)$
$15 \stackrel{?}{=} 21 - 6$	$99 \stackrel{?}{=} 21 + 78$
$15 = 15$	$99 = 99$ ∎

In solving word problems about geometric figures, make a drawing of the figure and write the given information on it. Lengths of sides of geometric figures cannot be negative; therefore, any value of the variable that would make a length negative must be rejected.

Example 4   The base of a triangle is 4 cm more than its altitude. The area is 30 sq. cm. Find the altitude and the base. (30 sq. cm $= 30$ cm^2)

Solution   Let $h =$ altitude (in centimeters)

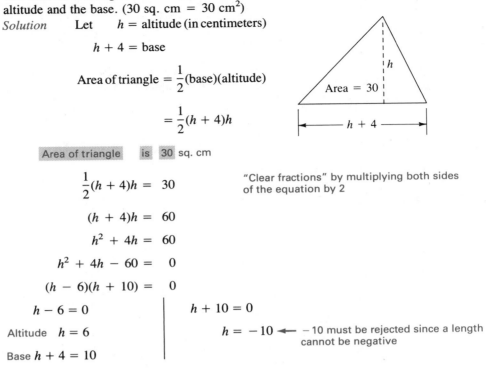

$$h + 4 = \text{base}$$

$$\text{Area of triangle} = \frac{1}{2}(\text{base})(\text{altitude})$$

$$= \frac{1}{2}(h + 4)h$$

Area of triangle   is   30 sq. cm

$$\frac{1}{2}(h + 4)h = 30$$

"Clear fractions" by multiplying both sides of the equation by 2

$$(h + 4)h = 60$$

$$h^2 + 4h = 60$$

$$h^2 + 4h - 60 = 0$$

$$(h - 6)(h + 10) = 0$$

$$h - 6 = 0 \qquad\qquad h + 10 = 0$$

Altitude $h = 6$ $h = -10$ ← −10 must be rejected since a length cannot be negative

Base $h + 4 = 10$

Therefore, the triangle has an altitude of 6 cm and a base of 10 cm.

Check The base is 4 cm more than the altitude. The area is $\frac{1}{2}(6\,\text{cm})(10\,\text{cm}) = 30$ cm^2. ■

Example 5   One square has a side 3 ft longer than the side of a second square. If the area of the larger square is 4 times as great as the area of the smaller square, find the length of the side of each square.

Solution   Let $x =$ length of side of smaller square

Area = x^2 x

Area = $(x + 3)^2$ $x + 3$

The area of the larger square	is	4 times as great as	the area of the smaller square

$$(x + 3)^2 = 4 \cdot x^2$$

$$(x + 3)^2 = 4x^2$$

$$x^2 + 6x + 9 = 4x^2$$

$$-3x^2 + 6x + 9 = 0$$

$$-3(x^2 - 2x - 3) = 0$$

$$-3(x - 3)(x + 1) = 0$$

$$-3 \neq 0 \;\mid\; x - 3 = 0 \qquad x + 1 = 0$$

Small square $x = 3$ $x = -1$ ← −1 cannot be a solution of the word statement

Large square $x + 3 = 6$

Therefore, the smaller square has a side of 3 ft and the larger square a side of 6 ft.
Check Area of smaller square is $(3 \text{ ft})^2 = 9 \text{ ft}^2$. Area of larger square is $(6 \text{ ft})^2 = 36 \text{ ft}^2$. The area of the larger square (36 sq. ft) is 4 times as great as the area of the smaller square (9 sq. ft). ■

Example 6 The width of a rectangle is 5 cm less than its length. Its area is 10 more (numerically*) than its perimeter. What are the dimensions of the rectangle?

Solution

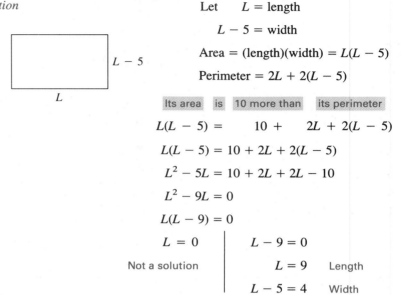

$$\text{Let} \quad L = \text{length}$$
$$L - 5 = \text{width}$$
$$\text{Area} = (\text{length})(\text{width}) = L(L - 5)$$
$$\text{Perimeter} = 2L + 2(L - 5)$$

Its area	is	10 more than	its perimeter

$$L(L - 5) = \quad 10 + \quad 2L + 2(L - 5)$$
$$L(L - 5) = 10 + 2L + 2(L - 5)$$
$$L^2 - 5L = 10 + 2L + 2L - 10$$
$$L^2 - 9L = 0$$
$$L(L - 9) = 0$$

$L = 0$ | $L - 9 = 0$
Not a solution | $L = 9$ Length
  | $L - 5 = 4$ Width

Therefore, the rectangle has a length of 9 cm and a width of 4 cm.
Check The width is 5 cm less than the length. The area is $(4 \text{ cm})(9 \text{ cm}) = 36 \text{ cm}^2$. The perimeter is $2(9 \text{ cm}) + 2(4 \text{ cm}) = 18 \text{ cm} + 8 \text{ cm} = 26 \text{ cm}$. 36 is 10 more than 26. ■

Example 7 The length of a rectangular solid is 2 cm more than its width. The height is 3 cm, and the volume is 72 cc (cubic centimeters). Find the width and the length.

Solution

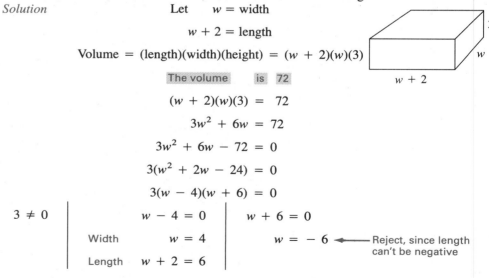

$$\text{Let} \quad w = \text{width}$$
$$w + 2 = \text{length}$$
$$\text{Volume} = (\text{length})(\text{width})(\text{height}) = (w + 2)(w)(3)$$

The volume	is	72

$$(w + 2)(w)(3) = \quad 72$$
$$3w^2 + 6w = 72$$
$$3w^2 + 6w - 72 = 0$$
$$3(w^2 + 2w - 24) = 0$$
$$3(w - 4)(w + 6) = 0$$

$3 \neq 0$ | $w - 4 = 0$ | $w + 6 = 0$
  | Width $w = 4$ | $w = -6$ ◄——— Reject, since length can't be negative
  | Length $w + 2 = 6$ |

Therefore, the width is 4 cm and the height is 6 cm.
Check The volume is $(6 \text{ cm})(4 \text{ cm})(3 \text{ cm}) = 72 \text{ cm}^3$. ■

---

*We say that the area is "numerically" 10 more than the perimeter because the area is measured in *square centimeters* whereas the perimeter is measured in *centimeters*.

EXERCISES 7.9

Set I Set up each problem algebraically and solve. Be sure to state what your variables represent.

1. The difference of two numbers is 5. Their product is 14. Find the numbers.

2. The difference of two numbers is 6. Their product is 27. Find the numbers.

3. The sum of two numbers is 12. Their product is 35. Find the numbers.

4. The sum of two numbers is −4. Their product is −12. Find the numbers.

5. The base of a triangle is 3 in. longer than the altitude. The area is 20 sq. in. Find the altitude and the base.

6. The base of a triangle is 3 cm longer than the altitude. The area is 90 sq. cm. Find the altitude and the base.

7. Find three consecutive integers such that the product of the first two plus the product of the last two is 8.

8. Find three consecutive integers such that the product of the first two plus the product of the first and third is 14.

9. Find three consecutive even integers such that twice the product of the first two is 16 more than the product of the last two.

10. Find three consecutive odd integers such that twice the product of the last two is 91 more than the product of the first two.

11. The length of a rectangle is 5 ft more than its width. Its area is 84 sq. ft. What are its dimensions?

12. The width of a rectangle is 3 ft less than its length. Its area is 28 sq. ft. What are its dimensions?

13. One square has a side 3 cm shorter than the side of a second square. The area of the larger square is 4 times as great as the area of the smaller square. Find the length of the side of each square.

14. One square has a side 4 ft longer than the side of a second square. The area of the larger square is 9 times as great as the area of the smaller square. Find the length of the side of each square.

15. The width of a rectangle is 4 yd less than its length. The area is 17 more (numerically) than its perimeter. What are the dimensions of the rectangle?

16. The area of a square is twice its perimeter (numerically). What is the length of its side?

17. The base of a triangle is 3 in. more than its altitude. Its area is 35 sq. in. Find the base and the altitude.

18. The base of a triangle is 5 m more than its altitude. Its area is 18 sq. m. Find the base and the altitude.

19. The sum of the base and the altitude of a triangle is 19 in. The area is 42 sq. in. Find the base and the altitude.

20. The sum of the base and the altitude of a triangle is 15 cm. The area of the triangle is 27 sq. cm. Find the base and the altitude.

21. The length of a rectangular solid is 4 cm more than its height. Its width is 5 cm, and its volume is 225 cc. Find its length and its height.

22. The length of a rectangular solid is 2 in. more than its width. Its height is 6 in. and its volume is 378 cu. in. Find its length and its width.

Set II Set up each problem algebraically and solve. Be sure to state what your variables represent.

1. The difference of two numbers is 12. Their product is 28. Find the numbers.

2. The difference of two numbers is 19. Their product is −84. Find the numbers.

3. The sum of two numbers is 10. Their product is −24. Find the numbers.

4. Find two consecutive integers whose product is 1 less than their sum.

5. The base of a triangle is 4 m longer than the altitude. The area is 48 sq. m. Find the altitude and the base.

6. The base of a triangle is 2 cm shorter than the altitude. The area is 40 sq. cm. Find the altitude and the base.

7. Find three consecutive integers such that the product of the first two minus the third is 7.

8. Find three consecutive integers such that the product of the first and third minus the second is 41.

9. Find three consecutive odd integers such that twice the product of the first two is 7 more than the product of the last two.

10. A 3-in.-wide mat surrounds a picture. The area of the picture itself is 176 sq. in. If the length of the outside of the mat is twice its width, what are the dimensions of the outside of the mat? What are the dimensions of the picture?

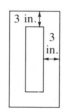

11. The length of a rectangle is 8 m more than its width. Its area is 48 sq. m. What are its dimensions?

12. The width of a rectangular box equals the length of the side of a certain cube. The length of the box is 3 cm more than its width, and the height of the box is 1 cm less than its width. The volume of the box is 9 cc more than the volume of the cube. Find the dimensions of the cube and of the box.

13. One square has a side 2 km shorter than the side of a second square. The area of the larger square is 9 times as great as the area of the smaller square. Find the length of the side of each square.

14. Bruce's vegetable garden is now square. If he forms a rectangle by increasing the length of one side by 3 ft and the length of the adjacent side by 6 ft, the area of the rectangle will be 3 times as great as the area of the square. What is the size of the vegetable garden now?

15. The width of a rectangle is 2 in. less than its length. The area is 4 more (numerically) than its perimeter. What are the dimensions of the rectangle?

16. A pieced quilt 5 ft by 6 ft is to be surrounded by a border of uniform width. How wide should the border be if the area of the border is to be 4 sq. ft less than the area of the pieced quilt?

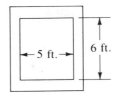

17. The base of a triangle is 5 cm more than its altitude. Its area is 7 sq. cm. What is the length of the altitude?

18. The difference of the base and altitude of a triangle is 4 ft. (The altitude is greater than the base.) The area of the triangle is 16 sq. ft. Find the base and the altitude.

19. The sum of the base and the altitude of a triangle is 17 m. The area of the triangle is 36 sq. m. Find the base and the altitude.

20. The length of a rectangular solid is 8 cm. Its width is 2 cm more than its height, and its volume is 280 cc. Find its height and its width.

21. The length of a rectangular solid is 5 cm more than its height. Its width is 3 cm, and its volume is 18 cc. Find its length and its height.

22. The length of a rectangular solid is 8 in. more than its width. Its height is 2 in. and its volume is 96 cu. in. Find its length and its width.

7.10 Review: 7.5–7.9

**Factoring by Grouping
7.5**

If a polynomial has four terms, it can sometimes be factored by grouping.

**Master Product Method
7.6**

The Master Product method can be used to factor a trinomial with a leading coefficient greater than 1.

**Factoring Completely
7.7**

1. Look for a greatest common factor first.

2. Look for a difference of two squares.

3. Look for a factorable trinomial.

4. Look for four terms that can be factored by grouping.

5. Check to see if any factor already obtained can be factored again.

**To Solve an Equation
by Factoring
7.8**

1. Write all nonzero terms on one side of the equation by adding the same expression to both sides. Only zero must remain on the other side. Then arrange the polynomial in descending powers.

2. Factor the polynomial.

3. Set each factor equal to zero, and solve for the variable.

4. Check apparent solutions in the *original* equation.

Review Exercises 7.10   Set I

In Exercises 1–14, factor the expression completely, or write "Not factorable."

1. $5 + 15a$

2. $50 - 2n^2$

3. $ab + 2b - a - 2$

4. $y^3 + 10y^2 + 16y$

5. $y^2 + 5y + 5$

6. $3b - 6b^2$

7. $5c^2 - 22c + 8$

8. $mn - 5n - m + 5$

9. $5x^2 - 35x - 150$

10. $x^2 + 81$

11. $x^3 + 5x^2 + 3x + 15$

12. $x^3 + 3x^2 - 9x - 27$

13. $x^2 - y^2 + x - y$

14. $15a^2 + 15ab - 30b^2$

In Exercises 15–24, solve each equation.

15. $(x - 5)(x + 3) = 0$

16. $x^2 - 5x - 24 = 0$

17. $m^2 = 18 + 3m$

18. $x^2 - 36 = 0$

19. $3z^2 = 12z$

20. $12u^2 = 47u - 45$

21. $3x^2 + 13x = 10$

22. $10x^2 + 13x = 3$

23. $2u(u + 6)(u - 2) = 0$

24. $39y^2 = 18y^3 + 18y$

In Exercises 25–30, set up each problem algebraically and solve. Be sure to state what your variables represent.

25. The difference of two numbers is 3. Their product is 28. Find the numbers.

26. One side of a square is 6 ft longer than the side of a second square. The area of the larger square is 16 times as great as the area of the smaller square. Find the length of the side of each square.

27. The width of a rectangle is 3 m less than its length. Its area is 40 sq. m. Find the length and width of the rectangle.

28. The length of a rectangle is 6 yd more than its width. The area is 12 more (numerically) than its perimeter. Find the dimensions of the rectangle.

29. Find two consecutive integers whose product is 1 more than their sum.

30. Find three consecutive odd integers such that 4 times the product of the first two is 3 less than the product of the last two.

Review Exercises 7.10   Set II

NAME _____

In Exercises 1–14, factor the expression completely, or write "Not factorable."

1. $8x + 10$

2. $6x^2 - 54$

3. $8x^2 + 32$

4. $4x^2y - 8xy^2 + 4xy$

5. $st - 3t - s + 3$

6. $4x^2 + 5x - 6$

7. $9 + x^2$

8. $w^2 - z^2 - w - z$

9. $x^3 + 2x^2 - 16x - 32$

10. $25x - 15 + 10x^2$

11. $a^2 - 3a + 3$

12. $6y^2 + 12y + 6$

13. $a^2 - b^2 - a + b$

14. $5x^2 + 10xy + 15$

In Exercises 15–24, solve each equation.

15. $(x + 8)(x - 6) = 0$

16. $x^2 - x - 20 = 0$

17. $z^2 = 4z + 21$

18. $m^2 - 81 = 0$

19. $5t^2 = 40t$

20. $12m^2 = 10 - 7m$

ANSWERS

1. _____

2. _____

3. _____

4. _____

5. _____

6. _____

7. _____

8. _____

9. _____

10. _____

11. _____

12. _____

13. _____

14. _____

15. _____

16. _____

17. _____

18. _____

19. _____

20. _____

21. $5h(h - 11)(h + 3) = 0$

22. $30x^3 = 87x^2 + 63x$

23. $14x^2 = 26x + 4$

24. $32x^2 + 96x + 72 = 0$

In Exercises 25–30, set up each problem algebraically and solve. Be sure to state what your variables represent.

25. The difference of two numbers is 6. Their product is 72. Find the numbers.

26. One side of a square is 8 cm longer than the side of a second square. The area of the larger square is 9 times as great as the area of the smaller square. Find the length of the side of each square.

27. The length of a rectangle is 7 yd more than its width. The area is 4 more (numerically) than its perimeter. Find the dimensions of the rectangle.

28. Find two consecutive integers whose product is 10 less than 4 times their sum.

29. Find three consecutive even integers such that twice the product of the first two is 10 more than the last integer.

30. The width of a rectangular solid is 3 cm more than its height. Its length is 18 cm and its volume is 324 cc. Find its height and its width.

21. _____

22. _____

23. _____

24. _____

25. _____

_____

26. _____

_____

27. _____

_____

28. _____

_____

29. _____

_____

_____

30. _____

_____

Chapter 7 Diagnostic Test

The purpose of this test is to see how well you understand factoring. We recommend that you work this diagnostic test *before* your instructor tests you on this chapter. Allow yourself about 50 minutes.

Complete solutions for all the problems on this test, together with section references, are given in the answer section in the back of this book. For the problems you do incorrectly, study the sections referred to.

In Problems 1–12, factor the polynomial completely, or write "Not factorable."

1. $8x + 12$

2. $5x^3 - 35x^2$

3. $25x^2 - 121y^2$

4. $8x^2 + 7$

5. $5a^2 - 180$

6. $z^2 + 9z + 8$

7. $m^2 + 5m - 6$

8. $11x^2 - 18x + 7$

9. $x^2 + 7x - 6$

10. $4y^2 + 19y - 5$

11. $5n - mn - 5 + m$

12. $6h^2k - 8hk^2 + 2k^3$

In Problems 13–16, solve each equation.

13. $3x^2 - 12x = 0$

14. $x^2 + 20 = 12x$

15. $3x^3 = x^2 + 10x$

16. $(x - 7)(x + 6) = -22$

In Problems 17 and 18, set up each problem algebraically and solve. Be sure to state what your variables represent.

17. Find three consecutive odd integers such that twice the product of the first two minus the product of the first and third is 49.

18. The length of a rectangle is 3 ft more than its width. Its area is 28 sq. ft. Find its width and its length.

Cumulative Review Exercises: Chapters 1–7

1. Evaluate $24 \div 2\sqrt{16} - 3^2 \cdot 5$.

2. Solve $5(2b - 4) = 8(3b + 5) - 4(9 + 6b)$.

3. Evaluate the formula using the values of the variables given with the formula.

$$C = \frac{a}{a + 12} \cdot A \qquad a = 8, \ A = 35$$

In Exercises 4 and 5, simplify each expression and write your answers using only positive exponents.

4. $(3y^{-3}z^2)^{-2}$

5. $\left(\dfrac{15x^2}{10x^3}\right)^3$

6. Write in scientific notation:

   a. 57,300,000

   b. 0.00351

In Exercises 7–10, perform the indicated operations and simplify.

7. $(2x^2 + 5x - 3) - (-4x^2 + 8x + 10) + (6x^2 + 3x - 8)$

8. $(y - 4)(3y^2 - 2y + 5)$

9. $\dfrac{12a^2 - 3a}{3a}$

10. $(8x^2 - 2x - 17) \div (2x - 3)$

In Exercises 11–16, factor each polynomial completely, or write "Not factorable."

11. $36x^2 - 18x$

12. $1 - 36t^2$

13. $x^2 - 8x + 15$

14. $7x^2 - x - 5$

15. $5k^2 - 34k - 7$

16. $3n^2 - 2n - 5$

In Exercises 17 and 18, set up the problem algebraically and solve.

17. How many pounds of cashews at \$7.50 per pound should be mixed with 20 lb of peanuts at \$3.50 per pound if the mixture is to be worth \$5.90 per pound?

18. The length and width of a rectangle are in the ratio of 3 to 2. The area is 150 sq. m. Find the width and the length.

8 Fractions

In this chapter, we define algebraic fractions and discuss how to perform necessary operations with them and how to solve equations and word problems involving them. A knowledge of the different methods of factoring discussed in Chapters 7 is essential to your work with algebraic fractions.

8.1 Algebraic Fractions

A simple **algebraic fraction** (also called a "rational expression") is an algebraic expression of the form $\frac{P}{Q}$ where P and Q are polynomials. We call P and Q the **terms** of the fraction. We call P the **numerator** and Q the **denominator** of the fraction.

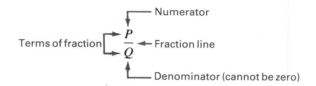

Excluded Values

Any value of the variable or variables that makes the denominator, Q, equal to zero *must be excluded* (see Section 1.6). We find excluded values by setting the denominator equal to zero and solving the resulting equation. The solutions of that equation are the values of the variable that must be excluded.

Example 1   Find the excluded values for each of these algebraic fractions:

a. $\frac{x}{5}$ If we set 5 equal to 0, we get a false statement; the equation has no solution. Therefore, no values of the variable must be excluded.

b. $\frac{7}{x}$ We set x equal to 0 and solve: $x = 0$. Therefore, 0 is an excluded value.

c. $\frac{4x + 1}{x + 2}$ We set $x + 2$ equal to 0 and solve:

$$x + 2 = 0$$
$$x = -2$$

Therefore, -2 is an excluded value.

d. $\frac{x^2 + 1}{x^2 + 3x - 4}$ We set $x^2 + 3x - 4$ equal to 0 and solve:

$$x^2 + 3x - 4 = 0$$
$$(x + 4)(x - 1) = 0$$

| $x + 4 = 0$ | $x - 1 = 0$ |
| $x = -4$ | $x = 1$ |

Therefore, -4 and 1 are excluded values.

e. $\frac{1}{2}$ No value of the variable is excluded. (If we set 2 equal to 0, we get a false statement.) Arithmetic fractions are also algebraic fractions; 1 and 2 are each polynomials of degree 0. ∎

NOTE   After this section, whenever a fraction is written, it is understood that the value(s) of the variable(s) that make the denominator zero are excluded.   ☑

A WORD OF CAUTION   If you are accustomed to writing fractions with a *slanted bar* (/), be sure to put parentheses around any numerator or denominator that contains more than one term.

$$(x - 5)/(x + 3) = \frac{x - 5}{x + 3}$$

but

$$x - 5/x + 3 = x - \frac{5}{x} + 3$$

Therefore, $(x - 5)/(x + 3) \neq x - 5/x + 3$.   ☑

Equivalent Fractions

Equivalent fractions are fractions that have the same value. In algebra, as in arithmetic, we get a fraction equivalent to the one we started with if we:

1. Multiply both numerator and denominator by the same nonzero number

2. Divide both numerator and denominator by the same nonzero number

A WORD OF CAUTION   We do *not* get a fraction equivalent to the one we started with if we *add* the same number to or *subtract* the same number from both the numerator and the denominator.   ☑

Example 2   Determine whether the pairs of fractions are equivalent:

a. $\dfrac{1}{2}, \dfrac{1 + x}{2 + x}$ No; we can't get the second fraction from the first by multiplying or dividing 1 and 2 both by the same number.

b. $\dfrac{3}{x}, \dfrac{6}{2x}$ Yes; if we multiply both 3 and x by 2, we get 6 and $2x$.

c. $\dfrac{8x}{12y}, \dfrac{2x}{3y}$ Yes; if we divide both $8x$ and $12y$ by 4, we get $2x$ and $3y$.

d. $\dfrac{3 + x}{4 + x}, \dfrac{3}{4}$ No; we can't get the second fraction from the first by multiplying or dividing both $(3 + x)$ and $(4 + x)$ by the same number.   ∎

The Three Signs of a Fraction

Every fraction has three signs associated with it, even if those signs are not visible: the sign of the fraction, the sign of the numerator, and the sign of the denominator. Consider the fraction $\frac{8}{4}$:

Let's compare three other fractions with $+\dfrac{+8}{+4}$:

$$+\,\frac{-8}{-4} = +\left(\frac{-8}{-4}\right) = +(+2) = 2$$

Sign of numerator and sign of denominator are different from $+\dfrac{+8}{+4}$

$$-\,\frac{-8}{+4} = -\left(\frac{-8}{+4}\right) = -(-2) = 2$$

Sign of fraction and sign of numerator are different from $+\dfrac{+8}{+4}$

$$-\,\frac{+8}{-4} = -\left(\frac{+8}{-4}\right) = -(-2) = 2$$

Sign of fraction and sign of denominator are different from $+\dfrac{+8}{+4}$

Because each of these fractions equals the same number, the fractions must all equal each other. Therefore,

$$+\,\frac{+8}{+4} = +\,\frac{-8}{-4} = -\,\frac{-8}{+4} = -\,\frac{+8}{-4} = 2$$

It can also be shown that

$$-\,\frac{+8}{+4} = +\,\frac{-8}{+4} = +\,\frac{+8}{-4} = -\,\frac{-8}{-4} = -2$$

RULE OF SIGNS FOR FRACTIONS

If any *two* of the three signs of a fraction are changed, the value of the fraction is unchanged. Fractions obtained in this way are *equivalent fractions*.

This rule of signs is helpful in reducing some fractions and in performing operations (adding, multiplying, and so on) on some fractions.

Example 3   Find the missing term:

a. $-\dfrac{-3x}{2y} = \dfrac{?}{2y}$

Because the signs of the *denominators* are the same in both fractions (they are understood to be $+$) and the signs of the *fractions* are different, the signs of the *numerators* must be different.

Therefore, $-\dfrac{-3x}{2y} = \dfrac{3x}{2y}$. The missing term is $3x$.

b. $\dfrac{x}{-5} = \dfrac{-x}{?}$

Because the signs of the *fractions* are the same in both fractions (they are understood to be $+$) and the signs of the *numerators* are different, the signs of the *denominators* must be different.

Therefore, $\dfrac{x}{-5} = \dfrac{-x}{5}$. The missing term is 5. ∎

Recall from Section 3.5 that

$$-(x - y) = -1(x - y) = -x + y = y - x$$

Therefore, $y - x$ can always be substituted for $-(x - y)$, and $-(x - y)$ can always be substituted for $y - x$. We use these facts in Example 4.

Example 4 Find the missing term:

a. $-\dfrac{1}{2 - x} = \dfrac{1}{?}$

The signs of the numerators are the same (both are understood to be $+$); the signs of the fractions are different. The signs of the denominators must be different. Therefore, the new denominator must be $-(2 - x)$. If we then substitute $x - 2$ for $-(2 - x)$, we have

 ┌─ This step need not be shown

$$-\frac{1}{2 - x} = \frac{1}{-(2 - x)} = \frac{1}{x - 2}$$

Therefore, the missing term is $x - 2$.

b. $\dfrac{a - 1}{-2} = \dfrac{?}{2}$

The signs of the fractions are both understood to be $+$; the signs of the denominators are different. The signs of the numerators must be different. Therefore,

 ┌─ This step need not be shown

$$\frac{a - 1}{-2} = \frac{-(a - 1)}{2} = \frac{1 - a}{2}$$

Therefore, the missing term is $1 - a$.

c. $\dfrac{x - y}{(u + v)(a - b)} = \dfrac{?}{(u + v)(b - a)}$

The signs of the fractions are both understood to be $+$. The signs of the denominators are different because $b - a$ is the negative of $a - b$. The signs of the numerators must be different. Therefore,

 ┌─ This step need not be shown

$$\frac{x - y}{(u + v)(a - b)} = \frac{-(x - y)}{(u + v)(b - a)} = \frac{y - x}{(u + v)(b - a)}$$

Therefore, the missing term is $y - x$. ∎

EXERCISES 8.1

Set I In Exercises 1–8, determine what value (or values) of the variable must be excluded.

1. $\dfrac{3x + 4}{x - 2}$

2. $\dfrac{5 - 4x}{x + 3}$

3. $\dfrac{x}{10}$

4. $\dfrac{y}{20}$

5. $\dfrac{x - 4}{3x^2 - 6x}$

6. $\dfrac{3x + 2}{4x^2 - 12x}$

7. $\dfrac{3 + x}{x^2 - x - 2}$

8. $\dfrac{x - 5}{x^2 + x - 12}$

In Exercises 9–14, determine whether the pairs of fractions are equivalent.

9. $\dfrac{x}{2y}, \dfrac{5x}{10y}$

10. $\dfrac{a}{7b}, \dfrac{2a}{14b}$

11. $\dfrac{x}{2y}, \dfrac{x+5}{2y+5}$

12. $\dfrac{3c}{5d}, \dfrac{3c+8}{5d+8}$

13. $\dfrac{6(x+1)}{12(3x-2)}, \dfrac{x+1}{2(3x-2)}$

14. $\dfrac{9(x+5)}{27(2x-7)}, \dfrac{x+5}{3(2x-7)}$

In Exercises 15–26, find the missing term.

15. $-\dfrac{5}{6} = \dfrac{?}{6}$

16. $-\dfrac{8}{9} = \dfrac{?}{9}$

17. $\dfrac{5}{-y} = \dfrac{-5}{?}$

18. $\dfrac{2}{-x} = \dfrac{?}{x}$

19. $\dfrac{2-x}{-9} = \dfrac{?}{9}$

20. $\dfrac{5-y}{-2} = \dfrac{?}{2}$

21. $\dfrac{6-y}{5} = \dfrac{y-6}{?}$

22. $\dfrac{8-x}{7} = \dfrac{x-8}{?}$

23. $-\dfrac{4}{x-5} = \dfrac{4}{?}$

24. $-\dfrac{3}{x-4} = \dfrac{3}{?}$

25. $\dfrac{a-b}{(c+d)(5-x)} = \dfrac{?}{(c+d)(x-5)}$

26. $\dfrac{x-y}{(u+v)(c-2)} = \dfrac{?}{(u+v)(2-c)}$

Set II   In Exercises 1–8, determine what value (or values) of the variable must be excluded.

1. $\dfrac{6x+4}{x-7}$

2. $\dfrac{x+8}{x}$

3. $\dfrac{x}{7}$

4. $\dfrac{7}{x}$

5. $\dfrac{x-7}{5x^2-10x}$

6. $\dfrac{5x-3}{x^2-2x-24}$

7. $\dfrac{8+x}{x^2-6x-27}$

8. $\dfrac{x+2}{x^2+6x+9}$

In Exercises 9–14, determine whether the pairs of fractions are equivalent.

9. $\dfrac{a}{5b}, \dfrac{10a}{50b}$

10. $\dfrac{x}{3y}, \dfrac{x+4}{3y+4}$

11. $\dfrac{x}{4y}, \dfrac{x+2}{4y+2}$

12. $\dfrac{7x}{3y}, \dfrac{7x(3+z)}{3y(3+z)}$

13. $\dfrac{5(2x-7)}{15(2x-5)}, \dfrac{2x-7}{3(2x-5)}$

14. $\dfrac{4(x+3)}{16(2x-5)}, \dfrac{4x+3}{32x-5}$

In Exercises 15–26, find the missing term.

15. $-\dfrac{4}{3} = \dfrac{?}{3}$

16. $-\dfrac{4}{3} = \dfrac{?}{-3}$

17. $\dfrac{8}{-b} = \dfrac{-8}{?}$

18. $\dfrac{9}{-a} = \dfrac{?}{a}$

19. $\dfrac{1-y}{-5} = \dfrac{?}{5}$

20. $\dfrac{5-y}{-2} = \dfrac{?}{2}$

21. $\dfrac{8-a}{3} = \dfrac{a-8}{?}$

22. $\dfrac{5-x}{a-b} = \dfrac{x-5}{?}$

23. $-\dfrac{7}{y-2} = \dfrac{7}{?}$

24. $-\dfrac{a-2}{x-3} = \dfrac{2-a}{?}$

25. $\dfrac{u-v}{(a+3)(7-x)} = \dfrac{?}{(a+3)(x-7)}$

26. $\dfrac{3-a}{(x+1)(y-3)} = \dfrac{?}{(3-y)(1+x)}$

8.2 Reducing Fractions to Lowest Terms

We reduce fractions to lowest terms in algebra for the same reason we do in arithmetic—it makes them simpler and easier to work with. After this section, it is understood that all fractions are to be reduced to lowest terms unless otherwise indicated.

The general rule we use for reducing fractions is

$$\frac{ak}{bk} = \frac{a}{b}, \text{ where } k \neq 0 \text{ and } b \neq 0$$

This means, of course, that if we divide both the numerator and the denominator of a fraction by the same nonzero number, the new fraction is equivalent to the original one (see Section 8.1).

In Arithmetic   We can reduce the fraction $\frac{6}{8}$ as follows:

$$\frac{\overset{3}{\cancel{6}}}{\underset{4}{\cancel{8}}} = \frac{3}{4}$$

This is possible only when the numerator and denominator have a common *factor.* For this example,

$$\frac{6}{8} = \frac{2 \cdot 3}{2 \cdot 4} = \frac{\overset{1}{\cancel{2}} \cdot 3}{\underset{1}{\cancel{2}} \cdot 4} = \frac{3}{4}$$

— 2 is a factor of the numerator

— 2 is a factor of the denominator

Both numerator and denominator were divided by 2

In Algebra   The procedure for reducing fractions in algebra is the same as in arithmetic.

TO REDUCE A FRACTION TO LOWEST TERMS

1. Factor the numerator and the denominator completely.

2. Divide numerator and denominator by all factors common to both.

The new fraction will, of course, be equivalent to the original fraction.

Example 1   Reduce to lowest terms:

a. $\dfrac{4x^2y}{2xy} = \dfrac{\overset{2}{\cancel{4}}}{\cancel{2}} \cdot \dfrac{x^2}{x} \cdot \dfrac{y}{y} = 2 \cdot x \cdot 1 = 2x$

   Here, the literal parts of the fraction are already factored; the common factors are 2, x, and y

b. $\dfrac{15ab^2c^3}{6a^4bc^2} = \dfrac{\overset{5}{\cancel{15}}}{\underset{2}{\cancel{6}}} \cdot \dfrac{a}{a^4} \cdot \dfrac{b^2}{b} \cdot \dfrac{c^3}{c^2} = \dfrac{5}{2} \cdot \dfrac{1}{a^3} \cdot \dfrac{b}{1} \cdot \dfrac{c}{1} = \dfrac{5bc}{2a^3}$

c. $\dfrac{x-3}{x^2-9} = \dfrac{\overset{1}{\cancel{(x-3)}}}{(x+3)\underset{1}{\cancel{(x-3)}}} = \dfrac{1}{x+3}$

d. $\dfrac{x^2-4x-5}{x^2+5x+4} = \dfrac{\overset{1}{\cancel{(x+1)}}(x-5)}{\underset{1}{\cancel{(x+1)}}(x+4)} = \dfrac{x-5}{x+4}$

e. $\dfrac{3x^2-5xy-2y^2}{6x^3y+2x^2y^2} = \dfrac{(x-2y)\overset{1}{\cancel{(3x+y)}}}{2x^2y\underset{1}{\cancel{(3x+y)}}} = \dfrac{x-2y}{2x^2y}$

f. $\dfrac{x-y}{y-x} = \dfrac{\boxed{-}}{\boxed{-}\,(y-x)}\cdot\dfrac{x-y}{} = -\dfrac{\overset{1}{\cancel{(x-y)}}}{\underset{1}{\cancel{(x-y)}}} = -1$

   Changing sign of fraction and denominator

g. $\dfrac{2b^2+ab-3a^2}{4a^2-9ab+5b^2} = \dfrac{(b-a)(2b+3a)}{(a-b)(4a-5b)}$

   $(-1)(b-a) = (a-b)$

   $= \dfrac{\boxed{(-1)(b-a)}\,(2b+3a)}{(-1)(a-b)(4a-5b)} = \dfrac{\overset{1}{\cancel{(a-b)}}\,(2b+3a)}{(-1)\underset{1}{\cancel{(a-b)}}(4a-5b)}$

   Changing the sign of both numerator and denominator is equivalent to multiplying each by -1

   $= \dfrac{(2b+3a)}{(-1)(4a-5b)}$

   $= \dfrac{2b+3a}{5b-4a}$

   or $\dfrac{(b-a)(2b+3a)}{(a-b)(4a-5b)} = \dfrac{(b-a)(2b+3a)}{-(b-a)(4a-5b)}$

   Substituting $-(b-a)$ for $a-b$

   $= \dfrac{2b+3a}{-(4a-5b)}$

   $= \dfrac{2b+3a}{5b-4a}$

h. $\dfrac{z}{2z} = \dfrac{\overset{1}{\cancel{z}}}{2\underset{1}{\cancel{z}}} = \dfrac{1}{2}$

NOTE   A factor of 1 will always remain in the numerator and denominator after they have been divided by factors common to both.   ☑

A WORD OF CAUTION   A common error made in reducing fractions is to forget that the number the numerator and denominator are divided by must be a factor of *both* (see Examples 1i and 1j).

3 is *not* a factor of the numerator

$$\frac{\cancel{3} + 2}{\cancel{3}} \ne 2 \qquad \text{Incorrect reduction}$$

The above reduction is incorrect because

$$\frac{3 + 2}{3} = \frac{5}{3} \ne 2 \qquad\qquad \boxed{\checkmark}$$

i. $\dfrac{x + 3}{x + 6}$

This fraction cannot be reduced; since neither x nor 3 is a *factor* of the numerator or the denominator.

j. x is not a factor of the numerator

$$\frac{x + y}{x} \qquad \text{This fraction cannot be reduced.} \quad \blacksquare$$

EXERCISES 8.2

Set I   Express each fraction in lowest terms.

1. $\dfrac{9}{12}$

2. $\dfrac{8}{14}$

3. $\dfrac{6ab^2}{3ab}$

4. $\dfrac{10m^2n}{5mn}$

5. $\dfrac{4x^2y}{2xy}$

6. $\dfrac{12x^3y}{4xy}$

7. $\dfrac{5x - 10}{x - 2}$

8. $\dfrac{3x + 12}{x + 4}$

9. $-\dfrac{5x - 6}{6 - 5x}$

10. $\dfrac{4 - 3z}{3z - 4}$

11. $\dfrac{5x^2 + 30x}{10x^2 - 40x}$

12. $\dfrac{4x^3 - 4x^2}{12x^2 - 12x}$

13. $\dfrac{2 + 4}{4}$

14. $\dfrac{3 + 9}{3}$

15. $\dfrac{5 + x}{5}$

16. $\dfrac{x + 8}{8}$

17. $\dfrac{x^2 - 1}{x + 1}$

18. $\dfrac{x^2 - 4}{x - 2}$

19. $\dfrac{6x^2 - x - 2}{10x^2 + 3x - 1}$

20. $\dfrac{8x^2 - 10x - 3}{12x^2 + 11x + 2}$

21. $\dfrac{x^2 - y^2}{(x + y)^2}$

22. $\dfrac{a^2 - 9b^2}{(a - 3b)^2}$

23. $\dfrac{2y^2 + xy - 6x^2}{3x^2 + xy - 2y^2}$

24. $\dfrac{10y^2 + 11xy - 6x^2}{4x^2 - 4xy - 15y^2}$

25. $\dfrac{8x^2 - 2y^2}{2ax - ay + 2bx - by}$

26. $\dfrac{3x^2 - 12y^2}{ax + 2by + 2ay + bx}$

27. $\dfrac{(-1)(z-8)}{8-z}$

28. $\dfrac{x-12}{(-1)(12-x)}$

29. $\dfrac{(-1)(a-2b)(b-a)}{(2a+b)(a-b)}$

30. $\dfrac{8(n-2m)}{(-1)(3n+m)(2m-n)}$

31. $\dfrac{9-16x^2}{16x^2-24x+9}$

32. $\dfrac{25-9x^2}{9x^2-30x+25}$

33. $\dfrac{10+x-3x^2}{2x^2+x-10}$

34. $\dfrac{12-19x+5x^2}{3x^2-5x-12}$

35. $\dfrac{18-3x-3x^2}{6x^2+6x-36}$

36. $\dfrac{16+4x-2x^2}{8x^2-16x-64}$

Set II Express each fraction in lowest terms.

1. $\dfrac{12}{16}$

2. $\dfrac{10}{2x}$

3. $\dfrac{8mn^3}{4n^2}$

4. $\dfrac{15x^3y}{10xy^2}$

5. $\dfrac{9x^2y}{3xy}$

6. $\dfrac{2ab}{8a^2b^3}$

7. $\dfrac{8x-12}{2x-3}$

8. $\dfrac{8x+5}{5+8x}$

9. $-\dfrac{7x-1}{1-7x}$

10. $-\dfrac{3x+4}{4+3x}$

11. $\dfrac{6x^2+9x}{12x^2+9x}$

12. $\dfrac{x^2-9}{3x+9}$

13. $\dfrac{5+10}{10}$

14. $\dfrac{5+10x}{10}$

15. $\dfrac{7+x}{7}$

16. $\dfrac{21+3x}{3}$

17. $\dfrac{x^2-16}{x+4}$

18. $\dfrac{x-1}{x^2-1}$

19. $\dfrac{4x^2-9x+2}{4x^2+7x-2}$

20. $\dfrac{x^2-9}{x^2+5x+6}$

21. $\dfrac{x^2-25}{(x+5)^2}$

22. $\dfrac{x^2-16}{x^2-x-12}$

23. $\dfrac{3y^2+5xy-2x^2}{3x^2-8xy-3y^2}$

24. $\dfrac{x^2+x-20}{x^2+2x-15}$

25. $\dfrac{8x^2-18y^2}{2ax-3ay+2bx-3by}$

26. $\dfrac{x^2-11x+30}{x^2-9x+20}$

27. $\dfrac{(-1)(a-5)}{5-a}$

28. $\dfrac{12a^3b+6a^2b^2}{18a^2b^2+9ab^3}$

29. $\dfrac{(-1)(x-y)(3y-x)}{(3y+x)(y-x)}$

30. $\dfrac{15m^2n^2-15mn^3}{10m^2-10mn}$

31. $\dfrac{4-25x^2}{(5x-2)^2}$

32. $\dfrac{16x^2-y^2}{y^2-8xy+16x^2}$

33. $\dfrac{8+10y-12y^2}{2y^2-3y-20}$

34. $\dfrac{60a^2-110a+30}{15a+5a^2-10a^3}$

35. $\dfrac{36-3x-3x^2}{9x^2+9x-108}$

36. $\dfrac{6x^3-54x}{108x^2-9x^3-9x^4}$

8.3 Multiplying and Dividing Fractions

Multiplying Fractions

Multiplication of fractions is defined as follows:

$$\frac{a}{b} \cdot \frac{c}{d} = \frac{a \cdot c}{b \cdot d}$$

In practice, however, we can often reduce the resulting fraction. Therefore, we use the following rules for multiplying fractions:

TO MULTIPLY FRACTIONS:

1. Factor the numerators and denominators of the fractions.

2. Divide the numerators and denominators by all factors common to both. (The common factors do not have to be in the same fraction.)

3. The answer is the product of the factors remaining in the numerator divided by the product of the factors remaining in the denominator. A factor of 1 will always remain in both numerator and denominator.

We can write the product as a single fraction *before* we divide by the common factors (see Examples 1 and 2a–c) or *after* we divide by the common factors (see Examples 2d–e).

Example 1

$$\frac{4}{9} \cdot \frac{3}{8} = \frac{\overset{1}{\cancel{4}} \cdot \overset{1}{\cancel{3}}}{\underset{3}{\cancel{9}} \cdot \underset{2}{\cancel{8}}} = \frac{1}{6} \quad \blacksquare$$

Example 2   Multiply the fractions:

a. $\dfrac{1}{m^2} \cdot \dfrac{m}{5} = \dfrac{1 \cdot \overset{1}{\cancel{m}}}{\underset{m}{\cancel{m^2}} \cdot 5} = \dfrac{1}{5m}$

b. $\dfrac{2y^3}{3x^2} \cdot \dfrac{12x}{5y^2} = \dfrac{2y^3 \cdot \overset{4}{\cancel{12}}x}{\underset{1}{\cancel{3}}x^2 \cdot 5y^2} = \dfrac{8y}{5x}$

c. $\dfrac{x}{2x-6} \cdot \dfrac{4x-12}{x^2} = \dfrac{x}{2(x-3)} \cdot \dfrac{4(x-3)}{x^2} = \dfrac{x \cdot \overset{2}{\cancel{4}}\overset{1}{\cancel{(x-3)}}}{\underset{1}{\cancel{2}}\underset{1}{\cancel{(x-3)}} \cdot x^2} = \dfrac{2}{x}$

d. $\dfrac{x+2}{6x^2} \cdot \dfrac{8x}{x^2-x-6} = \dfrac{\overset{1}{\cancel{(x+2)}}}{\underset{3}{\cancel{6}}x^2} \cdot \dfrac{\overset{4}{\cancel{8}}x}{\underset{1}{\cancel{(x+2)}}(x-3)} = \dfrac{4}{3x(x-3)}$

e. $\dfrac{10xy^3}{x^2-y^2} \cdot \dfrac{2x^2+xy-y^2}{15x^2y} = \dfrac{\overset{2}{\cancel{10}}xy^3}{\underset{1}{\cancel{(x+y)}}(x-y)} \cdot \dfrac{\overset{1}{\cancel{(x+y)}}(2x-y)}{\underset{3}{\cancel{15}}x^2y} = \dfrac{2y^2(2x-y)}{3x(x-y)} \quad \blacksquare$

Now that we have discussed multiplication of fractions, an alternate method for reducing fractions can be given that is based on the definition of multiplication and the identity property of multiplication. We can reduce the fraction $\frac{6}{8}$ as follows:

$$\frac{6}{8} = \frac{2 \cdot 3}{2 \cdot 4} = \frac{2}{2} \cdot \frac{3}{4} = 1 \cdot \frac{3}{4} = \frac{3}{4}$$

The Reciprocal of a Fraction To find the **reciprocal** of a fraction, interchange the numerator and the denominator; the reciprocal of a positive number is always positive, and the reciprocal of a negative number is always negative. Zero has no reciprocal. If $a \neq 0$ and $b \neq 0$, then the reciprocal of $\frac{a}{b}$ is $\frac{b}{a}$. Consider the product of $\frac{a}{b}$ and its reciprocal:

$$\frac{\overset{1}{\cancel{a}}}{\underset{1}{\cancel{b}}} \cdot \frac{\overset{1}{\cancel{b}}}{\underset{1}{\cancel{a}}} = 1$$

In fact, the product of any nonzero fraction and its reciprocal is always 1.

Example 3 Find the reciprocals of each of the following fractions:

a. $\frac{5}{6}$ The reciprocal of $\frac{5}{6}$ is $\frac{6}{5}$; notice that $\left(\frac{5}{6}\right)\left(\frac{6}{5}\right) = 1$.

b. -8 The reciprocal of -8 is $-\frac{1}{8}$; notice that $(-8)\left(-\frac{1}{8}\right) = 1$.

c. $-\frac{5}{7}$ The reciprocal of $-\frac{5}{7}$ is $-\frac{7}{5}$; notice that $\left(-\frac{5}{7}\right)\left(-\frac{7}{5}\right) = 1$.

d. $-\frac{1}{2}$ The reciprocal of $-\frac{1}{2}$ is -2; notice that $\left(-\frac{1}{2}\right)(-2) = 1$. ■

Multiplicative Inverse When the product of two numbers equals the multiplicative identity (that is, when the product is 1), we say that the numbers are **multiplicative inverses** of each other. Therefore, the reciprocal of a fraction can also be called its multiplicative inverse.

Dividing Fractions

Recall from Section 1.6 that in a division problem, the number we're dividing *by* is called the *divisor*, and the number we're dividing *into* is called the *dividend*.

In Arithmetic

Reciprocal of divisor

$$\frac{3}{5} \div \frac{4}{7} = \frac{3}{5} \cdot \frac{7}{4} = \frac{3 \cdot 7}{5 \cdot 4} = \frac{21}{20}$$

Dividend ⎯ ⎿⎯ Divisor

This method works because

$$\frac{3}{5} \div \frac{4}{7} = \frac{\dfrac{3}{5}}{\dfrac{4}{7}} = \frac{\dfrac{3}{5}}{\dfrac{4}{7}} \cdot \frac{\dfrac{7}{4}}{\dfrac{7}{4}} = \frac{\dfrac{3}{5} \cdot \dfrac{7}{4}}{\dfrac{\cancel{4}}{\cancel{7}} \cdot \dfrac{\cancel{7}}{\cancel{4}}} = \frac{\dfrac{3}{5} \cdot \dfrac{7}{4}}{1} = \frac{3}{5} \cdot \frac{7}{4}$$

└── The value of this fraction is 1

Therefore, $\dfrac{3}{5} \div \dfrac{4}{7} = \dfrac{3}{5} \cdot \dfrac{7}{4}$.

In Algebra We divide algebraic fractions the same way we divide fractions in arithmetic. The rule for dividing fractions is as follows:

TO DIVIDE FRACTIONS

Multiply the dividend by the multiplicative inverse (the reciprocal) of the divisor.

$$\frac{a}{b} \div \frac{c}{d} = \frac{a}{b} \cdot \frac{d}{c}$$

Dividend ──┘   └── Divisor

Example 4 Perform the indicated divisions:

a. $\dfrac{4}{3x} \div \dfrac{12}{x^3} = \dfrac{\overset{1}{\cancel{4}}}{3x} \cdot \dfrac{x^3}{\underset{3}{\cancel{12}}} = \dfrac{x^2}{9}$

b. $\dfrac{4r^3}{9s^2} \div \dfrac{8r^2s^4}{15rs} = \dfrac{\overset{1}{\cancel{4}}r^3}{\underset{3}{\cancel{9}}s^2} \cdot \dfrac{\overset{5}{\cancel{15}}rs}{\underset{2}{\cancel{8}}r^2s^4} = \dfrac{5r^2}{6s^5}$

c. $\dfrac{y^2 - x^2}{4xy - 2y^2} \div \dfrac{2x - 2y}{2x^2 + xy - y^2}$

$= \dfrac{(y + x)(y - x)}{2y\underset{1}{(2x - y)}} \cdot \dfrac{(x + y)(2x - y)}{2(x - y)} = \dfrac{(x + y)^2(y - x)}{4y(x - y)}$

$= \boxed{-} \dfrac{(x + y)^2 \overset{1}{(x - y)}}{4y\underset{1}{(x - y)}} = -\dfrac{(x + y)^2}{4y}$

└── Changing sign of fraction and numerator

d. $\dfrac{3y^3 - 3y^2}{16y^5 + 8y^4} \div \dfrac{3y^2 + 6y - 9}{4y + 12} = \dfrac{3y^2(y - 1)}{8y^4(2y + 1)} \cdot \dfrac{4(y + 3)}{3(y^2 + 2y - 3)}$

$= \dfrac{\overset{1}{\cancel{3}}y^2\overset{1}{(y - 1)}}{\underset{2}{\cancel{8}}y^4(2y + 1)} \cdot \dfrac{\overset{1}{\cancel{4}}\overset{1}{(y + 3)}}{\underset{1}{\cancel{3}}\underset{1}{(y + 3)}\underset{1}{(y - 1)}}$

$= \dfrac{1}{2y^2(2y + 1)}$ ∎

A WORD OF CAUTION $2y^2(2y + 1)$ is not an acceptable answer for Example 3d. The 1 in the numerator cannot be omitted. ☑

EXERCISES 8.3

Set I Perform the indicated operations.

1. $\dfrac{5}{6} \div \dfrac{5}{3}$

2. $\dfrac{3}{8} \div \dfrac{21}{12}$

3. $\dfrac{4a^3}{5b^2} \cdot \dfrac{10b}{8a^2}$

4. $\dfrac{6d^2}{8c} \cdot \dfrac{12c^2}{9d^3}$

5. $\dfrac{3x^2}{16} \div \dfrac{x}{8}$

6. $\dfrac{4y^3}{7} \div \dfrac{8y^2}{21}$

7. $\dfrac{3x^4y^2z}{18xy} \cdot \dfrac{15z}{x^3yz^2}$

8. $\dfrac{21a^2b^5}{4b^2c^2} \cdot \dfrac{6c^3}{7a^2b^3c}$

9. $\dfrac{x}{x+2} \cdot \dfrac{5x+10}{x^3}$

10. $\dfrac{b}{b+5} \cdot \dfrac{3b+15}{b^4}$

11. $\dfrac{y-2}{y} \cdot \dfrac{6}{3y-6}$

12. $\dfrac{a-1}{a} \cdot \dfrac{8}{4a-4}$

13. $\dfrac{b^3}{a+3} \div \dfrac{4b^2}{2a+6}$

14. $\dfrac{m^4}{m+2} \div \dfrac{6m^3}{3m+6}$

15. $\dfrac{5s-15}{30s} \div \dfrac{s-3}{45s^2}$

16. $\dfrac{3n-6}{15n} \div \dfrac{n-2}{20n^2}$

17. $\dfrac{a+4}{a-4} \div \dfrac{a^2+8a+16}{a^2-16}$

18. $\dfrac{x+8}{x-8} \div \dfrac{x^2+16x+64}{x^2-64}$

19. $\dfrac{5}{z+4} \cdot \dfrac{z^2-16}{(z-4)^2}$

20. $\dfrac{7}{x+3} \cdot \dfrac{x^2-9}{(x-3)^2}$

21. $\dfrac{3a-3b}{4c+4d} \cdot \dfrac{2c+2d}{b-a}$

22. $\dfrac{7x-7y}{8a+8b} \cdot \dfrac{4a+4b}{y-x}$

23. $\dfrac{4x-8}{4} \cdot \dfrac{x+2}{x^2-4}$

24. $\dfrac{5x-20}{5} \cdot \dfrac{x+4}{x^2-16}$

25. $\dfrac{4a+4b}{ab^2} \div \dfrac{3a+3b}{a^2b}$

26. $\dfrac{5x-5y}{x^2y} \div \dfrac{4x-4y}{xy^2}$

27. $\dfrac{a^2-9b^2}{a^2-6ab+9b^2} \div \dfrac{a+3b}{a-3b}$

28. $\dfrac{x^2-y^2}{x^2-2xy+y^2} \div \dfrac{x+y}{x-y}$

29. $\dfrac{x-y}{9x+9y} \div \dfrac{x^2-y^2}{3x^2+6xy+3y^2}$

30. $\dfrac{x-5y}{8x+40y} \div \dfrac{x^2-25y^2}{4x^2+40xy+100y^2}$

31. $\dfrac{a^2-9b^2}{6a^2-36ab+54b^2} \div \dfrac{a+3b}{2a-6b}$

32. $\dfrac{x^2-y^2}{8x^2-16xy+8y^2} \div \dfrac{x+y}{4x-4y}$

33. $\dfrac{2x^3y+2x^2y^2}{6x} \div \dfrac{x^2y^2-xy^3}{y-x}$

34. $\dfrac{5st^4+5s^2t^3}{10st^3} \div \dfrac{s^2t^2-s^3t}{s-t}$

Set II Perform the indicated operations.

1. $\dfrac{2}{3} \div \dfrac{1}{2}$

2. $\dfrac{5}{8} \div \dfrac{5}{14}$

3. $\dfrac{5x^3}{4y^2} \cdot \dfrac{8y}{10x^2}$

4. $\dfrac{8y^2}{12x} \cdot \dfrac{6x^2}{4y^3}$

5. $\dfrac{5a^2}{15} \div \dfrac{a}{3}$

6. $\dfrac{5x^3}{6} \div \dfrac{8x^2}{20}$

7. $\dfrac{6a^3b^2c^3}{24a^2b} \cdot \dfrac{18b}{a^2bc^2}$

8. $\dfrac{18x^2y^3}{10x^3y^2} \cdot \dfrac{5z^3}{9x^2y^2z}$

9. $\dfrac{s}{s+3} \cdot \dfrac{4s+12}{s^4}$

10. $\dfrac{x+8}{x+4} \cdot \dfrac{4x+16}{x^3}$

11. $\dfrac{x+7}{x} \cdot \dfrac{2}{3x+21}$

12. $\dfrac{y-3}{12y+12} \cdot \dfrac{9}{4y-12}$

13. $\dfrac{x^3}{y+5} \div \dfrac{4x^2}{3y+15}$

14. $\dfrac{x^5}{x-3} \div \dfrac{2x^7}{2x-6}$

15. $\dfrac{4a-16}{12a} \div \dfrac{a-4}{48a^3}$

16. $\dfrac{12b-6}{12b} \div \dfrac{2b-1}{24b^2}$

17. $\dfrac{x+1}{x-1} \div \dfrac{x^2+2x+1}{x^2-1}$

18. $\dfrac{x^2+x-2}{x-1} \div \dfrac{x^2+5x+6}{x^2}$

19. $\dfrac{9}{z+3} \cdot \dfrac{z^2-9}{(z-3)^2}$

20. $\dfrac{z^3}{z+4} \div \dfrac{z-1}{z^2+3z-4}$

21. $\dfrac{5x-5y}{3x+3y} \cdot \dfrac{4x+4y}{y-x}$

22. $\dfrac{3-y}{y+1} \cdot \dfrac{4+5y+y^2}{y^2+y-12}$

23. $\dfrac{2x-10}{8} \cdot \dfrac{x+5}{x^2-25}$

24. $\dfrac{x^2+10x+25}{x^2-25} \div \dfrac{5-x}{x+5}$

25. $\dfrac{5x+5y}{xy^3} \div \dfrac{8x+8y}{x^3y}$

26. $\dfrac{12f+16}{15f} \div \dfrac{6f^3+8f^2}{20f^4}$

27. $\dfrac{s^2-4t^2}{s^2-4st+4t^2} \div \dfrac{s+2t}{s-2t}$

28. $\dfrac{x^3-5x}{9x} \div \dfrac{4x^3-20x}{12x^2}$

29. $\dfrac{a-4b}{a+4b} \div \dfrac{a^2-16b^2}{2a^2+16ab+32b^2}$

30. $\dfrac{2b^2c-2bc^2}{b+c} \div \dfrac{4bc^2-4b^2c}{4b+4c}$

31. $\dfrac{x^2-25y^2}{3x^2-30xy+75y^2} \div \dfrac{x+5y}{4x-20y}$

32. $\dfrac{2x-1}{y} \div \dfrac{2x^2+x-1}{y^2}$

33. $\dfrac{8x^2y^2+16xy^3}{4y} \div \dfrac{2x^3y-2x^2y^2}{y-x}$

34. $\dfrac{a+2}{a+1} \cdot \dfrac{a^2+a}{a^2-4}$

8.4 Adding and Subtracting Like Fractions

Like Fractions **Like fractions** are fractions that have the same denominator.

Example 1 Examples of like fractions:

a. $\dfrac{2}{3}, \dfrac{5}{3}, \dfrac{1}{3}$ are like fractions.

Same denominator

b. $\dfrac{3}{x+2}, \dfrac{7}{x+2}, \dfrac{-5}{x+2}$ are like fractions.

Same denominator ■

Unlike Fractions   **Unlike fractions** are fractions that have different denominators.

Example 2   Examples of unlike fractions:

a. $\dfrac{1}{3}, \dfrac{2}{7}, \dfrac{3}{8}$ are unlike fractions.

Different denominators

b. $\dfrac{5}{3x}, \dfrac{5}{3x^2}, \dfrac{5}{3+x}$ are unlike fractions.

Different denominators ■

Adding Like Fractions
We know that

$$1 \text{ half-dollar} + 3 \text{ half-dollars} + 7 \text{ half-dollars} = (1 + 3 + 7)\,\text{half-dollars}$$
$$= 11 \text{ half-dollars}$$

In the same way, $\dfrac{1}{2} + \dfrac{3}{2} + \dfrac{7}{2} = \dfrac{11}{2}$. Similarly,

$$2 \text{ thirds} + 5 \text{ thirds} + 1 \text{ third} = (2 + 5 + 1) \text{ thirds} = 8 \text{ thirds}$$

$$\frac{2}{3} \;+\; \frac{5}{3} \;+\; \frac{1}{3} \;=\; \frac{2+5+1}{3} \;=\; \frac{8}{3}$$

TO ADD LIKE FRACTIONS

1. Add the numerators.

2. Write the sum of the numerators over the denominator of the like fractions.

$$\frac{a}{c} + \frac{b}{c} = \frac{a+b}{c}$$

3. Reduce the resulting fraction to lowest terms.

Example 3   Examples of adding like arithmetic fractions:

a. Add: $\dfrac{1}{9} + \dfrac{4}{9} + \dfrac{7}{9} = \dfrac{1+4+7}{9} = \dfrac{12}{9} = \dfrac{\overset{4}{\cancel{12}}}{\underset{3}{\cancel{9}}} = \dfrac{4}{3}$

b. Add: $\dfrac{11}{23} + \dfrac{5}{23} = \dfrac{11+5}{23} = \dfrac{16}{23}$ Already in lowest terms

c. Add: $\dfrac{3}{12} + \dfrac{1}{12} + \dfrac{4}{12} + \dfrac{7}{12} = \dfrac{3+1+4+7}{12} = \dfrac{\overset{5}{\cancel{15}}}{\underset{4}{\cancel{12}}} = \dfrac{5}{4}$ ■

This same method is used for adding like fractions in algebra.

Subtracting Like Fractions

Any subtraction of fractions can always be changed into an addition of fractions or the following definition can be used:

$$\frac{a}{c} - \frac{b}{c} = \frac{a-b}{c}$$

Example 4 Perform the indicated operations:

a. $\dfrac{2}{x} + \dfrac{5}{x} = \dfrac{2+5}{x} = \dfrac{7}{x}$

b. $\dfrac{7}{a-2} - \dfrac{4}{a-2} = \dfrac{7}{a-2} + \dfrac{-4}{a-2} = \dfrac{7+(-4)}{a-2} = \dfrac{3}{a-2}$

c. $\dfrac{3}{4a} - \dfrac{5}{4a} = \dfrac{3-5}{4a} = \dfrac{-2}{4a} = -\dfrac{\overset{1}{\cancel{2}}}{\underset{2}{\cancel{4}a}} = -\dfrac{1}{2a}$

d. $\dfrac{4x}{2x-y} - \dfrac{2y}{2x-y} = \dfrac{4x-2y}{2x-y} = \dfrac{2(2x-y)}{(2x-y)} = \dfrac{2\overset{1}{\cancel{(2x-y)}}}{\underset{1}{\cancel{(2x-y)}}} = 2$

e. $\dfrac{15}{d-5} + \dfrac{-3d}{d-5} = \dfrac{15-3d}{d-5} = \dfrac{3(5-d)}{d-5} = -\dfrac{3\overset{1}{\cancel{(d-5)}}}{\underset{1}{\cancel{d-5}}} = -3$

� *Changing sign of fraction and numerator* �   ■

In a subtraction problem, if the numerator of the fraction being subtracted contains more than one term, you *must* put parentheses around that numerator when you rewrite the problem as a single fraction (see Example 5).

Example 5 Subtract $\dfrac{5}{2x+3} - \dfrac{x+1}{2x+3}$.

Solution

$$\frac{5}{2x+3} - \frac{x+1}{2x+3} = \frac{5-(x+1)}{2x+3} = \frac{5-x-1}{2x+3} = \frac{4-x}{2x+3} \quad ■$$

A WORD OF CAUTION   It is *incorrect* to do Example 5 this way:

$$\cancel{\frac{5}{2x+3} - \frac{x+1}{2x+3} = \frac{5-x+1}{2x+3} = \frac{6-x}{2x+3}} \qquad ☑$$

When the denominators are not identical but are the *negatives* of each other, we can make the fractions like fractions by changing the signs of the numerator and denominator of one of the fractions (see Example 6).

Example 6 Add $\dfrac{9}{x-2} + \dfrac{5}{2-x}$.

Solution $x - 2$ and $2 - x$ are the negatives of each other.

Changing sign of numerator and denominator

$$\frac{9}{x-2} + \frac{5}{2-x} = \frac{9}{x-2} + \frac{-5}{-(2-x)} = \frac{9}{x-2} + \frac{-5}{x-2} = \frac{9-5}{x-2} = \frac{4}{x-2} \quad ■$$

In some problems, it is easier to change the sign of the denominator and the sign of the fraction (see Example 7).

Example 7 Subtract $\dfrac{8}{y-5} - \dfrac{3}{5-y}$.

Solution

Changing sign of fraction and denominator

$$\frac{8}{y-5} - \frac{3}{5-y} = \frac{8}{y-5} \boxed{+} \frac{3}{\boxed{-}(5-y)} = \frac{8}{y-5} + \frac{3}{y-5} = \frac{11}{y-5} \quad\blacksquare$$

A WORD OF CAUTION Students often confuse *addition of fractions* with *solving equations*, and they multiply both fractions by the same number. This is incorrect.

Correct method	*Incorrect method*
$\dfrac{1}{3x} + \dfrac{4}{3x} = \dfrac{5}{3x}$	$\dfrac{1}{3x} + \dfrac{4}{3x} = (3x)\left(\dfrac{1}{3x}\right) + (3x)\left(\dfrac{4}{3x}\right) = 1 + 4 = 5$

EXERCISES 8.4

Set I Perform the indicated operations.

1. $\dfrac{7}{a} + \dfrac{2}{a}$

2. $\dfrac{5}{b} + \dfrac{2}{b}$

3. $\dfrac{6}{x-y} - \dfrac{2}{x-y}$

4. $\dfrac{7}{m+n} - \dfrac{1}{m+n}$

5. $\dfrac{2}{3a} + \dfrac{4}{3a}$

6. $\dfrac{8}{5z} + \dfrac{2}{5z}$

7. $\dfrac{2y}{y+1} + \dfrac{2}{y+1}$

8. $\dfrac{10x}{2x+3} + \dfrac{15}{2x+3}$

9. $\dfrac{3}{x+3} + \dfrac{x}{x+3}$

10. $\dfrac{5}{y+5} + \dfrac{y}{y+5}$

11. $\dfrac{3x}{x-4} - \dfrac{12}{x-4}$

12. $\dfrac{7x}{x-2} - \dfrac{14}{x-2}$

13. $\dfrac{x-3}{y-2} - \dfrac{x+5}{y-2}$

14. $\dfrac{z-4}{a-b} - \dfrac{z+3}{a-b}$

15. $\dfrac{a+2}{2a+1} - \dfrac{1-a}{2a+1}$

16. $\dfrac{6x-1}{3x-2} - \dfrac{3x+1}{3x-2}$

17. $\dfrac{-x}{x-2} - \dfrac{2}{2-x}$

18. $\dfrac{-b}{2a-b} - \dfrac{2a}{b-2a}$

19. $\dfrac{-15w}{1-5w} - \dfrac{3}{5w-1}$

20. $\dfrac{-35}{6w-7} - \dfrac{30w}{7-6w}$

21. $\dfrac{7z}{8z-4} + \dfrac{6-5z}{4-8z}$

22. $\dfrac{5x}{9x-3} + \dfrac{4-7x}{3-9x}$

23. $\dfrac{31-8x}{12-8x} - \dfrac{5-16x}{8x-12}$

24. $\dfrac{13-30w}{15-10w} - \dfrac{10w+17}{10w-15}$

Set II   Perform the indicated operations.

1. $\dfrac{8}{y} + \dfrac{7}{y}$

2. $\dfrac{6}{x} - \dfrac{2}{x}$

3. $\dfrac{8}{x-y} - \dfrac{5}{x-y}$

4. $\dfrac{x+1}{a+b} + \dfrac{x-1}{a+b}$

5. $\dfrac{7}{5x} + \dfrac{8}{5x}$

6. $\dfrac{15}{7x} - \dfrac{1}{7x}$

7. $\dfrac{4x}{x+3} + \dfrac{12}{x+3}$

8. $\dfrac{2x+1}{x+4} - \dfrac{x-3}{x+4}$

9. $\dfrac{3}{2a+3} + \dfrac{2a}{2a+3}$

10. $\dfrac{9x}{3x-y} - \dfrac{3y}{3x-y}$

11. $\dfrac{5a}{a-2b} - \dfrac{10b}{a-2b}$

12. $\dfrac{8}{z-4} - \dfrac{2z}{z-4}$

13. $\dfrac{a-5}{x-4} - \dfrac{a+3}{x-4}$

14. $\dfrac{4x+y}{2x+y} + \dfrac{2x+2y}{2x+y}$

15. $\dfrac{7x-1}{3x-1} - \dfrac{x+1}{3x-1}$

16. $\dfrac{b}{b-2a} + \dfrac{2a}{2a-b}$

17. $\dfrac{-s}{s-5} - \dfrac{5}{5-s}$

18. $\dfrac{5}{2x+3} - \dfrac{x+4}{2x+3}$

19. $\dfrac{-3m}{5-m} - \dfrac{15}{m-5}$

20. $\dfrac{2x}{3x-8} + \dfrac{5x-8}{8-3x}$

21. $\dfrac{8x}{12x-15} + \dfrac{25-12x}{15-12x}$

22. $\dfrac{5x-2}{4x-5} - \dfrac{x+3}{4x-5}$

23. $\dfrac{11x-9y}{8x-6y} - \dfrac{13x-9y}{6y-8x}$

24. $\dfrac{18y-21}{10y-16} - \dfrac{3y+3}{10y-16}$

$\underline{8.5}$ Lowest Common Denominator (LCD)

We ordinarily use the **lowest common denominator (LCD)** when we add or subtract unlike fractions. The lowest common denominator is the smallest number that is exactly divisible by each of the denominators.

TO FIND THE LCD

1. Factor each denominator completely. Repeated factors should be expressed as powers.

2. Write down each different base that appears.

3. Raise each base to the *highest power* to which it occurs in *any* denominator.

4. The LCD is the product of all the factors found in step 3.

In Arithmetic   Consider the problem $\frac{7}{12} + \frac{4}{15}$. Find the LCD as follows:

Step 1: Find the prime factorization of each denominator.

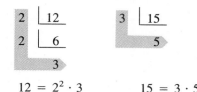

$$12 = 2^2 \cdot 3 \qquad\qquad 15 = 3 \cdot 5$$

Step 2: Write down each different base that appears in the prime factorizations.

$$2, 3, 5$$

Step 3: Raise each base to the *highest* power to which it occurs in any denominator.

$$2^2, 3, 5$$

Step 4: LCD $= 2^2 \cdot 3 \cdot 5 = 60$

In Algebra   This same method is used in finding the LCD for algebraic fractions.

Example 1   Find the LCD for $\dfrac{3}{2} + \dfrac{4}{y}$.
Solution

Step 1: The denominators are already factored.

Step 2: 2 and y are the different bases.

Step 3: 2^1, y^1 (highest power of each base)

Step 4: LCD $= 2^1 \cdot y^1 = 2y$ ∎

Example 2   Find the LCD for $\dfrac{2}{x} + \dfrac{5}{x^2}$.
Solution

Step 1: The denominators are already factored.

Step 2: x (x is the only base)

Step 3: x^2 (x^2 is the highest power in any denominator.)

Step 4: LCD $= x^2$ ∎

Example 3   Find the LCD for $\dfrac{7}{5b^3} + \dfrac{4}{15b^2}$.
Solution

Step 1: $5 \cdot b^3$, $3 \cdot 5 \cdot b^2$ Denominators in factored form

Step 2: 3, 5, b All the different bases

Step 3: 3^1, 5^1, b^3 Highest powers

Step 4: LCD $= 3^1 \cdot 5^1 \cdot b^3 = 15b^3$ ∎

Example 4 Find the LCD for $\dfrac{7}{18x^2y} + \dfrac{5}{8xy^4}$.

Solution

Step 1: $2 \cdot 3^2 \cdot x^2 \cdot y, \; 2^3 \cdot x \cdot y^4$ Denominators in factored form

Step 2: $2, \; 3, \; x, \; y$ All the different bases

Step 3: $2^3, \; 3^2, \; x^2, \; y^4$ Highest powers

Step 4: LCD $= 2^3 \cdot 3^2 \cdot x^2 \cdot y^4 = 72x^2y^4$ ■

Example 5 Find the LCD for $\dfrac{2}{x} + \dfrac{x}{x+2}$.

Solution

Step 1: The denominators are already factored.

Step 2: $x, \; (x+2)$

Step 3: $x^1, \; (x+2)^1$

Step 4: LCD $= x(x+2)$ ■

Example 6 Find the LCD for $\dfrac{16}{a^2b} + \dfrac{a-2}{2a(a-b)} - \dfrac{b+1}{4b^3(a-b)}$.

Solution

Step 1: $a^2b, \; 2a(a-b), \; 2^2b^3(a-b)$

Step 2: $2, \; a, \; b, \; (a-b)$ All the different bases

Step 3: $2^2, \; a^2, \; b^3, \; (a-b)^1$ Highest powers

Step 4: LCD $= 2^2a^2b^3(a-b)^1 = 4a^2b^3(a-b)$ ■

Example 7 Find the LCD for $\dfrac{8}{3x-3} - \dfrac{5}{x^2+2x+1}$.

Solution

Step 1: $3x - 3 = 3(x-1); \; x^2 + 2x + 1 = (x+1)^2$

Step 2: $3, \; (x-1), \; (x+1)$

Step 3: $3^1, \; (x-1)^1, \; (x+1)^2$

Step 4: LCD $= 3(x-1)(x+1)^2$ ■

Example 8 Find the LCD for $\dfrac{2x-3}{x^2+10x+25} - \dfrac{14}{4x^2+20x} + \dfrac{4x-3}{x^2+2x-15}$

Solution

Step 1: $x^2 + 10x + 25 = (x+5)^2$

$\qquad\qquad 4x^2 + 20x = 4x(x+5) = 2^2x(x+5)$

$\qquad x^2 + 2x - 15 = (x+5)(x-3)$

Step 2: $2, \; x, \; (x+5), \; (x-3)$

Step 3: $2^2, \; x, \; (x+5)^2, \; (x-3)$

Step 4: LCD $= 4x(x+5)^2(x-3)$ ■

EXERCISES 8.5

Set I   Find the LCD in each exercise. Do *not* add the fractions.

1. $\dfrac{x}{3} + \dfrac{2}{x}$

2. $\dfrac{3}{2y} - \dfrac{5}{y}$

3. $\dfrac{a-3}{12} - \dfrac{a+6}{4}$

4. $\dfrac{y-2}{3} + \dfrac{y+5}{9}$

5. $\dfrac{7}{5x} + \dfrac{3}{2x}$

6. $\dfrac{5}{3z} - \dfrac{9}{4z}$

7. $\dfrac{4}{x} + \dfrac{6}{x^3}$

8. $\dfrac{3}{a^2} + \dfrac{5}{a^4}$

9. $\dfrac{7}{12u^3v^2} - \dfrac{11}{18uv^3}$

10. $\dfrac{13}{50x^3y^4} - \dfrac{17}{20x^2y^5}$

11. $\dfrac{4}{a} + \dfrac{a}{a+3}$

12. $\dfrac{5}{b} + \dfrac{b}{b-5}$

13. $\dfrac{x}{2x+4} - \dfrac{5}{4x}$

14. $\dfrac{4}{3x} + \dfrac{2x}{3x+6}$

15. $\dfrac{3}{4z^2} + \dfrac{2z}{z^2+2z+1} - \dfrac{4z}{z+1}$

16. $\dfrac{2x}{x^2-2x+1} - \dfrac{5}{x-1} + \dfrac{11}{12x^3}$

17. $\dfrac{x-4}{x^2+3x+2} + \dfrac{3x+1}{x^2+2x+1}$

18. $\dfrac{2x+3}{x^2-x-12} + \dfrac{x-4}{x^2+6x+9}$

19. $\dfrac{x^2+1}{12x^3+24x^2} - \dfrac{4x+3}{x^2-4x+4} + \dfrac{1}{x^2-4}$

20. $\dfrac{2y+5}{y^2+6y+9} - \dfrac{7y}{y^2-9} - \dfrac{11}{8y^2-24y}$

21. $\dfrac{3x+1}{6x^2+x-2} + \dfrac{x^2+1}{9x^3+12x^2+4x} + \dfrac{5x^2-1}{4x^2-4x+1}$

22. $\dfrac{5x+1}{10x^2+13x-3} + \dfrac{3x^2-1}{4x^3+12x^2+9x} + \dfrac{x-4}{25x^2-10x+1}$

Set II   Find the LCD in each exercise. Do *not* add the fractions.

1. $\dfrac{y}{2} + \dfrac{5}{y}$

2. $\dfrac{5}{4y^2} + \dfrac{7}{6y}$

3. $\dfrac{x-5}{21} - \dfrac{x+3}{24}$

4. $\dfrac{x}{2} + \dfrac{x}{5}$

5. $\dfrac{3}{2x} + \dfrac{5}{3x}$

6. $\dfrac{x+1}{4} - \dfrac{x+3}{2}$

7. $\dfrac{8}{y^4} + \dfrac{3}{y}$

8. $\dfrac{6}{5x^3} + \dfrac{1}{10x^5}$

9. $\dfrac{5}{9x^2y} - \dfrac{7}{6xy^2}$

10. $\dfrac{5}{2 + x^2} + \dfrac{9}{2x^2}$

11. $\dfrac{2}{x} + \dfrac{x}{x + 5}$

12. $\dfrac{x + 1}{4} - \dfrac{x + 3}{12}$

13. $\dfrac{a}{3a + 6} - \dfrac{5}{3a}$

14. $\dfrac{9y}{x^2 - y^2} + \dfrac{6x}{x^2 + 2xy + y^2}$

15. $\dfrac{3}{2x} + \dfrac{x}{x^2 + 4x + 4} - \dfrac{1}{x + 2}$

16. $\dfrac{11}{x^2 + 16x + 64} - \dfrac{7}{2x^2 - 128} + \dfrac{1}{4x - 32}$

17. $\dfrac{x - 5}{x^2 + 4x + 3} + \dfrac{5x - 3}{x^2 + 7x + 12}$

18. $\dfrac{5}{8z^3} - \dfrac{8z}{z^2 - 4} + \dfrac{5z}{9z + 18}$

19. $\dfrac{x^2 + 3}{9x^3 + 45x^2} - \dfrac{5x + 2}{x^2 - 10x + 25} + \dfrac{1}{x^2 - 25}$

20. $\dfrac{a^2 + 1}{3a^2 + 5a - 2} - \dfrac{1}{3a^2 + 6a} + \dfrac{5a}{18a^4 - 6a^3}$

21. $\dfrac{5x + 1}{4x^2 - 11x - 3} + \dfrac{x + 3}{16x^3 + 8x^2 + x} + \dfrac{3x^2}{x^2 - 6x + 9}$

22. $\dfrac{6x^2}{6x^2 + 5x - 1} + \dfrac{3}{30x^3 - 5x^2} + \dfrac{5 + x}{10x^4 + 10x^3}$

8.6 Adding and Subtracting Unlike Fractions

Because the definition of addition of fractions is $\dfrac{a}{c} + \dfrac{b}{c} = \dfrac{a + b}{c}$, we can add fractions only when they are *like fractions*. The method of converting unlike fractions to like fractions is based on the definition of multiplication and the identity property of multiplication. Since $\dfrac{2}{2} = 1, \dfrac{x}{x} = 1, \dfrac{x + 2}{x + 2} = 1$, and so on, multiplying a fraction by a fraction whose numerator and denominator are equal produces a fraction equivalent to the original fraction. For example,

$$\frac{5}{6} = \frac{5}{6} \cdot \boxed{\frac{2}{2}} = \frac{10}{12}$$

Therefore, $\dfrac{10}{12}$ is equivalent to $\dfrac{5}{6}$.

$$\frac{x}{x + 2} = \frac{x}{x + 2} \cdot \boxed{\frac{x}{x}} = \frac{x^2}{x(x + 2)}$$

Therefore, $\dfrac{x^2}{x(x+2)}$ is equivalent to $\dfrac{x}{x+2}$.

$$\frac{2}{x} = \frac{2}{x} \cdot \boxed{\frac{x+2}{x+2}} = \frac{2x+4}{x(x+2)}$$

Therefore, $\dfrac{2x+4}{x(x+2)}$ is equivalent to $\dfrac{2}{x}$.

Consequently, the rules for adding unlike fractions are as follows:

TO ADD UNLIKE FRACTIONS

1. Find the LCD.

2. Convert all fractions to equivalent fractions that have the LCD as denominator by performing the following operations on each fraction:
   a. Divide the LCD by the denominator of the fraction.
   b. Multiply the numerator and the denominator of the fraction by the quotient from step 2a.

3. Add the resulting like fractions.

4. Reduce the resulting fraction to lowest terms.

In Arithmetic

Example 1   Find $\dfrac{1}{3} + \dfrac{5}{6}$.

Solution LCD $= 6$

Since $\dfrac{1}{3} = \dfrac{2}{6}$, then $\dfrac{1}{3} + \dfrac{5}{6} = \dfrac{2}{6} + \dfrac{5}{6} = \dfrac{2+5}{6} = \dfrac{7}{6}$. ∎

Example 2   Find $\dfrac{1}{2} + \dfrac{2}{3} + \dfrac{3}{4}$.

Solution LCD $= 12$

Since $\dfrac{1}{2} = \dfrac{6}{12}, \dfrac{2}{3} = \dfrac{8}{12}$, and $\dfrac{3}{4} = \dfrac{9}{12}$,

then $\dfrac{1}{2} + \dfrac{2}{3} + \dfrac{3}{4} = \dfrac{6}{12} + \dfrac{8}{12} + \dfrac{9}{12} = \dfrac{6+8+9}{12} = \dfrac{23}{12}$. ∎

In Algebra   This same method is used for adding unlike fractions in algebra.

Example 3   Add $\dfrac{2}{x} + \dfrac{5}{x^2}$.

Solution

Step 1: LCD $= x^2$ (Section 8.5, Example 2)

Step 2: $\dfrac{2}{x} = \dfrac{2}{x} \cdot \boxed{\dfrac{x}{x}} = \dfrac{2 \cdot x}{x \cdot x} = \dfrac{2x}{x^2}$

└── We multiply numerator and denominator by x in order to obtain an equivalent fraction whose denominator is the LCD x^2

Step 3: $\dfrac{2}{x} + \dfrac{5}{x^2} = \dfrac{2x}{x^2} + \dfrac{5}{x^2} = \dfrac{2x+5}{x^2}$

Step 4: The fraction cannot be reduced. ∎

A WORD OF CAUTION Students frequently reduce the fractions just after they have converted them to equivalent fractions with the LCD as the denominator (see step 2). The addition then cannot be done, because the fractions no longer have the same denominator. This gives us back what we started with:

$$\frac{2}{x} + \frac{5}{x^2} = \frac{2\overset{1}{\cancel{x}}}{\underset{x}{\cancel{x^2}}} + \frac{5}{x^2} = \underbrace{\frac{2}{x} + \frac{5}{x^2}}$$

└ We cannot add these fractions ☑

Example 4   Add $\dfrac{7}{18x^2y} + \dfrac{5}{8xy^4}$.

Solution

Step 1:   LCD $= 72x^2y^4$ (Section 8.5, Example 4)

$$4y^3 = \frac{72x^2y^4}{18x^2y} = \frac{\text{LCD}}{\text{denominator of fraction}}$$

Step 2: $\dfrac{7}{18x^2y} = \dfrac{7\,(4y^3)}{18x^2y\,(4y^3)} = \dfrac{28y^3}{72x^2y^4}$

$$\frac{5}{8xy^4} = \frac{5\,(9x)}{8xy^4\,(9x)} = \frac{45x}{72x^2y^4}$$

$$9x = \frac{72x^2y^4}{8xy^4} = \frac{\text{LCD}}{\text{denominator of fraction}}$$

Step 3: $\dfrac{7}{18x^2y} + \dfrac{5}{8xy^4} = \dfrac{28y^3}{72x^2y^4} + \dfrac{45x}{72x^2y^4} = \dfrac{28y^3 + 45x}{72x^2y^4}$

Step 4:   The fraction cannot be reduced.   ■

In applying step 2, if the multiplication involves any expression containing more than one term, you *must* put parentheses around that expression (see Example 5).

Example 5   Add $\dfrac{2}{x} + \dfrac{x}{x+2}$.

Solution

Step 1:   LCD $= x(x + 2)$ (Section 8.5, Example 5)

$$x + 2 = \frac{x(x+2)}{x} = \frac{\text{LCD}}{\text{denominator of fraction}}$$

Step 2: $\dfrac{2}{x} = \dfrac{2\,(x+2)}{x\,(x+2)}$

└ Note parentheses

$$\frac{x}{x+2} = \frac{x \cdot x}{(x+2)\,x} = \frac{x^2}{x(x+2)}$$

$$x = \frac{x(x+2)}{x+2} = \frac{\text{LCD}}{\text{denominator of fraction}}$$

Step 3: $\dfrac{2}{x} + \dfrac{x}{x+2} = \dfrac{2(x+2)}{x(x+2)} + \dfrac{x^2}{x(x+2)} = \dfrac{2x + 4 + x^2}{x(x+2)}$

Step 4: $\dfrac{2x + 4 + x^2}{x(x+2)} = \dfrac{x^2 + 2x + 4}{x(x+2)}$

$x^2 + 2x + 4$ does not factor. The fraction cannot be reduced.   ■

Example 6   Subtract $3 - \dfrac{2a}{a + 2}$.

Solution

Step 1: LCD $= a + 2$

Step 2: $3 = \dfrac{3\ (a + 2)}{1\ (a + 2)} = \dfrac{3a + 6}{a + 2}$

$\dfrac{2a}{a + 2} = \dfrac{2a}{a + 2}$

Step 3: $3 - \dfrac{2a}{a + 2} = \dfrac{3a + 6}{a + 2} - \dfrac{2a}{a + 2} = \dfrac{3a + 6 - 2a}{a + 2} = \dfrac{a + 6}{a + 2}$

Step 4: The fraction cannot be reduced. ∎

A WORD OF CAUTION   It is incorrect to cancel a numerator of one fraction against a denominator of a *different* fraction when adding or subtracting. For example,

$$\dfrac{x + 1}{x - 2} + \dfrac{5}{x + 1} = \dfrac{x + 1}{x - 2} + \dfrac{5}{x + 1} = \dfrac{1}{x - 2} + \dfrac{5}{1}$$ ☑

Example 7   Subtract $\dfrac{z - 1}{z + 3} - \dfrac{z + 3}{z - 1}$.

Solution We cannot "cancel" the $z - 1$'s or the $z + 3$'s because this is not a multiplication problem.

Step 1: LCD $= (z + 3)(z - 1)$

Step 2: $\dfrac{z - 1}{z + 3} = \dfrac{(z - 1)(z - 1)}{(z + 3)(z - 1)}$ ⟵

$\dfrac{z + 3}{z - 1} = \dfrac{(z + 3)(z + 3)}{(z - 1)(z + 3)}$ ⟵   — Note the parentheses

Step 3: $\dfrac{z - 1}{z + 3} - \dfrac{z + 3}{z - 1} = \dfrac{(z - 1)(z - 1)}{(z + 3)(z - 1)} - \dfrac{(z + 3)(z + 3)}{(z - 1)(z + 3)}$

$= \dfrac{z^2 - 2z + 1}{(z + 3)(z - 1)} - \dfrac{z^2 + 6z + 9}{(z + 3)(z - 1)}$

These parentheses are essential

$= \dfrac{(z^2 - 2z + 1) - (z^2 + 6z + 9)}{(z + 3)(z - 1)}$

$= \dfrac{z^2 - 2z + 1 - z^2 - 6z - 9}{(z + 3)(z - 1)}$

$= \dfrac{-8z - 8}{(z + 3)(z - 1)} = \dfrac{-8(z + 1)}{(z + 3)(z - 1)}$

Step 4: The fraction cannot be reduced. (Note: Other forms of the answer are also acceptable.) ∎

Example 8   Subtract $\dfrac{x-1}{2x^2+11x+12} - \dfrac{x-2}{2x^2+x-3}$.

Solution

Step 1: In order to find the LCD, we must factor each denominator.

$$2x^2 + 11x + 12 = (2x + 3)(x + 4)$$

$$2x^2 + x - 3 = (2x + 3)(x - 1)$$

The LCD is $(2x + 3)(x + 4)(x - 1)$.

Step 2: $\dfrac{x-1}{2x^2+11x+12} = \dfrac{x-1}{(2x+3)(x+4)} = \dfrac{(x-1)\,(x-1)}{(2x+3)(x+4)\,(x-1)}$

$\dfrac{x-2}{2x^2+x-3} = \dfrac{x-2}{(2x+3)(x-1)} = \dfrac{(x-2)\,(x+4)}{(2x+3)(x-1)\,(x+4)}$

Step 3: $\dfrac{x-1}{2x^2+11x+12} - \dfrac{x-2}{2x^2+x-3}$

$$= \dfrac{(x-1)(x-1)}{(2x+3)(x+4)(x-1)} - \dfrac{(x-2)(x+4)}{(2x+3)(x-1)(x+4)}$$

$$= \dfrac{x^2-2x+1}{(2x+3)(x+4)(x-1)} - \dfrac{x^2+2x-8}{(2x+3)(x+4)(x-1)}$$

$$= \dfrac{x^2-2x+1-(x^2+2x-8)}{(2x+3)(x+4)(x-1)} \qquad \text{The parentheses are essential}$$

$$= \dfrac{x^2-2x+1-x^2-2x+8}{(2x+3)(x+4)(x-1)}$$

$$= \dfrac{-4x+9}{(2x+3)(x+4)(x-1)}$$

Step 4: The fraction can't be reduced.   ■

A WORD OF CAUTION   It is incorrect to multiply all the fractions by the LCD. Example 8 cannot be done this way:

$$(2x+3)(x+4)(x-1)\cdot\dfrac{x-1}{(2x+3)(x+4)} - (2x+3)(x+4)(x-1)\cdot\dfrac{x-2}{(2x+3)(x-1)}$$

$$= (x-1)(x-1) - (x+4)(x-2) = x^2 - 2x + 1 - (x^2 + 2x - 8)$$

$$= -4x + 9 \qquad\qquad \boxed{}$$

EXERCISES 8.6

Set I   Perform the indicated operations.

1. $\dfrac{3}{a^2} + \dfrac{2}{a^3}$ **2.** $\dfrac{5}{u} + \dfrac{4}{u^3}$ **3.** $\dfrac{1}{2} + \dfrac{3}{x} - \dfrac{5}{x^2}$

4. $\dfrac{2}{3} - \dfrac{1}{y} + \dfrac{4}{y^2}$ **5.** $\dfrac{2}{xy} - \dfrac{3}{y}$ **6.** $\dfrac{5}{ab} - \dfrac{4}{a}$

7. $5 + \dfrac{2}{x}$

8. $3 + \dfrac{4}{y}$

9. $\dfrac{5}{4y^2} + \dfrac{9}{6y}$

10. $\dfrac{10}{5x} + \dfrac{3}{4x^2}$

11. $3x - \dfrac{3}{x}$

12. $4y - \dfrac{5}{y}$

13. $\dfrac{3}{a} + \dfrac{a}{a+3}$

14. $\dfrac{5}{b} + \dfrac{b}{b-5}$

15. $\dfrac{x}{2x+4} + \dfrac{-5}{4x}$

16. $\dfrac{4}{3x} + \dfrac{-2x}{3x+6}$

17. $x + \dfrac{2}{x} - \dfrac{3}{x-2}$

18. $m + \dfrac{3}{m} - \dfrac{2}{m-4}$

19. $\dfrac{a+b}{b} + \dfrac{b}{a-b}$

20. $\dfrac{y+x}{x} + \dfrac{x}{y-x}$

21. $\dfrac{3}{2e-2} - \dfrac{2}{3e-3}$

22. $\dfrac{2}{3f+6} - \dfrac{1}{5f+10}$

23. $\dfrac{3}{m-2} - \dfrac{5}{2-m}$

24. $\dfrac{7}{n-5} - \dfrac{2}{5-n}$

25. $\dfrac{x+1}{x-1} - \dfrac{x-1}{x+1}$

26. $\dfrac{x+4}{x-4} - \dfrac{x-4}{x+4}$

27. $\dfrac{y}{x^2-xy} + \dfrac{x}{y^2-xy}$

28. $\dfrac{b}{ab-a^2} + \dfrac{a}{ab-b^2}$

29. $\dfrac{2x}{x-3} - \dfrac{2x}{x+3} + \dfrac{36}{x^2-9}$

30. $\dfrac{x}{x+4} - \dfrac{x}{x-4} - \dfrac{32}{x^2-16}$

31. $\dfrac{x}{x^2+4x+4} + \dfrac{1}{x+2}$

32. $\dfrac{2x}{x^2-2x+1} - \dfrac{5}{x-1}$

33. $\dfrac{2x+3}{x^2+4x-5} - \dfrac{x+5}{2x^2+x-3}$

34. $\dfrac{3x+2}{x^2+6x-7} - \dfrac{x+7}{3x^2-x-2}$

35. $\dfrac{x^2}{x^2+8x+16} - \dfrac{x+1}{x^2-16}$

36. $\dfrac{x^2}{x^2+6x+9} - \dfrac{x+2}{x^2-9}$

37. $\dfrac{x+1}{2x^3+3x^2-5x} - \dfrac{x-1}{2x^3+7x^2+5x}$

38. $\dfrac{x+1}{3x^3+x^2-4x} - \dfrac{x-1}{3x^3+7x^2+4x}$

Set II Perform the indicated operations.

1. $\dfrac{3}{x} + \dfrac{4}{x^2}$

2. $\dfrac{5}{2x^3y} + \dfrac{3}{4xy^2}$

3. $\dfrac{1}{3} - \dfrac{1}{a} + \dfrac{2}{a^2}$

4. $\dfrac{1}{6} - \dfrac{1}{k} + \dfrac{3}{k^2}$

5. $\dfrac{5}{xy} - \dfrac{3}{x}$

6. $\dfrac{3}{b} - \dfrac{7}{ab}$

7. $2 + \dfrac{4}{z}$

8. $m - \dfrac{3}{m} + \dfrac{2}{m+4}$

9. $\dfrac{5}{6xy^2} + \dfrac{7}{8x^2y}$

10. $\dfrac{z+1}{z+2} - \dfrac{z-1}{z-2}$

11. $4m - \dfrac{5}{m}$

12. $\dfrac{2}{x+5} + \dfrac{3}{x-5}$

13. $\dfrac{2}{x} + \dfrac{x}{x+4}$

14. $\dfrac{2}{x+5} - \dfrac{3}{x-5}$

15. $\dfrac{x}{3x+6} + \dfrac{-3}{2x}$

16. $\dfrac{x+5}{x-2} - \dfrac{x-2}{x+5}$

17. $a + \dfrac{3}{a} - \dfrac{2}{a-2}$

18. $\dfrac{2}{a+3} - \dfrac{4}{a-1}$

19. $\dfrac{a-b}{b} + \dfrac{b}{a+b}$

20. $\dfrac{x+2}{x-3} - \dfrac{x+3}{x-2}$

21. $\dfrac{7}{4x-4} - \dfrac{4}{7x-7}$

22. $\dfrac{2x+3}{x+8} + \dfrac{x+8}{2x+3}$

23. $\dfrac{5}{y-7} - \dfrac{8}{7-y}$

24. $\dfrac{x-y}{x} - \dfrac{x}{x+y}$

25. $\dfrac{x-1}{x+2} - \dfrac{x-2}{x+1}$

26. $\dfrac{5}{x-3} - \dfrac{4}{3-x}$

27. $\dfrac{c}{d^2-cd} + \dfrac{d}{c^2-cd}$

28. $\dfrac{x-5}{x+5} - \dfrac{x+5}{x-5}$

29. $\dfrac{x}{x-1} - \dfrac{x}{x+1} + \dfrac{2}{x^2-1}$

30. $\dfrac{x+2}{x^2+x-2} + \dfrac{3}{x^2-1}$

31. $\dfrac{x}{x^2+10x+25} + \dfrac{1}{x+5}$

32. $\dfrac{b}{ab-a^2} - \dfrac{a}{b^2-ab}$

33. $\dfrac{4x+3}{x^2+3x-4} - \dfrac{x+4}{4x^2-x-3}$

34. $\dfrac{a^2}{4a^2+12a+9} - \dfrac{a+3}{4a^2-9}$

35. $\dfrac{x^2}{x^2+12x+36} - \dfrac{x+3}{x^2-36}$

36. $\dfrac{a^2}{9a^2-25} - \dfrac{a+1}{9a^2-30a+25}$

37. $\dfrac{x+1}{3x^3+x^2-4x} - \dfrac{x-1}{3x^3+7x^2+4x}$

38. $\dfrac{5}{3x^4-27x^2} + \dfrac{2}{x^3+6x^2+9x} - \dfrac{3}{6x-2x^2}$

8.7 Complex Fractions

A **simple fraction** is a fraction that has only one fraction line. Examples of simple fractions:

$$\frac{2}{x}, \quad \frac{3+y}{12}, \quad \frac{7a-7b}{ab^2}, \quad \frac{5}{x+y}$$

A **complex fraction** is a fraction that has more than one fraction line. Examples of complex fractions:

$$\frac{\dfrac{2}{x}}{3}, \quad \frac{a}{\dfrac{1}{c}}, \quad \frac{\dfrac{3}{z}}{\dfrac{5}{z}}, \quad \frac{\dfrac{3}{x}-\dfrac{2}{y}}{\dfrac{5}{x}+\dfrac{3}{y}}$$

The parts of a complex fraction are as follows:

$$\left.\begin{array}{c} \dfrac{1}{x} + \dfrac{3}{y} \\ \hline \dfrac{5}{x} - \dfrac{2}{y} \end{array}\right\} \begin{array}{l} \text{Primary numerator} \\ \leftarrow\text{Main fraction line} \\ \text{Primary denominator} \end{array}$$

Secondary fractions

$$\dfrac{\dfrac{1}{x} + \dfrac{3}{y}}{\dfrac{5}{x} - \dfrac{2}{y}}$$

Secondary fractions

TO SIMPLIFY COMPLEX FRACTIONS

Method 1: Multiply both numerator and denominator of the complex fraction by the LCD of the secondary fractions, then simplify the results.

Method 2: First simplify the numerator and denominator of the complex fraction, then divide the simplified numerator by the simplified denominator.

Note that in some of the following examples the solution by method 1 is easier than that by method 2. In others, the opposite is true.

Example 1   Simplify the following complex fraction:

$$\dfrac{\dfrac{1}{2} + \dfrac{3}{4}}{\dfrac{5}{6} - \dfrac{2}{3}}$$

Method 1 The LCD of the secondary denominators 2, 4, 6, and 3 is 12.

$$\underbrace{\dfrac{12}{12}}_{\text{This is 1}} \cdot \dfrac{\dfrac{1}{2} + \dfrac{3}{4}}{\dfrac{5}{6} - \dfrac{2}{3}} = \dfrac{12\left(\dfrac{1}{2} + \dfrac{3}{4}\right)}{12\left(\dfrac{5}{6} - \dfrac{2}{3}\right)} = \dfrac{\dfrac{12}{1}\left(\dfrac{1}{2}\right) + \dfrac{12}{1}\left(\dfrac{3}{4}\right)}{\dfrac{12}{1}\left(\dfrac{5}{6}\right) - \dfrac{12}{1}\left(\dfrac{2}{3}\right)} = \dfrac{6 + 9}{10 - 8} = \dfrac{15}{2} = 7\dfrac{1}{2}$$

Method 2

$$\dfrac{\dfrac{1}{2} + \dfrac{3}{4}}{\dfrac{5}{6} - \dfrac{2}{3}} = \left(\dfrac{1}{2} + \dfrac{3}{4}\right) \div \left(\dfrac{5}{6} - \dfrac{2}{3}\right) = \left(\dfrac{2}{4} + \dfrac{3}{4}\right) \div \left(\dfrac{5}{6} - \dfrac{4}{6}\right)$$

$$= \dfrac{5}{4} \div \dfrac{1}{6} = \dfrac{5}{4} \cdot \dfrac{\overset{3}{\cancel{6}}}{\underset{2}{\cancel{4}} \cdot 1} = \dfrac{15}{2} = 7\dfrac{1}{2} \quad \blacksquare$$

Example 2 Simplify $\dfrac{\dfrac{4b^2}{9a^2}}{\dfrac{8b}{3a^3}}$. ← Main fraction line

Method 1 The LCD of the secondary denominators $9a^2$ and $3a^3$ is $9a^3$.

$$\frac{9a^3}{9a^3}\left(\frac{\dfrac{4b^2}{9a^2}}{\dfrac{8b}{3a^3}}\right) = \frac{\dfrac{9a^3}{1}\left(\dfrac{4b^2}{9a^2}\right)}{\dfrac{9a^3}{1}\left(\dfrac{8b}{3a^3}\right)} = \frac{4ab^2}{24b} = \frac{ab}{6}$$

⌐— The value of this fraction is 1

Method 2

$$\frac{\dfrac{4b^2}{9a^2}}{\dfrac{8b}{3a^3}} = \frac{4b^2}{9a^2} \div \frac{8b}{3a^3} = \frac{\overset{1}{\cancel{4b^2}}}{\underset{3}{\cancel{9a^2}}} \cdot \frac{\overset{1}{\cancel{3a^3}}}{\underset{2}{\cancel{8b}}} = \frac{ab}{6} \quad \blacksquare$$

Example 3 Simplify $\dfrac{\dfrac{2}{x} - \dfrac{3}{x^2}}{5 + \dfrac{1}{x}}$.

Method 1 The LCD of the secondary denominators x and x^2 is x^2.

$$\frac{x^2}{x^2}\left(\frac{\dfrac{2}{x} - \dfrac{3}{x^2}}{5 + \dfrac{1}{x}}\right) = \frac{\dfrac{x^2}{1}\left(\dfrac{2}{x}\right) - \dfrac{x^2}{1}\left(\dfrac{3}{x^2}\right)}{\dfrac{x^2}{1}\left(\dfrac{5}{1}\right) + \dfrac{x^2}{1}\left(\dfrac{1}{x}\right)} = \frac{2x - 3}{5x^2 + x}$$

Method 2

$$\frac{\dfrac{2}{x} - \dfrac{3}{x^2}}{5 + \dfrac{1}{x}} = \left(\frac{2}{x} - \frac{3}{x^2}\right) \div \left(5 + \frac{1}{x}\right)$$

$$= \left(\frac{2x}{x^2} - \frac{3}{x^2}\right) \div \left(\frac{5x}{x} + \frac{1}{x}\right)$$

$$= \frac{2x - 3}{x^2} \div \frac{5x + 1}{x}$$

$$= \frac{2x - 3}{\underset{x}{\cancel{x^2}}} \cdot \frac{\overset{1}{\cancel{x}}}{5x + 1} = \frac{2x - 3}{5x^2 + x} \quad \blacksquare$$

Example 4   Simplify $\dfrac{1 + \dfrac{2}{a}}{1 - \dfrac{4}{a^2}}$.

Method 1 The LCD of the secondary denominators a and a^2 is a^2.

$$\frac{a^2}{a^2}\left(\frac{1 + \dfrac{2}{a}}{1 - \dfrac{4}{a^2}} \right) = \frac{\dfrac{a^2}{1}\left(\dfrac{1}{1}\right) + \dfrac{a^2}{1}\left(\dfrac{2}{a}\right)}{\dfrac{a^2}{1}\left(\dfrac{1}{1}\right) - \dfrac{a^2}{1}\left(\dfrac{4}{a^2}\right)} = \frac{a^2 + 2a}{a^2 - 4} = \frac{a\overset{1}{(a + 2)}}{\underset{1}{(a + 2)}(a - 2)}$$

$$= \frac{a}{a - 2}$$

Method 2

$$\frac{1 + \dfrac{2}{a}}{1 - \dfrac{4}{a^2}} = \left(1 + \frac{2}{a}\right) \div \left(1 - \frac{4}{a^2}\right)$$

$$= \left(\frac{a}{a} + \frac{2}{a}\right) \div \left(\frac{a^2}{a^2} - \frac{4}{a^2}\right)$$

$$= \frac{a + 2}{a} \div \frac{a^2 - 4}{a^2}$$

$$= \frac{a + 2}{a} \cdot \frac{a^2}{a^2 - 4}$$

$$= \frac{\overset{1}{a + 2}}{\underset{1}{a}} \cdot \frac{\overset{a}{a^2}}{\underset{1}{(a + 2)}(a - 2)}$$

$$= \frac{a}{a - 2} \quad \blacksquare$$

EXERCISES 8.7

Set I   Simplify each of the complex fractions.

1. $\dfrac{\dfrac{3}{4}}{\dfrac{5}{6}}$

2. $\dfrac{\dfrac{3}{5}}{\dfrac{3}{4}}$

3. $\dfrac{\dfrac{2}{3}}{\dfrac{4}{9}}$

4. $\dfrac{\dfrac{5}{6}}{\dfrac{5}{9}}$

5. $\dfrac{\dfrac{3}{4} - \dfrac{1}{2}}{\dfrac{5}{8} + \dfrac{1}{4}}$

6. $\dfrac{\dfrac{5}{6} - \dfrac{1}{3}}{\dfrac{2}{9} + \dfrac{1}{6}}$

7. $\dfrac{\dfrac{3}{5} + 2}{2 - \dfrac{3}{4}}$

8. $\dfrac{\dfrac{3}{16} + 5}{6 - \dfrac{7}{8}}$

9. $\dfrac{\dfrac{5x^3}{3y^4}}{\dfrac{10x}{9y}}$

10. $\dfrac{\dfrac{8a^4}{5b}}{\dfrac{4a^3}{15b^2}}$

11. $\dfrac{\dfrac{18cd^2}{5a^3b}}{\dfrac{12cd^2}{15ab^2}}$

12. $\dfrac{\dfrac{8x^2y}{7z^3}}{\dfrac{12xy^2}{21z^5}}$

13. $\dfrac{\dfrac{x+3}{5}}{\dfrac{2x+6}{10}}$ **14.** $\dfrac{\dfrac{a-4}{3}}{\dfrac{2a-8}{9}}$ **15.** $\dfrac{\dfrac{x+2}{2x}}{\dfrac{x+1}{4x^2}}$ **16.** $\dfrac{\dfrac{x-3}{3x^2}}{\dfrac{x-9}{9x}}$

17. $\dfrac{\dfrac{a}{b}+1}{\dfrac{a}{b}-1}$ **18.** $\dfrac{2+\dfrac{x}{y}}{2-\dfrac{x}{y}}$ **19.** $\dfrac{\dfrac{1}{x}+x}{\dfrac{1}{x}-x}$ **20.** $\dfrac{a-\dfrac{4}{a}}{a+\dfrac{4}{a}}$

21. $\dfrac{\dfrac{c}{d}+2}{\dfrac{c^2}{d^2}-4}$ **22.** $\dfrac{\dfrac{x^2}{y^2}-1}{\dfrac{x}{y}-1}$ **23.** $\dfrac{x+\dfrac{x}{y}}{1+\dfrac{1}{y}}$ **24.** $\dfrac{1-\dfrac{1}{b}}{3-\dfrac{3}{b}}$

25. $\dfrac{\dfrac{1}{x^2}-\dfrac{1}{y^2}}{\dfrac{1}{x}+\dfrac{1}{y}}$ **26.** $\dfrac{\dfrac{1}{a^2}-\dfrac{1}{4}}{\dfrac{1}{a}-\dfrac{1}{2}}$ **27.** $\dfrac{\dfrac{2}{x}-\dfrac{4}{x^2}}{\dfrac{1}{x}-\dfrac{2}{x^2}}$ **28.** $\dfrac{\dfrac{2}{y^2}+\dfrac{1}{y}}{\dfrac{8}{y^2}+\dfrac{4}{y}}$

Set II Simplify each of the complex fractions.

1. $\dfrac{\dfrac{5}{6}}{\dfrac{2}{9}}$ **2.** $\dfrac{\dfrac{3}{5}}{\dfrac{9}{10}}$ **3.** $\dfrac{\dfrac{5}{6}}{\dfrac{10}{18}}$ **4.** $\dfrac{\dfrac{8}{9}}{\dfrac{8}{27}}$

5. $\dfrac{\dfrac{7}{9}-\dfrac{1}{3}}{\dfrac{5}{2}-\dfrac{1}{4}}$ **6.** $\dfrac{\dfrac{5}{6}-\dfrac{1}{3}}{\dfrac{3}{2}-\dfrac{1}{4}}$ **7.** $\dfrac{\dfrac{3}{4}+2}{3-\dfrac{1}{2}}$ **8.** $\dfrac{\dfrac{1}{8}+3}{4-\dfrac{1}{2}}$

9. $\dfrac{\dfrac{3a^3}{5b^2}}{\dfrac{6a^2}{10b^3}}$ **10.** $\dfrac{\dfrac{4x^4}{5y^3}}{\dfrac{8x^2}{10y^4}}$ **11.** $\dfrac{\dfrac{16bc^2}{5a^2d}}{\dfrac{4ac^4}{10b^2d^3}}$ **12.** $\dfrac{\dfrac{9x^2y^3}{11wz^2}}{\dfrac{18xy^4}{22w^3z}}$

13. $\dfrac{\dfrac{x-2}{4}}{\dfrac{3x-6}{12}}$ **14.** $\dfrac{\dfrac{a}{b}-2}{2+\dfrac{a}{b}}$ **15.** $\dfrac{\dfrac{x+3}{3x}}{\dfrac{x+1}{6x^2}}$ **16.** $\dfrac{\dfrac{1}{x+2}}{\dfrac{4}{3x+6}}$

17. $\dfrac{\dfrac{4}{x}+1}{\dfrac{4}{x}-1}$ **18.** $\dfrac{2}{1+\dfrac{1}{x}}$ **19.** $\dfrac{\dfrac{3}{a}-a}{\dfrac{3}{a}+a}$ **20.** $\dfrac{1-\dfrac{1}{a^2}}{\dfrac{1}{a}-\dfrac{1}{a^2}}$

21. $\dfrac{\dfrac{x^2}{y^2}-4}{\dfrac{x}{y}+2}$ **22.** $\dfrac{\dfrac{1}{a}+\dfrac{1}{b}}{\dfrac{1}{ab}}$ **23.** $\dfrac{2+\dfrac{2}{b}}{a+\dfrac{a}{2}}$ **24.** $\dfrac{\dfrac{1}{x^2}-9}{\dfrac{1}{x}+3}$

25. $\dfrac{\dfrac{1}{x^2}-\dfrac{4}{y^2}}{\dfrac{1}{x}+\dfrac{2}{y}}$ **26.** $\dfrac{\dfrac{x+1}{6x^2}}{\dfrac{x+2}{2x}}$ **27.** $\dfrac{\dfrac{1}{y^2}-\dfrac{1}{9}}{\dfrac{1}{y}+\dfrac{1}{3}}$ **28.** $\dfrac{\dfrac{a^2}{9}-\dfrac{1}{b^2}}{\dfrac{a}{3}+\dfrac{1}{b}}$

8.8 Review: 8.1–8.7

**Algebraic Fractions
8.1**

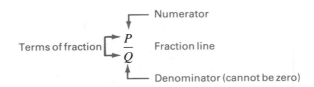

where P and Q are polynomials.

8.7 **A simple fraction** is a fraction that has only one fraction line.

8.7 **A complex fraction** is a fraction that has more than one fraction line.

**Equivalent Fractions
8.1**

Equivalent fractions are fractions that have the same value. If the numerator and denominator of a fraction are multiplied by or divided by the same nonzero number, the new fraction is equivalent to the original one.

**Excluded Values
8.1**

Any value of the variable(s) that would make any denominator be zero must be excluded.

**The Three Signs
of a Fraction
8.1**

Every fraction has three signs associated with it: the sign of the entire fraction, the sign of the numerator, and the sign of the denominator. If any two of the three signs of a fraction are changed, the value of the fraction is unchanged.

**To Reduce a Fraction to
Lowest Terms
8.2**

1. Factor the numerator and the denominator completely.

2. Divide the numerator and the denominator by all factors common to both.

**To Multiply Fractions
8.3**

1. Factor the numerators and the denominators of the fractions.

2. Divide the numerators and the denominators by all factors common to both.

3. The answer is the product of factors remaining in the numerator divided by the product of factors remaining in the denominator.

**To Divide Fractions
8.3**

Multiply the dividend by the multiplicative inverse (the reciprocal) of the divisor.

$$\frac{a}{b} \div \frac{c}{d} = \frac{a}{b} \cdot \frac{d}{c}$$

Dividend ⟶   ⟵ Divisor

**To Find the LCD
8.5**

1. Factor each denominator completely. Repeated factors should be expressed as powers.

2. Write down each different base that appears.

3. Raise each base to the highest power to which it occurs in any denominator.

4. The LCD is the product of all the factors found in step 3.

**To Add Like Fractions
8.4**

1. Add the numerators.

2. Write the sum of the numerators over the denominator of the like fractions.

3. Reduce the resulting fraction to lowest terms.

To Add Unlike Fractions
8.6

1. Find the LCD.

2. Convert all fractions to equivalent fractions having the LCD as denominator.

3. Add the resulting like fractions.

4. Reduce the resulting fraction to lowest terms.

To Simplify
Complex Fractions
8.7

Method 1: Multiply both numerator and denominator of the complex fraction by the LCD of the secondary fractions, then simplify the results.

Method 2: First simplify the numerator and denominator of the complex fraction, then divide the simplified numerator by the denominator.

Review Exercises 8.8   Set I

In Exercises 1–4, determine what value (or values) of the variable must be excluded.

1. $\dfrac{2x-1}{x+4}$

2. $8y - \dfrac{4}{5y}$

3. $\dfrac{x-2}{x^2-9}$

4. $\dfrac{x+4}{x^3+x^2-2x}$

In Exercises 5 and 6, determine whether the pairs of fractions are equivalent.

5. $\dfrac{x+1}{2x+2}, \dfrac{1}{2}$

6. $\dfrac{4x+5}{3x+5}, \dfrac{4x}{3x}$

In Exercises 7–9, find the missing term.

7. $\dfrac{x-6}{7} = -\dfrac{?}{-7}$

8. $\dfrac{2}{a-b} = -\dfrac{?}{b-a}$

9. $\dfrac{3-x}{4} = \dfrac{x-3}{?}$

In Exercises 10–15, reduce each fraction to lowest terms.

10. $\dfrac{6ab^4}{3a^3b^3}$

11. $\dfrac{2+4m}{2}$

12. $\dfrac{3-6n}{3}$

13. $\dfrac{a^2-4}{a+2}$

14. $\dfrac{x-5}{x^2-3x-10}$

15. $\dfrac{a-b}{ax+ay-bx-by}$

In Exercises 16–28, perform the indicated operations.

16. $\dfrac{8}{k} + \dfrac{2}{k}$

17. $5 - \dfrac{3}{2x}$

18. $\dfrac{4m}{2m-3} - \dfrac{2m+3}{2m-3}$

19. $\dfrac{-5a^2}{3b} \div \dfrac{10a}{9b^2}$

20. $\dfrac{21x^3}{4y^2} \div \dfrac{-7x}{8y^4}$

21. $\dfrac{3x + 6}{6} \cdot \dfrac{2x^2}{4x + 8}$

22. $\dfrac{y - 3}{2} - \dfrac{y + 4}{3}$

23. $\dfrac{a - 2}{a - 1} + \dfrac{a + 1}{a + 2}$

24. $\dfrac{k + 3}{k - 2} - \dfrac{k + 2}{k - 3}$

25. $\dfrac{2x^2 - 6x}{x + 2} \div \dfrac{x}{4x + 8}$

26. $\dfrac{4z^2}{z - 5} \div \dfrac{z}{2z - 10}$

27. $\dfrac{3x + 4}{2x^2 + 7x + 6} - \dfrac{x + 2}{3x^2 + 10x + 8}$

28. $\dfrac{5x - 20}{3x^3 + 24x^2 + 48x} - \dfrac{x + 4}{5x^2 - 80}$

In Exercises 29–32, simplify the complex fractions.

29. $\dfrac{\dfrac{5k^2}{3m^2}}{\dfrac{10k}{9m}}$

30. $\dfrac{3 - \dfrac{a}{b}}{2 + \dfrac{a}{b}}$

31. $\dfrac{\dfrac{a^2}{b^2} - 1}{\dfrac{a}{b} - 1}$

32. $\dfrac{1 - \dfrac{16}{x^2}}{1 - \dfrac{4}{x}}$

Review Exercises 8.8 Set II

NAME _____

In Exercises 1–4, determine what value (or values) of the variable must be excluded.

ANSWERS

1. $\dfrac{x + 2}{x - 1}$

2. $3x + \dfrac{5}{2x}$

1. _____

2. _____

3. $\dfrac{x + 2}{x^2 + 2x - 3}$

4. $\dfrac{x - 2}{2x^2 + x - 3}$

3. _____

4. _____

5. _____

In Exercises 5 and 6, determine whether the pairs of fractions are equivalent.

6. _____

5. $\dfrac{5x}{3y}, \ \dfrac{5x + 1}{3y + 1}$

6. $\dfrac{3x(x - 1)}{2x(x - 1)}, \ \dfrac{3x}{2x}$

7. _____

8. _____

9. _____

In Exercises 7–9, find the missing term.

7. $\dfrac{6}{x - 3} = \dfrac{-6}{?}$

8. $\dfrac{4 - x}{7} = -\dfrac{?}{7}$

10. _____

11. _____

9. $\dfrac{x - 3}{(2 + x)(5 - 2x)} = \dfrac{?}{(x + 2)(2x - 5)}$

12. _____

13. _____

14. _____

In Exercises 10–15, reduce each fraction to lowest terms.

15. _____

10. $\dfrac{6ab^3}{2ab}$

11. $\dfrac{x + 2}{x^2 - 2x - 8}$

12. $\dfrac{9 - y^2}{3 - y}$

13. $\dfrac{x + 3}{x^2 - x - 12}$

14. $\dfrac{4z^3 + 4z^2 - 24z}{2z^2 + 4z - 6}$

15. $\dfrac{6k^3 - 12k^2 - 18k}{3k^2 + 3k - 36}$

In Exercises 16–28, perform the indicated operations.

16. $\dfrac{3}{x} + \dfrac{5}{x}$

17. $4 - \dfrac{3}{2x}$

18. $\dfrac{2a}{3a + 1} - \dfrac{3a - 1}{3a + 1}$

19. $\dfrac{x + 1}{2} - \dfrac{x - 3}{5}$

20. $\dfrac{y - 2}{y + 1} - \dfrac{y - 1}{y + 2}$

21. $\dfrac{15x^3}{4y^2} \div \dfrac{5x^2}{8y}$

22. $\dfrac{4x - 4}{2} \cdot \dfrac{6x^2}{3x - 3}$

23. $\dfrac{z^2 + 3z + 2}{z^2 + 2z + 1} \div \dfrac{z^2 + 2z - 3}{z^2 - 1}$

24. $\dfrac{5x - 5}{10} \cdot \dfrac{4x^3}{2x - 2}$

25. $\dfrac{x + 4}{5} - \dfrac{x - 2}{3}$

26. $\dfrac{x^2 + 2x - 3}{2x^2 - x - 1} \cdot \dfrac{6x + 3}{3x^2 + 11x + 6}$

27. $\dfrac{15x + 3}{50x^2 - 20x + 2} - \dfrac{4}{75x^2 - 3}$

28. $\dfrac{5}{6x^2 + 5x - 1} - \dfrac{x + 1}{36x^2 - 12x + 1}$

16. _____

17. _____

18. _____

19. _____

20. _____

21. _____

22. _____

23. _____

24. _____

25. _____

26. _____

27. _____

28. _____

In Exercises 29–32, simplify the complex fractions.

29. $\dfrac{\dfrac{10a^2b}{12a^4b^3}}{\dfrac{5ab^2}{16a^2b^3}}$

30. $\dfrac{2 + \dfrac{a}{b}}{\dfrac{a}{b} - 2}$

31. $\dfrac{\dfrac{1}{x^2} - \dfrac{9}{y^2}}{\dfrac{1}{x} - \dfrac{3}{y}}$

32. $\dfrac{\dfrac{x}{y} + 2}{\dfrac{x}{y} - 2}$

29. _____

30. _____

31. _____

32. _____

8.9 Solving Equations That Have Fractions

In Section 4.4, we removed fractions from simple equations by multiplying both sides of the equation by the same number. All those equations contained only one denominator. If an equation has *more* than one denominator, we can "clear fractions" by multiplying both sides of the equation by the lowest common denominator (LCD).

Extraneous Roots   If any of the denominators have variables in them, the LCD will also have variables in it. If we multiply both sides of an equation by an expression containing the variable, we may get solutions that are not solutions of the original equation. We call such solutions **extraneous roots**. For this reason, when the LCD has variables in it, we must check all solutions in the original equation. Any value of the variable that makes any denominator in the equation zero is *not* a solution; it is an extraneous root. Any value of the variable that gives a false statement (such as $3 = 0$) is also an extraneous root.

8.9A Solving Equations That Simplify to First-Degree Equations

In this section, after we remove fractions and grouping symbols, the equation that remains will be a *first-degree equation* (a polynomial equation whose highest-degree term is first-degree).

TO SOLVE AN EQUATION WITH FRACTIONS THAT SIMPLIFIES TO A FIRST-DEGREE EQUATION

1. Remove fractions by multiplying each term by the LCD.

2. Remove grouping symbols.

3. Collect and combine like terms, getting all terms with the variable on one side, all other terms on the other side.

4. Divide both sides by the coefficient of the variable.

5. Check apparent solutions in the original equation. Reject any extraneous roots.

Example 1   Solve $\dfrac{x}{2} + \dfrac{x}{3} = 5$.

Solution The LCD of the fractions is 6. Multiply both sides by the LCD, 6.

$$6\left(\frac{x}{2} + \frac{x}{3}\right) = 6(5)$$ Use the distributive rule on the left side

$$6\left(\frac{x}{2}\right) + 6\left(\frac{x}{3}\right) = 6(5)$$ This results in *each term* of the equation being multiplied by the LCD, 6

$$\frac{\overset{3}{\cancel{6}}}{1}\left(\frac{x}{\cancel{2}}\right) + \frac{\overset{2}{\cancel{6}}}{1}\left(\frac{x}{\cancel{3}}\right) = \frac{6}{1}\left(\frac{5}{1}\right)$$

$$3x + 2x = 30$$

$$5x = 30$$

$$x = 6$$

Check

$$\frac{x}{2} + \frac{x}{3} = 5$$

$$\frac{6}{2} + \frac{6}{3} \overset{?}{=} 5$$

$$3 + 2 = 5 \quad \blacksquare$$

Example 2   Solve $\dfrac{x-4}{2} - \dfrac{x}{5} = \dfrac{1}{10}$.

Solution LCD $= 10$

$$10\left(\frac{x-4}{2} - \frac{x}{5}\right) = 10\left(\frac{1}{10}\right) \qquad \text{Multiply both sides by the LCD}$$

$$\frac{\overset{5}{\cancel{10}}}{1}\left(\frac{x-4}{\underset{1}{\cancel{2}}}\right) - \frac{\overset{2}{\cancel{10}}}{1}\left(\frac{x}{\underset{1}{\cancel{5}}}\right) = \frac{\overset{1}{\cancel{10}}}{1}\left(\frac{1}{\underset{1}{\cancel{10}}}\right)$$

$$5(x-4) - 2x = 1$$

$$5x - 20 - 2x = 1$$

$$3x = 21$$

$$x = 7$$

Check

$$\frac{x-4}{2} - \frac{x}{5} = \frac{1}{10}$$

$$\frac{7-4}{2} - \frac{7}{5} \overset{?}{=} \frac{1}{10}$$

$$\frac{3}{2} - \frac{7}{5} \overset{?}{=} \frac{1}{10}$$

$$\frac{15}{10} - \frac{14}{10} = \frac{1}{10} \quad \blacksquare$$

NOTE   There could not have been any extraneous roots in Examples 1 and 2 because the LCD did not contain a variable.   ☑

Before you begin to solve an equation, note all excluded values of the variable. Any excluded value cannot be a solution of the equation.

Example 3   Solve $\dfrac{x}{x-3} = \dfrac{3}{x-3} + 4$.

NOTE   3 is an excluded value because it makes the denominators $x - 3$ zero. Therefore, 3 cannot be a solution of this equation.   ☑

Solution LCD $= x - 3$

$$(x - 3)\left(\frac{x}{x - 3}\right) = (x - 3)\left(\frac{3}{x - 3} + 4\right)$$

$$\frac{\overset{1}{(x - 3)}}{1} \cdot \frac{x}{\underset{1}{(x - 3)}} = \frac{\overset{1}{(x - 3)}}{1} \cdot \frac{3}{\underset{1}{(x - 3)}} + \frac{(x - 3)}{1} \cdot \frac{4}{1}$$

$$x \qquad = \qquad 3 \qquad + \quad 4(x - 3)$$

$$x = 3 + 4x - 12$$

$$x = 4x - 9$$

$$-3x = -9$$

$$x = 3$$

Since 3 is an excluded value, this equation has *no solution*.

Check If we had failed to look for excluded values, the check would show there is no solution.

$$\frac{x}{x - 3} = \frac{3}{x - 3} + 4$$

$$\frac{3}{3 - 3} = \frac{3}{3 - 3} + 4$$

$$\frac{3}{0} = \frac{3}{0} + 4$$

Not possible ■

EXERCISES 8.9A

Set I   Solve the equations.

1. $\dfrac{x}{3} + \dfrac{x}{4} = 7$ 　　　　　　　　　　2. $\dfrac{x}{5} + \dfrac{x}{3} = 8$

3. $\dfrac{10}{c} = 2$ 　　　　　　　　　　　　　4. $\dfrac{6}{z} = 4$

5. $\dfrac{a}{2} - \dfrac{a}{5} = 6$ 　　　　　　　　　6. $\dfrac{b}{3} - \dfrac{b}{7} = 12$

7. $\dfrac{9}{2x} = 3$ 　　　　　　　　　　　　8. $\dfrac{14}{3x} = 7$

9. $\dfrac{M - 2}{5} + \dfrac{M}{3} = \dfrac{1}{5}$ 　　　　　10. $\dfrac{y + 2}{4} + \dfrac{y}{5} = \dfrac{1}{4}$

11. $\dfrac{7}{x + 4} = \dfrac{3}{x}$ 　　　　　　　　12. $\dfrac{5}{x + 6} = \dfrac{2}{x}$

13. $\dfrac{3x}{x - 2} = 5$ 　　　　　　　　　14. $\dfrac{4}{x + 3} = \dfrac{2}{x}$

15. $\dfrac{3x - 1}{x - 2} = 3 + \dfrac{2x + 1}{x - 2}$ 　　　16. $\dfrac{5x - 2}{x - 4} = 7 + \dfrac{4x + 2}{x - 4}$

17. $\dfrac{x}{x^2 + 1} = \dfrac{2}{1 + 2x}$ 18. $\dfrac{3x}{3x^2 + 2} = \dfrac{1}{x + 1}$

19. $\dfrac{2x - 1}{3} + \dfrac{3x}{4} = \dfrac{5}{6}$ 20. $\dfrac{3z - 2}{4} + \dfrac{3z}{8} = \dfrac{3}{4}$

21. $\dfrac{2(m - 3)}{5} - \dfrac{3(m + 2)}{2} = \dfrac{7}{10}$ 22. $\dfrac{5(x - 4)}{6} - \dfrac{2(x + 4)}{9} = \dfrac{5}{18}$

23. $\dfrac{x}{x - 2} = \dfrac{2}{x - 2} + 5$ 24. $\dfrac{x}{x + 5} = 4 - \dfrac{5}{x + 5}$

Set II   Solve the equations.

1. $\dfrac{x}{5} + \dfrac{x}{2} = 7$ 2. $\dfrac{x}{6} + \dfrac{x - 6}{3} = 2$

3. $\dfrac{8}{z} = 4$ 4. $\dfrac{15}{x - 1} = 3$

5. $\dfrac{a}{4} - \dfrac{a}{3} = 1$ 6. $\dfrac{x}{3} - \dfrac{x}{5} = 2$

7. $\dfrac{12}{5x} = 2$ 8. $\dfrac{3}{x} + \dfrac{1}{2} = \dfrac{6}{x}$

9. $\dfrac{y - 1}{2} + \dfrac{y}{5} = \dfrac{3}{10}$ 10. $\dfrac{24}{x} = \dfrac{5}{2} + \dfrac{4}{x}$

11. $\dfrac{6}{y - 2} = \dfrac{3}{y}$ 12. $\dfrac{5}{x - 3} = \dfrac{2}{x}$

13. $\dfrac{6x}{x - 2} = 8$ 14. $\dfrac{3}{x + 7} + \dfrac{17}{2x} = \dfrac{1}{x}$

15. $\dfrac{7x - 2}{x - 1} = 5 + \dfrac{6x - 1}{x - 1}$ 16. $\dfrac{3}{x + 1} - \dfrac{2}{x} = \dfrac{5}{2x}$

17. $\dfrac{x}{3x^2 + 5} = \dfrac{1}{1 + 3x}$ 18. $\dfrac{4x - 1}{x - 6} = 2 + \dfrac{3x + 5}{x - 6}$

19. $\dfrac{2z - 4}{3} + \dfrac{3z}{2} = \dfrac{5}{6}$ 20. $\dfrac{5}{1 + x} - \dfrac{5}{x} = \dfrac{1}{x}$

21. $\dfrac{2(m - 1)}{5} - \dfrac{3(m + 1)}{2} = \dfrac{3}{10}$ 22. $\dfrac{5x}{5x^2 + 2} = \dfrac{1}{x + 2}$

23. $\dfrac{y}{y - 3} + 10 = \dfrac{3}{y - 3}$ 24. $\dfrac{x}{x + 3} = 2 + \dfrac{3}{x + 3}$

8.9B   Solving Proportions

A **proportion** is a statement that two ratios or two rates are equal. That is, a proportion is an equation of the form $\dfrac{a}{b} = \dfrac{c}{d}$.

When we multiply both sides of the equation $\dfrac{a}{b} = \dfrac{c}{d}$ by the LCD (which is bd), we obtain the new equation $ad = bc$. When we rewrite $\dfrac{a}{b} = \dfrac{c}{d}$ as $ad = bc$, we often say we are "cross-multiplying." We can cross-multiply as the first step in solving a proportion; however, we must be sure to check all solutions in the original equation in case there are any extraneous roots.

In the proportion $\dfrac{a}{b} = \dfrac{c}{d}$, a is called the *first term*, b the *second term*, c the *third term*, and d the *fourth term*. Also, we sometimes call a and d the **extremes** of the proportion, and b and c the **means** of the proportion. In this case, the cross-multiplication rule may be stated as follows:

In any proportion, the product of the means equals the product of the extremes.

Example 4 Given the proportion $\dfrac{4}{7} = \dfrac{3}{z}$, identify the first, second, third, and fourth terms, the means, and the extremes. Then solve the proportion for z.

Solution The first term is 4, the second term is 7, the third term is 3, the fourth term is z; the means are 7 and 3, and the extremes are 4 and z. To find z, we can first cross-multiply.

$$4z = 3(7)$$
$$4z = 21$$
$$z = \frac{21}{4} \quad \blacksquare$$

Example 5 Solve $\dfrac{3}{a} = 4$.

Solution The LCD is a. If we multiply both sides of $\dfrac{3}{a} = 4$ by a, we have

$$\overset{1}{\cancel{a}}\left(\frac{3}{\cancel{a}}\right)_1 = a(4)$$

$$3 = 4a$$

These equations are equivalent

Alternate solution If we rewrite 4 as $\dfrac{4}{1}$, we have $\dfrac{3}{a} = \dfrac{4}{1}$, which is a proportion.

$$3 \cdot 1 = 4 \cdot a \qquad \text{Cross-multiplying}$$

Then, using either method,

$$a = \frac{3}{4}$$

Check

$$\frac{3}{a} = 4$$

$$\frac{3}{\frac{3}{4}} \overset{?}{=} 4 \qquad \text{Remember: } \frac{3}{\frac{3}{4}} = 3 \div \frac{3}{4} = 3 \cdot \frac{4}{3} = 4$$

$$4 = 4 \quad \blacksquare$$

Example 6   Solve $\dfrac{9x}{x-3} = 6$.

Solution   The LCD is $x - 3$.   *Alternate solution*

$$\frac{9x}{x-3} = 6$$

$$(x-3)\left(\frac{9x}{x-3}\right) = (x-3)(6)$$

$$9x = 6(x-3)$$

$$\frac{9x}{x-3} = \frac{6}{1}$$ Writing $\frac{6}{1}$ makes the equation a proportion

$$9x \cdot 1 = 6(x-3)$$ Cross-multiplying

└─These equations are equivalent─┘

Then, using either method,

$$9x = 6x - 18 \qquad \text{Removing ( )}$$
$$\underline{-6x \quad -6x}$$
$$3x = \qquad -18 \qquad \text{Getting } x\text{'s on one side}$$
$$x = -6 \qquad \text{Dividing both sides by 3}$$

Check

$$\frac{9x}{x-3} = 6$$

$$\frac{9(-6)}{-6-3} \overset{?}{=} 6$$

$$\frac{-54}{-9} \overset{?}{=} 6$$

$$6 = 6 \quad \blacksquare$$

Example 7   Solve $\dfrac{x+1}{x-2} = \dfrac{7}{4}$.

Solution

$$\frac{x+1}{x-2} = \frac{7}{4} \qquad \text{This is a proportion}$$

$$7(x-2) = 4(x+1) \qquad \text{Cross-multiplying } or \text{ multiplying both sides by the LCD}$$

$$7x - 14 = 4x + 4 \qquad \text{Removing ( )}$$

$$3x = 18$$

$$x = 6 \qquad \text{Dividing both sides by 3}$$

Check

$$\frac{x+1}{x-2} = \frac{7}{4}$$

$$\frac{6+1}{6-2} \overset{?}{=} \frac{7}{4}$$

$$\frac{7}{4} = \frac{7}{4} \quad \blacksquare$$

The terms of a proportion can be any kind of number except that no denominator can be zero. (The checking will not be shown in Examples 8–10.)

Example 8   Solve for P: $\dfrac{P}{3} = \dfrac{\frac{5}{6}}{5}$

Solution

$$\frac{P}{3} = \frac{\frac{5}{6}}{5}$$

$$5 \cdot P = \frac{\overset{1}{\cancel{3}}}{1} \cdot \frac{5}{\underset{2}{\cancel{6}}} \qquad \text{Cross-multiplying}$$

$$5 \cdot P = \frac{5}{2}$$

$$\frac{\overset{1}{\cancel{5}} \cdot P}{\underset{1}{\cancel{5}}} = \frac{\frac{5}{2}}{5}$$

$$P = \frac{5}{2} \div 5 = \frac{\overset{1}{\cancel{5}}}{2} \cdot \frac{1}{\underset{1}{\cancel{5}}} = \frac{1}{2} \quad \blacksquare$$

Example 9   Solve for x: $\dfrac{3\frac{1}{2}}{5\frac{1}{4}} = \dfrac{x}{4}$

Solution

$$\frac{3\frac{1}{2}}{5\frac{1}{4}} = \frac{x}{4}$$

$$\left(5\frac{1}{4}\right)x = \left(3\frac{1}{2}\right)4 \qquad \text{Cross-multiplying}$$

$$\frac{21}{4} \cdot x = \frac{7}{\underset{1}{\cancel{2}}} \cdot \frac{\overset{2}{\cancel{4}}}{1} = 14$$

$$\frac{\frac{\overset{1}{\cancel{21}}}{\cancel{4}} \cdot x}{\underset{1}{\frac{\cancel{21}}{\cancel{4}}}} = \frac{14}{\frac{21}{4}}$$

$$x = 14 \div \frac{21}{4}$$

$$x = \frac{\overset{2}{\cancel{14}}}{1} \cdot \frac{4}{\underset{3}{\cancel{21}}} = \frac{8}{3} = 2\frac{2}{3} \quad \blacksquare$$

Example 10   Solve for B: $\dfrac{0.24}{2.7} = \dfrac{4}{B}$.

Solution

$$\frac{0.24}{2.7} = \frac{4}{B}$$

$$\frac{\overset{4}{\cancel{24}}}{\underset{45}{\cancel{270}}} = \frac{4}{B}$$
First multiply numerator and denominator by 100 to eliminate both decimal points; then reduce.

$$\frac{4}{45} = \frac{4}{B}$$

$$4 \cdot B = 45 \cdot 4 \qquad \text{Cross-multiplying}$$

$$\frac{\overset{1}{\cancel{4}} \cdot B}{\underset{1}{\cancel{4}}} = \frac{45 \cdot \overset{1}{\cancel{4}}}{\underset{1}{\cancel{4}}}$$

$$B = 45 \quad \blacksquare$$

A WORD OF CAUTION   Because students can solve a proportion by cross-multiplication, they sometimes use cross-multiplication incorrectly in a *sum* or *product* of fractions. It is incorrect to cross-multiply when adding or multiplying fractions.

Incorrect application of cross-multiplication	*Correct application of cross-multiplication*	*Incorrect application of cross-multiplication*
This is a *sum*	This is an *equation*	This is a *product*
$\dfrac{3}{4} + \dfrac{2}{3} \neq 9 + 8$	If $\dfrac{16}{6} = \dfrac{8}{3}$,	$\dfrac{16}{6} \cdot \dfrac{8}{3} \neq \dfrac{16 \cdot 3}{6 \cdot 8}$
Correct sum:	then $16 \cdot 3 = 6 \cdot 8$	*Correct product:*
$\dfrac{3}{4} + \dfrac{2}{3} = \dfrac{9}{12} + \dfrac{8}{12} = \dfrac{17}{12}$	$48 = 48$	$\dfrac{16}{6} \cdot \dfrac{8}{3} = \dfrac{\overset{8}{\cancel{16}} \cdot 8}{\underset{3}{\cancel{6}} \cdot 3} = \dfrac{64}{9}$ ☑

EXERCISES 8.9B

Set I   **1.** In the proportion $\dfrac{8}{14} = \dfrac{16}{x}$, find the following:

    a. the first term   b. the second term

    c. the third term   d. the fourth term

    e. the means   f. the extremes

2. In the proportion $\dfrac{3}{5} = \dfrac{x}{20}$, find the following:

    a. the first term   b. the second term

    c. the third term   d. the fourth term

    e. the means   f. the extremes

In Exercises 3–26, solve for the variable.

3. $\dfrac{x}{4} = \dfrac{2}{3}$ **4.** $\dfrac{x}{5} = \dfrac{6}{4}$ **5.** $\dfrac{8}{x} = \dfrac{4}{5}$

6. $\dfrac{10}{x} = \dfrac{15}{4}$

7. $\dfrac{4}{7} = \dfrac{x}{21}$

8. $\dfrac{15}{12} = \dfrac{x}{9}$

9. $\dfrac{100}{x} = \dfrac{40}{30}$

10. $\dfrac{144}{36} = \dfrac{96}{x}$

11. $\dfrac{x+1}{x-1} = \dfrac{3}{2}$

12. $\dfrac{x+1}{5} = \dfrac{x-1}{3}$

13. $\dfrac{2x+7}{9} = \dfrac{2x+3}{5}$

14. $\dfrac{2x+7}{3x+10} = \dfrac{3}{4}$

15. $\dfrac{5x-10}{10} = \dfrac{3x-5}{7}$

16. $\dfrac{8x-2}{3x+4} = \dfrac{3}{2}$

17. $\dfrac{\frac{3}{4}}{6} = \dfrac{P}{16}$

18. $\dfrac{\frac{2}{5}}{4} = \dfrac{P}{25}$

19. $\dfrac{A}{9} = \dfrac{3\frac{1}{3}}{5}$

20. $\dfrac{A}{8} = \dfrac{2\frac{1}{4}}{18}$

21. $\dfrac{7.7}{B} = \dfrac{3.5}{5}$

22. $\dfrac{6.8}{B} = \dfrac{17}{57.4}$

23. $\dfrac{P}{100} = \dfrac{\frac{3}{2}}{15}$

24. $\dfrac{P}{100} = \dfrac{\frac{7}{5}}{35}$

25. $\dfrac{12\frac{1}{2}}{100} = \dfrac{A}{48}$

26. $\dfrac{16\frac{2}{3}}{100} = \dfrac{9}{B}$

Set II   **1.** In the proportion $\dfrac{5}{6} = \dfrac{15}{x}$, find the following:

   a. the first term

   b. the second term

   c. the third term

   d. the fourth term

   e. the means

   f. the extremes

   2. In the proportion $\dfrac{81}{x} = \dfrac{9}{5}$, find the following:

   a. the first term

   b. the second term

   c. the third term

   d. the fourth term

   e. the means

   f. the extremes

In Exercises 3–26, solve for the variable.

3. $\dfrac{x}{15} = \dfrac{6}{5}$

4. $\dfrac{x+2}{5} = \dfrac{5-x}{8}$

5. $\dfrac{81}{x} = \dfrac{9}{5}$

6. $\dfrac{4}{13} = \dfrac{16}{x}$

7. $\dfrac{18}{28} = \dfrac{x}{14}$

8. $\dfrac{x}{100} = \dfrac{75}{125}$

9. $\dfrac{26}{x} = \dfrac{39}{14}$

10. $\dfrac{x}{18} = \dfrac{24}{30}$

11. $\dfrac{x+5}{x-5} = \dfrac{27}{10}$

12. $\dfrac{x+8}{x-8} = 5$

13. $\dfrac{4x+3}{2} = \dfrac{4x+1}{3}$

14. $\dfrac{3x-7}{x-5} = 2$

15. $\dfrac{2x+5}{3} = \dfrac{3x-1}{2}$

16. $\dfrac{15}{22} = \dfrac{x}{33}$

17. $\dfrac{\frac{3}{4}}{6} = \dfrac{P}{16}$

18. $\dfrac{\frac{5}{6}}{\frac{1}{6}} = \dfrac{z}{2}$

19. $\dfrac{A}{16} = \dfrac{2\frac{1}{2}}{10}$

20. $\dfrac{1.2}{2} = \dfrac{x}{2.4}$

21. $\dfrac{P}{100} = \dfrac{\frac{3}{4}}{15}$

22. $\dfrac{8}{\frac{1}{2}} = \dfrac{x}{4}$

23. $\dfrac{P}{100} = \dfrac{12\frac{1}{2}}{100}$

24. $\dfrac{9}{y} = \dfrac{3\frac{1}{3}}{\frac{1}{6}}$

25. $\dfrac{6\frac{1}{4}}{100} = \dfrac{1}{B}$

26. $\dfrac{3.1}{x} = \dfrac{10}{2.5}$

8.9C Solving Equations That Simplify to Second-Degree Equations

In this section, the equation that remains after we remove fractions and grouping symbols will usually be a *second-degree equation* (a polynomial whose highest-degree term is second-degree), and it can be solved by factoring. If the new equation is a first-degree equation, solve it by using the methods discussed in Section 8.9A.

TO SOLVE AN EQUATION WITH FRACTIONS THAT SIMPLIFIES TO A SECOND-DEGREE EQUATION

1. Remove fractions by multiplying each term by the LCD.

2. Remove grouping symbols.

3. Combine like terms.

4. If there are second-degree terms, get all nonzero terms to one side by adding the same expression to both sides. Only zero must remain on the other side. Then arrange the terms in descending powers.

5. Factor the polynomial.

6. Set each factor equal to zero, and solve each resulting equation for the variable.

7. Check apparent solutions in the original equation. Reject any extraneous roots.

Example 11 Solve $\dfrac{2}{x} + \dfrac{3}{x^2} = 1$.

Solution LCD $= x^2$. (Note that 0 is an excluded value.)

$$x^2 \left(\frac{2}{x} + \frac{3}{x^2} \right) = x^2 (1) \qquad \text{Multiply both sides by } x^2$$

$$\frac{x^2}{1}\left(\frac{2}{x}\right) + \frac{x^2}{1}\left(\frac{3}{x^2}\right) = \frac{x^2}{1}\left(\frac{1}{1}\right)$$

Second-degree term

$$2x + 3 = x^2$$

$$0 = x^2 - 2x - 3$$

$$0 = (x - 3)(x + 1)$$

$$x - 3 = 0 \quad | \quad x + 1 = 0$$

$$x = 3 \quad | \quad x = -1$$

Check for x = 3	*Check for x = −1*
$\dfrac{2}{x} + \dfrac{3}{x^2} = 1$	$\dfrac{2}{x} + \dfrac{3}{x^2} = 1$
$\dfrac{2}{3} + \dfrac{3}{3^2} \overset{?}{=} 1$	$\dfrac{2}{-1} + \dfrac{3}{(-1)^2} \overset{?}{=} 1$
$\dfrac{2}{3} + \dfrac{1}{3} \overset{?}{=} 1$	$-2 + 3 \overset{?}{=} 1$
$1 = 1$	$1 = 1$ ∎

Example 12 Solve $\dfrac{8}{x} = \dfrac{3}{x + 1} + 3$.

Solution LCD $= x(x + 1)$. (Note that 0 and -1 are excluded values.)

$$x(x + 1)\left(\frac{8}{x}\right) = x(x + 1)\left(\frac{3}{x + 1} + 3\right)$$

$$\frac{\overset{1}{\cancel{x}}(x + 1)}{1}\frac{8}{\underset{1}{\cancel{x}}} = \frac{x(\overset{1}{\cancel{x + 1}})}{1}\frac{3}{\underset{1}{(\cancel{x + 1})}} + \frac{x(x + 1)}{1}\frac{3}{1}$$

$$8(x + 1) = 3x + 3x(x + 1)$$

Second-degree term

$$8x + 8 = 3x + 3x^2 + 3x$$

$$0 = 3x^2 - 2x - 8$$

$$0 = (3x + 4)(x - 2)$$

$$3x + 4 = 0 \qquad \bigg| \qquad x - 2 = 0$$

$$3x = -4 \qquad \bigg| \qquad x = 2$$

$$x = -\frac{4}{3} \qquad \bigg|$$

You should check to see that both $x = 2$ and $x = -\frac{4}{3}$ make the two sides of the equation equal. ∎

NOTE   When we multiply both sides of an equation by the LCD, the denominators cancel completely. When we add or subtract fractions, the denominators do *not* cancel.

☑

A WORD OF CAUTION   A common mistake students make is to confuse an *equation* such as $\frac{2}{x} + \frac{3}{x^2} = 1$ with an *addition problem* such as $\frac{2}{x} + \frac{3}{x^2}$.

The equation	*The addition problem*
Both sides are multiplied by the LCD to remove fractions.	Each fraction is changed into an equivalent fraction with the LCD for a denominator.

The equation

Both sides are multiplied by the LCD to remove fractions.

$$\frac{2}{x} + \frac{3}{x^2} = 1 \qquad \text{LCD} = x^2$$

$$\frac{x^2}{1} \cdot \frac{2}{x} + \frac{x^2}{1} \cdot \frac{3}{x^2} = \frac{x^2}{1} \cdot \frac{1}{1}$$

$$2x + 3 = x^2$$

This equation is then solved by factoring (see Example 11). Here the result is two numbers (-1 and 3) that make both sides of the given equation equal.

The addition problem

Each fraction is changed into an equivalent fraction with the LCD for a denominator.

$$\frac{2}{x} + \frac{3}{x^2} \qquad \text{LCD} = x^2$$

This is 1

$$= \frac{2}{x} \cdot \frac{x}{x} + \frac{3}{x^2}$$

$$= \frac{2x}{x^2} + \frac{3}{x^2} = \frac{2x + 3}{x^2}$$

Here the result is a fraction that represents the sum of the given fractions.

The usual mistake made is to multiply both terms of *the sum* by the LCD.

$$\frac{x^2}{1} \cdot \frac{2}{x} + \frac{x^2}{1} \cdot \frac{3}{x^2} = 2x + 3$$

$$\neq \frac{2}{x} + \frac{3}{x^2}$$

The sum has been multiplied by x^2 and therefore is no longer equal to its original value.

☑

EXERCISES 8.9C

Set I   Solve the equations.

1. $z + \frac{1}{z} = \frac{17}{z}$

2. $y + \frac{3}{y} = \frac{12}{y}$

3. $\frac{2}{x} - \frac{2}{x^2} = \frac{1}{2}$

4. $\frac{3}{x} - \frac{4}{x^2} = \frac{1}{2}$

5. $\frac{x}{x + 1} = \frac{4x}{3x + 2}$

6. $\frac{x}{3x - 4} = \frac{3x}{2x + 2}$

7. $\frac{5}{x} - 1 = \frac{x + 11}{x}$

8. $\frac{7}{x} - 1 = \frac{x + 15}{x}$

9. $\frac{1}{x - 1} + \frac{2}{x + 1} = \frac{5}{3}$

10. $\frac{2}{3x + 1} + \frac{1}{x - 1} = \frac{7}{10}$

11. $\frac{3}{2x + 5} + \frac{x}{4} = \frac{3}{4}$

12. $\frac{5}{2x - 1} - \frac{x}{6} = \frac{4}{3}$

Set II   Solve the equations.

1. $x + \dfrac{1}{x} = \dfrac{10}{x}$

2. $y + \dfrac{2}{y} = \dfrac{18}{y}$

3. $\dfrac{5}{x} - \dfrac{1}{x^2} = \dfrac{9}{4}$

4. $\dfrac{1}{2} - \dfrac{1}{2x} = \dfrac{6}{x^2}$

5. $\dfrac{2x}{3x + 1} = \dfrac{4x}{5x + 1}$

6. $\dfrac{x}{8} + \dfrac{x}{2} = \dfrac{10}{x}$

7. $\dfrac{8}{x} - 1 = \dfrac{x + 18}{x}$

8. $\dfrac{5}{x - 3} + \dfrac{1}{6} = \dfrac{7}{x - 2}$

9. $\dfrac{4}{x + 1} = \dfrac{3}{x} + \dfrac{1}{15}$

10. $\dfrac{6}{x + 2} - 1 = \dfrac{x - 4}{x + 2}$

11. $\dfrac{1}{x - 2} - \dfrac{4}{x + 2} = \dfrac{1}{5}$

12. $\dfrac{1}{x - 5} + \dfrac{3}{x + 2} = \dfrac{5}{6}$

8.10 Literal Equations

Literal equations are equations that contain more than one variable.

Example 1   Examples of literal equations:

a. $3x + 4y = 12$

This is an equation in two variables. We might be asked to solve it for x or for y.

b. $\dfrac{4ab}{d} = 15$

This is an equation in three variables. We might be asked to solve it for a, for b, or for d.

c. $A = P(1 + rt)$

This is an equation in four variables. We might be asked to solve it for P, for r, or for t. (It has already been solved for A.)   ■

Generally, when we solve a literal equation for one of its variables, the solution will contain the other variables as well as constants. The variable we are solving for must appear only once all by itself on one side of the equal sign. All other variables and all constants must be on the other side.

TO SOLVE A LITERAL EQUATION

1. Remove fractions (if there are any) by multiplying both sides by the LCD.

2. Remove grouping symbols (if there are any).

3. Collect like terms. All terms containing the variable you are solving for should be on one side, all other terms on the other side.

4. Factor out the variable you are solving for (if it appears in more than one term).

5. Divide both sides by the coefficient of the variable you are solving for.

Example 2 Solve the given equation for the indicated variable:

a. Solve $3x + 4y = 12$ for x.
   Solution

$$3x + 4y = 12 \qquad \text{Subtract } 4y \text{ from both sides}$$

$$3x = 12 - 4y \qquad \text{Divide both sides by 3}$$

$$x = \frac{12 - 4y}{3} \qquad \text{Solution for } x$$

b. Solve $3x + 4y = 12$ for y.
   Solution

$$3x + 4y = 12 \qquad \text{Subtract } 3x \text{ from both sides}$$

$$4y = 12 - 3x \qquad \text{Divide both sides by 4}$$

$$y = \frac{12 - 3x}{4} \qquad \text{Solution for } y$$

c. Solve $\dfrac{4ab}{d} = 15$ for a.
   Solution The LCD is d.

$$\frac{4ab}{d} = 15$$

$$\overset{1}{\cancel{d}}\left(\frac{4ab}{\cancel{d}_1}\right) = (15)d \qquad \text{Multiply both sides by } d$$

$$4ab = 15d \qquad \text{Simplify}$$

$$\frac{4ab}{4b} = \frac{15d}{4b} \qquad \text{Divide both sides by } 4b$$

$$a = \frac{15d}{4b} \qquad \text{Solution}$$

d. Solve $A = P(1 + rt)$ for t.
   Solution

$$A = P(1 + rt)$$

$$A = P + Prt \qquad \text{Remove ( ) by using the distributive rule}$$

$$A - P = Prt \qquad \begin{array}{l}\text{Collect terms with the letter you are solving for } (t) \text{ on}\\ \text{one side and all other terms on the other side}\end{array}$$

$$\frac{A - P}{Pr} = t \qquad \text{Divide both sides by } Pr$$

$$t = \frac{A - P}{Pr} \qquad \text{Solution}$$

e. Solve $I = \dfrac{nE}{R + nr}$ for n.

Solution The LCD is $R + nr$.

$$I = \frac{nE}{R + nr}$$

$$(R + nr)(I) = (\overset{1}{\cancel{R + nr}})\left(\frac{nE}{\underset{1}{\cancel{R + nr}}}\right) \qquad \text{Multiply both sides by the LCD}$$

$$IR + Inr = nE \qquad\qquad \text{Cancel and use the distributive rule}$$

$$\underline{ -Inr \qquad -Inr}$$

$$IR \qquad\;\; = nE - Inr \qquad \text{Get all terms with } n \text{ on one side}$$

$$IR = n(E - Ir) \qquad\qquad \text{Factor out } n$$

$$\frac{IR}{E - Ir} = n \qquad\qquad \text{Divide both sides by } E - Ir$$

$$n = \frac{IR}{E - Ir} \qquad\qquad \text{Solution} \quad \blacksquare$$

EXERCISES 8.10

Set I   Solve each equation for the variable listed after each equation.

1. $2x + y = 4; \; x$

2. $x + 3y = 6; \; y$

3. $y - z = -8; \; z$

4. $m - n = -5; \; n$

5. $2x - y = -4; \; y$

6. $3y - z = -5; \; z$

7. $2x - 3y = 6; \; x$

8. $3x - 2y = 6; \; x$

9. $2(x - 3y) = x + 4; \; x$

10. $3x - 14 = 2(y - 2x); \; x$

11. $PV = k; \; V$

12. $IR = E; \; R$

13. $I = prt; \; p$

14. $V = \ell wh; \; \ell$

15. $p = 2\ell + 2w; \; \ell$

16. $P = 2\ell + 2w; \; w$

17. $y = mx + b; \; x$

18. $V = k + gt; \; t$

19. $S = \dfrac{a}{1 - r}; \; r$

20. $I = \dfrac{E}{R + r}; \; R$

21. $C = \dfrac{5}{9}(F - 32); \; F$

22. $A = \dfrac{h}{2}(B + b); \; B$

23. $L = a + (n - 1)d; \; n$

24. $A = 2\pi rh + 2\pi r^2; \; h$

25. $z = \dfrac{Rr}{R + r};\ R$

26. $c = \dfrac{ab}{a + b};\ b$

27. $\dfrac{1}{F} = \dfrac{1}{u} + \dfrac{1}{v};\ u$

28. $\dfrac{1}{c} = \dfrac{1}{a} + \dfrac{1}{b};\ a$

Set II Solve each equation for the variable listed after each equation.

1. $x + 2y = 5;\ x$

2. $3x - 5y = 8;\ y$

3. $x - y = -4;\ y$

4. $8x + 3y = 1;\ x$

5. $2x - y = -4;\ x$

6. $y - 3x = -4;\ x$

7. $3x - 4y = 12;\ y$

8. $5y - 2x = 10;\ x$

9. $3(x + 2y) = x + 2;\ x$

10. $4(3x + y) = y - 3;\ y$

11. $d = rt;\ t$

12. $\dfrac{3xy}{z} = 10;\ z$

13. $I = prt;\ r$

14. $\dfrac{x}{a} + \dfrac{y}{b} = 1;\ y$

15. $x = 3y + 4z;\ y$

16. $A = \dfrac{1}{2}bh;\ h$

17. $y = mx + b;\ m$

18. $A = P(1 + rt);\ r$

19. $T = \dfrac{s}{2 - x};\ x$

20. $\dfrac{mn}{m + n} = 1;\ m$

21. $A = \dfrac{h}{2}(B + b);\ h$

22. $\dfrac{ab}{c} = a + b;\ b$

23. $N = x + (y + 2)z;\ y$

24. $5xy - 3 = 2(3x + y);\ y$

25. $I = \dfrac{E}{R + r};\ r$

26. $S = \dfrac{n}{2}(A + L);\ L$

27. $\dfrac{1}{F} = \dfrac{1}{u} + \dfrac{1}{v};\ v$

28. $a = \dfrac{b}{2 + c};\ c$

8.11 Word Problems Involving Fractions

In this section, we will discuss word problems that involve fractions. All the types of word problems discussed in Chapter 5 can lead to equations that have fractions.

Example 1   The denominator of a fraction exceeds the numerator by 8. If 2 is added to the numerator and 4 is subtracted from the denominator, the value of the resulting fraction is $\frac{5}{6}$. What is the original fraction?

Solution In this example, we must let x equal the numerator *or* the denominator of the fraction—*not* the original fraction.

Let x = the numerator of the original fraction

$x + 8$ = the denominator of the original fraction

$\dfrac{x + 2}{(x + 8) - 4}$ = the new fraction

| The value of the resulting fraction | is | $\frac{5}{6}$ |

$$\frac{x+2}{(x+8)-4} = \frac{5}{6}$$

$$\frac{x+2}{x+4} = \frac{5}{6} \qquad \text{This is a proportion}$$

$$
\begin{array}{ll}
6(x+2) = 5(x+4) & \text{Cross-multiplying } or \\
6x + 12 = 5x + 20 & \text{multiplying both sides by the LCD} \\
\underline{-5x - 12 \qquad -5x - 12} &
\end{array}
$$

$$x = 8 \qquad \text{The numerator}$$

$$x + 8 = 16 \qquad \text{The denominator}$$

The original fraction is $\frac{8}{16}$.

Check The denominator exceeds the numerator by 8.

$$\frac{8+2}{16-4} = \frac{10}{12} = \frac{5}{6} \quad \blacksquare$$

Example 2 The sum of a number and its reciprocal is $\frac{25}{12}$. Find the number.

Solution

$$\text{Let } x = \text{the number}$$

$$\frac{1}{x} = \text{its reciprocal}$$

| The sum of a number and its reciprocal | is | $\frac{25}{12}$ |

$$x + \frac{1}{x} = \frac{25}{12}$$

$$(12x)\left(x + \frac{1}{x}\right) = (\overset{1}{\cancel{12}}x)\left(\frac{25}{\underset{1}{\cancel{12}}}\right) \qquad \text{Multiplying both sides by the LCD}$$

$$12x^2 + 12 = 25x \qquad \text{Second-degree equation}$$

$$12x^2 - 25x + 12 = 0$$

$$(3x - 4)(4x - 3) = 0$$

$$
\begin{array}{c|c}
3x - 4 = 0 & 4x - 3 = 0 \\
3x = 4 & 4x = 3 \\
x = \dfrac{4}{3} & x = \dfrac{3}{4} \\
\dfrac{1}{x} = \dfrac{3}{4} & \dfrac{1}{x} = \dfrac{4}{3}
\end{array}
$$

There are two answers: (1) The number is $\frac{4}{3}$ and its reciprocal is $\frac{3}{4}$; (2) the number is $\frac{3}{4}$ and its reciprocal is $\frac{4}{3}$.

$$\textit{Check } \frac{4}{3} + \frac{3}{4} = \frac{16}{12} + \frac{9}{12} = \frac{25}{12} \text{ and } \frac{3}{4} + \frac{4}{3} = \frac{25}{12} \quad \blacksquare$$

Solving Word Problems by Using Proportions

Some word problems are easily solved by using proportions. The method for doing so is described in the following box.

SOLVING WORD PROBLEMS THAT LEAD TO PROPORTIONS

1. Represent the unknown quantity by a letter.

2. Be sure to put the units next to the numbers when you write the proportion.

3. Be sure the same units occupy corresponding positions in the two ratios (or rates) of the proportion.

Correct arrangements

$$\frac{\text{miles}}{\text{hours}} = \frac{\text{miles}}{\text{hours}}$$

$$\frac{\text{hours}}{\text{miles}} = \frac{\text{hours}}{\text{miles}}$$

$$\frac{\text{miles}}{\text{miles}} = \frac{\text{hours}}{\text{hours}}$$

Incorrect arrangements

$$\frac{\text{dollars}}{\text{weeks}} = \frac{\text{weeks}}{\text{dollars}}$$

$$\frac{\text{dollars}}{\text{weeks}} = \frac{\text{dollars}}{\text{days}}$$

4. Once the numbers have been correctly entered into the proportion by using the units as a guide, drop the units before cross-multiplying to solve for the unknown.

Example 3 The scale on an architectural drawing is stated as "1 inch equals 8 feet." What are the dimensions of a room that measures $3\frac{1}{2}$ by 4 in. on the drawing?

Solution There are really two variables. We must find the *width* of the room (it corresponds to $3\frac{1}{2}$ in.) and the *length* of the room (it corresponds to 4 in.)

Let x = the width of the room (in feet).

$$\frac{3\frac{1}{2}\text{ in.}}{x\text{ ft}} = \frac{1\text{ in.}}{8\text{ ft}} \qquad \text{We set the two ratios equal to each other}$$

$$\frac{3\frac{1}{2}}{x} = \frac{1}{8}$$

$$x = 8\left(3\frac{1}{2}\right) = 8\left(\frac{7}{2}\right) = 28\text{ ft}$$

Let y = the length of the room (in feet).

$$\frac{4\text{ in.}}{y\text{ ft}} = \frac{1\text{ in.}}{8\text{ ft}}$$

$$\frac{4}{y} = \frac{1}{8}$$

$$y = 8(4) = 32\text{ ft}$$

Therefore, the room is 28 by 32 ft. ∎

Example 4 Jim knows he can drive 406 mi on 14 gal of gasoline. At this rate, how far can he expect to drive on 35 gal?

Solution Let x = the number of miles on 35 gal.

$$\frac{406 \text{ mi}}{14 \text{ gal}} = \frac{x \text{ mi}}{35 \text{ gal}} \qquad \text{We set the two rates equal to each other}$$

$$\frac{406}{14} = \frac{x}{35}$$

$$14x = (406)(35)$$

$$x = \frac{(406)(35)}{14} = 1{,}015 \text{ mi} \quad \blacksquare$$

Work Problems

Work problems are similar to rate-time-distance problems. One basic relationship used to solve work problems is the following:

Rate × Time = Amount of work

In symbols: $r \cdot t = w$

If we know the rate and the time, we can find the amount of work done. If we know the amount of work done in a certain length of time, we can find the rate. If we know the rate and the amount of work done, we can find the time.

Example 5 If Machine A produces brackets at the rate of thirty-five brackets per hour for 7 hr, how many brackets has it produced?

Solution We use the formula $w = rt$.

$$w = 35 \frac{\text{brackets}}{\text{hr}} \times 7 \text{ hr} = 245 \text{ brackets} \quad \blacksquare$$

Example 6 Machine B produces sprocket wheels at the rate of 175 sprocket wheels per hour. How long does it take the machine to produce 805 sprocket wheels?

Solution Because $r \cdot t = w$, $t = \dfrac{w}{r}$. Therefore,

$$t = \frac{805 \text{ sprocket wheels}}{175 \dfrac{\text{spr. wh.}}{\text{hr}}} = 4.6 \text{ hr, or } 4 \text{ hr } 36 \text{ min} \quad \blacksquare$$

The other basic relationship used to solve work problems is as follows:

$$\begin{pmatrix} \text{Amount } A \\ \text{does} \\ \text{in time } x \end{pmatrix} + \begin{pmatrix} \text{Amount } B \\ \text{does} \\ \text{in time } x \end{pmatrix} = \begin{pmatrix} \text{Amount done} \\ \text{together} \\ \text{in time } x \end{pmatrix}$$

Example 7 George can build a fence in 6 days. Brian can do the same job in 4 days. (a) What is George's rate? (b) What is Brian's rate? (c) How long would it take them to build the same fence if they worked together?

Solution

a. Because $r \cdot t = w$, $r = \dfrac{w}{t}$. Therefore, George's rate is

$$\frac{1 \text{ fence}}{6 \text{ days}} = \frac{1}{6} \frac{\text{fence}}{\text{day}}$$

b. Brian's rate is $\dfrac{1 \text{ fence}}{4 \text{ days}} = \dfrac{1}{4} \dfrac{\text{fence}}{\text{day}}$.

c. Let $x =$ the number of days to build the fence together.

George's rate	$\cdot$	George's time	$=$	Amount George builds
$\dfrac{1}{6}$	$\cdot$	x	$=$	$\dfrac{x}{6}$

Brian's rate	$\cdot$	Brian's time	$=$	Amount Brian builds
$\dfrac{1}{4}$	$\cdot$	x	$=$	$\dfrac{x}{4}$

Amount George builds in x days	$+$	Amount Brian builds in x days	$=$	Amount they build together in x days
$\dfrac{x}{6}$	$+$	$\dfrac{x}{4}$	$=$	1 1 fence is built

$\text{LCD} = 12$

$$12\left(\frac{x}{6}\right) + 12\left(\frac{x}{4}\right) = 12(1)$$

$$2x + 3x = 12$$

$$5x = 12$$

$$x = \frac{12}{5} = 2\frac{2}{5} \text{ days}$$

Check

George's work: $\dfrac{1 \text{ fence}}{6 \text{ day}} \cdot \dfrac{12}{5} \text{ days} = \dfrac{2}{5} \text{ fence}$

Brian's work: $\dfrac{1 \text{ fence}}{4 \text{ day}} \cdot \dfrac{12}{5} \text{ days} = \dfrac{3}{5} \text{ fence}$

Together: $\dfrac{2}{5} \text{ fence} + \dfrac{3}{5} \text{ fence} = 1 \text{ fence}$ ∎

Example 8   In a film-processing lab, machine A can process 5,400 ft of film in 60 min. (a) What is the rate of machine A? (b) How long does it take machine B to process 4,400 ft of film if the two machines working together can process 14,240 ft in 80 min?

Solution

a. The rate of machine A is $\dfrac{5,400 \text{ ft}}{60 \text{ min}} = 90 \dfrac{\text{ft}}{\text{min}}$.

b. Let x = the number of minutes for machine B to process 4,400 ft of film

$\dfrac{4,400 \text{ ft}}{x \text{ min}}$ = rate of machine B

Number of feet A processes in 80 min		Number of feet B processes in 80 min		Number of feet they process together in 80 min
	+		=	

$$\left(90\,\frac{\text{ft}}{\text{min}}\right)(80\text{ min}) + \left(\frac{4400}{x}\,\frac{\text{ft}}{\text{min}}\right)(80\text{ min}) = 14{,}240 \text{ ft}$$

$$7{,}200 + \frac{352{,}000}{x} = 14{,}240$$

LCD is x

$$7{,}200x + 352{,}000 = 14{,}240x$$

$$352{,}000 = 7{,}040x$$

$$x = 50 \text{ min}$$

Check If machine B processes 4,400 ft of film in 50 min, its rate is $\dfrac{4,400}{50}\,\dfrac{\text{ft}}{\text{min}} = 88\,\dfrac{\text{ft}}{\text{min}}$. The amount of film it processes in 80 min is $\left(88\,\dfrac{\text{ft}}{\text{min}}\right)(80\text{ min}) = 7{,}040$ ft.

$$7{,}040 \text{ ft} + 7{,}200 \text{ ft} = 14{,}240 \text{ ft}$$

Machine B ⟶   ⟵ Machine A ∎

EXERCISES 8.11

Set I   Set up each problem algebraically, solve, and check. Be sure to state what your variables represent. (Note: These word problems do not necessarily lead to equations involving fractions.)

1. The denominator of a fraction exceeds the numerator by 6. If 4 is added to the numerator and subtracted from the denominator, the resulting fraction equals $\frac{11}{10}$. What is the original fraction?

2. The denominator of a fraction exceeds the numerator by 4. If 6 is subtracted from the numerator and added to the denominator, the resulting fraction equals $\frac{5}{13}$. What is the original fraction?

3. The denominator of a fraction is twice the numerator. If 5 is added to the numerator and subtracted from the denominator, the value of the resulting fraction is $\frac{4}{5}$. What is the original fraction?

4. The denominator of a fraction is 3 times the numerator. If 5 is added to the numerator and subtracted from the denominator, the value of the resulting fraction is $\frac{1}{2}$. What is the original fraction?

5. The sum of a fraction and its reciprocal is $\frac{29}{10}$. Find the number.

6. The sum of a fraction and its reciprocal is $\frac{130}{33}$. Find the number.

For Exercises 7 and 8, use this statement: The scale in an architectural drawing is 1 inch equals 8 feet.

7. Find the dimensions of a room that measures $2\frac{1}{2}$ by 3 in. on the drawing.

8. Find the dimensions of a room that measures $3\frac{1}{4}$ by $4\frac{1}{4}$ in. on the drawing.

9. The ratio of a woman's weight on earth compared to her weight on the moon is 6 : 1. How much would a 150-lb woman weigh on the moon?

10. The ratio of a man's weight on Mars compared to his weight on earth is 2 : 5. How much would a 196-lb man weigh on Mars?

11. The ratio of the weight of lead to the weight of an equal volume of aluminum is 21 : 5. If an aluminum bar weighs 150 lb, what would a lead bar of the same size weigh?

12. The ratio of the weight of platinum to the weight of an equal volume of copper is 12 : 5. If a platinum bar weighs 18 lb, what would a copper bar of the same size weigh?

13. On the first $6\frac{1}{2}$ hr of their shift, a fire crew built twenty-six chains of fire line. How much fire line can they build in the remainder of a 10-hr day if they work at the same pace?

14. The IRS informed a local businessman that for every $5,000 worth of sales he made, his tax would increase by $75. If the man's sales totaled $125,000 for the year, how much was his tax?

15. A meat packer paid a cattle producer $3,420 for nine steers. Assuming all of the animals are of the same approximate weight, how much will the producer receive for sixteen steers?

16. A hog producer fed 1,320 hogs until they reached a marketing weight of 100 kg each. He received a check for their sale amounting to $59,400. How much did the producer receive per hog? What was the price paid for the hogs per kg of live weight?

17. Bill can do a job in 8 hr. Mike can do the same job in 10 hr. How long will it take them to do the same job if they work together?

18. Jeff can paint a house in 5 days. Fred can paint the same house in 7 days. How long will it take them to paint the same house if they work together?

19. The sum of the numerator and denominator of a fraction is 42. If 2 is added to the numerator and 6 is added to the denominator, the resulting fraction is $\frac{2}{3}$. What is the original fraction?

20. The sum of the numerator and denominator of a fraction is 40. If 3 is added to the numerator and 2 is added to the denominator, the resulting fraction is $\frac{4}{5}$. What is the original fraction?

21. Ruth can proofread 220 pages of a deposition in 4 hr. How long does it take Sandra to proofread 210 pages if, when they work together, they can proofread 291 pages in 3 hr?

22. Barbara can type 95 pages of a manuscript in 3 hr. How long does it take Tricia to type 110 pages if she and Barbara working together can type 322 pages in 6 hr?

23. Jim and Jerry live 63 mi apart. Both leave their homes at 7 A.M. by bicycle, riding toward one another. They meet at 10 A.M. If Jim's average speed is three-fourths of Jerry's, how fast does each cycle?

24. Rebecca and Jill live 75 mi apart. Both leave their homes at 8 A.M. by bicycle, riding toward one another. They meet at 10:30 A.M. If Jill's average speed is seven-eighths of Rebecca's, how fast does each cycle?

25. The speed of the current of a river is 5 mph. If a boat can travel 999 mi with the current in the same time it could travel 729 mi against the current, what is the speed of the boat in still water?

26. The speed of the current in a river is 4 mph. If a boat can travel 288 mi with the current in the same time it could travel 224 mi against the current, what is the speed of the boat in still water?

27. Two numbers differ by 12. One-sixth the larger exceeds one-fifth the smaller by 2. Find the numbers.

28. Two numbers differ by 8. One-seventh the larger exceeds one-eighth the smaller by 2. Find the numbers.

Set II   Set up each problem algebraically, solve, and check. Be sure to state what your variables represent. (Note: These word problems do not necessarily lead to equations involving fractions.)

1. The denominator of a fraction exceeds the numerator by 20. If 7 is added to the numerator and subtracted from the denominator, the resulting fraction equals $\frac{6}{7}$. Find the fraction.

2. Suppose the wind speed is 30 mph. If an airplane can fly 240 mi against the wind in the same time that it can fly 420 mi with the wind, what is the speed of the plane in still air?

3. The denominator of a fraction is twice the numerator. If 2 is added to the numerator and subtracted from the denominator, the value of the resulting fraction is $\frac{5}{7}$. What is the original fraction?

4. When a number is subtracted from its reciprocal, the difference is $\frac{21}{10}$. What is the number?

5. The sum of a fraction and its reciprocal is $\frac{34}{15}$. Find the number.

6. Mr. Summers drove his motorboat upstream a certain distance while pulling his son Brian on a water ski. He returned to the starting point pulling his other son Derek. The round trip took 25 min of skiing time. On both legs of the trip, the speedometer read 30 mph. If the speed of the current is 6 mph, how far upstream did he travel?

For Exercises 7 and 8, use this statement: The scale in an architectural drawing is 1 inch equals 8 feet.

7. Find the dimensions of a room that measures $2\frac{1}{4}$ by $3\frac{3}{4}$ in. on the drawing.

8. Find the dimensions of a room that measures $3\frac{1}{8}$ by $4\frac{3}{8}$ in. on the drawing.

9. An apartment house manager spent 22 hr painting three apartments. How long can he expect to take painting the remaining fifteen apartments?

10. Ralph drove 420 mi in $\frac{3}{4}$ of a day. About how far can he drive in $2\frac{1}{2}$ days?

11. The ratio of the width of a rectangle to its length is 4 : 9. If its area is 900 sq. m, find the width and the length.

12. A crew of ten men takes a week to overhaul 25 trucks in a fleet of 100 trucks. How many men would it take to complete the fleet overhaul in one week?

13. A car burns $2\frac{1}{2}$ qt of oil on a 1,800-mi trip. How many quarts of oil can the owner expect to use on a 12,000-mi trip?

14. Fifteen defective axles were found in 100,000 cars of a particular model. How many defective axles would you expect to find in the 2 million cars made of that same model?

15. Mr. Sanders has 300 Leghorn hens and needs 22.86 cm of roosting space per hen. How many meters of roosting space does he need?

16. The Forest Service must determine how many acres will be needed to make an addition of twenty-two campsites. If the existing 24-acre campground accommodates fifty-five campsites, how many additional acres will they need?

17. David can do a job in 9 hr. Mark can do the same job in 11 hr. How long will it take them to do the same job if they work together?

18. Ben bought 120 stamps consisting of 20¢, 15¢, and 3¢ stamps at a total cost of $12.80. He bought twice as many 20¢ stamps as 15¢ stamps and 20 more 3¢ stamps than 20¢ stamps. How many of each kind did he buy?

19. The sum of the numerator and denominator of a fraction is 68. If 1 is added to the numerator and 12 is added to the denominator, the resulting fraction is $\frac{2}{7}$. What is the original fraction?

20. The sum of two numbers is 32. One-half the smaller exceeds one-third the larger by 1. Find the numbers.

21. Karla can type 65 pages of a manuscript in 2 hr. How long does it take Susan to type 450 pages if, when they work together, they can type 210 pages in 3 hr?

22. The sum of the first two of three consecutive odd integers added to the sum of the last two is 60. Find the integers.

23. Jim and Jerry live 80 mi apart. Both leave their homes at 6 A.M. by bicycle, riding toward one another. They meet at 10 A.M. If Jim's average speed is two-thirds of Jerry's, how fast does each cycle?

24. It takes Jim 3 times as long as Jeff to paint a certain house. Working together, Jim and Jeff could paint the same house in 3 days.

    a. How long would it take Jeff working alone to paint the house?

    b. How long would it take Jim working alone to paint the house?

25. The speed of the current in a river is 5 mph. If a boat can travel 198 mi with the current in the same time it could travel 138 mi against the current, what is the speed of the boat in still water?

26. Colin paddles a kayak downstream for 3 hr. After having lunch, he paddles upstream for 5 hr. At that time he is still 6 mi short of getting back to his starting point. If the speed of the stream is 2 mph, how fast does Colin paddle in still water? How far downstream did he travel?

27. Two numbers differ by 3. One-fifth the larger exceeds one-sixth the smaller by 1. Find the numbers.

28. Machine A takes 3 times as long as machine B to do a certain job. Both machines running together can do this same job in 4 hr. How long does it take each machine working alone to do the job?

8.12 Review: 8.9 – 8.11

To Solve an Equation with Fractions 8.9

1. Remove fractions by multiplying each term by the LCD. (If the equation is a proportion, we can cross-multiply.)

2. Remove grouping symbols

First-degree equations	*Quadratic equations*
3. Collect and combine like terms, getting all terms with the variable on one side, all other terms on the other side.	3. Get *all* nonzero terms to one side by adding the same expression to both sides. *Only zero must remain on the other side*. Then arrange the terms in descending powers.
4. Divide both sides by the coefficient of the variable.	4. Factor the polynomial.
	5. Set each factor equal to zero, then solve for the unknown.

Check apparent solutions in the original equation. Any value of the variable that makes any denominator in the equation zero is not a solution.

Literal Equations 8.10

Literal equations are equations that have more than one variable.

To solve a literal equation, proceed in the same way used to solve an equation with a single variable. The solution will be expressed in terms of the other variable given in the literal equation, as well as in numbers.

Solving Word Problems Using Proportions 8.11

1. Represent the unknown quantity by a variable.

2. Write a proportion; be sure the same units occupy corresponding positions in the two ratios (or rates) of the proportion. Put the units next to the numbers when writing the proportion.

3. Solve the proportion by cross-multiplying, dropping the units before cross-multiplying.

Review Exercises 8.12   Set I

In Exercises 1–7, solve for the variable.

1. $\dfrac{6}{m} = 5$

2. $x - \dfrac{3x}{5} = 2$

3. $\dfrac{z}{5} - \dfrac{z}{8} = 3$

4. $\dfrac{3}{2} = \dfrac{3x + 4}{5x - 1}$

5. $\dfrac{4}{2z} + \dfrac{2}{z} = 1$

6. $\dfrac{4}{x^2} - \dfrac{3}{x} = \dfrac{5}{2}$

7. $\dfrac{7}{2x - 1} + \dfrac{1}{18} = \dfrac{x}{6}$

In Exercises 8–11, solve for the variable listed after the equation.

8. $2x - 7y = 14;\ y$

9. $\dfrac{2m}{n} = P;\ n$

10. $V = \dfrac{1}{3}Bh;\ B$

11. $\dfrac{F - 32}{C} = \dfrac{9}{5};\ C$

In Exercises 12–17, set up each problem algebraically, solve, and check. Be sure to state what your variables represent.

12. The numerator of a fraction exceeds the denominator by 4. If 3 is added to the denominator and subtracted from the numerator, the value of the resulting fraction is $\frac{7}{8}$. Find the original fraction.

13. Mr. Maxwell takes 30 min to drive to work in the morning, but he takes 45 min to return home over the same route during the evening rush hour. If his average morning speed is 10 mph faster than his average evening speed, how far is it from his home to his work?

14. The sum of a number and its reciprocal is $\frac{13}{6}$. Find the number.

15. Machine A can do a job in 6 hr. How long does it take machine B to do the same job if, when the two machines work together, they get the job done in 4 hr?

16. Karla drove 324 miles in $\frac{3}{5}$ of a day. At that same rate, about how far can she drive in a day and a half?

17. Rebecca weaves $2\frac{1}{2}$ yards of fabric in $\frac{5}{8}$ of a day. At that same rate, how many yards of fabric can she weave in half a day?

Review Exercises 8.12 Set II

NAME _____

In Exercises 1–7, solve for the variable.

ANSWERS

1. $\dfrac{3}{x} = 4$

2. $\dfrac{x}{3} - \dfrac{x}{2} = 2$

3. $\dfrac{x+2}{5} + \dfrac{2x}{3} = 3$

4. $\dfrac{3x}{7} = \dfrac{x-1}{5}$

5. $\dfrac{x+2}{-2} = \dfrac{3}{x-3}$

6. $\dfrac{17}{6x} + \dfrac{5}{2x^2} = \dfrac{2}{3}$

7. $\dfrac{3}{x} - \dfrac{8}{x^2} = \dfrac{1}{4}$

1. _____

2. _____

3. _____

4. _____

5. _____

6. _____

7. _____

8. _____

9. _____

10. _____

11. _____

12. _____

In Exercises 8–11, solve for the variable listed after the equation.

8. $\dfrac{P}{V} = C; V$

9. $V = LWH; H$

10. $V^2 = 2gS; S$

11. $5(x - 2y) = 14 + 3(2x - y); y$

In Exercises 12–17, set up each problem algebraically, solve, and check. Be sure to state what your variables represent.

12. When a number is subtracted from its reciprocal, the difference is $\frac{40}{21}$. Find the number.

13. When the speed of the wind is 25 mph, a certain airplane can fly only 480 mi against the wind in the same time it can fly 780 mi with the wind. Find the speed of the plane in still air.

14. The denominator of a fraction exceeds the numerator by 7. If 10 is added to the numerator and 15 is subtracted from the denominator, the value of the resulting fraction is $\frac{11}{5}$. Find the original fraction.

15. Alan can paint a house in 5 days. How long would it take Justin alone to paint the same house if the two men working together can paint the house in 3 days?

16. Jill drove 220 miles in $\frac{2}{5}$ of a day. At that same rate, about how far can she drive in half a day?

17. Susan crochets $3\frac{1}{2}$ scarfs in $2\frac{1}{3}$ days. At that same rate, how many scarfs can she crochet in 6 days?

13. _____

14. _____

15. _____

16. _____

17. _____

Chapter 8 Diagnostic Test

The purpose of this test is to see how well you understand operations with fractions. We recommend that you work this diagnostic test *before* your instructor tests you on this chapter. Allow yourself about 50 minutes.

Complete solutions for all the problems on this test, together with section references, are given in the answer section in the back of this book. For the problems you do incorrectly, study the sections referred to.

1. What value(s) of the variable must be excluded, if any, in each of the following expressions?

   a. $\dfrac{3x}{x + 4}$

   b. $\dfrac{5x - 4}{x^2 + 4x}$

2. Find the missing term in each of the following expressions.

   a. $-\dfrac{-8}{3} = \dfrac{8}{?}$

   b. $\dfrac{3}{x - 5} = \dfrac{?}{5 - x}$

In Problems 3 and 4, reduce each fraction to lowest terms.

3. $\dfrac{x^2 - 9}{x^2 - 6x + 9}$

4. $\dfrac{4x^2 - 23xy + 15y^2}{20x^2y - 15xy^2}$

In Problems 5–10, perform the indicated operations. Be sure to reduce fractions to lowest terms.

5. $\dfrac{a}{a^2 + 5a + 6} \cdot \dfrac{4a + 8}{6a^3 + 18a^2}$

6. $\dfrac{6x}{x - 2} - \dfrac{3}{x - 2}$

7. $\dfrac{2x}{x^2 - 9} \div \dfrac{4x^2}{x^2 - 6x + 9}$

8. $\dfrac{b}{b - 1} - \dfrac{b + 1}{b}$

9. $\dfrac{x + 3}{x^2 - 10x + 25} + \dfrac{x}{2x^2 - 7x - 15}$

10. $\dfrac{x}{x + 5} - \dfrac{x}{x - 5} - \dfrac{50}{x^2 - 25}$

In Problems 11 and 12, simplify each complex fraction.

11. $\dfrac{\dfrac{6x^4}{11y^2}}{\dfrac{9x}{22y^4}}$

12. $\dfrac{\dfrac{6}{x} + \dfrac{15}{x^2}}{2 + \dfrac{5}{x}}$

In Problems 13–16, solve each equation.

13. $\dfrac{x + 7}{3} = \dfrac{2x - 1}{4}$

14. $x - \dfrac{4}{x} = 3$

15. $\dfrac{3}{4x + 2} + \dfrac{x + 1}{6} = 1$

16. $\dfrac{3x - 5}{x} = \dfrac{x + 1}{3}$

17. Solve for y: $5x + 7y = 18$

In Problems 18–20, set up each problem algebraically, solve, and check. Be sure to state what your variables represent.

18. The denominator of a fraction is 6 more than the numerator. If 1 is subtracted from the numerator and 7 is added to the denominator, the value of the resulting fraction is $\frac{1}{3}$. What is the original fraction?

19. Mr. Ames drove 21 mi in $\frac{2}{5}$ hr. At that same rate, how far can he drive in $2\frac{2}{3}$ hr?

20. Sherry can type 128 pages of a manuscript in 4 hr. How long does it take Barbara to type 111 pages if both women working together can type 345 pages in 5 hr?

Cumulative Review Exercises: Chapters 1–8

1. Evaluate $10 - (3\sqrt{4} - 5^2)$.

2. Evaluate the formula using the values of the variables given with the formula.

$$V = \frac{25}{8}\left(\frac{H}{D} - \frac{A}{R}\right) \quad H = 10, D = 18, A = 1, R = 5$$

3. Simplify. Write your answer using only positive exponents.

$$\left(\frac{24y^{-3}}{8y^{-1}}\right)^{-2}$$

In Exercises 4–9, perform the indicated operations and simplify.

4. $\dfrac{6x^2 - 2x}{2x}$

5. $(15z^2 + 11z + 4) \div (3z - 2)$

6. $\dfrac{5x - 5}{10} \cdot \dfrac{4x^3}{2x - 2}$

7. $\dfrac{2x}{x + 1} - \dfrac{2x - 1}{x + 1}$

8. $5 + \dfrac{3}{2x^3}$

9. $\dfrac{2x - 1}{x^2 + 9x + 20} - \dfrac{x + 5}{2x^2 + 7x - 4}$

In Exercises 10 and 11, solve each equation.

10. $\dfrac{14}{3x} + \dfrac{42}{x} = 1$

11. $\dfrac{3}{x} - \dfrac{8}{x^2} = \dfrac{1}{4}$

In Exercises 12–15, factor each expression.

12. $x^2 - x - 42$

13. $3w^2 - 48$

14. $20a^2 - 7ab - 3b^2$

15. $3x^2 - 6x - 2xy + 4y$

In Exercises 16–20, set up the problem algebraically, solve, and check. Be sure to state what your variables represent.

16. The sum of two consecutive integers is 33. What are the integers?

17. The sum of two numbers is 5. Their product is -24. What are the numbers?

18. A dealer makes up a 15-lb mixture of oranges. One kind costs 78¢ per pound, and the other costs 99¢ per pound. How many pounds of each kind must be used in order for the mixture to cost 85¢ per pound?

19. Manny has twenty coins with a total value of $1.65. If all the coins are nickels or dimes, how many of each does he have?

20. Margaret has $2.60 in nickels, dimes, and quarters. If she has 1 more dime than quarters and 3 times as many nickels as quarters, how many of each kind of coin does she have?

9 Graphing

Many algebraic relationships are easier to understand if a picture called a graph is drawn. In this chapter, we graph ordered pairs and equations and inequalities in two variables. We also discuss writing the equation of a straight line.

9.1 The Rectangular Coordinate System

In Chapter 4, we graphed the solution of an equation in one variable on the real number line. Solutions of equations and inequalities in two variables usually do not lie on a single number line. Instead, they lie in a *plane*, which is a flat surface. (In this book, we work in the plane only and not in three-dimensional space.)

The **rectangular coordinate system** in the plane consists of a horizontal number line called the **horizontal axis**, or **x-axis**, and a vertical number line called the **vertical axis** or **y-axis**. These lines intersect at a point called the **origin**, and they separate the plane into four regions called **quadrants**, numbered as shown in Figure 9.1.1.

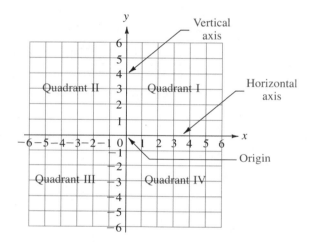

FIGURE 9.1.1   Rectangular Coordinate System

Ordered Pairs   While points on the number line can be represented by a single real number, points in the plane must be represented by an *ordered pair* of real numbers. Two ordered pairs are equal to each other if they contain the same elements *in the same order*. We enclose the elements of an ordered pair within parentheses, never braces or brackets. Thus, $(a, b) \neq (b, a)$.

The first number of the ordered pair is called the *x-coordinate* (or horizontal coordinate, first coordinate, or abscissa). The second number of the ordered pair is called the *y-coordinate* (or vertical coordinate, second coordinate, or ordinate). Consider the ordered pair $(3, 2)$. We call 3 and 2 the *coordinates* of the point $(3, 2)$. The first number, 3, is called the *x*-coordinate of the point $(3, 2)$. The second number, 2, is called the *y*-coordinate of the point $(3, 2)$.

Graph of a Point   There is exactly one point in the plane corresponding to each ordered pair of real numbers, and there is exactly one ordered pair of real numbers corresponding to each point in the plane. The **origin** is the point that corresponds to $(0, 0)$. See Figure 9.1.3 for the graph of the point $(3, 2)$.

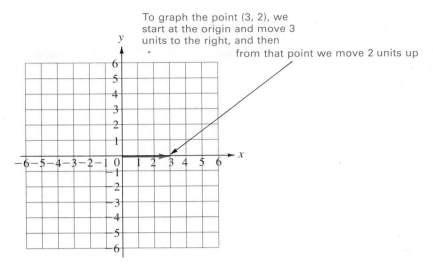

FIGURE 9.1.2

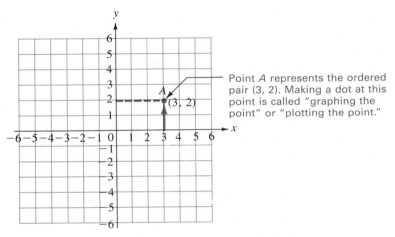

FIGURE 9.1.3   Graph of an Ordered Pair

The *x*-coordinate tells us how far the point is from the *vertical* axis (the *y*-axis). A positive *x*-coordinate indicates that the point is to the right of the *y*-axis. (See Figure 9.1.4.) A negative *x*-coordinate indicates that the point is to the left of the *y*-axis.

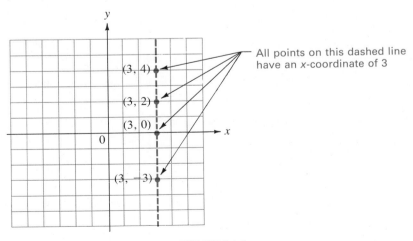

FIGURE 9.1.4

The *y*-coordinate tells us how far the point is from the *horizontal* axis (the *x*-axis). A positive *y*-coordinate indicates that the point is above the *x*-axis. (See Figure 9.1.5.) A negative *y*-coordinate indicates that the point is below the *x*-axis.

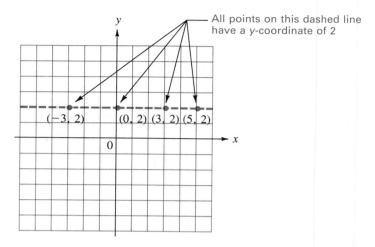

FIGURE 9.1.5

NOTE   When the order is changed in an ordered pair, we get a different point. For example, (1, 4) and (4, 1) are two different points (see Figure 9.1.6). ☑

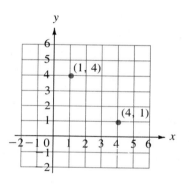

FIGURE 9.1.6

Example 1   Graph the following points:

a. (3, 5)   Start at the origin and move *right* 3 units; then move *up* 5 units (point *A* in Figure 9.1.7).

b. (−5, 2)   Start at the origin and move *left* 5 units; then move *up* 2 units (point *B* in Figure 9.1.7).

c. (−5, −4)   Start at the origin and move *left* 5 units; then move *down* 4 units (point *C* in Figure 9.1.7).

d. (0, −3)   Start at the origin, but because the first number is zero, do not move either right or left. Just move *down* 3 units (point *D* in Figure 9.1.7).

e. (4, −6)   Start at the origin and move *right* 4 units; then move *down* 6 units (point *E* in Figure 9.1.7).

The phrase "plot the points" means the same as "graph the points."

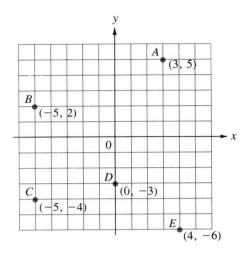

FIGURE 9.1.7 ∎

Example 2   Name the coordinates of points B, C, D, and E in Figure 9.1.8. Also name the quadrant in which each point lies.

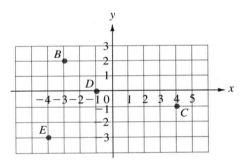

FIGURE 9.1.8

Solution The coordinates of point B are $(-3, 2)$, because B lies 3 units to the *left* of the y-axis and 2 units *above* the x-axis. B lies in quadrant II.

The coordinates of point C, which lies in quadrant IV, are $(4, -1)$, because C lies 4 units to the *right* of the y-axis and 1 unit *below* the x-axis.

The coordinates of point D are $(-1, 0)$, because D lies 1 unit to the *left* of the y-axis and *on* the x-axis. D does not lie in any quadrant; it lies on the x-axis.

The coordinates of point E are $(-4, -3)$, because E lies 4 units to the *left* of the y-axis and 3 units *below* the x-axis. E lies in quadrant III. ∎

The points at which the sides of a triangle meet are called the vertices of the triangle (see Example 3).

Example 3   Draw the triangle whose vertices have the following coordinates:

$$A\ (-1, 2)\quad B\ (3, -2)\quad C\ (-3, -4)$$

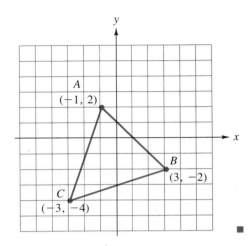

NOTE   The scales on the two axes do not have to be equal. If we needed to graph the point (3, 200), for example, it would be better if the scales were *not* equal.   ☑

EXERCISES 9.1

Set I   **1.** Graph each of the following points.

 a. (3, 1)   b. (−4, −2)   c. (0, 3)

 d. (5, −4)   e. (4, 0)   f. (−2, 4)

2. Graph each of the following points.

 a. (2, 4)   b. (2, −4)   c. (3, 0)

 d. (−3, −2)   e. (0, 0)   f. (0, −4)

In Exercises 3 and 4, use Figure 9.1.9. Give the coordinates of each point and name the quadrant in which the point lies.

3. a. R   b. N   c. U   d. S

4. a. M   b. P   c. Q   d. T

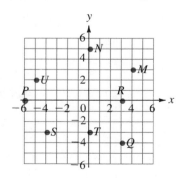

FIGURE 9.1.9

In Exercises 5–8, use Figure 9.1.10.

5. Write the x-coordinate of each of these points.

 a. A   b. C   c. E   d. F

6. Write the *y*-coordinate of each of these points.

    a. *B*   b. *D*   c. *E*   d. *F*

7. a. What is the abscissa of point *F*?

    b. What is the ordinate of point *C*?

8. a. What is the ordinate of point *B*?

    b. What is the abscissa of point *D*?

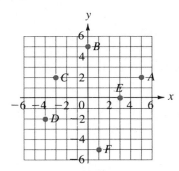

FIGURE 9.1.10

9. What is the *y*-coordinate of the origin?

10. What is the *x*-coordinate of the origin?

11. Draw the triangle whose vertices have the following coordinates:

$$A\ (0, 0) \quad B\ (3, 2) \quad C\ (-4, 5)$$

12. Draw the triangle whose vertices have the following coordinates:

$$A\ (-2, -3) \quad B\ (-2, 4) \quad C\ (3, 5)$$

Set II   **1.** Graph each of the following points.

    a. $(-2, 4)$ b. $(2, -5)$ c. $(-1, -3)$

    d. $(3, 6)$ e. $(-5, -1)$ f. $(5, -2)$

2. Graph each of the following points.

    a. $(-3, 5)$ b. $(0, 4)$ c. $(-2, 0)$

    d. $(2, -4)$ e. $(-4, -3)$ f. $(0, -2)$

In Exercises 3 and 4, use Figure 9.1.11. Give the coordinates of each point and name the quadrant in which the point lies.

3. a. *C*   b. *D*   c. *E*   d. *F*

4. a. *K*   b. *L*   c. *W*   d. *U*

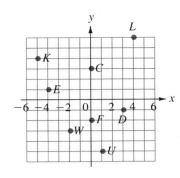

FIGURE 9.1.11

In Exercises 5–8, use Figure 9.1.12.

5. Write the *x*-coordinate of each of these points.

    a. *S*   b. *T*   c. *U*   d. *V*

6. Write the *y*-coordinate of each of these points.

    a. *S*   b. *T*   c. *U*   d. *V*

7. a. What is the abscissa of point *U*?

    b. What is the ordinate of point *S*?

8. a. What is the ordinate of point *V*?

    b. What is the abscissa of point *T*?

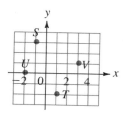

FIGURE 9.1.12

9. What name is given to the point $(0, 0)$?

10. Does every point on the *y*-axis have the same *x*-coordinate? If so, what is it?

11. Draw the triangle whose vertices have the following coordinates:

$$A\,(-6,\,-4) \quad B\,(2,\,-5) \quad C\,(5,\,3)$$

12. Does every point on the *x*-axis have the same *x*-coordinate? If so, what is it?

9.2   Linear Equations in Two Variables

If an equation can be put into the form

$$Ax + By + C = 0$$

where A, B, and C are any real numbers and A and B are not both zero, it is called a **linear equation in two variables**. (Note the word *line* in the word *line*ar.)

Example 1   Examples of linear equations in two variables:

    a. $7x = 5y + 3$ The equation can be rewritten $7x - 5y - 3 = 0$.

    b. $2y = 4$ The equation can be rewritten $0x + 2y - 4 = 0$.

    c. $x = -5$ The equation can be rewritten $1x + 0y + 5 = 0$.   ∎

The Solutions of Equations in Two Variables

An ordered pair (a, b) is a *solution* of an equation in x and y if we get a true statement when we substitute a for x and b for y in that equation. There are always many ordered pairs that are solutions for such an equation and many others that are not solutions.

Example 2   Determine whether each ordered pair is a solution of the equation $2x - 3y = 6$:

a. $(2, -3)$

Substituting 2 for x and -3 for y, we have

$$2x - 3y = 6$$
$$2(2) - 3(-3) \stackrel{?}{=} 6$$
$$4 + 9 \stackrel{?}{=} 6$$
$$13 = 6 \quad \text{False}$$

$(2, -3)$ is not a solution.

b. $(3, 0)$

Substituting 3 for x and 0 for y, we have

$$2x - 3y = 6$$
$$2(3) - 3(0) \stackrel{?}{=} 6$$
$$6 - 0 \stackrel{?}{=} 6$$
$$6 = 6 \quad \text{True}$$

$(3, 0)$ is a solution.

c. $(0, 0)$

Substituting 0 for x and 0 for y, we have

$$2x - 3y = 6$$
$$2(0) - 3(0) \stackrel{?}{=} 6$$
$$0 + 0 \stackrel{?}{=} 6$$
$$0 = 6 \quad \text{False}$$

$(0, 0)$ is not a solution.

d. $(0, -2)$

Substituting 0 for x and -2 for y, we have

$$2x - 3y = 6$$
$$2(0) - 3(-2) \stackrel{?}{=} 6$$
$$0 + 6 \stackrel{?}{=} 6$$
$$6 = 6 \quad \text{True}$$

$(0, -2)$ is a solution.   ∎

Finding Solutions for Equations in Two Variables

We can find a solution for an equation in two variables when we are given a value for one of the variables (see Example 3).

Example 3 Find the solution for the equation $2x - 3y = 6$ given that (a) $x = 0$, (b) $y = 0$, (c) $x = 2$.

Solution a. If we substitute 0 for x, we have

$$2x - 3y = 6$$
$$2(0) - 3y = 6 \qquad \text{We can now solve for } y$$
$$-3y = 6$$
$$y = -2 \qquad \text{We often say "}-2 \text{ is the } y\text{-value that corresponds to } x = 0\text{"}$$

Therefore, $(0, -2)$ is a solution for $2x - 3y = 6$.

b. If we substitute 0 for y, we have

$$2x - 3y = 6$$
$$2x - 3(0) = 6 \qquad \text{We can now solve for } x$$
$$2x = 6$$
$$x = 3 \qquad 3 \text{ is the } x\text{-value that corresponds to } y = 0$$

Therefore, $(3, 0)$ is a solution for $2x - 3y = 6$.

c. If we substitute 2 for x, we have

$$2x - 3y = 6$$
$$2(2) - 3y = 6 \qquad \text{We can now solve for } y$$
$$4 - 3y = 6$$
$$-3y = 2$$
$$y = -\frac{2}{3}$$

Therefore, $(2, -\frac{2}{3})$ is a solution for $2x - 3y = 6$. ■

If we aren't given a value for one of the variables, we can usually choose any value we wish for either variable (see Example 4).

Example 4 Find three ordered pairs that are solutions of $y = x + 1$.

Solution We can let x have any value whatever and solve the equation for y. We can also let y have any value whatever and solve the equation for x.

If $x = 5$, $y = 5 + 1 = 6$. Therefore, $(5, 6)$ is a solution.

If $x = 402$, $y = 402 + 1 = 403$. Therefore, $(402, 403)$ is a solution.

If $y = -23$, we have

$$-23 = x + 1$$
$$x = -24$$

Therefore, $(-24, -23)$ is a solution.

NOTE Infinitely many other answers are possible. ■

Example 5   Complete the following ordered pairs so they will be solutions of the equation $x + 3 = 0$: ($\quad$, 5), ($\quad$, -2), ($\quad$, 0).

Solution   The equation $x + 3 = 0$ is equivalent to the equation $x + 0y + 3 = 0$. When we substitute *any* value for y in this equation, we have

$$x + 3 = 0$$

Therefore, $x = -3$ for *all* y-values. The ordered pairs are $(-3, 5)$, $(-3, -2)$, $(-3, 0)$. ∎

In Example 5, we cannot let x have any value whatever, since x must always equal -3.

EXERCISES 9.2

Set I   In Exercises 1–4, determine whether the given ordered pairs are solutions of the given equation.

1. $3x + 2y = 6$

    a. $(0, 0)$ b. $(0, 3)$ c. $(3, 0)$ d. $(2, 0)$

2. $5x + 2y = 10$

    a. $(0, 0)$ b. $(5, 0)$ c. $(0, 5)$ d. $(2, 0)$

3. $3x + 4y = 0$

    a. $(0, 0)$ b. $(0, 3)$ c. $(4, -3)$ d. $(-4, 3)$

4. $x - 2y = 0$

    a. $(0, 0)$ b. $(0, 2)$ c. $(2, 0)$ d. $(-4, -2)$

In Exercises 5–10, complete the ordered pairs so they will be solutions of the given equation.

5. $4x - 3y = 12$

    a. $(0, \quad)$ b. $(\quad, 0)$ c. $(3, \quad)$ d. $(\quad, -4)$

6. $3x + y = 3$

    a. $(0, \quad)$ b. $(\quad, 0)$ c. $(3, \quad)$ d. $(\quad, -3)$

7. $2x - 5y = 0$

    a. $(0, \quad)$ b. $(\quad, 0)$ c. $(2, \quad)$ d. $(\quad, 2)$

8. $3x + y = 0$

    a. $(0, \quad)$ b. $(\quad, 0)$ c. $(3, \quad)$ d. $(\quad, -3)$

9. $x - 5 = 0$

    a. $(\quad, 0)$ b. $(\quad, 5)$ c. $(\quad, 2)$ d. $(\quad, -2)$

10. $y + 3 = 0$

    a. $(0, \quad)$ b. $(3, \quad)$ c. $(-3, \quad)$ d. $(5, \quad)$

Set II   In Exercises 1–4, determine whether the given ordered pairs are solutions of the given equation.

1. $2x + 3y = 6$

    a. $(0, 0)$ b. $(0, 3)$ c. $(3, 0)$ d. $(2, 0)$

2. $2x - 5y = 10$

    a. (0, 0)   b. (5, 0)   c. (0, 5)   d. (2, 0)

3. $2x + 5y = 0$

    a. (0, 0)   b. (0, 5)   c. (5, −2)   d. (−5, 2)

4. $2x - y = 0$

    a. (0, 0)   b. (0, 2)   c. (2, 0)   d. (−4, −2)

In Exercises 5–10, complete the ordered pairs so they will be solutions of the given equation.

5. $5x - 2y = 10$

    a. (0,)   b. (, 0)   c. (2,)   d. (, −5)

6. $y = -2$

    a. (0,)   b. (−2,)   c. (3,)   d. (−10,)

7. $3x - y = 0$

    a. (0,)   b. (, 0)   c. (3,)   d. (, 3)

8. $x = -4$

    a. (, 0)   b. (, 2)   c. (, −4)   d. (, −30)

9. $y + 2 = 0$

    a. (0,)   b. (5,)   c. (−2,)   d. (3,)

10. $y = 0$

    a. (0,)   b. (3,)   c. (−3,)   d. (5,)

9.3 Graphing Straight Lines

In Section 9.1, we showed how to graph points. In this section, we show how to graph straight lines. As we mentioned in Section 9.2, there are many ordered pairs that will be solutions for a linear equation in two variables.

Example 1   Graph four points (ordered pairs) that lie on the graph of $y = x + 1$.

Solution By choosing four values for x and finding the corresponding values for y, we obtain a set of points (ordered pairs) on the graph of $y = x + 1$. These ordered pairs are listed in the following table of values.

Table of values

Equation: $y = x + 1$

When $x = 0$, $y = (0) + 1 = 1$

When $x = 2$, $y = (2) + 1 = 3$

When $x = 5$, $y = (5) + 1 = 6$

When $x = -3$, $y = (-3) + 1 = -2$

x	y
0	1
2	3
5	6
−3	−2

Each of these ordered pairs, called "a pair of corresponding values," represents a point on the line ⟶

In Figure 9.3.1, we plot the four points from the table of values.

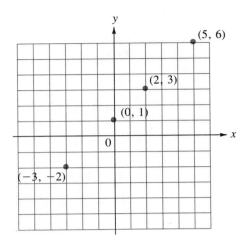

FIGURE 9.3.1

Notice that the points appear to lie in a straight line. This is *not* yet the graph of $y = x + 1$. ∎

There are infinitely many points whose coordinates satisfy an equation in two variables. The following statement must be memorized; it will not be proved.

> The graph of any first-degree (linear) equation in one or two variables is a straight line.

Example 2   Graph the line whose equation is $y = x + 1$.
Solution In Example 1, we graphed four points that lie on the graph of $y = x + 1$. We must now connect these points with a straight line. The line actually extends infinitely far in both directions; to indicate this, we put arrowheads on both ends of it (see Figure 9.3.2).

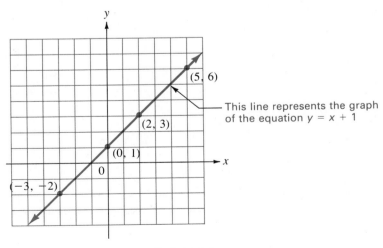

FIGURE 9.3.2 ∎

A straight line can be drawn if we know any two points that lie on that line. Although only two points are necessary to draw the line, it is advisable to plot a third point as a *checkpoint*. If the three points do not lie in a straight line, some mistake has been made in the calculation of the coordinates of the points or in graphing the points.

Intercepts   Students often ask which points to use to plot the graph of a straight line. We usually start by choosing zeros for the *x*- and *y*-values. The points found by this method are called the *x*-intercept and the *y*-intercept.

The **x-intercept** of an equation is the point where its graph meets the *x*-axis. The *y*-value at this point is zero (Figure 9.3.3).

The **y-intercept** of an equation is the point where its graph meets the *y*-axis. The *x*-value at this point is zero (Figure 9.3.3).

Example 3   Graph $3x - 2y = -6$.
Solution We first let *x* equal zero and *y* equal zero.

x	y
0	
	0

y-intercept: If $x = 0$, $3(0) - 2y = -6$
$$-2y = -6$$
$$y = 3$$
Therefore, the *y*-intercept is (0, 3).

x	y
0	3
	0

x-intercept: If $y = 0$, $3x - 2(0) = -6$
$$3x = -6$$
$$x = -2$$
Therefore, the *x*-intercept is $(-2, 0)$.

x	y
0	3
-2	0

Checkpoint: If $x = 2$, $3(2) - 2y = -6$
$$6 - 2y = -6$$
$$-2y = -12$$
$$y = 6$$
Therefore, this checkpoint is (2, 6).

x	y
0	3
-2	0
2	6

We graph these three points and then draw the straight line through them (Figure 9.3.3).

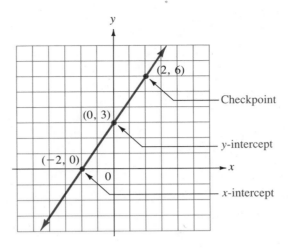

FIGURE 9.3.3 ∎

NOTE   Some students find it helpful to first solve the equation for y and then let x have three different values. In Example 3, the equation becomes $y = \frac{3}{2}x + 3$ (verify this!) and the chart of values remains unchanged.   ☑

Example 4   Graph $4x + 3y = 12$.
Solution

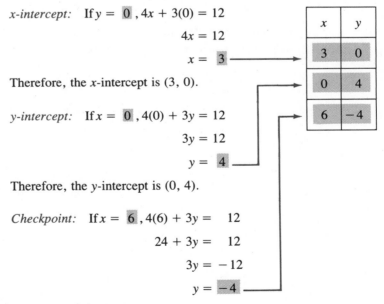

x-intercept: If $y = \boxed{0}$, $4x + 3(0) = 12$

$$4x = 12$$

$$x = \boxed{3}$$

Therefore, the x-intercept is $(3, 0)$.

y-intercept: If $x = \boxed{0}$, $4(0) + 3y = 12$

$$3y = 12$$

$$y = \boxed{4}$$

Therefore, the y-intercept is $(0, 4)$.

Checkpoint: If $x = \boxed{6}$, $4(6) + 3y = 12$

$$24 + 3y = 12$$

$$3y = -12$$

$$y = \boxed{-4}$$

Therefore, this checkpoint is $(6, -4)$.

Plot the x-intercept $(3, 0)$, the y-intercept $(0, 4)$, and the checkpoint $(6, -4)$; then draw the straight line through them (Figure 9.3.4).

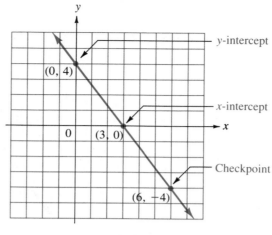

FIGURE 9.3.4 ∎

Example 5   **Graph** $3x - 4y = 0$.
Solution

x-intercept: If $y = \boxed{0}$, $3x - 4(0) = 0$

$$3x = 0$$

$$x = \boxed{0}$$

Therefore, the *x*-intercept is (0, 0). Since the line goes through the origin, the *y*-intercept is also (0, 0).

   We have found only one point on the line: (0, 0). Therefore, we must find another point on the line. To find another point, we must set either variable equal to a number and then solve the equation for the other variable. For example,

$$\text{if } y = \boxed{3}, \; 3x - 4(3) = \; 0$$

$$3x = 12$$

$$x = \boxed{4}$$

This gives the point (4, 3) on the line.

Checkpoint: If $x = \boxed{-4}$, $3(-4) - 4y = \; 0$

$$-12 - 4y = \; 0$$

$$-4y = \; 12$$

$$y = \boxed{-3}$$

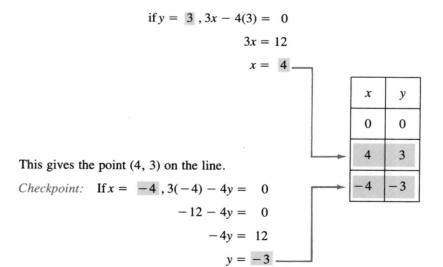

Plot the points (0, 0), (4, 3), and (−4, −3); then draw the straight line through them (Figure 9.3.5).

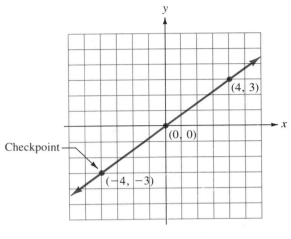

FIGURE 9.3.5 ∎

NOTE Sometimes the x- and y- intercepts are so close together that drawing the line through them accurately would be very difficult. In this case, find another point on the line far enough away from the intercepts that it is easy to draw an accurate line. To find the other point, set either variable equal to a number and then solve the equation for the other variable. ☑

Example 6   Graph $3x - 5y = 15$.
Solution Substitute three values for x, and find the corresponding values for y.

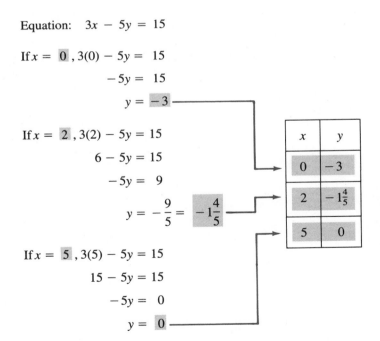

Equation: $3x - 5y = 15$

If $x =$ [0] , $3(0) - 5y = 15$

$-5y = 15$

$y = [-3]$

If $x =$ [2] , $3(2) - 5y = 15$

$6 - 5y = 15$

$-5y = 9$

$y = -\dfrac{9}{5} = [-1\tfrac{4}{5}]$

If $x =$ [5] , $3(5) - 5y = 15$

$15 - 5y = 15$

$-5y = 0$

$y = [0]$

x	y
0	-3
2	$-1\frac{4}{5}$
5	0

Plot the three points from the table of values, and draw a straight line through them (Figure 9.3.6).

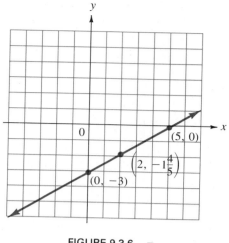

FIGURE 9.3.6 ∎

Some equations of a line have only one variable. The graphs of such equations are either vertical or horizontal lines (see Examples 7 and 8).

Example 7   Graph $x = 3$.

Solution The equation $x = 3$ is equivalent to $0y + x = 3$.
If $y = \boxed{5}$, $0(5) + x = 3$

$$0 + x = 3$$

$$x = \boxed{3}$$

If $y = \boxed{-2}$, $0(-2) + x = 3$

$$0 + x = 3$$

$$x = \boxed{3}$$

x	y
3	5
3	-2

You can see that no matter what value y has in this equation, x is always 3. Therefore, all the points with an x-value of 3 lie in a vertical line whose x-intercept is (3, 0) (Figure 9.3.7).

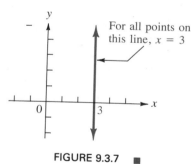

FIGURE 9.3.7 ∎

The graph of $x = a$, where a is any real number, is always a *vertical line*.

Example 8   Graph $y + 4 = 0$.
Solution

$$y + 4 = 0$$

$$y = -4$$

In the equation $y + 4 = 0$, no matter what value x has, y is always -4. Therefore, all the points with a y-value of -4 lie in a horizontal line whose y-intercept is $(0, -4)$ (Figure 9.3.8).

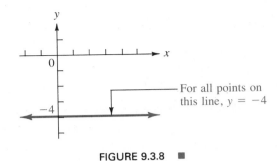

FIGURE 9.3.8 ■

The graph of $y = b$, where b is any real number, is always a *horizontal line*. The methods used in these examples are summarized below.

TO GRAPH A STRAIGHT LINE

Method 1 (Intercept method)

1. Find the x-intercept: Set $y = 0$; then solve for x.

2. Find the y-intercept: Set $x = 0$; then solve for y.

3. If both intercepts are $(0, 0)$, an additional point must be found.

4. Find one other point by letting x or y have any value.

5. Draw a straight line through the three points.

Method 2

1. In the equation, substitute a number for one of the variables and solve for the corresponding value of the other variable. Find *three* points in this way, and list them in a table of values.

2. Graph the points listed in the table of values.

3. Draw a straight line through the points.

Exceptions:

The graph of $x = a$ is a vertical line that passes through $(a, 0)$.

The graph of $y = b$ is a horizontal line that passes through $(0, b)$.

EXERCISES 9.3

Set I   In Exercises 1–26, graph the line whose equation is given.

1. $x + y = 3$

2. $x + y = 5$

3. $2x - 3y = 6$

4. $3x - 4y = 12$

5. $y = 8$

6. $y = -3$

7. $x = -2$

8. $x = 9$

9. $x + 5 = 0$

10. $x + 1 = 0$

11. $y + 5 = 0$

12. $y + 2 = 0$

13. $3x = 5y + 15$

14. $2x = 5y + 10$

15. $x = -2y$

16. $x = -5y$

17. $y = -4x$

18. $y = -2x$

19. $4 - y = x$

20. $6 - y = x$

21. $y = x$

22. $y = -x$

23. $y = \frac{1}{2}x + 2$

24. $y = \frac{3}{4}x + 1$

25. $3x = 24 + 4y$

26. $9x = 18 + 2y$

In Exercises 27 and 28, (a) graph the two equations for each exercise on the same set of axes and (b) give the coordinates of the point where the two lines cross.

27. $x - y = 5$

$x + y = 1$

28. $3x - 4y = -12$

$3x + y = 18$

Set II In Exercises 1–26, graph the line whose equation is given.

1. $x + y = 1$

2. $x = \frac{2}{3}y + 4$

3. $4x - 3y = 12$

4. $x - y = 5$

5. $y = 1$

6. $y = 0$

7. $x = -4$

8. $x = 0$

9. $x + 7 = 0$

10. $x + y = -3$

11. $y + 8 = 0$

12. $3 - y = 2x$

13. $2x = 4y + 8$

14. $5y = x - 5$

15. $x = -3y$

16. $x + y = 0$

17. $y = -3x$

18. $x - y = 0$

19. $2 - y = x$

20. $5x - y = 0$

21. $y = x + 3$

22. $y = x + 5$

23. $y = \frac{2}{3}x + 1$

24. $y = \frac{2}{3}x - 2$

25. $y = \frac{2}{3}x$

26. $7x = 3y + 21$

In Exercises 27 and 28, (a) graph the two equations on the same set of axes and (b) give the coordinates of the point where the two lines cross.

27. $x - 2y = 4$

$x + 2y = 2$

28. $2x + 3y = 12$

$2x = 3y$

9.4 The Slope of a Line

Subscripts A small number written below and to the right of a variable is used to indicate a particular value of that variable. It is called a **subscript**. For example,

$$x_1$$

x_1 is read "x sub one"

x_2 ← A different subscript indicates a different *value* of that variable

x_2 is read "x sub two"

Example 1 Examples of subscripted variables:

a. y_1 and y_2 represent two different values of y.

b. P_1 and P_2 represent two different values of P.

c. $P_1(x_1, y_1)$ and $P_2(x_2, y_2)$ represent two different *points*. ■

The Slope of a Line

If we imagine a line as representing a hill, then the **slope** of the line is a measure of the steepness of the hill. To measure the slope of a line, we choose any two points on the line, $P_1(x_1, y_1)$ and $P_2(x_2, y_2)$ (Figure 9.4.1).

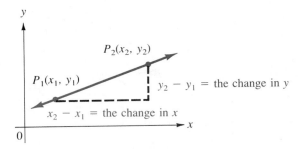

FIGURE 9.4.1

The letter m is used to represent the slope of a line. The slope is defined as follows:

SLOPE OF A LINE

$$\text{Slope} = \frac{\text{the change in } y}{\text{the change in } x}$$

$$m = \frac{y_2 - y_1}{x_2 - x_1}$$

$$or \quad m = \frac{y_1 - y_2}{x_1 - x_2}$$

A WORD OF CAUTION   It is *incorrect* to write $\dfrac{x_2 - x_1}{y_2 - y_1}$, $\dfrac{x_1 - x_2}{y_1 - y_2}$, $\dfrac{y_2 - y_1}{x_1 - x_2}$, or $\dfrac{y_1 - y_2}{x_2 - x_1}$ ☑
as the slope of a line.

Example 2   Find the slope of the line through the points $(-2, -4)$ and $(6, 0)$ (see Figure 9.4.2).
Solution

$$\text{Let } P_1 = (-2, -4)$$

$$P_2 = (6, 0)$$

That is, $x_1 = -2$, $y_1 = -4$,
$x_2 = 6$, and $y_2 = 0$. Therefore,

$$m = \frac{y_2 - y_1}{x_2 - x_1}$$

$$= \frac{0 - (-4)}{6 - (-2)} = \frac{4}{8} = \frac{1}{2}$$

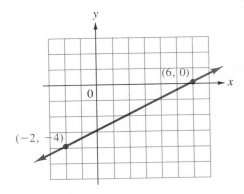

FIGURE 9.4.2

The slope is not changed if the points P_1 and P_2 are interchanged.

$$\text{Let } P_1 = (6, 0)$$

$$P_2 = (-2, -4)$$

That is, $x_1 = 6$, $y_1 = 0$, $x_2 = -2$, and $y_2 = -4$. Therefore,

$$m = \frac{y_2 - y_1}{x_2 - x_1}$$

$$= \frac{-4 - 0}{-2 - 6} = \frac{-4}{-8} = \frac{1}{2} \quad \blacksquare$$

Notice that in Example 2, a point moving along the line in the positive x-direction *rises*, and the slope of the line is *positive*.

Example 3   Find the slope of the line through the points $A(-1, 3)$ and $B(2, -3)$ (see Figure 9.4.3).
Solution

$$m = \frac{y_2 - y_1}{x_2 - x_1}$$

$$= \frac{-3 - (3)}{2 - (-1)} = \frac{-6}{3} = -2$$

or

$$m = \frac{y_1 - y_2}{x_1 - x_2}$$

$$= \frac{3 - (-3)}{-1 - 2} = \frac{6}{-3} = -2$$

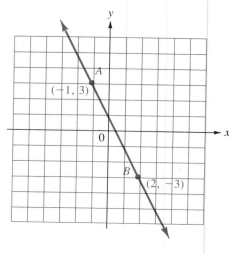

FIGURE 9.4.3 $\blacksquare$

Notice that in Example 3, a point moving along the line in the positive x-direction *falls*, and the slope of the line is *negative*.

Example 4   Find the slope of the line through the points $E(-4, -3)$ and $F(2, -3)$ (see Figure 9.4.4).

$$m = \frac{y_2 - y_1}{x_2 - x_1}$$

$$= \frac{(-3) - (-3)}{(2) - (-4)} = \frac{0}{6} = 0$$

Whenever the slope is zero, the line is horizontal.

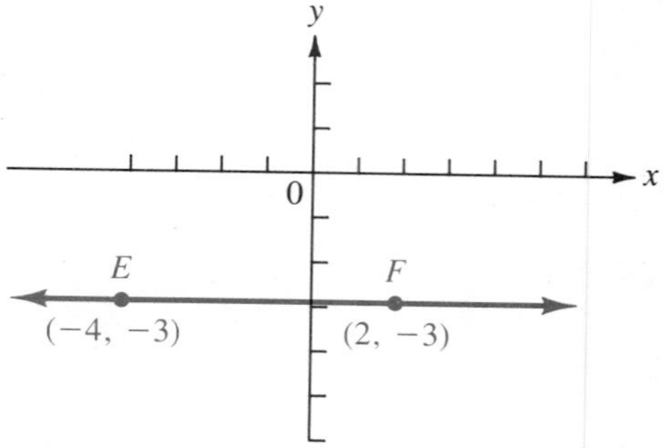

FIGURE 9.4.4 $\blacksquare$

Example 5 Find the slope of the line through the points $R(4, 5)$ and $S(4, -2)$ (see Figure 9.4.5).

$$m = \frac{y_2 - y_1}{x_2 - x_1}$$

$$= \frac{(-2) - (5)}{(4) - (4)} = \frac{-7}{0}$$

Note that $\dfrac{-7}{0}$ is not a real number. Therefore, the slope does not exist when the line is vertical.

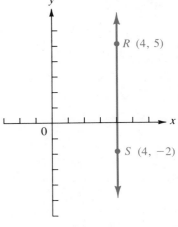

FIGURE 9.4.5

The facts about the slope of a line are summarized as follows:

The slope of a line is positive if a point moving along the line in the positive x-direction (to the right) rises (Figure 9.4.2).

The slope of a line is negative if a point moving along the line in the positive x-direction falls (Figure 9.4.3).

The slope is zero if the line is horizontal (Figure 9.4.4).

The slope does not exist if the line is vertical (Figure 9.4.5).

Given the equation of a line, we can find its slope by finding the coordinates of any two points on the line (see Example 6).

Example 6 For each of the following equations, graph the line and find its slope:

a. $y = 3x$

x	y
-2	-6
0	0
1	3

Point A

Point B

Point C

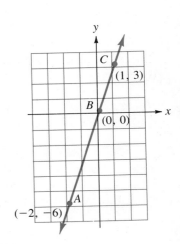

For the slope, if we use points A and B, we have

$$m = \frac{-6 - (0)}{-2 - (0)} = \frac{-6}{-2} = 3$$

If we use points B and C, we have

$$m = \frac{0 - (3)}{0 - (1)} = \frac{-3}{-1} = 3$$

b. $y = 3x + 2$

x	y	
-2	-4	Point D
0	2	Point E
1	5	Point F

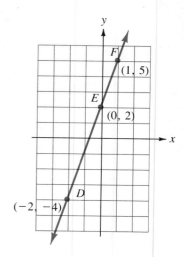

For the slope, if we use points D and E, we have

$$m = \frac{-4 - (2)}{-2 - (0)} = \frac{-6}{-2} = 3$$

If we use points E and F, we have

$$m = \frac{2 - (5)}{0 - (1)} = \frac{-3}{-1} = 3$$

You can verify that, for each of the lines, if any other points on the lines had been used, the slope would still have been 3.

c. $y = -\frac{1}{3}x$

x	y	
-3	1	Point G
0	0	Point H
1	$-\frac{1}{3}$	Point I

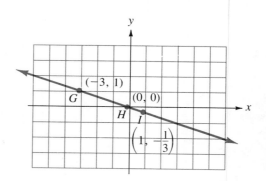

For the slope, if we use points G and H, we have

$$m = \frac{1 - (0)}{-3 - (0)} = \frac{1}{-3} = -\frac{1}{3}$$

If we use points H and I, we have

$$m = \frac{0 - (-\frac{1}{3})}{0 - (1)} = \frac{\frac{1}{3}}{-1} = -\frac{1}{3}$$

d. Graph the lines from Examples 6a and 6b on the same axes.

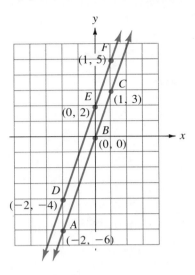

e. Graph the lines from Examples 6a and 6c on the same axes.

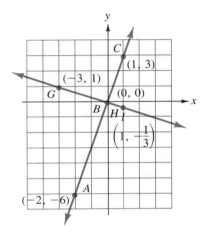

Parallel Lines   Two or more lines in the same plane are *parallel* if they will not meet even if they are extended indefinitely.

Perpendicular Lines   Two lines are *perpendicular* if they intersect in a right (90°) angle.

The following rules about the slopes of parallel and perpendicular lines are stated here without proof.

Parallel lines have the same slope, and any lines that have the same slope are parallel.

Perpendicular lines (if neither is vertical) have slopes whose product is -1. Or, if one line with slope m_1 is perpendicular to another line with slope m_2, then

$$m_1 = -\frac{1}{m_2}$$

A vertical line is parallel to any other vertical line and is perpendicular to any horizontal line.

Notice that in Examples 6a and 6b, the lines were *parallel* to each other and the *slopes* of the two lines were equal. Notice also that the lines in Examples 6a and 6c appear to be perpendicular to each other and the *product* of their slopes is $(3)(-\frac{1}{3}) = -1$.

EXERCISES 9.4

Set I   In Exercises 1–14, find the slope of the line through the given pair of points.

1. $(-2, 1)$ and $(7, 3)$ **2.** $(-1, -3)$ and $(5, 2)$

3. $(-1, -1)$ and $(5, -3)$ **4.** $(-2, -1)$ and $(6, -5)$

5. $(-5, -3)$ and $(4, -3)$ **6.** $(-2, -4)$ and $(3, -4)$

7. $(-7, 5)$ and $(8, -3)$ **8.** $(12, -9)$ and $(-5, -4)$

9. $(-6, -15)$ and $(4, -5)$ **10.** $(-7, 8)$ and $(-4, 5)$

11. $(-2, 5)$ and $(-2, 8)$ **12.** $(6, -4)$ and $(6, -11)$

13. $(-4, 3)$ and $(-4, -2)$ **14.** $(-5, 2)$ and $(-5, 7)$

In Exercises 15–18, graph each set of lines on the same axes and find the slope of each line.

15. a. $y = \frac{1}{2}x$ b. $y = \frac{1}{2}x + 3$ c. $y = \frac{1}{2}x - 2$

16. a. $y = 2x$ b. $y = 2x + 1$ c. $y = 2x - 3$

17. a. $y = -3x$ b. $y = -3x + 4$ c. $y = -3x - 1$

18. a. $y = -\frac{3}{4}x$ b. $y = -\frac{3}{4}x + 2$ c. $y = -\frac{3}{4}x - 3$

Set II   In Exercises 1–14, find the slope of the line through the given pair of points.

1. $(-3, 5)$ and $(2, -1)$ **2.** $(4, 0)$ and $(7, 0)$

3. $(-3, 2)$ and $(6, 4)$ **4.** $(-2, -2)$ and $(4, -1)$

5. $(-4, -3)$ and $(6, -3)$ **6.** $(-10, 2)$ and $(-4, -8)$

7. $(-3, 8)$ and $(9, 12)$ **8.** $(-7, 3)$ and $(-7, -2)$

9. $(4, -2)$ and $(4, 3)$ **10.** $(3, -1)$ and $(0, 0)$

11. $(5, -2)$ and $(1, 1)$ **12.** $(-3, -1)$ and $(2, -1)$

13. $(-3, 7)$ and $(-3, -5)$ **14.** $(8, -2)$ and $(-2, 8)$

In Exercises 15–18, graph each set of lines on the same axes and find the slope of each line.

15. a. $y = \frac{1}{4}x$ b. $y = \frac{1}{4}x + 2$ c. $y = \frac{1}{4}x - 3$

16. a. $y = -4x$ b. $y = -4x + 3$ c. $y = -4x - 1$

17. a. $y = -x$ b. $y = -x + 2$ c. $y = -x - 3$

18. a. $y = -\frac{2}{3}x$ b. $y = -\frac{2}{3}x + 4$ c. $y = -\frac{2}{3}x - 2$

9.5 Deriving Equations of Straight Lines

In Section 9.3, we discussed the graph of a straight line. In this section, we show how to write the *equation* of a line when certain facts about the line are known.

Three forms of the equation of a straight line are especially useful: the *general form*, the *point-slope form*, and the *slope-intercept form*.

General Form of the Equation of a Line

The general form of the equation of a line, as used in this text, is defined as follows:

GENERAL FORM OF THE EQUATION OF A LINE

$$Ax + By + C = 0$$

where A, B, and C are integers, $A \geq 0$, and A and B are not both 0.

A is the coefficient of x, and the x-term is written first on the left side of the equation; B is the coefficient of y, and the y-term is written next; C is the constant, and it is the last term on the left side. Zero must be the only term on the right side of the equation.

To write the equation of a straight line in the *general form*, multiply both sides of the equation by the LCD and move all terms to the left side of the equal sign. If A is then negative, multiply both sides of the equation by -1.

Example 1 Write $-\frac{2}{3}x + \frac{1}{2}y = 1$ in general form.

Solution LCD $= 6$

$$\frac{6}{1}\left(-\frac{2}{3}x\right) + \frac{6}{1}\left(\frac{1}{2}y\right) = \frac{6}{1}\left(\frac{1}{1}\right)$$

$$-4x + 3y = 6$$

$$4x - 3y + 6 = 0 \qquad \text{General form} \qquad \blacksquare$$

Point-Slope Form of the Equation of a Line

Let $P_1(x_1, y_1)$ be a known point on a line whose slope is m. Let $P(x, y)$ represent *any* other point on that line. Then, using the definition of slope, we have

$$m = \frac{y - y_1}{x - x_1}$$

$$\frac{m}{1} = \frac{y - y_1}{x - x_1}$$

$$1(y - y_1) = m(x - x_1)$$

$$y - y_1 = m(x - x_1)$$

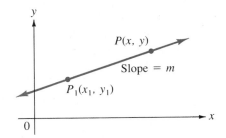

427

POINT-SLOPE FORM OF THE EQUATION OF A LINE

$$y - y_1 = m(x - x_1)$$

where m = slope of the line and $P_1(x_1, y_1)$ is a known point on the line.

When we are given the *slope* of a line and a *point* through which it passes, we use the *point-slope form* of the equation of a line when we write the equation of the line (see Examples 2 and 3).

Example 2   Write the general form of the equation of the line that passes through $(2, -3)$ and has a slope of 4.

Solution Into the point-slope form of the equation of a line, substitute 2 for x_1, -3 for y_1, and 4 for m.

$$y - y_1 = m(x - x_1) \quad \text{Point-slope form}$$
$$y - (-3) = 4(x - 2)$$
$$y + 3 = 4x - 8$$
$$4x - y - 11 = 0 \quad \text{General form} \quad \blacksquare$$

Example 3   Write the general form of the equation of the line that passes through $(-1, 4)$ and has a slope of $-\frac{2}{3}$.

Solution

$$y - y_1 = m(x - x_1) \quad \text{Point-slope form}$$
$$y - 4 = -\frac{2}{3}[x - (-1)]$$
$$3y - 12 = -2(x + 1) \quad \text{Multiplying both sides by 3}$$
$$3y - 12 = -2x - 2$$
$$2x + 3y - 10 = 0 \quad \text{General form} \quad \blacksquare$$

Slope-Intercept Form of the Equation of a Line

Let $(0, b)$ be the *y-intercept* of a line whose slope is m. Then

$$y - y_1 = m(x - x_1)$$
$$y - b = m(x - 0)$$
$$y - b = mx$$
$$y = mx + b$$

Slope   y-intercept

SLOPE-INTERCEPT FORM OF THE EQUATION OF A LINE

$$y = mx + b$$

where m = slope of the line and b = y-intercept of the line.

When we are given the *slope* of a line and its *y-intercept* and are asked to write the equation of the line, we can use the *slope-intercept form* of the equation (see Example 4). The point-slope form could also be used.

Example 4 Write the general form of the equation of the line that has a slope of $-\frac{1}{2}$ and a *y*-intercept of -3.

Solution Into the slope-intercept form, substitute $-\frac{1}{2}$ for m and -3 for b.

$$y = mx + b \qquad \text{Slope-intercept form}$$

$$y = -\frac{1}{2}x + (-3)$$

$$2y = 2\left(-\frac{1}{2}x - 3\right) \qquad \text{Multiplying both sides by 2}$$

$$2y = -x - 6$$

$$x + 2y + 6 = 0 \qquad \text{General form} \quad \blacksquare$$

The Equation of a Horizontal Line

Because a horizontal line has a slope, its equation can be written using the slope-intercept form or the point-slope form, *or* the following statement can be memorized:

> The equation of a horizontal line is of the form $y = b$ or $y - b = 0$, where b is the *y*-coordinate of every point on the line.

Example 5 Find the equation of the horizontal line that passes through $(-2, 4)$.

Solution

Method 1 The slope of every horizontal line is 0. Therefore,

$$y - 4 = 0(x - [-2]) \qquad \text{Point-slope form}$$

$$y - 4 = 0 \qquad \text{General form}$$

Method 2 The *y*-coordinate of the given point is 4. Therefore, the equation of the line must be

$$y = 4 \quad \blacksquare$$

The Equation of a Vertical Line

Because a vertical line has no slope, we cannot use the point-slope or the slope-intercept form when we write its equation. The only way to write the equation of a vertical line is to memorize the following statement:

> The equation of a vertical line is of the form $x = a$ or $x - a = 0$, where a is the *x*-coordinate of every point on the line.

Example 6 Write the equation of the vertical line that passes through $(-2, 4)$.

Solution Because the x-coordinate of the given point is -2, the equation of the line must be

$$x = -2$$

or $\qquad\qquad x + 2 = 0$ General form ■

In Example 7, we show how to write the equation of a line given two points on the line.

Example 7 Find the equation of the line passing through the points $(-5, 3)$ and $(-15, -9)$.

Solution

1. Find the slope from the two given points.

$$m = \frac{(-9) - (3)}{(-15) - (-5)} = \frac{-12}{-10} = \frac{6}{5}$$

2. Use this slope with *either* given point to find the equation of the line.

Using the point $(-5, 3)$	*Using the point* $(-15, -9)$
$y - y_1 = m(x - x_1)$	$y - y_1 = m(x - x_1)$
$y - 3 = \dfrac{6}{5}[x - (-5)]$	$y - (-9) = \dfrac{6}{5}[x - (-15)]$
$5y - 15 = 6(x + 5)$	$5(y + 9) = 6(x + 15)$
$5y - 15 = 6x + 30$	$5y + 45 = 6x + 90$
$\boxed{0 = 6x - 5y + 45}$	$\boxed{0 = 6x - 5y + 45}$

This result shows that the same equation is obtained no matter which of the two given points is used. ■

We sometimes need to change an equation into the slope-intercept form. This means, of course, that the equation must be solved for y.

Example 8 Write $2x + 3y + 6 = 0$ in the slope-intercept form, find the slope of the line, and find the y-intercept of the line.

Solution Solve the given equation for y.

$$2x + 3y + 6 = 0 \qquad \text{General form}$$
$$3y = -2x - 6$$
$$y = -\frac{2}{3}x - 2 \qquad \text{Slope-intercept form}$$

Slope $\longrightarrow$ $\qquad$ $\longleftarrow$ y-intercept

The slope is $-\frac{2}{3}$, and the y-intercept is -2. ■

EXERCISES 9.5

Set I In Exercises 1–6, write each equation in general form.

1. $3x = 2y - 4$ $\qquad\qquad\qquad$ **2.** $2x = 3y + 7$

3. $y = -\frac{3}{4}x - 2$ $\qquad\qquad\qquad$ **4.** $y = -\frac{3}{5}x - 4$

5. $2(3x + y) = 5(x - y) + 4$

6. $3(2x - y) = 2(x + 3y) - 5$

In Exercises 7–10, write the general form of the equation of the line through the given point and with the indicated slope.

7. $(3, 4), m = \frac{1}{2}$

8. $(5, 6), m = \frac{1}{3}$

9. $(-1, -2), m = -\frac{2}{3}$

10. $(-2, -3), m = -\frac{5}{4}$

In Exercises 11–14, write the general form of the equation of the line with the indicated slope and y-intercept.

11. $m = \frac{3}{4}$, y-intercept $= -3$

12. $m = \frac{2}{7}$, y-intercept $= -2$

13. $m = -\frac{2}{5}$, y-intercept $= \frac{1}{2}$

14. $m = -\frac{5}{3}$, y-intercept $= \frac{3}{4}$

In Exercises 15–18, find the general form of the equation of the line that passes through the given points.

15. $(4, -1)$ and $(2, 4)$

16. $(5, -2)$ and $(3, 1)$

17. $(0, 0)$ and $(3, 4)$

18. $(0, 0)$ and $(-2, -5)$

19. Write the equation of the horizontal line that passes through the point $(-3, 5)$.

20. Write the equation of the horizontal line that passes through the point $(-2, -3)$.

21. Write the equation of the vertical line that passes through the point $(-3, 5)$.

22. Write the equation of the vertical line that passes through the point $(-2, -3)$.

In Exercises 23–26, (a) write the given equation in the slope-intercept form, (b) give the slope of the line, and (c) give the y-intercept of the line.

23. $3x + 4y + 12 = 0$

24. $2x + 5y + 10 = 0$

25. $2x + 3y - 9 = 0$

26. $2x + 5y - 15 = 0$

Set II In Exercises 1–6, write each equation in general form.

1. $5x = 4y - 7$

2. $3 - 2y = 5x$

3. $y = -\frac{4}{5}x - 3$

4. $\frac{2}{3}x = 3y - \frac{1}{6}$

5. $3(2x - y) = 4(x + y) - 6$

6. $x = 4$

In Exercises 7–10, write the general form of the equation of the line through the given point and with the indicated slope.

7. $(-6, 3), m = \frac{1}{2}$

8. $(5, -7), m = -\frac{3}{4}$

9. $(3, -5), m = -\frac{3}{5}$

10. $(0, 0), m = 0$

In Exercises 11–14, write the general form of the equation of the line with the indicated slope and y-intercept.

11. $m = -\frac{2}{3}$, y-intercept $= -4$

12. $m = \frac{5}{4}$, y-intercept $= -3$

13. $m = -\frac{1}{3}$, y-intercept $= \frac{2}{5}$

14. $m = 0$, y-intercept $= 0$

In Exercises 15–18, find the general form of the equation of the line that passes through the given points.

15. $(-3, 4)$ and $(5, -2)$

16. $(5, -3)$ and $(-2, -4)$

17. $(3, -5)$ and $(-3, 2)$

18. $(4, 6)$ and $(-3, 6)$

19. Write the equation of the horizontal line that passes through the point $(-4, -2)$.

20. Write the equation of the horizontal line that passes through the point $(2, -5)$.

21. Write the equation of the vertical line that passes through the point $(-4, -2)$.

22. Write the equation of the vertical line that passes through the point $(2, -5)$.

In Exercises 23–26, (a) write the given equation in the slope-intercept form, (b) give the slope of the line, and (c) give the y-intercept of the line.

23. $5x + 3y + 15 = 0$

24. $\frac{3}{4}x - 2y - 3 = 0$

25. $4x + 2y - 7 = 0$

26. $x = \frac{1}{2}y + 3$

9.6 Graphing Curves

In Section 9.3, we showed how to graph straight lines. In this section, we show how to graph curves. We need only two points to draw a straight line. To draw a curve, however, we must find more than two points. In this book, you will be told what values to use for one of the variables.

The *x-intercepts* of a curve are the points at which the curve crosses the x-axis. There are sometimes no x-intercepts, sometimes one, and sometimes more than one. We find the x-intercepts by setting y equal to zero and trying to solve the resulting equation for x. (We won't always be able to solve the equation; this doesn't necessarily mean there are no x-intercepts.)

The y-intercepts of a curve are the points where the curve intersects the y-axis. We find the y-intercepts by setting x equal to zero and trying to solve the equation for y. There are sometimes no y-intercepts, sometimes one, and sometimes more than one.

TO GRAPH A CURVE

1. Use the equation to make a table of values.

2. Plot the points from the table of values.

3. Draw a smooth curve through the points. In this book we will join them in order from left to right.

Example 1   Find the x- and y-intercepts for $y = x^2 - x - 2$, and graph the curve.
Solution To find the x-intercepts, we let $y = 0$ and solve for x.

$$0 = x^2 - x - 2$$

$$0 = (x - 2)(x + 1)$$

$$x - 2 = 0 \quad | \quad x + 1 = 0$$

$$x = 2 \quad | \quad x = -1$$

Therefore, the x-intercepts are 2 and -1.

To find the y-intercept, we let $x = 0$ and solve for y.

$$y = 0^2 - 0 - 2 = -2$$

Therefore, the y-intercept is -2.

Make a table of values by finding the values of y that correspond to $x = -2$, $x = -1$, $x = 0$, $x = \frac{1}{2}$, $x = 1$, $x = 2$, and $x = 3$.

			x	y
If $x = -2$, $y = (-2)^2 - (-2) - 2 = 4 + 2 - 2 = 4$			-2	4
If $x = -1$, $y = (-1)^2 - (-1) - 2 = 1 + 1 - 2 = 0$			-1	0
If $x = 0$, $y = (0)^2 - (0) - 2 = 0 + 0 - 2 = -2$			0	-2
If $x = \frac{1}{2}$, $y = (\frac{1}{2})^2 - (\frac{1}{2}) - 2 = \frac{1}{4} - \frac{1}{2} - 2 = -2\frac{1}{4}$			$\frac{1}{2}$	$-2\frac{1}{4}$
If $x = 1$, $y = (1)^2 - (1) - 2 = 1 - 1 - 2 = -2$			1	-2
If $x = 2$, $y = (2)^2 - (2) - 2 = 4 - 2 - 2 = 0$			2	0
If $x = 3$, $y = (3)^2 - (3) - 2 = 9 - 3 - 2 = 4$			3	4

In Figure 9.6.1, we graph these points and draw a smooth curve through them. In drawing the smooth curve, start with the point in the table of values with the smallest x-value. Draw to the point having the next larger x-value. Continue in this way through all the points. The graph of the equation $y = x^2 - x - 2$ is called a *parabola*.

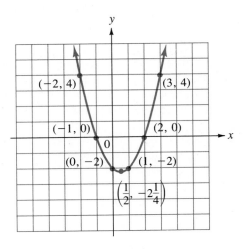

FIGURE 9.6.1 ■

The graph of any equation that can be expressed in the form $y = ax^2 + bx + c$ is called a parabola.

Example 2 Try to find the x- and y-intercepts for $y = x^2 - 2x - 1$ and graph the curve.

Solution This is a parabola, because the equation is in the form

$$y = ax^2 + bx + c$$

x-intercepts: We cannot factor $x^2 - 2x - 1$. Therefore, we cannot find the x-intercepts.

y-intercept: Solve $y = (0)^2 - 2(0) - 1 = -1$. Therefore, the y-intercept is -1.

Make a table of values, letting x have integral values from -2 to 4.

x	y
-2	7
-1	2
0	-1
1	-2
2	-1
3	2
4	7

If $x = -2$, $y = (-2)^2 - 2(-2) - 1 = 7$

If $x = -1$, $y = (-1)^2 - 2(-1) - 1 = 2$

If $x = 0$, $y = (0)^2 - 2(0) - 1 = -1$

If $x = 1$, $y = (1)^2 - 2(1) - 1 = -2$

If $x = 2$, $y = (2)^2 - 2(2) - 1 = -1$

If $x = 3$, $y = (3)^2 - 2(3) - 1 = 2$

If $x = 4$, $y = (4)^2 - 2(4) - 1 = 7$

In Figure 9.6.2, we graph these points and draw a smooth curve through them.

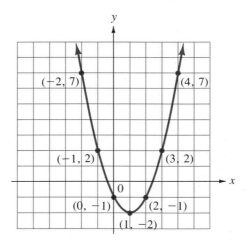

FIGURE 9.6.2

We see from the graph that there were x-intercepts even though we couldn't find them algebraically. ∎

Example 3   Find the x- and y-intercepts for $y = x^3 - 4x$, and graph the curve.

Solution

x-intercepts: Solve: $0 = x^3 - 4x$

$$0 = x(x^2 - 4)$$

$$0 = x(x - 2)(x + 2)$$

$x = 0$	$x - 2 = 0$	$x + 2 = 0$
	$x = 2$	$x = -2$

Therefore, the x-intercepts are 0, 2, and -2.

y-intercepts: Solve $y = (0)^3 - 4(0) = 0$. Therefore, the y-intercept is 0.

Make a table of values, letting x have integral values from -3 to 3.

If $x = -3$, $y = (-3)^3 - 4(-3) = -27 + 12 = -15$

If $x = -2$, $y = (-2)^3 - 4(-2) = -8 + 8 = 0$

If $x = -1$, $y = (-1)^3 - 4(-1) = -1 + 4 = 3$

If $x = 0$, $y = (0)^3 - 4(0) = 0 - 0 = 0$

If $x = 1$, $y = (1)^3 - 4(1) = 1 - 4 = -3$

If $x = 2$, $y = (2)^3 - 4(2) = 8 - 8 = 0$

If $x = 3$, $y = (3)^3 - 4(3) = 27 - 12 = 15$

x	y
-3	-15
-2	0
-1	3
0	0
1	-3
2	0
3	15

In Figure 9.6.3, we graph these points and draw a smooth curve through them.

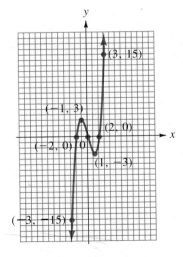

FIGURE 9.6.3   ■

EXERCISES 9.6

Set I In Exercises 1–4, find the x- and y-intercepts and complete the table of values for each equation; then draw its graph.

1. $y = x^2$

x	y
-2	
-1	
0	
1	
2	

2. $y = \dfrac{x^2}{4}$

x	y
-3	
-2	
-1	
0	
1	
2	
3	

3. $y = x^2 - 2x$

x	y
-2	
-1	
0	
1	
2	
3	
4	

4. $y = 3x - x^2$

x	y
-2	
-1	
0	
1	
2	
3	
4	

5. Use integer values of x from -4 to $+2$ to make a table of values for the equation $y = x^2 + 2x - 2$. Graph the points and draw a smooth curve through them.

6. Use integer values of x from -4 to $+2$ to make a table of values for the equation $y = x^2 + 2x + 2$. Graph the points and draw a smooth curve through them.

7. Use integer values of x from -2 to $+4$ to make a table of values for the equation $y = 2x - x^2$. Graph the points and draw a smooth curve through them.

8. Use integer values of x from -3 to $+1$ to make a table of values for the equation $y = 2x + x^2$. Graph the points and draw a smooth curve through them.

9. Use integer values of x from -2 to $+2$ to make a table of values for the equation $y = x^3$. Graph the points and draw a smooth curve through them.

10. Use integer values of x from -2 to $+2$ to make a table of values for the equation $y = x^3 - 3x + 4$. Graph the points and draw a smooth curve through them.

Set II   In Exercises 1–4, find the x- and y-intercepts and complete the table of values for each equation; then draw its graph.

1. $y = \dfrac{x^2}{2}$

x	y
-3	
-2	
-1	
0	
1	
2	
3	

2. $y = x^2 + 4x$

x	y
-5	
-4	
-3	
-2	
-1	
0	
1	

3. $y = x^2 - 4$

x	y
-3	
-2	
-1	
0	
1	
2	
3	

4. $y = 4 - x^2$

x	y
-3	
-2	
-1	
0	
1	
2	
3	

5. Use integer values of x from -1 to $+5$ to make a table of values for the equation $y = x^2 - 4x + 2$. Graph the points and draw a smooth curve through them.

6. Use integer values of x from -1 to $+5$ to make a table of values for the equation $y = x^2 - 4x - 2$. Graph the points and draw a smooth curve through them.

7. Use integer values of x from -1 to $+6$ to make a table of values for the equation $y = 5x - x^2$. Graph the points and draw a smooth curve through them.

8. Use integer values of x from -3 to $+1$ to make a table of values for the equation $y = -2x - x^2$. Graph the points and draw a smooth curve through them.

9. Use integer values of x from -2 to $+2$ to make a table of values for the equation $y = -x^3$. Graph the points and draw a smooth curve through them.

10. Use integer values of x from -2 to $+2$ to make a table of values for the equation $y = 1 - 2x - x^3$. Graph the points and draw a smooth curve through them.

9.7 Graphing First-Degree Inequalities in Two Variables

In Section 4.7, we solved inequalities in one variable and graphed the solution on the real number line. In this section, we graph solutions of inequalities in one or two variables *in the plane* of the rectangular coordinate system.

Half-planes   Any line in a plane separates that plane into two half-planes. For example, in Figure 9.7.1, the line *AB* separates the plane into the two half-planes shown.

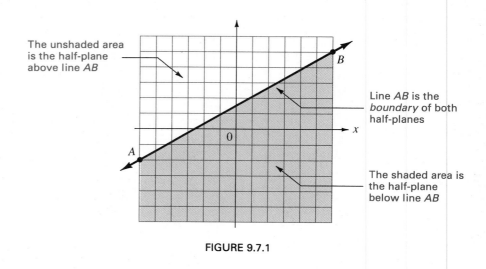

The unshaded area is the half-plane above line *AB*

Line *AB* is the *boundary* of both half-planes

The shaded area is the half-plane below line *AB*

FIGURE 9.7.1

The following statement must be memorized; it will not be proved.

> Any first-degree inequality in one or two variables has a graph that is a half-plane.

The equation of the boundary line of the half-plane is obtained by replacing the inequality sign with an equal sign.

How to Determine Whether the Boundary is a Dashed or Solid Line

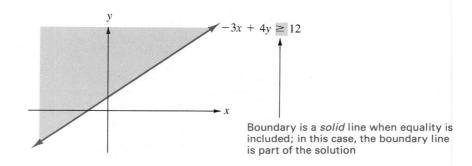

$-3x + 4y \geq 12$

Boundary is a *solid* line when equality is included; in this case, the boundary line is part of the solution

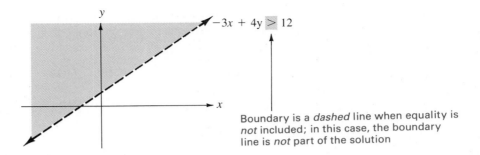

Boundary is a *dashed* line when equality is *not* included; in this case, the boundary line is *not* part of the solution

How to Determine the Correct Half-plane

1. *If the boundary does not go through the origin,* substitute the coordinates of the origin $(0, 0)$ into the inequality.

   If the resulting statement is *true,* the solution is the half-plane containing $(0, 0)$.

   If the resulting statement is *false,* the solution is the half-plane *not* containing $(0, 0)$.

2. *If the boundary goes through the origin,* select a point *not* on the boundary. Substitute the coordinates of this point into the inequality.

   If the resulting statement is *true,* the solution is the half-plane containing the point selected.

   If the resulting statement is *false,* the solution is the half-plane *not* containing the point selected.

TO GRAPH A FIRST-DEGREE INEQUALITY (IN THE PLANE)

1. The boundary line is *solid* if equality is included $(\geq, \leq)$.
The boundary line is *dashed* if equality is *not* included $(>, <)$.

2. Graph the boundary line.

3. Select and shade the correct half-plane.

Example 1   Graph the inequality $2x - 3y < 6$.
Solution Change $<$ to $=$ to find the boundary line.
Boundary line: $2x - 3y = 6$

Step 1: The boundary is a *dashed* line because the equality is *not* included.

$$2x - 3y < 6$$

Step 2: Plot the graph of the boundary line: $2x - 3y = 6$ (Figure 9.7.2).

x-intercept: If $y = 0, 2x - 3(0) = 6$

$$2x = 6$$

$$x = 3$$

y-intercept: If $x = 0, 2(0) - 3y = 6$

$$-3y = 6$$

$$y = -2$$

x	y
3	0
0	-2

Therefore, the boundary line goes through $(3, 0)$ and $(0, -2)$.

Step 3: Select the correct half-plane. The solution of the inequality is only one of the two half-planes determined by the boundary line. Substitute the coordinates of the origin (0, 0) into the inequality:

$$2x - 3y < 6$$

$$2(0) - 3(0) < 6$$

$$0 < 6 \quad \text{True}$$

Therefore, the half-plane containing the origin is the solution. The solution is the shaded area in Figure 9.7.2.

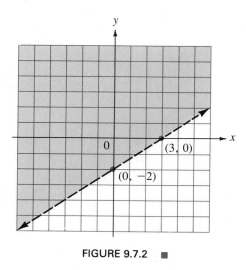

FIGURE 9.7.2 ■

Example 2   Graph the inequality $3x + 4y \leq -12$.
Solution

Step 1: The boundary is a *solid* line because the equality is included.

$$3x + 4y \leq -12$$

Step 2: Plot the graph of the boundary line: $3x + 4y = -12$ (Figure 9.7.3).

x-intercept: If $y = 0$, $3x + 4(0) = -12$

$$3x = -12$$

$$x = -4$$

y-intercept: If $x = 0$, $3(0) + 4y = -12$

$$4y = -12$$

$$y = -3$$

x	y
-4	0
0	-3

Therefore, the boundary line goes through $(-4, 0)$ and $(0, -3)$.

Step 3: Select the correct half-plane. Substitute the coordinates of the origin (0, 0):

$$3x + 4y \leq -12$$

$$3(0) + 4(0) \leq -12$$

$$0 \leq -12 \quad \text{False}$$

Therefore, the solution is the half-plane *not* containing (0, 0). The solution is the shaded area in Figure 9.7.3.

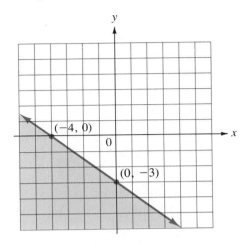

FIGURE 9.7.3 ■

Some inequalities have equations with only one variable. Such inequalities have graphs whose boundaries are either *vertical* or *horizontal* lines.

Example 3   Graph the inequality $x + 4 < 0$.
Solution

Step 1: The boundary is a *dashed* line because the equality is *not* included.

$$x + 4 < 0$$

Step 2: Plot the graph of the boundary line: $x + 4 = 0$ or $x = -4$ (see Figure 9.7.4).

Step 3: Select the correct half-plane. Substitute the coordinates of the origin (0, 0):

$$x + 4 < 0$$
$$(0) + 4 < 0$$
$$4 < 0 \qquad \text{False}$$

Therefore, the solution is the half-plane *not* containing (0, 0). The solution is the shaded area in Figure 9.7.4.

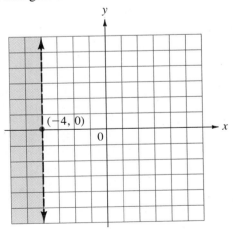

FIGURE 9.7.4 ■

In Section 4.7, we discussed how to graph the solution of an inequality such as $x + 4 < 0$ on a *single number line*. Since $x + 4 < 0$, then $x < -4$.

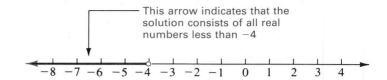

This arrow indicates that the solution consists of all real numbers less than -4

Example 3 of this section shows that the solution of this same inequality, $x + 4 < 0$, represents an entire half-plane when it is graphed in the rectangular coordinate system (see Figure 9.7.4).

Example 4 Graph the inequality $2y - 5x \geq 0$.
Solution

Step 1: The boundary is a *solid* line because the equality is included.

$$2y - 5x \ \geq\ 0$$

Step 2: Plot the graph of the boundary line: $2y - 5x \ =\ 0$ (Figure 9.7.5).

x-intercept: If $y = 0$, $2(0) - 5x = 0$

$$-5x = 0$$

$$x = 0$$

y-intercept: If $x = 0$, $2y - 5(0) = 0$

$$2y = 0$$

$$y = 0$$

Therefore, the boundary line passes through the origin (0, 0).

Find another point on the line:

If $x = 2$, $2y - 5(2) = 0$

$$2y - 10 = 0$$

$$2y = 10$$

$$y = 5$$

This gives the point (2, 5) on the line.

x	y
0	0
0	0
2	5

Step 3: Select the correct half-plane. Since the boundary goes through the origin (0, 0), select a point *not* on the boundary, say (1, 0). Substitute the coordinates of (1, 0) into $2y - 5x \geq 0$:

$$2(0) - 5(1) \ \geq\ 0$$

$$-5 \geq 0 \qquad \text{False}$$

Therefore, the solution is the half-plane *not* containing (1, 0) (Figure 9.7.5).

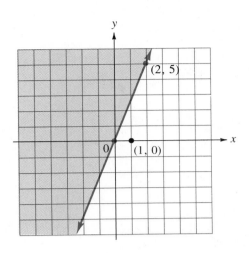

FIGURE 9.7.5 ∎

EXERCISES 9.7

Set I   Graph each of the following inequalities in the plane.

1. $x + 2y < 4$

2. $3x + y < 6$

3. $2x - 3y > 6$

4. $5x - 2y > 10$

5. $x \geq -2$

6. $x \leq 3$

7. $y \leq 4$

8. $y \geq -3$

9. $3y - 4x \geq 12$

10. $2y - 3x \geq 6$

11. $3x + 2y \geq -6$

12. $2x + y \geq -4$

13. $3x - 4y \geq 10$

14. $2x - 3y \leq 11$

15. $2(x + 1) + 3 \leq 3(2x - 1)$

16. $4(2x - 1) + 6 \leq 3(4x + 2)$

17. $4(2y + 3) - 1 \leq 5(2y - 1)$

18. $5(y - 1) + 4 \leq 2(3y + 2)$

19. $\dfrac{x}{5} - \dfrac{y}{3} > 1$

20. $\dfrac{x}{2} - \dfrac{y}{2} < 1$

21. $\dfrac{y}{4} - \dfrac{x}{5} > 1$

22. $\dfrac{y}{3} - \dfrac{x}{2} > 1$

23. $\dfrac{2x + y}{3} - \dfrac{x - y}{2} \geq \dfrac{5}{6}$

24. $\dfrac{4x - 3y}{5} - \dfrac{2x - y}{2} \geq \dfrac{2}{5}$

Set II   Graph each of the following inequalities in the plane.

1. $4x + 2y < 8$

2. $2x - 7y \geq 14$

3. $5x + 2y > 10$

4. $x + y > 0$

5. $x \geq -4$

6. $y \geq -4$

7. $y \leq 1$

8. $x \leq 1$

9. $2y - 5x \geq 10$

10. $x - y > 0$

11. $x + 3y \geq -3$

12. $2x + 7y < 14$

13. $5x - 4y > 12$

14. $y \leq -2$

15. $4(x + 2) - 3 \leq 2(x - 3)$

16. $x > 0$

17. $3(y - 4) + 4 \leq 2(3y + 2)$

18. $3x - 2(2x - 7) \leq 2(3 + x) - 4$

19. $\dfrac{x}{5} - \dfrac{y}{2} > 1$

20. $\dfrac{x}{4} - \dfrac{y}{2} \geq 1$

21. $\dfrac{y}{3} - \dfrac{x}{2} > 1$

22. $\dfrac{y}{2} + \dfrac{x}{4} < 1$

23. $\dfrac{3x + y}{2} - \dfrac{x + y}{3} \geq \dfrac{7}{6}$

24. $\dfrac{x - y}{2} - \dfrac{x - 4y}{7} \leq \dfrac{5}{14}$

9.8 Review: 9.1–9.7

**Ordered Pairs
9.1**

An ordered pair of numbers is used to represent a point in the plane.

x-coordinate		y-coordinate
Abscissa		Ordinate
Horizontal coordinate	(a, b)	Vertical coordinate
First coordinate		Second coordinate

**Linear Equation in
Two Variables
9.2**

An equation that can be expressed in the form $Ax + By + C = 0$ is a **linear equation in two variables**. An ordered pair (a, b) is a solution of the equation $Ax + By + C = 0$ if we get a true statement when we substitute a for x and b for y in that equation.

**Graphing Straight Lines
9.3**

Method 1 (Intercept method)	*Method 2*
1. Find the x-intercept: Set $y = 0$; then solve for x.	1. In the equation, substitute a number for one of the variables and solve for the corresponding value of the other variable. Find *three* points in this way, and list them in a table of values.
2. Find the y-intercept: Set $x = 0$; then solve for y.	
3. If both intercepts are $(0, 0)$, an additional point must be found.	
4. Find one other point by letting x or y have any value.	2. Graph the points listed in the table of values.
5. Draw a straight line through the three points.	3. Draw a straight line through the points.

Exceptions:

The graph of $x = a$ is a vertical line that passes through $(a, 0)$.

The graph of $y = b$ is a horizontal line that passes through $(0, b)$.

**Slope
9.4**

The slope of a line through the points $P_1(x_1, y_1)$ and $P_2(x_2, y_2)$ is

$$m = \frac{y_2 - y_1}{x_2 - x_1}$$

The slope of a horizontal line is 0.

The slope of a vertical line *does not exist*.

Parallel lines have the same slope.

The product of the slopes of two perpendicular lines is -1.

Equations of Lines 9.5

1. *General Form:* $Ax + By + C = 0$
   where A, B, and C are integers and A and B are not both 0.

2. *Point-Slope Form:* $y - y_1 = m(x - x_1)$
   where $m = $ slope and $P_1(x_1, y_1)$ is a known point.

3. *Slope-Intercept Form:* $y = mx + b$
   where $m = $ slope and $b = y$-intercept.

4. The equation of any *horizontal line* is $y = b$, where b is the y-coordinate of every point on the line.

5. The equation of any *vertical line* is $x = a$, where a is the x-coordinate of every point on the line.

To Graph a Curve 9.6

1. Use the equation to make a table of values.

2. Plot the points from the table of values.

3. Draw a smooth curve through the points.

To Graph a First-Degree Inequality in the Plane 9.7

Find the boundary line by replacing the inequality symbol with an equal sign.
1. The boundary line is *solid* if equality is included ($\geq$, $\leq$).

   The boundary line is *dashed* if equality is *not* included ($>$, $<$).

2. Graph the boundary line.

3. Select and shade the correct half-plane.

Review Exercises 9.8   Set I

1. Draw the four-sided figure whose sides are straight lines and whose vertices have the following coordinates: $A(-4, -5)$, $B(2, -4)$, $C(5, 1)$, $D(-2, 3)$.

2. Given the point $B(-2, 4)$, do the following.

   a. Find the y-coordinate of B.   b. Find the abscissa of B.

   c. Find the x-coordinate of B.   d. Graph B.

In Exercises 3–9, graph each line.

3. $x - y = 5$ 4. $y - x = 3$

5. $x = -2$ 6. $y = -3$

7. $4y - 5x = 20$ 8. $5y - 4x = 20$

9. $x + 2y = 0$

In Exercises 10–12, find the slope of line through the given pair of points.

10. $(4, -1)$ and $(4, 3)$ 11. $(2, -6)$ and $(-3, 5)$

12. $(3, -5)$ and $(-2, 4)$

In Exercises 13–17, write the general form of the equation of the given line.

13. The line through $(3, -5)$ with slope $m = -\frac{2}{3}$.

14. The line with a slope of $-\frac{1}{3}$ and a y-intercept of 9.

15. The line that passes through the points $(-3, 4)$ and $(1, -2)$.

16. The horizontal line through $(-2, 5)$.

17. The vertical line through $(-2, 5)$.

In Exercises 18–20, complete the table of values and graph the curve.

18. $y = x^2 - 3x$

x	y
-2	
-1	
0	
1	
2	
3	
4	

19. $y = x^3 - 3x$

x	y
-3	
-2	
-1	
0	
1	
2	
3	

20. $y = x^3 + 3x^2 - 2$

x	y
-4	
-3	
-2	
-1	
0	
1	
2	

In Exercises 21–25, graph the inequalities in the plane.

21. $3x - 7y < 21$

22. $x + 4y \leq 0$

23. $y > 3$

24. $\frac{x}{2} - \frac{y}{6} \geq 1$

25. $\frac{y}{3} - \frac{x}{4} \leq 1$

Review Exercises 9.8   Set II

NAME _____

1. Draw the four-sided figure whose sides are straight lines and whose vertices have the following coordinates: $A(4, -5)$, $B(-2, 4)$, $C(3, 2)$, $D(5, -1)$.

ANSWERS

1.

2. Given the point $P(2, -4)$, do the following.

   a. Find the x-coordinate of P.

   b. Find the ordinate of P.

   c. Graph P.

2a. _____

b. _____

c.

In Exercises 3–9, graph each line.

3. $x - y = -4$

3.

4. $y + 2 = 0$

4.

5. $x + 3y = 6$

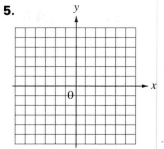

6. $x = \frac{3}{5}y$

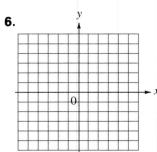

7. $2x - 3y = 7$

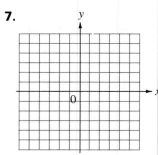

8. $y - 2(x - 3) = 0$

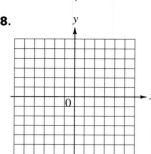

9. $x = 4$

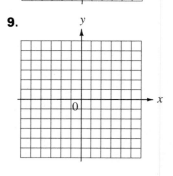

In Exercises 10–12, find the slope of the line through the given pair of points.

10. $(4, -3)$ and $(2, -3)$ **11.** $(-13, -10)$ and $(-8, 7)$

10. _____

11. _____

12. $(-5, 3)$ and $(-5, 0)$

In Exercises 13–17, write the general form of the equation of the given line.

13. The line through $(4, -3)$ with slope $m = -\frac{2}{5}$.

14. The line with a slope of $-\frac{4}{3}$ and a y-intercept of 5.

15. The line that passes through the points $(-4, 3)$ and $(2, -1)$.

16. The horizontal line through $(6, -5)$.

17. The vertical line through $(6, -5)$.

In Exercises 18–20, complete the table of values and graph the curve.

18. $y = x^2 + 1$

x	y
-3	
-2	
-1	
0	
1	
2	
3	

19. $y = 1 - x^2$

x	y
-3	
-2	
-1	
0	
1	
2	
3	

20. $y = 2x - x^3$

x	y
-2	
-1	
0	
1	
2	

12. _____

13. _____

14. _____

15. _____

16. _____

17. _____

18.

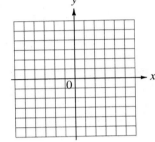

19.

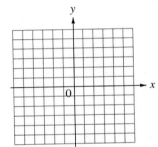

20.

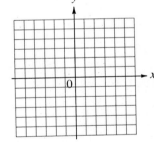

In Exercises 21–25, graph the inequalities in the plane.

21. $2y - 5x \leq 11$

22. $2x - 3y \leq 0$

23. $x - 3 < 0$

24. $\dfrac{x}{5} - \dfrac{y}{4} > 1$

25. $\dfrac{y}{3} - \dfrac{x}{2} \geq 1$

21.

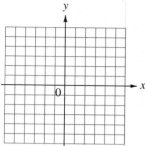

22.

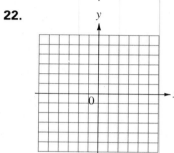

23.

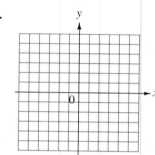

24.

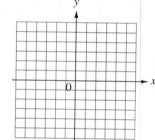

25.

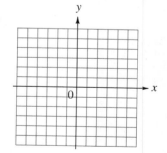

Chapter 9 Diagnostic Test

The purpose of this test is to see how well you understand graphing. We recommend that you work this diagnostic test *before* your instructor tests you on this chapter. Allow yourself about 50 minutes.

Complete solutions for all the problems on this test, together with section references, are given in the answer section in the back of this book. For the problems you do incorrectly, study the sections referred to.

1. a. Draw the triangle whose vertices have the following coordinates: $A(4, -3)$, $B(2, 0)$, $C(-3, 2)$.

b. Read the coordinates of the points D, E, F, and G from the graph below.

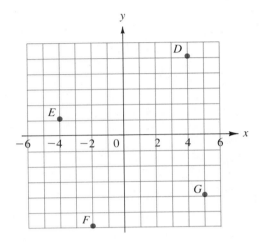

2. Determine whether the given ordered pairs are solutions of the equation $5x - 3y = 15$.

a. $(0, 0)$ b. $(3, 0)$ c. $(0, 3)$ d. $(0, -5)$

In Problems 3–6, graph each line.

3. $y = -2$

4. $3x + y = 3$

5. $2x - y = 4$

6. $4x + 9y = 18$

7. Find the slopes of the lines through the given points.

a. $(5, 2)$ and $(-3, 1)$ b. $(-3, 2)$ and $(-3, 0)$

c. $(2, -4)$ and $(0, -4)$

8. Write the equation of the given line.

a. The vertical line that passes through the point $(-3, 4)$.

b. The horizontal line that passes through the point $(-3, 4)$.

9. Write the equation of the line with a slope of 5 and a y-intercept of 3.

10. Given the points $A(-3, 2)$ and $B(1, -4)$, find the following.

a. The slope of the line through A and B.

b. The general form of the equation of the line through A and B.

11. Graph the curve $y = x^2 - x - 6$, using integral values of x from -3 to 4.

12. Graph the inequality $3x - 2y \le 6$ in the plane.

Cumulative Review Exercises: Chapters 1–9

1. Evaluate the formula using the values of the variables given with the formula.

$$A = \frac{h}{2}(B + b) \quad h = 5, B = 11, b = 7$$

2. Simplify. Write your answer using only positive exponents.

$$\left(\frac{35a}{15a^{-2}b}\right)^{-2}$$

In Exercises 3–7, perform the indicated operations and simplify.

3. $(6z^2 - z - 7) \div (3z - 5)$

4. $\dfrac{2x - 1}{x - 2} - \dfrac{1}{x + 1}$

5. $\dfrac{b + 1}{a^2 + ab} \div \dfrac{b^2 - b - 2}{a^3 - ab^2}$

6. $\dfrac{x^2 - \dfrac{4}{y^2}}{x + \dfrac{2}{y}}$

7. $\dfrac{3}{x^2 - 7x + 6} - \dfrac{4}{2x^2 + x - 3}$

8. Solve for x: $\dfrac{2x - 4}{3} = \dfrac{4 - 6x}{5}$

9. Solve for x: $\dfrac{2x}{9} - \dfrac{11}{3} = \dfrac{5x}{6}$

10. Solve for k: $C = \dfrac{5}{2h} - \dfrac{h}{k}$

11. Graph $y = 3x - 2$.

12. Graph $y = x^2 - 4x$, using integral values of x from -1 to 5.

13. Write the equation of the straight line that passes through $(7, 1)$ and $(2, -4)$.

In Exercises 14–22, set up the problem algebraically, solve, and check. Be sure to state what your variables represent.

14. One number is 4 more than another. Their product is equal to $-\frac{7}{4}$. Find the two numbers.

15. A 50-lb mixture of Delicious and Jonathan apples costs $44.58. If the Delicious apples cost 98¢ a pound and the Jonathan apples cost 85¢ a pound, how many pounds of each kind are there?

16. The base of a triangle is 3 cm more than its altitude. Its area is 44 sq. cm. Find the lengths of the altitude and the base.

17. The sum of a number and its reciprocal is $\frac{53}{14}$. Find the number.

18. Find two consecutive integers whose product is 5 more than their sum.

19. How many cubic centimeters of water must be added to 10 cc of a 17% solution of a disinfectant to reduce it to a 0.2% solution?

20. Terri types 473 words in 8.6 min. What is her rate of typing?

21. A camera shop buys eighteen lenses. If it had bought fifteen lenses of higher quality, it would have paid $23 more per lens for the same total cost. Find the cost of each type of lens.

22. A hardware store buys twelve saws. If it had bought eight saws of higher quality, it would have paid $2.75 more per saw for the same total cost. Find the cost of each type of saw.

Critical Thinking

Each of the following problems has an error. Can you find it?

1. Simplify $\dfrac{4}{x} + \dfrac{2}{x+2}$.

LCD $= x(x + 2)$

$$\overset{1}{\cancel{x}}(x+2)\dfrac{4}{\underset{1}{\cancel{x}}} + x\cancel{(x+2)}\overset{1}{}\dfrac{2}{\underset{1}{\cancel{x+2}}} = 4x + 8 + 2x = 6x + 8$$

2. Simplify $\dfrac{1 - \dfrac{1}{x^2}}{1 - \dfrac{1}{x}}$.

$$\dfrac{x^2\left(1 - \dfrac{1}{x^2}\right)}{x\left(1 - \dfrac{1}{x}\right)} = \dfrac{x^2 - 1}{x - 1}$$

$$= \dfrac{(x+1)\cancel{(x-1)}^{1}}{\underset{1}{\cancel{x-1}}}$$

$$= x + 1$$

3. Solve $x^2 - 3x = 4$.

$$x^2 - 3x = 4$$
$$x(x - 3) = 4$$

$x = 4$ $\bigg|$ $\begin{array}{c} x - 3 = 4 \\ x = 7 \end{array}$

4. Solve $4x - 2(x - 3) < 3(x + 3)$.

$$4x - 2(x - 3) < 3(x + 3)$$
$$4x - 2x + 6 < 3x + 9$$
$$2x + 6 < 3x + 9$$
$$-x + 6 < 9$$
$$-x < 3$$
$$x < -3$$

10 Systems of Equations

In previous chapters, we showed how to solve a single equation for a single variable. In this chapter, we show how to solve systems of two linear equations in two variables. Systems that have *more* than two equations and variables are not discussed in this book.

10.1 Basic Definitions

A **system of two equations** in (the same) two variables is a set of two equations considered together, or simultaneously. The *solution*, if one exists, of such a system is the set of ordered pairs that satisfies both equations.

Example 1

Examples of systems of two equations in two variables and their solutions:

a. $\begin{cases} x - y = 6 \\ x + y = 2 \end{cases}$ is a system of two equations in x and y.

$(4, -2)$ is a solution of the system, because when we substitute 4 for x and -2 for y into the first equation, we have

$$4 - (-2) = 6 \quad \text{True}$$

and when we substitute 4 for x and -2 for y in the second equation, we have

$$4 + (-2) = 2 \quad \text{True}$$

This can be stated as "$(4, -2)$ satisfies the system $\begin{cases} x - y = 6 \\ x + y = 2 \end{cases}$."

b. $\begin{cases} x + y = 4 \\ 2x - y = 2 \end{cases}$ is a system of two equations in x and y.

$(2, 2)$ is a solution of that system, because when we substitute 2 for x and 2 for y in the first equation, we have

$$2 + 2 = 4 \quad \text{True}$$

Substituting 2 for x and 2 for y in the second equation gives

$$2(2) - 2 = 2 \quad \text{True} \quad \blacksquare$$

Example 2

Determine whether $(0, 3)$ is a solution of the system $\begin{cases} x + y = 3 \\ 3x - y = 5 \end{cases}$.

$0 + 3 = 3$	Substituting 0 for x and 3 for y in the first equation
$3 = 3$	True
$3(0) - 3 = 5$	Substituting 0 for x and 3 for y in the second equation
$-3 = 5$	False

Because a false statement results when we substitute the ordered pair into the second equation, $(0, 3)$ is *not* a solution of the given system. $\blacksquare$

EXERCISES 10.1

Set I   Determine whether each ordered pair is a solution of the given system of equations.

1. $\begin{cases} 3x + y = 8 \\ 2x - y = 2 \end{cases}$ a. $(0, 8)$ b. $(1, 0)$ c. $(2, 2)$

2. $\begin{cases} 4x - y = 3 \\ 3x + y = 11 \end{cases}$ a. $(0, -3)$ b. $(2, 5)$ c. $(1, 1)$

3. $\begin{cases} 2x + 3y = 6 \\ 4x + 6y = 12 \end{cases}$ a. $(3, 2)$ b. $(0, 2)$ c. $(3, 0)$

4. $\begin{cases} 3x - 2y = 6 \\ 6x - 4y = 12 \end{cases}$ a. $(2, 0)$ b. $(0, -3)$ c. $(-2, -6)$

5. $\begin{cases} x + y = 2 \\ x + y = 4 \end{cases}$ a. $(0, 4)$ b. $(0, 2)$ c. $(4, 0)$

6. $\begin{cases} 2x - y = 4 \\ 2x - y = 2 \end{cases}$ a. $(0, -4)$ b. $(2, 0)$ c. $(1, 0)$

Set II   Determine whether each ordered pair is a solution of the given system of equations.

1. $\begin{cases} 2x + y = 6 \\ 3x - y = -1 \end{cases}$ a. $(0, 6)$ b. $(1, 4)$ c. $(3, 0)$

2. $\begin{cases} 4x - 2y = 8 \\ 2x - y = 4 \end{cases}$ a. $(0, -4)$ b. $(2, 0)$ c. $(6, 8)$

3. $\begin{cases} x + 3y = 6 \\ 2x + 6y = 12 \end{cases}$ a. $(6, 0)$ b. $(0, 2)$ c. $(1, 3)$

4. $\begin{cases} 3x - y = 3 \\ 6x - 2y = 3 \end{cases}$ a. $(1, 0)$ b. $(0, -3)$ c. $(\frac{1}{2}, 0)$

5. $\begin{cases} 4x + y = 4 \\ 8x + 2y = 4 \end{cases}$ a. $(0, 4)$ b. $(1, 0)$ c. $(\frac{1}{2}, 0)$

6. $\begin{cases} 2x + y = 6 \\ 2x - y = 2 \end{cases}$ a. $(0, 6)$ b. $(2, 2)$ c. $(1, 0)$

10.2 Graphical Method for Solving a Linear System

The graphical method for solving a system of equations is not an exact method of solution, but it can sometimes be used successfully in solving such systems. Recall from Chapter 9 that the graph of a linear equation in two variables is a straight line.

TO SOLVE A SYSTEM OF LINEAR EQUATIONS BY THE GRAPHICAL METHOD

1. Graph each straight line on the same set of axes.

2. Three outcomes are possible:
   a. The lines intersect in exactly one point (Figure 10.2.1). The solution is the ordered pair that represents the point of intersection.
   b. The lines will never cross, even if they are extended indefinitely (Figure 10.2.2). Such lines are called *parallel lines*. There is no solution for the system of equations.
   c. Both equations have the same line for their graph (Figure 10.2.3). Infinitely many ordered pairs are solutions; any ordered pair that represents a point on the line is a solution.

3. If the lines graphed in step 1 intersect in exactly one point, check the coordinates of that point in *both* equations.

Example 1   Solve the system $\begin{Bmatrix} x + 2y = 8 \\ x - y = 2 \end{Bmatrix}$ graphically.

Solution We draw the graph of each line on the same set of axes (Figure 10.2.1).

Line 1: $x + 2y = 8$

x	y
0	4
8	0
2	3

y-intercept: If $x = 0$, then $y = 4$.

x-intercept: If $y = 0$, then $x = 8$.

Checkpoint: If $x = 2$, then $y = 3$.

Line 2: $x - y = 2$

x	y
0	-2
2	0
-2	-4

y-intercept: If $x = 0$, then $y = -2$.

x-intercept: If $y = 0$, then $x = 2$.

Checkpoint: If $x = -2$, then $y = -4$.

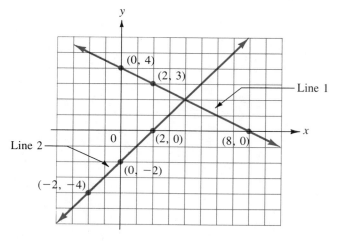

FIGURE 10.2.1

The lines appear to intersect at the point (4, 2). We check that solution by substituting 4 for x and 2 for y in *both* equations.

$$(1) \quad x + 2y = 8 \qquad\qquad (2) \quad x - y = 2$$
$$4 + 2(2) \stackrel{?}{=} 8 \qquad\qquad 4 - 2 \stackrel{?}{=} 2$$
$$8 = 8 \quad \text{True} \qquad\qquad 2 = 2 \quad \text{True}$$

Therefore, the solution of the system is (4, 2).

The coordinates of any point on line 1 satisfy the equation of line 1. The coordinates of any point on line 2 satisfy the equation of line 2. The only point that lies on *both* lines is (4, 2). Therefore, it is the only point whose coordinates satisfy *both* equations. ∎

When a system of equations has a solution, it is called a **consistent system**. When each equation in the system has a different graph, the system is called an **independent system**. Therefore, the system in Example 1 is a consistent, independent system.

Example 2 Solve the system $\begin{cases} 3x + 2y = 6 \\ 6x + 4y = 24 \end{cases}$ graphically.

Solution We draw the graph of each line on the same set of axes (Figure 10.2.2).

Line 1: $3x + 2y = 6$

y-intercept: If $x = 0$, then $y = 3$.

x-intercept: If $y = 0$, then $x = 2$.

Checkpoint: If $x = -2$, then $y = 6$.

x	y
0	3
2	0
-2	6

Line 2: $6x + 4y = 24$

y-intercept: If $x = 0$, then $y = 6$.

x-intercept: If $y = 0$, then $x = 4$.

Checkpoint: If $x = 2$, then $y = 3$.

x	y
0	6
4	0
2	3

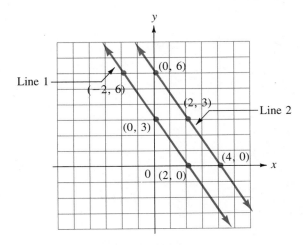

FIGURE 10.2.2

The lines will never meet; they are parallel. There is *no solution* for the system of equations. ∎

A system (such as that in Example 2) that has no solution is called an **inconsistent system**. Since each equation in Example 2 has a different graph, the system is independent.

Example 3   Solve the system $\begin{Bmatrix} x - 2y = -4 \\ 3x - 6y = -12 \end{Bmatrix}$ graphically.

Solution We draw the graph of each line on the same set of axes (Figure 10.2.3).

Line 1: $x - 2y = -4$

y-intercept: If $x = 0$, then $y = 2$.

x-intercept: If $y = 0$, then $x = -4$.

Checkpoint: If $x = 4$, then $y = 4$.

x	y
0	2
-4	0
4	4

Line 2: $3x - 6y = -12$

y-intercept: If $x = 0$, then $y = 2$.

x-intercept: If $y = 0$, then $x = -4$.

Checkpoint: If $x = 2$, then $y = 3$.

x	y
0	2
-4	0
2	3

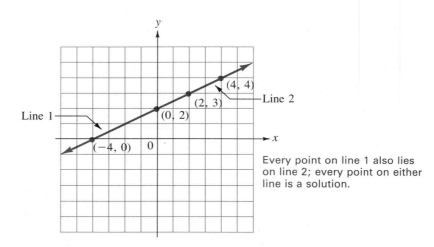

Every point on line 1 also lies on line 2; every point on either line is a solution.

FIGURE 10.2.3

Since both lines go through $(0, 2)$ and $(-4, 0)$, they must be the same line (Figure 10.2.3). There are infinitely many solutions for this system of equations; four of them are $(0, 2)$, $(-4, 0)$, $(4, 4)$, and $(2, 3)$. ■

When both equations in the system have the same graph, the system is called a **dependent system**. Therefore, the system of equations in Example 3 is a consistent, dependent system.

EXERCISES 10.2

Set I   Find the solution of each system of equations graphically. If the lines are parallel, write
"No solution." If both equations have the same line for their graph, write "Many solutions."

1. $\begin{cases} 2x + y = 6 \\ 2x - y = -2 \end{cases}$ **2.** $\begin{cases} 2x - y = -4 \\ x + y = 1 \end{cases}$

3. $\begin{cases} x - 2y = -6 \\ 4x + 3y = 20 \end{cases}$ **4.** $\begin{cases} x - 3y = 6 \\ 4x + 3y = 9 \end{cases}$

5. $\begin{cases} x + 2y = 0 \\ x - 2y = -2 \end{cases}$ **6.** $\begin{cases} 2x + y = 0 \\ 2x - y = -6 \end{cases}$

7. $\begin{cases} 3x - 2y = -9 \\ x + y = 2 \end{cases}$ **8.** $\begin{cases} x + 2y = -4 \\ 2x - y = -3 \end{cases}$

9. $\begin{cases} 10x - 4y = 20 \\ 6y - 15x = -30 \end{cases}$ **10.** $\begin{cases} 12x - 9y = 36 \\ 6y - 8x = -24 \end{cases}$

11. $\begin{cases} 2x - 4y = -2 \\ -3x + 6y = 12 \end{cases}$ **12.** $\begin{cases} 4x - 2y = 8 \\ -6x + 3y = 6 \end{cases}$

Set II   Find the solution of each system of equations graphically. If the lines are parallel, write
"No solution." If both equations have the same line for their graph, write "Many solutions."

1. $\begin{cases} 2x + y = -4 \\ x - y = -5 \end{cases}$ **2.** $\begin{cases} 3x + 2y = 0 \\ 3x - 2y = 12 \end{cases}$

3. $\begin{cases} 3x - 4y = -3 \\ 3x + 2y = 6 \end{cases}$ **4.** $\begin{cases} 3x + 2y = 3 \\ 5x - 4y = 16 \end{cases}$

5. $\begin{cases} x + y = 1 \\ x - y = -3 \end{cases}$ **6.** $\begin{cases} x - 4y = 4 \\ 8y - 2x = -8 \end{cases}$

7. $\begin{cases} 4x - 3y = 6 \\ x - y = 1 \end{cases}$ **8.** $\begin{cases} 3x - y = 6 \\ y - 3x = 3 \end{cases}$

9. $\begin{cases} 4x - 6y = 12 \\ 9y - 6x = -18 \end{cases}$ **10.** $\begin{cases} 4x + 3y = 3 \\ x + y = -1 \end{cases}$

11. $\begin{cases} x - 2y = 10 \\ 2y - x = 2 \end{cases}$ **12.** $\begin{cases} x - 4y = 2 \\ x + 2y = -4 \end{cases}$

10.3 Addition Method for Solving a Linear System

The graphical method for solving a system of equations has two disadvantages: (1) It is
slow, and (2) it is not an exact method of solution. The method we discuss in this section
has neither of these disadvantages. We need one definition before we proceed.

EQUIVALENT SYSTEMS OF EQUATIONS

Two systems of equations are **equivalent** if the systems both have the same
solution set.

When we solve a system of two equations in x and y algebraically, we must try to find an equivalent system in which one of the equations is in the form $x = a$ and the other is in the form $y = b$. That is, we try to find an equivalent system in which each equation contains only one variable.

In using the addition method for solving a linear system, we use the following property of equality:

$$\text{If } a = b \text{ and } c = d, \text{ then } a + c = b + d$$

Before we use this property, we may need to use the multiplication property of equality to make the coefficients of one of the variables become the *negatives* of each other. Then, when we add the equations, at least one of the variables will be eliminated because the sum of its coefficients will be zero.

The general method for solving a system of equations by the addition method is as follows:

TO SOLVE A SYSTEM OF EQUATIONS BY THE ADDITION METHOD

1. Multiply the equations by numbers that make the coefficients of one of the variables become the negatives of each other.

2. Add the equations.

3. Three outcomes are possible:
   a. *One variable remains.* Solve the resulting equation for that variable. Then substitute this known value into either of the system's equations to solve for the other variable. There is one solution for the system. The system is consistent and independent.
   b. *Both variables are eliminated and a false statement results.* There is no solution; the system is inconsistent and independent.
   c. *Both variables are eliminated and a true statement results.* There are many solutions; the system is consistent and dependent.

4. If one solution is found in step 3, check it in both equations.

We can always make the coefficients of one of the variables equal numerically by simply multiplying the equations by each other's coefficients (see Example 1). Sometimes, however, smaller numbers can be used (see Example 2).

Solving Systems with Only One Solution by the Addition Method

Example 1   Solve the system $\begin{Bmatrix} 3x + 4y = 6 \\ 2x + 3y = 5 \end{Bmatrix}$ using the addition method (a) eliminating the x's and (b) eliminating the y's.

Solution

a. Equation 1: $3x + 4y = 6$
   Equation 2: $2x + 3y = 5$

   If Equation 1 is multiplied by 2 and Equation 2 is multiplied by -3, the coefficients of x will add to zero in the resulting equations.

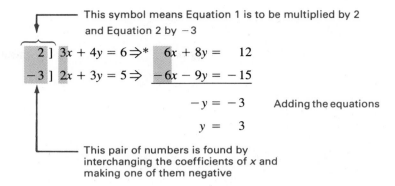

This symbol means Equation 1 is to be multiplied by 2 and Equation 2 by -3

$$2\;]\;3x + 4y = 6 \Rightarrow^* \quad 6x + 8y = 12$$
$$-3\;]\;2x + 3y = 5 \Rightarrow \quad \underline{-6x - 9y = -15}$$
$$-y = -3 \qquad \text{Adding the equations}$$
$$y = 3$$

This pair of numbers is found by interchanging the coefficients of x and making one of them negative

The value $y = 3$ can be substituted into either Equation 1 or Equation 2 to find the value of x. It is usually easier to substitute into the equation that has the smaller coefficients.

We substitute $y = 3$ into Equation 2:

$$2x + 3y = 5$$
$$2x + 3(3) = 5$$
$$2x + 9 = 5$$
$$2x = -4$$
$$x = -2$$

If we substitute $y = 3$ into Equation 1, we will still get -2 as the value for x:

$$3x + 4y = 6$$
$$3x + 4(3) = 6$$
$$3x + 12 = 6$$
$$3x = -6$$
$$x = -2$$

Therefore, the solution of the system is $\left\{ \begin{array}{l} x = -2 \\ y = 3 \end{array} \right\}$, or $(-2, 3)$ if written as an ordered pair.

The coefficients of y add to zero

b. $\quad 3\;]\;3x + 4y = 6 \Rightarrow \quad 9x + 12y = 18$
$\quad -4\;]\;2x + 3y = 5 \Rightarrow \quad \underline{-8x - 12y = -20}$
$$x = -2$$

This pair of numbers is found by interchanging the coefficients of y and making one of them negative

---

*The symbol $\Rightarrow$, read "implies," means that the second statement is true if the first statement is true. For example,

$$x - 2 = 0 \Rightarrow x = 2$$

is read "$x - 2 = 0$ implies $x = 2$" and means that $x = 2$ is true if $x - 2 = 0$ is true.

The value $x = -2$ can be substituted into either Equation 1 or Equation 2 to find the value of y. Substituting into Equation 2, we have

$$2x + 3y = 5$$
$$2(-2) + 3y = 5$$
$$-4 + 3y = 5$$
$$3y = 9$$
$$y = 3$$

The solution, $(-2, 3)$, is, of course, the same solution we obtained in Example 1a. The check will not be shown. ■

Example 2 Find the smallest numbers that can be used to eliminate (a) the x's and (b) the y's for the system $\begin{Bmatrix} 10x - 9y = 5 \\ 15x + 6y = 4 \end{Bmatrix}$. (Do not solve the system.)

Solution

a.

This pair is found by interchanging the coefficients of x and making one of them negative

| 15] | 3] | $10\,x - 9y = 5 \Rightarrow$ | $30\,x - 27y = 15$ |
| -10] | -2] | $15\,x + 6y = 4 \Rightarrow$ | $-30\,x - 12y = -8$ |

This pair is found by reducing the ratio $\frac{15}{10}$ to $\frac{3}{2}$

b.

This pair is found by interchanging the coefficients of y

| 6] | 2] | $10x - 9\,y = 5 \Rightarrow 20x - 18\,y = 10$ |
| 9] | 3] | $15x + 6\,y = 4 \Rightarrow 45x + 18\,y = 12$ |

This pair is found by reducing the ratio $\frac{6}{9}$ to $\frac{2}{3}$ ■

In Example 2b, it was not necessary to make one of the numbers negative, because the coefficients of y already had different signs.

Example 3 Solve the system $\begin{Bmatrix} x - y = 6 \\ y + x = 2 \end{Bmatrix}$ by the addition method.

Solution In this example, the coefficients of the y's are already the negatives of each other. Therefore, we can omit step 1; when we add the equations, the y's will be eliminated. We must, however, be sure to line up like terms when we write the equations.

Equation 1 $x - y = 6$

Equation 2 $\underline{x + y = 2}$ Rearranging so like terms are lined up

$2x \qquad = 8$ Adding the equations

$x = 4$

Substituting $x = 4$ into Equation 1, we have

$$x - y = 6$$
$$4 - y = 6$$
$$-y = 2$$
$$y = -2$$

The system $\begin{Bmatrix} x = 4 \\ y = -2 \end{Bmatrix}$ is equivalent to the system $\begin{Bmatrix} x - y = 6 \\ y + x = 2 \end{Bmatrix}$, because both systems have the same solution set.

Check See Example 1a, Section 10.1.

Therefore, the solution is $\begin{Bmatrix} x = 4 \\ y = -2 \end{Bmatrix}$, or $(4, -2)$. ∎

Example 4 Solve the system $\begin{Bmatrix} x - 2y = -6 \\ 4x + 3y = 20 \end{Bmatrix}$.

Solution

Equation 1: $x - 2y = -6$
Equation 2: $4x + 3y = 20$

If both sides of Equation 1 are multiplied by -4, the coefficients of x will add to zero.

$$-4]\ x - 2y = -6 \Rightarrow -4x + 8y = +24$$
$$4x + 3y = 20 \Rightarrow \underline{4x + 3y = 20}$$
$$11y = 44 \qquad \text{Adding the equations}$$
$$y = 4$$

Then, substituting $y = 4$ into Equation 1, we have

$$x - 2y = -6$$
$$x - 2(4) = -6$$
$$x - 8 = -6$$
$$x = 2$$

Therefore, the solution of the system is $\begin{Bmatrix} x = 2 \\ y = 4 \end{Bmatrix}$ or $(2, 4)$.

The check will not be shown. ∎

Example 5 Solve the system $\begin{Bmatrix} 3y + 2x = 4 \\ 5x + 6y = 11 \end{Bmatrix}$.

Solution We first arrange the equations so like terms are lined up.

$$5]\ 2x + 3y = 4 \Rightarrow 10x + 15y = 20$$
$$-2]\ 5x + 6y = 11 \Rightarrow \underline{-10x - 12y = -22}$$
$$3y = -2$$
$$y = -\frac{2}{3}$$

Substituting $y = -\frac{2}{3}$ into $2x + 3y = 4$, we have

$$2x + 3y = 4$$
$$2x + \frac{\cancel{3}}{1}\left(\frac{-2}{\cancel{3}}\right) = 4$$
$$2x - 2 = 4$$
$$2x = 6$$
$$x = 3$$

Therefore, the solution of the system is $(3, -\frac{2}{3})$.

The check will not be shown. ∎

In all the examples so far, when we attempted to solve the system by the addition method, we had one variable left after we added the equations. Therefore, there was exactly one solution for each system. Each system was consistent and independent, and the lines representing the equations intersected in exactly one point.

Solving Systems with No Solution by the Addition Method

Sometimes when we attempt to solve a system of equations by the addition method, both variables drop out and we're left with a *false statement*. In this case, there is no solution for the system. The system is inconsistent, and the graphs of the equations are parallel lines.

Example 6   Solve the system $\left\{\begin{matrix} 4x - 2y = 5 \\ 2x - y = 2 \end{matrix}\right\}$ by the addition method.

Solution

Equation 1: $-2\,]$ $-1\,]\,4x - 2y = 5 \Rightarrow$ $-4\,x\ +2\,y = -5$

Equation 2: $4\,]$ $2\,]\,2x - \ y = 2 \Rightarrow$ $\underline{\quad 4\,x\ -2\,y = \ \ \ 4}$

$$0 = -1 \qquad \text{False}$$

Because both variables were eliminated and we were left with a *false* statement, there is no solution for the system. ∎

In Example 6, the system was independent and inconsistent.

Solving Systems with More Than One Solution by the Addition Method

Sometimes when we attempt to solve a system of equations by the addition method, both variables drop out and we're left with a *true statement*. In this case, there are many solutions for the system; the graphs of the equations are lines that coincide.

Example 7   Solve the system $\left\{\begin{matrix} 4x - 2y = 4 \\ 2x - y = 2 \end{matrix}\right\}$ by the addition method.

Solution

Equation 1: $-2\,]$ $-1\,]\,4x - 2y = 4 \Rightarrow$ $-4\,x\ +2\,y = -4$

Equation 2: $4\,]$ $2\,]\,2x - \ y = 2 \Rightarrow$ $\underline{\quad 4\,x\ -2\,y = \ \ \ 4}$

$$0 = \ \ 0 \qquad \text{True}$$

Because both variables were eliminated and we were left with a *true* statement, there are many solutions for the system. Any ordered pair that satisfies Equation 1 or Equation 2 is a solution of the system. ∎

In Example 7, the system of equations was consistent and dependent.

EXERCISES 10.3

Set I   Find the solution of each system of equations by the addition method. Check your solutions. Write "Inconsistent" if no solution exists. Write "Dependent" if many solutions exist.

1. $\left\{\begin{matrix} 2x - y = -4 \\ x + y = -2 \end{matrix}\right\}$

2. $\left\{\begin{matrix} 2x + y = 6 \\ x - y = 0 \end{matrix}\right\}$

3. $\begin{cases} x - 2y = 10 \\ x + y = 4 \end{cases}$

4. $\begin{cases} x + 4y = 4 \\ x - 2y = -2 \end{cases}$

5. $\begin{cases} x - 3y = 6 \\ 4x + 3y = 9 \end{cases}$

6. $\begin{cases} 2x + 5y = 2 \\ 3x - 5y = 3 \end{cases}$

7. $\begin{cases} x + y = 2 \\ 3x - 2y = -9 \end{cases}$

8. $\begin{cases} x + 2y = -4 \\ 2x - y = -3 \end{cases}$

9. $\begin{cases} x + 2y = 0 \\ y - 2x = 0 \end{cases}$

10. $\begin{cases} x - 3y = 0 \\ 3y - x = 0 \end{cases}$

11. $\begin{cases} 4x + 3y = 2 \\ 3x + 5y = -4 \end{cases}$

12. $\begin{cases} 5x + 7y = 1 \\ 3x + 4y = 1 \end{cases}$

13. $\begin{cases} 6x - 10y = 6 \\ 9x - 15y = -4 \end{cases}$

14. $\begin{cases} 7x - 2y = 7 \\ 21x - 6y = 6 \end{cases}$

15. $\begin{cases} 3x - 5y = -2 \\ 10y - 6x = 4 \end{cases}$

16. $\begin{cases} 15x - 9y = -3 \\ 6y - 10x = 2 \end{cases}$

Set II Find the solution of each system of equations by the addition method. Check your solutions. Write "Inconsistent" if no solution exists. Write "Dependent" if many solutions exist.

1. $\begin{cases} x - y = -5 \\ 3x + y = -3 \end{cases}$

2. $\begin{cases} x - 2y = 4 \\ -2x + 4y = 3 \end{cases}$

3. $\begin{cases} x + y = 5 \\ x - 3y = 3 \end{cases}$

4. $\begin{cases} x - 2y = 3 \\ 3x + 7y = -4 \end{cases}$

5. $\begin{cases} x - 2y = 6 \\ 3x - 2y = 12 \end{cases}$

6. $\begin{cases} 3x + 5y = 4 \\ 2x + 3y = 4 \end{cases}$

7. $\begin{cases} 2x + 6y = 2 \\ 3x + 9y = 3 \end{cases}$

8. $\begin{cases} 4x - 3y = 7 \\ 2x + 7y = 12 \end{cases}$

9. $\begin{cases} y - 5x = 0 \\ x - 5y = 0 \end{cases}$

10. $\begin{cases} 5x + 2y = 7 \\ 10x - 4y = 14 \end{cases}$

11. $\begin{cases} 7x + 2y = 14 \\ 2x - 3y = 4 \end{cases}$

12. $\begin{cases} 6x + 3y = 4 \\ y + 2x = 2 \end{cases}$

13. $\begin{cases} 2x - 5y = 4 \\ 6x - 15y = 8 \end{cases}$

14. $\begin{cases} 5x + 3y = -12 \\ 3y - 6x = 10 \end{cases}$

15. $\begin{cases} 6x + 4y = 2 \\ 3x - 2y = 3 \end{cases}$

16. $\begin{cases} 9x + 3y = 0 \\ 2x - 5y = 0 \end{cases}$

10.4 Substitution Method for Solving a Linear System

All linear systems of two equations in two variables can be solved by the addition method shown in Section 10.3. However, it is important that you also learn the **substitution method** of solution so you can apply it later in solving more complicated systems.

TO SOLVE A LINEAR SYSTEM OF TWO EQUATIONS BY THE SUBSTITUTION METHOD

1. Solve one equation for one of the variables in terms of the other.

2. Substitute the expression obtained in step 1 into the other equation. Simplify both sides of the resulting equation.

3. Three outcomes are possible:
   a. *One variable remains*. Solve the equation resulting from step 2 for its variable, and substitute that value into either equation to find the value of the other variable. There is one solution for the system. The system is consistent and independent.
   b. *Both variables are eliminated and a false statement results*. There is no solution; the system is inconsistent and independent.
   c. *Both variables are eliminated and a true statement results*. There are many solutions; the system is consistent and dependent.

4. If one solution was found in step 3, check it in both equations.

When students use the substitution method, they often worry about which equation to start with and which variable to solve for. Sometimes one of the equations is already solved for a variable (see Examples 1 and 4); if so, use that equation and variable. Sometimes, in one of the equations the coefficient of one of the variables is 1 (see Examples 2 and 5); if so, solve that equation for the variable with a coefficient of 1. Otherwise, solve for the variable with the smallest coefficient (see Examples 3 and 6).

Example 1 Examples of systems with one equation already solved for one variable:

a. $\begin{cases} x + y = 4 \\ y = x + 2 \end{cases}$

       └─ Already solved for y

       ┌─ Already solved for x

b. $\begin{cases} x = -3 - y \\ 2x + 5y = 0 \end{cases}$ ∎

Example 2 Examples of systems in which one variable has a coefficient of 1:

       ┌─ y has a coefficient of 1

a. $\begin{cases} 2x + y = 5 \\ 3x - 2y = 18 \end{cases} \Rightarrow y = 5 - 2x$

b. $\begin{cases} 2x + 6y = 3 \\ x - 4y = 2 \end{cases} \Rightarrow x = 4y + 2$

     x has a coefficient of 1 ∎

Example 3   Examples of choosing the variable with the smallest coefficient:

Smallest of the four coefficients

a. $\begin{Bmatrix} 6x + \boxed{3}\,y = -1 \\ 4x + 9y = 4 \end{Bmatrix} \Rightarrow y = \dfrac{-1-6x}{3}$

Smallest possible denominator

Smallest of the four coefficients

b. $\begin{Bmatrix} \boxed{3}\,x + 9y = 7 \\ 12x - 8y = 15 \end{Bmatrix} \Rightarrow x = \dfrac{7-9y}{3}$

Smallest possible denominator ■

Solving Systems with One Solution by the Substitution Method

Example 4   Solve the system $\begin{Bmatrix} x + y = 4 \\ y = x + 2 \end{Bmatrix}$ by the substitution method.

Solution

Step 1: Equation 2 has already been solved for y: $y = x + 2$.

Step 2: Substitute $\boxed{x+2}$ for y in Equation 1:

$$x + y \qquad\quad = 4$$
$$x + \boxed{(x+2)} = 4$$

Step 3: $2x + 2 = 4$ Both sides are simplified; the variable x remains

        $2x = 2$ Solve for x

         $x = 1$

To find y, substitute 1 for x in Equation 2:

$$y = x + 2$$
$$y = \boxed{1} + 2$$
$$y = 3$$

Therefore, $(1, 3)$ is the solution for this system.

Step 4: (1) $x + y = 4$ (2) $y = x + 2$

          $1 + 3 \overset{?}{=} 4$ $3 \overset{?}{=} 1 + 2$

            $4 = 4$ True $3 = 3$ True ■

Example 5   Solve the system $\begin{Bmatrix} 2x + y = 5 \\ 3x - 2y = 18 \end{Bmatrix}$ by the substitution method.

Solution The coefficient of y in Equation 1 is 1.

Step 1: Solve Equation 1 for y:

$$2x + \boxed{y} = 5$$
$$\underline{-2x} \qquad\quad \underline{-2x}$$
$$y = \boxed{5 - 2x}$$

Step 2: Substitute $5 - 2x$ in place of y in Equation 2:

$$3x - 2y = 18$$
$$3x - 2\,(5 - 2x) = 18$$
$$3x - 10 + 4x = 18$$

Step 3: $\qquad\qquad 7x - 10 = 18$ Both sides are simplified; the variable x remains

$$7x = 28 \qquad \text{Solve for } x$$
$$x = 4$$

To find y, substitute 4 for x in Equation 1:

$$2x + y = 5$$
$$2(\,4\,) + y = 5$$
$$8 + y = 5$$
$$y = -3$$

Therefore, $(4, -3)$ is the solution for this system.

The check will not be shown. ∎

— Smallest of the four coefficients

Example 6   Solve the system $\begin{Bmatrix} 6x + 3y = -1 \\ 4x + 9y = 4 \end{Bmatrix}$.

Solution No variable has a coefficient of 1. Solve Equation 1 for y because it has the smallest coefficient.

Step 1: Solve Equation 1 for y:

$$6x + 3y = -1$$
$$\underline{-6x \qquad\qquad -6x}$$
$$3y = -1 - 6x$$
$$y = \frac{-1 - 6x}{3}$$

Step 2: Substitute $\dfrac{-1 - 6x}{3}$ in place of y in Equation 2:

$$4x + 9y = 4$$
$$4x + \overset{3}{\cancel{9}}\left(\frac{-1 - 6x}{\underset{1}{\cancel{3}}} \right) = 4$$
$$4x + 3(-1 - 6x) = 4$$
$$4x - 3 - 18x = 4$$
$$-14x = 7$$
$$x = \frac{7}{-14} = -\frac{1}{2}$$

Step 3: Substitute $x = -\dfrac{1}{2}$ into $y = \dfrac{-1 - 6x}{3}$:

$$y = \frac{-1 - \overset{3}{\cancel{6}}\left(-\dfrac{1}{\underset{1}{\cancel{2}}}\right)}{3}$$

$$y = \frac{-1 - 3(-1)}{3} = \frac{-1 + 3}{3}$$

$$y = \frac{2}{3}$$

Therefore, $\left(-\dfrac{1}{2}, \dfrac{2}{3}\right)$ is the solution for this system.

The check will not be shown. ∎

Solving Systems with No Solution by the Substitution Method

Example 7 shows how to identify systems that have no solution when you are using the substitution method.

Coefficient is 1

Example 7   Solve the system $\begin{Bmatrix} x + 3y = 4 \\ 2x + 6y = 4 \end{Bmatrix}$ using the substitution method.

Solution

Step 1: Solve Equation 1 for x:

$$x + 3y = 4 \Rightarrow x = \boxed{4 - 3y}$$

Step 2: Substitute $\boxed{4 - 3y}$ in place of x in Equation 2:

$$2x + 6y = 4$$
$$2(\,4 - 3y\,) + 6y = 4$$
$$8 - 6y + 6y = 4$$

Step 3: $8 = 4$ A false statement

Because both variables were eliminated and we were left with a *false* statement, there is *no solution* for this system of equations. ∎

Solving Systems with More Than One Solution by the Substitution Method

Example 8 shows how to identify systems that have more than one solution when you are using the substitution method.

Example 8   Solve the system $\begin{Bmatrix} 6x + 3y = 6 \\ 2x + y = 2 \end{Bmatrix}$ using the substitution method.

Coefficient is 1

Solution

Step 1: Solve Equation 2 for y:

$$2x + y = 2 \Rightarrow y = \boxed{2 - 2x}$$

Step 2: Substitute $2 - 2x$ in place of y in Equation 1:

$$6x + 3y = 6$$

$$6x + 3(2 - 2x) = 6$$

$$6x + 6 - 6x = 6$$

Step 3: $\qquad\qquad\qquad\qquad\qquad\qquad 6 = 6 \qquad$ A true statement

Because both variables were eliminated and we were left with a *true* statement, this system of equations has *many* solutions. Any ordered pair that satisfies Equation 1 also satisfies Equation 2. ■

EXERCISES 10.4

Set I Find the solution of each system of equations. Write "Inconsistent" if no solution exists. Write "Dependent" if many solutions exist. In Exercises 1–10, use the substitution method.

1. $\begin{cases} 2x - 3y = 1 \\ x = y + 2 \end{cases}$
2. $\begin{cases} y = 2x + 3 \\ 3x + 2y = 20 \end{cases}$

3. $\begin{cases} 3x + 4y = 2 \\ y = x - 3 \end{cases}$
4. $\begin{cases} 2x + 3y = 11 \\ x = y - 2 \end{cases}$

5. $\begin{cases} 4x + y = 2 \\ 7x + 3y = 1 \end{cases}$
6. $\begin{cases} 5x + 7y = 1 \\ x + 4y = -5 \end{cases}$

7. $\begin{cases} 4x - y = 3 \\ 8x - 2y = 6 \end{cases}$
8. $\begin{cases} x - 3y = 2 \\ 3x - 9y = 6 \end{cases}$

9. $\begin{cases} x + 3 = 0 \\ 3x - 2y = 6 \end{cases}$
10. $\begin{cases} y - 4 = 0 \\ 3y - 5x = 15 \end{cases}$

In Exercises 11–18, use any convenient method.

11. $\begin{cases} 8x + 4y = 7 \\ 3x + 6y = 6 \end{cases}$
12. $\begin{cases} 5x - 4y = 2 \\ 15x + 12y = 12 \end{cases}$

13. $\begin{cases} 3x - 2y = 8 \\ 2y - 3x = 4 \end{cases}$
14. $\begin{cases} 4x - 5y = 15 \\ 5y - 4x = 10 \end{cases}$

15. $\begin{cases} 8x + 5y = 2 \\ 7x + 4y = 1 \end{cases}$
16. $\begin{cases} 4x - 9y = 7 \\ 3x - 8y = 4 \end{cases}$

17. $\begin{cases} 4x + 4y = 3 \\ 6x + 12y = -6 \end{cases}$
18. $\begin{cases} 4x + 9y = -11 \\ 10x + 6y = 11 \end{cases}$

Set II Find the solution of each system of equations. Write "Inconsistent" if no solution exists. Write "Dependent" if many solutions exist. In Exercises 1–10, use the substitution method.

1. $\begin{cases} x - y = 1 \\ y = 2x - 3 \end{cases}$
2. $\begin{cases} x + 3y = 4 \\ x = 2 - y \end{cases}$

3. $\begin{cases} x + 2y = -1 \\ x = 5 + y \end{cases}$
4. $\begin{cases} x + y = 1 \\ y = x + 7 \end{cases}$

5. $\begin{cases} x - 4y = 9 \\ 3x + 8y = 7 \end{cases}$
6. $\begin{cases} 3x + 4y = 18 \\ 5x - y = 7 \end{cases}$

7. $\begin{cases} 5x + y = 6 \\ 10x + 2y = 12 \end{cases}$ **8.** $\begin{cases} 2x + 5y = -5 \\ 3y + x = -2 \end{cases}$

9. $\begin{cases} 5x + y = 0 \\ 3x + 2y = 7 \end{cases}$ **10.** $\begin{cases} x + y = 4 \\ 2x - y = 2 \end{cases}$

In Exercises 11–18, use any convenient method.

11. $\begin{cases} 7x + 5y = -4 \\ 4x + 3y = -2 \end{cases}$ **12.** $\begin{cases} 4x + 7y = 9 \\ 6x + 5y = -3 \end{cases}$

13. $\begin{cases} 2x - y = 3 \\ y = 2x + 1 \end{cases}$ **14.** $\begin{cases} 3x + 6y = 9 \\ 4x + 8y = 12 \end{cases}$

15. $\begin{cases} 8x + 3y = 4 \\ 2x - 3y = 6 \end{cases}$ **16.** $\begin{cases} x + 3y = 0 \\ x - 2y = 10 \end{cases}$

17. $\begin{cases} 3x + 2y = 7 \\ 6x - 4y = 7 \end{cases}$ **18.** $\begin{cases} 7x - 2y = 5 \\ 5x + 3y = 8 \end{cases}$

10.5 Using Systems of Equations to Solve Word Problems

In solving word problems involving more than one unknown, it is sometimes difficult to represent each unknown in terms of a single variable. In this section, we eliminate that difficulty by using a different variable for each unknown. When we use two variables, we must find two equations to represent the given facts; we then use a system of equations to solve the problem.

TO SOLVE A WORD PROBLEM USING A SYSTEM OF EQUATIONS

1. Read the problem completely and determine how many unknown numbers there are.

2. Draw a diagram showing the relationships in the problem whenever possible.

3. Represent each unknown number by a different variable.

4. Use the word statement to write a system of equations. *There must be as many equations as variables.*

5. Solve the system of equations using one of the following:
   a. Addition method (Section 10.3)
   b. Substitution method (Section 10.4)
   c. Graphical method (Section 10.2)

6. Check the solutions in the word problem.

In Example 1, the word problem is solved on the left by using a single variable and a single equation. On the right, it is solved by using two variables and a system of equations.

Example 1 The sum of two numbers is 20. Their difference is 6. What are the numbers?
Solution

Using one variable	*Using two variables*

Using one variable

Let x = larger number

$20 - x$ = smaller number

Their difference is 6

$$x - (20 - x) = 6$$

$$x - 20 + x = 6$$

$$2x = 26$$

Larger number $x = 13$

Smaller number $20 - x = 7$

The difficulty in using the one-variable method to solve this problem is that some students cannot decide whether to represent the second unknown number by $x - 20$ or by $20 - x$.

Using two variables

Let x = larger number

y = smaller number

The sum of two numbers is 20

(1) $x + y = 20$

Their difference   is 6

(2) $x - y = 6$

Using the addition method:

$$\begin{array}{ll}(1) & x + y = 20 \\ (2) & x - y = 6\end{array}$$

$$2x = 26 \quad \text{Adding (1) and (2)}$$

$$x = 13 \quad \text{Larger number}$$

Substituting $x = 13$ into Equation 1, we have

$$x + y = 20$$

$$13 + y = 20$$

$$y = 7 \quad \text{Smaller number}$$

Check

$$13 + 7 = 20 \quad \text{The sum of the numbers is 20.}$$

$$13 - 7 = 6 \quad \text{The difference of the numbers is 6.} \quad \blacksquare$$

Example 2 Kevin spent $3.76 on eighteen stamps, buying only 18¢ and 22¢ stamps. How many of each kind did he buy?
Solution

Let x = the number of 18¢ stamps

y = the number of 22¢ stamps

Number of 18¢ stamps	+	number of 22¢ stamps	=	total number of stamps
x	+	y	=	18

Value of 18¢ stamps	+	value of 22¢ stamps	=	total value of stamps
$18x$	+	$22y$	=	376 Value in cents

Using the addition method:

$$\begin{array}{l}(1)\\(2)\end{array}\left\{\begin{array}{rcl}x + y &=& 18\\18x + 22y &=& 376\end{array}\right\}$$

$$-18x - 18y = -324 \qquad \text{Multiplying Equation 1 by } -18$$

$$\underline{18x + 22y = 376} \qquad \text{Equation 2}$$

$$4y = 52 \qquad \text{Adding}$$

$$y = 13 \qquad \text{The number of 22¢ stamps}$$

Substituting $y = 13$ into Equation 1, we have

$$x + y = 18$$

$$x + 13 = 18$$

$$x = 5 \qquad \text{The number of 18¢ stamps}$$

Check $13 + 5 = 18$ There are 18 stamps altogether.

$13(\$0.22) + 5(\$0.18) = \$2.86 + \$0.90 = \$3.76$ The total value of the stamps is \$3.76. ■

Example 3   A fraction has the value $\frac{2}{3}$. If 3 is added to the numerator and 2 is added to the denominator, the resulting fraction has the value $\frac{3}{4}$. Find the original fraction.
Solution

Let $x =$ the numerator of the original fraction

$y =$ the denominator of the original fraction

$$\begin{array}{l}(1)\\(2)\end{array}\left\{\begin{array}{l}\dfrac{x}{y} = \dfrac{2}{3}\\[2mm]\dfrac{x+3}{y+2} = \dfrac{3}{4}\end{array}\right\} \qquad \begin{array}{l}\text{The value of the original fraction is } \dfrac{2}{3}\\[2mm]\text{The value of the new fraction is } \dfrac{3}{4}\end{array}$$

Before we can solve the system of equations, we must clear fractions and line up the like terms:

$$(1) \qquad 3x = 2y \qquad\qquad \text{Multiplying both sides by the LCD}$$

$$(2) \quad 4(x + 3) = 3(y + 2) \qquad \text{Multiplying both sides by the LCD}$$

$$(1) \qquad 3x = 2y \qquad\qquad \Rightarrow 3x - 2y = 0$$

$$(2) \quad 4(x + 3) = 3(y + 2) \Rightarrow 4x + 12 = 3y + 6 \Rightarrow 4x - 3y = -6$$

Therefore, the system is

$$\begin{array}{l}(1)\\(2)\end{array}\left\{\begin{array}{rcl}3x - 2y &=& 0\\4x - 3y &=& -6\end{array}\right\}$$

We now solve the system:

$$(1) \quad 3]\ 3x - 2y = 0 \quad \Rightarrow \quad 9x - 6y = 0$$

$$(2) \quad -2]\ 4x - 3y = -6 \Rightarrow \underline{-8x + 6y = 12}$$

$$x = 12$$

Substituting 12 for x in Equation 1, we have

$$3x - 2y = 0$$
$$3(12) - 2y = 0$$
$$36 - 2y = 0$$
$$-2y = -36$$
$$y = 18$$

Therefore, the original fraction is $\dfrac{12}{18}$.

Check $\dfrac{12}{18} = \dfrac{2}{3}$ The value of the original fraction is $\dfrac{2}{3}$.

$\dfrac{12 + 3}{18 + 2} = \dfrac{15}{20} = \dfrac{3}{4}$ If 3 is added to the numerator and 2 is added to the denominator, the value of the resulting fraction is $\dfrac{3}{4}$.   ■

Some word problems should be solved using one unknown, while others are best solved using a system of equations.

HOW TO CHOOSE WHICH METHOD TO USE FOR SOLVING A WORD PROBLEM

1. Read the problem completely and determine how many unknown numbers there are.

2. If there is only one unknown number, use the one-variable method.

3. If there is more than one unknown number, try to represent all of them in terms of one variable. If this is too difficult, represent each unknown number by a different variable and then solve using a system of equations.

Systems that have more than two equations or more than two variables are not discussed in this book.

EXERCISES 10.5

Set I   Set up each of the following problems algebraically *using two variables*, solve, and check. Be sure to state what your variables represent.

1. The sum of two numbers is 80. Their difference is 12. What are the numbers?

2. The sum of two numbers is 90. Their difference is 4. What are the numbers?

3. The sum of two angles is 90°. Their difference is 60°. What are the angles?

4. The sum of two angles is 180°. Their difference is 32°. What are the angles?

5. Find two numbers such that twice the smaller plus 3 times the larger is 34, and 5 times the smaller minus twice the larger is 9.

6. Find two numbers such that 5 times the larger plus 3 times the smaller is 47, and 4 times the larger minus twice the smaller is 20.

7. Jason paid $16.88 for a 6-lb mixture of granola and dried apple chunks. If the granola cost $2.10 per pound and the dried apple chunks cost $4.24 per pound, how many pounds of each did he buy?

8. A 100-lb mixture of two different grades of coffee costs $351.50. If grade A costs $3.80 per pound and grade B costs $3.30 per pound, how many pounds of each grade were used?

9. Don spent $4.48 for twenty-two stamps. If he bought only 22¢ and 18¢ stamps, how many of each kind did he buy?

10. Sue spent $12.38 for fifty stamps. If she bought only 22¢ and 45¢ stamps, how many of each type did she buy?

11. The length of a rectangle is 1 ft 6 in. longer than its width. Its perimeter is 19 ft. Find its dimensions.

12. The length of a rectangle is 2 ft 6 in. longer than its width. Its perimeter is 25 ft. Find its dimensions.

13. A fraction has the value $\frac{2}{3}$. If 4 is added to the numerator and the denominator is decreased by 2, the resulting fraction has the value $\frac{6}{7}$. What is the original fraction?

14. A fraction has the value $\frac{3}{4}$. If 4 is added to its numerator and 8 is subtracted from its denominator, the value of the resulting fraction is 1. What is the original fraction?

15. A boat takes 7 hr to travel 252 mi with the current and takes 9 hr to travel the same distance against the same current. Find the average speed of the boat in still water and the average speed of the current.

16. A pilot takes 5 hr to fly 450 mi against the wind and only 3 hr to return with the wind. Find the average speed of the plane in still air and the average speed of the wind.

17. A tie and a pin cost $1.10. The tie costs $1 more than the pin. What is the cost of each?

18. A number of birds are resting on two limbs of a tree. One limb is above the other. A bird on the lower limb says to the birds on the upper limb, "If one of you will come down here, we will have an equal number on each limb." A bird from above replies, "If one of you will come up here we will have twice as many up here as you will have down there." How many birds are sitting on each limb?

Set II   Set up each of the following problems algebraically *using two variables*, solve, and check. Be sure to state what your variables represent.

1. The sum of two numbers is 42. Their difference is 12. What are the numbers?

2. Half the sum of two numbers is 15. Half their difference is 8. Find the numbers.

3. The sum of two angles is 90°. Their difference is 16°. Find the angles.

4. A 20-lb mixture of Product A and Product B costs $19.75. If Product A costs 85¢ per pound and Product B costs $1.40 per pound, find the number of pounds of each.

5. Find two numbers such that twice the smaller plus 4 times the larger is 66, and 6 times the smaller minus 3 times the larger is 3.

6. Beatrice has seventeen coins with a total value of $3.05. If these coins are all nickels and quarters, how many of each kind does she have?

7. A 20-lb mixture of almonds and hazelnuts costs $53.00. If almonds cost $2.50 per pound and hazelnuts cost $3.00 per pound, find the number of pounds of each.

8. A fraction has the value $\frac{1}{2}$. If 2 is added to the numerator and 1 is subtracted from the denominator, the value of the resulting fraction is $\frac{4}{7}$. Find the original fraction.

9. Rachelle spent $4.55 for fifteen stamps. If she bought only 25¢ and 35¢ stamps, how many of each kind did she buy?

10. A class received $233 for selling 200 tickets to the school play. If student tickets cost $1 each and nonstudent tickets cost $2 each, how many nonstudent tickets were sold?

11. The length of a rectangle is 3 ft 6 in. longer than its width. Its perimeter is 13 ft. Find its dimensions.

12. Jill spent $59.50 for fourteen tickets to a movie. Some were for children and some for adults. If children's tickets cost $3.50 and adults' tickets cost $5.25, how many of each kind did she buy?

13. A fraction has the value $\frac{1}{2}$. If 4 is added to the numerator and 2 is subtracted from the denominator, the resulting fraction has the value $\frac{2}{3}$. What is the original fraction?

14. Several families went to a school play together. They spent $10.60 for eight tickets. If adults' tickets cost $1.95 and children's tickets cost 95¢, how many of each kind of ticket were bought?

15. Jerry takes 6 hr to ride his bicycle 36 mi against the wind and only 2 hr to return with the wind. Find the average riding speed in still air and the average speed of the wind.

16. A 100-lb mixture of Product A and Product B costs $22. If Product A costs 20¢ per pound and Product B costs 25¢ per pound, how many pounds of each kind are there?

17. A bracelet and a pin cost $11. One costs $10 more than the other. What is the cost of each?

18. A pilot takes 7 hr to fly 875 mi against the wind and only 5 hr to return with the wind. Find the average speed of the plane in still air and the average speed of the wind.

10.6 Review: 10.1–10.5

Solution of a System of Equations
10.1

A solution of a system of two equations in two unknowns is an ordered pair that, when substituted into each equation, makes them both true.

Solving a System of Equations
10.2, 10.3, 10.4

In solving a system of equations, there are three possibilities:

1. There is only one solution.
   a. Graphical method: The lines intersect at one point.
   b. Algebraic method:
      Addition } The equations can be solved for
      Substitution } a single ordered pair.

2. There is no solution.
   a. Graphical method: The lines are parallel.
   b. Algebraic method:
      Addition } Both variables drop out and
      Substitution } a false statement results.

3. There are many solutions.
   a. Graphical method: Both equations have the same line for a graph.
   b. Algebraic method:

$$\left.\begin{matrix} \text{Addition} \\ \text{Substitution} \end{matrix}\right\} \quad \begin{matrix} \text{Both variables drop out and} \\ \text{a true statement results.} \end{matrix}$$

To Solve a Word Problem Using a System of Equations 10.5

1. Read the problem completely and determine how many unknown numbers there are.

2. Draw a diagram showing the relationships in the problem, whenever possible.

3. Represent each unknown number by a different variable.

4. Use the word statement to write a system of two equations in two unknowns.

5. Solve the system of equations using:
   a. Addition method (Section 10.3)
   b. Substitution method (Section 10.4)
   c. Graphical method (Section 10.2)

6. Check the solutions in the word problem.

Review Exercises 10.6   Set I

In Exercises 1–3, find the solution of each system graphically. Write "Inconsistent" if no solution exists. Write "Dependent" if many solutions exist.

1. $\begin{cases} x + y = 6 \\ x - y = 4 \end{cases}$
2. $\begin{cases} x + y = 5 \\ x - y = -3 \end{cases}$

3. $\begin{cases} 3x - y = 6 \\ 6x - 2y = 12 \end{cases}$

In Exercises 4–6, find the solution of each system, using the addition method. Write "Inconsistent" if no solution exists. Write "Dependent" if many solutions exist.

4. $\begin{cases} 4x + 5y = 22 \\ 3x + y = 11 \end{cases}$
5. $\begin{cases} 4x - 8y = 4 \\ 3x - 6y = 3 \end{cases}$

6. $\begin{cases} 4x - 7y = 28 \\ 7y - 4x = 20 \end{cases}$

In Exercises 7–9, find the solution of each system, using the substitution method. Write "Inconsistent" if no solution exists. Write "Dependent" if many solutions exit.

7. $\begin{cases} x = y + 2 \\ 4x - 5y = 3 \end{cases}$
8. $\begin{cases} 7x - 3y = 1 \\ y = x + 5 \end{cases}$

9. $\begin{cases} x + y = 4 \\ 2x - y = 2 \end{cases}$

In Exercise 10, solve the system by any convenient method. Write "Inconsistent" if no solution exists. Write "Dependent" if many solutions exist.

10. $\begin{cases} 5x - 4y = -7 \\ -6x + 8y = 2 \end{cases}$

In Exercises 11–14, set up each problem algebraically *using two variables*, solve, and check. Be sure to state what your variables represent.

11. The sum of two numbers is 84. Their difference is 22. What are the numbers?

12. A fraction has the value $\frac{4}{5}$. If 3 is subtracted from the numerator and 5 is added to the denominator, the value of the resulting fraction is $\frac{3}{5}$. What is the original fraction?

13. An office manager paid $107.75 for twenty boxes of paper for the copy machine and the laser printer. If paper for the laser printer costs $7.50 per box and paper for the copy machine costs $4.25 per box, how many boxes of each kind were purchased?

14. A boat takes 7 hr to travel 231 mi with the current and 11 hr to travel the same distance against the same current. Find the average speed of the boat in still water and the average speed of the current.

Review Exercises 10.6   Set II

NAME _____

In Exercises 1 and 2, find the solution of each system graphically. Write "Inconsistent" if no solution exists. Write "Dependent" if many solutions exist.

ANSWERS

1. $\begin{cases} 3x + y = -9 \\ 3x - 2y = 0 \end{cases}$

1.

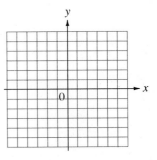

2. $\begin{cases} 2x - 3y = 3 \\ 3y - 2x = 6 \end{cases}$

2.

In Exercises 3–5, find the solution of each system using the addition method. Write "Inconsistent" if no solution exists. Write "Dependent" if many solutions exist.

3. $\begin{cases} 3x - 2y = 10 \\ 5x + 4y = 24 \end{cases}$

4. $\begin{cases} 6x + 4y = -1 \\ 4x + 6y = -9 \end{cases}$

3. _____

4. _____

5. _____

6. _____

5. $\begin{cases} 6x + 4y = 13 \\ 8x + 10y = 1 \end{cases}$

In Exercises 6 and 7, solve each system using the substitution method. Write "Inconsistent" if no solution exists. Write "Dependent" if many solutions exist.

6. $\begin{cases} x - 3y = 15 \\ 5x + 7y = -13 \end{cases}$

7. $\begin{cases} 4x - 6y = 2 \\ 6x - 9y = 3 \end{cases}$

In Exercises 8–10, solve each system by any convenient method. Write "Inconsistent" if no solution exists. Write "Dependent" if many solutions exist.

8. $\begin{cases} 5x + 4y = 2 \\ 2x + 5y = 11 \end{cases}$

9. $\begin{cases} 3x - 5y = 2 \\ 2x + 7y = 22 \end{cases}$

10. $\begin{cases} 5x - 7y = 17 \\ 4x - 5y = 13 \end{cases}$

In Exercises 11–14, set up each problem algebraically *using two variables*, solve, and check. Be sure to state what your variables represent.

11. The sum of two numbers is 95. Their difference is 19. What are the numbers?

12. The width of a rectangle is 5 m shorter than its length. Its perimeter is 98 m. Find its dimensions.

13. An office manager paid $730 for a total of eighty rolls of two different kinds of tape. If one kind costs $10 per roll and the other kind costs $8 per roll, how many rolls of each kind were bought?

14. A fraction has the value $\frac{1}{3}$. If 5 is added to the numerator and 5 is subtracted from the denominator, the value of the resulting fraction is $\frac{1}{2}$. What is the original fraction?

Chapter 10 Diagnostic Test

The purpose of this test is to see how well you understand systems of equations. We recommend that you work this diagnostic test *before* your instructor tests you on this chapter. Allow yourself about 50 minutes.

Complete solutions for all the problems on this test, together with section references, are given in the answer section in the back of this book. For problems you do incorrectly, study the sections referred to.

In Problems 1–7, write "Inconsistent" if no solution exists, and write "Dependent" if many solutions exist.

1. Solve graphically: $\begin{cases} 2x + 3y = 6 \\ x + 3y = 0 \end{cases}$

2. Solve by addition: $\begin{cases} x + 3y = 5 \\ -x + 4y = 2 \end{cases}$

3. Solve by addition: $\begin{cases} 4x + 3y = 5 \\ 5x - 2y = 12 \end{cases}$

4. Solve by substitution: $\begin{cases} 2x - 5y = -20 \\ x = y - 7 \end{cases}$

5. Solve by any convenient method: $\begin{cases} 3x + 4y = -1 \\ 2x - 5y = -16 \end{cases}$

6. Solve by any convenient method: $\begin{cases} 5x + y = 10 \\ 10x + 2y = 9 \end{cases}$

7. Solve by any convenient method: $\begin{cases} 6x - 9y = 3 \\ 15y - 10x = -5 \end{cases}$

In Problems 8–10, set up each problem algebraically *using two variables*, solve, and check. Be sure to state what your variables represent.

8. The sum of two numbers is 13. Their difference is 37. What are the numbers?

9. A pilot takes 11 hr to fly 1,650 mi against the wind and only 7.5 hr to return with the wind. Find the average speed of the plane in still air and the average speed of the wind.

10. The length of a rectangle is 3 cm more than its width. Its perimeter is 98 cm. Find the dimensions of the rectangle.

Cumulative Review Exercises: Chapters 1–10

1. Simplify. Write your answer using only positive exponents.

$$\left(\frac{24w^{-2}z^4}{16w}\right)^{-3}$$

2. Evaluate the formula using the values of the variables given with the formula.

$$A = P(1 + r)^n \qquad P = 1,000, r = 0.15, n = 2$$

3. Solve for x: $\dfrac{2 + 5x}{6x - 1} = \dfrac{3}{4}$

4. Solve for x: $\dfrac{3x - 1}{4} + 6 = \dfrac{5 - x}{5}$

5. Add and simplify:

$$\frac{1 + a}{2 - 3a} + \frac{2 + a}{4a}$$

6. Graph the equation $5x - 7y = 15$.

7. Draw the graph of $4y = x^2$. Use the following values for x in your table of values: $-4, -2, -1, 0, 1, 2, 4$.

8. Solve the inequality $2(x - 3) \le 4$ and graph its solution on the number line.

9. Graph the inequality $3x - 4y > 12$ in the plane.

10. Write the general form of the equation of the line that passes through the points $A(-3, 2)$ and $B(4, -1)$.

In Exercises 11–14, solve each system of equations by any convenient method. Write "Inconsistent" if no solution exists. Write "Dependent" if many solutions exist.

11. $\begin{cases} 2x - 3y = 3 \\ 3y - 2x = 5 \end{cases}$

12. $\begin{cases} 8x - 4y = 12 \\ 6x - 3y = 9 \end{cases}$

13. $\begin{cases} x + 5y = 11 \\ 3x + 4y = 11 \end{cases}$

14. $\begin{cases} 6x + 8y = 15 \\ 4x + 2y = -5 \end{cases}$

In Exercises 15 and 16, factor the polynomial completely, or write "Not factorable."

15. $5x^2 - 2x - 13$

16. $7x^2 - 3x - 11$

In Exercises 17–20, set up each problem algebraically, solve, and check. Be sure to state what your variables represent.

17. A drugstore buys eighteen cameras. If it had bought ten cameras of a higher quality, it would have paid $48 more per camera for the same total expenditure. Find the price of each type of camera.

18. Mr. McAllister invested part of $27,300 at 14% and the remainder at 21%. His total yearly income from these investments is $4,648. How much is invested at each rate?

In Exercises 19 and 20, *use two variables*.

19. The sum of two numbers is 6. Their difference is 40. What are the numbers?

20. The width of a rectangle is 5 m less than its length. Its perimeter is 58 m. Find the dimensions of the rectangle.

11 Radicals

Roots of numbers were introduced in Section 1.10, and square roots of algebraic terms were introduced in Section 7.2. In this chapter, we discuss the simplification of square roots and operations involving square roots. We also solve radical equations and some word problems that lead to such equations.

11.1 Review of Square Roots

We learned in Section 1.10 that the *square root* of a number p is a number that, when squared, gives p. We also learned that every positive real number has both a positive and a negative square root and that the *positive* square root is called the *principal square root*. Recall that the symbol $\sqrt{p}$ represents the principal square root of p and that p is called the *radicand* (Figure 11.1.1).

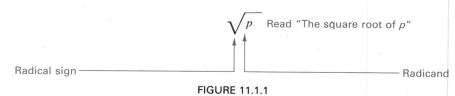

FIGURE 11.1.1

Example 1   Identify the radicand in each expression:

a. $\sqrt{7}$ 7 is the radicand.

b. $\sqrt{2x}$ $2x$ is the radicand.

c. $\sqrt{\dfrac{3x}{2y}}$ $\dfrac{3x}{2y}$ is the radicand. ■ ·

Just as the number 4 has two square roots, 2 and -2, the algebraic expression x^2 has two square roots, x and $-x$. Since x itself can be either a positive or a negative number, we don't know which is the principal square root. However, we do know that $|x|$ must be positive (or zero); therefore, $|x|$ is the principal square root of x^2; in symbols, $\sqrt{x^2} = |x|$. However, in the remainder of this chapter, we assume that all variables represent positive numbers. For this reason, the absolute value symbol need not be used, and we will write $\sqrt{x^2} = x$.

The Square of a Square Root   As stated in Rule 11.1, when you square the square root of a number, you get that number.

RULE 11.1   THE SQUARE OF A SQUARE ROOT

$$(\sqrt{a})^2 = a$$

Example 2   Examples of squaring the square root of a number:

a. $(\sqrt{25})^2 = (5)^2 = 25$; therefore, $(\sqrt{25})^2 = 25$.

   Replacing $\sqrt{25}$ with 5

b. $(\sqrt{16})^2 = (4)^2 = 16$; therefore, $(\sqrt{16})^2 = 16$.

   Replacing $\sqrt{16}$ with 4

Replacing $\sqrt{121}$ with 11

c. $\left(\sqrt{121}\right)^2 = \left(11\right)^2 = 121$; therefore, $\left(\sqrt{121}\right)^2 = 121$. ∎

Perfect Square When a factor has an even exponent or is the square of an integer, it can be called a **perfect square**. In Example 2, all the radicands were perfect squares. However, Rule 11.1 applies even when a is not a perfect square.

Example 3 Simplify each expression by using the rule $\left(\sqrt{a}\right)^2 = a$:

a. $\left(\sqrt{7}\right)^2 = 7$ 　　　　　　　　　b. $\left(\sqrt{13}\right)^2 = 13$

c. $\left(\sqrt{x}\right)^2 = x$ 　　　　　　　　　d. $\left(\sqrt{cd^4}\right)^2 = cd^4$

e. $\left(\sqrt{x+3}\right)^2 = x + 3$ ∎

EXERCISES 11.1

Set I　In Exercises 1–6, identify the radicand.

1. $\sqrt{17}$ 　　　　　　**2.** $\sqrt{23}$ 　　　　　　**3.** $\sqrt{4xy^2}$

4. $\sqrt{9a^2b}$ 　　　　　**5.** $\sqrt{x+1}$ 　　　　　**6.** $\sqrt{m-n}$

In Exercises 7–14, simplify each expression.

7. $\left(\sqrt{9}\right)^2$ 　　　　　**8.** $\left(\sqrt{100}\right)^2$ 　　　　　**9.** $\left(\sqrt{15}\right)^2$

10. $\left(\sqrt{29}\right)^2$ 　　　　**11.** $\left(\sqrt{3a^2}\right)^2$ 　　　　**12.** $\left(\sqrt{7y^2}\right)^2$

13. $\left(\sqrt{x+y}\right)^2$ 　　　**14.** $\left(\sqrt{u+v}\right)^2$

Set II　In Exercises 1–6, identify the radicand.

1. $\sqrt{35}$ 　　　　　　**2.** $\sqrt{82}$ 　　　　　　**3.** $\sqrt{25ab^2}$

4. $\sqrt{49c^2d}$ 　　　　　**5.** $\sqrt{y+5}$ 　　　　　**6.** $\sqrt{s+t}$

In Exercises 7–14, simplify each expression.

7. $\left(\sqrt{49}\right)^2$ 　　　　　**8.** $\left(\sqrt{225}\right)^2$ 　　　　　**9.** $\left(\sqrt{31}\right)^2$

10. $\left(\sqrt{53}\right)^2$ 　　　　**11.** $\left(\sqrt{5r^2}\right)^2$ 　　　　**12.** $\left(\sqrt{2m^2}\right)^2$

13. $\left(\sqrt{e+4}\right)^2$ 　　　**14.** $\left(\sqrt{c+d}\right)^2$

11.2　Simplifying Square Roots

We will consider square roots of two kinds:

1. Square roots that do not involve fractions

2. Square roots involving fractions

11.2A　Square Roots That Do Not Involve Fractions

A number of conditions must be satisfied for a square root to be in *simplest radical form*. The *first* of these conditions is that no factor of the radicand can have an exponent greater than 1 when the radicand is expressed in prime factored form.

In simplifying radicals, we may make use of the following property:

RULE 11.2

$$\sqrt{ab} = \sqrt{a}\sqrt{b}, \text{ if } a \geq 0 \text{ and } b \geq 0$$

When the radicand contains a factor with an *even* exponent, we remove that factor from the radical by dividing its exponent by 2, as was done in Section 7.2 (see Example 1).

Example 1   Simplify the radicals:

a. $\sqrt{5^2 x^2 y^4} = 5^{2 \div 2} x^{2 \div 2} y^{4 \div 2} = 5xy^2$

b. $\sqrt{3^4 z^6} = 3^{4 \div 2} z^{6 \div 2} = 3^2 z^3$ ∎

When some of the factors have odd exponents, we use the following procedure:

TO SIMPLIFY THE PRINCIPAL SQUARE ROOT OF A PRODUCT

1. Express the radicand in prime factored form.

2. Find the square root of each factor as follows:
   a. If the exponent of a factor is 1, that factor must remain under the radical sign.
   b. If the exponent of a factor is an even number, remove that factor from the radical by dividing its exponent by 2.
   c. If the exponent of a factor is an odd number, write that factor as the product of two factors—one factor with an even exponent, and the other factor with an exponent of 1. Then remove from the radical the factor that has an even exponent by dividing its exponent by 2.

3. The simplified radical is the product of all the factors found in step 2.

Example 2   Simplify the radicals:

a. $\sqrt{3^5} = \sqrt{3^4 \cdot 3^1}$ ◄——— $3^5 = 3^4 \cdot 3^1$

$= \sqrt{3^4}\sqrt{3}$ └── The factor 3^4 is the highest power of 3 whose exponent (4) is exactly divisible by 2

$= 3^2\sqrt{3}$

$= 9\sqrt{3}$

b. $\sqrt{x^7} = \sqrt{x^6 \cdot x^1}$ ◄——— $x^7 = x^6 \cdot x^1$

$= \sqrt{x^6}\sqrt{x}$ └── The factor x^6 is the highest power of x whose exponent (6) is exactly divisible by 2

$= x^3\sqrt{x}$ ∎

A convenient arrangement of the work involved in simplifying the principal square root of a number is shown in Example 3.

Example 3   Simplify the radicals:

a. $\sqrt{48} = \sqrt{2^4 \cdot 3}$ ←— Prime factored form of 48

$= \sqrt{2^4}\sqrt{3}$

$= 2^2\sqrt{3}$

$= 4\sqrt{3}$

$$
\begin{array}{r|l}
2 & 48 \\
\hline
2 & 24 \\
\hline
2 & 12 \\
\hline
2 & 6 \\
\hline
 & 3
\end{array}
$$

$48 = 2^4 \cdot 3$

b. $\sqrt{75} = \sqrt{3 \cdot 5^2}$ ←— Prime factored form of 75

$= \sqrt{3}\sqrt{5^2}$

$= \sqrt{3}(5)$

$= 5\sqrt{3}$

$$
\begin{array}{r|l}
5 & 75 \\
\hline
5 & 15 \\
\hline
 & 3
\end{array}
$$

$75 = 3 \cdot 5^2$

c. $\sqrt{360} = \sqrt{2^3 \cdot 3^2 \cdot 5}$

$= \sqrt{2^2 \cdot 2 \cdot 3^2 \cdot 5}$

$= 2 \cdot 3\sqrt{2 \cdot 5}$

$= 6\sqrt{10}$

$$
\begin{array}{r|l}
2 & 360 \\
\hline
2 & 180 \\
\hline
2 & 90 \\
\hline
3 & 45 \\
\hline
3 & 15 \\
\hline
 & 5
\end{array}
$$

$360 = 2^3 \cdot 3^2 \cdot 5$ ■

In Example 4, the radicand is a product of factors.

Example 4   Simplify the radicals:

a. $\sqrt{12x^4y^3} = \sqrt{2^2 \cdot 3 \cdot x^4 \cdot y^2 \cdot y}$

$= 2x^2y\sqrt{3y}$

$$
\begin{array}{r|l}
2 & 12 \\
\hline
2 & 6 \\
\hline
 & 3
\end{array}
$$

$12 = 2^2 \cdot 3$

b. $\sqrt{24a^5b^7} = \sqrt{2^3 \cdot 3 \cdot a^5b^7}$

$= \sqrt{2^2 \cdot 2 \cdot 3 \cdot a^4 \cdot a \cdot b^6 \cdot b}$

$= 2a^2b^3\sqrt{2 \cdot 3 \cdot ab}$

$= 2a^2b^3\sqrt{6ab}$ ■

$$
\begin{array}{r|l}
2 & 24 \\
\hline
2 & 12 \\
\hline
2 & 6 \\
\hline
 & 3
\end{array}
$$

$24 = 2^3 \cdot 3$

The square root of a number can be simplified if by inspection you can see that it has a factor that is a perfect square (see Example 5).

Example 5   Simplify the radicals.

a. $\sqrt{12} = \sqrt{4 \cdot 3}$ 4 is a factor of 12 and is a perfect square.

$= \sqrt{4} \cdot \sqrt{3}$

$= 2 \cdot \sqrt{3}$

$= 2\sqrt{3}$

b. $\sqrt{50} = \sqrt{2 \cdot \boxed{25}}$ 25 is a factor of 50 and is a perfect square.

$$= \sqrt{2} \cdot \sqrt{25}$$

$$= \sqrt{2} \cdot 5$$

$$= 5\sqrt{2} \quad \blacksquare$$

EXERCISES 11.2A

Set I   Simplify each of the square roots.

1. $\sqrt{25x^2}$ **2.** $\sqrt{100y^2}$ **3.** $\sqrt{81s^4}$ **4.** $\sqrt{64t^6}$

5. $\sqrt{4x^2}$ **6.** $\sqrt{9y^2}$ **7.** $\sqrt{16z^4}$ **8.** $\sqrt{25b^6}$

9. $\sqrt{12}$ **10.** $\sqrt{20}$ **11.** $\sqrt{18}$ **12.** $\sqrt{45}$

13. $\sqrt{8}$ **14.** $\sqrt{32}$ **15.** $\sqrt{x^3}$ **16.** $\sqrt{y^5}$

17. $\sqrt{m^7}$ **18.** $\sqrt{n^9}$ **19.** $\sqrt{24}$ **20.** $\sqrt{54}$

21. $\sqrt{32x^2y^3}$ **22.** $\sqrt{98ab^3}$ **23.** $\sqrt{8s^4t^5}$ **24.** $\sqrt{44x^3y^6}$

25. $\sqrt{18a^2b^3}$ **26.** $\sqrt{27c^5d^2}$ **27.** $\sqrt{40x^6y^2}$ **28.** $\sqrt{135a^6b^8}$

29. $\sqrt{60h^5k^4}$ **30.** $\sqrt{90m^7n^6}$ **31.** $\sqrt{280y^5z^6}$

32. $\sqrt{270g^8h^9}$ **33.** $\sqrt{500x^7y^9}$ **34.** $\sqrt{216x^{11}y^{13}}$

Set II   Simplify each of the square roots.

1. $\sqrt{36z^2}$ **2.** $\sqrt{120a^2}$ **3.** $\sqrt{25w^2}$ **4.** $\sqrt{64u^6}$

5. $\sqrt{16t^2}$ **6.** $\sqrt{12b^2}$ **7.** $\sqrt{25a^4}$ **8.** $\sqrt{84y^8}$

9. $\sqrt{75}$ **10.** $\sqrt{72}$ **11.** $\sqrt{125}$ **12.** $\sqrt{400}$

13. $\sqrt{432}$ **14.** $\sqrt{200}$ **15.** $\sqrt{a^5}$ **16.** $\sqrt{u^7}$

17. $\sqrt{z^5}$ **18.** $\sqrt{a^9}$ **19.** $\sqrt{28}$ **20.** $\sqrt{96u^6}$

21. $\sqrt{12a^2b^5}$ **22.** $\sqrt{8x^5y^3}$ **23.** $\sqrt{18c^6d^7}$ **24.** $\sqrt{50s^3t^5}$

25. $\sqrt{84c^2d^4}$ **26.** $\sqrt{125xy^2}$ **27.** $\sqrt{28ab^5}$ **28.** $\sqrt{150x^3y}$

29. $\sqrt{80a^7b^4}$ **30.** $\sqrt{121x^4y^2}$ **31.** $\sqrt{88s^6t}$

32. $\sqrt{40p^3q^4}$ **33.** $\sqrt{54a^3b^4}$ **34.** $\sqrt{250x^5y^2z}$

11.2B   Square Roots Involving Fractions

Another basic rule for working with square roots follows:

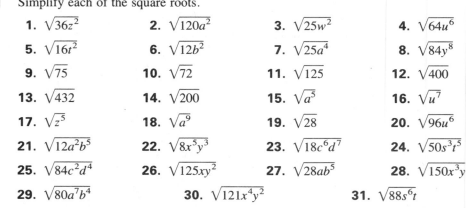

RULE 11.3

$$\sqrt{\frac{a}{b}} = \frac{\sqrt{a}}{\sqrt{b}}, \text{ if } a \geq 0, b > 0$$

A *second* condition that must be satisfied for a square root to be in *simplest radical form* is that the radicand cannot be a fraction. Rule 11.3 can be used in simplifying radicals in which the radicand is a fraction (see Example 6).

Example 6   Examples of simplifying the principal square root of a fraction (assume $y \neq 0$ and $k \neq 0$):

a. $\sqrt{\dfrac{4}{9}} = \dfrac{\sqrt{4}}{\sqrt{9}} = \dfrac{2}{3}$

b. $\sqrt{\dfrac{25}{36}} = \dfrac{\sqrt{25}}{\sqrt{36}} = \dfrac{5}{6}$

c. $\sqrt{\dfrac{x^4}{y^6}} = \dfrac{\sqrt{x^4}}{\sqrt{y^6}} = \dfrac{x^2}{y^3}$

d. $\sqrt{\dfrac{50h^2}{2k^4}} = \sqrt{\dfrac{\overset{25}{\cancel{50}}h^2}{\underset{1}{\cancel{2}}k^4}} = \dfrac{\sqrt{25h^2}}{\sqrt{k^4}} = \dfrac{5h}{k^2}$

       ⌐— Simplify fraction first

e. $\sqrt{\dfrac{3x^2}{4}} = \dfrac{\sqrt{3x^2}}{\sqrt{4}} = \dfrac{x\sqrt{3}}{2}$ ∎

In Example 6, all the denominators were (or became, after the fraction was simplified) perfect squares.

Rationalizing the Denominator

If the denominator is *not* a perfect square, it is necessary to multiply the numerator and denominator by a number that makes the resulting denominator a perfect square. This procedure is called **rationalizing the denominator**.

Example 7   Simplify the radicals:

a. $\sqrt{\dfrac{4}{5}} = \sqrt{\dfrac{4 \cdot 5}{5 \cdot 5}} = \dfrac{\sqrt{4 \cdot 5}}{\sqrt{5^2}} = \dfrac{\sqrt{4}\sqrt{5}}{5} = \dfrac{2\sqrt{5}}{5}$

       ⌐— Multiply $\dfrac{4}{5}$ by $\dfrac{5}{5}$ in order to make the new denominator a perfect square

b. $\sqrt{\dfrac{1}{7}} = \sqrt{\dfrac{1 \cdot 7}{7 \cdot 7}} = \dfrac{\sqrt{7}}{\sqrt{7^2}} = \dfrac{\sqrt{7}}{7}$

c. $\sqrt{\dfrac{3}{20}} = \sqrt{\dfrac{3 \cdot 5}{20 \cdot 5}} = \dfrac{\sqrt{15}}{\sqrt{100}} = \dfrac{\sqrt{15}}{10}$

       ⌐— Multiply $\dfrac{3}{20}$ by $\dfrac{5}{5}$ in order to make the new denominator a perfect square

(We could have multiplied $\frac{3}{20}$ by $\frac{20}{20}$ instead of by $\frac{5}{5}$.) ∎

A *third* condition that must be satisfied for an algebraic expression to be in *simplest radical form* is that no denominator can contain a square root. Example 8 shows how to rationalize the denominator when it contains a square root.

Example 8   Rationalize the denominator of each of the following fractions (assume $x > 0$):

a. $\dfrac{2}{\sqrt{5}} = \dfrac{2}{\sqrt{5}} \cdot \dfrac{\sqrt{5}}{\sqrt{5}} = \dfrac{2\sqrt{5}}{\sqrt{5}\sqrt{5}} = \dfrac{2\sqrt{5}}{(\sqrt{5}\,)^2} = \dfrac{2\sqrt{5}}{5}$

Denominator is now a rational number

Multiply numerator and denominator by $\sqrt{5}$; the value of this fraction is 1, and multiplying $\dfrac{2}{\sqrt{5}}$ by 1 does not change its value

Denominator is *not* a rational number

b. $\dfrac{6}{\sqrt{3}} = \dfrac{6 \cdot \sqrt{3}}{\sqrt{3} \cdot \sqrt{3}} = \dfrac{6\sqrt{3}}{(\sqrt{3}\,)^2} = \dfrac{\overset{2}{\cancel{6}}\sqrt{3}}{\underset{1}{\cancel{3}}} = 2\sqrt{3}$

c. $\dfrac{3xy}{\sqrt{x}} = \dfrac{3xy \cdot \sqrt{x}}{\sqrt{x} \cdot \sqrt{x}} = \dfrac{3xy\sqrt{x}}{(\sqrt{x}\,)^2} = \dfrac{3xy\sqrt{x}}{x} = 3y\sqrt{x}$ ∎

THE SIMPLIFIED FORM OF AN EXPRESSION THAT HAS SQUARE ROOTS

1. No prime factor of a radicand has an exponent greater than 1.

2. No radicand contains a fraction.

3. No denominator contains a square root.

EXERCISES 11.2B

Set I   Simplify each of the following expressions; assume all variables in denominators are nonzero.

1. $\sqrt{\dfrac{9}{25}}$ 2. $\sqrt{\dfrac{36}{49}}$ 3. $\sqrt{\dfrac{16}{25}}$ 4. $\sqrt{\dfrac{81}{100}}$

5. $\sqrt{\dfrac{y^4}{x^2}}$ 6. $\sqrt{\dfrac{x^6}{y^8}}$ 7. $\sqrt{\dfrac{4x^2}{9}}$ 8. $\sqrt{\dfrac{16a^2}{49}}$

9. $\sqrt{\dfrac{2}{8k^2}}$ 10. $\sqrt{\dfrac{2m^2}{18}}$ 11. $\sqrt{\dfrac{4x^3y}{xy^3}}$ 12. $\sqrt{\dfrac{x^5y}{9xy^3}}$

13. $\sqrt{\dfrac{1}{5}}$ 14. $\sqrt{\dfrac{1}{3}}$ 15. $\sqrt{\dfrac{7}{13}}$ 16. $\sqrt{\dfrac{5}{17}}$

17. $\sqrt{\dfrac{1}{18}}$ 18. $\sqrt{\dfrac{1}{50}}$ 19. $\sqrt{\dfrac{7}{27}}$ 20. $\sqrt{\dfrac{5}{8}}$

21. $\dfrac{3}{\sqrt{7}}$ 22. $\dfrac{2}{\sqrt{3}}$ 23. $\dfrac{10}{\sqrt{5}}$ 24. $\dfrac{14}{\sqrt{2}}$

25. $\sqrt{\dfrac{m^2}{3}}$ 26. $\sqrt{\dfrac{k^2}{5}}$ 27. $\sqrt{\dfrac{x^5z^4}{36y^2}}$ 28. $\sqrt{\dfrac{a^3c^6}{25b^2}}$

29. $\sqrt{\dfrac{3a^2b}{4b^3}}$ **30.** $\sqrt{\dfrac{10uv^2}{8u}}$ **31.** $\sqrt{\dfrac{b^2c^4}{16d^3}}$ **32.** $\sqrt{\dfrac{h^4k^8}{49p^5}}$

33. $\sqrt{\dfrac{8m^2n}{2n^2}}$ **34.** $\sqrt{\dfrac{18xy^2}{2x^2}}$

Set II Simplify each of the following expressions; assume all variables in denominators are nonzero.

1. $\sqrt{\dfrac{16}{49}}$ **2.** $\sqrt{\dfrac{64}{81}}$ **3.** $\sqrt{\dfrac{25}{64}}$ **4.** $\sqrt{\dfrac{81}{121}}$

5. $\sqrt{\dfrac{a^2}{b^6}}$ **6.** $\sqrt{\dfrac{c^8}{d^4}}$ **7.** $\sqrt{\dfrac{36m^2}{25}}$ **8.** $\sqrt{\dfrac{16a^4}{49}}$

9. $\sqrt{\dfrac{3}{27x^2}}$ **10.** $\sqrt{\dfrac{18c^2}{2d^4}}$ **11.** $\sqrt{\dfrac{h^3k^3}{16hk^5}}$ **12.** $\sqrt{\dfrac{50a^4b}{2b^3}}$

13. $\sqrt{\dfrac{1}{11}}$ **14.** $\sqrt{\dfrac{1}{2}}$ **15.** $\sqrt{\dfrac{2}{19}}$ **16.** $\sqrt{\dfrac{2}{18}}$

17. $\sqrt{\dfrac{1}{45}}$ **18.** $\sqrt{\dfrac{1}{32}}$ **19.** $\sqrt{\dfrac{2}{75}}$ **20.** $\sqrt{\dfrac{3}{28}}$

21. $\dfrac{5}{\sqrt{10}}$ **22.** $\dfrac{3}{\sqrt{75}}$ **23.** $\dfrac{6}{\sqrt{15}}$ **24.** $\dfrac{7}{\sqrt{28}}$

25. $\sqrt{\dfrac{e^2}{7}}$ **26.** $\sqrt{\dfrac{f^2}{2}}$ **27.** $\sqrt{\dfrac{r^3s^6}{9t^2}}$ **28.** $\sqrt{\dfrac{50a}{2a^5b^4}}$

29. $\sqrt{\dfrac{15zw^4}{18z}}$ **30.** $\sqrt{\dfrac{12ab^2}{5a^5b}}$ **31.** $\sqrt{\dfrac{d^4e^6}{100f^3}}$ **32.** $\sqrt{\dfrac{125x^3y^4}{3xy^2}}$

33. $\sqrt{\dfrac{45tu^2}{5t^2}}$ **34.** $\sqrt{\dfrac{7cd^3}{5c^2d}}$

11.3 Products and Quotients of Square Roots

11.3A Multiplying Square Roots

In Section 11.2A, we used Rule 11.2 in the following direction:

$$\overleftarrow{\phantom{\sqrt{ab} = \sqrt{a}\sqrt{b}}}$$
$$\sqrt{ab} = \sqrt{a}\sqrt{b}$$

$$\sqrt{4 \cdot 3} = \sqrt{4}\sqrt{3} \qquad \text{Using Rule 11.2 to find the square root of a product}$$

$$= 2\sqrt{3}$$

In this section, we use Rule 11.2 in the opposite way:

$$\overrightarrow{\sqrt{a}\sqrt{b} = \sqrt{ab}}$$

$$\sqrt{2}\sqrt{8} = \sqrt{2 \cdot 8} \qquad \text{Using Rule 11.2 to find the product of square roots}$$

$$= \sqrt{16} = 4$$

In words,

Product of square roots	=	square root of product
$\sqrt{a}\sqrt{b}$	=	$\sqrt{ab}$

A WORD OF CAUTION   While the product of square roots equals the square root of the product, the *sum* of square roots does not equal the square root of the sum. That is,

$$\sqrt{a} + \sqrt{b} \neq \sqrt{a + b}$$

To see that this is so, let $a = 9$ and $b = 16$. Then

$$\sqrt{a} + \sqrt{b} = \sqrt{9} + \sqrt{16} = 3 + 4 = 7$$

$$\sqrt{a + b} = \sqrt{9 + 16} = \sqrt{25} = 5$$

Since $7 \neq 5$, $\sqrt{a} + \sqrt{b} \neq \sqrt{a + b}$.   ☑

Example 1   Examples of multiplying square roots:

Product of square roots = square root of product

a. $\sqrt{2}\sqrt{2} = \sqrt{2 \cdot 2} = \sqrt{2^2} = 2$

b. $\sqrt{2}\sqrt{8} = \sqrt{2 \cdot 8} = \sqrt{16} = 4$

c. $\sqrt{4x}\sqrt{x} = \sqrt{4x \cdot x} = \sqrt{4x^2} = 2x$

d. $\sqrt{3y}\sqrt{12y^3} = \sqrt{3y \cdot 12y^3} = \sqrt{36y^4} = 6y^2$

e. $\sqrt{3}\sqrt{6}\sqrt{2} = \sqrt{3 \cdot 6 \cdot 2} = \sqrt{36} = 6$

f. $\sqrt{5z}\sqrt{10}\sqrt{2z^3} = \sqrt{5z \cdot 10 \cdot 2z^3} = \sqrt{100z^4} = 10z^2$ ■

Multiplying a Square Root by Itself   By Rule 11.2,

$$\sqrt{ab} = \sqrt{a}\sqrt{b}$$

If $a = b$, then

$$\sqrt{aa} = \sqrt{a}\sqrt{a}$$

$$\sqrt{a^2} = \sqrt{a}\sqrt{a}$$

$$a = \sqrt{a}\sqrt{a} \quad \text{or} \quad \sqrt{a}\sqrt{a} = a$$

RULE 11.4   MULTIPLYING A SQUARE ROOT BY ITSELF

$$\sqrt{a}\sqrt{a} = a, \text{ if } a \geq 0$$

Example 2   Examples of multiplying a square root by itself:

a. $\sqrt{16}\sqrt{16} = 16$

b. $\sqrt{7}\sqrt{7} = 7$

c. $\sqrt{2x}\sqrt{2x} = 2x$ ∎

An expression that contains radicals is not simplified if any term has more than one radical sign in it.

Example 3   Find the products and simplify the result:

a. $\sqrt{3x}\sqrt{6x^2} = \sqrt{3x \cdot 6x^2} = \sqrt{18x^3}$ Multiplying the square roots

$$= \sqrt{2 \cdot 9 \cdot x^2 \cdot x} \left.\rule{0pt}{20pt}\right\} \text{ Simplifying}$$

$$= 3x\sqrt{2x}$$

b. $\sqrt{3w^2}\sqrt{2}\sqrt{8w^3} = \sqrt{3w^2 \cdot 2 \cdot 8w^3} = \sqrt{48w^5}$ Multiplying the square roots

$$= \sqrt{3 \cdot 16 \cdot w^4 \cdot w} \left.\rule{0pt}{20pt}\right\} \text{ Simplifying}$$

$$= 4w^2\sqrt{3w}$$

c. $5\sqrt{2y^3} \cdot 3\sqrt{6y^4} = (5 \cdot 3)\sqrt{2y^3 \cdot 6y^4} = 15\sqrt{12y^7}$ Multiplying the square roots

$$= 15\sqrt{3 \cdot 4 \cdot y^6 \cdot y}$$

$$= 15 \cdot 2 \cdot y^3\sqrt{3y} \left.\rule{0pt}{30pt}\right\} \text{ Simplifying}$$

$$= 30y^3\sqrt{3y}$$ ∎

EXERCISES 11.3A

Set I   Find the following products and simplify the results.

1. $\sqrt{3}\sqrt{3}$ 2. $\sqrt{7}\sqrt{7}$ 3. $\sqrt{4}\sqrt{4}$

4. $\sqrt{9}\sqrt{9}$ 5. $\sqrt{3}\sqrt{12}$ 6. $\sqrt{2}\sqrt{32}$

7. $\sqrt{9x}\sqrt{x}$ 8. $\sqrt{25y}\sqrt{y}$ 9. $\sqrt{5}\sqrt{10}\sqrt{2}$

10. $\sqrt{6}\sqrt{12}\sqrt{2}$ 11. $\sqrt{5ab^2}\sqrt{20ab}$ 12. $\sqrt{3x^2y}\sqrt{27xy}$

13. $\sqrt{2a}\sqrt{6}\sqrt{3a}$ 14. $\sqrt{2}\sqrt{h^3}\sqrt{8h}$

15. $5\sqrt{2x} \cdot \sqrt{8x^3} \cdot 2\sqrt{3x^5}$ 16. $4\sqrt{2M^3} \cdot \sqrt{3M} \cdot 3\sqrt{12M^3}$

Set II   Find the following products and simplify the results.

1. $\sqrt{11}\sqrt{11}$ 2. $\sqrt{6}\sqrt{6}$ 3. $\sqrt{8}\sqrt{8}$

4. $\sqrt{3}\sqrt{3}$ 5. $\sqrt{18}\sqrt{2}$ 6. $\sqrt{4x}\sqrt{4x}$

7. $\sqrt{16a}\sqrt{a}$ 8. $\sqrt{49y}\sqrt{2y}$ 9. $\sqrt{3}\sqrt{6}\sqrt{2}$

10. $\sqrt{6x}\sqrt{2x}\sqrt{8x}$ **11.** $\sqrt{18x^3}\sqrt{2xy^2}$ **12.** $\sqrt{30ab^5}\sqrt{3a^2b}$

13. $\sqrt{12mn}\sqrt{mn^3}\sqrt{3}$ **14.** $\sqrt{b^3}\sqrt{32}\sqrt{2b}$

15. $6\sqrt{5z^3}\cdot\sqrt{2z}\cdot4\sqrt{8z^3}$ **16.** $2\sqrt{6x^5}\cdot\sqrt{3x}\cdot5\sqrt{20x^3}$

11.3B   Dividing Square Roots

In Section 11.2B, we used Rule 11.3 in the following direction:

$$\sqrt{\frac{a}{b}} = \frac{\sqrt{a}}{\sqrt{b}} \qquad (b \neq 0)$$

Using Rule 11.3 to find the square root of a quotient

$$\sqrt{\frac{4}{9}} = \frac{\sqrt{4}}{\sqrt{9}} = \frac{2}{3}$$

In this section, we use Rule 11.3 in the opposite way:

$$\frac{\sqrt{a}}{\sqrt{b}} = \sqrt{\frac{a}{b}} \qquad (b \neq 0)$$

Using Rule 11.3 to find the quotient of square roots

$$\frac{\sqrt{50}}{\sqrt{2}} = \sqrt{\frac{50}{2}}$$
$$= \sqrt{25} = 5$$

In words,

> | Quotient of square roots | = | square root of quotient |
>
> $$\frac{\sqrt{a}}{\sqrt{b}} = \sqrt{\frac{a}{b}} \quad (b \neq 0)$$

Example 4   Examples of dividing square roots (assume $a > 0$, $x > 0$, $y > 0$):

Quotient of square roots = square root of quotient

a. $\dfrac{\sqrt{8}}{\sqrt{2}} = \sqrt{\dfrac{8}{2}} = \sqrt{4} = 2$

b. $\dfrac{\sqrt{a^5}}{\sqrt{a^3}} = \sqrt{\dfrac{a^5}{a^3}} = \sqrt{a^2} = a$

c. $\dfrac{\sqrt{27x}}{\sqrt{3x^3}} = \sqrt{\dfrac{27x}{3x^3}} = \sqrt{\dfrac{9}{x^2}} = \dfrac{3}{x}$

d. $\dfrac{\sqrt{28xy^3}}{\sqrt{7xy}} = \sqrt{\dfrac{28xy^3}{7xy}} = \sqrt{4y^2} = 2y$ ■

In Example 5, we rationalize the denominator after the division has been performed.

Example 5   Simplify each of the following expressions (assume $x > 0$):

a. $\dfrac{\sqrt{5x}}{\sqrt{10x^2}} = \sqrt{\dfrac{5x}{10x^2}} = \sqrt{\dfrac{1}{2x}} = \dfrac{\sqrt{1}}{\sqrt{2x}} = \dfrac{1}{\sqrt{2x}} \cdot \dfrac{\sqrt{2x}}{\sqrt{2x}} = \dfrac{\sqrt{2x}}{2x}$ $\sqrt{2x}\,\sqrt{2x} = 2x$ by Rule 11.4

b. $\dfrac{6\sqrt{25x^3}}{5\sqrt{3x}} = \dfrac{6}{5}\sqrt{\dfrac{25x^3}{3x}} = \dfrac{6\sqrt{25x^2}}{5\,\sqrt{3}} = \dfrac{6}{5}\cdot\dfrac{\overset{x}{\cancel{5}x}}{\sqrt{3}} \cdot \dfrac{\sqrt{3}}{\sqrt{3}} = \dfrac{\overset{2}{\cancel{6}}x\sqrt{3}}{\cancel{3}} = 2x\sqrt{3}$ ∎

EXERCISES 11.3B

Set I   Simplify each of the following expressions; assume all variables in denominators are nonzero.

1. $\dfrac{\sqrt{20}}{\sqrt{5}}$

2. $\dfrac{\sqrt{7}}{\sqrt{28}}$

3. $\dfrac{\sqrt{32}}{\sqrt{2}}$

4. $\dfrac{\sqrt{98}}{\sqrt{2}}$

5. $\dfrac{\sqrt{4}}{\sqrt{5}}$

6. $\dfrac{\sqrt{9}}{\sqrt{7}}$

7. $\dfrac{\sqrt{15x}}{\sqrt{5x}}$

8. $\dfrac{\sqrt{18y}}{\sqrt{3y}}$

9. $\dfrac{\sqrt{72x^3y^2}}{\sqrt{2xy^2}}$

10. $\dfrac{\sqrt{27x^2y^3}}{\sqrt{3x^2y}}$

11. $\dfrac{\sqrt{x^4y}}{\sqrt{5y}}$

12. $\dfrac{\sqrt{m^6n}}{\sqrt{3n}}$

13. $\dfrac{4\sqrt{45m^3}}{3\sqrt{10m}}$

14. $\dfrac{6\sqrt{400x^4}}{5\sqrt{6x}}$

Set II   Simplify each of the following expressions; assume all variables in denominators are nonzero.

1. $\dfrac{\sqrt{27}}{\sqrt{3}}$

2. $\dfrac{\sqrt{2}}{\sqrt{50}}$

3. $\dfrac{\sqrt{75}}{\sqrt{3}}$

4. $\dfrac{\sqrt{7}}{\sqrt{28}}$

5. $\dfrac{\sqrt{25}}{\sqrt{2}}$

6. $\dfrac{\sqrt{36}}{\sqrt{11}}$

7. $\dfrac{\sqrt{35x}}{\sqrt{5x}}$

8. $\dfrac{\sqrt{18m}}{\sqrt{6m}}$

9. $\dfrac{\sqrt{75a^5b^3}}{\sqrt{3a^3b^3}}$

10. $\dfrac{\sqrt{18x^3y}}{\sqrt{5xy^5}}$

11. $\dfrac{\sqrt{x^7y}}{\sqrt{3x}}$

12. $\dfrac{\sqrt{c^3d^4}}{\sqrt{7c^2d}}$

13. $\dfrac{9\sqrt{20a^6}}{2\sqrt{15a^3}}$

14. $\dfrac{12\sqrt{15b^3c}}{7\sqrt{3bc^5}}$

11.4 Sums and Differences of Square Roots

Like square roots are square roots that have the same radicand.

Example 1   Examples of like square roots:

a. $3\sqrt{5},\ 2\sqrt{5},\ -7\sqrt{5}$

b. $2\sqrt{x},\ -9\sqrt{x},\ 11\sqrt{x}$ ∎

Unlike square roots are square roots that have different radicands.

Example 2 Examples of unlike square roots:

a. $2\sqrt{15}, -6\sqrt{11}, 8\sqrt{24}$

b. $5\sqrt{y}, 3\sqrt{x}, -4\sqrt{13}$ ■

Combining Like Square Roots

The addition of like square roots is an application of the distributive property. For example,

$$3\sqrt{7} + 2\sqrt{7} = (3 + 2)\sqrt{7} = 5\sqrt{7}$$

Therefore, we add like square roots by adding their coefficients and then multiplying that sum by the square root.

Example 3 Combine all like radicals:

a. $5\sqrt{2} + 3\sqrt{2} = (5 + 3)\sqrt{2} = 8\sqrt{2}$

b. $6\sqrt{3} - 4\sqrt{3} = (6 - 4)\sqrt{3} = 2\sqrt{3}$

c. $7\sqrt{x} - 3\sqrt{x} = (7 - 3)\sqrt{x} = 4\sqrt{x}$

d. $\dfrac{3}{2}\sqrt{5} + \dfrac{\sqrt{5}}{2} = \left(\dfrac{3}{2} + \dfrac{1}{2}\right)\sqrt{5} = 2\sqrt{5}$

$\dfrac{\sqrt{5}}{2} = \dfrac{1}{2}\sqrt{5}$, because $\dfrac{1}{2}\sqrt{5} = \dfrac{1}{2} \cdot \dfrac{\sqrt{5}}{1} = \dfrac{\sqrt{5}}{2}$ ■

Combining Unlike Square Roots

When two or more *unlike* radicals are connected with *addition* or *subtraction* symbols, we usually cannot express the sum or difference with just one radical sign. We can, however, find an *approximation* to the sum or difference by using a calculator or Table I (found inside the back cover).

Example 4 Find the approximate value of (a) $\sqrt{5} + \sqrt{7}$ and (b) $\sqrt{12}$. Determine whether $\sqrt{5} + \sqrt{7}$ equals $\sqrt{12}$.

a. Using a calculator and rounding off each approximation to three decimal places, we have

$$\sqrt{5} + \sqrt{7} \doteq 2.236 + 2.646 = 4.882$$

b. $\sqrt{12} \doteq 3.464$

Because $4.882 \neq 3.464$, $\sqrt{5} + \sqrt{7} \neq \sqrt{12}$. ■

While most unlike radicals cannot be combined, some *can* be combined. Sometimes terms that were not like radicals become like radicals after they have been expressed in simplest radical form; if this is the case, they can be combined (see Example 5).

TO COMBINE UNLIKE RADICALS

1. Express each term in simplest radical form.

2. Combine any terms that have like radicals by adding their coefficients and multiplying that sum by the like radical.

Example 5 Simplify:

a. $\sqrt{8} + \sqrt{18}$

$= \sqrt{4 \cdot 2} + \sqrt{9 \cdot 2} = \boxed{2}\sqrt{2} + \boxed{3}\sqrt{2} = 5\sqrt{2}$ $2\sqrt{2}$ and $3\sqrt{2}$ are like radicals

b. $\sqrt{12} - \sqrt{27} + 5\sqrt{3}$

$= \sqrt{4 \cdot 3} - \sqrt{9 \cdot 3} + 5\sqrt{3} = \boxed{2}\sqrt{3} \boxed{-3}\sqrt{3} + \boxed{5}\sqrt{3} = 4\sqrt{3}$

c. $\sqrt{20} - 2\sqrt{45} - \sqrt{15}$

$= \sqrt{4 \cdot 5} - 2\sqrt{9 \cdot 5} - \sqrt{3 \cdot 5}$

$= 2\sqrt{5} - 2 \cdot 3\sqrt{5} - \sqrt{15}$ $2\sqrt{5}$ and $-2 \cdot 3\sqrt{5}$ are like radicals

$= \boxed{2}\sqrt{5} \boxed{-6}\sqrt{5} - \sqrt{15} = -4\sqrt{5} - \sqrt{15}$

d. $2\sqrt{\dfrac{1}{2}} - 6\sqrt{\dfrac{1}{8}} - 10\sqrt{\dfrac{4}{5}}$

$= 2\sqrt{\dfrac{1 \cdot 2}{2 \cdot 2}} - 6\sqrt{\dfrac{1 \cdot 2}{8 \cdot 2}} - 10\sqrt{\dfrac{4 \cdot 5}{5 \cdot 5}}$ Rationalizing the denominators

$= \dfrac{2}{1} \cdot \dfrac{\sqrt{2}}{\sqrt{4}} - \dfrac{6}{1} \cdot \dfrac{\sqrt{2}}{\sqrt{16}} - \dfrac{10}{1} \cdot \dfrac{\sqrt{4}\sqrt{5}}{\sqrt{25}}$

$= \dfrac{\overset{1}{\cancel{2}}}{1} \cdot \dfrac{\sqrt{2}}{\underset{1}{\cancel{2}}} - \dfrac{\overset{3}{\cancel{6}}\sqrt{2}}{\underset{2}{\cancel{4}}} - \dfrac{\overset{2}{\cancel{10}}}{1} \cdot \dfrac{2\sqrt{5}}{\underset{1}{\cancel{5}}}$

$= \sqrt{2} - \dfrac{3}{2}\sqrt{2} - 4\sqrt{5}$ $\sqrt{2}$ and $-\frac{3}{2}\sqrt{2}$ are like radicals

$= \left(1 - \dfrac{3}{2}\right)\sqrt{2} - 4\sqrt{5} = -\dfrac{1}{2}\sqrt{2} - 4\sqrt{5}$ ∎

Although we cannot add the unlike square roots in the expression $-\frac{1}{2}\sqrt{2} - 4\sqrt{5}$, we can find the approximate value by using a calculator or Table I (see Example 6).

Example 6 Approximate $-\dfrac{1}{2}\sqrt{2} - 4\sqrt{5}$.

Using Table I

$$\sqrt{2} \doteq 1.414$$

$$\sqrt{5} \doteq 2.236$$

Therefore,

$$-\frac{1}{2}\sqrt{2} - 4\sqrt{5} \doteq -\frac{1}{2}(1.414) - 4(2.236)$$

$$= -0.707 - 8.944$$

$$= \boxed{-9.651}$$

Using a calculator

$$\sqrt{2} \doteq 1.414213562$$

$$\sqrt{5} \doteq 2.236067977$$

Therefore,

$$-\frac{1}{2}\sqrt{2} - 4\sqrt{5} \doteq -\frac{1}{2}(1.414213562) - 4(2.236067977)$$

$$= -0.707106781 - 8.94427191$$

$$= -9.651378691$$

$$\doteq \boxed{-9.651} \qquad \text{Rounded off to three decimal places}$$

Notice that when the calculator answer is rounded off to the same number of decimal places as the answer obtained using Table I, we get the same number, $\boxed{-9.651}$. However, answers sometimes differ slightly because of the rounding off process. ■

An expression that contains square roots is not in simplest form unless all like radicals have been combined.

EXERCISES 11.4

Set I   In Exercises 1–32, simplify each expression.

1. $2\sqrt{3} + 5\sqrt{3}$ **2.** $4\sqrt{2} + 3\sqrt{2}$ **3.** $3\sqrt{x} - \sqrt{x}$

4. $5\sqrt{a} - \sqrt{a}$ **5.** $\dfrac{3}{2}\sqrt{2} - \dfrac{\sqrt{2}}{2}$ **6.** $\dfrac{4}{3}\sqrt{3} - \dfrac{\sqrt{3}}{3}$

7. $5 \cdot 8\sqrt{5} + \sqrt{5}$ **8.** $3 \cdot 4\sqrt{7} + \sqrt{7}$ **9.** $\sqrt{25} + \sqrt{5}$

10. $\sqrt{16} + \sqrt{6}$ **11.** $\sqrt{50} + \sqrt{50}$ **12.** $\sqrt{32} + \sqrt{32}$

13. $\sqrt{68} - \sqrt{17}$ **14.** $\sqrt{52} - \sqrt{13}$ **15.** $\sqrt{45} - \sqrt{28}$

16. $\sqrt{90} - \sqrt{54}$ **17.** $2\sqrt{3} + \sqrt{12}$ **18.** $3\sqrt{2} + \sqrt{8}$

19. $2\sqrt{50} - \sqrt{32}$ **20.** $3\sqrt{24} - \sqrt{54}$ **21.** $3\sqrt{32} - \sqrt{8}$

22. $4\sqrt{27} - 3\sqrt{12}$ **23.** $\sqrt{\dfrac{1}{2}} + \sqrt{8}$ **24.** $\sqrt{\dfrac{1}{3}} + \sqrt{12}$

25. $\sqrt{24} - \sqrt{\dfrac{2}{3}}$ **26.** $\sqrt{45} - \sqrt{\dfrac{4}{5}}$ **27.** $10\sqrt{\dfrac{3}{5}} + \sqrt{60}$

28. $8\sqrt{\dfrac{3}{16}} + \sqrt{48}$ **29.** $\sqrt{\dfrac{25}{2}} - \dfrac{3}{\sqrt{2}}$ **30.** $5\sqrt{\dfrac{1}{5}} - \sqrt{\dfrac{9}{20}}$

31. $3\sqrt{\dfrac{1}{6}} + \sqrt{12} - 5\sqrt{\dfrac{3}{2}}$ **32.** $3\sqrt{\dfrac{5}{2}} + \sqrt{20} - 5\sqrt{\dfrac{1}{10}}$

In Exercises 33 and 34, approximate each expression using Table I or a calculator. Round off answers to three decimal places.

33. $\dfrac{1}{3}\sqrt{7} + 2\sqrt{3}$ **34.** $3\sqrt{6} + \dfrac{1}{4}\sqrt{5}$

Set II   In Exercises 1–32, simplify each expression.

1. $3\sqrt{7} + 2\sqrt{7}$ **2.** $8\sqrt{14} + 7\sqrt{14}$ **3.** $4\sqrt{m} - \sqrt{m}$

4. $12\sqrt{c} - \sqrt{c}$ **5.** $\dfrac{2}{3}\sqrt{5} - \dfrac{\sqrt{5}}{3}$ **6.** $\dfrac{3}{2}\sqrt{6} - \dfrac{\sqrt{6}}{2}$

7. $3 \cdot 6\sqrt{3} + \sqrt{3}$　　　　**8.** $5 \cdot 8\sqrt{11} + \sqrt{11}$　　　　**9.** $\sqrt{9} + \sqrt{17}$

10. $\sqrt{25} + \sqrt{26}$　　　　**11.** $\sqrt{18} + \sqrt{18}$　　　　**12.** $\sqrt{48} + \sqrt{48}$

13. $\sqrt{44} - \sqrt{11}$　　　　**14.** $\sqrt{117} - \sqrt{13}$　　　　**15.** $\sqrt{52} - \sqrt{44}$

16. $\sqrt{160} - \sqrt{63}$　　　　**17.** $5\sqrt{2} + \sqrt{18}$　　　　**18.** $7\sqrt{7} + \sqrt{28}$

19. $\sqrt{45} - 3\sqrt{20}$　　　　**20.** $\sqrt{48} - 5\sqrt{27}$　　　　**21.** $4\sqrt{28} - \sqrt{63}$

22. $8\sqrt{45} - 3\sqrt{20}$　　　　**23.** $\sqrt{20} + \sqrt{\dfrac{1}{5}}$　　　　**24.** $\sqrt{18} + \sqrt{\dfrac{1}{2}}$

25. $\sqrt{75} - \sqrt{\dfrac{4}{3}}$　　　　**26.** $\sqrt{\dfrac{1}{3}} - \sqrt{108}$　　　　**27.** $6\sqrt{\dfrac{5}{9}} + \sqrt{45}$

28. $5\sqrt{8} + 3\sqrt{\dfrac{1}{2}}$　　　　**29.** $\sqrt{\dfrac{16}{3}} - \dfrac{6}{\sqrt{3}}$　　　　**30.** $\dfrac{4}{\sqrt{5}} - \sqrt{\dfrac{36}{5}}$

31. $10\sqrt{\dfrac{1}{15}} + \sqrt{60} - 8\sqrt{\dfrac{3}{5}}$　　　　**32.** $7\sqrt{40} - 3\sqrt{\dfrac{5}{2}} + 8\sqrt{\dfrac{1}{10}}$

In Exercises 33 and 34, approximate each expression using Table I or a calculator. Round off answers to three decimal places.

33. $\dfrac{1}{4}\sqrt{13} + 3\sqrt{6}$　　　　**34.** $5\sqrt{5} + \dfrac{1}{5}\sqrt{7}$

11.5 Products and Quotients Involving More Than One Term

Example 1 illustrates the use of the distributive property when radicals are involved.

Example 1

$$\sqrt{2}(3\sqrt{2} - 5) = \sqrt{2} \cdot (3\sqrt{2}) + \sqrt{2}(-5)$$
$$= 3\sqrt{2 \cdot 2} - 5\sqrt{2}$$
$$= 3 \cdot 2 - 5\sqrt{2}$$
$$= 6 - 5\sqrt{2} \quad \blacksquare$$

The FOIL method can be used when we multiply two expressions that contain square roots and have two terms (see Example 2).

Example 2　$(2\sqrt{3} - 5)(4\sqrt{3} - 6)$

Solution

$$2\sqrt{3} \cdot 4\sqrt{3} = 8 \cdot 3 = 24$$

$$(-5)(-6) = 30$$

$$(2\sqrt{3} - 5)(4\sqrt{3} - 6)$$

$$-20\sqrt{3}$$

$$-12\sqrt{3} \qquad \text{Adding like radicals}$$

$$= 24 - 32\sqrt{3} + 30$$
$$= 54 - 32\sqrt{3} \quad \blacksquare$$

Example 3 illustrates squaring an expression that contains square roots and has two terms.

Example 3 $(3\sqrt{5} - 7x)^2$

Solution Using the formula $(a - b)^2 = a^2 - 2ab + b^2$, we have

$$(3\sqrt{5} - 7x)^2 = (3\sqrt{5})^2 - 2(3\sqrt{5})(7x) + (7x)^2$$

$$= 3^2(\sqrt{5})^2 - 42\sqrt{5}\,x + 7^2x^2 \qquad \text{Notice that } x \text{ is } \textit{not} \text{ under the radical sign}$$

$$= 9 \cdot 5 - 42\sqrt{5}\,x + 49x^2$$

$$= 45 - 42\sqrt{5}\,x + 49x^2$$

We obtain the same answer if we write $(3\sqrt{5} - 7x)^2$ as $(3\sqrt{5} - 7x)(3\sqrt{5} - 7x)$ and use the FOIL method. ■

In a division problem, when the dividend contains more than one term but the divisor contains only one term, we divide *each term* of the dividend by the divisor (see Example 4).

Example 4

$$\frac{3\sqrt{14} - \sqrt{8}}{\sqrt{2}} = \frac{3\sqrt{14}}{\sqrt{2}} - \frac{\sqrt{8}}{\sqrt{2}} = 3\sqrt{\frac{14}{2}} - \sqrt{\frac{8}{2}}$$

$$= 3\sqrt{7} - \sqrt{4} = 3\sqrt{7} - 2 \quad ■$$

Rationalizing a Denominator That Contains Two Terms

In order to discuss division problems in which the divisor contains square roots and has two terms, we need a new definition.

CONJUGATE

The conjugate of the algebraic expression $a + b$ is the algebraic expression $a - b$.

Example 5   Examples of conjugates of algebraic expressions that contain square roots:

a. The conjugate of $1 - \sqrt{2}$ is $1 + \sqrt{2}$.

b. The conjugate of $2\sqrt{3} + 5$ is $2\sqrt{3} - 5$.

c. The conjugate of $\sqrt{x} - \sqrt{y}$ is $\sqrt{x} + \sqrt{y}$. ■

Example 6   Find the product of $1 - \sqrt{2}$ and its conjugate.

Solution The conjugate of $1 - \sqrt{2}$ is $1 + \sqrt{2}$. Therefore, the product is

$$(1 - \sqrt{2})(1 + \sqrt{2}) = 1^2 - (\sqrt{2})^2 = 1 - 2 = -1$$

Notice that -1 is a *rational number*. ■

The product of an algebraic expression that contains square roots and its conjugate is *always* a rational number. Because of this fact, the following procedure should be used when a denominator contains square roots and has two terms:

> **TO RATIONALIZE A DENOMINATOR THAT CONTAINS SQUARE ROOTS AND HAS TWO TERMS**
>
> Multiply the numerator and the denominator by the conjugate of the denominator.

Example 7 Examples of rationalizing denominators that contain square roots and have two terms:

a.
$$\frac{2}{1 + \sqrt{3}} = \frac{2}{1 + \sqrt{3}} \cdot \boxed{\frac{1 - \sqrt{3}}{1 - \sqrt{3}}} = \frac{2(1 - \sqrt{3})}{(1)^2 - (\sqrt{3})^2} = \frac{2(1 - \sqrt{3})}{1 - 3}$$

$$= \frac{2(1 - \sqrt{3})}{-2} = \frac{\overset{1}{\cancel{2}}(1 - \sqrt{3})}{\underset{-1}{\cancel{-2}}}$$

$$= \sqrt{3} - 1$$

Multiply the numerator and denominator by $1 - \sqrt{3}$ (the conjugate of the denominator $1 + \sqrt{3}$); the value of this fraction is 1, and multiplying $\frac{2}{1 + \sqrt{3}}$ by 1 does not change its value

b.
$$\frac{6}{\sqrt{5} - \sqrt{3}} = \frac{6}{\sqrt{5} - \sqrt{3}} \cdot \boxed{\frac{\sqrt{5} + \sqrt{3}}{\sqrt{5} + \sqrt{3}}} = \frac{6(\sqrt{5} + \sqrt{3})}{(\sqrt{5})^2 - (\sqrt{3})^2} = \frac{6(\sqrt{5} + \sqrt{3})}{5 - 3}$$

$$= \frac{\overset{3}{\cancel{6}}(\sqrt{5} + \sqrt{3})}{\underset{1}{\cancel{2}}} = 3\sqrt{5} + 3\sqrt{3}$$

Multiply numerator and denominator by $\sqrt{5} + \sqrt{3}$ (the conjugate of the denominator $\sqrt{5} - \sqrt{3}$) ∎

EXERCISES 11.5

Set I In Exercises 1–18, simplify each expression.

1. $\sqrt{2}(\sqrt{2} + 1)$ 2. $\sqrt{3}(\sqrt{3} + 1)$ 3. $\sqrt{3}(2\sqrt{3} + 1)$

4. $\sqrt{5}(3\sqrt{5} + 1)$ 5. $\sqrt{x}(\sqrt{x} - 3)$ 6. $\sqrt{y}(4 - \sqrt{y})$

7. $(\sqrt{7} + 2)(\sqrt{7} + 3)$ 8. $(\sqrt{3} + 2)(\sqrt{3} + 4)$

9. $(\sqrt{8} - 3\sqrt{2})(\sqrt{8} + 2\sqrt{5})$ 10. $(\sqrt{2} + 4\sqrt{3})(\sqrt{12} - 2\sqrt{3})$

11. $(\sqrt{2} + \sqrt{14})^2$ 12. $(\sqrt{5} + \sqrt{11})^2$

13. $(\sqrt{2x} + 3)^2$ 14. $(\sqrt{7x} + 4)^2$

15. $\dfrac{\sqrt{8} + \sqrt{18}}{\sqrt{2}}$ 16. $\dfrac{\sqrt{12} + \sqrt{27}}{\sqrt{3}}$

17. $\dfrac{\sqrt{20} + 5\sqrt{10}}{\sqrt{5}}$ 18. $\dfrac{2\sqrt{6} + \sqrt{14}}{\sqrt{6}}$

In Exercises 19 and 20, write the conjugate for each expression.

19. a. $2 + \sqrt{3}$ b. $2\sqrt{5} - 7$

20. a. $3\sqrt{2} - 5$ b. $\sqrt{7} + 4$

In Exercises 21–28, rationalize the denominators and simplify.

21. $\dfrac{3}{\sqrt{2} - 1}$ **22.** $\dfrac{5}{\sqrt{2} - 1}$ **23.** $\dfrac{6}{\sqrt{3} - \sqrt{2}}$

24. $\dfrac{8}{\sqrt{2} - \sqrt{3}}$ **25.** $\dfrac{6}{\sqrt{5} + \sqrt{2}}$ **26.** $\dfrac{4}{\sqrt{7} + \sqrt{5}}$

27. $\dfrac{x - 4}{\sqrt{x} + 2}$ **28.** $\dfrac{y - 9}{\sqrt{y} - 3}$ $(y \neq 9)$

In Exercises 29–32, approximate each expression using Table I or a calculator. Round off answers to two decimal places.

29. $\dfrac{1}{3}\sqrt{7} - 2\sqrt{3}$ **30.** $3\sqrt{6} - \dfrac{1}{4}\sqrt{5}$

31. $\dfrac{3 + 2\sqrt{11}}{6}$ **32.** $\dfrac{3 - 2\sqrt{11}}{6}$

Set II   In Exercises 1–18, simplify each expression.

1. $\sqrt{5}\left(\sqrt{5} + 1\right)$ **2.** $\sqrt{7}\left(\sqrt{7} - 5\right)$ **3.** $\sqrt{2}\left(3\sqrt{2} + 1\right)$

4. $\sqrt{5}\left(8\sqrt{5} - \sqrt{3}\right)$ **5.** $\sqrt{z}\left(\sqrt{3} - \sqrt{z}\right)$ **6.** $\sqrt{8}\left(\sqrt{5} + \sqrt{2}\right)$

7. $\left(\sqrt{6} + 5\right)\left(\sqrt{6} + 2\right)$ **8.** $\left(\sqrt{11} - 2\right)\left(\sqrt{11} - 3\right)$

9. $\left(\sqrt{2} - 5\sqrt{8}\right)\left(\sqrt{2} + 4\sqrt{8}\right)$ **10.** $\left(\sqrt{7} + 3\sqrt{2}\right)\left(3\sqrt{7} + 2\sqrt{2}\right)$

11. $\left(\sqrt{7} + \sqrt{18}\right)^2$ **12.** $\left(\sqrt{13} + \sqrt{12}\right)^2$

13. $\left(3\sqrt{y} + 5\right)^2$ **14.** $\left(5\sqrt{z} + 2\right)^2$

15. $\dfrac{\sqrt{15} + \sqrt{10}}{\sqrt{5}}$ **16.** $\dfrac{\sqrt{21} - \sqrt{14}}{\sqrt{3}}$

17. $\dfrac{5\sqrt{32} + \sqrt{24}}{\sqrt{8}}$ **18.** $\dfrac{8\sqrt{48} - \sqrt{16}}{\sqrt{3}}$

In Exercises 19 and 20, write the conjugate for each expression.

19. a. $5 - \sqrt{7}$ b. $3\sqrt{11} - 2$

20. a. $-2 + \sqrt{3}$ b. $-8 + \sqrt{5}$

In Exercises 21–28, rationalize the denominators and simplify.

21. $\dfrac{10}{1 + \sqrt{2}}$ **22.** $\dfrac{8}{2 + \sqrt{12}}$ **23.** $\dfrac{13}{\sqrt{6} - \sqrt{5}}$

24. $\dfrac{21}{\sqrt{7} + \sqrt{3}}$ **25.** $\dfrac{15}{\sqrt{3} + \sqrt{8}}$ **26.** $\dfrac{12}{\sqrt{5} - 2}$

27. $\dfrac{z - 25}{\sqrt{z} + 5}$ **28.** $\dfrac{m^2 - 16}{2 - \sqrt{m}}$ $(m \neq 4)$

In Exercises 29–32, approximate each expression using Table I or a calculator. Round off answers to two decimal places.

29. $\dfrac{-5 + 2\sqrt{5}}{7}$

30. $\dfrac{-1 - 5\sqrt{3}}{3}$

31. $\dfrac{5 + 6\sqrt{7}}{11}$

32. $\dfrac{\sqrt{2} + \sqrt{3}}{5}$

11.6 Radical Equations

A **radical equation** is an equation in which the variable appears in a radicand. In this text, we will consider only radical equations that have square roots.

Example 1   Examples of radical equations:

a. $\sqrt{x} = 7$

b. $\sqrt{x + 2} = 3$

c. $\sqrt{2x - 3} = \sqrt{x + 7} - 2$ ∎

The following property of real numbers is used in solving radical equations:

RULE 11.5

If two numbers are equal, then their squares are equal.

$$\text{If} \qquad a = b$$
$$\text{then} \quad a^2 = b^2$$

We can remove square root signs from equations by using Rule 11.5; that is, we square both sides of the equation. However, we sometimes introduce *extraneous roots* (see Section 8.9) when we do this. Therefore, all apparent solutions to the equation must be checked in the original equation.

TO SOLVE A RADICAL EQUATION

1. Arrange the terms so that one term with a radical is by itself on one side of the equation.

2. Square both sides of the equation.

3. Combine like terms.

4. If a radical still remains, repeat steps 1, 2, and 3.

5. Solve the resulting equation for the variable.

6. Check apparent solutions in the original equation.

Example 2   Solve $\sqrt{x} = 7$.

Solution   *Check*

$$\sqrt{x} = 7$$ $$\sqrt{x} = 7$$

$$\left(\sqrt{x}\right)^2 = (7)^2$$ $$\sqrt{49} \stackrel{?}{=} 7$$

$$x = 49$$ $$7 = 7 \quad \blacksquare$$

Example 3   Solve $\sqrt{x + 2} = 3$.

Solution   *Check*

$$\sqrt{x + 2} = 3$$ $$\sqrt{x + 2} = 3$$

$$\left(\sqrt{x + 2}\right)^2 = (3)^2$$ $$\sqrt{7 + 2} \stackrel{?}{=} 3$$

$$x + 2 = 9$$ $$\sqrt{9} \stackrel{?}{=} 3$$

$$x = 7$$ $$3 = 3 \quad \blacksquare$$

A WORD OF CAUTION   If one side of the equation has *more than one term*, squaring each *term* is not the same as squaring both sides of the equation. That is,

$$\text{if} \qquad a = b + c$$

$$\text{then} \quad a^2 = (b + c)^2$$

$$\text{but} \quad a^2 \neq b^2 + c^2$$

To verify this with numbers, suppose that $a = 7$, $b = 2$, and $c = 5$.

$$7 = 2 + 5 \qquad \text{True statement}$$

$$7^2 = (2 + 5)^2 \qquad 7^2 = 7^2$$

$$7^2 \neq 2^2 + 5^2 \qquad 49 \neq 4 + 25 \qquad \boxed{\checkmark}$$

In Examples 4 and 5, we must be sure to square the *entire* right side of the equation.

Example 4   Solve $\sqrt{2x + 1} = x - 1$.

Solution

$$\sqrt{2x + 1} = x - 1$$

$$\left(\sqrt{2x + 1}\right)^2 = (x - 1)^2$$

When squaring $(x - 1)$, do not forget this middle term

$$2x + 1 = x^2 - 2x + 1$$

$$0 = x^2 - 4x \qquad \text{Adding} -2x - 1 \text{ to both sides}$$

$$0 = x(x - 4)$$

$$x = 0 \qquad \bigg| \qquad x - 4 = 0$$

$$x = 4$$

Check for x = 0

$$\sqrt{2x + 1} \overset{?}{=} x - 1$$

$$\sqrt{2(0) + 1} \overset{?}{=} (0) - 1$$

$$\sqrt{1} \overset{?}{=} -1 \qquad \text{The symbol } \sqrt{1} \text{ } \textit{always} \text{ stands for}$$
$$\qquad\qquad\qquad \text{the } \textit{principal} \text{ square root of 1, which is 1 (} \textit{not} - 1)$$

$$1 \neq -1$$

Therefore, 0 is not a solution of $\sqrt{2x + 1} = x - 1$ because it does not satisfy the equation.

Check for x = 4

$$\sqrt{2x + 1} = x - 1$$

$$\sqrt{2(4) + 1} \overset{?}{=} (4) - 1$$

$$\sqrt{9} \overset{?}{=} 3$$

$$3 = 3 \qquad \text{True}$$

Therefore, 4 is a solution because it does satisfy the equation. ∎

Example 5 Solve $\sqrt{2x - 3} = \sqrt{x + 7} - 2$.

Solution

$$\sqrt{2x - 3} = \sqrt{x + 7} - 2$$

$$(\sqrt{2x - 3})^2 = (\sqrt{x + 7} - 2)^2$$

When squaring $(\sqrt{x + 7} - 2)$, do not forget this middle term

$$2x - 3 = x + 7 \quad \boxed{-4\sqrt{x + 7}} + 4$$

$$2x - 3 = x + 11 - 4\sqrt{x + 7} \qquad \text{We must repeat steps 1, 2, and 3}$$

$$x - 14 = -4\sqrt{x + 7} \qquad \text{Adding } -x - 11 \text{ to both sides}$$

$$(x - 14)^2 = (-4\sqrt{x + 7})^2$$

$$x^2 - 28x + 196 = 16(x + 7)$$

$$x^2 - 28x + 196 = 16x + 112$$

$$x^2 - 44x + 84 = 0 \qquad \text{Adding } -16x - 112 \text{ to both sides}$$

$$(x - 2)(x - 42) = 0$$

$$x - 2 = 0 \qquad\qquad x - 42 = 0$$

$$x = 2 \qquad\qquad\quad x = 42$$

Check for x = 2

$$\sqrt{2x - 3} = \sqrt{x + 7} - 2$$

$$\sqrt{2(2) - 3} \overset{?}{=} \sqrt{(2) + 7} - 2$$

$$\sqrt{4 - 3} \overset{?}{=} \sqrt{9} - 2$$

$$\sqrt{1} \overset{?}{=} 3 - 2$$

$$1 = 1 \qquad \text{True}$$

Therefore, 2 is a solution because it does satisfy the equation.
Check for x = 42

$$\sqrt{2x - 3} = \sqrt{x + 7} - 2$$

$$\sqrt{2(42) - 3} \overset{?}{=} \sqrt{(42) + 7} - 2$$

$$\sqrt{84 - 3} \overset{?}{=} \sqrt{49} - 2$$

$$\sqrt{81} \overset{?}{=} 7 - 2$$

$$9 = 5 \qquad \text{False}$$

Therefore, 42 is not a solution because it does not satisfy the equation. ∎

Word Problems Involving Radical Equations

Many applications of algebra in the sciences, business, engineering, and so forth involve radical equations.

Example 6 Find the amount of power, P, consumed if an appliance has a resistance of 16 ohms and draws 5 amps (amperes) of current, using the following formula from electricity:

$$I = \sqrt{\frac{P}{R}}$$

where I, the current, is measured in amps (amperes); P, the power, is measured in watts; and R, the resistance, is measured in ohms.

Solution We must find the power, P, when $R = 16$ and $I = 5$.

$$I = \sqrt{\frac{P}{R}}$$

$$5 = \sqrt{\frac{P}{16}}$$

$$(5)^2 = \left(\sqrt{\frac{P}{16}}\right)^2$$

$$25 = \frac{P}{16}$$

$$P = 400$$

Check

$$I = \sqrt{\frac{P}{R}}$$

$$5 \overset{?}{=} \sqrt{\frac{400}{16}}$$

$$5 \overset{?}{=} \sqrt{25}$$

$$5 = 5 \qquad \text{True} \quad ∎$$

EXERCISES 11.6

Set I   In Exercises 1–22, solve each equation.

1. $\sqrt{x} = 5$

2. $\sqrt{x} = 4$

3. $\sqrt{2x} = 4$

4. $\sqrt{3x} = 6$

5. $\sqrt{x - 3} = 2$

6. $\sqrt{x + 4} = 6$

7. $\sqrt{2x + 1} = 9$

8. $\sqrt{5x - 4} = 4$

9. $\sqrt{3x + 1} = 5$

10. $\sqrt{7x + 8} = 6$

11. $\sqrt{x + 1} = \sqrt{2x - 7}$

12. $\sqrt{3x - 2} = \sqrt{x + 4}$

13. $\sqrt{3x - 2} = x$

14. $\sqrt{5x - 6} = x$

15. $\sqrt{4x - 1} = 2x$

16. $\sqrt{6x - 1} = 3x$

17. $\sqrt{x - 3} + 5 = x$

18. $\sqrt{4x + 5} + 5 = 2x$

19. $\sqrt{2x + 2} = 1 + \sqrt{x + 2}$

20. $\sqrt{3x + 1} = 1 + \sqrt{x + 4}$

21. $\sqrt{3x - 5} = \sqrt{x + 6} - 1$

22. $\sqrt{5x - 1} = \sqrt{x + 14} - 1$

In Exercises 23–26, solve each problem for the required unknown, using the formula $I = \sqrt{\dfrac{P}{R}}$, as given in Example 6.

23. Find the amount of power, P, consumed if an appliance has a resistance of 9 ohms and draws 7 amps of current.

24. Find the amount of power, P, consumed if an appliance has a resistance of 25 ohms and draws 2 amps of current.

25. Find R, the resistance in ohms, for an electrical system that consumes 500 watts and draws 5 amps of current.

26. Find R, the resistance in ohms, for an electrical system that consumes 600 watts and draws 5 amps of current.

In Exercises 27 and 28, use this formula from statistics: $\sigma = \sqrt{npq}$, where σ (*sigma*) is the *standard deviation*, n is the number of trials, p is the probability of success, and q is the probability of failure.

27. Find the probability of success, p, if the standard deviation is 3, the probability of failure is $\frac{3}{4}$, and the number of trials is 48.

28. Find the probability of success, p, if the standard deviation is $3\frac{1}{5}$, the probability of failure is $\frac{4}{5}$, and the number of trials is 64.

In Exercises 29 and 30, use this formula from geometry: $c = \sqrt{a^2 + b^2}$, where c is the length of the longest side of a right triangle and a and b are the lengths of the other two sides.

29. Find the length of one side of a right triangle if the length of the longest side is 5 and the length of one of the other sides is 3.

30. Find the length of one side of a right triangle if the length of the longest side is 13 and the length of one of the other sides is 5.

Set II   In Exercises 1–22, solve each equation.

1. $\sqrt{x} = 8$

2. $\sqrt{x} = 11$

3. $\sqrt{5x} = 10$

4. $\sqrt{3x} = 12$

5. $\sqrt{x-6}=3$

6. $\sqrt{x-5}=2$

7. $\sqrt{6x+1}=5$

8. $\sqrt{3x-2}=5$

9. $\sqrt{9x-5}=7$

10. $\sqrt{5x+6}=6$

11. $\sqrt{9-2x}=\sqrt{5x-12}$

12. $\sqrt{6x-1}=\sqrt{4x-5}$

13. $\sqrt{3x+10}=x$

14. $x=\sqrt{8x-15}$

15. $\sqrt{3x+2}=3x$

16. $2x=\sqrt{40-12x}$

17. $\sqrt{x-6}+8=x$

18. $1+\sqrt{3x-3}=x$

19. $\sqrt{3x+1}=2+\sqrt{x+1}$

20. $\sqrt{2x+7}=1+\sqrt{x+3}$

21. $\sqrt{2x-1}=1+\sqrt{x-1}$

22. $\sqrt{2x-5}=\sqrt{x+9}-1$

In Exercises 23–26, solve each problem for the required unknown, using the formula $I=\sqrt{\dfrac{P}{R}}$, as given in Example 6.

23. Find the amount of power, P, consumed if an appliance has a resistance of 12 ohms and draws 5 amps of current.

24. Find the amount of power, P, consumed if an appliance has a resistance of 20 ohms and draws 6 amps of current.

25. Find R, the resistance in ohms, for an electrical system that consumes 700 watts and draws 5 amps of current.

26. Find R, the resistance in ohms, for an electrical system that consumes 400 watts and draws 4 amps of current.

In Exercises 27 and 28, use this formula from statistics: $\sigma=\sqrt{npq}$, where σ (*sigma*) is the *standard deviation*, n is the number of trials, p is the probability of success, and q is the probability of failure.

27. Find the probability of success, p, if the standard deviation is $2\frac{2}{3}$, the probability of failure is $\frac{2}{3}$, and the number of trials is 32.

28. Find the probability of success, p, if the standard deviation is 5, the probability of failure is $\frac{5}{6}$, and the number of trials is 180.

In Exercises 29 and 30, use this formula from geometry: $c=\sqrt{a^2+b^2}$, where c is the length of the longest side of a right triangle and a and b are the lengths of the other two sides.

29. Find the length of one side of a right triangle if the length of the longest side is 17 and the length of one of the other sides is 8.

30. Find the length of one side of a right triangle if the length of the longest side is 10 and the length of one of the other sides is 6.

11.7 Review: 11.1–11.6

Square Roots
11.1

The **square root** of a number N is a number that when squared gives N. A positive real number N has two square roots, a positive root called the **principal square root**, written $\sqrt{N}$, and a negative square root, written $-\sqrt{N}$.

Square Root Rules 11.1–11.3	It is understood that all variables in radicands represent positive numbers.

Rule 11.1 $(\sqrt{a})^2 = a$

Rule 11.2 $\sqrt{ab} = \sqrt{a}\sqrt{b}$

Rule 11.3 $\sqrt{\dfrac{a}{b}} = \dfrac{\sqrt{a}}{\sqrt{b}}$ $(b \neq 0)$

Rule 11.4 $\sqrt{a}\sqrt{a} = a$

To Simplify the Square Root of a Product 11.2

1. Express the radicand in prime factored form.

2. Find the square root of each factor as follows:
   a. If the exponent of a factor is 1, that factor must remain under the radical sign.
   b. If the exponent of a factor is an even number, remove that factor from the radical by dividing its exponent by 2.
   c. If the exponent of a factor is an odd number, write that factor as the product of two factors—one factor with an even exponent, and the other factor with an exponent of 1. Then remove from the radical the factor that has an even exponent by dividing its exponent by 2.

3. The simplified radical is the product of all the factors found in step 2.

The Simplified Form of an Expression That Has Square Roots 11.2
11.3
11.4

1. No prime factor of a radicand has an exponent greater than 1.

2. No radicand contains a fraction.

3. No denominator contains a square root.

4. No term contains more than one radical sign.

5. All like radicals are combined.

To Multiply Square Roots 11.3, 11.5

Use Rule 11.2: $\sqrt{a}\sqrt{b} = \sqrt{ab}$. Simplify the result. If one of the factors has more than one term, use the distributive rule. If both factors have two terms, use the FOIL method.

To Divide Square Roots 11.3

Use Rule 11.3: $\dfrac{\sqrt{a}}{\sqrt{b}} = \sqrt{\dfrac{a}{b}}$. You may, however, need to rationalize the denominator.

Conjugates 11.5

The **conjugate** of $a + b$ is $a - b$.

To Rationalize a Denominator 11.3

A. With one term: Multiply numerator and denominator by a number that makes the denominator a perfect square.

11.5

B. With two terms: Multiply numerator and denominator by the conjugate of the denominator.

To Combine Square Roots 11.4

A. Like radicals: Add their coefficients and multiply that sum by the like radical.

B. Unlike radicals:

1. Express each term in simplest radical form.

2. Combine any terms that have like radicals by adding their coefficients and multiplying that sum by the like radical.

To Solve a Radical Equation
11.6

1. Arrange the terms so that one term with a radical is by itself on one side of the equation.

2. Square both sides of the equation.

3. Combine like terms.

4. If a radical still remains, repeat steps 1, 2, and 3.

5. Solve the resulting equation for the variable.

6. Check the solutions in the original equation. There may be extraneous roots.

Review Exercises 11.7   Set I

1. Identify the radicand in each expression.

   a. $\sqrt{9x}$ b. $\sqrt{\dfrac{1}{2}}$ c. $\sqrt{x-5}$

In Exercises 2–26, simplify each expression.

2. $\sqrt{100}$ **3.** $(\sqrt{3})^2$ **4.** $(\sqrt{a+b})^2$

5. $\sqrt{x^3}$ **6.** $\sqrt{36a^4b^2}$ **7.** $\sqrt{a^3b^3}$

8. $\sqrt{11}\sqrt{11}$ **9.** $\sqrt{2}\sqrt{32}$ **10.** $\sqrt{50}$

11. $\sqrt{180}$ **12.** $4\sqrt{2}-\sqrt{2}$ **13.** $\sqrt{18}-\sqrt{8}$

14. $\sqrt{27}-\sqrt{12}$ **15.** $\dfrac{1}{\sqrt{5}}$ **16.** $\sqrt{\dfrac{1}{7}}$

17. $\dfrac{6}{\sqrt{3}}$ **18.** $\sqrt{\dfrac{9}{2}}$ **19.** $\sqrt{8}-\sqrt{\dfrac{1}{2}}$

20. $\sqrt{54}-\sqrt{\dfrac{2}{3}}$ **21.** $\sqrt{2}(\sqrt{8}+\sqrt{18})$ **22.** $(5+\sqrt{2})(5-\sqrt{2})$

23. $(2\sqrt{3}+1)^2$ **24.** $(3\sqrt{2}-1)^2$ **25.** $\dfrac{8}{\sqrt{3}-2}$

26. $\dfrac{6}{2+\sqrt{5}}$

In Exercises 27–34, solve each equation.

27. $\sqrt{x}=4$ **28.** $\sqrt{3a}=6$

29. $\sqrt{2x-1}=5$ **30.** $\sqrt{5a-4}=\sqrt{3a+2}$

31. $\sqrt{7x-6}=x$ **32.** $\sqrt{9x-8}=x$

33. $\sqrt{2x+7}=\sqrt{x}+2$ **34.** $\sqrt{3x+1}=\sqrt{x+8}-1$

In Exercise 35, use the formula $c=\sqrt{a^2+b^2}$ (where c is the length of the longest side of a right triangle).

35. Find the length of one side of a right triangle if the length of the longest side is 13 and the length of one of the other sides is 12.

Review Exercises 11.7   Set II

NAME _____

1. Identify the radicand in each expression.

  a. $\sqrt{4-z}$

  b. $\sqrt{\dfrac{2}{3}}$

  c. $\sqrt{5a}$

ANSWERS

1a. _____

  b. _____

  c. _____

In Exercises 2–26, simplify each expression.

2. $\sqrt{64}$

3. $\left(\sqrt{11}\right)^2$

4. $\left(\sqrt{m+n}\right)^2$

5. $\sqrt{w^7}$

6. $\sqrt{25t^4s^2}$

7. $\sqrt{u^3v^5}$

8. $\sqrt{6}\sqrt{6}$

9. $\sqrt{18}\sqrt{2}$

10. $\sqrt{68}$

11. $\sqrt{175}$

12. $5\sqrt{6}-\sqrt{6}$

13. $\sqrt{45}-\sqrt{20}$

14. $\sqrt{63}-\sqrt{28}$

15. $\dfrac{1}{\sqrt{13}}$

16. $\sqrt{\dfrac{1}{17}}$

17. $\dfrac{12}{\sqrt{6}}$

18. $\sqrt{\dfrac{16}{5}}$

19. $\sqrt{60}-\sqrt{\dfrac{3}{5}}$

2. _____

3. _____

4. _____

5. _____

6. _____

7. _____

8. _____

9. _____

10. _____

11. _____

12. _____

13. _____

14. _____

15. _____

16. _____

17. _____

18. _____

19. _____

20. $\sqrt{24} - \sqrt{\dfrac{2}{3}}$

21. $\sqrt{5}(\sqrt{5} + \sqrt{20})$

22. $(\sqrt{6} - 2)(\sqrt{6} + 2)$

23. $(3 - 2\sqrt{7})^2$

24. $(\sqrt{2} - 5)^2$

25. $\dfrac{3}{1 - \sqrt{2}}$

26. $\dfrac{12}{\sqrt{5} + 3}$

In Exercises 27–34, solve each equation.

27. $\sqrt{z} = 5$

28. $\sqrt{5h} = 10$

29. $\sqrt{4x - 3} = 5$

30. $\sqrt{7 - 3m} = \sqrt{m + 3}$

31. $x = \sqrt{x + 20}$

32. $\sqrt{4x - 7} = \sqrt{x} + 1$

33. $\sqrt{5x - 4} = 2 + \sqrt{x}$

34. $\sqrt{2x + 5} = \sqrt{x + 14} - 1$

20. _____

21. _____

22. _____

23. _____

24. _____

25. _____

26. _____

27. _____

28. _____

29. _____

30. _____

31. _____

32. _____

33. _____

34. _____

35. _____

In Exercise 35, use the formula $c = \sqrt{a^2 + b^2}$ (where c is the length of the longest side of a right triangle).

35. Find the length of one side of a right triangle if the length of the longest side is 17 and the length of one of the other sides is 15.

Chapter 11 Diagnostic Test

The purpose of this test is to see how well you understand operations with radicals. We recommend that you work this diagnostic test *before* your instructor tests you on this chapter. Allow yourself about 50 minutes.

    Complete solutions for all the problems on this test, together with section references, are given in the answer section in the back of this book. For the problems you do incorrectly, study the sections referred to.

1. Identify the radicand in each of the following.

   a. $\sqrt{3x^3y}$
                       b. $\sqrt{2x + 5}$

In Problems 2–22, simplify the expression; assume all variables in denominators are nonzero.

2. $\sqrt{16z^2}$
                      3. $\sqrt{50}$

4. $\sqrt{x^6y^5}$
                     5. $\sqrt{\dfrac{3}{14}}$

6. $\sqrt{\dfrac{48}{3n^2}}$
                   7. $\sqrt{60}$

8. $\sqrt{13}\sqrt{13}$
                   9. $\sqrt{3}\sqrt{75y^2}$

10. $\sqrt{5}(2\sqrt{5} - 3)$
        11. $(\sqrt{13} + \sqrt{3})(\sqrt{13} - \sqrt{3})$

12. $(4\sqrt{x} + 3)(3\sqrt{x} + 4)$
    13. $(8 + \sqrt{10})^2$

14. $\dfrac{\sqrt{5}}{\sqrt{80}}$
                   15. $\dfrac{\sqrt{3}}{\sqrt{6}}$

16. $\dfrac{\sqrt{18} - \sqrt{36}}{\sqrt{3}}$
       17. $\dfrac{\sqrt{x^4y}}{\sqrt{3y}}$

18. $\dfrac{6}{1 + \sqrt{5}}$
             19. $8\sqrt{t} - \sqrt{t}$

20. $7\sqrt{3} - \sqrt{27}$
         21. $\sqrt{54} + 4\sqrt{24}$

22. $\sqrt{45} + 6\sqrt{\dfrac{4}{5}}$

In Problems 23–25, solve each equation.

23. $\sqrt{8x} = 12$
             24. $\sqrt{5x + 11} = 6$

25. $\sqrt{6x - 5} = x$

Cumulative Review Exercises: Chapters 1–11

1. Simplify. Write your answer using only positive exponents.

$$\left(\frac{30x^4y^{-3}}{12y^{-1}}\right)^{-2}$$

2. Simplify the following expression. Then factor if possible.

$$(2h - 6)^2 - 2[-4h(3 - h)]$$

3. Divide and simplify.

$$\frac{x^2 + x - 2}{x^2 - 1} \div \frac{x^2 - 2x - 8}{x^2 - 4x}$$

4. Solve the inequality $3(x - 2) \le 6$ and graph its solution on the number line.

5. Solve for x: $\dfrac{2x - 3}{2} = \dfrac{5x + 4}{6} - \dfrac{5}{3}$

6. Subtract.

$$\frac{2x}{x + 2} - \frac{5}{x - 1}$$

7. Solve graphically: $\begin{Bmatrix} 4x - 3y = -9 \\ 3y + x = -6 \end{Bmatrix}$

8. Solve by addition: $\begin{Bmatrix} 3x + 2y = 8 \\ 2x - y = 17 \end{Bmatrix}$

9. Solve by substitution: $\begin{Bmatrix} 2x - 5y = 2 \\ 3x - 7y = 2 \end{Bmatrix}$

10. What is the multiplicative inverse of $-\frac{4}{7}$?

In Exercises 11–15, simplify each expression.

11. $\left(\sqrt{2x}\right)^2$

12. $\left(\sqrt{11} + \sqrt{14}\right)^2$

13. $\sqrt{5}(3 + \sqrt{80})$

14. $\sqrt{175} - \sqrt{28}$

15. $\dfrac{8}{\sqrt{6} - 2}$

16. Solve the equation $\sqrt{3x + 2} = 4$.

In Exercises 17–20, set up each problem algebraically, solve, and check. Be sure to state what your variables represent.

17. The sum of two numbers is $\frac{1}{2}$. Their difference is $7\frac{1}{2}$. Find the two numbers.

18. Nadya takes 20 min to drive to work in the morning, but she takes 36 min to return home over the same route during the evening rush hour. If her average morning speed is 16 mph faster than her average evening speed, how far is it from her apartment to her office?

19. Sylvia has thirty-two coins with a total value of $2.45. If all the coins are nickels and dimes, how many of each does she have?

20. The sum of the first two of three consecutive odd integers added to the sum of the last two is 140. Find the integers.

12 Quadratic Equations

In this chapter, we discuss several methods for solving quadratic equations. We have already solved quadratic equations by factoring in Sections 7.8, 8.9C, and 11.6.

12.1 The General Form for Quadratic Equations

A *quadratic equation* is a polynomial equation whose highest degree term is a second-degree term.

Example 1 Examples of quadratic equations:

a. $8x^2 + 2x - 4 = 0$ b. $x^2 + 1 = 0$

c. $\frac{1}{3}x - 3x^2 = \frac{4}{5}$ d. $3x^2 = 2x$ ■

The *general form* of a quadratic equation, as used in this text, is as follows:

THE GENERAL FORM OF A QUADRATIC EQUATION

$$ax^2 + bx + c = 0$$

where a, b, and c are integers and $a > 0$.

Notice that a is the coefficient of x^2, b is the coefficient of x, and c is the constant.

TO CHANGE A QUADRATIC EQUATION INTO GENERAL FORM

1. Remove fractions by multiplying each term by the LCD.

2. Remove grouping symbols.

3. Collect and combine like terms.

4. Arrange all nonzero terms in descending powers on one side, leaving only zero on the other side.

Example 2 Change each of the following quadratic equations into the general form and identify a, b, and c:

a. $8x = 3 - 2x^2$
   Solution Add $2x^2 - 3$ to both sides.

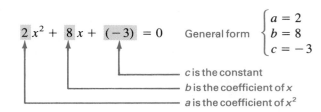

$$2\,x^2 + 8\,x + (-3) = 0 \qquad \text{General form} \quad \begin{cases} a = 2 \\ b = 8 \\ c = -3 \end{cases}$$

c is the constant
b is the coefficient of x
a is the coefficient of x^2

b. $3x^2 = 5$

Solution Add -5 to both sides.

$$3x^2 - 5 = 0 \qquad \text{General form} \qquad \begin{cases} a = 3 \\ b = 0 \\ c = -5 \end{cases}$$

The coefficient of x is an understood 0.

c. $8x = 7x^2$

Solution Add $-8x$ to both sides.

Writing the equation so a is positive

$$0 = 7x^2 - 8x \qquad \text{General form} \qquad \begin{cases} a = 7 \\ b = -8 \\ c = 0 \end{cases}$$

The constant is an understood 0.

d. $\dfrac{1}{2}x^2 - 3x = \dfrac{2}{3}$

Solution

$$6\left(\dfrac{1}{2}x^2 - 3x\right) = 6\left(\dfrac{2}{3}\right) \qquad \text{Multiply both sides by the LCD, 6}$$

$$3x^2 - 18x = 4$$

$$3x^2 - 18x - 4 = 0 \qquad \text{General form} \qquad \begin{cases} a = 3 \\ b = -18 \\ c = -4 \end{cases}$$

e. $x(x - 2) = 5$

Solution Remove parentheses first.

$$x^2 - 2x = 5$$

$$x^2 - 2x - 5 = 0 \qquad \text{General form} \qquad \begin{cases} a = 1 \\ b = -2 \\ c = -5 \end{cases}$$

f. $(x + 3)(x - 2) = 2x + 1$

Solution

$$x^2 + x - 6 = 2x + 1$$

$$x^2 - x - 7 = 0 \qquad \text{General form} \qquad \begin{cases} a = 1 \\ b = -1 \\ c = -7 \end{cases} \qquad \blacksquare$$

EXERCISES 12.1

Set I   Write each of the following quadratic equations in the general form and identify a, b, and c.

1. $2x^2 = 5x + 3$

2. $3x^2 = 4 - 2x$

3. $6x^2 = x$

4. $2x - 3x^2 = 0$

5. $\dfrac{3x}{2} + 5 = x^2$

6. $4 - x^2 = \dfrac{2x}{3}$

7. $x^2 - \dfrac{5x}{4} + \dfrac{2}{3} = 0$

8. $2x^2 + \dfrac{3x}{5} = \dfrac{1}{3}$

9. $x(x - 3) = 4$

10. $2x(x + 1) = 12$

11. $3x(x + 1) = (x + 1)(x + 2)$

12. $(x + 1)(x - 3) = 4x(x - 1)$

Set II Write each of the following quadratic equations in the general form and identify a, b, and c.

1. $5x = 4x^2 + 1$

2. $3 = 5x - x^2$

3. $5x^2 = x$

4. $(x - 2)(2x + 4) = 0$

5. $\dfrac{4x}{3} + 2 = x^2$

6. $7x^2 = 3$

7. $3x^2 + \dfrac{x}{4} = \dfrac{1}{5}$

8. $\left(\dfrac{x}{2} - 3\right)\left(x + \dfrac{1}{4}\right) = 2$

9. $(x - 1)(3x) = 7$

10. $2 = 3x^2$

11. $3x(x + 1) = (x + 2)(x - 3)$

12. $(4x - 1)(x + 2) = x(x - 4)$

12.2 Solving Quadratic Equations by Factoring

We will review the method of solving quadratic equations by factoring first used in Section 7.8.

TO SOLVE A QUADRATIC EQUATION BY FACTORING

1. Arrange the equation in general form (Section 12.1).

$$ax^2 + bx + c = 0$$

2. Factor the polynomial.

3. Set each factor equal to zero and solve for the variable.

4. Check apparent solutions in the original equation.

Example 1   Solve $x^2 - 2x - 8 = 0$.
Solution

$$x^2 - 2x - 8 = 0 \qquad \text{Equation is already in general form}$$

$$(x + 2)(x - 4) = 0$$

$$\begin{array}{c|c} x + 2 = 0 & x - 4 = 0 \\ x = -2 & x = 4 \end{array}$$

Check for $x = -2$	*Check for* $x = 4$
$x^2 - 2x - 8 = 0$	$x^2 - 2x - 8 = 0$
$(-2)^2 - 2(-2) - 8 = 0$	$(4)^2 - 2(4) - 8 = 0$
$4 + 4 - 8 = 0$ True	$16 - 8 - 8 = 0$ True

Therefore, -2 and 4 are solutions. ∎

We can also solve equations of this type by getting all terms to the *right* side, leaving only zero on the *left* side (see Example 2).

Example 2   Solve $5 - 2x^2 = 3x$.
Solution

$$5 - 2x^2 = 3x$$
$$\underline{-5 + 2x^2 \qquad\qquad -5 + 2x^2}$$
$$0 = 3x - 5 + 2x^2$$
$$0 = \boxed{2}\,x^2 + 3x - 5 \qquad \text{The coefficient of } x^2 \text{ is } positive$$
$$0 = (x - 1)(2x + 5)$$

$$\begin{array}{c|c} x - 1 = 0 & 2x + 5 = 0 \\ x = 1 & 2x = -5 \\ & x = -\dfrac{5}{2} \end{array}$$

The checking is left to the student. ∎

Example 3   Solve $x^2 = \dfrac{3 - 5x}{2}$.
Solution

$$\frac{x^2}{1} = \frac{3 - 5x}{2}$$
$$2 \cdot x^2 = 1 \cdot (3 - 5x) \qquad \text{Cross-multiplying}$$
$$2x^2 = 3 - 5x$$
$$2x^2 + 5x - 3 = 0$$
$$(2x - 1)(x + 3) = 0$$

$$\begin{array}{c|c} 2x - 1 = 0 & x + 3 = 0 \\ 2x = 1 & x = -3 \\ x = \dfrac{1}{2} & \end{array}$$

The checking is left to the student. ∎

Example 4 Solve $\dfrac{x-1}{x-3} = \dfrac{12}{x+1}$.

Solution

$$\frac{x-1}{x-3} = \frac{12}{x+1}$$

$$(x-1)(x+1) = 12(x-3) \qquad \text{Cross-multiplying}$$

$$x^2 - 1 = 12x - 36$$

$$x^2 - 12x + 35 = 0$$

$$(x-7)(x-5) = 0$$

$$x - 7 = 0 \quad \bigg| \quad x - 5 = 0$$

$$x = 7 \quad \bigg| \quad x = 5$$

The checking is left to the student. ■

Example 5 Solve $(6x+2)(x-4) = 2 - 11x$.

Solution

$$(6x+2)(x-4) = 2 - 11x$$

$$6x^2 - 22x - 8 = 2 - 11x$$

$$6x^2 - 11x - 10 = 0$$

$$(3x+2)(2x-5) = 0$$

$$3x + 2 = 0 \quad \bigg| \quad 2x - 5 = 0$$

$$3x = -2 \quad \bigg| \quad 2x = 5$$

$$x = -\frac{2}{3} \quad \bigg| \quad x = \frac{5}{2}$$

The checking is left to the student. ■

Example 6 Michelle bicycled from her home to a beach and back, a total distance of 120 mi. Her average speed returning was 3 mph slower than her average speed going to the beach. If her total bicycling time was 9 hr, what was her average speed going to the beach?

Solution

Let $\quad x = $ average speed going to the beach

$x - 3 = $ average speed returning from the beach

The distance each way is 60 mi. We solve the distance-rate-time formula $d = rt$ for t:

$$t = \frac{d}{r}$$

We then use this formula to find the time going to the beach and the time returning from the beach:

$$\frac{60}{x} = \text{time going to beach}$$

$$\frac{60}{x-3} = \text{time returning from beach}$$

The sum of the two times equals 9 hr:

$$\frac{60}{x} + \frac{60}{x-3} = 9$$

$$x(x-3)\left(\frac{60}{x} + \frac{60}{x-3}\right) = x(x-3)\,(9) \qquad \text{Multiply both sides by the LCD}$$

$$\overset{1}{\cancel{x}}(x-3)\left(\frac{60}{\cancel{x}}\right) + x(x\overset{1}{\cancel{-3}})\left(\frac{60}{\underset{1}{\cancel{x-3}}}\right) = x(x-3)(9)$$

$$60(x-3) + 60x = 9x(x-3)$$

$$60x - 180 + 60x = 9x^2 - 27x$$

$$120x - 180 = 9x^2 - 27x$$

$$0 = 9x^2 - 147x + 180$$

$$0 = 3(3x^2 - 49x + 60)$$

$$\frac{0}{3} = \frac{\overset{1}{\cancel{3}}(3x^2 - 49x + 60)}{\underset{1}{\cancel{3}}}$$

$$0 = (3x-4)(x-15)$$

$3x - 4 = 0$	$x - 15 = 0$
$3x = 4$	$x = 15$
$x = \dfrac{4}{3}$	$x = 15$ The average speed going to the beach

Check for $x = \frac{4}{3}$ If $x = \frac{4}{3}$ is the speed going to the beach, then

$$x - 3 = \frac{4}{3} - 3 = -\frac{5}{3} \qquad \text{The speed returning}$$

But this is not possible, since a speed cannot be negative.
Check for $x = 15$ If $x = 15$ is the speed going to the beach, then

$$x - 3 = 15 - 3 = 12 \qquad \text{The speed returning}$$

$$t_1 = \frac{60}{15} = 4 \qquad \text{The time going to the beach}$$

$$t_2 = \frac{60}{12} = 5 \qquad \text{The time returning}$$

The sum of the two times is 9 hr. ■

EXERCISES 12.2

Set I   In Exercises 1–22, solve each equation by factoring.

1. $x^2 + x - 6 = 0$ **2.** $x^2 - x - 6 = 0$

3. $x^2 + x = 12$ **4.** $x^2 + 2x = 15$

5. $2x^2 - x = 1$ **6.** $2x^2 + x = 1$

7. $\dfrac{x}{8} = \dfrac{2}{x}$

8. $\dfrac{x}{3} = \dfrac{12}{x}$

9. $\dfrac{x+2}{3} = \dfrac{-1}{x-2}$

10. $\dfrac{x+3}{2} = \dfrac{-4}{x-3}$

11. $x^2 + 9x + 8 = 0$

12. $x^2 + 7x + 6 = 0$

13. $2x^2 + 4x = 0$

14. $3x^2 - 9x = 0$

15. $x^2 = x + 2$

16. $x^2 = x + 6$

17. $\dfrac{x}{2} + \dfrac{2}{x} = \dfrac{5}{2}$

18. $\dfrac{x}{3} + \dfrac{2}{x} = \dfrac{7}{3}$

19. $\dfrac{x-1}{4} + \dfrac{6}{x+1} = 2$

20. $\dfrac{3x+1}{5} + \dfrac{8}{3x-1} = 3$

21. $2x^2 = \dfrac{2-x}{3}$

22. $4x^2 = \dfrac{14x-3}{2}$

In Exercises 23–28, set up each problem algebraically and solve. Be sure to state what your variables represent.

23. The length of a rectangle is 5 more than its width. If its area is 24, find its dimensions.

24. The length of a rectangle is 4 more than its width. If its area is 77, find its dimensions.

25. If the product of two consecutive even integers is increased by 4, the result is 84. Find the integers.

26. If the product of two consecutive integers is decreased by 6, the result is 36. Find the integers.

27. Bruce drove from Los Angeles to the Mexican border and back to Los Angeles, a total distance of 240 mi. His average speed returning to Los Angeles was 20 mph faster than his average speed going to Mexico. If his total driving time was 5 hr, what was his average speed driving from Los Angeles to Mexico?

28. Nick drove from his home to San Diego and back, a total distance of 360 mi. His average speed returning to his home was 15 mph faster than his average speed going to San Diego. If his total driving time was 7 hr, what was his average speed driving from his home to San Diego?

Set II   In Exercises 1–22, solve each equation by factoring.

1. $x^2 + x - 12 = 0$

2. $x^2 + 2x - 8 = 0$

3. $x^2 + 4x = 5$

4. $x^2 + 2x = 63$

5. $3x^2 - 2x = 1$

6. $2x^2 = 3 - 5x$

7. $\dfrac{x}{27} = \dfrac{3}{x}$

8. $\dfrac{x-2}{5} = \dfrac{1}{x+2}$

9. $\dfrac{x+1}{1} = \dfrac{3}{x-1}$

10. $2x^2 + 7x + 3 = 0$

11. $x^2 + 6x + 5 = 0$

12. $x^2 + x = 0$

13. $2x^2 - 6x = 0$

14. $x^2 + 2x = 15$

15. $x^2 = x + 12$

16. $x + \dfrac{1}{4} = \dfrac{3}{4x}$

17. $\dfrac{x}{2} - \dfrac{3}{x} = \dfrac{5}{4}$

18. $\dfrac{x + 3}{4} + \dfrac{3}{x + 3} = \dfrac{61}{28}$

19. $\dfrac{2x + 1}{5} + \dfrac{10}{2x + 1} = 3$

20. $\dfrac{x}{2} = \dfrac{19}{10} + \dfrac{3}{x}$

21. $3x^2 = \dfrac{11x + 10}{2}$

22. $2x^2 = \dfrac{1 - 5x}{3}$

In Exercises 23–28, set up each problem algebraically and solve. Be sure to state what your variables represent.

23. The length of a rectangle is 5 more than its width. If its area is 36, find its dimensions.

24. The length of a rectangle is 3 times its width. If the numerical sum of its area and perimeter is 80, find its dimensions.

25. If the product of two consecutive even integers is increased by 8, the result is 16. Find the integers.

26. If the product of two consecutive odd integers is decreased by 3, the result is 12. Find the integers.

27. Leon drove from his home to Lake Shasta and back, a total distance of 600 mi. His average speed returning to his home was 10 mph faster than his average speed going to Lake Shasta. If his total driving time was 11 hr, what was his average speed driving from his home to Lake Shasta?

28. Ruth drove from Creston to Des Moines, a distance of 90 mi. Then she continued on from Des Moines to Omaha, a distance of 120 mi. Her average speed was 10 mph faster on the second part of the journey than on the first part. If the total driving time was 6 hr, what was her average speed on the first leg of the journey?

12.3 **Incomplete Quadratic Equations**

An **incomplete quadratic equation** is one in which b or c (or both) is zero. If a were zero, the equation would not be a quadratic equation. Therefore, in these problems, a cannot be zero.

Example 1   Examples of incomplete quadratic equations:

a. $12x^2 + 5 = 0$ $(b = 0)$

b. $7x^2 - 2x = 0$ $(c = 0)$

c. $3x^2 = 0$ $(b = 0 \text{ and } c = 0)$ ∎

We don't need any special rules or methods for solving an incomplete quadratic equation when $c = 0$ (see Examples 2 and 3).

Example 2   Solve $12x^2 = 3x$.
Solution

$$12x^2 - 3x = 0 \qquad \text{General form}$$

$$3x(4x - 1) = 0 \qquad \text{GCF} = 3x$$

$$3x = 0 \qquad \bigg| \qquad 4x - 1 = 0$$

$$x = 0 \qquad \bigg| \qquad 4x = 1$$

$$\qquad\qquad\qquad x = \frac{1}{4}$$

The checking is left to the student.   ∎

A WORD OF CAUTION   In doing Example 2, a common mistake students make is to divide both sides of the equation by x.

$$12x^2 = 3x$$

$$12x = 3 \qquad \text{Dividing both sides by } x$$

$$x = \frac{1}{4} \qquad \text{Using this method, only the solution } x = \frac{1}{4} \text{ is found, } not\, x = 0$$

By dividing both sides of the equation by x, we lost the solution $x = 0$. Do not divide both sides of an equation by an expression containing the variable, because you may lose solutions.   ☑

Example 3   Solve $\frac{2}{5}x = 3x^2$.

Solution LCD = 5

$$\frac{5}{1} \cdot \frac{2}{5}x = \boxed{5} \cdot (3x^2)$$

$$2x = 15x^2$$

$$0 = 15x^2 - 2x \qquad \text{General form}$$

$$0 = x(15x - 2)$$

$$x = 0 \qquad \bigg| \qquad 15x - 2 = 0$$

$$\qquad\qquad 15x = 2$$

$$\qquad\qquad x = \frac{2}{15}$$

The checking is left to the student.   ∎

Some incomplete quadratic equations in which $b = 0$ can be solved by factoring. Other such equations can be solved by using the following rule:

RULE 12.1

If $x^2 = p$, then $x = \sqrt{p}$ or $x = -\sqrt{p}$.

We can combine both answers by using the $\pm$ symbol (read "plus or minus") that was introduced in Section 1.11. Using Rule 12.1, an incomplete quadratic equation in which $b = 0$ (that is, an equation of the form $ax^2 + c = 0$) can be solved by the following method:

TO SOLVE A QUADRATIC EQUATION WHEN $b = 0$

1. Arrange the equation so the second-degree term is on one side of the equal sign and the constant term is on the other side.

2. Divide both sides by a, the coefficient of x^2.

3. Take the square root of both sides, inserting $\pm$ in front of the constant.

4. Express the constant in simplest form.

5. When the radicand is *positive* or *zero*, the square roots are real numbers; when it is *negative*, the square roots are *not* real numbers.

In Example 4, we demonstrate the above method and also show that the given equation can be solved by factoring.

Example 4   Solve $x^2 - 4 = 0$.
Solution $x^2 - 4 = 0$ is in the general form, and $b = 0$.

$$x^2 = 4$$
$$\sqrt{x^2} = \pm\sqrt{4} \qquad \text{Using Rule 12.1}$$
$$x = \pm 2$$

This equation can also be solved by factoring.

$$x^2 - 4 = 0$$
$$(x + 2)(x - 2) = 0$$

$$x + 2 = 0 \qquad \bigg| \qquad x - 2 = 0$$
$$x = -2 \qquad \bigg| \qquad x = 2$$

Notice that we do obtain both 2 and -2 as solutions when we solve by factoring. The checking is left to the student. ∎

Example 5 can be solved by factoring; however, we show only the method described in the box on this page.

Example 5 Solve $4x^2 - 9 = 0$.

Solution

$$4x^2 - 9 = 0$$

$$4x^2 = 9$$

$$x^2 = \frac{9}{4}$$

$$x = \pm\sqrt{\frac{9}{4}}$$

$$x = \pm\frac{3}{2}$$

The checking is left to the student. ∎

Examples 6 and 7 *cannot* be solved by factoring over the integers.

Example 6 Solve $3x^2 - 5 = 0$.

Solution

$$3x^2 - 5 = 0$$

$$3x^2 = 5$$

$$x^2 = \frac{5}{3}$$

$$x = \pm\sqrt{\frac{5}{3}} = \pm\frac{\sqrt{5}}{\sqrt{3}} \cdot \boxed{\frac{\sqrt{3}}{\sqrt{3}}} \longleftarrow \text{To rationalize the denominator}$$

$$x = \pm\frac{\sqrt{15}}{3}$$

We can use Table I or a calculator to approximate $\sqrt{15}$ and express the answers as approximate decimals.

$$x = \pm\frac{\sqrt{15}}{3} = \pm\frac{3.873}{3} = \pm1.291 \doteq \pm1.29 \qquad \text{Rounded off to two decimal places}$$

Check for $x = +\dfrac{\sqrt{15}}{3}$	*Check for* $x = -\dfrac{\sqrt{15}}{3}$
$3x^2 - 5 = 0$	$3x^2 - 5 = 0$
$3\left(\dfrac{\sqrt{15}}{3}\right)^2 - 5 \stackrel{?}{=} 0$	$3\left(-\dfrac{\sqrt{15}}{3}\right)^2 - 5 \stackrel{?}{=} 0$
$3\left(\dfrac{15}{9}\right) - 5 \stackrel{?}{=} 0$	$3\left(\dfrac{15}{9}\right) - 5 \stackrel{?}{=} 0$
$5 - 5 = 0$	$5 - 5 = 0$ ∎

In Example 7, the solutions *do exist*. However, they are not real numbers; they are *imaginary numbers*, and imaginary numbers are not discussed in this text.

Example 7　Solve $x^2 + 25 = 0$.
Solution

$$x^2 + 25 = 0$$
$$x^2 = -25$$
$$x = \pm \boxed{\sqrt{-25}} \qquad \text{Solutions are not real numbers because radicand is negative}$$

Because the radicand is negative, our answer is "The solutions are not real numbers." ■

EXERCISES 12.3

Set I　Solve each equation, or write "The solutions are not real numbers."

1. $8x^2 = 4x$ **2.** $15x^2 = 5x$

3. $x^2 - 9 = 0$ **4.** $x^2 - 36 = 0$

5. $x^2 - 4x = 0$ **6.** $x^2 - 16x = 0$

7. $x^2 + 16 = 0$ **8.** $x^2 + 1 = 0$

9. $x^2 - 4 = 0$ **10.** $x^2 - 16 = 0$

11. $5x^2 = 4$ **12.** $3x^2 = 25$

13. $8 - 2x^2 = 0$ **14.** $27 - 3x^2 = 0$

15. $x(x + 3) = 3x - 4$ **16.** $x(x + 5) = 5x - 9$

17. $2(x + 3) = 6 + x(x + 2)$

18. $5(2x - 3) - x(2 - x) = 8(x - 1) - 7$

19. $2x(3x - 4) = 2(3 - 4x)$

20. $5x(2x - 3) = 3(4 - 5x)$

Set II　Solve each equation, or write "The solutions are not real numbers."

1. $8x^2 = 12x$ **2.** $3x = 6x^2$

3. $x^2 - 25 = 0$ **4.** $x^2 = 100$

5. $x^2 - 49x = 0$ **6.** $x^2 = 9x$

7. $x^2 + 81 = 0$ **8.** $x^2 + 4x = 0$

9. $x^2 - 121 = 0$ **10.** $5x^2 = 25$

11. $3x^2 = 7$ **12.** $x^2 - 25x = 0$

13. $64 - 4x^2 = 0$ **14.** $x(x + 7) = 7x - 1$

15. $x(x + 1) = x - 9$ **16.** $x^2 + 3(x - 1) = -3$

17. $3(x + 4) = 12 + x(x + 2)$ **18.** $x(x + 3) - 7 = 3(x - 1)$

19. $4x(5x - 2) = 2(40 - 4x)$ **20.** $5x(2x - 3) = 3(10 - 5x)$

12.4　The Quadratic Formula

The methods we have shown in previous sections can be used to solve only *some* quadratic equations. The method we show in this section can be used to solve *all* quadratic equations.

In Example 1, we use a method called *completing the square* to solve a quadratic equation.

Example 1 Solve $x^2 - 4x + 1 = 0$.
Solution

$$x^2 - 4x + 1 = 0$$

$$x^2 - 4x = -1 \qquad \text{Adding } -1 \text{ to both sides}$$

Take $\dfrac{1}{2}$ of -4: $\dfrac{1}{2}(-4) = -2$

Then $(-2)^2 = 4$

$$x^2 - 4x + 4 = -1 + 4 \qquad \text{Adding } 4 \text{ to both sides makes the left side a perfect square}$$

$$(x - 2)^2 = 3 \qquad \text{Factoring the left side}$$

$$\sqrt{(x - 2)^2} = \pm\sqrt{3} \qquad \text{Taking the square root of both sides}$$

$$x - 2 = \pm\sqrt{3} \qquad \text{Simplifying the radical}$$

$$x = 2 \pm \sqrt{3} \qquad \text{Adding 2 to both sides}$$

Check for $x = 2 + \sqrt{3}$

$$
\begin{array}{rcl}
x^2 - 4x + 1 &=& 0 \\
(2 + \sqrt{3})^2 - 4(2 + \sqrt{3}) + 1 &\overset{?}{=}& 0 \\
4 + 4\sqrt{3} + 3 - 8 - 4\sqrt{3} + 1 &\overset{?}{=}& 0 \\
0 &=& 0
\end{array}
$$

We leave the check for $x = 2 - \sqrt{3}$ to the student. ■

The method of completing the square can be used to solve *any* quadratic equation. We now use it to solve the general form of the quadratic equation, and, in this way, we derive the *quadratic formula*.

Derivation of the Quadratic Formula

$$ax^2 + bx + c = 0 \qquad \text{General form}$$

$$ax^2 + bx = -c \qquad \text{Adding } -c \text{ to both sides}$$

$$x^2 + \frac{b}{a}x = -\frac{c}{a} \qquad \text{Dividing both sides by } a$$

Take $\dfrac{1}{2}$ of $\dfrac{b}{a}$: $\dfrac{1}{2}\left(\dfrac{b}{a}\right) = \dfrac{b}{2a}$

Then $\left(\dfrac{b}{2a}\right)^2 = \dfrac{b^2}{4a^2}$

$$x^2 + \frac{b}{a}x + \frac{b^2}{4a^2} = \frac{b^2}{4a^2} - \frac{c}{a} = \frac{b^2}{4a^2} - \frac{4ac}{4a^2} \qquad \text{Adding } \dfrac{b^2}{4a^2} \text{ to both sides to make the left side a perfect square}$$

$$\left(x + \frac{b}{2a}\right)^2 = \frac{b^2 - 4ac}{4a^2} \qquad \text{Factoring the left side and adding the fractions on the right side}$$

$$x + \frac{b}{2a} = \pm\sqrt{\frac{b^2 - 4ac}{4a^2}} \qquad \text{Taking the square root of both sides}$$

$$x + \frac{b}{2a} = \pm \frac{\sqrt{b^2 - 4ac}}{\sqrt{4a^2}} = \pm \frac{\sqrt{b^2 - 4ac}}{2a} \qquad \text{Simplifying radicals}$$

$$x = -\frac{b}{2a} \pm \frac{\sqrt{b^2 - 4ac}}{2a} \qquad \text{Adding } -\frac{b}{2a} \text{ to both sides}$$

Therefore, $\qquad x = \frac{-b \pm \sqrt{b^2 - 4ac}}{2a} \qquad$ The *quadratic formula*

The procedure for using the quadratic formula can be summarized as follows:

TO SOLVE A QUADRATIC EQUATION BY FORMULA

1. Arrange the equation in general form (Section 12.1):

$$ax^2 + bx + c = 0$$

2. Substitute the values of a, b, and c into the *quadratic formula*:

$$x = \frac{-b \pm \sqrt{b^2 - 4ac}}{2a} \qquad (a \neq 0)$$

3. Simplify your answers.

4. Check your answers by substituting them into the original equation.

A WORD OF CAUTION   Be sure that your fraction bar is long enough and that your radical sign is long enough.

$$x = \frac{-b}{2a} \pm \sqrt{b^2 - 4ac} \quad \text{is incorrect,}$$

$$x = \frac{-b \pm \sqrt{b^2} - 4ac}{2a} \quad \text{is incorrect,}$$

and

$$x = \frac{-b \pm \sqrt{b^2} - 4ac}{2a} \quad \text{is incorrect.} \qquad \boxed{\checkmark}$$

Example 2   Solve $x^2 - 5x + 6 = 0$ by using the quadratic formula.

Solution

Substitute $\begin{cases} a = 1 \\ b = -5 \\ c = 6 \end{cases}$ into the formula $x = \frac{-b \pm \sqrt{b^2 - 4ac}}{2a}$.

$$x = \frac{-(-5) \pm \sqrt{(-5)^2 - 4(1)(6)}}{2(1)}$$

$$= \frac{5 \pm \sqrt{25 - 24}}{2} = \frac{5 \pm \sqrt{1}}{2}$$

$$= \frac{5 \pm 1}{2} = \begin{cases} \dfrac{5 + 1}{2} = \dfrac{6}{2} = 3 \\[2mm] \dfrac{5 - 1}{2} = \dfrac{4}{2} = 2 \end{cases}$$

We obtain the same answers if we solve the equation by factoring.

$$x^2 - 5x + 6 = 0$$

$$(x - 2)(x - 3) = 0$$

$$x - 2 = 0 \quad | \quad x - 3 = 0$$

$$x = 2 \quad | \quad x = 3$$

The checking is left to the student. ■

Solving a quadratic equation by factoring is ordinarily faster than using the formula. Therefore, first check to see if the equation can be solved by factoring. If it cannot, use the formula. The equations in Examples 3–6 cannot be solved by factoring.

Example 3   Solve $x^2 - 6x - 3 = 0$.
Solution

Substitute $\begin{cases} a = 1 \\ b = -6 \\ c = -3 \end{cases}$ into the formula $x = \dfrac{-b \pm \sqrt{b^2 - 4ac}}{2a}$.

$$x = \frac{-(-6) \pm \sqrt{(-6)^2 - 4(1)(-3)}}{2(1)}$$

$$= \frac{6 \pm \sqrt{36 + 12}}{2} = \frac{6 \pm \sqrt{48}}{2}$$

$$= \frac{6 \pm 4\sqrt{3}}{2} = \frac{\overset{1}{\cancel{2}}(3 \pm 2\sqrt{3})}{\underset{1}{\cancel{2}}} = 3 \pm 2\sqrt{3}$$

The checking is left to the student. ■

Example 4   Solve $\frac{1}{4}x^2 = 1 - x$.
Solution LCD $= 4$

$$\boxed{\frac{4}{1}} \cdot \frac{1}{4}x^2 = \boxed{4} \cdot (1 - x) \qquad \text{First, change the equation to general form}$$

$$x^2 = 4 - 4x$$

$$x^2 + 4x - 4 = 0 \qquad \text{General form}$$

Substitute $\begin{cases} a = 1 \\ b = 4 \\ c = -4 \end{cases}$ into $x = \dfrac{-b \pm \sqrt{b^2 - 4ac}}{2a}$.

$$x = \frac{-(4) \pm \sqrt{(4)^2 - 4(1)(-4)}}{2(1)}$$

$$= \frac{-4 \pm \sqrt{16 + 16}}{2} = \frac{-4 \pm \sqrt{32}}{2}$$

$$= \frac{-4 \pm 4\sqrt{2}}{2} = \frac{\overset{1}{\cancel{2}}(-2 \pm 2\sqrt{2})}{\underset{1}{\cancel{2}}} = -2 \pm 2\sqrt{2}$$

The checking is left to the student. ■

Example 5   Solve $4x^2 - 5x + 2 = 0$.
Solution

Substitute $\begin{cases} a = 4 \\ b = -5 \\ c = 2 \end{cases}$ into $x = \dfrac{-b \pm \sqrt{b^2 - 4ac}}{2a}$.

$$x = \frac{-(-5) \pm \sqrt{(-5)^2 - 4(4)(2)}}{2(4)}$$

$$= \frac{5 \pm \sqrt{25 - 32}}{8} = \frac{5 \pm \sqrt{-7}}{8} \longleftarrow \text{Solutions are not real numbers because the radicand is negative}$$

There are no real solutions for the equation. ■

In some practical applications, we need decimal approximations to the exact answers. In such cases, we can use a calculator or Table I to approximate the square roots; we can then express the answers as approximate decimals (see Example 6).

Example 6   Solve $x^2 - 5x + 3 = 0$. Round off the answers to two decimal places.
Solution

Substitute $\begin{cases} a = 1 \\ b = -5 \\ c = 3 \end{cases}$ into $x = \dfrac{-b \pm \sqrt{b^2 - 4ac}}{2a}$.

$$x = \frac{-(-5) \pm \sqrt{(-5)^2 - 4(1)(3)}}{2(1)}$$

$$= \frac{5 \pm \sqrt{25 - 12}}{2} = \frac{5 \pm \sqrt{13}}{2} \qquad (\sqrt{13} \doteq 3.606 \text{ from Table I})$$

$$\doteq \frac{5 \pm 3.606}{2} = \begin{cases} \dfrac{5 + 3.606}{2} = \dfrac{8.606}{2} = 4.303 = 4.30 \\[2mm] \dfrac{5 - 3.606}{2} = \dfrac{1.394}{2} = 0.697 = 0.70 \end{cases}$$

Check We suggest that you use a calculator to check these answers in the original equation. The answers will not check exactly, since they are approximations and not exact answers. ■

EXERCISES 12.4

Set I   In Exercises 1–16, use the quadratic formula to solve each equation. If the radicand is negative, write "The answers are not real numbers." Express all other answers in simplest radical form.

1. $3x^2 - x - 2 = 0$ **2.** $2x^2 + 3x - 2 = 0$

3. $x^2 - 4x + 1 = 0$ **4.** $x^2 - 4x - 1 = 0$

5. $\dfrac{x}{2} + \dfrac{2}{x} = \dfrac{5}{2}$ **6.** $\dfrac{x}{3} + \dfrac{2}{x} = \dfrac{7}{3}$

7. $2x^2 = 8x - 5$ **8.** $3x^2 = 6x - 2$

9. $\dfrac{1}{x} + \dfrac{x}{x - 1} = 3$ **10.** $\dfrac{2}{x} - \dfrac{x}{x + 1} = 4$

11. $x^2 + x + 1 = 0$

12. $x^2 + x + 2 = 0$

13. $4x^2 + 4x = 1$

14. $9x^2 - 6x = 2$

15. $3x^2 + 2x + 1 = 0$

16. $4x^2 + 3x + 2 = 0$

In Exercises 17 and 18, set up each problem algebraically and solve. Express all answers in simplest radical form.

17. A number less its reciprocal is $\frac{2}{3}$. What is the number?

18. If a number is subtracted from twice its reciprocal, the result is $\frac{5}{4}$. What is the number?

In Exercises 19–24, find the solutions and express them in simplest radical form. Also use a calculator or Table I to approximate the solutions; round off answers to two decimal places. (In Exercises 21–24, set up each problem algebraically.)

19. $4x^2 = 1 - 4x$

20. $9x^2 = 2 + 6x$

21. The length of a rectangle is 2 more than its width. If its area is 2, find its dimensions.

22. The length of a rectangle is 4 more than its width. If its area is 6, find its dimensions.

23. The perimeter of a square is numerically 6 less than its area. Find the length of a side.

24. The area of a square is numerically 2 more than its perimeter. Find the length of a side.

Set II   In Exercises 1–16, use the quadratic formula to solve each equation. If the radicand is negative, write "The answers are not real numbers." Express all other answers in simplest radical form.

1. $3x^2 - 8x - 3 = 0$

2. $x^2 - 3x + 3 = 0$

3. $x^2 - 4x + 2 = 0$

4. $2x^2 = 2x - 5$

5. $\dfrac{x}{3} + \dfrac{1}{x} = \dfrac{7}{6}$

6. $\dfrac{x}{2} + \dfrac{1}{x} = \dfrac{33}{8}$

7. $3x^2 + 4 = 2x$

8. $5x^2 + 3x + 1 = 0$

9. $\dfrac{2}{x} = \dfrac{3x}{x + 2} - 1$

10. $\dfrac{3}{x} = \dfrac{x}{x + 2} + 1$

11. $x^2 + 3x + 5 = 0$

12. $x^2 - 3x + 4 = 0$

13. $9x^2 + 9x = 1$

14. $5x(x + 5) = 1$

15. $3x^2 + 4 = 2x$

16. $x^2 = \dfrac{3 - 5x}{2}$

In Exercises 17 and 18, set up each problem algebraically and solve. Express all answers in simplest radical form.

17. A number less its reciprocal is $\frac{1}{4}$. What is the number?

18. A number is equal to the sum of its reciprocal and $\frac{2}{5}$. What is the number?

In Exercises 19–24, find the solutions and express them in simplest radical form. Also use a calculator or Table I to approximate the solutions; round off answers to two decimal places. (In Exercises 21–24, set up each problem algebraically.)

19. $5x^2 + 4x = 3$

20. $3x^2 = 5 + 6x$

21. The length of a rectangle is 2 more than its width. If its area is 6, find its dimensions.

22. The length of a rectangle is 5 more than its width. If its area is 3, find its dimensions.

23. The perimeter of a square is numerically 3 more than its area. Find the length of a side.

24. The area of a square is numerically 3 more than its perimeter. Find the length of a side.

12.5 The Pythagorean Theorem

Right Triangles   A triangle that has a right angle (a 90° angle) is called a **right triangle**. The side opposite the right angle of a right triangle is called the **hypotenuse**; it is always the longest side of the triangle. The other two sides are often called the **legs** of the right triangle.

The diagonal of a rectangle divides the rectangle into two right triangles. The parts of a right triangle and a rectangle and one of its diagonals are shown in Figure 12.5.1.

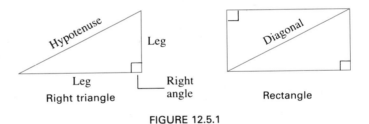

FIGURE 12.5.1

The Pythagorean theorem, which follows, applies only to right triangles.

THE PYTHAGOREAN THEOREM

The square of the hypotenuse of a right triangle is equal to the sum of the squares of the other sides.

$$c^2 = a^2 + b^2$$

The Pythagorean theorem can be used to determine whether or not a given triangle is a right triangle; to do this, we use the rule as follows: If the square of the longest side of a triangle equals the sum of the squares of the two shorter sides, the triangle is a right triangle.

Example 1   Determine whether each of the following triangles is a right triangle:

a. 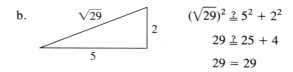 $5^2 \overset{?}{=} 3^2 + 4^2$

$25 \overset{?}{=} 9 + 16$

$25 = 25$

Therefore, the given triangle *is* a right triangle.

b. 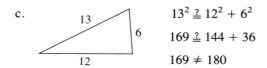 $(\sqrt{29})^2 \overset{?}{=} 5^2 + 2^2$

$29 \overset{?}{=} 25 + 4$

$29 = 29$

Therefore, the given triangle *is* a right triangle.

c. $13^2 \overset{?}{=} 12^2 + 6^2$

$169 \overset{?}{=} 144 + 36$

$169 \neq 180$

Therefore, this triangle *is not* a right triangle. ∎

The Pythagorean theorem can also be used to find one side of a right triangle when the other two sides are known.

Example 2   Find the hypotenuse of a right triangle whose legs are 8 cm and 6 cm.
Solution

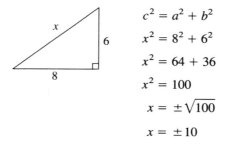

$c^2 = a^2 + b^2$

$x^2 = 8^2 + 6^2$

$x^2 = 64 + 36$

$x^2 = 100$

$x = \pm\sqrt{100}$

$x = \pm 10$

NOTE $x = 10$ and $x = -10$ are solutions of this equation. However, -10 cannot be a solution of this *geometric problem*, because we usually consider lengths in geometric figures to be positive numbers. For this reason, in the problems of this section we will take only the positive (*principal*) square root.   ☑

Therefore, the length of the hypotenuse is 10 cm. ∎

While lengths in geometric figures cannot be negative, they *can* be irrational. In fact, in a square whose sides measure 1 in., the *exact* length of the diagonal is $\sqrt{2}$ in. The *decimal approximation* of this length is 1.41 in.

Example 3   Use the Pythagorean theorem to find x in the given right triangle.
Solution

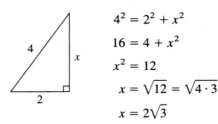

$$4^2 = 2^2 + x^2$$
$$16 = 4 + x^2$$
$$x^2 = 12$$
$$x = \sqrt{12} = \sqrt{4 \cdot 3}$$
$$x = 2\sqrt{3}$$

The *exact* length of the side is $2\sqrt{3}$; the *approximate* length is 3.46. ■

Example 4   Find the length of the diagonal of a rectangle that has a length of 6 ft and a width of 4 ft.
Solution Let $x =$ the length of the diagonal.

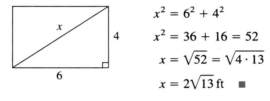

$$x^2 = 6^2 + 4^2$$
$$x^2 = 36 + 16 = 52$$
$$x = \sqrt{52} = \sqrt{4 \cdot 13}$$
$$x = 2\sqrt{13} \text{ ft} ■$$

Example 5   The length of a rectangle is 2 m more than its width. If the length of its diagonal is 10 m, find the dimensions of the rectangle.
Solution

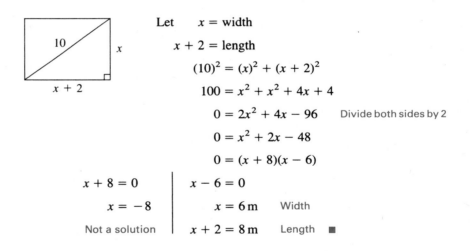

Let $x =$ width
      $x + 2 =$ length
$$(10)^2 = (x)^2 + (x + 2)^2$$
$$100 = x^2 + x^2 + 4x + 4$$
$$0 = 2x^2 + 4x - 96 \quad \text{Divide both sides by 2}$$
$$0 = x^2 + 2x - 48$$
$$0 = (x + 8)(x - 6)$$

$x + 8 = 0$	$x - 6 = 0$	
$x = -8$	$x = 6 \text{ m}$	Width
Not a solution	$x + 2 = 8 \text{ m}$	Length ■

EXERCISES 12.5

Set I   In Exercises 1–14, use the Pythagorean theorem to find x in the given right triangle.

1.

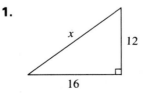

12

16

2.

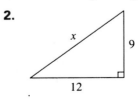

9

12

3.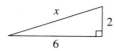

4.

5.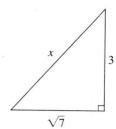

6.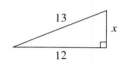

7.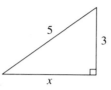

8.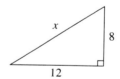

9.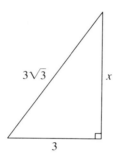

10.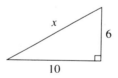

11.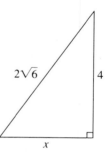

12.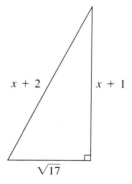

13.

14.

In Exercises 15–22, set up each problem algebraically and solve. If the answers are not rational numbers, express them in simplest radical form.

15. Find the width of a rectangle with a diagonal of 25 cm and a length of 24 cm.

16. Find the width of a rectangle with a diagonal of 41 in. and a length of 40 in.

17. One leg of a right triangle is 4 m less than twice the other leg. If its hypotenuse is 10 m, how long are the two legs?

18. The length of a rectangle is 3 yd more than its width. If the length of its diagonal is 15 yd, find the dimensions of the rectangle.

19. Find the diagonal of a square whose side is $\sqrt{6}$ in.

20. Find the diagonal of a square whose side is $\sqrt{14}$ m.

21. Find the side of a square with a diagonal of $4\sqrt{2}$ cm.

22. Find the side of a square with a diagonal of $5\sqrt{2}$ in.

In Exercises 23 and 24, set up each problem algebraically and solve. Use a calculator or Table I and round off answers to two decimal places.

23. Find the diagonal of a square whose side is 41.6 cm.

24. Find the width of a rectangle with a diagonal of 4.53 m and a length of 3.67 m.

Set II   In Exercises 1–14, use the Pythagorean theorem to find x in the given right triangle.

1.

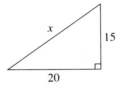

2.

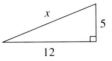

3.

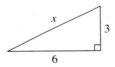

4.

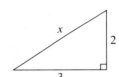

5.

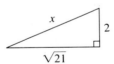

6.

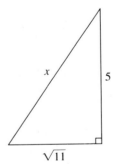

7.

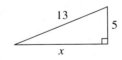

8.

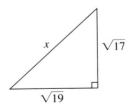

9.

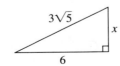

10.

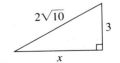

11.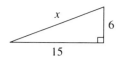

12.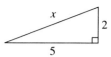

13.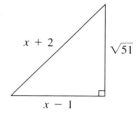

14.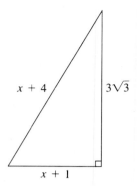

In Exercises 15–22, set up each problem algebraically and solve. If the answers are not rational numbers, express them in simplest radical form.

15. Find the width of a rectangle with a diagonal of 17 cm and a length of 15 cm.

16. Find the width of a rectangle with a diagonal of $10\sqrt{3}$ ft and a length of 15 ft.

17. One leg of a right triangle is 2 cm less than twice the other leg. If the hypotenuse is 5 cm, how long are the two legs?

18. The width of a rectangle is 4 cm less than its length. If the length of the diagonal is $4\sqrt{5}$ cm, find the dimensions of the rectangle.

19. Find the diagonal of a square whose side is $\sqrt{3}$ cm.

20. Find the diagonal of a square whose side is $\sqrt{15}$ ft.

21. Find the side of a square with a diagonal of $7\sqrt{2}$ cm.

22. Find the side of a square with a diagonal of 24 in.

In Exercises 23 and 24, set up each problem algebraically and solve. Use a calculator or Table I and round off answers to two decimal places.

23. Find the diagonal of a square whose side is 7.13 in.

24. Find the width of a rectangle with a diagonal of 7.81 cm and a length of 2.68 cm.

12.6 Review: 12.1–12.5

**Quadratic Equations
12.1**

A **quadratic equation** is a polynomial equation whose highest-degree term is a second-degree term.

The **general form** of a quadratic equation as used in this text is

$$ax^2 + bx + c = 0$$

where a, b, and c are integers, and $a > 0$.

Methods of Solving Quadratic Equations

12.2 Factoring

1. Arrange in general form.

2. Factor the polynomial.

3. Set each factor equal to zero and solve for the variable.

4. Check answers in the original equation.

12.4 Formula

1. Arrange in general form.

2. Substitute the values of a, b, and c into the quadratic formula

$$x = \frac{-b \pm \sqrt{b^2 - 4ac}}{2a}$$

3. Simplify the answers.

12.3 Incomplete quadratic

a. *When $c = 0$:* Find the greatest common factor (GCF), then solve by factoring.

b. *When $b = 0$:* 1. Write the equation as $ax^2 = -c$.

    2. Divide both sides by a.

    3. Take the square root of both sides, inserting $\pm$ in front of the constant, and simplify.

The Pythagorean Theorem

12.5 The square of the hypotenuse of a right triangle is equal to the sum of the squares of the other sides.

$$c^2 = a^2 + b^2$$

Review Exercises 12.6   Set I

In Exercises 1–14, solve each equation by any convenient method, or write "The solutions are not real numbers."

1. $x^2 + x = 6$

2. $x^2 - 49x = 0$

3. $x^2 - 2x - 4 = 0$

4. $x^2 - 4x + 1 = 0$

5. $x^2 = 5x$

6. $\dfrac{3x}{5} = \dfrac{5}{12x}$

7. $\dfrac{x + 2}{3} = \dfrac{1}{x - 2} + \dfrac{2}{3}$

8. $x^2 + 144 = 0$

9. $5(x + 2) = x(x + 5)$

10. $3(x + 4) = x(x + 3)$

11. $\dfrac{2}{x} + \dfrac{x}{x + 1} = 5$

12. $\dfrac{3}{x} - \dfrac{x}{x + 2} = 2$

13. $3x^2 + 2x + 1 = 0$

14. $(2x - 1)(3x + 5) = x(x + 7) + 4$

15. Use the Pythagorean theorem to solve for x.

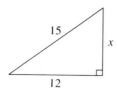

In Exercises 16–18, set up each problem algebraically and solve. Express each answer in simplest radical form.

16. One leg of a right triangle is 7 in. more than the other. The hypotenuse is 13 in. Find the length of the longer leg.

17. Find the diagonal of a square whose side is $\sqrt{10}$ m.

18. Find the diagonal of a square whose side is 8 yd.

In Exercises 19 and 20, set up each problem algebraically and solve. Express each answer in simplest radical form, and use a calculator to approximate each answer; round off answers to two decimal places.

19. A rectangle is 8 in. long and 4 in. wide. Find the length of its diagonal.

20. The width of a rectangle is 6 ft less than its length. Its area is 6 sq. ft. Find its dimensions.

Review Exercises 12.6   Set II

NAME _____

In Exercises 1–14, solve each equation by any convenient method, or write "The solutions are not real numbers."

1. $x^2 + x = 12$

2. $x^2 - 36 = 0$

3. $x^2 - 2x - 2 = 0$

4. $x^2 = 11x$

5. $x^2 + 1 = 0$

6. $x^2 + x + 1 = 0$

7. $\dfrac{2x}{3} = \dfrac{3}{8x}$

8. $\dfrac{x+2}{4} = \dfrac{1}{x-2} + 1$

9. $2(x + 1) = x(x - 4)$

10. $\dfrac{3}{x} + \dfrac{x}{x-1} = 2$

11. $3x^2 - x = 0$

12. $4x^2 = 2x - 1$

13. $x^2 + 2x + 5 = 0$

14. $x^2 + 2x - 5 = 0$

ANSWERS

1. _____
2. _____
3. _____
4. _____
5. _____
6. _____
7. _____
8. _____
9. _____
10. _____
11. _____
12. _____
13. _____
14. _____
15. _____

15. Use the Pythagorean theorem to solve for x.

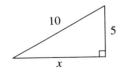

In Exercises 16–18, set up each problem algebraically and solve. Express each answer in simplest radical form.

16. The length of the diagonal of a square is $\sqrt{32}$ yd. What is the length of a side?

17. Find the width of a rectangle with a diagonal of 13 in. and a length of 12 in.

18. A rectangle is 9 ft long and 3 ft wide. Find the length of its diagonal.

16. _____

17. _____

18. _____

19. _____

20. _____

_____

In Exercises 19 and 20, set up each problem algebraically and solve. Express each answer in simplest radical form, and use a calculator to approximate each answer; round off answers to two decimal places.

19. Find the diagonal of a rectangle with a width of 8 cm and a length of 10 cm.

20. One leg of a right triangle is 2 m longer than the other. The hypotenuse is $2\sqrt{2}$ m. Find the length of each leg.

Chapter 12 Diagnostic Test

The purpose of this test is to see how well you understand quadratic equations. We recommend that you work this diagnostic test *before* your instructor tests you on this chapter. Allow yourself about 50 minutes.

Complete solutions for all the problems on this test, together with section references, are given in the answer section in the back of this book. For the problems you do incorrectly, study the sections referred to.

In Problems 1 and 2, write each quadratic equation in the general form, and identify a, b, and c in each equation.

1. $5 + 3x^2 = 7x$

2. $3(x + 5) = 3(4 - x^2)$

In Problems 3–10, solve each equation, or write "The solutions are not real numbers." Use any convenient method. If necessary, use the quadratic formula, which is as follows:

If $ax^2 + bx + c = 0$, then $x = \dfrac{-b \pm \sqrt{b^2 - 4ac}}{2a}$.

3. $x^2 = 8x + 15$

4. $3x^2 + 2x = 8$

5. $5x^2 = 17x$

6. $2x^2 - 9x - 5 = 0$

7. $x^2 + 2x + 6 = 0$

8. $6x^2 = 54$

9. $3(x + 1) = x^2$

10. $x(x + 3) = x$

11. Use the Pythagorean theorem to find x in the right triangle.

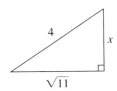

In Problems 12 and 13, set up each problem algebraically and solve. If answers are not rational numbers, express them in simplest radical form.

12. Find the diagonal of a square whose side is 7 m.

13. The length of a rectangle is 6 ft more than its width. Its area is 78 sq. ft. Find its length and its width.

Cumulative Review Exercises: Chapters 1–12

1. Is division associative?

2. Is subtraction commutative?

3. What is the additive identity?

4. Simplify. Write your answer using only positive exponents.

$$\left(\frac{18s^{-1}t^{-3}}{12s}\right)^{-2}$$

In Exercises 5–14, perform the indicated operation, or write "Not possible." Express all answers in simplest form.

5. $73 \div 0$

6. $(2\sqrt{3} - 3)(4\sqrt{3} + 5)$

7. $(4x^2 + 3x - 2)(5x - 3)$

8. $(\sqrt{21} - \sqrt{5})^2$

9. $\dfrac{x^2 + 3x}{2x^2 + 7x + 5} \div \dfrac{x^2 - 9}{x^2 - 2x - 3}$

10. $(x^2 - 14x - 5) \div (x - 3)$ (Use long division)

11. $(x^2 + 2x - 8) \div \dfrac{x^2 - 4}{x^2 + 2x}$

12. $\dfrac{x + 3}{x - 3} - \dfrac{x^2 + 9}{x^2 - 9}$

13. $\dfrac{x^2 + 3x - 2}{x^2 + 5x + 6} - \dfrac{x - 1}{2x^2 + 7x + 3}$

14. $5\sqrt{18} - \sqrt{\dfrac{9}{2}}$

In Exercises 15–25, solve for all variables, or write "Identity," or write "No solution."

15. $7 - (2x + 3) = 3x - 2 - (4x - 3)$

16. $\begin{cases} 2x + 3y = 11 \\ 4x + 9y = 4 \end{cases}$

17. $\begin{cases} 3x + 4y = 3 \\ 2x - 5y = 25 \end{cases}$

18. $\sqrt{6x + 13} = 3$

19. $\sqrt{5x + 3} = 10x$

20. $\dfrac{3x - 5}{4} + 2 = \dfrac{3 + 4x}{3}$

21. $12x - 3 \le 7x - 13$

22. $x^2 + 5x = 14$

23. $3 + 4x^2 - 12x = 0$

24. $x(3x - 1) = 2x + 6$

25. $7x^2 - 1 = 0$

26. Rationalize the denominator and simplify: $\dfrac{14}{3 - \sqrt{2}}$

In 27–30, factor completely, or write "Not factorable."

27. $3x^2 - 12x - 15$

28. $5x^2 - 45$

29. $6x^2 - 12x + 5$

30. $7x^2 - 13x - 2$

In Exercises 31–34, set up each problem algebraically, solve, and check. Be sure to state what your variables represent.

31. On a scale drawing, $\frac{1}{4}$ in. represents 1 ft. What will be the size of the drawing of a room that is 12 ft by 18 ft?

32. Find two consecutive integers whose product is 11 more than their sum.

33. A 5 lb mixture of beef back ribs and beef stew meat costs $7.35. If the stew meat is $2.10 per pound and the back ribs are $1.20 per pound, how many pounds of each kind are there?

34. One leg of a right triangle is 7 cm longer than the other leg. The length of the hypotenuse is 13 cm. Find the lengths of the two legs.

Critical Thinking

Each of the following problems has an error. Can you find it?

1. Simplify $\sqrt{18} + 3\sqrt{32}$.

$$\sqrt{18} + 3\sqrt{32} = \sqrt{9 \cdot 2} + 3\sqrt{16 \cdot 2}$$
$$= 3\sqrt{2} + 7\sqrt{2}$$
$$= 10\sqrt{2}$$

2. Simplify $\dfrac{b}{a} - \dfrac{a-b}{a+b}$.

$$\frac{b}{a} - \frac{a-b}{a+b} = \frac{b}{a} \cdot \frac{\boxed{a+b}}{\boxed{a+b}} - \frac{a-b}{a+b} \cdot \frac{\boxed{a}}{\boxed{a}}$$

$$= \frac{ab + b^2}{a(a+b)} - \frac{a^2 - ab}{a(a+b)}$$

$$= \frac{ab + b^2 - a^2 - ab}{a(a+b)}$$

$$= \frac{b^2 - a^2}{a(a+b)}$$

$$= \frac{\overset{1}{\cancel{(b+a)}}(b-a)}{a\underset{1}{\cancel{(a+b)}}}$$

$$= \frac{b-a}{a}$$

3. Solve $\sqrt{3x-2} = \sqrt{x} + 2$.

$$\sqrt{3x-2} = \sqrt{x} + 2$$
$$(\sqrt{3x-2})^2 = (\sqrt{x}+2)^2$$
$$3x - 2 = x + 4$$
$$2x - 2 = 4$$
$$2x = 6$$
$$x = 3$$

Problem 4 has two errors. Can you find both?

4. Solve $2x^2 - 4x - 1 = 0$.
   Using the quadratic formula,

$$\left.\begin{array}{l} a = 2 \\ b = -4 \\ c = -1 \end{array}\right\} \quad x = \frac{-(-4) \pm \sqrt{(-4)^2 - 4(2)(-1)}}{2(2)}$$

$$= \frac{4 \pm \sqrt{16 - 8}}{4}$$

$$= \frac{4 \pm \sqrt{8}}{4} = \frac{\overset{1}{\cancel{4}} \pm 2\sqrt{2}}{\underset{1}{\cancel{4}}}$$

$$= \pm 2\sqrt{2}$$

APPENDIX A
Sets

Ideas in all branches of mathematics, such as arithmetic, algebra, geometry, calculus, and statistics, can be explained in terms of sets. For this reason, you will find a basic understanding of sets helpful. We have already used sets in the text and in writing answers and solutions for the exercises.

The sets of numbers mentioned in this appendix are defined in Chapter 1, and a few of the definitions that follow also appear in Chapter 1.

A.1 Basic Definitions

SET

A set is a collection of objects or things.

Example 1   Examples of sets:

a. The first five letters of our alphabet

b. The set of whole numbers between 3 and 8

c. The set of students in your math class ■

ELEMENT OF A SET

The objects or things that make up a set are called its elements (or members).

Roster Method of Representing a Set   A **roster** is a list of the members of a group. The method of representing a set by listing its elements is called the **roster method**.

Whenever the roster method is used, it is customary to put the elements inside a pair of braces { }. We never use parentheses for sets. It is customary, though not necessary, to arrange numbers and letters in sets in numerical and alphabetical order (to make reading easier) and to represent elements of a set with lowercase letters.

Example 2   Examples showing the elements of sets and the roster method:

a. Set $\{a, b, c\}$ has elements a, b, and c.

b. Set $\{1, 5, 12, 23\}$ has elements 1, 5, 12, and 23.

c. Set $\{1, 2, 3, \ldots\}$ has elements 1, 2, 3, and so on. The three dots to the right of the 3 indicate that the numbers go on forever, following the pattern that has been established. ■

Naming a Set   A set is usually named by a capital letter such as A, N, W, and so on. The expression "$P = \{1, 5, 7\}$" is read "P is the set whose elements are 1, 5, and 7."

The Symbol $\in$ If we wish to show that a number or object is a member of a given set, we use the symbol $\in$, which is read "is an element of." Thus, the expression $2 \in A$ is read "2 is an element of set A" (or "2 is a member of set A"). If $A = \{2, 3, 4\}$, we can say $2 \in A$, $3 \in A$, and $4 \in A$. If we wish to show that a number or object is *not* a member of a given set, we use the symbol $\notin$, which is read "is not an element of" or "is not a member of." If $A = \{2, 3, 4\}$, then $5 \notin A$ (read "5 is not an element of set A"). To help you remember this symbol, notice that $\in$ looks like the first letter of the word *element*.

Example 3   Examples showing the use of $\in$ and $\notin$:

a. If $B = \{5, 9\}$, then $5 \in B$, $9 \in B$, $3 \notin B$, and $1 \notin B$.

b. If $C = \{b, e, g, m\}$, then $g \in C$, $e \in C$, and $k \notin C$.

c. If $D = \{\text{Abe, Helen, John}\}$, then Helen $\in D$ but Mary $\notin D$.   ■

Cardinal Number of a Set   The **cardinal number** of a set is the number of elements in that set. The symbol $n(A)$ is read "the cardinal number of set A" and means the number of elements in set A. When a set is represented by the roster method, its cardinal number is found by counting its elements.

Example 4   Examples of the cardinal number of a set:

a. If $A = \{5, 8, 6, 9\}$, then $n(A) = 4$.

b. If $H = \{\text{Ed, Mabel}\}$, then $n(H) = 2$.

c. If $Q = \{a, h, l, s, t, v\}$, then $n(Q) = 6$.

d. If $E = \{2, 4, 3, 2\}$, then $n(E) = 3$. The cardinal number is 3 instead of 4 because the set has only 3 *different* elements.   ■

Equal Sets   Two sets are equal if they both have exactly the same members.

Example 5   Examples of equal sets and unequal sets:

a. $\{1, 5, 7\} = \{5, 1, 7\}$. Notice that both sets have exactly the same elements, even though they are not listed in the same order.

b. $\{1, 5, 5, 5\} = \{5, 1\}$. Notice that both sets have exactly the same elements. It is not necessary to write the same element more than once when writing the roster of a set.

c. $\{7, 8, 11\} \neq \{7, 11\}$. These sets are *not* equal because they do not both have exactly the same elements.   ■

The Empty Set (or Null Set)   If we think of the braces $\{\ \}$ as a basket, then the set $A = \{1, 5, 7\}$ has three elements in its basket: 1, 5, and 7. Set $B = \{1, 5\}$ has two elements in its basket, and set $C = \{5\}$ has only one element. Set $D = \{\ \}$ is empty, having no elements in its basket. A set with no elements is called the **empty set** (or **null set**). We use the symbols $\{\ \}$ or $\varnothing$ to represent the empty set. Whenever either symbol appears, read it as "the empty set."

Example 6   Examples of the empty set:

a. The set of all people in this classroom who are 10 ft tall $= \varnothing$.

b. The set of all digits greater than $10 = \{\ \}$.   ■

A WORD OF CAUTION $\{0\}$ is *not* the correct symbol for the empty set, and $\{\varnothing\}$ is *not* the correct symbol for the empty set. Use either the symbol $\{\ \}$ or the symbol $\varnothing$. Do not combine the two symbols.   ☑

Universal Set   A **universal set** is a set containing all the elements being considered in a particular problem. We use the symbol U to represent the universal set under consideration.

Example 7   Examples of universal sets:

a. Suppose we are going to consider only digits. Then $U = \{0, 1, 2, 3, 4, 5, 6, 7, 8, 9\}$, since U contains *all* digits.

b. Suppose we are going to consider whole numbers. Then $U = \{0, 1, 2, 3, \ldots\}$, since U contains *all* whole numbers.

c. Suppose we are going to consider sets of football players at East Los Angeles College in 1978. Then U is the set of football players at East Los Angeles College in 1978.

d. Suppose we are going to consider different sets of cats. Then U is the set of *all* cats. ∎

Notice that there can be different universal sets. We might want to consider the set of all students in this class as the universal set if we were going to deal only with class members. On the other hand, we might use the set of all fifty of the United States as our universal set if we were going to consider only sets of states.

Finite Set   If, in counting the elements of a set, the counting comes to an end, the set is called a **finite set**. This means that the number of elements in a finite set must be a particular whole number (which we call its cardinal number).

Example 8   Examples of finite sets:

a. $A = \{5, 9, 10, 13\}$; $n(A) = 4$.

b. $D = $ the set of digits; $n(D) = 10$.

c. $C = $ the set of whole numbers less than 100; $n(C) = 100$.

d. $\varnothing = \{\ \ \}$; $n(\varnothing) = 0$. ∎

Infinite Set   If, in counting the elements of a set, the counting never comes to an end, the set is called an **infinite set**. A set is infinite if it is not finite. The cardinal number of an infinite set is not a whole number.

Example 9   Examples of infinite sets:

a. $N = \{1, 2, 3, \ldots\}$

b. $W = \{0, 1, 2, 3, \ldots\}$

c. The set of all fractions ∎

EXERCISES A.1

1. Can the collection $\{\square, X, =, 5\}$ be called a set?

2. Are $\{2, 7\}$ and $\{7, 2\}$ equal sets?

3. Are $\{2, 2, 7, 7\}$ and $\{7, 2\}$ equal sets?

4. Write the set of digits < 3 (see Section 1.1 for the meaning of $<$ and $>$).

5. Write the set of digits > 9.

6. Write the set of whole numbers < 3.

7. Write the set of whole numbers > 9.

8. Write the set of whole numbers > 4 and < 6.

9. Write the set of whole numbers > 4 and < 5.

10. Write the set of whole numbers < 4 and < 5.

11. Write all the elements of the set $\{2, a, 3\}$.

12. Write all the elements in the set of digits.

13. Write the cardinal number of each of the following sets.

   a. $\{1, 1, 3, 5, 5, 5\}$

   b. $\{0\}$

   c. $\{a, b, g, x\}$

   d. The set of whole numbers < 8

   e. $\emptyset$

14. The empty set has no elements. Therefore, the statement $\emptyset = \{\ \ \}$ is true. The statement $\emptyset = \{0\}$ is not true. Explain.

15. State which of the following sets are finite and which are infinite.

   a. The set of digits

   b. The set of whole numbers

   c. The set of days in the week

   d. The set of books in the East Los Angeles College library

16. State which of the following statements are true and which are false.

   a. If $A = \{5, 11, 19\}$, then $11 \in A$.

   b. If $B = \{x, y, z, w\}$, then $a \notin B$.

   c. If $C = \{$Ann, Bill, Charles$\}$, then Dan $\in C$.

   d. $0 \in \emptyset$.

17. Given the universal set $U = \{7, 12, 15, 20, 23\}$, which of the following statements are true and which are false?

   a. $12 \in U$ b. $8 \in U$

   c. $15 \notin U$ d. $11 \notin U$

18. Given the universal set $U = \{5, 7, 11, 14, 20\}$, which of the following statements are true and which are false?

   a. $12 \in U$ b. $5 \in U$

   c. $15 \notin U$ d. $11 \notin U$

A.2 Subsets

Subset   A set A is called a **subset** of set B if every member of A is also a member of B. "A is a subset of B" is written $A \subseteq B$.

Example 1   Examples of subsets:

a. $A = \{3, 5\}$ is a subset of $B = \{3, 5, 7\}$ because every member of A is also a member of B. Therefore, $A \subseteq B$.

b. $P = \{a, c, g, f\}$ is a subset of $Q = \{d, f, a, g, h, c\}$ because every member of P is also a member of Q. Therefore, $P \subseteq Q$.

c. $X = \{$Joe, Betty$\}$ is a subset of $Y = \{$Mary, Betty, Jack, Joe$\}$ because every member of X is also a member of Y. Therefore, $X \subseteq Y$.

d. $D = \{4, 7\}$ is *not* a subset of $E = \{7, 8, 5\}$ because $4 \in D$, but $4 \notin E$. Therefore, $D \nsubseteq E$, read "D is not a subset of E."

e. $K = \{d, f, m\}$ is *not* a subset of $L = \{f, m\}$ because $d \in K$, but $d \notin L$. Therefore, $K \nsubseteq L$, read "K is not a subset of L." ■

Proper Subset   Subset A is called a **proper subset** of B if there is at least one member of B that is not a member of subset A. "A is a proper subset of B" is written $A \subset B$. The empty set is a proper subset of every set except itself.

Example 2   Examples of proper subsets:

a. Subset $A = \{3, 5\}$ is a proper subset of $B = \{3, 5, 7\}$ because $7 \in B$, but $7 \notin A$. Therefore, $A \subset B$.

b. Subset $X = \{$Joe, Betty$\}$ is a proper subset of $Y = \{$Mary, Betty, Jack, Joe$\}$ because Mary $\in Y$, but Mary $\notin X$. Therefore, $X \subset Y$. ■

Improper Subset   Subset A is called an **improper subset** of B if there is no member of B that is not also a member of A.

Example 3   Examples of improper subsets:

a. Subset $A = \{10, 12\}$ is an improper subset of $B = \{12, 10\}$, because no member of B is not also a member of A. Note that $A = B$. This means that A is an improper subset of A. In other words, *any set is an improper subset of itself.*

b. D is an improper subset of D. ■

By considering set $A = \{3, 5, 7\}$, we find that all its subsets are $\varnothing$, $\{3\}$, $\{5\}$, $\{7\}$, $\{3, 5\}$, $\{3, 7\}$, $\{5, 7\}$, and $\{3, 5, 7\}$. Of these eight subsets, only $A = \{3, 5, 7\}$ is an improper subset of A; the seven others are proper subsets of A.

Example 4   Examples of finding all the subsets of a set:

a. All the subsets of $\{6, 8\}$ are $\varnothing$, $\{6\}$, $\{8\}$, and $\{6, 8\}$. Of these four subsets, only $\{6, 8\}$ is an improper subset.

b. All the subsets of $\{a\}$ are $\varnothing$ and $\{a\}$. Of these two subsets, only $\{a\}$ is an improper subset. ■

EXERCISES A.2

1. $M = \{1, 2, 3, 4, 5\}$. State whether each of the following sets is a proper subset, an improper subset, or not a subset of M.

a. $A = \{3, 5\}$ b. $B = \{0, 1, 7\}$

c. $\varnothing$ d. $C = \{2, 4, 1, 3, 5\}$

2. $P = \{x, z, w, r, t, y\}$. State whether each of the following sets is a proper subset, an improper subset, or not a subset of P.

a. $D = \{x, y, r\}$ b. P

c. $\varnothing$ d. $E = \{x, y, z, s, t\}$

3. Write all the subsets of $\{R, G, Y\}$.

4. Write all the subsets of $\{\square, \triangle\}$.

A.3 Union and Intersection of Sets

Venn Diagrams A useful tool for helping us understand set concepts is the Venn diagram. A simple Venn diagram is shown in Figure A.3.1.

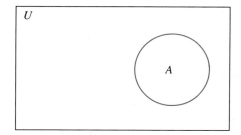

FIGURE A.3.1   Venn Diagram

It is customary to represent the universal set U by a rectangle and the other sets by circles enclosed within the U-rectangle. Just as we enclose all members of a particular set within braces in conventional set notation, in Venn diagrams we think of all members of a set A as being enclosed in a circle marked by the A. Using the same reasoning, we see that all elements in the universe under consideration are enclosed in the U-rectangle.

Union of Sets The **union** of sets A and B, written $A \cup B$, is the set that contains all the elements of A as well as all the elements of B.

In Figure A.3.2, $A \cup B$ is represented by the shaded area. In terms of set notation, suppose $A = \{b, c, g\}$ and $B = \{1, 2, 5, 7\}$. Then $A \cup B = \{b, c, g, 1, 2, 5, 7\}$.

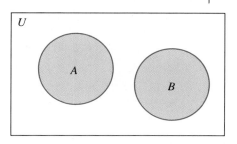

FIGURE A.3.2 $A \cup B$

Example 1 Examples of the union of sets:

a. If $C = \{x, y\}$ and $D = \{4\}$, then $C \cup D = \{x, y, 4\}$.

b. If $E = \{\text{Mary, Helen}\}$ and $G = \{\text{high, low}\}$, then $E \cup G = \{\text{Mary, Helen, high, low}\}$.

c. If $H = \{4, 8, 26\}$ and $K = \{15, 17\}$, then $H \cup K = \{4, 8, 26, 15, 17\}$.

d. If $S = \{2, 5, 9\}$ and $T = \{9, 2, 7\}$, then $S \cup T = \{2, 5, 7, 9\}$.

e. If $P = \{a, 4, 3, c\}$ and $Q = \{3, k, a\}$, then $P \cup Q = \{a, 4, 3, c, k\}$.

f. If $A = \{2, 3, 4\}$ and $B = \{\ \}$, then $A \cup B = \{2, 3, 4\} = A$. ∎

Intersection of Sets The **intersection** of sets C and D, written $C \cap D$, is the set that contains only elements in both C and D. Consider $C = \{g, f, m\}$ and $D = \{g, m, t, z\}$. Then $C \cap D = \{g, m\}$, because g and m are the only elements in both C and D. In the Venn diagram (Figure A.3.3), the shaded area represents $C \cap D$ because that area lies in both circles.

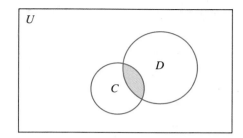

FIGURE A.3.3 $C \cap D$

To be certain that you can distinguish between the union and intersection of sets, two Venn diagrams (Figures A.3.4 and A.3.5) are shown for the same sets, P and Q.

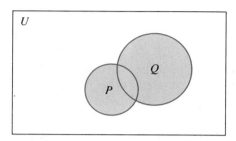

FIGURE A.3.4 Shaded Area is $P \cup Q$ **FIGURE A.3.5** Shaded Area is $P \cap Q$

Example 2   Examples of the intersection and the union of sets:

a. If $A = \{$John, Henry$\}$ and $B = \{$Bill, Henry, Tom$\}$, then $A \cap B = \{$Henry$\}$ and $A \cup B = \{$John, Henry, Bill, Tom$\}$.

b. If $G = \{1, 5, 7\}$ and $H = \{2, 5, 6, 7, 8\}$, then $G \cap H = \{5, 7\}$ and $G \cup H = \{1, 5, 7, 2, 6, 8\}$.

c. If $K = \{a, 6, b\}$ and $M = \{c, 5, 4\}$, then $K \cap M = \{\ \ \}$ and $K \cup M = \{a, 6, b, c, 5, 4\}$. ∎

Disjoint Sets   **Disjoint sets** are sets whose intersection is the empty set. A Venn diagram showing disjoint sets is Figure A.3.2, in which sets A and B are disjoint.

Example 3   Examples of disjoint sets and of sets that are not disjoint:

a. If $A = \{1, 2\}$ and $B = \{3, 4\}$, then $A \cap B = \varnothing$. Therefore, A and B are disjoint sets.

b. If $P = \{a, b, c\}$ and $Q = \{2, 8\}$, then $P \cap Q = \varnothing$. Therefore, P and Q are disjoint sets.

c. If $R = \{5, 7, 9\}$ and $T = \{9, 10, 12\}$, then $R \cap T = \{9\} \neq \varnothing$. Therefore, R and T are *not* disjoint sets. ∎

EXERCISES A.3   **1.** Write the union and intersection of each pair of given sets.

     a. $\{1, 5, 7\}, \{2, 4\}$

     b. $\{a, b\}, \{x, y, z, a\}$

     c. $\{\ \ \}, \{k, 2\}$

     d. $\{$river, boat$\}, \{$boat, streams, down$\}$

2. Given that $A = \{1, 3, 5, 7\}$, $B = \{2, 4, 6\}$, $C = \{1, 2, 3, 4\}$, and $D = \{5, 6, 7\}$, find the following.

   a. $A \cup B$ b. $C \cup D$

   c. $A \cap B$ d. $B \cap D$

3. Given that $A = \{$Bob, John$\}$, $B = \{$Charles, Tom, Bob$\}$, $C = \{$Tom, John, Dick$\}$, and $D = \{$Ray, Bob$\}$, find any two sets that are disjoint.

4. Given the Venn diagram below, write each of the following sets in roster notation. (The small letters within a particular circle in the diagram are elements of that set.)

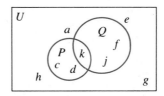

   a. P b. Q c. $P \cup Q$

   d. $P \cap Q$ e. U

5. Given: $X = \{2, 5, 6, 11\}$ and $Y = \{7, 5, 11, 13\}$.

   a. Find $X \cap Y$.

   b. Find $Y \cap X$.

   c. Is $X \cap Y = Y \cap X$?

6. Given: $K = \{a, 4, 7, b\}$, $L = \{m, 4, 6, b\}$, and $M = \{n, 4, 7, t\}$.

   a. Write $K \cap L$ in roster notation.

   b. Find $n(K \cap L)$.

   c. Write $L \cup M$ in roster notation.

   d. Find $n(L \cup M)$.

7. Shade in $A \cup B$ in the Venn diagram given here.

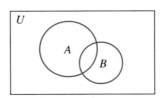

8. Shade in $P \cap Q$.

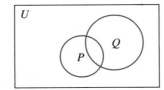

9. Write the name of the set representing the shaded area.

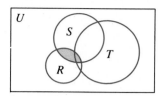

10. Write the name of the set representing the shaded area.

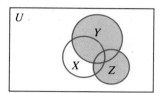

APPENDIX B
Brief Review of Arithmetic

This brief review is given as an aid for students who need to refresh their knowledge of arithmetic. For a more complete treatment, see Johnston, Willis, and Lazaris, *Essential Arithmetic, 5th ed.* (Belmont, Calif.: Wadsworth Publishing Company, 1988).

Whole Numbers

All whole numbers can be considered as decimals.

$$15 = 15.$$

Whole number ⎡⎤   Decimal point
                Decimal

Therefore, all operations with whole numbers can be done in the same way as operations with decimals.

Decimals

Decimal Places The number of decimal places in a number is the number of digits written to the right of the decimal point.

$$75.14 \text{ (two decimal places)} \quad 1.086 \text{ (three decimal places)}$$

Approximately Equal The symbol $\doteq$ (read "approximately equal to") is used to show that two numbers are approximately equal to each other.

$78 \doteq 80$ Read "78 is approximately equal to 80"

$\pi \doteq 3.14$ Read "π is approximately equal to 3.14"

$0.99 \doteq 1$ Read "0.99 is approximately equal to 1"

Rounding Off

$876. \doteq 880$ Rounded to tens

$44.62 \doteq 44.6$ Rounded to one decimal place

$0.16504 \doteq 0.17$ Rounded to two decimal places

Addition and Subtraction To add or subtract decimals, write the numbers with their decimal points in the same vertical line.

$$\text{Add: } 17.6 + 2.65 + 439 + 0.015$$

$$
\begin{array}{r}
17.6 \\
2.65 \\
439. \\
0.015 \\
\hline
459.265
\end{array}
$$

$$\text{Subtract: } 26.54 \text{ from } 518.3$$

$$
\begin{array}{r}
518.30 \\
26.54 \\
\hline
491.76
\end{array}
$$

Multiplication The number of decimal places in a product is the sum of the number of decimal places in the numbers being multiplied.

$$
\begin{array}{r}
56.75 \quad \text{(two decimal places)} \\
\times\ 3.8 \quad \text{(one decimal place)} \\
\hline
45\ 400 \\
170\ 25 \\
\hline
215.650 \quad \text{(three decimal places)}
\end{array}
$$

Division

$$
\begin{array}{r}
2\ 4.88 \quad \doteq 24.9 \quad \text{Rounded to tenths} \\
3.8_\wedge\ )\overline{94.5_\wedge 60} \longleftarrow \text{Decimal points of divisor and dividend were} \\
76 \qquad\qquad \text{both moved } one \text{ place to the right} \\
\hline
18\ 5 \\
15\ 2 \\
\hline
3\ 36 \\
3\ 04 \\
\hline
320 \\
304 \\
\hline
16
\end{array}
$$

Multiplying a Decimal by a Power of 10 Move the decimal point to the *right* as many places as the number of zeros in the power of ten.

$$23.4 \times 100 = 2{,}340\underset{2}{\longrightarrow} \qquad 1.56 \times 10^3 = 1{,}560\underset{3}{\longrightarrow}$$

Dividing a Decimal by a Power of 10 Move the decimal point to the *left* as many places as the number of zeros in the power of ten.

$$\frac{72.8}{10} = 7.\underset{1}{2}8 \qquad \frac{875}{10^2} = 8.\underset{2}{75}$$

Order of Operations

A. If there are any parentheses in the expression, that part of the expression within a pair of parentheses is evaluated first.

B. The evaluation then proceeds in this order:
   1. Powers and roots are done first.
   2. Multiplication and division are done in order from left to right.
   3. Addition and subtraction are done in order from left to right.

a. $12 \div 3 \times 4$

   $4 \quad \times 4 = 16$

b. $4 \times 3^2 \div 6 - 2$

   $4 \times 9 \div 6 - 2$

   $36 \quad \div 6 - 2$

   $6 \quad - 2 = 4$

c. $5(8 - 2) + 6(4)$

   $5(6) \quad + 6(4)$

   $30 \quad + 24 = 54$

d. $5\sqrt{16} \div 4 - 2$

   $5(4) \div 4 - 2$

   $20 \div 4 - 2$

   $5 \quad - 2 = 3$

Fractions

Raising a Fraction to Higher Terms   Multiply both numerator and denominator by the same number (not zero).

$$\frac{2}{3} = \frac{2 \cdot 2}{3 \cdot 2} = \frac{4}{6}$$

$\frac{2}{3}$ and $\frac{4}{6}$ are called **equivalent fractions**.

Reducing a Fraction to Lower Terms   Divide both numerator and denominator by a number that is a divisor of both.

$$\frac{\overset{3}{\cancel{15}}}{\underset{15}{\cancel{75}}} = \frac{\overset{1}{\cancel{3}}}{\underset{5}{\cancel{15}}} = \frac{1}{5} \quad \begin{cases} \dfrac{15}{75} \quad \text{and} \quad \dfrac{1}{5} \quad \text{are } \textit{equivalent fractions}. \\[2ex] \dfrac{1}{5} \quad \text{is} \quad \dfrac{15}{75} \quad \text{reduced to } \textit{lowest terms}. \end{cases}$$

Changing an Improper Fraction into a Mixed Number   Divide the numerator by the denominator and express the remainder as a fraction.

$$\frac{17}{3} = 3\overline{)17}^{\;5\,R2} = 5\frac{2}{3}$$

Changing a Mixed Number into an Improper Fraction

$$\frac{\text{whole number} \times \text{denominator} + \text{numerator}}{\text{denominator}}$$

$$2\frac{3}{5} = \frac{2 \times 5 + 3}{5} = \frac{10 + 3}{5} = \frac{13}{5}$$

Adding

$$\begin{array}{ll}
\dfrac{3}{8} \;=\; \dfrac{9}{24} & 8 = 2^3 \\[2ex]
& 6 = 2 \cdot 3 \\[1ex]
+\dfrac{5}{6} \;=\; \dfrac{20}{24} & \text{LCD} = 2^3 \cdot 3 \\[1ex]
\overline{\phantom{+\dfrac{5}{6}}} \quad \overline{\phantom{\dfrac{20}{24}}} & \qquad = 8 \cdot 3 \\[1ex]
\dfrac{29}{24} = 1\dfrac{5}{24} & \qquad = 24
\end{array}$$

Subtracting

$$\begin{array}{lll}
13\dfrac{1}{2} \;=\; & 13\dfrac{3}{6} = 12 + 1 + \dfrac{3}{6} = 12 + \dfrac{6}{6} + \dfrac{3}{6} = & 12\dfrac{9}{6} \\[2ex]
-\,9\dfrac{2}{3} \;=\; & -\,9\dfrac{4}{6} & = \;-\,9\dfrac{4}{6} \\[2ex]
& & \overline{\phantom{-\,9\dfrac{4}{6}}} \\[1ex]
& & \quad 3\dfrac{5}{6}
\end{array}$$

Multiplying

$$\frac{\overset{2}{\cancel{6}}}{\underset{1}{\cancel{7}}} \times \frac{\overset{2}{\cancel{14}}}{\underset{5}{\cancel{15}}} = \frac{4}{5}$$

Dividing

$$\frac{5}{8} \div 3\frac{3}{4} = \frac{5}{8} \div \frac{15}{4} = \frac{\overset{1}{\cancel{5}}}{\underset{2}{\cancel{8}}} \times \frac{\overset{1}{\cancel{4}}}{\underset{3}{\cancel{15}}} = \frac{1}{6}$$

Simplifying a Complex Fraction   Divide the numerator of the complex fraction by the denominator.

$$\frac{\dfrac{5}{12}}{\dfrac{15}{16}} = \frac{5}{12} \div \frac{15}{16} = \frac{\overset{1}{\cancel{5}}}{\underset{3}{\cancel{12}}} \times \frac{\overset{4}{\cancel{16}}}{\underset{3}{\cancel{15}}} = \frac{4}{9}$$

$$\frac{3 - \dfrac{1}{5}}{\dfrac{1}{3} + \dfrac{1}{5}} = \frac{\dfrac{15}{5} - \dfrac{1}{5}}{\dfrac{5}{15} + \dfrac{3}{15}} = \frac{\dfrac{14}{5}}{\dfrac{8}{15}} = \frac{14}{5} \div \frac{8}{15} = \frac{\overset{7}{\cancel{14}}}{\underset{1}{\cancel{5}}} \times \frac{\overset{3}{\cancel{15}}}{\underset{4}{\cancel{8}}} = \frac{21}{4} = 5\frac{1}{4}$$

Changing a Fraction to a Decimal   Divide the numerator by the denominator.

$$\frac{5}{12} = \begin{array}{r} 0.416 \doteq 0.42 \qquad \text{Rounded to two decimal places} \\ 12\overline{)5.000} \\ \underline{4\ 8} \\ 20 \\ \underline{12} \\ 80 \\ \underline{72} \\ 8 \end{array}$$

Changing a Decimal to a Fraction

$$0.48 = 48 \text{ hundredths} = \frac{48}{100} = \frac{12}{25}$$

Percents

The Meaning of a Percent   *Percent* means hundredths. For example, 5% of a quantity is $\frac{5}{100}$, or 0.05 of that quantity.

Changing a Decimal to a Percent   Move the decimal point two places to the *right* and attach the percent symbol to the right of the number.

$$0.13 = 13\% \qquad 0.003 = 0.3\% \qquad 2.5 = 250\%$$

Changing a Percent to a Decimal   Move the decimal point two places to the *left* and drop the percent symbol.

$$83\% = 0.83 \qquad 3\% = 0.03 \qquad 115\% = 1.15$$

Changing a Fraction to a Percent First change the fraction to a decimal (rounding off, if necessary); then move the decimal point two places to the right and attach the percent symbol to the right of the number.

$$\frac{1}{4} = 0.25 = 25\%$$

$$\frac{5}{12} \doteq 0.4167 \doteq 41.67\%$$

The decimal was rounded to four decimal places. The percent was rounded to two decimal places.

Changing a Percent to a Fraction Write the number with a denominator of 100 and drop the percent symbol. Reduce the resulting fraction if possible.

$$32\% = \frac{32}{100} = \frac{8}{25} \qquad 105\% = \frac{105}{100} = \frac{21}{20} \qquad 0.07\% = \frac{0.07}{100} = \frac{7}{10,000}$$

EXERCISES B.1 In Exercises 1–4, reduce each fraction to lowest terms.

1. $\dfrac{6}{9}$ **2.** $\dfrac{12}{20}$ **3.** $\dfrac{49}{24}$ **4.** $\dfrac{210}{240}$

In Exercises 5–8, change each improper fraction into a mixed number and reduce if possible.

5. $\dfrac{5}{2}$ **6.** $\dfrac{11}{8}$ **7.** $\dfrac{47}{25}$ **8.** $\dfrac{78}{33}$

In Exercises 9–12, change each mixed number into an improper fraction.

9. $3\dfrac{7}{8}$ **10.** $2\dfrac{7}{9}$ **11.** $2\dfrac{5}{6}$ **12.** $8\dfrac{3}{5}$

In Exercises 13–38, perform the indicated operation.

13. $\dfrac{2}{3} + \dfrac{1}{4}$ **14.** $\dfrac{2}{7} + \dfrac{3}{14}$ **15.** $\dfrac{7}{10} - \dfrac{2}{5}$

16. $\dfrac{3}{4} - \dfrac{2}{5}$ **17.** $\dfrac{3}{16} \times \dfrac{20}{9}$ **18.** $\dfrac{6}{35} \times \dfrac{15}{8}$

19. $\dfrac{4}{3} \div \dfrac{8}{9}$ **20.** $\dfrac{5}{2} \div \dfrac{5}{8}$ **21.** $2\dfrac{2}{3} + 1\dfrac{3}{5}$

22. $3\dfrac{1}{4} + 4\dfrac{1}{2}$ **23.** $5\dfrac{4}{5} - 3\dfrac{7}{10}$ **24.** $4\dfrac{5}{6} - 2\dfrac{2}{3}$

25. $4\dfrac{1}{5} \times 2\dfrac{1}{7}$ **26.** $2\dfrac{1}{3} \times 4\dfrac{1}{2}$ **27.** $2\dfrac{2}{5} \div 1\dfrac{1}{15}$

28. $2\dfrac{1}{4} \div 3\dfrac{3}{8}$ **29.** $\dfrac{\frac{5}{8}}{\frac{5}{6}}$ **30.** $\dfrac{\frac{3}{5}}{\frac{6}{5}}$

31. Add $34.5 + 1.74 + 18 + 0.016$.

32. Add $74.1 + 19 + 1.55 + 0.095$.

33. Subtract 34.67 from 356.4.

34. Subtract 81.94 from 418.5.

35. 100×7.45

36. $1,000 \times 3.54$

37. $\dfrac{46.8}{100}$

38. $\dfrac{89.5}{1,000}$

In Exercises 39–42, perform the indicated operations. Then round off your answer to the indicated place.

39. 70.9×94.78 (one decimal place)

40. 40.8×68.59 (one decimal place)

41. $6.007 \div 7.25$ (two decimal places)

42. $8.009 \div 4.67$ (two decimal places)

43. Change $5\frac{3}{4}$ to a decimal.

44. Change $4\frac{3}{5}$ to a decimal.

45. Change 0.65 to a fraction.

46. Change 0.25 to a fraction.

47. Change 5.9 to a mixed number.

48. Change 4.3 to a mixed number.

In Exercises 49–52, change each number to a percent.

49. 4.7
50. 0.001
51. 0.258
52. 12

In Exercises 53–56, change each percent to a decimal.

53. 3.5%
54. 0.02%
55. 157%
56. 17.8%

In Exercises 57–60, change each fraction to a percent. In Exercises 59 and 60, round off to two decimal places.

57. $\dfrac{1}{8}$
58. $\dfrac{3}{5}$
59. $\dfrac{7}{12}$
60. $\dfrac{1}{7}$

In Exercises 61–64, change each percent to a fraction reduced to lowest terms.

61. 35%
62. 80%
63. 12%
64. 18%

APPENDIX C
Functions

C.1 Definitions

We live in a world of functions. If we are paid at a fixed hourly rate, our weekly salary is a function of the number of hours we work each week. If the cost of sending a letter first-class is 25¢, then the cost of mailing several letters is a function of the number of letters being mailed.

Function   A **function** is a set of ordered pairs in which no two of the ordered pairs have the *same first coordinate* and *different second coordinates*. A function can also be thought of as a rule that assigns one and only one value to y for each value of x. Because the value of y depends on the value of x, we often call y the **dependent variable** and x the **independent variable**.

Domain   The **domain** of a function is the set of all the values that x can have so that y is a real number; it is the set of all the *first coordinates*, or x-values, of the ordered pairs.

Range   The **range** of a function is the set of all the values that y can have; it is the set of all the *second coordinates*, or y-values, of the ordered pairs. It is often more difficult to find the range than the domain of a function.

Example 1   Examples of functions and their domains:

a. $\{(3, 7), (-2, 7), (0, -5)\}$

(Recall from Section 1.1 that we enclose the elements of a set within braces.) This is a set of three ordered pairs; they all have different first coordinates. Therefore, it is a function. The domain of $\{(\boxed{3}, 7), (\boxed{-2}, 7), (\boxed{0}, -5)$ is the set $\{3, -2, 0\}$.

b. $y = x + 1$

For any value of x we choose, there will be one and only one corresponding value for y. Therefore, this is a function. Because we can choose *any* value for x, the domain of this function is the set of all real numbers.

c. $y = \dfrac{3}{x - 1}$

Excluded values were introduced in Section 8.1. For the fraction $\dfrac{3}{x - 1}$, 1 is an excluded value. Therefore, the domain is the set of all real numbers except 1. For any value of x from the domain we choose, there will be one and only one corresponding value for y. Therefore, it is a function.

d. $y = \sqrt{x + 4}$

We learned in Section 1.10A that square roots of negative numbers are not real numbers. Therefore, if y is to be a real number, the radicand $x + 4$ must be greater than or equal to zero. We must then solve the inequality $x + 4 \geq 0$.

$$x + 4 \geq 0$$
$$x \geq -4 \qquad \text{Adding } -4 \text{ to both sides}$$

Therefore, the domain is the set of all real numbers greater than or equal to -4. If we let x equal any number greater than or equal to -4, we get one and only one corresponding value for y. Therefore, this is a function.   ∎

The rules for finding the domain of a function when an *equation* for the function has been given are as follows:

The domain of the function will be the set of all real numbers unless:

1. The domain is restricted by some statement accompanying the equation.

2. There are variables in a denominator or variables with negative exponents.

3. Variables occur under a radical sign when the index of the radical is an even number.

The Graph of a Function   The graph of a function is simply the graph of all the ordered pairs in the function.

Example 2   Find the domain of each of the following functions and graph each function:

a. $\{(2, 4), (-1, 3), (0, 3), (3, 1)\}$

The domain of $\{(\boxed{2}, 4), (\boxed{-1}, 3), (\boxed{0}, 3), (\boxed{3}, 1)\}$ is $\{2, -1, 0, 3\}$.

The graph is as follows:

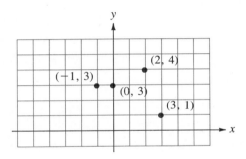

b. $y = x + 1, x \leq 2$

In this example, there *are* restrictions on the domain, namely, $x \leq 2$. Therefore, the domain is the set of all real numbers less than or equal to 2.

For the graph, we make a table of values, taking care to choose x-values only from the domain.

x	y
-2	-1
0	1
2	3

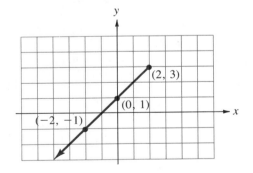

We then plot those points and connect them with a straight line that does not extend to the right of the point where $x = 2$. ■

NOTE   It is possible to graph *any* set of ordered pairs, whether it is a function or not.

The following statement is true because of the way in which a function is defined:

> No vertical line can meet the graph of a function in more than one point.

Example 3   Graph these sets of ordered pairs; they are *not* functions:

a. {(3 , −2), (4, 1), (3 , 3)}

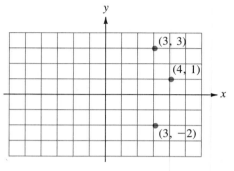

The set is *not* a function because the ordered pairs (3 , −2) and (3 , 3) have two different second elements but the same first element (3). The vertical line $x = 3$ would meet the graph in more than one point.

b. $x = 1$

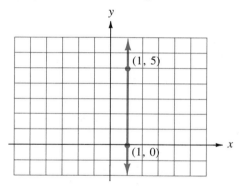

$x = 1$ is *not* a function because there are ordered pairs that satisfy the equation $x = 1$ that have the same first element but different second elements. (1, 5) and (1, 0) are two such pairs. ▪

Example 4   Determine from the graph whether each of the following is a function:

a. This *is* the graph of a function, because no vertical line can meet the graph in more than one point.

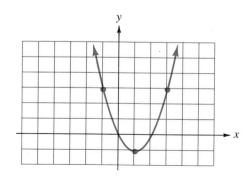

b. This is *not* the graph of a function, because a vertical line through (2, 5) would also pass through (2, 2).

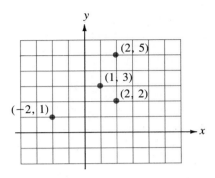

c. This *is* the graph of a function because no vertical line can meet the graph in more than one point.

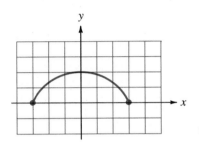

EXERCISES C.1   In Exercises 1–6, determine whether the set of ordered pairs is a function, and find the domain of each of the functions.

1. {(4, −1), (2, 5), (3, 0), (0, 4)}

2. {(6, 3), (−2, 5), (4, −1), (0, 0)}

3. {(2, −5), (3, 4), (2, 0), (7, −4), (−1, 3)}

4. {(3, 7), (−1, 4), (5, −2), (3, 0), (−8, 2)}

5. {(−8, −2), (3, −4), (6, −2), (9, −4)}

6. {(−3, −7), (−1, −7), (0, −2), (3, 0)}

In Exercises 7–14, find the domain of each function.

7. $y = \dfrac{7}{x - 12}$

8. $y = \dfrac{5}{x - 2}$

9. $y = 3x + 4$

10. $y = 2x + 5$

11. $y = \sqrt{x - 5}$

12. $y = \sqrt{x - 10}$

13. $y = x + 3, x \geq 0$

14. $y = x + 7, x \geq -3$

15. Which of the following are graphs of functions?

a.

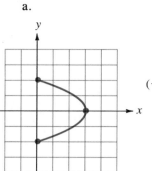

b.

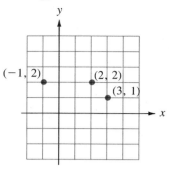

c.

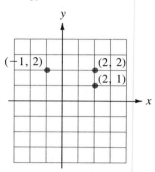

16. Which of the following are graphs of functions?

a.

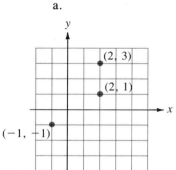

b.

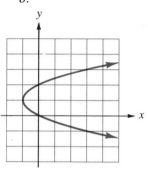

c.

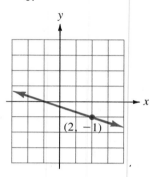

C.2 Functional Notation

Functional Notation   When we have a rule that assigns one and only one value to y for each value of x, we say that y is a function of x. This can be written $y = f(x)$, which is read "y equals f of x."

A WORD OF CAUTION $f(x)$ does *not* mean f times x. It is simply a different way of writing y when y is a function of x. ☑

The equations $y = x + 1$ and $f(x) = x + 1$ mean exactly the same thing, and we often combine the equations in the following manner:

$$y = f(x) = x + 1$$

Evaluating a Function   To *evaluate* a function means to determine what value $f(x)$ or y has for a particular x-value. The notation $f(a)$ (read "f of a") means that the function $f(x)$ is to be evaluated at $x = a$.

Example 1 If $f(x) = 3x + 2$, find $f(4)$.
Solution   Since $f(x) = 3x + 2$,

$$f(4) = 3(4) + 2 \qquad \text{Substituting 4 for } x$$
$$f(4) = 12 + 2$$
$$f(4) = 14$$

The statement $f(4) = 14$ means that $y = 14$ when $x = 4$. ■

Example 2 If $f(x) = \dfrac{3x^2 + 1}{x - 2}$, find $f(1), f(-3)$, and $f(0)$.
Solution

$$f(1) = \frac{3(1)^2 + 1}{1 - 2} = \frac{3 + 1}{-1} = \frac{4}{-1} = -4$$

$$f(-3) = \frac{3(-3)^2 + 1}{-3 - 2} = \frac{3(9) + 1}{-5} = \frac{27 + 1}{-5} = -\frac{28}{5}$$

$$f(0) = \frac{3(0)^2 + 1}{0 - 2} = \frac{0 + 1}{-2} = \frac{1}{-2} = -\frac{1}{2} \quad ■$$

Letters other than f can be used to name functions.

Variables other than x can be used for the independent variable.

Variables other than y can be used for the dependent variable.

Example 3 Examples in which letters other than f, x, and y are used to represent functions and variables:

a. $A = F(r) = \pi r^2$ A is a function of r.

b. $s = g(t) = 4t - 1$ s is a function of t.

c. $V = h(s) = s^3$ V is a function of s. ■

EXERCISES C.2 Evaluate the functions.

1. If $f(x) = (x + 2)^2$, find the following.

   a. $f(0)$ b. $f(2)$ c. $f(-3)$ d. $f(1)$

2. If $f(x) = (2x + 3)^2$, find the following.

   a. $f(0)$ b. $f(1)$ c. $f(-2)$ d. $f(2)$

3. If $g(x) = x^2 + 4x + 4$, find the following.

   a. $g(0)$ b. $g(2)$ c. $g(-3)$ d. $g(1)$

4. If $h(x) = 4x^2 + 12x + 9$, find the following.

   a. $h(0)$ b. $h(1)$ c. $h(-2)$ d. $h(2)$

5. If $F(x) = x^2 + 4$, find the following.

   a. $F(0)$ b. $F(2)$ c. $F(-3)$ d. $F(1)$

6. If $G(x) = 4x^2 + 9$, find the following.

 a. $G(0)$ b. $G(1)$ c. $G(-2)$ d. $G(2)$

7. If $f(x) = 3x^2 + x - 1$, find the following.

 a. $f(0)$ b. $f(-2)$ c. $f(5)$ d. $f(1)$

8. If $f(x) = 2x^2 + 3x - 5$, find the following.

 a. $f(0)$ b. $f(-3)$ c. $f(4)$ d. $f(1)$

9. If $f(x) = x^3 - 1$, find the following.

 a. $f(0)$ b. $f(-1)$ c. $f(2)$ d. $f(1)$

10. If $f(x) = x^3 + 1$, find the following.

 a. $f(0)$ b. $f(-2)$ c. $f(1)$ d. $f(-1)$

11. If $g(t) = \dfrac{4t^2 + 5t - 1}{t + 2}$, find the following.

 a. $g(0)$ b. $g(-1)$ c. $g(2)$ d. $g(4)$

12. If $h(t) = \dfrac{-2t^2 - 3t + 1}{t + 3}$, find the following.

 a. $h(0)$ b. $h(-2)$ c. $h(1)$ d. $h(-1)$

Answers

To Set I Exercises, Diagnostic Tests, and
Cumulative Review Exercises

Exercises 1.1 (page 6)

1. 5   **2.** 5   **3.** 1   **4.** 0   **5.** 9   **6.** 99   **7.** 10
8. 100   **9.** $>$ **10.** $<$ **11.** $>$ **12.** $<$ **13.** Yes
14. Yes   **15.** Yes   **16.** Yes   **17.** No   **18.** No   **19.** No
20. Yes   **21.** Three   **22.** Four

Exercises 1.2 (page 10)

1. Negative seventy-five (or minus seventy-five)
2. Negative forty-nine (or minus forty-nine)   **3.** -54 **4.** -109
5. -2, because -2 is to the right of -4 on the number line   **6.** 0
7. -5, because -5 is to the right of -10 on the number line
8. -15 **9.** -1 **10.** Yes   **11.** Yes   **12.** Yes   **13.** Yes
14. There is none   **15.** 1, 2, 3, 4   **16.** 7, 8, 9   **17.** 0, 2, 4
18. 1, 3, 5, 7
19. $>$, because 0 is to the right of -3 on the number line   **20.** $>$
21. $<$, because -5 is to the left of 2 on the number line   **22.** $<$
23. $>$, because -2 is to the right of -10 on the number line   **24.** $<$
25. -62 **26.** $-45°$ F

Exercises 1.3 (page 16)

1. Since they have the same sign, add their absolute values: $4 + 5 = 9$. Sum has sign of both numbers: $+$. $(4) + (5) = 9$.
2. 8
3. Since they have the same sign, add their absolute values: $3 + 4 = 7$. Sum has sign of both numbers: $-$. $(-3) + (-4) = -7$.
4. -8
5. Since they have different signs, subtract their absolute values: $6 - 5 = 1$. Sum has sign of number with larger absolute value: $-$. Therefore, $(-6) + (5) = -1$.
6. -5
7. Since their signs are different, subtract their absolute values: $7 - 3 = 4$. Sum has sign of number with larger absolute value: $+$. Therefore, $(7) + (-3) = 4$.
8. 5
9. Since they have the same sign, add their absolute values: $8 + 4 = 12$. Sum has sign of both numbers: $-$. $(-8) + (-4) = -12$.
10. -11
11. Since they have different signs, subtract their absolute values: $9 - 3 = 6$. Sum has sign of number with larger absolute value: $-$. Therefore, $(3) + (-9) = -6$.
12. -4 **13.** $-5 + 0 = -5$, because 0 is the additive identity
14. -17
15. Since they have different signs, subtract their absolute values: $9 - 7 = 2$. Sum has sign of number with larger absolute value: $+$. Therefore, $(-7) + (9) = 2$.
16. 3
17. Since they have the same sign, add their absolute values: $2 + 11 = 13$. Sum has sign of both numbers: $-$. $(-2) + (-11) = -13$.
18. -9
19. Since they have different signs, subtract their absolute values: $15 - 5 = 10$. Sum has sign of number with larger absolute value: $-$. Therefore, $(5) + (-15) = -10$.
20. -8

21. Since they have different signs, subtract their absolute values: $9 - 8 = 1$. Sum has sign of number with larger absolute value: $+$. Therefore, $(-8) + (9) = 1$.
22. 6
23. Since they have different signs, subtract their absolute values: $4 - 4 = 0$. Zero is neither positive nor negative. The sum $(-4) + (4) = 0$.
24. 0
25. Since they have the same sign, add their absolute values: $27 + 13 = 40$. Sum has sign of both numbers: $-$. $(-27) + (-13) = -40$.
26. -54
27. Since their signs are different, subtract their absolute values: $121 - 80 = 41$. Sum has sign of number with larger absolute value: $+$. Therefore, $(-80) + (121) = 41$.
28. 65
29. Since their signs are different, subtract their absolute values: $105 - 73 = 32$. Sum has sign of number with larger absolute value: $+$. Therefore, $(105) + (-73) = 32$.
30. 105
31. Since their signs are the same, add their absolute values. Sum has sign of both numbers: $-$

$$\left(-1\frac{1}{2}\right) = \left(-1\frac{5}{10}\right)$$
$$+ \left(-3\frac{2}{5}\right) = \left(-3\frac{4}{10}\right)$$
$$\overline{}$$
$$-4\frac{9}{10}$$

32. $-7\frac{3}{4}$
33. Since their signs are different, subtract their absolute values. Sum has sign of number with larger absolute value: $+$

$$\left(4\frac{5}{6}\right) = \left(4\frac{5}{6}\right)$$
$$+ \left(-1\frac{1}{3}\right) = \left(-1\frac{2}{6}\right)$$
$$\overline{}$$
$$3\frac{3}{6} = 3\frac{1}{2}$$

34. $4\frac{5}{8}$
35. Since the signs are different, subtract their absolute values: $6.075 - 3.146 = 2.929$. Sum has sign of number with larger absolute value: $+$. Therefore, the sum is 2.929.
36. 88.373
37. $|-73| = 73$. The absolute value of a number is never negative.
38. 55   **39.** $|48| = 48$. Therefore, $-|48| = -48$.   **40.** -26
41. $|-6| = 6$. Therefore, $|-6| + (-2) = 6 + (-2)$. Since their signs are different, subtract their absolute values: $6 - 2 = 4$. Sum has sign of number with larger absolute value: $+$. Therefore, $|-6| + (-2) = 6 + (-2) = 4$.
42. 3
43. $|-17| = 17$. Therefore, $-8 + |-17| = -8 + 17$. Since their signs are different, subtract their absolute values: $17 - 8 = 9$. Sum has sign of number with larger absolute value: $+$. Therefore, $-8 + |-17| = -8 + 17 = 9$.
44. 21

45. $|23| = 23$. Therefore, $-9 + |23| = -9 + 23$. Since their signs are different, subtract their absolute values: $23 - 9 = 14$. Sum has sign of number with larger absolute value: $+$. Therefore, $-9 + |23| = -9 + 23 = 14$.

46. 31

47. We must add $53°$ to $-35°$. Since they have different signs, subtract their absolute values: $53 - 35 = 18$. Sum has sign of number with larger absolute value: $+$. Therefore, the sum is $18°$ F.

48. $17°$ F

Exercises 1.4 (page 20)

1. 6   **2.** $-\frac{2}{3}$ **3.** $(-3) - (-2) = (-3) + (+2) = -1$

4. -1 **5.** $(-6) - (2) = (-6) + (-2) = -8$ **6.** -13

7. $(9) - (-5) = (9) + (+5) = 14$ **8.** 10

9. $(2) - (-7) = (2) + (+7) = 9$ **10.** 8

11. $(-5) - (-9) = (-5) + (+9) = 4$ **12.** 6

13. $(-4) - (3) = (-4) + (-3) = -7$ **14.** -15

15. $(6) - (11) = (6) + (-11) = -5$ **16.** -4

17. $(-9) - (-4) = -9 + (+4) = -5$ **18.** -3

19. $(4) - (-7) = (4) + (+7) = 11$ **20.** 17

21. $(-15) - (11) = (-15) + (-11) = -26$ **22.** -40

23. $0 - 7 = -7$ **24.** -15 **25.** $0 - (-9) = -(-9) = 9$

26. 25   **27.** $16 - 0 = 16$ **28.** 10

29. $(156) - (-97) = (156) + (+97) = 253$ **30.** 373

31. $(-354) - (-286) = (-354) + (+286) = -68$ **32.** -109

33. $(-7) - (-2.009) = (-7.000) + (+2.009) = -4.991$

34. -8.11 **35.** $(+5) - (-2) = (+5) + (+2) = 7$ **36.** -5

37. $\left(-5\frac{1}{4}\right) - \left(2\frac{1}{2}\right) = \left(-5\frac{1}{4}\right) + \left(-2\frac{1}{2}\right)$
$= \left(-5\frac{1}{4}\right) + \left(-2\frac{2}{4}\right) = -7\frac{3}{4}$

38. $-10\frac{1}{2}$

39. $(\$473.29) - (\$238.43) = (\$473.29) + (-\$238.43) = \$234.86$

40. \$578.89

41. $42 - (-7) = 42 + (+7) = 49$. Therefore, the rise in temperature was $49°$ F.

42. $43.1°$ F   **43.** $(-141) - (68) = (-141) + (-68) = -209$ ft

44. 9800 ft

45. $(29,028) - (-36,198) = (29,028) + (+36,198) = 65,226$ ft (This is 12.35 miles.)

46. 3997 ft

Exercises 1.5 (page 25)

1. $3(-2) = -(3 \times 2) = -6$ **2.** -24

3. $(-5)(2) = -(5 \times 2) = -10$ **4.** -35

5. $(-8)(-2) = +(8 \times 2) = 16$ **6.** 42

7. $8(-4) = -(8 \times 4) = -32$ **8.** -45

9. $(-7)(9) = -(7 \times 9) = -63$ **10.** -48

11. $(-10)(-10) = +(10 \times 10) = 100$ **12.** 81

13. $(8)(-7) = -(8 \times 7) = -56$ **14.** -72

15. $(-26)(10) = -(26 \times 10) = -260$ **16.** -132

17. $(-20)(-10) = +(20 \times 10) = 200$ **18.** 600

19. $(75)(-15) = -(75 \times 15) = -1125$ **20.** -1118

21. $(-30)(5) = -(30 \times 5) = -150$ **22.** -300

23. $(-7)(-20) = +(7 \times 20) = 140$

24. 360   **25.** $(-5\frac{1}{2})(0) = 0$ **26.** 0

27. $\left(6\frac{3}{4}\right)\left(-8\frac{1}{4}\right) = \left(\frac{27}{4}\right)\left(-\frac{33}{4}\right) = -\frac{891}{16}$ or $-55\frac{11}{16}$

28. $-\frac{1826}{27}$ or $-67\frac{17}{27}$

29. $\left(-3\frac{3}{5}\right)\left(-8\frac{1}{5}\right) = \left(-\frac{18}{5}\right)\left(-\frac{41}{5}\right) = \frac{738}{25}$ or $29\frac{13}{25}$

30. $\frac{684}{35}$ or $19\frac{19}{35}$ **31.** $(-3.5)(-1.4) = +(3.5 \times 1.4) = 4.90$

32. 7.52   **33.** $(2.74)(-100) = -(2.74 \times 100) = -274$

34. -304 **35.** $\left(2\frac{1}{3}\right)\left(-3\frac{1}{2}\right) = \left(\frac{7}{3}\right)\left(-\frac{7}{2}\right) = -\frac{49}{6} = -8\frac{1}{6}$

36. $13\frac{13}{20}$ **37.** $(0)\left(-\frac{2}{3}\right) = 0$ **38.** 0   **39.** 12   **40.** $-\frac{5}{7}$

Exercises 1.6 (page 28)

1. 2. Positive because numbers have same signs.   **2.** 3

3. -4. Negative because numbers have different signs.   **4.** -2

5. -5 **6.** -2 **7.** 2   **8.** 5   **9.** -5 **10.** -6

11. -4 **12.** -5 **13.** 3   **14.** 3   **15.** -3 **16.** -4

17. 9   **18.** 7   **19.** -15 **20.** -2.5 **21.** -3

22. -7 **23.** -3 **24.** -3 **25.** Not possible

26. Not possible   **27.** Cannot be determined

28. Cannot be determined   **29.** $0 \div 7 = 0$ **30.** 0

31. Not possible   **32.** Not possible   **33.** $\frac{-15}{6} = \frac{-5}{2} = -2\frac{1}{2}$

34. $-2\frac{1}{4}$ **35.** $\frac{7.5}{-0.5} = \frac{75}{-5} = \frac{15}{-1} = -15$ **36.** -5

37. $\frac{-6.3}{-0.9} = \frac{-63}{-9} = 7$ **38.** 8   **39.** $\frac{-367}{100} = -3.67$

40. -4.86 **41.** $\frac{78.5}{-96.5} \doteq -0.813$ **42.** $\doteq -1.168$

Review Exercises 1.7 (page 32)

1. 8, 9   **2.** 10   **3.** -9 **4.** -1

5a. $<$, because -3 is to the left of 8 on the number line

b. $>$, because 5 is to the right of -2 on the number line

6a. $>$ **b.** $>$ **7.** 1   **8.** -4

9. Since they have different signs, subtract their absolute values: $3 - 2 = 1$. Sum has same sign as number with larger absolute value: $+$. Therefore, sum is 1.

10. -1 **11.** 3. Positive because numbers have same sign.

12. 2   **13.** $(-5) - (-3) = (-5) + (+3) = -2$

14. -5 **15.** $(+5) - |-2| = (+5) - (2) = (+5) + (-2) = 3$

16. 5   **17.** 12. Positive because numbers have same sign.

18. 20   **19.** $(-7) - (3) = (-7) + (-3) = -10$

20. -6 **21.** $(4) + (-12) = -8$ **22.** -2

23. -120.

24. -90 **25.** -8. Negative because they have different signs.

26. -6 **27.** $(9) - (-4) = (9) + (+4) = 13$ **28.** 11

29. $\left(-2\frac{2}{3}\right)\left(2\frac{1}{2}\right) = \left(-\frac{8}{3}\right)\left(\frac{5}{2}\right) = -\left(\frac{\overset{4}{\cancel{8}}}{3} \times \frac{5}{\cancel{2}}\right) = -\frac{20}{3} = -6\frac{2}{3}$

30. $4\frac{3}{10}$ **31.** -3 **32.** -3

33. Since they have the same sign, find sum of their absolute values: 10 + 2 = 12. Sum has same sign: $-$. Therefore, the sum is -12.

34. -11 **35.** -24. Negative because numbers have opposite signs.

36. -35 **37.** 5. Positive because both numbers have the same sign.

38. 8   **39.** 0. Zero divided by any number other than zero is zero.

40. 0   **41.** $(-10) - (-6) = -10 + (+6) = -4$ **42.** 3

Exercises 1.8 (page 40)

1. True. Commutative property of addition.

2. True. Commutative property of addition.

3. True. Associative property of addition.

4. True. Associative property of addition.

5. $6 - 2 \overset{?}{=} 2 - 6$

$4 \neq -4$

False.

6. False.   **7.** True. Associative property of multiplication.

8. True. Associative property of multiplication.

9. $8 \div 4 \overset{?}{=} 4 \div 8$

$2 \neq \frac{1}{2}$

False.

10. False.   **11.** True. Commutative property of multiplication.

12. True. Commutative property of multiplication.

13. True. Commutative property of addition.

14. True. Commutative property of addition.

15. True. Commutative property of addition.

16. True. Commutative property of addition.

17. True. Commutative property of addition.

18. True. Commutative property of addition.

19. True. 1 is the multiplicative identity.

20. True. 0 is the additive identity.   **21.** True. Additive inverse.

22. True. Additive inverse.   **23.** False   **24.** False

25. True. Associative and commutative properties of addition.

26. True. Commutative and associative properties of multiplication.

27. False.

28. False.

29. True. Associative property of multiplication. (Grouping changed.)

30. True. Associative property of multiplication. (Grouping changed.)

31. True. Commutative property of addition. (Order changed.)

32. True. Commutative property of addition. (Order changed.)

33. $8 + (-11) = -3$ **34.** -12

35. $(-5)(-4)(-2) = -(5 \times 4 \times 2) = -40$ **36.** -48

37. $(2)(-5)(-9) = +(2 \times 5 \times 9) = 90$ **38.** 24

39. $(-2)(-3)(-5)(-4) = +(2 \times 3 \times 5 \times 4) = 120$ **40.** 56

Exercises 1.9 (page 43)

1. $3^3 = 3 \cdot 3 \cdot 3 = 27$ **2.** 16   **3.** $(-5)^2 = (-5)(-5) = 25$

4. -216 **5.** $7^2 = 7 \cdot 7 = 49$ **6.** 81

7. $0^3 = 0 \cdot 0 \cdot 0 = 0$ **8.** 0   **9.** $(-10)^1 = -10$ **10.** 100

11. $10^3 = 10 \cdot 10 \cdot 10 = 1000$ **12.** 10,000

13. $(-10)^5 = (-10)(-10)(-10)(-10)(-10) = -100,000$

14. 1,000,000   **15.** $2^1 = 2$ **16.** 32

17. $(-2)^6 = (-2)(-2)(-2)(-2)(-2)(-2) = 64$ **18.** -128

19. $2^8 = 2 \cdot 2 \cdot 2 \cdot 2 \cdot 2 \cdot 2 \cdot 2 \cdot 2 = 256$ **20.** 625

21. $40^3 = 40 \cdot 40 \cdot 40 = 64,000$ **22.** 0

23. $(-12)^3 = (-12)(-12)(-12) = -1728$ **24.** 225

25. $(-1)^5 = (-1)(-1)(-1)(-1)(-1) = -1$ **26.** -1

27. $-2^2 = -(2 \cdot 2) = -4$ **28.** -9 **29.** $(-1)^{99} = -1$

30. 1   **31.** 161.29   **32.** 237.16   **33.** $0^8 = 0$ **34.** 0

Exercises 1.10A (page 45)

1. 4, because $4^2 = 16$ **2.** 5   **3.** $-(\sqrt{4}) = -(2) = -2$

4. -3 **5.** 9, because $9^2 = 81$ **6.** 6

7. 10, because $10^2 = 100$ **8.** 12   **9.** $-(\sqrt{81}) = -(9) = -9$

10. -11 **11.** 8, because $8^2 = 64$ **12.** 13

Exercises 1.10B (page 46)

1. Try 20: $20^2 = 400$, therefore, 20 is too small. Try 23: $23^2 = 529$, therefore, $\sqrt{529} = 23$.

2. 19

3. Try 25: $25^2 = 625$, therefore, 25 is too large. Try 22: $22^2 = 484$, therefore, 22 is too large. Try 21: $21^2 = 441$, therefore, $\sqrt{441} = 21$.

4. 25

5. Try 18: $18^2 = 324$, therefore, 18 is too large. Try 17: $17^2 = 289$, therefore, $\sqrt{289} = 17$.

6. 18

7. If the number we are taking the square root of is an odd number, then its square root must be an odd number. Try 29: $29^2 = 841$, therefore, 29 is too large. Try 27: $27^2 = 729$, therefore $\sqrt{729} = 27$.

8. 36

Exercises 1.10C (page 47)

1. 3.606   **2.** 4.243   **3.** 6.083   **4.** 7.071   **5.** 8.888

6. 7.746   **7.** 9.274   **8.** 9.592   **9.** 683   **10.** 821

11. 522   **12.** 299

Exercises 1.10D (page 50)

1. By trial, we find: $2^3 = 2 \cdot 2 \cdot 2 = 8$; $3^3 = 3 \cdot 3 \cdot 3 = 27$; $4^3 = 4 \cdot 4 \cdot 4 = 64$. Therefore, $\sqrt[3]{64} = 4$.

2. 3   **3.** 3 (See the trial in the solution for Exercise 1.)

4. 5   **5.** -3, because $\sqrt[3]{27} = 3$, $-\sqrt[3]{27} = -3$.

6. -5 **7.** $-\sqrt[4]{1} = -(\sqrt[4]{1}) = -(1) = -1$

8. 2   **9.** -5, because $(-5)(-5)(-5) = -125$

10. -2 **11.** Not a real number   **12.** Not a real number

13. -10, because $(-10)(-10)(-10) = -1000$ **14.** -4

15. -1. Any odd root of -1 is -1.   **16.** -2

17. 3, because $3 \cdot 3 \cdot 3 \cdot 3 \cdot 3 \cdot 3 = 729$.   **18.** 6

19. Not a real number   **20.** Not a real number

21. The rational numbers are $\frac{1}{2}$, -12, and $0.26262626\ldots$. The irrational numbers are $\sqrt[3]{13}$ and $0.196732468\ldots$. The real numbers are $\sqrt[3]{13}$, $\frac{1}{2}$, -12, $0.26262626\ldots$, and $0.196732468\ldots$. The number that is not real is $\sqrt{-15}$.

22. The rational numbers are 18, $\frac{11}{32}$, and $0.37373737\ldots$. The irrational numbers are $\sqrt[3]{12}$ and $0.67249713\ldots$. The real numbers are $0.67249713\ldots$, 18, $\frac{11}{32}$, $0.37373737\ldots$, and $\sqrt[3]{12}$. The number that is not real is $\sqrt{-27}$.

Exercises 1.11 (page 54)

1. ± 1, ± 2, ± 4 **2.** ± 1, ± 3, ± 9

3. ± 1, ± 2, ± 5, ± 10 **4.** ± 1, ± 2, ± 7, ± 14

5. ± 1, ± 3, ± 5, ± 15 **6.** ± 1, ± 2, ± 4, ± 8, ± 16

7. ± 1, ± 2, ± 3, ± 6, ± 9, ± 18

8. ± 1, ± 2, ± 4, ± 5, ± 10, ± 20 **9.** ± 1, ± 3, ± 7, ± 21

10. ± 1, ± 2, ± 11, ± 22 **11.** ± 1, ± 3, ± 9, ± 27

12. ± 1, ± 2, ± 4, ± 7, ± 14, ± 28 **13.** ± 1, ± 3, ± 11, ± 33

14. ± 1, ± 2, ± 17, ± 34

15. ± 1, ± 2, ± 4, ± 11, ± 22, ± 44

16. ± 1, ± 3, ± 5, ± 9, ± 15, ± 45

17. Prime; 1, 5   **18.** Composite; 1, 2, 4, 8   **19.** Prime; 1, 13

20. Composite; 1, 3, 5, 15   **21.** Composite; 1, 2, 3, 4, 6, 12

22. Prime; 1, 11   **23.** Composite; 1, 3, 7, 21   **24.** Prime; 1, 23

25. Composite; 1, 5, 11, 55   **26.** Prime; 1, 41

27. Composite; 1, 7, 49   **28.** Prime; 1, 31

29. Composite; 1, 3, 17, 51

30. Composite; 1, 2, 3, 6, 7, 14, 21, 42

31. Composite; 1, 3, 37, 111

32. Prime; 1, 101

33.
$$2\,\underline{|\,14}$$
$$7$$
$$14 = 2 \cdot 7$$

34. $15 = 3 \cdot 5$

35.
$$3\,\underline{|\,21}$$
$$7$$
$$21 = 3 \cdot 7$$

36. $22 = 2 \cdot 11$

37.
$$2\,\underline{|\,26}$$
$$13$$
$$26 = 2 \cdot 13$$

38. $27 = 3^3$ **39.** 29 is prime   **40.** 31 is prime

41.
$$2\,\underline{|\,32}$$
$$2\,\underline{|\,16}$$
$$2\,\underline{|\,8}$$
$$2\,\underline{|\,4}$$
$$2$$
$$32 = 2^5$$

42. $33 = 3 \cdot 11$

43.
$$2\,\underline{|\,34}$$
$$17$$
$$34 = 2 \cdot 17$$

44. $35 = 5 \cdot 7$

45.
$$2\,\underline{|\,84}$$
$$2\,\underline{|\,42}$$
$$3\,\underline{|\,21}$$
$$7$$
$$84 = 2^2 \cdot 3 \cdot 7$$

46. $75 = 3 \cdot 5^2$

47.
$$2\,\underline{|\,144}$$
$$2\,\underline{|\,72}$$
$$2\,\underline{|\,36}$$
$$2\,\underline{|\,18}$$
$$3\,\underline{|\,9}$$
$$3$$
$$144 = 2^4 \cdot 3^2$$

48. $180 = 2^2 \cdot 3^2 \cdot 5$

Review Exercises 1.12 (page 56)

1a. True. Associative property of multiplication.
b. True. Commutative property of addition.
c. False
d. True. Associative property of addition.
e. True. Commutative property of multiplication.
f. True. Commutative property of addition.

2. 23   **3.** 0   **4.** 1   **5.** $0^3 = 0 \cdot 0 \cdot 0 = 0$ **6.** 16

7. $-5^2 = -5 \cdot 5 = -25$ **8.** 11

9. $-\sqrt[4]{16} = -(\sqrt[4]{16}) = -(2) = -2$ **10.** Not a real number

11. -2, because $(-2)^3 = -8$ **12.** 2   **13.** $16^2 = 16 \cdot 16 = 256$

14. The rational numbers are -31, $\frac{7}{12}$, and $4.15151515\ldots$. The irrational numbers are $-\sqrt{5}$, $5.3892531\ldots$, and $\sqrt[3]{71}$. The real numbers are $-\sqrt{5}$, -31, $5.3892531\ldots$, $\frac{7}{12}$, $4.15151515\ldots$, and $\sqrt[3]{71}$. There are no numbers that are not real.

15. 5.568

16. The prime factorization of 270 is $2 \cdot 3^3 \cdot 5$. The integral factors of 270 are ± 1, ± 2, ± 3, ± 5, ± 6, ± 9, ± 10, ± 15, ± 18, ± 27, ± 30, ± 45, ± 54, ± 90, ± 135, and ± 270.

Chapter 1 Diagnostic Test (page 59)

Following each problem number is the textbook section number (in parentheses) where that kind of problem is discussed.

1. (1.1) $<$ 2. (1.2) $>$ 3. (1.2) $<$ 4. (1.1) 99
5. (1.2) -9 6. (1.1) 0, 1, 2 7. (1.3) 0
8. (1.4) $38° - (-20°) = 38° + 20° = 58°$
9. (1.8) True. Associative property of multiplication.
10. (1.8) False. 11. (1.8) True. Commutative property of addition.
12. (1.4) $5 - 24 = 5 + (-24) = -19$
13. (1.4) $-17 - 8 = -17 + (-8) = -25$
14. (1.3) $4 + (-18) = -14$
15. (1.4) $5 - 24 = 5 + (-24) = -19$
16. (1.6) $3 \div 0$ is not possible. (We cannot divide by zero.)
17. (1.5) $(-10)(-5) = +(10)(5) = 50$
18. (1.4) $(-10) - 5 = -10 + (-5) = -15$
19. (1.9) $-7^2 = -7 \cdot 7 = -49$
20. (1.9) $(-7)^2 = (-7)(-7) = 49$
21. (1.3) $15 + (-8) = 7$ 22. (1.5) $-6(0) = 0$
23. (1.6) $\dfrac{15}{-21} = -\dfrac{5}{7}$ 24. (1.9) $(-4)^3 = (-4)(-4)(-4) = -64$
25. (1.9) $-4^3 = -4 \cdot 4 \cdot 4 = -64$
26. (1.6) $(-24) \div (-4) = 6$
27. (1.6) $\dfrac{-33}{11} = -3$ 28. (1.6) $42 \div (-14) = -3$
29. (1.6) $\dfrac{0}{0}$ cannot be determined.
30. (1.4) $-8 - (-17) = -8 + (+17) = 9$
31. (1.3) $-13 + 5 = -8$
32. (1.5) $(-6)(-5)(-2) = -6 \times 5 \times 2 = -60$
33. (1.3) $-12 + (-9) = -21$ 34. (1.9) $0^4 = 0 \cdot 0 \cdot 0 \cdot 0 = 0$
35. (1.4) $-7 - 8 = -7 + (-8) = -15$
36. (1.6) $0 \div (-12) = 0$ 37. (1.5) $6(-9) = -54$
38. (1.4) $6 - 9 = 6 + (-9) = -3$ 39. (1.5) $-7(-8) = 56$
40. (1.4) $-11 - (-4) = -11 + (+4) = -7$
41. (1.6) $-56 \div (-7) = 8$
42. (1.4) $3 - (-12) = 3 + (+12) = 15$
43. (1.5) $0(-6) = 0$ 44. (1.4) $0 - 6 = -6$
45. (1.3) $|-4| = 4$ 46. (1.3) $|6| = 6$
47. (1.10) $\sqrt{49} = 7$, because $7^2 = 49$
48. (1.10) $\sqrt{81} = 9$, because $9^2 = 81$
49. (1.10) $\sqrt[3]{64} = 4$, because $4^3 = 64$
50. (1.10) $\sqrt[3]{-8} = -2$, because $(-2)^3 = -8$
51. (1.10) The rational numbers are -63, $0.38383838\ldots$, and $\frac{2}{23}$. The irrational numbers are $-\sqrt[3]{19}$ and $0.25375913\ldots$. The real numbers are $-\sqrt[3]{19}$, -63, $0.38383838\ldots$, $\frac{2}{23}$, and $0.25375913\ldots$. The number that is not real is $\sqrt{-121}$.
52. (1.11) **a.** The prime factorization of 180 is $2^2 \cdot 3^2 \cdot 5$.
  b. The integral factors of 180 are ± 1, ± 2, ± 3, ± 4, ± 5, ± 6, ± 9, ± 10, ± 12, ± 15, ± 18, ± 20, ± 30, ± 36, ± 45, ± 60, ± 90, and ± 180.

Exercises 2.1 (page 64)

1. $12 - 8 - 6 = 4 - 6 = -2$ 2. 2
3. $17 - 11 + 13 - 9 = 6 + 13 - 9 = 19 - 9 = 10$
4. 12 5. $7 + 2 \cdot 4 = 7 + 8 = 15$ 6. 28
7. $9 - 3 \cdot 2 = 9 - 6 = 3$ 8. -10
9. $10 \div 2 \cdot 5 = 5 \cdot 5 = 25$ 10. 60
11. $12 \div 6 \div 2 = 2 \div 2 = 1$ 12. $\frac{1}{3}$
13. $(-12) \div 2 \cdot (-3) = (-6) \cdot (-3) = 18$ 14. -36
15. $8 \cdot 5^2 = 8 \cdot 25 = 200$ 16. 96
17. $(-485)^2 \cdot 0 \cdot (-5)^2 = 0 \cdot (-5)^2 = 0$ 18. 0
19. $12 \cdot 4 + 16 \div 8 = 48 + 16 \div 8 = 48 + 2 = 50$
20. 15 21. $28 \div 4 \cdot 2(6) = 7 \cdot 2(6) = 14(6) = 84$ 22. -48
23. $(-2)^2 + (-4)(5) - (-3)^2 = 4 + (-20) - (9)$
  $= -16 + (-9) = -25$ 24. -3
25. $2 \cdot 3 + 3^2 - 4 \cdot 2 = 2 \cdot 3 + 9 - 4 \cdot 2 = 6 + 9 - 8$
  $= 15 + (-8) = 7$ 26. 624
27. $(10^2)\sqrt{16} + 5(4) - 80 = (100)(4) + 5(4) - 80$
  $= 400 + 20 - 80 = 420 - 80 = 340$
28. 39

Exercises 2.2 (page 67)

1. $2 \cdot (-6) \div 3 \cdot (8 - 4) = 2 \cdot (-6) \div 3 \cdot (4) = -12 \div 3 \cdot 4 =$
  $-4 \cdot 4 = -16$
2. -50 3. $24 - [(-6) + 18] = 24 - [12] = 12$ 4. 11
5. $[12 - (-19)] - 16 = [31] - 16 = 15$ 6. 6
7. $[11 - (5 + 8)] - 24 = [11 - (13)] - 24 = [-2] - 24 = -26$
8. -25
9. $20 - [5 - (7 - 10)] = 20 - [5 - (-3)] = 20 - [8] = 12$
10. 3 11. $\dfrac{7 + (-12)}{8 - 3} = \dfrac{-5}{5} = -1$ 12. -4
13. $15 - \{4 - [2 - 3(6 - 4)]\} = 15 - \{4 - [2 - 3(2)]\}$
  $= 15 - \{4 - [2 - 6]\} = 15 - \{4 - [-4]\} = 15 - \{8\} = 7$
14. 30
15. $32 \div (-2)^3 - 5\left\{7 - \dfrac{6 - 2}{5}\right\} = 32 \div (-2)^3 - 5\left\{7 - \dfrac{4}{5}\right\}$
  $= 32 \div (-2)^3 - \cancel{5}\left\{\dfrac{31}{\cancel{5}}\right\} = 32 \div (-8) - 31 = (-4) - 31$
  $= -35$
16. -16 17. $\sqrt{3^2 + 4^2} = \sqrt{9 + 16} = \sqrt{25} = 5$ 18. 12
19. $\sqrt{16.3^2 - 8.35^2} = \sqrt{265.69 - 69.7225} = \sqrt{195.9675} \doteq 13.999$
20. $\doteq 45.400$
21. $(1.5)^2 \div (-2.5) + \sqrt{35} \doteq 2.25 \div (-2.5) + 5.916$
  $= -0.90 + 5.916 = 5.016$
22. $\doteq -55.152$
23. $18.91 - [64.3 - (8.6^2 + 14.2)] = 18.91 - [64.3 - (73.96 + 14.2)]$
  $= 18.91 - [64.3 - 88.16] = 18.91 - [-23.86]$
  $= 18.91 + (+23.86) = 42.77$ 24. $\doteq 8.396$

Exercises 2.3 (page 71)

1a. 2, 4 **b.** x, y **2a.** 7, 3 **b.** a, b

3a. 7, -8, 2 **b.** u, v **4a.** 3, -5, -2 **b.** x, y

5. $3b = 3(-5) = -15$ **6.** -75

7. $9 - 6b = 9 - 6(-5) = 9 + 30 = 39$

8. -27

9. $b^2 = (-5)^2 = (-5)(-5) = 25$ **10.** -49

11. $2a - 3b = 2(3) - 3(-5) = 6 + 15 = 21$ **12.** 26

13. $x - y - 2b = (4) - (-7) - 2(-5) = 4 + 7 + 10 = 21$

14. 29

15. $3b - ab + xy = 3(-5) - (3)(-5) + (4)(-7)$
$= -15 + 15 - 28 = -28$

16. -27 **17.** $x^2 - y^2 = (4)^2 - (-7)^2 = 16 - 49 = -33$

18. 24

19. $4 + a(x + y) = 4 + (3)[(4) + (-7)] = 4 + 3[-3]$
$= 4 - 9 = -5$

20. 15

21. $2(a - b) - 3c = 2[(3) - (-5)] - 3(-1) = 2[8] - 3(-1)$
$= 16 + 3 = 19$

22. 17

23. $3x^2 - 10x + 5 = 3(4)^2 - 10(4) + 5 = 3(16) - 10(4) + 5$
$= 48 - 40 + 5 = 13$

24. 156

25. $a^2 - 2ab + b^2 = (3)^2 - 2(3)(-5) + (-5)^2$
$= 9 - 2(3)(-5) + 25 = 9 + 30 + 25 = 64$

26. 121   **27.** $\dfrac{3x}{y + b} = \dfrac{3(4)}{(-7) + (-5)} = \dfrac{12}{-12} = -1$ **28.** 3

29. $\dfrac{E + F}{EF} = \dfrac{(-1) + (3)}{(-1)(3)} = \dfrac{2}{-3} = -\dfrac{2}{3}$

30. $-\dfrac{9}{20}$

31. $\dfrac{(1 + G)^2 - 1}{H} = \dfrac{[1 + (-5)]^2 - 1}{-4} = \dfrac{[-4]^2 - 1}{-4}$
$= \dfrac{16 - 1}{-4} = \dfrac{15}{-4} = -3\dfrac{3}{4}$

32. $\dfrac{1}{3}$

33. $2E - [F - (3K - H)] = 2(-1) - [(3) - \{3(0) - (-4)\}]$
$= 2(-1) - [(3) - \{0 + 4\}] = 2(-1) - [3 - \{4\}]$
$= 2(-1) - [-1] = -2 + 1 = -1$

34. 1

35. $G - \sqrt{G^2 - 4EH} = (-5) - \sqrt{(-5)^2 - 4(-1)(-4)}$
$= (-5) - \sqrt{25 - 4(-1)(-4)} = (-5) - \sqrt{25 - 16}$
$= (-5) - \sqrt{9} = -5 - 3 = -8$

36. -8

37. $\dfrac{\sqrt{2H - 5G}}{0.2F^2} = \dfrac{\sqrt{2(-4) - 5(-5)}}{0.2(3)^2} = \dfrac{\sqrt{-8 + 25}}{0.2(9)}$
$= \dfrac{\sqrt{17}}{1.8} \doteq \dfrac{4.123}{1.8} \doteq 2.291$

38. 20

Exercises 2.4 (page 75)

1. $A = \dfrac{1}{2}bh = \dfrac{1}{\underset{1}{\cancel{2}}}\left(\dfrac{15}{1}\right)\left(\dfrac{\overset{7}{\cancel{14}}}{1}\right) = 105$ **2.** 486

3. $I = \dfrac{E}{R} = \dfrac{110}{22} = 5$ **4.** $6\dfrac{2}{3}$

5. $I = prt = 600(0.09)(4.5) = 243$ **6.** 140

7. $F = \dfrac{9}{5}C + 32$

$F = \dfrac{9}{5}(25) + 32$

$F = 45 + 32 = 77$

8. -13

9. $A = \pi R^2 \doteq 3.14(10)^2 = 3.14(100) = 314$ **10.** $\doteq 1256$

11. $s = \dfrac{1}{2}gt^2 = \dfrac{1}{2}(32)(3)^2$

$s = \dfrac{1}{2}(32)(9)$

$s = 144$

12. 400

13. $V = C - Crt$
$V = 500 - (500)(0.1)(2)$
$V = 500 - 100 = 400$

14. 600

15. $\sigma = \sqrt{npq}$
$\sigma = \sqrt{(100)(0.9)(0.1)}$
$\sigma = \sqrt{9} = 3$

16. 4

17. $C = \dfrac{5}{9}(F - 32)$

$C = \dfrac{5}{9}(-4 - 32)$

$C = \dfrac{5}{9}(-36) = -20$

18. -15

19. $C = \dfrac{a}{a + 12} \cdot A$

$C = \dfrac{6}{6 + 12} \cdot 30$

$C = \dfrac{6}{18} \cdot 30 = 10$

20. 12

21. $V = \dfrac{4}{3}\pi R^3 \doteq \dfrac{4}{3}(3.14)(3)^3$

$V \doteq \dfrac{4}{3}(3.14)(27)$

$V \doteq 113.04$

22. $\doteq 904.32$

23. $A = P(1 + i)^n$
$A = 1000(1 + 0.06)^2$
$A = 1123.60$

24. 2519.42

Review Exercises 2.5 (page 78)

1. $26 - 14 + 8 - 11 = 12 + 8 - 11 = 20 - 11 = 9$

2. -13 **3.** $11 - 7 \cdot 3 = 11 - 21 = -10$ **4.** 12

5. $15 \div 5 \cdot 3 = 3 \cdot 3 = 9$ **6.** 58

7. $6 - [8 - (3 - 4)] = 6 - [8 - (-1)] = 6 - [9] = -3$

8. -3

9. $\dfrac{6 + (-14)}{3 - 7} = \dfrac{-8}{-4} = 2$ **10.** 6

11. $3x - y + z = 3(-2) - 3 + (-4) = -6 - 3 - 4 = -13$

12. 25

13. $5x^2 - 3x + 10 = 5(-2)^2 - 3(-2) + 10 = 5(4) - 3(-2) + 10$
$= 20 + 6 + 10 = 36$

14. -17

15. $x - 2[y - x(y + z)] = -2 - 2[3 - (-2)(3 + (-4))]$
$= -2 - 2[3 - 2] = -2 - 2(1) = -2 - 2 = -4$

16. $6\dfrac{5}{6}$

17. $C = \dfrac{a}{a + 12}(35)$

$C = \dfrac{8}{8 + 12}\left(\dfrac{35}{1}\right)$

$C = \dfrac{8}{20}\left(\dfrac{35}{1}\right) = 14$

18. 110

19. $C = \dfrac{5}{9}(F - 32) = \dfrac{5}{9}\left(15\dfrac{1}{2} - 32\right)$

$C = \dfrac{5}{9}\left(-\dfrac{33}{2}\right) = -9\dfrac{1}{6}$

20. 665.50

21. $\sigma = \sqrt{npq} = \sqrt{(100)(0.5)(0.5)}$
$\sigma = \sqrt{25} = 5$

22. 1100

Chapter 2 Diagnostic Test (page 83)

Following each problem is the textbook section number (in parentheses) where that kind of problem is discussed.

1. (2.1) $23 - 19 - 3 + 11 = 4 - 3 + 11 = 1 + 11 = 12$

2. (2.1) $8 + 9 \cdot 7 = 8 + 63 = 71$

3. (2.1) $54 \div 9 \cdot 6 = 6 \cdot 6 = 36$

4. (2.1) $-5^2 + (-4)^2 = -25 + 16 = -9$

5. (2.1) $6 \cdot 4^2 - 4 = 6 \cdot 16 - 4 = 96 - 4 = 92$

6. (2.1) $5\sqrt{36} - 6(-5) = 5(6) - 6(-5) = 30 + 30 = 60$

7. (2.2) $\dfrac{7 - 15}{-2 + 6} = \dfrac{-8}{4} = -2$

8. (2.2) $(2^3 - 8)(8^2 - 9^2) = (8 - 8)(64 - 81) = (0)(-17) = 0$

9. (2.2) $10 - [7 - (4 - 9)] = 10 - [7 - (-5)]$
$= 10 - [7 + (+5)] = 10 - 12 = -2$

10. (2.2) $\{-10 - [5 + (4 - 7)]\} - 3 = \{-10 - [5 + (-3)]\} - 3$
$= \{-10 - 2\} - 3 = -12 - 3 = -15$

11. (2.2) $\sqrt{13^2 - 5^2} = \sqrt{169 - 25} = \sqrt{144} = 12$

12. (2.2) $48 \div (-4)^2 - 3\left(10 - \dfrac{10}{-2}\right) = 48 \div 16 - 3(10 - [-5])$
$= 3 - 3(15) = 3 - 45 = -42$

13. (2.3) $3c - by + cx = 3(-2) - (-7)(5) + (-2)(-6)$
$= -6 - (-35) + 12 = -6 + 35 + 12 = 29 + 12 = 41$

14. (2.3) $4x - [a - (3c - b)] = 4(-6) - \{(-4) - (3[-2] - [-7])\}$
$= -24 - \{-4 - (-6 + 7)\}$
$= -24 - \{-4 - (+1)\} = -24 - \{-4 + (-1)\}$
$= -24 - \{-5\} = -24 + \{+5\} = -19$

15. (2.3) $x^2 + 2xy + y^2 = (-6)^2 + 2(-6)(5) + (5)^2$
$= 36 + (-12)(5) + 25 = 36 + (-60) + 25 = -24 + 25 = 1$

16. (2.3) $(x + y)^2 = ([-6] + [5])^2 = (-1)^2 = 1$

17. (2.4) $C = \dfrac{5}{9}(F - 32); F = 68$

$C = \dfrac{5}{9}([68] - 32) = \dfrac{5}{9}(36) = 20$

18. (2.4) $A = \pi R^2; \pi \doteq 3.14, R = 3$
$A \doteq (3.14)(3)^2 \doteq (3.14)(9) \doteq 28.26$

19. (2.4) $A = P(1 + rt); P = 500, r = 0.10, t = 3.5$
$A = 500(1 + [0.1][3.5]) = 500(1 + 0.35) = 500(1.35) = 675$

20. (2.4) $V = C - Crt; C = 800, r = 0.06, t = 10$
$V = 800 - 800(0.06)(10) = 800 - 480 = 320$

Cumulative Review Exercises: Chapters 1 and 2 (page 84)

1. $(-16)(-2) = 32$ **2.** -21

3. $(-5) - (-11) = (-5) + (11) = 6$ **4.** -4

5. $(-15)(0)(4) = 0$ **6.** -8 **7.** $(-27) + (10) = -17$

8. 4 **9.** $\dfrac{-48}{-12} = 4$ **10.** 0 **11.** $(25)^2 = (25)(25) = 625$

12. -45 **13.** $-2^4 = -(2^4) = -16$ **14.** -4

15. $\dfrac{-12}{0}$ is not possible because we cannot divide by zero.

16. 2 **17.** $\sqrt{64} = 8$ because $8^2 = 64$ **18.** 0

19. $(8) - (17) = (8) + (-17) = -9$ **20.** -29

21. $24 \div 12 \cdot 2 = 2 \cdot 2 = 4$ **22.** 24

23. $8 - 6 \cdot 5 = 8 - 30 = -22$ **24.** 1

25. $17 - 9 - 2 = 8 - 2 = 6$ **26.** 38 **27.** $0 \div 5 = 0$

28. $C = \dfrac{5}{9}(F - 32);\quad F = 21\dfrac{1}{2} = \dfrac{43}{2}$

$C = \dfrac{5}{9}\left(\dfrac{43}{2} - \dfrac{64}{2}\right)$

$C = \dfrac{5}{9}\left(-\dfrac{21}{2}\right)$

$C = -\dfrac{35}{6}$ or $-5\dfrac{5}{6}$

29. True   **30.** True   **31.** True   **32.** False   **33.** True

34. True   **35.** False

Exercises 3.1 (page 88)

1a. Factor
  b. Term
  c. Factor
  d. Factor

2a. Term
  b. Factor
  c. Factor
  d. Factor

3a. Term
  b. Term
  c. Factor
  d. Term

4a. Factor
  b. Factor
  c. Term
  d. Term

5. One term
  No second term

6. One term
  No second term

7. Three terms
  Second term: $-5F$

8. Three terms
  Second term: $-2T$

9. Three terms
  Second term: $\dfrac{2x+y}{3xy}$

10. Three terms
  Second term: $\dfrac{5x-y}{7xy}$

11. Two terms
  Second term: $-6u(2u+v^2)$

12. Two terms
  Second term: $-2E(8E+F^2)$

13. One term
  No second term

14. One term
  No second term

15. 3   **16.** 4   **17.** 1   **18.** 1   **19.** -1 **20.** -1
21. $\dfrac{4}{5}$ **22.** $\dfrac{1}{2}$

Exercises 3.2 (page 90)

1. $x^3 \cdot x^4 = x^{3+4} = x^7$ **2.** x^{11} **3.** $y \cdot y^3 = y^{1+3} = y^4$
4. z^5 **5.** $m^2 \cdot m = m^{2+1} = m^3$ **6.** a^4
7. $10^2 \cdot 10^3 = 10^{2+3} = 10^5$ **8.** 10^7
9. $2 \cdot 2^3 \cdot 2^2 = 2^{1+3+2} = 2^6$ **10.** 3^6
11. $x \cdot x^3 \cdot x^4 = x^{1+3+4} = x^8$ **12.** y^9 **13.** $x^2 y^5$ **14.** $a^3 b^2$
15. $3^2 \cdot 5^3 = 9 \cdot 125 = 1125$ **16.** 72
17. $a^4 + a^2$ (cannot be rewritten)   **18.** $x^3 + x^4$
19. $a^x \cdot a^w = a^{x+w}$ **20.** x^{a+b} **21.** $x^y y^x$ (cannot be combined)
22. $a^b b^a$ **23.** $x^2 y^3 x^5 = x^{2+5} y^3 = x^7 y^3$ **24.** $z^7 w^2$
25. $a^2 b^3 a^5 = a^{2+5} b^3 = a^7 b^3$ **26.** $x^{12} y$
27. $s^7 + s^4$ (cannot be rewritten)   **28.** $t^3 + t^8$

Exercises 3.3 (page 92)

1. $-(2 \cdot 4)(a \cdot a^2) = -8a^3$ **2.** $-15x^4$
3. $+(5 \cdot 6)(h^2 h^3) = 30h^5$ **4.** $48k^4$
5. $(-5x^3)^2 = (-5x^3)(-5x^3) = +(5 \cdot 5)(x^3 x^3) = 25x^6$
6. $49y^8$ **7.** $+(2 \cdot 4 \cdot 3)(a^3 a a^4) = 24a^8$ **8.** $-48b^6$
9. $+(9 \cdot 2)(mm^5 m^2) = 18m^8$ **10.** $28n^7$
11. $-(5 \cdot 7)x^2 y = -35x^2 y$ **12.** $-12x^3 y$
13. $+(6 \cdot 4)(m^3 m)(n^2 n^2) = 24m^4 n^4$ **14.** $-40h^6 k^4$

15. $-(2 \cdot 3)(x^{10} \cdot x^{12})(y^2 \cdot y^7) = -6x^{22} y^9$ **16.** $10a^{12} b^{15}$
17. $(3xy^2)^2 = (3xy^2)(3xy^2) = (3 \cdot 3)(x \cdot x)(y^2 \cdot y^2) = 9x^2 y^4$
18. $16x^4 y^6$ **19.** $-(5)(x^4)(y^5 \cdot y^4)(z \cdot z^7) = -5x^4 y^9 z^8$
20. $-21E^2 F^{11} G^{18}$ **21.** $-(2^3 \cdot 2^2)(R \cdot R^5)S^2 T^4 = -32R^6 S^2 T^4$
22. $-243x^9 yz^5$ **23.** $+(5 \cdot 4)(c^2 \cdot c^5)(d \cdot d)(e^3 \cdot e^2) = 20c^7 d^2 e^5$
24. $24m^4 n^7 r^5$ **25.** $-(2 \cdot 3 \cdot 7)(x^2 \cdot x)(y^2 \cdot y)(z \cdot z) = -42x^3 y^3 z^2$
26. $420x^4 y^3 z^5$ **27.** $-(3 \cdot 5)(x \cdot x^2 \cdot x^3)(y \cdot y^2 \cdot y^3) = -15x^6 y^6$
28. $-6x^2 y^2 z^2$ **29.** $-(2 \cdot 5)(a \cdot a)(b \cdot b)(c \cdot c) = -10a^2 b^2 c^2$
30. $+6x^3 y^3 z^3$ **31.** $(5 \cdot 2)(x^2 \cdot x)(y \cdot y)(z^3 \cdot z) = 10x^3 y^2 z^4$
32. $6a^5 b^7$
33. $-(5 \cdot 7)(x^2 \cdot x^5 \cdot x)(y^2 \cdot y^3 \cdot y)(z \cdot z \cdot z^5) = -35x^8 y^6 z^7$
34. $-32R^9 S^7 T^{13}$
35. $-(3 \cdot 7)(h^2 \cdot h^4)(k^1 \cdot k^5)(m^3 \cdot m^1) = -21h^6 k^6 m^4$
36. $-48k^3 m^3 n^3$

Exercises 3.4 (page 95)

1. $5a + 30$ **2.** $4x + 40$ **3.** $7x + 7y$ **4.** $5m + 5n$
5. $(3)(m) + (3)(-4) = 3m - 12$ **6.** $3a - 15$
7. $(4)(x) + (4)(-y) = 4x - 4y$ **8.** $9m - 9n$
9. $(a)(6) + (a)(x) = 6a + ax$ **10.** $7b + by$
11. $(-2)(x) + (-2)(-3) = -2x + 6$ **12.** $-3x + 15$
13. $(-3)(2x^2) + (-3)(-4x) + (-3)(5) = -6x^2 + 12x - 15$
14. $-15x^2 + 10x + 35$ **15.** $(4x)(3x^2) + (4x)(-6) = 12x^3 - 24x$
16. $15x^3 - 30x$
17. $(-2x)(5x^2) + (-2x)(3x) + (-2x)(-4) = -10x^3 - 6x^2 + 8x$
18. $-8x^3 + 20x^2 - 12x$
19. $(x)(6) + (-4)(6) = 6x - 24$ **20.** $-15 + 10x$
21. $(y^2)(7) + (-4y)(7) + (3)(7) = 7y^2 - 28y + 21$
22. $63 - 7z + 14z^2$
23. $(2x^2)(4x) + (-3x)(4x) + (5)(4x) = 8x^3 - 12x^2 + 20x$
24. $15w^3 + 10w^2 - 40w$ **25.** $(x)(xy) + (x)(-3) = x^2 y - 3x$
26. $a^2 b - 4a$ **27.** $(3a)(ab) + (3a)(-2a^2) = 3a^2 b - 6a^3$
28. $12x^2 - 8xy^2$
29. $(-2x)(-3y) + (4x^2 y)(-3y) = 6xy - 12x^2 y^2$
30. $6az - 4a^2 z^2$
31. $(-2xy)(x^2 y) + (-2xy)(-y^2 x) + (-2xy)(-y) + (-2xy)(-5)$
  $= -2x^3 y^2 + 2x^2 y^3 + 2xy^2 + 10xy$
32. $-24ab + 3a^3 b + 3ab^3 - 3a^2 b^2$
33. $(-3)(x) + (-3)(-2y) + (-3)(2) = -3x + 6y - 6$
34. $-2x + 6y - 8$
35. $(3x^3)(-2xy) + (-2x^2 y)(-2xy) + (y^3)(-2xy)$
  $= -6x^4 y + 4x^3 y^2 - 2xy^4$
36. $-8yz^4 + 2y^2 z^3 + 2y^4 z$
37. $(2xy^2 z)(-5xz^3) + (-7x^2 z^2)(-5xz^3) = -10x^2 y^2 z^4 + 35x^3 z^5$
38. $-9a^3 bc^4 + 12a^2 b^3 c^3$
39. $(5x^2 y^3 z)(-4xz^2) + (-2xz^3)(-4xz^2) + (y^4)(-4xz^2)$
  $= -20x^3 y^3 z^3 + 8x^2 z^5 - 4xy^4 z^2$
40. $-12x^3 y^3 z^2 + 6xy^3 z^3 + 8x^2 y^2 z^4$

Exercises 3.5 (page 99)

1. $8 + (a - b) = 8 + 1(a - b) = 8 + a - b$ **2.** $7 + m - n$
3. $5 - (x - y) = 5 - 1(x - y) = 5 - x + y$ **4.** $6 - a + b$

5. $12 - 3(m - n) = 12 - 3m + 3n$ **6.** $14 - 5x + 5y$

7. $(R - S) - 8 = R - S - 8$ **8.** $x - y - 2$

9. $(R - S)(-8) = -8R + 8S$ **10.** $-2x + 2y$

11. $(3x - 2y)(-5z) = -15xz + 10yz$ **12.** $-10ac + 14bc$

13. $(3x - 2y) - 5z = 3x - 2y - 5z$ **14.** $5a - 7b - 2c$

15. $10 - 2(a - b) = 10 - 2a + 2b$ (Using the distributive rule)

16. $12 - 5x + 10y$

17. $2(x - y) + 3 = 2x - 2y + 3$ (Using the distributive rule)

18. $4a - 8b + 5$

19. $2a - 3(x - y) = 2a - 3x + 3y$ (Using the distributive rule)

20. $5x - 2a + 2b$

21. $a - [x - 1\,(b - c)]$

$= a - 1\,[x - b + c]$ (Removing innermost grouping symbols first)

$= a - x + b - c$

22. $a - x - b + c$

23. $4 - 2[a - 3(x - y)] = 4 - 2[a - 3x + 3y]$

$= 4 - 2a + 6x - 6y$

24. $8 - 3x + 18a - 6b$

25. $3(a - 2x) - 2(y - 3b) = 3a - 6x - 2y + 6b$

26. $4x - 2b - 3y + 15a$

27. $-10[-2(x - 3y) + a] - b = -10[-2x + 6y + a] - b$

$= 20x - 60y - 10a - b$

28. $30x - 15y - 5a - c$

29. $(a - b) - \{[x - 1\,(3 - y)] - R\}$

$= (a - b) - \{[x - 3 + y] - R\}$

$= (a - b) - 1\,\{x - 3 + y - R\} = a - b - x + 3 - y + R$

30. $x - y - a + c - 5 + b$

Exercises 3.6 (page 102)

1. $(15 - 3)x = 12x$ **2.** $3a$ **3.** $(5 - 12)a = -7a$

4. $-14x$ **5.** $(2 - 5 + 6)a = 3a$ **6.** $4y$

7. $(5 - 8 + 1)x = -2x$ **8.** $-a$

9. $3x - 3x + 2y = (3 - 3)x + 2y = 0x + 2y = 2y$ **10.** $4a$

11. $(4 + 1 - 10)y = -5y$ **12.** $-3x$

13. $(3 - 5 + 2)mn = (0)mn = 0$ **14.** 0

15. $2xy - 5xy + xy = (2 - 5 + 1)xy = -2xy$ **16.** $4mn$

17. $(8 - 2)x^2y = 6x^2y$ **18.** $7ab^2$ **19.** $(1 - 3)a^2b = -2a^2b$

20. $-4x^2y^2$

21. $5ab - 2ab + 2c = (5 - 2)ab + 2c = 3ab + 2c$

22. $3xy - 3z$ **23.** $(5 - 2 - 4)xyz^2 = -xyz^2$ **24.** $2a^2bc$

25. $(5 - 2)u + 10v = 3u + 10v$ **26.** $4w + 5v$

27. $8x - 4x - 2y = (8 - 4)x - 2y = 4x - 2y$ **28.** $11x - 8y$

29. $7x^2y - 4x^2y - 2xy^2 = (7 - 4)x^2y - 2xy^2 = 3x^2y - 2xy^2$

30. $2xy^2 - 5x^2y$

31. $(5x^2 - 2x^2) + (-3x + 8x) + (7 - 9) = 3x^2 + 5x - 2$

32. $-2y^2 + 2y + 1$ **33.** $(12.67 + 9.08 - 6.73)\text{sec} = 15.02 \text{ sec}$

34. 83.00 ft

35. $a + 5(b + 3c + a) = \underline{a} + 5b + 15c + \underline{5a} = 6a + 5b + 15c$

36. $10x + 9y + 63z$

37. $7x + 4x(3y - 5 + 8x) = \underline{7x} + 12xy - \underline{20x} + 32x^2$

$= 32x^2 + 12xy - 13x$

38. $65y + 32yz - 72y^2$

39. $4x - 1(5 - 2y + 8x) = \underline{4x} - 5 + 2y - \underline{8x} = 2y - 5 - 4x$

40. $1 - 2a - 3b - 4c$

41. $6x^2 + 5 - 1(3x + 4 - x^2) = \underline{6x^2} + \underset{\sim}{5} - 3x - \underset{\sim}{4} + \underline{x^2}$

$= 7x^2 - 3x + 1$

42. $-3a^2 + 5a - 3$

43. $8x^2(7x^2 - 3x + 1) - 5x(6x^2 - 8x + 9)$

$= 56x^4 - \underline{24x^3} + \underset{\sim}{8x^2} - \underline{30x^3} + \underset{\sim}{40x^2} - 45x$

$= 56x^4 - 54x^3 + 48x^2 - 45x$

44. $42a^5 + 18a^4 + 33a^3 - 15a^2$

Exercises 3.7A (page 105)

1. y^{10} **2.** N^{12} **3.** x^{16} **4.** z^{28} **5.** 2^8 **6.** 3^8

7. x^5y^5 **8.** a^4b^4 **9.** 2^6c^6 **10.** 3^4x^4 **11.** x^{28}

12. v^{24} **13.** 10^6 **14.** 10^{14} **15.** $(-2 \cdot 3)^2 = (-6)^2 = 36$

16. 144 **17.** $2(-3)^3 = 2(-27) = -54$ **18.** -24

19. $-4 \cdot 5^2 = -4 \cdot 25 = -100$ **20.** -48

Exercises 3.7B (page 109)

1. $\dfrac{x^7}{x^2} = x^{7-2} = x^5$ **2.** y^2 **3.** $x^4 - x^2$ (cannot be rewritten)

4. $s^8 - s^3$ **5.** $\dfrac{a^5}{a} = a^{5-1} = a^4$ **6.** b^6

7. $\dfrac{10^{11}}{10} = 10^{11-1} = 10^{10}$ **8.** 5^5 **9.** $\dfrac{x^8}{y^4}$ (cannot be rewritten)

10. $\dfrac{a^4}{b^3}$ **11.** $\dfrac{6x^2}{2x} = \dfrac{\overset{3}{\cancel{6}}}{\underset{1}{\cancel{2}}} \cdot \dfrac{x^2}{x} = \dfrac{3}{1} \cdot \dfrac{x}{1} = 3x$ **12.** $3y^2$

13. $\dfrac{a^3}{b^2}$ cannot be simplified because the bases are different.   **14.** $\dfrac{x^5}{y^3}$

15. $\dfrac{10x^4}{5x^3} = \dfrac{\overset{2}{\cancel{10}}}{\underset{1}{\cancel{5}}} \cdot \dfrac{x^4}{x^3} = \dfrac{2}{1} \cdot \dfrac{x}{1} = 2x$ **16.** $\dfrac{5y^3}{3}$

17. $\dfrac{12h^4k^3}{8h^2k} = \dfrac{12}{8} \cdot \dfrac{h^4}{h^2} \cdot \dfrac{k^3}{k} = \dfrac{3}{2} \cdot \dfrac{h^2}{1} \cdot \dfrac{k^2}{1} = \dfrac{3h^2k^2}{2}$ **18.** $\dfrac{4a^4b}{3}$

19. $\dfrac{x^{5\alpha}}{x^{3\alpha}} = x^{5\alpha - 3\alpha} = x^{2\alpha}$ **20.** M^{4x}

21. $\dfrac{a^4 - b^3}{a^2}$ (cannot be rewritten)   **22.** $\dfrac{x^6 + y^4}{y^2}$ **23.** $\dfrac{s^7}{t^7}$

24. $\dfrac{x^9}{y^9}$ **25.** $\dfrac{2^4}{x^4}$ **26.** $\dfrac{3^2}{z^2}$ **27.** $\dfrac{x^6}{2^6}$ **28.** $\dfrac{c^3}{5^3}$

29. $(a^2b^3)^2 = a^{2 \cdot 2}b^{3 \cdot 2} = a^4b^6$ **30.** $x^{12}y^{15}$

31. $(2z^3)^2 = 2^{1 \cdot 2}z^{3 \cdot 2} = 2^2z^6$ **32.** 3^3w^6

33. $\left(\dfrac{xy^4}{z^2}\right)^2 = \dfrac{x^{1 \cdot 2}y^{4 \cdot 2}}{z^{2 \cdot 2}} = \dfrac{x^2y^8}{z^4}$ **34.** $\dfrac{a^9b^3}{c^6}$

35. $\left(\dfrac{5y^3}{2x^2}\right)^4 = \dfrac{5^{1 \cdot 4}y^{3 \cdot 4}}{2^{1 \cdot 4}x^{2 \cdot 4}} = \dfrac{5^4y^{12}}{2^4x^8}$ **36.** $\dfrac{6^3b^{12}}{7^3c^6}$

37. $\left(\dfrac{x^3y^7}{xy^2}\right)^3 = (x^2y^5)^3 = x^{2 \cdot 3}y^{5 \cdot 3} = x^6y^{15}$ **38.** a^6b^{15}

39. $\dfrac{(-4)^2}{-4^2} = \dfrac{16}{-16} = -1$ **40.** -1

Exercises 3.8 (page 117)

1. $x^{-4} = \dfrac{1}{x^4}$ **2.** $\dfrac{1}{y^7}$ **3.** $\dfrac{1}{a^{-4}} = a^4$ **4.** b^5

5. $r^{-4}st^{-2} = \dfrac{r^{-4}}{1} \cdot \dfrac{s}{1} \cdot \dfrac{t^{-2}}{1} = \dfrac{1}{r^4} \cdot \dfrac{s}{1} \cdot \dfrac{1}{t^2} = \dfrac{s}{r^4 t^2}$ **6.** $\dfrac{t}{r^5 s^3}$

7. $(xy)^{-2} = \dfrac{1}{(xy)^2} = \dfrac{1}{x^2 y^2}$ **8.** $\dfrac{1}{a^4 b^4}$

9. $\dfrac{h^2}{k^{-4}} = \dfrac{h^2}{1} \cdot \dfrac{1}{k^{-4}} = \dfrac{h^2}{1} \cdot \dfrac{k^4}{1} = h^2 k^4$ **10.** $m^3 n^2$

11. $\dfrac{x^{-4}}{y} = \dfrac{x^{-4}}{1} \cdot \dfrac{1}{y} = \dfrac{1}{x^4} \cdot \dfrac{1}{y} = \dfrac{1}{x^4 y}$ **12.** $\dfrac{1}{a^5 b}$

13. $ab^{-2}c^0 = \dfrac{a}{1} \cdot \dfrac{b^{-2}}{1} \cdot \dfrac{c^0}{1} = \dfrac{a}{1} \cdot \dfrac{1}{b^2} \cdot \dfrac{1}{1} = \dfrac{a}{b^2}$ **14.** $\dfrac{z}{x^3}$

15. $x^{-3} \cdot x^4 = x^{-3+4} = x^1 = x$ **16.** y^4

17. $10^3 \cdot 10^{-2} = 10^{3-2} = 10^1 = 10$ **18.** $\frac{1}{2}$

19. $(x^2)^{-4} = x^{2(-4)} = x^{-8} = \dfrac{1}{x^8}$ **20.** $\dfrac{1}{z^6}$

21. $(a^{-2})^3 = a^{-2(3)} = a^{-6} = \dfrac{1}{a^6}$ **22.** $\dfrac{1}{b^{10}}$

23. $\dfrac{y^{-2}}{y^5} = y^{(-2)-(5)} = y^{-7} = \dfrac{1}{y^7}$ **24.** $\dfrac{1}{z^4}$

25. $\dfrac{10^2}{10^{-5}} = 10^{2-(-5)} = 10^7$ **26.** 2^5 **27.** $\left(\dfrac{x}{y}\right)^{-3} = \left(\dfrac{y}{x}\right)^3 = \dfrac{y^3}{x^3}$

28. $\dfrac{t^2}{s^2}$ **29.** $\dfrac{1}{x^{+2}} = \dfrac{x^{-2}}{1} = x^{-2}$ **30.** y^{-3}

31. $\dfrac{h}{k} = \dfrac{h}{1} \cdot \dfrac{1}{k} = \dfrac{h}{1} \cdot \dfrac{k^{-1}}{1} = hk^{-1}$ **32.** mn^{-1}

33. $\dfrac{x^2}{yz^5} = \dfrac{x^2}{1} \cdot \dfrac{1}{y} \cdot \dfrac{1}{z^5} = \dfrac{x^2}{1} \cdot \dfrac{y^{-1}}{1} \cdot \dfrac{z^{-5}}{1} = x^2 y^{-1} z^{-5}$ **34.** $a^3 b^{-2} c^{-1}$

35. $10^4 \cdot 10^{-2} = 10^{4-2} = 10^2 = 100$ **36.** 3

37. $10^{-4} = \dfrac{1}{10^4} = \dfrac{1}{10,000}$ **38.** $\dfrac{1}{8}$ **39.** $5^0 \cdot 7^2 = 1 \cdot 49 = 49$

40. 64 **41.** $\dfrac{10^0}{10^2} = \dfrac{1}{100}$ **42.** 25

43. $\dfrac{10^{-3} \cdot 10^2}{10^5} = 10^{-3+2-5} = 10^{-6} = \dfrac{1}{10^6} = \dfrac{1}{1,000,000}$ **44.** $\dfrac{1}{8}$

45. $(10^2)^{-1} = 10^{2(-1)} = 10^{-2} = \dfrac{1}{10^2} = \dfrac{1}{100}$ **46.** $\dfrac{1}{64}$

47. $(-5)^{-3} = \dfrac{1}{(-5)^3} = -\dfrac{1}{125}$ **48.** $-\dfrac{1}{64}$

49. $(-12)^{-2} = \dfrac{1}{(-12)^2} = \dfrac{1}{144}$ **50.** $\dfrac{1}{169}$

51. $\dfrac{a^3 b^0}{c^{-2}} = \dfrac{a^3}{1} \cdot \dfrac{b^0}{1} \cdot \dfrac{1}{c^{-2}} = \dfrac{a^3}{1} \cdot \dfrac{1}{1} \cdot \dfrac{c^2}{1} = a^3 c^2$ **52.** $e^2 f^3$

53. $\dfrac{p^4 r^{-1}}{t^{-2}} = \dfrac{p^4}{1} \cdot \dfrac{r^{-1}}{1} \cdot \dfrac{1}{t^{-2}} = \dfrac{p^4}{1} \cdot \dfrac{1}{r^1} \cdot \dfrac{t^2}{1} = \dfrac{p^4 t^2}{r}$ **54.** $\dfrac{u^5 w^3}{v^2}$

55. $\dfrac{8x^{-3}}{12x} = \dfrac{\overset{2}{\cancel{8}}}{\underset{3}{\cancel{12}}} \cdot \dfrac{x^{-3}}{x} = \dfrac{2}{3} \cdot \dfrac{x^{-3-1}}{1} = \dfrac{2}{3} \cdot \dfrac{x^{-4}}{1} = \dfrac{2}{3} \cdot \dfrac{1}{x^4} = \dfrac{2}{3x^4}$

56. $\dfrac{3}{2y^3}$ **57.** $\dfrac{20h^{-2}}{35h^{-4}} = \dfrac{\overset{4}{\cancel{20}}}{\underset{7}{\cancel{35}}} \cdot \dfrac{h^{-2}}{h^{-4}} = \dfrac{4}{7} \cdot \dfrac{h^{(-2)-(-4)}}{1} = \dfrac{4}{7} \cdot \dfrac{h^2}{1} = \dfrac{4h^2}{7}$

58. $\dfrac{5k^3}{4}$

59. $\dfrac{15m^0 n^{-2}}{5m^{-3}n^4} = \dfrac{\overset{3}{\cancel{15}}}{\underset{1}{\cancel{5}}} \cdot \dfrac{m^0}{m^{-3}} \cdot \dfrac{n^{-2}}{n^4} = \dfrac{3}{1} \cdot \dfrac{m^{0-(-3)}}{1} \cdot \dfrac{n^{-2-4}}{1}$

$= \dfrac{3}{1} \cdot \dfrac{m^3}{1} \cdot \dfrac{n^{-6}}{1} = \dfrac{3}{1} \cdot \dfrac{m^3}{1} \cdot \dfrac{1}{n^6} = \dfrac{3m^3}{n^6}$

60. $\dfrac{7x^2 y}{6}$ **61.** $x^{3m} \cdot x^{-m} = x^{3m-m} = x^{2m}$ **62.** y^{3n}

63. $(x^{3b})^{-2} = x^{3b(-2)} = x^{-6b} = \dfrac{1}{x^{6b}}$ **64.** $\dfrac{1}{y^{6a}}$

65. $\dfrac{x^{2a}}{x^{-5a}} = x^{2a-(-5a)} = x^{7a}$ **66.** a^{8x}

67. $(m^{-2}n)^4 = m^{(-2)4}n^{1\cdot4} = m^{-8}n^4 = \dfrac{n^4}{m^8}$ **68.** $\dfrac{r^5}{p^{15}}$

69. $(x^{-2}y^3)^{-4} = x^{(-2)(-4)}y^{3(-4)} = x^8 y^{-12} = \dfrac{x^8}{y^{12}}$ **70.** $\dfrac{w^6}{z^8}$

71. $\left(\dfrac{M^{-2}}{N^3}\right)^4 = \dfrac{M^{(-2)4}}{N^{3\cdot4}} = \dfrac{M^{-8}}{N^{12}} = \dfrac{1}{M^8 N^{12}}$ **72.** $R^{15}S^{12}$

73. $\left(\dfrac{a^2 b^{-4}}{b^{-5}}\right)^2 = (a^2 b^{5-4})^2 = a^{2\cdot2}b^{1\cdot2} = a^4 b^2$ **74.** $x^3 y^6$

75. $\left(\dfrac{mn^{-1}}{m^3}\right)^{-2} = (m^{1-3}n^{-1})^{-2} = m^{(-2)(-2)}n^{(-1)(-2)} = m^4 n^2$

76. $a^3 b^6$

77. $\left(\dfrac{x^4}{x^{-1}y^{-2}}\right)^{-1} = (x^{4+1}y^2)^{-1} = x^{5(-1)}y^{2(-1)} = x^{-5}y^{-2} = \dfrac{1}{x^5 y^2}$

78. $\dfrac{1}{x^5 y^4}$ **79.** $(10^0 k^{-4})^{-2} = k^{(-4)(-2)} = k^8$ **80.** z^{10}

81. $\left(\dfrac{r^7 s^8}{r^9 s^6}\right)^0 = 1$ **82.** 1

Exercises 3.9 (page 122)

1. 8060 **2.** $31,400$ **3.** 0.00132 **4.** 0.00082 **5.** 5.26

6. 9.11 **7.** $3{,}5{,}300 = 3.53 \times 10^4$ **8.** 8.25×10^5

9. $0.003.12 = 3.12 \times 10^{-3}$ **10.** 1.45×10^{-4}

11. $8.97 = 8.97 \times 10^0$ **12.** 2.497×10^0

13. $0.8.15 = 8.15 \times 10^{-1}$ **14.** 2.74×10^{-1}

15. $0.0002. = 2 \times 10^{-4}$ **16.** 6×10^{-3}

17. $4.5 = 4.5 \times 10^1$ **18.** 1.2×10^1 **19.** 5.418×10^{11}

20. 3×10^{-10} **21.** 9×10^{-4} **22.** 1.5×10^{-3}

Review Exercises 3.10 (page 124)

1a. 3 terms
  b. $2ab$

2a. 2 terms
  b. $-2(x^2 + y^2)$

3a. 1 term
  b. No second term

4a. 2 terms
  b. $-2(x^3 + 1)$

5. Term **6.** Factor **7.** $m^2 m^3 = m^{2+3} = m^5$ **8.** y^8

9. $2 \cdot 2^2 = 2^{1+2} = 2^3$ **10.** x^{7y} **11.** $10 \cdot 10^y = 10^{1+y}$

12. a^2 **13.** $41x$ **14.** $-54xy^2$ **15.** $-9ab - 3a + 5b$

16. $2x - y - 3$ **17.** $8(3x - y) = 24x - 8y$ **18.** $-5a - 5b$

19. $(3s - 2t)(-2) = -6s + 4t$ **20.** $3x - 2t - 2$

21. $5x(x^2 + 7) = 5x^3 + 35x$ **22.** $-10x^3y - 6x^2 + 2x$

23. $6 + 2(5x^2y + 3x - 1) = \underline{6} + 10x^2y + 6x - \underline{2}$
   $= 4 + 10x^2y + 6x$

24. $-63x^{12}y^8$ **25.** $100e^6f^{10}g^5$ **26.** $27x^3$

27. $(x^2y^3)^4 = x^{2\cdot4}y^{3\cdot4} = x^8y^{12}$ **28.** n^{15}

29. $(p^{-3})^5 = p^{-3\cdot5} = p^{-15} = \dfrac{1}{p^{15}}$ **30.** $\dfrac{1}{16c^4}$ **31.** $(k^{-7})^0 = 1$

32. $\dfrac{t^3}{s^{12}}$ **33.** $(-2x^4)^2 = 4x^8$ **34.** $-\dfrac{125}{a^{12}}$

35. $(-10)^{-3} = \dfrac{1}{(-10)^3} = -\dfrac{1}{1000}$ **36.** r^2

37. $\dfrac{x^{-4}}{x^5} = \dfrac{1}{x^5x^4} = \dfrac{1}{x^9}$ **38.** $\dfrac{c^4}{d^4}$ **39.** $\dfrac{m^0}{m^{-3}} = \dfrac{1 \cdot m^3}{1} = m^3$

40. $x^4 + x^2$ (cannot be simplified)   **41.** $\dfrac{n^{-6}}{n^0} = \dfrac{1}{n^0n^6} = \dfrac{1}{n^6}$

42. $\dfrac{x^{10}y^{15}}{z^{20}}$ **43.** $\left(\dfrac{a^{-4}}{b^3c^0}\right)^{-5} = \left(\dfrac{b^3c^0}{a^{-4}}\right)^5 = (b^3a^4)^5 = b^{15}a^{20}$

44. $\dfrac{1}{x^8}$ **45.** $\left(\dfrac{r^{-6}}{s^5t^{-3}}\right)^4 = \left(\dfrac{t^3}{s^5r^6}\right)^4 = \dfrac{t^{12}}{s^{20}r^{24}}$ **46.** 1

47. $(5a^3b^{-4})^{-2} = 5^{-2}a^{-6}b^8$
   $= \dfrac{b^8}{5^2a^6}$ or $\dfrac{b^8}{25a^6}$

48. $5 + n$

49. $2x(3x^2 - x) - 1(3x^2 - 4) = 6x^3 - 2x^2 - 3x^2 + 4$
   $= 6x^3 - 5x^2 + 4$

50. $21m^3n^3 - 14m^2n^2$ **51.** $\dfrac{a^3}{b^2} = a^3b^{-2}$ **52.** m^2n^3

53. $\dfrac{u^{-4}v^3}{10^2w^{-5}} = 10^{-2}u^{-4}v^3w^5$ **54.** $\dfrac{1}{16}$

55. $10^{-4} = \dfrac{1}{10^4}$ or $\dfrac{1}{10,000}$ **56.** 8   **57.** $4^0 \cdot 3^2 = 1 \cdot 9 = 9$

58. -1 **59.** $4.\underrightarrow{5,300} = 4.53 \times 10^4$ **60.** 3.156×10^{-2}

Chapter 3 Diagnostic Test (page 131)

Following each problem number is the textbook section number (in parentheses) where that kind of problem is discussed.

1. (3.1) **a.** $5x$ is a factor of $5xy$.
   b. $5x$ is a term of $5x + y$.
   c. $5x$ is a term of $3 + 5x$.
   d. $5x$ is a term of $(3 + 2y) + 5x$.
   e. $5x$ is a factor of $(3 + 2y)(+5x)$.

2. (3.3) $(-3xy)(5x^3y)(-2xy^4)$. Because two factors are negative, the answer is positive.
   $(3 \cdot 5 \cdot 2)(xx^3x)(yyy^4) = 30x^5y^6$

3. (3.4) $2(x - 3y)$
   $= (2)x + (2)(-3y)$
   $= 2x - 6y$

4. (3.4) $(x - 4)(-5)$
   $= x(-5) + (-4)(-5)$
   $= -5x + 20$

5. (3.4) $2xy^2(x^2 - 3y - 4)$
   $= 2xy^2(x^2) + 2xy^2(-3y) + 2xy^2(-4)$
   $= 2x^3y^2 - 6xy^3 - 8xy^2$

6. (3.5) $5 - (x - y) = 5 - 1(x - y) = 5 - x + y$

7. (3.5) $-2[-4(3c - d) + a] - b = -2[-12c + 4d + a] - b$
   $= 24c - 8d - 2a - b$

8. (3.6) $4x - 3x + 5x = (4 - 3 + 5)x = 6x$

9. (3.6) $2a - 5b - 7 - 3b + 4 - 5a = -3a - 8b - 3$

10. (3.2) $x^3 \cdot x^4 = x^{3+4} = x^7$ **11.** (3.7) $(x^2)^3 = x^{2\cdot3} = x^6$

12. (3.7) $\dfrac{x^5}{x^2} = x^{5-2} = x^3$ **13.** (3.8) $x^{-4} = \dfrac{1}{x^4}$

14. (3.8) $x^2y^{-3} = \dfrac{x^2}{1} \cdot \dfrac{1}{y^3} = \dfrac{x^2}{y^3}$

15. (3.8) $\dfrac{a^{-3}}{b} = \dfrac{a^{-3}}{1} \cdot \dfrac{1}{b} = \dfrac{1}{a^3} \cdot \dfrac{1}{b} = \dfrac{1}{a^3b}$

16. (3.7) $\dfrac{x^{5a}}{x^{3a}} = x^{5a-3a} = x^{2a}$ **17.** (3.8) $(4^{3x})^0 = 1$

18. (3.7) $(x^2y^4)^3 = x^{2\cdot3}y^{4\cdot3}$
   $= x^6y^{12}$

19. (3.8) $(a^{-3}b)^2 = a^{-3\cdot2}b^{1\cdot2} = a^{-6}b^2$
   $= \dfrac{1}{a^6} \cdot \dfrac{b^2}{1} = \dfrac{b^2}{a^6}$

20. (3.7) $\left(\dfrac{p^3}{q^2}\right)^2 = \dfrac{p^{3\cdot2}}{q^{2\cdot2}} = \dfrac{p^6}{q^4}$

21. (3.8) $\left(\dfrac{x}{y^2}\right)^{-3} = \dfrac{x^{1(-3)}}{y^{2(-3)}} = \dfrac{x^{-3}}{y^{-6}} = \dfrac{1}{x^3} \cdot \dfrac{y^6}{1} = \dfrac{y^6}{x^3}$

22. (3.8) $\left(\dfrac{\cancel{4}x^{-2}}{\cancel{2}x^{-3}}\right)^{-1}_1 = (2x^{-2-(-3)})^{-1} = (2^1x^1)^{-1}$
   $= 2^{1(-1)}x^{1(-1)} = 2^{-1}x^{-1}$
   $= \dfrac{1}{2^1} \cdot \dfrac{1}{x^1} = \dfrac{1}{2x}$

23. (3.6) $5x - 3(y - x) = 5x - 3y + 3x = 8x - 3y$

24. (3.6) $3h(2k^2 - 5h) - h(2h - 3k^2) = 6hk^2 - 15h^2 - 2h^2 + 3hk^2$
   $= 9hk^2 - 17h^2$

25. (3.6) $x(x^2 + 2x + 4) - 2(x^2 + 2x + 4) = x^3 + 2x^2 + 4x - 2x^2 - 4x - 8$
   $= x^3 - 8$

26. (3.8) $\dfrac{a^3}{b} = a^3b^{-1}$

27. (3.8) $\dfrac{h^{-2}}{k^{-3}h^{-4}} = h^{-2-(-4)}k^3$
   $= h^2k^3$

28. (3.8) $2^3 \cdot 2^2 = 2^{3+2}$
   $= 2^5 = 32$

29. (3.8) $10^{-4} \cdot 10^2 = 10^{-2} = \dfrac{1}{10^2} = \dfrac{1}{100}$

30. (3.8) $5^{-2} = \dfrac{1}{5^2} = \dfrac{1}{25}$

31. (3.8) $(2^{-3})^2 = 2^{(-3)2} = 2^{-6} = \dfrac{1}{2^6} = \dfrac{1}{64}$

32. (3.8) $(4^{-2})^{-1} = 4^{(-2)(-1)}$
   $= 4^2 = 16$

33. (3.8) $\dfrac{10^{-3}}{10^{-4}} = 10^{-3-(-4)} = 10^1$

34. (3.8) $\dfrac{-3^2}{(-3)^2} = \dfrac{-(3)(3)}{(-3)(-3)} = \dfrac{-9}{9} = -1$

35. (3.8) $(5^0)^2 = 1^2 = 1$

36. (3.9) **a.** $1.326 = 1.326 \times 10^0$
   b. $0.\underleftarrow{5.27} = 5.27 \times 10^{-1}$

ANSWERS

Cumulative Review Exercises: Chapters 1–3 (page 132)

1. False   **2.** True   **3.** False   **4.** False   **5.** False

6. False   **7.** $\dfrac{14 - 23}{-8 + 11} = \dfrac{-9}{3} = -3$ **8.** 12

9. $-4^2 + (-4)^2 = -16 + 16 = 0$ **10.** 0

11. $6(-3) - 4\sqrt{36} = 6(-3) - 4(6) = -18 - 24 = -42$

12. -28 **13.** $\dfrac{0}{-11} = 0$ **14.** -25

15. $A = P(1 + rt)$

$A = 1200[1 + 0.15(4)]$

$A = 1200[1 + 0.6]$

$A = 1200[1.6] = 1920$

16. $\doteq 1256$ **17.** $(-5p)(4p^3) = -20p^4$ **18.** $200h^9 j^{10} k^5$

19. $7 - 4(3xy - 3) = \underline{7} - 12xy + \underline{12} = 19 - 12xy$

20. $-26x^2y^2 + 9x^2y^3$

Exercises 4.1 (page 135)

1. $(-4) + 2 \neq 3$. Therefore, -4 is not a solution of $x + 2 = 3$.

2. No

3. $(4) + 1 = 5$. Therefore, 4 is a solution of $x + 1 = 5$.   **4.** Yes

5. $2 + (-3) = -1$. Therefore, -3 is in the solution set.   **6.** Yes

7. $5 + (3) \neq 1$. Therefore, 3 is not in the solution set.   **8.** No

Exercises 4.2 (page 138)

1. $x + 5 = 8$ $Check: x + 5 = 8$

$\quad\ \ -5 \quad -5$ $\qquad 3 + 5 = 8$

$\quad x \quad\ \ = \quad 3$ $\qquad\quad\ 8 = 8$

2. 5

3. $x - 3 = 4$ $Check: x - 3 = 4$

$\quad\ \ +3 \quad +3$ $\qquad 7 - 3 = 4$

$\quad x \quad\ \ = \quad 7$ $\qquad\quad\ 4 = 4$

4. 9

5. $3 + x = -4$ $Check:\quad 3 + x = -4$

$\ -3 \qquad -3$ $\qquad\ 3 + (-7) = -4$

$\qquad x = -7$ $\qquad\qquad -4 = -4$

6. -7

7. $x + 4 = 21$ $Check:\ x + 4 = 21$

$\quad\ \ -4 \quad -4$ $\qquad 17 + 4 = 21$

$\quad x \quad\ \ = \quad 17$ $\qquad\quad 21 = 21$

8. 9

9. $x - 35 = 7$ $Check:\ x - 35 = 7$

$\quad +35 \quad +35$ $\qquad 42 - 35 = 7$

$\quad x \quad\ \ = \quad 42$ $\qquad\qquad 7 = 7$

10. 51

11. $9 = x + 5$ $Check: 9 = x + 5$

$\ -5 \qquad -5$ $\qquad 9 = 4 + 5$

$\quad 4 = x$ $\qquad\ 9 = 9$

$\quad x = 4$

12. 3

13. $12 = x - 11$ $Check: 12 = x - 11$

$\ +11 \qquad +11$ $\qquad 12 = 23 - 11$

$\quad 23 = x$ $\qquad\quad 12 = 12$

$\quad x = 23$

14. 29

15. $-17 + x = 28$ $Check: -17 + x = 28$

$\ +17 \qquad +17$ $\qquad -17 + 45 = 28$

$\qquad x = 45$ $\qquad\qquad 28 = 28$

16. 47

17. $-28 = -15 + x$ $Check: -28 = -15 + x$

$\ +15 \quad +15$ $\qquad -28 = -15 + (-13)$

$\ -13 = \qquad x$ $\qquad -28 = -28$

$\quad x = -13$

18. -29

19. $x + \dfrac{1}{2} = 2\dfrac{1}{2}$ $Check: x + \dfrac{1}{2} = 2\dfrac{1}{2}$

$\quad -\dfrac{1}{2} \quad -\dfrac{1}{2}$ $\qquad 2 + \dfrac{1}{2} = 2\dfrac{1}{2}$

$\quad x \quad\ = \quad 2$ $\qquad\quad 2\dfrac{1}{2} = 2\dfrac{1}{2}$

20. 5

21.

$$5.6 + x = 2.8$$
$$\underline{-5.6 \qquad -5.6}$$
$$x = -2.8$$

Check:
$$5.6 + x = 2.8$$
$$5.6 + (-2.8) = 2.8$$
$$2.8 = 2.8$$

22. -0.08

23.

$$7.84 = x - 3.98$$
$$\underline{+3.98 \qquad +3.98}$$
$$11.82 = x$$
$$x = 11.82$$

Check: $7.84 = x - 3.98$
$$7.84 = 11.82 - 3.98$$
$$7.84 = 7.84$$

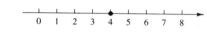

24. 7.07

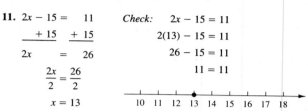

Exercises 4.3 (page 142)

1. $2x = 8$

$$\frac{2x}{2} = \frac{8}{2}$$
$$x = 4$$

Check: $2x = 8$
$$2(4) = 8$$
$$8 = 8$$

2. 5

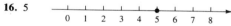

3. $21 = 7x$

$$\frac{21}{7} = \frac{7x}{7}$$
$$3 = x$$
$$x = 3$$

Check: $21 = 7x$
$$21 = 7(3)$$
$$21 = 21$$

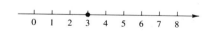

4. 7

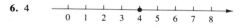

5. $11x = 33$

$$\frac{11x}{11} = \frac{33}{11}$$
$$x = 3$$

Check: $11x = 33$
$$11(3) = 33$$
$$33 = 33$$

6. 4

7. $4x + 1 = 9$

$$\underline{-1 \qquad -1}$$
$$4x = 8$$
$$\frac{4x}{4} = \frac{8}{4}$$
$$x = 2$$

Check: $4x + 1 = 9$
$$4(2) + 1 = 9$$
$$8 + 1 = 9$$
$$9 = 9$$

8. 2

9. $6x - 2 = 10$

$$\underline{+2 \qquad +2}$$
$$6x = 12$$
$$\frac{6x}{6} = \frac{12}{6}$$
$$x = 2$$

Check: $6x - 2 = 10$
$$6(2) - 2 = 10$$
$$12 - 2 = 10$$
$$10 = 10$$

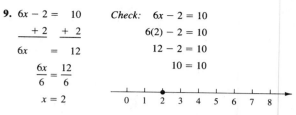

10. 1

11. $2x - 15 = 11$

$$\underline{+15 \qquad +15}$$
$$2x = 26$$
$$\frac{2x}{2} = \frac{26}{2}$$
$$x = 13$$

Check: $2x - 15 = 11$
$$2(13) - 15 = 11$$
$$26 - 15 = 11$$
$$11 = 11$$

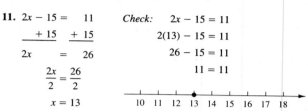

12. 6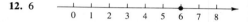

The checks will not be shown for Exercises 13–30.

13. $4x + 2 = -14$

$$\underline{-2 \qquad -2}$$
$$4x = -16$$
$$\frac{4x}{4} = \frac{-16}{4}$$
$$x = -4$$

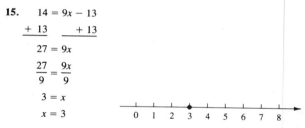

14. -3

15.

$$14 = 9x - 13$$
$$\underline{+13 \qquad +13}$$
$$27 = 9x$$
$$\frac{27}{9} = \frac{9x}{9}$$
$$3 = x$$
$$x = 3$$

16. 5

17. $12x + 17 = 65$

$$\underline{-17 \qquad -17}$$
$$12x = 48$$
$$\frac{12x}{12} = \frac{48}{12}$$
$$x = 4$$

18. 2

19. $8x - 23 = 31$

$$\underline{+23 \qquad +23}$$
$$8x = 54$$
$$\frac{8x}{8} = \frac{54}{8}$$
$$x = 6\frac{3}{4}$$

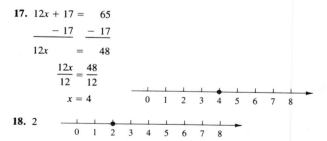

20. $10\frac{1}{3}$

21. $14 - 4x = -28$

$$\underline{-14 \qquad -14}$$

$$-4x = -42$$

$$\frac{-4x}{-4} = \frac{-42}{-4}$$

$$x = 10\frac{1}{2}$$

22. $10\frac{1}{3}$

23. $8 = 25 - 3x$

$$\underline{-25 \qquad -25}$$

$$-17 = \qquad -3x$$

$$\frac{-17}{-3} = \frac{-3x}{-3}$$

$$5\frac{2}{3} = x$$

$$x = 5\frac{2}{3}$$

24. $8\frac{1}{2}$

25. $-73 = 24x + 31$

$$\underline{-31 \qquad -31}$$

$$-104 = 24x$$

$$\frac{-104}{24} = \frac{24x}{24}$$

$$-4\frac{1}{3} = x$$

$$x = -4\frac{1}{3}$$

26. $-2\frac{1}{2}$

27. $18x - 4.8 = 6$

$$\underline{+4.8 \qquad +4.8}$$

$$18x = 10.8$$

$$\frac{18x}{18} = \frac{10.8}{18}$$

$$x = 0.6$$

28. $1.03\frac{1}{3}$

29. $2.5x - 3.8 = -7.9$

$$\underline{+3.8 \qquad +3.8}$$

$$2.5x = -4.1$$

$$\frac{2.5x}{2.5} = \frac{-4.1}{2.5}$$

$$x = -1.64$$

30. $-0.066\frac{2}{3}$

Exercises 4.4 (page 148)

1. $\frac{x}{3} = 4$ 　　*Check:* $\frac{x}{3} = 4$

$$\overset{1}{\cancel{3}}\left(\frac{x}{\cancel{3}}\right)_1 = 3(4) \qquad \frac{12}{3} = 4$$

$$x = 12 \qquad\qquad 4 = 4$$

2. 15

3. $\frac{x}{5} = -2$ 　　*Check:* $\frac{x}{5} = -2$

$$\overset{1}{\cancel{5}}\left(\frac{x}{\cancel{5}}\right)_1 = 5(-2) \qquad \frac{-10}{5} = -2$$

$$x = -10 \qquad\qquad -2 = -2$$

4. -24

5. $4 - \frac{x}{7} = 0$ 　　*Check:* $4 - \frac{x}{7} = 0$

$$\underline{-4 \qquad -4} \qquad\qquad 4 - \frac{28}{7} = 0$$

$$-\frac{x}{7} = -4 \qquad\qquad 4 - 4 = 0$$

$$\overset{1}{\cancel{7}}\left(-\frac{x}{\cancel{7}}\right)_1 = 7(-4)$$

$$-x = -28$$

$$x = 28$$

6. 24

The checks will not be shown for Exercises 7–24.

7. $-13 = \frac{x}{9}$

$$9(-13) = \overset{1}{\cancel{9}}\left(\frac{x}{\cancel{9}}\right)_1$$

$$-117 = x$$

$$x = -117$$

8. -120

9. $\frac{x}{10} = 3.14$

$$\overset{1}{\cancel{10}}\left(\frac{x}{\cancel{10}}\right)_1 = 10(3.14)$$

$$x = 31.4$$

10. 39

11. $\dfrac{x}{4} + 6 = 9$

$\quad\ \underline{-6 \quad -6}$

$\quad \dfrac{x}{4} = 3$

$\overset{1}{\cancel{4}}\left(\dfrac{x}{\cancel{4}}\right) = 4(3)$

$\quad x = 12$

10 11 12 13 14 15 16 17 18

12. 25

20 21 22 23 24 25 26 27 28

13. $\dfrac{x}{10} - 5 = 13$

$\quad\ \underline{+5 \quad +5}$

$\quad \dfrac{x}{10} = 18$

$\overset{1}{\cancel{10}}\left(\dfrac{x}{\cancel{10}}\right) = 10(18)$

$\quad x = 180$

175 176 177 178 179 180 181 182 183

14. 320

315 316 317 318 319 320 321 322 323

15. $-14 = \dfrac{x}{6} - 7$

$\quad \underline{+7 \qquad +7}$

$\quad -7 = \dfrac{x}{6}$

$6(-7) = \overset{1}{\cancel{6}}\left(\dfrac{x}{\cancel{6}}\right)$

$\quad -42 = x$

$\quad x = -42$

−50 −49 −48 −47 −46 −45 −44 −43 −42 −41

16. −88

−92 −91 −90 −89 −88 −87 −86 −85 −84

17. $7 = \dfrac{2x}{5} + 3$

$\quad \underline{-3 \qquad -3}$

$\quad 4 = \dfrac{2x}{5}$

$5(4) = \overset{1}{\cancel{5}}\left(\dfrac{2x}{\cancel{5}}\right)$

$\quad 20 = 2x$

$\quad \dfrac{20}{2} = \dfrac{2x}{2}$

$\quad 10 = x$

$\quad x = 10$

4 5 6 7 8 9 10 11 12

18. 4

0 1 2 3 4 5 6 7 8

19. $4 - \dfrac{7x}{5} = 11$

$\quad \underline{-4 \qquad -4}$

$\quad -\dfrac{7x}{5} = 7$

$\overset{1}{\cancel{5}}\left(\dfrac{-7x}{\cancel{5}}\right) = 5(7)$

$\quad -7x = 35$

$\quad \dfrac{-7x}{-7} = \dfrac{35}{-7}$

$\quad x = -5$

−6 −5 −4 −3 −2 −1 0 1 2

20. −45

−50 −49 −48 −47 −46 −45 −44 −43 −42

21. $-24 + \dfrac{5x}{8} = 41$

$\quad \underline{+24 \qquad +24}$

$\quad \dfrac{5x}{8} = 65$

$\overset{1}{\cancel{8}}\left(\dfrac{5x}{\cancel{8}}\right) = 8(65)$

$\quad 5x = 8(65)$

$\quad \dfrac{5x}{5} = \dfrac{8\overset{13}{\cancel{(65)}}}{\cancel{5}}$

$\quad x = 8(13) = 104$

101 102 103 104 105 106 107

22. 20

16 17 18 19 20 21 22 23 24

23. $41 = 25 - \dfrac{4x}{5}$

$\quad \underline{-25 \quad -25}$

$\quad 16 = -\dfrac{4x}{5}$

$5(16) = \overset{1}{\cancel{5}}\left(\dfrac{-4x}{\cancel{5}}\right)$

$\quad 5(16) = -4x$

$\quad \dfrac{5(16)}{-4} = \dfrac{-4x}{-4}$

$\quad -20 = x$

$\quad x = -20$

−23 −22 −21 −20 −19 −18 −17

24. −35

−40 −39 −38 −37 −36 −35 −34 −33 −32

Exercises 4.5A (page 151)

1. $3x + 11 = 14x$

$\quad \underline{-3x \qquad -3x}$

$\quad 11 = 11x$

$\quad \dfrac{11}{11} = \dfrac{11x}{11}$

$\quad 1 = x$

$\quad x = 1$

0 1 2 3 4

2. 1

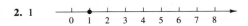

3. $9x - 7 = 2x$

$\underline{-2x + 7 \quad -2x + 7}$

$7x \quad = \quad 7$

$\dfrac{7x}{7} = \dfrac{7}{7}$

$x = 1$

4. 1

5. $2x - 7 = x$

$\underline{-x \quad -x}$

$x - 7 = 0$

$\underline{+7 \quad +7}$

$x \quad = \quad 7$

6. 2

7. $5x = 3x - 4$

$\underline{-3x \quad -3x}$

$2x = \quad -4$

$\dfrac{2x}{2} = \dfrac{-4}{2}$

$x = -2$

8. -3

9. $9 - 2x = x$

$\underline{+2x \quad +2x}$

$9 \quad = \quad 3x$

$\dfrac{9}{3} = \dfrac{3x}{3}$

$3 = x$

$x = 3$

10. 1

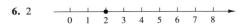

11. $3x - 4 = 2x + 5$

$\underline{-2x \quad -2x}$

$x - 4 = \quad 5$

$\underline{+4 \quad +4}$

$x \quad = \quad 9$

12. 6

13. $6x + 7 = 3 + 8x$

$\underline{-6x \quad -6x}$

$7 = 3 + 2x$

$\underline{-3 \quad -3}$

$4 = \quad 2x$

$\dfrac{4}{2} = \dfrac{2x}{2}$

$2 = x$

$x = 2$

14. -7

15. $7x - 8 = 8 - 9x$

$\underline{+9x \quad +9x}$

$16x - 8 = 8$

$\underline{+8 \quad +8}$

$16x \quad = 16$

$\dfrac{16x}{16} = \dfrac{16}{16}$

$x = 1$

16. 1

17. $3x - 7 - x = 15 - 2x - 6$

$2x - 7 = 9 - 2x$

$\underline{+2x \quad +2x}$

$4x - 7 = 9$

$\underline{+7 \quad +7}$

$4x \quad = 16$

$\dfrac{4x}{4} = \dfrac{16}{4}$

$x = 4$

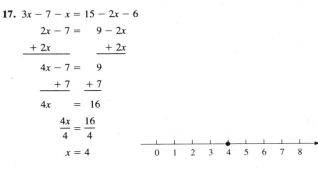

18. -3

19. $8x - 13 + 3x = 12 + 5x - 7$

$11x - 13 = 5 + 5x$

$\underline{-5x \quad -5x}$

$6x - 13 = 5$

$\underline{+13 \quad +13}$

$6x \quad = 18$

$\dfrac{6x}{6} = \dfrac{18}{6}$

$x = 3$

20. 2

21. $7 - 9x - 12 = 3x + 5 - 8x$

$-9x - 5 = -5x + 5$

$\underline{+9x \quad +9x}$

$-5 = 4x + 5$

$\underline{-5 \quad -5}$

$-10 = 4x$

$\dfrac{-10}{4} = \dfrac{4x}{4}$

$-\dfrac{5}{2} = x$

$x = -2\dfrac{1}{2}$

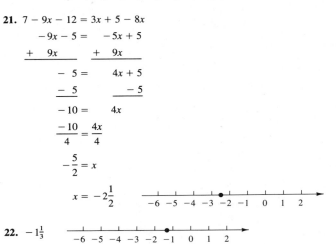

22. $-1\frac{1}{3}$

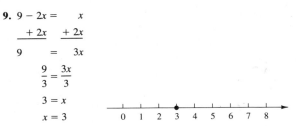

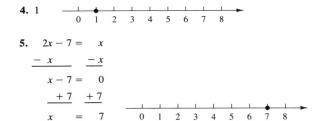

23.
$$7.84 - 1.15x = 2.45$$
$$\underline{-7.84 \qquad\qquad -7.84}$$
$$-1.15x = -5.39$$
$$\frac{-1.15x}{-1.15} = \frac{-5.39}{-1.15}$$
$$x \doteq 4.69$$

24. $\doteq 0.662$

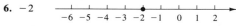

Exercises 4.5B (page 153)

1.
$$5x - 3(2 + 3x) = 6$$
$$5x - 6 - 9x = 6$$
$$-4x = 12$$
$$\frac{-4x}{-4} = \frac{12}{-4}$$
$$x = -3$$

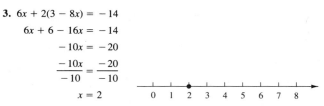

2. -18

3.
$$6x + 2(3 - 8x) = -14$$
$$6x + 6 - 16x = -14$$
$$-10x = -20$$
$$\frac{-10x}{-10} = \frac{-20}{-10}$$
$$x = 2$$

4. 2

5.
$$7x + 5 = 3(3x + 5)$$
$$7x + 5 = 9x + 15$$
$$-10 = 2x$$
$$-5 = x$$
$$x = -5$$

6. -2

7.
$$9 - 4x = 5(9 - 8x)$$
$$9 - 4x = 45 - 40x$$
$$36x = 36$$
$$x = 1$$

8. 2

9.
$$3y - 2(2y - 7) = 2(3 + y) - 4$$
$$3y - 4y + 14 = 6 + 2y - 4$$
$$-y + 14 = 2 + 2y$$
$$-3y = -12$$
$$y = 4$$

10. 1

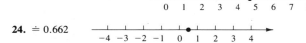

11.
$$6(3 - 4x) + 12 = 10x - 2(5 - 3x)$$
$$18 - 24x + 12 = 10x - 10 + 6x$$
$$30 - 24x = 16x - 10$$
$$40 = 40x$$
$$1 = x$$
$$x = 1$$

12. 1

13.
$$2(3x - 6) - 3(5x + 4) = 5(7x - 8)$$
$$6x - 12 - 15x - 12 = 35x - 40$$
$$-9x - 24 = 35x - 40$$
$$16 = 44x$$
$$\frac{16}{44} = x$$
$$\frac{4}{11} = x$$
$$x = \frac{4}{11}$$

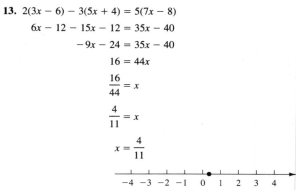

14. $\dfrac{1}{18}$

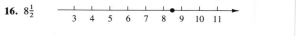

15.
$$6(5 - 4h) = 3(4h - 2) - 7(6 + 8h)$$
$$30 - 24h = 12h - 6 - 42 - 56h$$
$$30 - 24h = -48 - 44h$$
$$20h = -78$$
$$h = -3\frac{9}{10}$$

16. $8\frac{1}{2}$

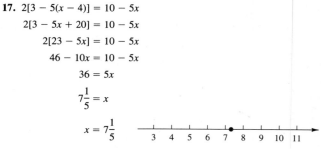

17.
$$2[3 - 5(x - 4)] = 10 - 5x$$
$$2[3 - 5x + 20] = 10 - 5x$$
$$2[23 - 5x] = 10 - 5x$$
$$46 - 10x = 10 - 5x$$
$$36 = 5x$$
$$7\frac{1}{5} = x$$
$$x = 7\frac{1}{5}$$

18. 16

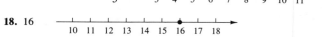

19.
$$3[2h - 6] = 2\{2(3 - h) - 5\}$$
$$6h - 18 = 2\{6 - 2h - 5\}$$
$$6h - 18 = 2\{1 - 2h\}$$
$$6h - 18 = 2 - 4h$$
$$10h = 20$$
$$h = 2$$

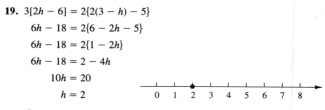

20. $\dfrac{7}{10}$

21. $5(3 - 2x) - 10 = 4x + [-(2x - 5) + 15]$

$15 - 10x - 10 = 4x + [-2x + 5 + 15]$

$15 - 10x - 10 = 4x - 2x + 20$

$-15 = 12x$

$-\dfrac{15}{12} = x$

$-1\dfrac{1}{4} = x$

$x = -1\dfrac{1}{4}$

22. -1

23. $9 - 3(2x - 7) - 9x = 5x - 2[6x - (4 - x) - 20]$

$9 - 6x + 21 - 9x = 5x - 2[6x - 4 + x - 20]$

$9 - 6x + 21 - 9x = 5x - 2[7x - 24]$

$9 - 6x + 21 - 9x = 5x - 14x + 48$

$-6x = 18$

$x = -3$

24. 21

25. $-2\{5 - [6 - 3(4 - x)] - 2x\} = 13 - [-(2x - 1)]$

$-2\{5 - [6 - 12 + 3x] - 2x\} = 13 - [-2x + 1]$

$-2\{5 - 6 + 12 - 3x - 2x\} = 13 + 2x - 1$

$-2\{11 - 5x\} = 12 + 2x$

$-22 + 10x = 12 + 2x$

$8x = 34$

$x = \dfrac{34}{8}$

$x = 4\dfrac{1}{4}$

26. 10

27. $5.073x - 2.937(8.622 + 7.153x) = 6.208$

$5.073x - 25.322814 - 21.008361x = 6.208$

$-15.935361x = 31.530814$

$x \doteq -1.979$

28. $\doteq 0.502$

29. $8.23x - 4.07(6.75x - 5.59) = 3.84(9.18 - x) - 2.67$

$8.23x - 27.4725x + 22.7513 = 35.2512 - 3.84x - 2.67$

$-19.2425x + 22.7513 = 32.5812 - 3.84x$

$-15.4025x = 9.8299$

$x = -0.6382015907$

$x \doteq -0.638$

30. $\doteq 1.011$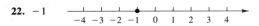

Exercises 4.6 (page 157)

1. $x + 3 = 8$

$\dfrac{-3}{x} \quad \dfrac{-3}{5}$

$x = 5$

Conditional

2. -2; Conditional

3. $2x + 5 = 7 + 2x$

$\dfrac{-2x}{5 = 7} \quad \dfrac{-2x}{} \qquad$ False

No solution

4. No solution

5. $6 + 4x = 4x + 6$

$\dfrac{-4x}{6} \quad \dfrac{-4x}{6}$ True

Identity

6. Identity

7. $5x - 2(4 - x) = 6$

$5x - 8 + 2x = 6$

$7x = 14$

$x = 2$

Conditional

8. 2; Conditional

9. $6x - 3(5 + 2x) = -15$

$6x - 15 - 6x = -15$

$-15 = -15$ True

Identity

10. Identity

11. $4x - 2(6 + 2x) = -15$

$4x - 12 - 4x = -15$

$0 = -3$ False

No solution

12. No solution

13. $7(2 - 5x) - 32 = 10x - 3(6 + 15x)$

$14 - 35x - 32 = 10x - 18 - 45x$

$-35x - 18 = -35x - 18$

$-18 = -18$ True

Identity

14. $\dfrac{34}{41}$; Conditional

15. $2(2x - 5) - 3(4 - x) = 7x - 20$

$4x - 10 - 12 + 3x = 7x - 20$

$-22 = -20$ False

No solution

16. Identity

17. $2[3 - 4(5 - x)] = 2(3x - 11)$

$2[3 - 20 + 4x] = 6x - 22$

$6 - 40 + 8x = 6x - 22$

$2x = 12$

$x = 6$

Conditional

18. No solution

19. $460.2x - 23.6(19.5x - 51.4) = 1213.04$

$460.2x - 460.2x + 1213.04 = 1213.04$

$1213.04 = 1213.04$ True

Identity

20. $\doteq 2.415$ Conditional

Exercises 4.7 (page 164)

1. $x - 5 < \quad 2$

$\underline{\quad + 5 \quad + 5}$

$x \quad < \quad 7$

2. $x < 11$

3. $5x + 4 \leq \quad 19$

$\underline{\quad - 4 \quad - 4}$

$5x \quad \leq \quad 15$

$\dfrac{\overset{1}{\cancel{5}}x}{\underset{1}{\cancel{5}}} \leq \dfrac{15}{5}$

$x \leq 3$

4. $x \leq 3$

5. $6x + 7 > 3 + 8x$

$7 - 3 > 8x - 6x$

$4 > 2x$

$\dfrac{4}{2} > \dfrac{\overset{1}{\cancel{2}}x}{\underset{1}{\cancel{2}}}$

$2 > x$

$x < 2$

6. $x > -7$

7. $2x - 9 > 3(x - 2)$

$2x - 9 > 3x - 6$

$2x - 3x > 9 - 6$

$-x > 3$

$x < -3$

8. $x < -3$

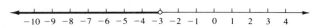

9. $6(3 - 4x) + 12 \geq 10x - 2(5 - 3x)$

$18 - 24x + 12 \geq 10x - 10 + 6x$

$18 + 12 + 10 \geq 10x + 6x + 24x$

$40 \geq 40x$

$1 \geq x$

$x \leq 1$

10. $x \leq 1$

11. $4(6 - 2x) \neq 5x - 2$

$24 - 8x \neq 5x - 2$

$26 \neq 13x$

$2 \neq x$

$x \neq 2$

12. $x \neq 4$

13. $2[3 - 5(x - 4)] < 10 - 5x$

$2[3 - 5x + 20] < 10 - 5x$

$2[23 - 5x] < 10 - 5x$

$46 - 10x < 10 - 5x$

$46 - 10 < 10x - 5x$

$36 < 5x$

$\dfrac{36}{5} < \dfrac{\overset{1}{\cancel{5}}x}{\underset{1}{\cancel{5}}}$

$7\dfrac{1}{5} < x$

$x > 7\dfrac{1}{5}$

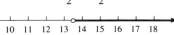

14. $x > 16$

15. $7(x - 5) - 4x > x - 8$

$7x - 35 - 4x > x - 8$

$3x - 35 > x - 8$

$2x > 27$

$x > \dfrac{27}{2} \text{ or } 13\dfrac{1}{2}$

16. $x \neq 1$

17. $3x - 5(x + 2) \le 4x + 8$

$\quad 3x - 5x - 10 \le 4x + 8$

$\qquad -2x - 10 \le 4x + 8$

$\qquad\qquad -18 \le 6x$

$\qquad\qquad\quad -3 \le x$

$\qquad\qquad\qquad x \ge -3$

18. $x < 1$

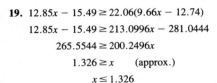

19. $12.85x - 15.49 \ge 22.06(9.66x - 12.74)$

$\quad 12.85x - 15.49 \ge 213.0996x - 281.0444$

$\qquad\quad 265.5544 \ge 200.2496x$

$\qquad\qquad\quad 1.326 \ge x \quad \text{(approx.)}$

$\qquad\qquad\qquad x \le 1.326$

20. $x > -8.410 \quad \text{(approx.)}$

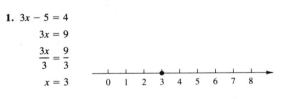

Review Exercises 4.8 (page 166)

The checks will not be shown.

1. $3x - 5 = 4$

$\quad 3x = 9$

$\quad \dfrac{3x}{3} = \dfrac{9}{3}$

$\quad x = 3$

2. 2

3. $\quad 2 = 20 - 9x$

$\quad -18 = -9x$

$\quad \dfrac{-18}{-9} = \dfrac{-9x}{-9}$

$\quad 2 = x$

$\quad x = 2$

4. No solution

5. $\quad 7.5 = \dfrac{A}{10}$

$\quad 10(7.5) = \overset{1}{\cancel{10}}\left(\dfrac{A}{\cancel{10}}\right)_1$

$\quad 75 = A$

$\quad A = 75$

6. 180

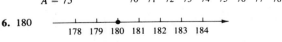

7. $7 - 2(M - 4) = 5$

$\quad 7 - 2M + 8 = 5$

$\quad -2M + 15 = 5$

$\qquad -2M = -10$

$\qquad \dfrac{-2M}{-2} = \dfrac{-10}{-2}$

$\qquad\quad M = 5$

8. Identity

9. $\quad 6R - 8 = 6(2 - 3R)$

$\quad 6R - 8 = 12 - 18R$

$\quad \underline{+\,18R \qquad\quad +\,18R}$

$\quad 24R - 8 = 12$

$\quad \underline{\quad +\,8 \quad +\,8\quad}$

$\quad 24R = 20$

$\quad \dfrac{24R}{24} = \dfrac{20}{24}$

$\qquad R = \dfrac{5}{6}$

10. $1\frac{5}{7}$

11. $\quad 56T - 18 = 7(8T - 4)$

$\quad 56T - 18 = 56T - 28$

$\quad \underline{-\,56T \qquad -\,56T}$

$\qquad\quad -18 = -28 \quad \text{False}$

No solution

12. No solution

13. $15(4 - 5V) = 16(4 - 6V) + 10$

$\quad 60 - 75V = 64 - 96V + 10$

$\quad 60 - 75V = 74 - 96V$

$\qquad\quad 21V = 14$

$\qquad\quad \dfrac{21V}{21} = \dfrac{14}{21}$

$\qquad\qquad V = \dfrac{2}{3}$

14. Identity

15. $5x - 7(4 - 2x) + 8 = 10 - 9(11 - x)$

$\quad 5x - 28 + 14x + 8 = 10 - 99 + 9x$

$\qquad 19x - 20 = -89 + 9x$

$\qquad\qquad 10x = -69$

$\qquad\qquad\quad x = -6.9$

16. -8.3

17. $2[-7y - 3(5 - 4y) + 10] = 10y - 12$

$\quad 2[-7y - 15 + 12y + 10] = 10y - 12$

$\qquad\qquad 2[5y - 5] = 10y - 12$

$\qquad\qquad 10y - 10 = 10y - 12$

$\qquad\qquad\quad -10 = -12 \quad \text{False}$

No solution

18. No solution

19. $4[-24 - 6(3x - 5) + 22x] = 0$

$4[-24 - 18x + 30 + 22x] = 0$

$4[6 + 4x] = 0$

$24 + 16x = 0$

$16x = -24$

$x = -1\frac{1}{2}$

20. 1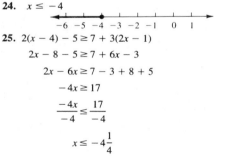

21. $x + 7 > 2$

$x > 2 - 7$

$x > -5$

22. $x > -4$

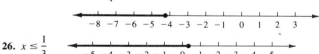

23. $x - 3 \leq -8$

$x \leq 3 - 8$

$x \leq -5$

24. $x \leq -4$

25. $2(x - 4) - 5 \geq 7 + 3(2x - 1)$

$2x - 8 - 5 \geq 7 + 6x - 3$

$2x - 6x \geq 7 - 3 + 8 + 5$

$-4x \geq 17$

$\dfrac{-4x}{-4} \leq \dfrac{17}{-4}$

$x \leq -4\frac{1}{4}$

26. $x \leq \dfrac{1}{3}$

Chapter 4 Diagnostic Test (page 171)

Following each problem number is the textbook section number (in parentheses) where that kind of problem is discussed.

1. (4.2) $x - 5 = 3$ *Check:* $x - 5 = 3$

$\underline{+5 \quad +5}$ $8 - 5 \overset{?}{=} 3$

$x = 8$ $3 = 3$

2. (4.3) $5y + 7 = 22$ *Check:* $5y + 7 = 22$

$\underline{-7 \quad -7}$ $5(3) + 7 \overset{?}{=} 22$

$5y = 15$ $15 + 7 \overset{?}{=} 22$

$\dfrac{5y}{5} = \dfrac{15}{5}$ $22 = 22$

$y = 3$

3. (4.3) $17 - 3z = -1$ *Check:* $17 - 3z = -1$

$\underline{-17 \qquad -17}$ $17 - 3(6) \overset{?}{=} -1$

$-3z = -18$ $17 - 18 \overset{?}{=} -1$

$\dfrac{-3z}{-3} = \dfrac{-18}{-3}$ $-1 = -1$

$z = 6$

4. (4.4) $\dfrac{x}{6} = 5.1$ *Check:* $\dfrac{x}{6} = 5.1$

$6\left(\dfrac{x}{6}\right) = (5.1)(6)$ $\dfrac{30.6}{6} \overset{?}{=} 5.1$

$x = 30.6$ $5.1 = 5.1$

5. (4.4) $-6 = \dfrac{w}{7}$ *Check:* $-6 = \dfrac{w}{7}$

$7(-6) = \left(\dfrac{w}{7}\right)(7)$ $-6 \overset{?}{=} \dfrac{-42}{7}$

$-42 = w$ $-6 = -6$

$w = -42$

6. (4.4) $\dfrac{x}{4} - 5 = 3$ *Check:* $\dfrac{x}{4} - 5 = 3$

$\underline{+5 \quad +5}$ $\dfrac{32}{4} - 5 \overset{?}{=} 3$

$\dfrac{x}{4} = 8$ $8 - 5 \overset{?}{=} 3$

$4\left(\dfrac{x}{4}\right) = (8)(4)$ $3 = 3$

$x = 32$

7. (4.5) $6x + 1 = 17 - 2x$ *Check:* $6x + 1 = 17 - 2x$

$\underline{+2x - 1 \quad -1 + 2x}$ $6(2) + 1 \overset{?}{=} 17 - 2(2)$

$8x = 16$ $12 + 1 \overset{?}{=} 17 - 4$

$\dfrac{8x}{8} = \dfrac{16}{8}$ $13 = 13$

$x = 2$

8. (4.5) $3z - 21 + 5z = 4 - 6z + 17$

$8z - 21 = 21 - 6z$

$\underline{+6z + 21 \quad +21 + 6z}$

$14z = 42$

$\dfrac{14z}{14} = \dfrac{42}{14}$

$z = 3$

Check: $3z - 21 + 5z = 4 - 6z + 17$

$3(3) - 21 + 5(3) \overset{?}{=} 4 - 6(3) + 17$

$9 - 21 + 15 \overset{?}{=} 4 - 18 + 17$

$3 = 3$

9. (4.5) $5k - 9(7 - 2k) = 6$

$5k - 63 + 18k = 6$

$23k - 63 = \quad 6$

$\underline{\quad + 63 \quad + 63 \quad}$

$23k \quad = \quad 69$

$\dfrac{23k}{23} = \dfrac{69}{23}$

$k = 3$

Check: $5k - 9(7 - 2k) = 6$

$5(3) - 9(7 - 2[3]) \overset{?}{=} 6$

$15 - 9(7 - 6) \overset{?}{=} 6$

$15 - 9(1) \overset{?}{=} 6$

$15 - 9 \overset{?}{=} 6$

$6 = 6$

10. (4.6) $10x - 2(5x - 7) = 14$

$10x - 10x + 14 = 14$

$14 = 14 \quad True$

Identity (The variable dropped out and we were left with a true statement.)

11. (4.6) $2y - 4(3y - 2) = 5(6 + y) - 7$

$2y - 12y + 8 = 30 + 5y - 7$

$-10y + 8 = \quad 23 + 5y$

$\underline{\quad + 10y - 23 \quad - 23 + 10y \quad}$

$-15 = \quad\quad 15y$

$\dfrac{-15}{15} = \dfrac{15y}{15}$

$y = -1$

Check: $2y - 4(3y - 2) = 5(6 + y) - 7$

$2(-1) - 4(3[-1] - 2) \overset{?}{=} 5(6 + [-1]) - 7$

$-2 - 4(-3 - 2) \overset{?}{=} 5(5) - 7$

$-2 - 4(-5) \overset{?}{=} 25 - 7$

$-2 + 20 \overset{?}{=} 25 - 7$

$18 = 18$

12. (4.6) $7(3z + 4) = 14 + 3(7z - 1)$

$21z + 28 = 14 + 21z - 3$

$21z + 28 = \quad 11 + 21z$

$\underline{\quad - 21z - 28 \quad - 28 - 21z \quad}$

$0 = -17 \quad\quad False$

No solution (The variable dropped out and we were left with a false statement.)

13. (4.5) $3[7 - 6(x - 2)] = -3 + 2x$

$3[7 - 6x + 12] = -3 + 2x$

$3[19 - 6x] = -3 + 2x$

$57 - 18x = -3 + 2x$

$\underline{\quad + 3 + 18x \quad + 3 + 18x \quad}$

$60 \quad = \quad\quad 20x$

$\dfrac{60}{20} = \dfrac{20x}{20}$

$x = 3$

Check: $3[7 - 6(x - 2)] = -3 + 2x$

$3[7 - 6(3 - 2)] \overset{?}{=} -3 + 2[3]$

$3[7 - 6(1)] \overset{?}{=} -3 + 6$

$3[7 - 6] \overset{?}{=} 3$

$3[1] \overset{?}{=} 3$

$3 = 3$

14. (4.7) $4x + 5 > -3$

$\underline{\quad - 5 \quad\quad - 5 \quad}$

$4x \quad > \quad -8$

$\dfrac{4x}{4} > \dfrac{-8}{4}$

$x > -2$

15. (4.7) $5x - 2 \le 10 - x$

$5x - 2 \le \quad 10 - x$

$\underline{\quad + x + 2 \quad + 2 + x \quad}$

$6x \quad \le \quad 12$

$\dfrac{6x}{6} \le \dfrac{12}{6}$

$x \le 2$

Cumulative Review Exercises: Chapters 1–4 (page 172)

1. $\dfrac{-7}{0}$ Not possible

2. 10

3. $7\sqrt{16} - 5(-4)$

$= 7(4) - 5(-4)$

$= 28 + 20 = 48$

4. 37

5. $V = \dfrac{4}{3}\pi R^3$

$\doteq \dfrac{4}{\overset{}{3}}(3.14)(\overset{3}{9})(9)(9)$

$\doteq 4(3.14)(3)(9)(9)$

$\doteq 3052.08$

6. 3630

7. $15x - [9y - (7x - 10y)]$

$= 15x - [9y - 7x + 10y]$

$= 15x - 9y + 7x - 10y$

$= 22x - 19y$

8. $-14a - 57b$

9. $x(2x^2 - 5x + 12) - 6(3x^2 + 2x - 5)$

$= 2x^3 - 5x^2 + 12x - 18x^2 - 12x + 30$

$= 2x^3 - 23x^2 + 30$

10. $10h^2k^2 - 3h^2k^3$

11. $2x + 1 = \quad 2x + 7$

$\underline{\quad - 2x - 1 \quad - 2x - 1 \quad}$

$0 = 6 \quad\quad False$

No solution

12. $x = 3$

13. $10 - 4(2 - 3x) = 2 + 12x$

$10 - 8 + 12x = 2 + 12x$

$2 + 12x = \quad 2 + 12x$

$\underline{\quad - 2 - 12x \quad - 2 - 12x \quad}$

$0 = 0 \quad\quad True$

Identity

14. $N = 4$

15. $\dfrac{C}{7} - 15 = \quad 13$

$\quad\quad \underline{+ 15 \quad\quad + 15}$

$\quad\dfrac{C}{7} \quad = \quad 28$

$\quad 7\left(\dfrac{C}{7}\right) = 28(7)$

$\quad\quad C = 196$

16. $W = \dfrac{5}{3}$ or $1\dfrac{2}{3}$

17. $9 - 3(x - 2) = 3(5 - x)$

$\quad 9 - 3x + 6 = 15 - 3x$

$\quad\quad 15 - 3x = \quad 15 - 3x$

$\quad\quad \underline{-15 + 3x \quad -15 + 3x}$

$\quad\quad\quad\quad 0 = 0 \quad\quad \text{True}$

Identity

18. $x < 4$

19. $\quad 3 - x \geq \quad 4$

$\quad \underline{-4 + x \quad -4 + x}$

$\quad\quad -1 \quad \geq \quad\quad x$

$\quad\quad\quad\quad x \leq -1$

20. $x \geq -2$

Exercises 5.1A (page 176)

1. $x + 10$ **2.** $A + B$ **3.** $A - 5$ **4.** $B - C$ **5.** $6z$

6. AB **7.** $x - 7$ **8.** $9 + A$ **9.** $3x - 4$ **10.** $6 + 2x$

11. $x - uv$ **12.** $PQ - x$ **13.** $5x^2$ **14.** $10x^3$

15. $(A + B)^2$ **16.** $\left(\dfrac{A}{B}\right)^2$ **17.** $\dfrac{x + 7}{y}$ **18.** $\dfrac{T}{x + 9}$

19. $x(y - 6)$ **20.** $A(3 + B)$

Exercises 5.1B (page 181)

1. Fred's salary is the unknown. Let S represent Fred's salary. Then $S + 75$ is the algebraic expression for "Fred's salary plus seventy-five dollars."

2. $S - 42$, where S is Jaime's salary

3. The number of children is the unknown. Let N represent the number of children. Then $N - 2$ is the algebraic expression for "two less than the number of children in Mr. Moore's family."

4. $N + 2$, where $N =$ the number of players on Jerry's team

5. Joyce's age is the unknown. Let x represent Joyce's age. Then $4x$ is the algebraic expression for "four times Joyce's age."

6. $\dfrac{x}{4}$ or $\dfrac{1}{4}x$, where x is Rene's age

7. The cost of a record is the unknown. Let C represent the cost of a record (in cents). Then $20C + 89$ is the algebraic expression for "twenty times the cost of a record increased by eighty-nine cents."

8. $5C - 17$, where C is the cost of the pen

9. The cost of a hamburger is the unknown. Let C represent the cost of a hamburger. Then $\dfrac{1}{5}C$ or $\dfrac{C}{5}$ is the algebraic expression for "one-fifth the cost of a hamburger."

10. $\dfrac{L}{8}$, where L is the length of the building

11. Let $x =$ speed of car. Then $5x + 100$ is the algebraic expression for "five times the speed of the car plus one hundred miles per hour."

12. $2x - 40$, where x is the speed of the car

13. Let $x =$ the unknown. Then $x^2 =$ the square of the unknown, and $5x^2 - 10$ is the algebraic expression that represents "ten less than five times the square of the unknown."

14. $4x^3 + 8$, where x is the unknown number

15. *First method:* Let $\quad W =$ width

$\quad\quad W + 12 =$ length

Second method: Let $\quad L =$ length

$\quad\quad L - 12 =$ width

16. *First method:* Let $\quad B =$ base

$\quad\quad B - 7 =$ altitude

Second method: Let $\quad A =$ altitude

$\quad\quad A + 7 =$ base

17. Let $x =$ unknown number. Then $50 - \dfrac{7}{x}$ represents the statement in the problem.

18. $15 + \dfrac{x}{9}$, where x is the unknown number

19. Let $L =$ length of rectangle. Then $2L + 11$ represents the statement in the problem.

20. $D - 1$, where D is the diameter

21. If $W =$ Walter's weight, then $320 - W =$ Carlos' weight.

If $C =$ Carlos's weight, then $320 - C =$ Walter's weight.

22. *First method:* Let $\quad T =$ Teresa's weight

$\quad\quad 224 - T =$ Lucy's weight

Second method: Let $\quad L =$ Lucy's weight

$\quad\quad 224 - L =$ Teresa's weight

23. Let $w =$ weight in kilograms. Then $2.2w$ represents the statement in the problem.

24. $0.62D$, where D is the distance in kilometers

25. Let $x =$ the unknown number. Then $\dfrac{8 + x}{x^2}$ represents the statement.

26. $\dfrac{x^2 + 11}{x}$, where x is the unknown number

27. Let $C =$ Celsius temperature. Then $32 + \dfrac{9}{5}C$ represents the statement.

28. $\dfrac{5}{9}(F - 32)$, where F is the Fahrenheit temperature

29. Let $\quad x =$ the smallest integer

$\quad\quad x + 1 =$ the next integer

$\quad\quad x + 2 =$ the largest integer

The sum is $x + (x + 1) + (x + 2)$.

30. $x + (x + 1)$, where

$\quad x =$ the smaller integer and

$\quad x + 1 =$ the larger integer

31. Let $\quad x =$ first odd integer

$\quad\quad x + 2 =$ second odd integer

$\quad\quad x + 4 =$ third odd integer

$\quad\quad x + 6 =$ fourth odd integer

The sum is $x + (x + 2) + (x + 4) + (x + 6)$.

32. $x + (x + 2) + (x + 4)$, where

x = first even integer

$x + 2$ = second even integer

$x + 4$ = third even integer

33. Let r = radius

$6 + r$

34. $5 + r$, where r = radius

35. Let r = radius

$2r$ = height

36. $4h$ is radius, where h = height

37. Let h = height

$h - 4$ = radius

Volume = $\pi(h - 4)^2 h$

38. Area = $\frac{1}{2}(h - 5)h$, where

h = altitude and

$h - 5$ = base

39. Let r = radius

$4r$ = height

Volume = $\frac{1}{3}\pi r^2 (4r)$

40. Volume = $\pi(6h)^2 h$, where

h = height and

$6h$ = radius

Exercises 5.2 (page 185)

1. Let x = unknown number
$13 + 2x = 25$

2. Let x = unknown number
$25 + 3x = 34$

3. Let x = unknown number
$5x - 8 = 22$

4. Let x = unknown number
$4x - 5 = 15$

5. Let x = unknown number
$7 - x = x + 1$

6. Let x = unknown number
$6 + x = 12 - x$

7. Let x = unknown number
$\frac{1}{5}x = 4$

8. Let x = unknown number
$\frac{x}{12} = 6$

9. Let x = unknown number
$\frac{1}{2}x - 4 = 6$

10. Let x = unknown number
$\frac{1}{3}x - 5 = 4$

11. Let x = unknown number
$2(5 + x) = 26$

12. Let x = unknown number
$4(9 + x) = 18$

13. Let x = unknown number
$(x + x)(3) = 24$

14. Let x = unknown number
$5(x + x) = 40$

15. Let x = length of shorter piece of rope (in meters)

$x + 13$ = length of longer piece (in meters)

$x + (x + 13) = 75$

16. Let x = length of shorter piece of wire (in meters)

$x + 8$ = length of longer piece (in meters)

$x + (x + 8) = 46$

17. Let x = length of shorter piece of wire (in centimeters)

$2x$ = length of longer piece of wire (in centimeters)

$x + 2x = 72$

18. Let x = length of shorter piece of wire (in centimeters)

$3x$ = length of longer piece of wire (in centimeters)

$x + 3x = 72$

19. Let x = number of cans of pears

$x + 8$ = number of cans of peaches

$x + (x + 8) = 42$

20. Let x = number of cans of corn

$x + 11$ = number of cans of peas

$x + (x + 11) = 49$

21. Let x = first integer

$x + 1$ = second integer

$x + 2$ = third integer

$x + 3$ = fourth integer

$x + (x + 1) + (x + 2) + (x + 3) = 106$

22. Let x = smallest integer

$x + 1$ = next integer

$x + 2$ = third integer

$x + (x + 1) + (x + 2) = -72$

23. Let w = width (in cm)

$w + 4$ = length (in cm)

$2w + 2(w + 4) = 36$

24. Let w = width (in feet)   or   Let ℓ = length (in feet)

$w + 6$ = length (in feet) $\ell - 6$ = width (in feet)

$2w + 2(w + 6) = 64$ $2\ell + 2(\ell - 6) = 64$

Exercises 5.3 (page 190)

(The checks will not be shown.)

1. Let x = unknown number
$$13 + 2x = 25$$
$$2x = 12$$
$$x = 6$$

2. 3

3. Let x = unknown number
$$5x - 8 = 22$$
$$5x = 30$$
$$x = 6$$

4. 5

5. Let x = unknown number
$$7 - x = x + 1$$
$$6 = 2x$$
$$x = 3$$

6. 3

7. Let x = unknown number

$$\frac{1}{2}x - 4 = 6$$

$$\underline{+4 \quad +4}$$

$$\frac{1}{2}x = 10$$

$$2\left(\frac{1}{2}x\right) = 2(10)$$

$$x = 20$$

8. 27

9. Let x = unknown number

$$2(5 + x) = 26$$

$$10 + 2x = 26$$

$$2x = 16$$

$$x = 8$$

10. -4

11. Let x = length of shorter piece of wire (in cm)

$x + 10$ = length of longer piece (in cm)

$$x + (x + 10) = 36$$

$$x + x + 10 = 36$$

$$2x + 10 = 36$$

$$2x = 26$$

$$x = 13 \text{ cm}$$

$$x + 10 = 23 \text{ cm}$$

12. 4 yd and 6 yd

13. Let x = number of packages of apricots

$x + 7$ = number of packages of apples

$$x + (x + 7) = 15$$

$$x + x + 7 = 15$$

$$2x + 7 = 15$$

$$2x = 8$$

$$x = 4 \text{ packages of apricots}$$

$$x + 7 = 11 \text{ packages of apples}$$

14. 5 skeins of green yarn; 8 skeins of blue yarn

15. $A = \frac{1}{2}bh$

$$A = \frac{1}{2}(26 \text{ cm})(13 \text{ cm}) = 169 \text{ sq. cm}$$

16. $V = 30$ cu. m; $S = 62$ sq. m

17. $V \doteq \frac{4}{3}(3.14)(6 \text{ yd})^3 = 904.32$ cu. yd

$S \doteq 4(3.14)(6 \text{ yd})^2 = 452.16$ sq. yd

18. $A \doteq 50.24$ sq. ft $\quad C \doteq 25.12$ ft

19. Let x = first odd integer

$x + 2$ = second one

$x + 4$ = third one

$$x + (x + 2) + (x + 4) = 177$$

$$x + x + 2 + x + 4 = 177$$

$$3x + 6 = 177$$

$$3x = 171$$

$$x = 57, \text{ first odd integer}$$

$$x + 2 = 59, \text{ second odd integer}$$

$$x + 4 = 61, \text{ third odd integer}$$

20. -50, -48, and -46

21. Let x = first integer

$x + 1$ = second integer

$x + 2$ = third integer

$$4[x + (x + 2)] = 140 + (x + 1)$$

$$4[x + x + 2] = 140 + x + 1$$

$$4[2x + 2] = 141 + x$$

$$8x + 8 = 141 + x$$

$$\underline{-x - 8 \qquad -8 - x}$$

$$7x = 133$$

$$x = 19, \text{ first integer}$$

$$x + 1 = 20, \text{ second integer}$$

$$x + 2 = 21, \text{ third integer}$$

22. 14, 15, and 16

23. Let x = number of degrees in third angle

$$62 + 47 + x = 180$$

$$109 + x = 180$$

$$x = 71°$$

24. $125°$

25. Let w = width of rectangle (in ft)

$2w$ = length (in ft)

$$2w + 2(2w) = 102$$

$$2w + 4w = 102$$

$$6w = 102$$

$$w = 17 \text{ ft, width}$$

$$2w = 34 \text{ ft, length}$$

26. Length = 64 cm; width = 16 cm

27. Let x = length of rectangle (in ft)

$$2x + 2(5) = 44$$

$$2x + 10 = 44$$

$$\underline{-10 \qquad -10}$$

$$2x = 34$$

$$x = 17 \text{ ft}$$

28. 12 yd

29. Let x = unknown number

$$2(4 + x) + x = x + 10$$

$$8 + 2x + x = x + 10$$

$$2x = 2$$

$$x = 1$$

30. 5

31. Let x = unknown number

$$3(8 + 2x) = 4(3x + 8)$$

$$24 + 6x = 12x + 32$$

$$-6x = 8$$

$$\frac{-6x}{-6} = \frac{8}{-6}$$

$$x = -1\frac{1}{3}$$

32. $\frac{10}{11}$

33. Let x = unknown number

$$10x - 3(4 + x) = 5(9 + 2x)$$
$$10x - 12 - 3x = 45 + 10x$$
$$-3x = 57$$
$$x = -19$$

34. $-2\frac{1}{2}$

35. Let x = unknown number

$$8.66x - 5.75(6.94 + x) = 4.69(8.55 + 3.48x)$$
$$8.66x - 39.905 - 5.75x = 40.0995 + 16.3212x$$
$$2.91x - 39.905 = 40.0995 + 16.3212x$$
$$-80.0045 = 13.4112x$$
$$x \doteq -5.97$$

36. $\doteq -32.6$

37. Let x = unknown number

$$\begin{array}{rcl} x + 18 & \geq & 5 \\ -18 & & -18 \\ \hline x & \geq & -13 \end{array}$$

(Any number greater than or equal to -13)

38. The unknown number can be any number greater than or equal to 35.

39. Let x = unknown length

$$\begin{array}{rcl} 37 + x & < & 80 \\ -37 & & -37 \\ \hline x & < & 43 \end{array}$$

The second piece must be less than 43 m.

40. Any length greater than 17 m

Exercises 5.4A (page 195)

1.

$$\begin{array}{ccc} \text{number} & & \text{value} \\ \text{of} & \cdot & \text{of each} = \text{total value} \\ \text{nickels} & & \text{nickel} \\ \\ 7 & \cdot & (5) = 35¢ \end{array}$$

2. $2.75

3.

$$\begin{array}{ccc} \text{number} & & \text{value} \\ \text{of} & \cdot & \text{of each} = \text{total value} \\ \text{50-cent coins} & & \text{coin} \\ \\ 9 & \cdot & (50) = 450¢ \\ & & = \$4.50 \end{array}$$

4. $1.20

5.

$$\begin{array}{ccc} \text{number} & & \text{value} \\ \text{of} & \cdot & \text{of each} = \text{total value} \\ \text{quarters} & & \text{quarter} \\ \\ x & \cdot & (25¢) = 25x¢ \end{array}$$

6. $5y¢$

7.

$$\begin{array}{ccc} \text{cost of} & & \text{cost of} \\ \text{adults'} & + & \text{children's} = \text{total cost} \\ \text{tickets} & & \text{tickets} \\ \\ 7(1.50) & + & 5(0.75) = 14.25 \\ 10.50 & + & 3.75 = \$14.25 \end{array}$$

8. $21.25

9.

$$\begin{array}{ccc} \text{cost of} & & \text{cost of} \\ \text{adults'} & + & \text{children's} = \text{total cost} \\ \text{tickets} & & \text{tickets} \\ \\ 3.5x & + & 1.9y = 3.5x + 1.9y \end{array}$$

10. $2.75x + 1.50y$

11.

$$\begin{array}{ccc} \text{cost of} & & \text{cost of} \\ \text{25¢ stamps} & + & \text{6¢ stamps} = \text{total cost in cents} \\ \\ 25x & + & 6y = 25x + 6y \end{array}$$

12. $4x + 2y$

Exercises 5.4B (page 197)

The checks will not be shown.

1. Let D = number of dimes

$$13 - D = \text{number of nickels}$$
$$10D + 5(13 - D) = 95$$
$$10D + 65 - 5D = 95$$
$$5D = 30$$
$$D = 6 \text{ dimes}$$
$$13 - D = 13 - 6 = 7 \text{ nickels}$$

2. 5 nickels, 6 dimes

3. Let N = number of nickels

$$12 - N = \text{number of quarters}$$
$$5N + 25(12 - N) = 220$$
$$5N + 300 - 25N = 220$$
$$-20N = -80$$
$$N = 4 \text{ nickels}$$
$$12 - N = 12 - 4 = 8 \text{ quarters}$$

4. 10 nickels, 8 quarters

5. Let x = number of nickels

$$x + 4 = \text{number of quarters}$$
$$3x = \text{number of dimes}$$
$$5x + 25(x + 4) + 10(3x) = 400$$
$$5x + 25x + 100 + 30x = 400$$
$$60x = 300$$
$$x = 5 \text{ nickels}$$
$$x + 4 = 5 + 4 = 9 \text{ quarters}$$
$$3x = 3(5) = 15 \text{ dimes}$$

6. 2 nickels, 9 dimes, 18 quarters

7. Let x = number of quarters

$$x - 3 = \text{number of dimes}$$
$$2x - 3 = \text{number of nickels}$$
$$25x + 10(x - 3) + 5(2x - 3) = 225$$
$$25x + 10x - 30 + 10x - 15 = 225$$
$$45x = 270$$
$$x = 6 \text{ quarters}$$
$$x - 3 = 6 - 3 = 3 \text{ dimes}$$
$$2x - 3 = 12 - 3 = 9 \text{ nickels}$$

8. 7 pennies, 12 nickels, 19 dimes

9. Let x = number of box seats sold

$5x$ = number of balcony seats sold

This leaves $1080 - x - 5x = 1080 - 6x$ for the number of orchestra seats

$$10x + 4(5x) + 7(1080 - 6x) = 6600$$
$$10x + 20x + 7560 - 42x = 6600$$
$$-12x = -960$$
$$x = 80 \text{ box seats}$$
$$5x = 5(80) = 400 \text{ balcony seats}$$
$$1080 - 6x = 1080 - 6(80) = 600 \text{ orchestra seats}$$

10. 4400 box seats, 17,600 reserved seats, 35,200 general admission seats

11. Let x = number of 12-cent stamps

$2x$ = number of 10-cent stamps

This leaves $60 - x - 2x = 60 - 3x$ for the number of 2-cent stamps

$$12x + 10(2x) + 2(60 - 3x) = 380$$
$$12x + 20x + 120 - 6x = 380$$
$$26x = 260$$
$$x = 10 \quad \text{12-cent stamps}$$
$$2x = 2(10) = 20 \quad \text{10-cent stamps}$$
$$60 - 3x = 60 - 3(10) = 30 \quad \text{2-cent stamps}$$

12. Ten 8-cent stamps, thirty 6-cent stamps, sixty 12-cent stamps.

Exercises 5.5 (page 203)

The checks will not be shown.

1. Let $4x$ = larger number

$3x$ = smaller number

$$3x + 4x = 35$$
$$7x = 35$$
$$x = 5$$
$$\text{smaller number} = 3x = 15$$
$$\text{larger number} = 4x = 20$$

2. 91, 39

3. Let $4x$ = width

$9x$ = length

$$2(9x) + 2(4x) = 78$$
$$18x + 8x = 78$$
$$26x = 78$$
$$x = 3$$
$$\text{length} = 9x = 27$$
$$\text{width} = 4x = 12$$

4. 28, 8

5. Let $4x$ = amount spent for food

$5x$ = amount spent for rent

x = amount spent for clothing

$$4x + 5x + x = 460$$
$$10x = 460$$
$$x = \$46 \text{ spent for clothing}$$
$$4x = \$184 \text{ spent for food}$$
$$5x = \$230 \text{ spent for rent}$$

6. $2700 for tuition, $2700 for housing, $675 for food

7. Let $6x$ = one part

$5x$ = other part

$$6x + 5x = 88$$
$$11x = 88$$
$$x = 8$$
$$6x = 48$$
$$5x = 40$$

8. 22, 77

9. Let $4x$ = shortest side

$5x$ = next side

$6x$ = longest side

$$4x + 5x + 6x = 90$$
$$15x = 90$$
$$x = 6$$
$$4x = 24 \text{ ft}$$
$$5x = 30 \text{ ft}$$
$$6x = 36 \text{ ft}$$

10. 20 yd, 24 yd, 28 yd

11. Let $2x$ = amount received by first nephew

$3x$ = amount received by second nephew

$4x$ = amount received by third nephew

$$2x + 3x + 4x = \$27,000$$
$$9x = \$27,000$$
$$x = \$3000$$
$$2x = \$6000$$
$$3x = \$9000$$
$$4x = \$12,000$$

12. $75,000; $15,000; $60,000

13. Let $4x$ = cu. yd of gravel

$2\frac{1}{2}x$ = cu. yd of sand

$1x$ = cu. yd of cement

$$4x = 5$$
$$x = \frac{5}{4} = 1\frac{1}{4} \text{ cu. yd of cement}$$
$$2\frac{1}{2}x = \frac{5}{2}\left(\frac{5}{4}\right) = \frac{25}{8} = 3\frac{1}{8} \text{ cu. yd of sand}$$

14. 519 seafood dinners

15. Let $7x$ = length

$4x$ = width

$$2(7x) + 2(4x) < 88$$
$$14x + 8x < 88$$
$$22x < 88$$
$$x < 4$$
$$4x < 16$$

The width must be less than 16.

16. The length must be greater than 24.

17. Let $2x$ = length of shortest side

    $4x$ = length of next side

    $7x$ = length of longest side

$$2x + 4x + 7x > 117$$
$$13x > 117$$
$$x > 9$$
$$7x > 63$$

The longest side must be greater than 63.

18. The length of the shortest side must be less than 24.

19. $\dfrac{360 \text{ miles}}{7 \text{ hours}}$

20. $\dfrac{35 \text{ miles}}{4 \text{ hours}}$

21. $\dfrac{3 \text{ afghans}}{25 \text{ days}}$

22. $\dfrac{500 \text{ words}}{11 \text{ minutes}}$

23. Let x = number of square yards

$$x \, \cancel{\text{sq yd}} \left(23 \, \frac{\text{dollars}}{\cancel{\text{sq yd}}} \right) = 805 \text{ dollars}$$
$$23x = 805$$
$$x = 35 \text{ sq. yd}$$

24. 36 hours

25. Let x = number of gallons of gasoline

$$x \, \cancel{\text{gallons}} \left(23 \, \frac{\text{miles}}{\cancel{\text{gallon}}} \right) = 368 \text{ miles}$$
$$23x = 368$$
$$x = 16 \text{ gal}$$

26. 12 rolls

Exercises 5.6 (page 209)

1. Let x = number

    $0.30x = 15$

       $x = 50$

2. 80

3. Let x = percent

    $250x = 115$

       $x = 0.46 = 46\%$

4. $146\frac{2}{3}\%$

5. Let x = unknown number

    $x = 0.25(40) = 10$

6. 29.25

7. Let x = number

    $0.15x = 127.5$

       $x = 850$

8. 800

9. Let x = percent

    $8x = 17$

      $x = 2.125 = 212.5\%$

10. 200%

11. Let x = unknown number

    $x = 0.63(48) = 30.24$

12. 42.63

13. Let x = number

    $1.25x = 750$

       $x = 600$

14. 250

15. Let x = percent

    $16x = 23$

      $x = 1.4375 = 143.75\%$

16. $\doteq 247.8\%$

17. Let x = unknown number

    $x = 2.00(12) = 24$

18. 27

19. Let x = number

    $0.15x = 37.5$

      $x = 250$

20. \$36.45

21. Let x = number

$$0.66\tfrac{2}{3}x = 42 \quad \left(0.66\tfrac{2}{3} = \frac{66\tfrac{2}{3}}{100} = \frac{\tfrac{200}{3}}{100} = \frac{2}{3} \right)$$
$$\tfrac{2}{3}x = 42$$
$$x = 63$$

22. 216

23. Sixty-eight is 80% of what number?

Let x = number

    $0.80x = 68$

      $x = 85$

24. 32,200

25. Seven is what percent of 42?

Let x = percent

    $42x = 7$

      $x = 0.16\tfrac{2}{3} = 16\tfrac{2}{3}\%$

26. \$25.30

27. Fifty-four is what percent of 210?

Let x = percent

    $210x = 54$

      $x \doteq 0.257 = 25.7\%$

28. 11%

29. The markup is 35% of \$125.

Let x = number

    $x = 0.35(\$125) = \$43.75$ (Markup)

Selling price = Cost + Markup
Selling price = \$125 + \$43.75 = \$168.75

30. \$86.40

31. Total weight of steers $= 15 \times 1027 = 15,405$ lb
Total weight of heifers $= 18 \times 956 = 17,208$ lb
If 3% of the weight is lost in shipping, then 97% of the weight remains at time of sale.
Let $x =$ amount of money from sale of steers
$x = 0.97 \times 15,405 \times 0.84 = 12,551.9940$
Let $y =$ amount of money from sale of heifers
$y = 0.97 \times 17,208 \times 0.78 = 13,019.5728$
$$x + y = 25,571.5668$$
Amount of check $= \$25,571.57$

32. $\doteq 38.6\%$

Exercises 5.7 (page 215)

The checks will not be shown.

1. Let $x =$ number of hours
$$\left(51\frac{\text{miles}}{\text{hour}}\right)(x\ \text{hours}) = 408\ \text{miles}$$
$$51x = 408$$
$$x = 8\ \text{hours}$$

2. 9 hours

3. Let $\quad x =$ speed with no traffic
$x - 10 =$ speed with traffic
$$4x = 5(x - 10)$$
$$4x = 5x - 50$$
$$\underline{50 - 4x \quad -4x + 50}$$
$$50 \quad = \quad x$$
a. $x = 50$ mph, with no traffic
b. distance $= \left(50\frac{\text{miles}}{\text{hour}}\right)(4\ \text{hours}) = 200\ \text{miles}$

4. a. 60 mph **b.** 300 mi

5. Let $\quad x =$ Mr. Robinson's speed
$x + 9 =$ Mr. Reid's speed
$$5x = 4(x + 9)$$
$$5x = 4x + 36$$
$$\underline{-4x \quad -4x}$$
a. $x \quad = 36$ mph, Mr. Robinson's speed
b. $x + 9 \quad = 45$ mph, Mr. Reid's speed
c. distance $= \left(\frac{36\ \text{miles}}{\text{hour}}\right)(5\ \text{hours}) = 180\ \text{mi}$

6a. 45 mph
b. 54 mph
c. 270 mi

7. Let $\quad x =$ speed of boat in still water
$x - 3 =$ speed upstream
$x + 3 =$ speed downstream
$$4(x - 3) = 3(x + 3)$$
$$4x - 12 = 3x + 9$$
$$\underline{-3x + 12 \quad -3x + 12}$$
$$x \quad = \quad 21\ \text{mph in still water}$$

8. 24 mph

9. Let $\quad x =$ speed of boat in still water
$x - 4 =$ speed upstream
$x + 4 =$ speed downstream
$$4(x - 4) + 6 = 3(x + 4)$$
$$4x - 16 + 6 = 3x + 12$$
$$4x - 10 = 3x + 12$$
$$4x - 10 = \quad 3x + 12$$
$$\underline{-3x + 10 \quad -3x + 10}$$
$$x \quad = \quad 22\ \text{mph in still water}$$

10. 24 mph

11. Let $\quad t =$ number of hours driven on Saturday
$17 - t =$ number of hours driven the next day
$54t =$ distance driven on Saturday
$48(17 - t) =$ distance driven the next day
$$54t = 48(17 - t) \quad \text{(The distances are equal.)}$$
$$54t = 816 - 48t$$
$$\underline{48t \quad\quad +48t}$$
$$102t = 816$$
a. $t = 8$ hours driven on Saturday
b. distance $= \left(\frac{54\ \text{miles}}{\text{hour}}\right)(8\ \text{hours}) = 432\ \text{mi}$

12a. 3 hours
b. 48 miles

Exercises 5.8 (page 220)

1. Let $\quad x =$ number of lb of apple chunks
$10 - x =$ number of lb of granola
$$4.2x + 2.2(10 - x) = 3(10)$$
$$42x + 22(10 - x) = 30(10)$$
$$42x + 220 - 22x = 300$$
$$20x + 220 = 300$$
$$\underline{-220 \quad -220}$$
$$20x = 80$$
$$x = 4\ \text{lb of apple chunks}$$
$$10 - x = 6\ \text{lb of granola}$$

2. 23 lb of macadamia nuts and 27 lb of peanuts

3. Let $\quad x =$ number of lb of apples at 95 cents per pound
$30 - x =$ number of lb of apples at 65 cents per pound
$$95x + 65(30 - x) = 78(30)$$
$$95x + 1950 - 65x = 2340$$
$$30x + 1950 = 2340$$
$$\underline{-1950 \quad -1950}$$
$$30x = 390$$
$$x = 13\ \text{lb at 95 cents per pound}$$
$$30 - x = 17\ \text{lb at 65 cents per pound}$$

4. 35 lb of English toffee, 25 lb of peanut brittle

5. Let $\quad x =$ number of lb of gumdrops

$27 - x =$ number of lb of caramels

$2.8x + 2.2(27 - x) = 66$

$28x + 22(27 - x) = 660$

$28x + 594 - 22x = 660$

$6x + 594 = \quad 660$

$\underline{\quad - 594 \quad - 594}$

$6x \quad\quad = \quad\quad 66$

$x = 11$ lb of gumdrops

$27 - x = 16$ lb of caramels

6. 23 lb of Brand C and 17 lb of Brand D

7. Let $x =$ number of lb of candy

(There will be $50 + x$ pounds of the mixture.)

$3x + 2.4(50) = 2.5(50 + x)$

$30x + 24(50) = 25(50 + x)$

$30x + 1200 = \quad\quad 1250 + 25x$

$\underline{- 25x - 1200 \quad - 1200 - 25x}$

$5x \quad\quad = \quad\quad 50$

$x = 10$ lb of candy

8. 20 lb of cashews

Exercises 5.9 (page 224)

1. What is 20% of 16 mℓ?

Let $x =$ unknown number

$x = (0.20)(16\,\text{m}\ell) = 3.2\,\text{m}\ell$

2. 0.4 ℓ

3. Forty-three is what percent of 500?

Let $x =$ percent

$500x = 43$

$x = 0.086 = 8.6\%$

4. 5.4%

5. Twenty-four is 40% of what number?

Let $x =$ unknown number

$0.40x = 24$

$x = 60\,\text{m}\ell$

6. 50 mℓ

7. Let $x =$ number of cc of 20% solution

$0.50(100) + 0.20(x) = 0.25(x + 100)$

$50(100) + 20x = 25(x + 100)$

$5000 + 20x = 25x + 2500$

$2500 = 5x$

$500 = x$

$x = 500$ cc of 20% solution

8. 20 pints

9. Let $x =$ number of mℓ of water

(There will be $[500 + x]$ mℓ altogether.)

$0.25(x + 500) = 0.40(500)$

$25(x + 500) = 40(500)$

$25x + 12{,}500 = 20{,}000$

$25x = 7500$

$x = 300\,\text{m}\ell$

10. 120 mℓ

11. Let $x =$ number of ℓ of pure alcohol

(There will be $[10 + x]$ ℓ altogether.)

$1x + 0.2(10) = 0.5(10 + x)$

$10x + 2(10) = 5(10 + x)$

$10x + 20 = \quad 50 + 5x$

$\underline{- \quad 5x - 20 \quad - 20 - 5x}$

$5x \quad\quad = \quad\quad 30$

$x = 6$ ℓ of pure alcohol

12. 5 ℓ

Review Exercises 5.10 (page 226)

1. Let $6x =$ one number

$7x =$ other number

$6x + 7x = 52$

$13x = 52$

$x = 4$

$6x = 24$, one number

$7x = 28$, other number

2. $1102.50

3. Let $x =$ number

$0.31x = 77.5$

$x = 250$

4. 37.5%

5. The amount of increase is 5% of $380.

Let $x =$ increase

$x = 0.05(\$380) = \$19$

The new rent $= \$380 + \$19 = \$399$.

6. 9 oz lead and 15 oz zinc

7. Let $3x =$ smallest side

$4x =$ next side

$5x =$ longest side

$3x + 4x + 5x = 108$

$12x = 108$

$x = 9$

$3x = 27$, shortest side

$4x = 36$, next side

$5x = 45$, longest side

8. 15 dimes; 7 50¢ pieces

9. Let $\quad x =$ number of 10-cent stamps

$2x =$ number of 12-cent stamps

$2(2x) =$ number of 2-cent stamps

$10x + 12(2x) + 2(4x) = 210$

$10x + 24x + 8x = 210$

$42x = 210$

$x = 5 \quad$ 10-cent stamps

$2x = 10 \quad$ 12-cent stamps

$4x = 20 \quad$ 2-cent stamps

10. 1000 box seats, 5500 reserved seats, 22,000 general admission seats

11. We first find 24% of 1800.

Let x = that number

$$x = 0.24(1,800) = 432$$

The number of minority students needed is $432 - 256 = 176$.

12. 5 mph

13. Let x = Mrs. Koontz's average speed

$x + 5$ = Mrs. Fowler's average speed

$$\left(\frac{x \text{ miles}}{\text{hour}}\right)(10 \text{ hours}) = \left(\frac{[x + 5] \text{ miles}}{\text{hour}}\right)(9 \text{ hours}) \quad \text{(The distances are equal.)}$$

$$10x = 9(x + 5)$$

$$10x = 9x + 45$$

$$\underline{-9x \quad -9x}$$

a. $\quad x = 45$ mph, Mrs. Koontz's average speed

b. $\quad x + 5 = 50$ mph, Mrs. Fowler's average speed

c. The distance is $\left(\dfrac{45 \text{ miles}}{\text{hour}}\right)(10 \text{ hours}) = 450$ miles.

14. 12 lb at 85 cents per pound and 18 lb at 95 cents per pound

15. Let x = cc of water added

$$0.25(500) + 0 = 0.05(500 + x)$$

$$25(500) + 0 = 5(500 + x)$$

$$12,500 = 2500 + 5x$$

$$10,000 = 5x$$

$$x = 2000 \text{ cc of water}$$

Chapter 5 Diagnostic Test (page 233)

Following each problem number is the textbook section number (in parentheses) where that kind of problem is discussed.

1. (5.9)   Twenty-eight is 40% of what number?

Let x = unknown number

$$0.40x = 28 \qquad \qquad \textit{Check:} \quad 40\% \text{ of } 70 \text{ m}\ell = (0.40)(70 \text{ m}\ell)$$

$$x = 70 \text{ m}\ell \qquad \qquad \qquad \qquad \qquad \doteq 28 \text{ m}\ell$$

2. (5.9)   Seventeen is what percent of 20?

Let x = percent $\qquad \textit{Check:} \quad 85\% \text{ of } 20 = (0.85)(20)$

$$20x = 17 \qquad \qquad \qquad \qquad \qquad = 17$$

$$x = 0.85 = 85\%$$

3. (5.3)   Let x = unknown number

$$16 + 3x = 37 \qquad \textit{Check:} \quad 16 + 3x = 37$$

$$\underline{-16 \qquad = -16} \qquad \qquad 16 + 3(7) \doteq 37$$

$$3x = 21 \qquad \qquad \qquad 16 + 21 \doteq 37$$

$$x = 7 \qquad \qquad \qquad 37 = 37$$

4. (5.4)   Let $\quad x$ = number of nickels

$25 - x$ = number of dimes

Then amount of money in nickels = $5x$

and amount of money in dimes = $10(25 - x)$

$$5x + 10(25 - x) = 165$$

$$5x + 250 - 10x = 165$$

$$\underline{-5x = -85} \qquad \textit{Check:} \quad \text{nickels} \quad 17(5\text{¢}) = 85\text{¢}$$

$$x = 17 \qquad \qquad \qquad \text{dimes} \quad 8(10\text{¢}) = \underline{80\text{¢}}$$

$$25 - x = 8 \qquad \qquad \qquad \qquad \qquad 165\text{¢}$$

5. (5.8)   Let $\quad x$ = number of lb of apricots

$50 - x$ = number of lb of granola

$$2.70x + 2.20(50 - x) = 2.34(50)$$

$$270x + 220(50 - x) = 234(50)$$

$$270x + 11,000 - 220x = 11,700$$

$$50x + 11,000 = 11,700$$

$$\underline{-11,000 \qquad -11,000}$$

$$50x \qquad = \qquad 700$$

$$x = 14 \text{ lb of apricots}$$

$$50 - x = 36 \text{ lb of granola}$$

Check: $\$2.70(14) + \$2.20(36) \doteq \$2.34(50)$

$$\$37.80 + \$79.20 \doteq \$117$$

$$\$117 = \$117$$

6. (5.7)   Let $\quad x$ = Kevin's average speed

$x + 9$ = Jason's average speed

$6x$ = Kevin's distance

$5(x + 9)$ = Jason's distance

$$6x = 5(x + 9) \quad \text{(The distances are equal.)}$$

$$6x = 5x + 45$$

$$\underline{-5x \qquad -5x}$$

a. $\qquad x = 45$, Kevin's average speed (in mph)

b. $\qquad x + 9 = 54$, Jason's average speed (in mph)

c. distance = $\left(\dfrac{45 \text{ miles}}{\text{hour}}\right)(6 \text{ hours}) = 270$ mi

Check: $45\dfrac{\text{mi}}{\text{hr}}(6 \text{ hr}) \doteq 54\dfrac{\text{mi}}{\text{hr}}(5 \text{ hr})$

$$270 \text{ mi} = 270 \text{ mi}$$

7. (5.8)   Let x = number of lb of nuts to be added

$$3.50x + 8.00(30) = 4.85(x + 30)$$

$$350x + 800(30) = 485(x + 30)$$

$$350x + 24,000 = 485x + 14,550$$

$$\underline{-350x - 14,550 \qquad -350x - 14,550}$$

$$9450 = 135x$$

$$70 = x$$

$$x = 70 \text{ lb of nuts to be added}$$

(The mixture will contain 70 lb + 30 lb, or 100 lb.)

Check: $3.50\dfrac{\text{dollars}}{\text{lb}}(70 \text{ lb}) + 8\dfrac{\text{dollars}}{\text{lb}}(30 \text{ lb}) \doteq 4.85\dfrac{\text{dollars}}{\text{lb}}(100 \text{ lb})$

$$\$245 + \$240 \doteq \$485$$

$$\$485 = \$485$$

8. (5.9) Let $\quad x =$ number of mℓ of 30% solution

$200 - x =$ number of mℓ of 80% solution

$0.30x + 0.80(200 - x) = 0.45(200)$

$30x + 80(200 - x) = 45(200)$

$30x + 16{,}000 - 80x = 9000$

$16{,}000 - 50x = \quad 9000$

$\underline{\quad -9000 + 50x \quad -9000 + 50x}$

$7000 \qquad = \qquad 50x$

$140 = x$

$x = 140\,\text{mℓ}$ of the 30% solution

$200 - x = 60\ \text{mℓ}$ of the 80% solution

Check: $\quad 0.30(140) + 0.80(60) \overset{?}{=} 0.45(200)$

$42 + 48 \overset{?}{=} 90$

$90 = 90$

9. (5.5) Let $7x =$ length of smallest side (in m)

$9x =$ length of next side

$11x =$ length of longest side

$7x + 9x + 11x = 135$

$27x = 135$

$x = 5$

$7x = 35\,\text{m},$ smallest side

$9x = 45\,\text{m},$ next side

$11x = 55\,\text{m},$ longest side

Check: $\quad 35\,\text{m} + 45\,\text{m} + 55\,\text{m} \overset{?}{=} 135\,\text{m}$

$135\,\text{m} = 135\,\text{m}$

10. (5.6) We must first find the markup, which is 30% of $240.

Let $x =$ markup

$x = 0.30(\$240) = \$72$

The selling price is $\$240 + \$72 = \$312.$

Cumulative Review Exercises: Chapters 1–5 (page 234)

1. Cannot be determined **2.** 98

3. $\quad 46 - 2\{4 - [3(5 - 8) - 10]\}$

$= 46 - 2\{4 - [3(-3) - 10]\}$

$= 46 - 2\{4 - [-9 - 10]\}$

$= 46 - 2\{4 - [-19]\}$

$= 46 - 2\{23\}$

$= 46 - 46 = 0$

4. 45

5. $S = 4\pi R^2$

$S \doteq 4(3.14)5^2$

$S \doteq 4(3.14)25$

$S \doteq 4(25)3.14$

$S \doteq 100(3.14) = 314$

6. $10^5 = 100{,}000$ **7.** $\dfrac{x^{3c}}{x^c} = x^{3c-c} = x^{2c}$ **8.** $\dfrac{8a^6}{b^3}$

9. $\left(\dfrac{\cancel{8}^2 y^{-1}}{\cancel{8}_1 y^3}\right)^2 = (2y^{-1-3})^2 = (2y^{-4})^2 = 2^2 y^{-8} = \dfrac{4}{y^8}$ **10.** $\dfrac{x^4}{9}$

11. $\quad 32 - 4x = \quad 15$

$\underline{\quad -15 + 4x \quad -15 + 4x}$

$17 \qquad = \qquad 4x$

$\dfrac{17}{4} = x$

$x = \dfrac{17}{4}\quad \text{or}\quad 4\dfrac{1}{4}$

12. -33

13. $5w - 12 + 7w = 6 - 8w - 3$

$-12 + 12w = \quad 3 - 8w$

$\underline{+12 + \ 8w \qquad +12 + 8w}$

$20w = \quad 15$

$w = \dfrac{15}{20} = \dfrac{3}{4}$

14. $y = -3$

15. $6z - 22 = 2[8 - 4(5z - 1)]$

$6z - 22 = 2[8 - 20z + 4]$

$6z - 22 = 2[12 - 20z]$

$6z - 22 = 24 - 40z$

$\underline{40z + 22 \qquad 22 + 40z}$

$46z \qquad = 46$

$z = 1$

16. Yes **17.** 1 **18.** 88 and 72

19. Let $\quad x =$ number of nickels

$15 - x =$ number of quarters

$5x + 25(15 - x) = 175$

$5x + 375 - 25x = 175$

$-20x = -200$

$x = 10$ nickels

$15 - x = 5$ quarters

20. 60 mph

Exercises 6.1 (page 238)

1a. First degree
b. Second degree

2a. First degree
b. Third degree

3. Not a polynomial   **4.** Not a polynomial

5a. Third degree
b. Sixth degree

6a. Third degree
b. Third degree

7a. Second degree
b. Fourth degree

8a. Second degree
b. Third degree

9. Not a polynomial   **10.** Not a polynomial
11. Not a polynomial   **12.** Not a polynomial

13a. Second degree
b. Second degree

14a. Third degree
b. Third degree

15. Not a polynomial   **16.** Not a polynomial

17. $8x^5 + 7x^3 - 4x - 5$ Leading coefficient is 8
18. $-3y^5 - 2y^3 + 4y^2 + 10$ Leading coefficient is -3
19. $xy^3 + 8xy^2 - 4x^2y$ Leading coefficient is 1
20. $x^4y^2 + 3x^3y - 3xy^3$ Leading coefficient is 1

Exercises 6.2 (page 242)

1. $\underline{2m^2} - \underset{\sim}{m} + 4 + \underline{3m^2} + \underset{\sim}{m} - \underset{\sim}{5} = 5m^2 - 1$
2. $11n^2 + 2n + 3$
3. $2x^3 \underline{-4} + 4x^2 \underline{+8x} - 9x \underline{+7} = 2x^3 + 4x^2 - x + 3$
4. $9z^2 + 9$
5. $\underline{3x^2} + \underline{4x} - 10 - \underline{5x^2} + \underline{3x} - \underset{\sim}{7}$
   $= -2x^2 + 7x - 17$
6. $-a^2 - 7a + 14$
7. $(8b^2 + 2b - 14) - (-5b^2 + 4b + 8)$
   $= 8b^2 + 2b - 14 + 5b^2 - 4b - 8 = 13b^2 - 2b - 22$
   Or $\quad\quad 8b^2 + 2b - 14 \quad\quad\quad 8b^2 + 2b - 14$
   $\quad\quad \underline{-(-5b^2 + 4b + 8)} \Rightarrow \underline{5b^2 - 4b - 8}$
   $\quad\quad\quad\quad\quad\quad\quad\quad\quad\quad\quad 13b^2 - 2b - 22$
8. $19c^2 + 5c + 1$
9. $\underline{6a} - \underline{5a^2} + 6 + \underline{4a^2} + \underset{\sim}{6} - \underline{3a} = -a^2 + 3a + 12$
10. $11b^2 + 3$
11. $(4a^2 + 6 - 3a) - (5a + 3a^2 - 4) = 4a^2 + 6 - 3a - 5a - 3a^2 + 4$
   $\quad\quad\quad\quad\quad\quad\quad\quad\quad\quad\quad\quad = a^2 - 8a + 10$
   Or $\quad\quad 4a^2 - 3a + 6 \quad\quad\quad 4a^2 - 3a + 6$
   $\quad\quad \underline{-(3a^2 + 5a - 4)} \Rightarrow \underline{-3a^2 - 5a + 4}$
   $\quad\quad\quad\quad\quad\quad\quad\quad\quad\quad\quad a^2 - 8a + 10$
12. $2b^2 - 9b + 15$
13. $17a^3 \quad\quad\quad + 4a - 9$
   $\quad\quad \underline{8a^2 - 6a + 9}$
   $17a^3 + 8a^2 - 2a$
14. $-20b^4 + b^3 + 5b^2 - 1$
15. $14x^2y^3 - 11xy^2 + 8xy$
   $-9x^2y^3 + 6xy^2 - 3xy$
   $\underline{7x^2y^3 - 4xy^2 - 5xy}$
   $12x^2y^3 - 9xy^2$
16. $9a^2b + 2ab^2 - 10ab$
17. $15x^3 - 4x^2 \quad\quad + 12 \quad\quad 15x^3 - 4x^2 \quad\quad + 12$
   $\underline{-(8x^3 \quad\quad + 9x - 5)} \Rightarrow \underline{-8x^3 \quad\quad - 9x + 5}$
   $\quad\quad\quad\quad\quad\quad\quad\quad\quad\quad 7x^3 - 4x^2 \quad - 9x + 17$
18. $-7y^3 - 28y^2 + 19y - 24$
19. $10a^2b - 6ab + 5ab^2 \quad\quad 10a^2b - 6ab + 5ab^2$
   $\underline{-(3a^2b + 6ab - 7ab^2)} \Rightarrow \underline{-3a^2b - 6ab + 7ab^2}$
   $\quad\quad\quad\quad\quad\quad\quad\quad\quad\quad 7a^2b - 12ab + 12ab^2$
20. $22m^3n^2 - 4m^2n^2 - 9mn$
21. $\underline{7m^8} - \underset{\sim}{4m^4} + \underset{\sim}{4m^4} + \underline{m^5} + \underline{8m^8} - \underline{m^5} = 15m^8$
22. $-h^6 + 17h$
23. $\underline{6r^3t} + \underline{14r^2t} - \underset{\sim}{11} + 19 - \underline{8r^2t} + \underline{r^3t} + \underset{\sim}{8} - \underline{6r^2t} = 7r^3t + 16$
24. $4m^2n^2 + 11$

25. $(7x^2y^2 - 3x^2y + xy + 7) - (3x^2y^2 - 5xy + 4 + 7x^2y)$
   $= \underline{7x^2y^2} - \underline{3x^2y} + xy + \underset{\sim}{7} - \underline{3x^2y^2} + \underline{5xy} - \underset{=}{4} - \underline{7x^2y}$
   $= 4x^2y^2 - 10x^2y + 6xy + 3$
26. $9x^2y^2 - 2x^2y - 4xy - 13$
27. $(x^2 + 4) - [(x^2 - 5) - (3x^2 + 1)]$
   $= (x^2 + 4) - [x^2 - 5 - 3x^2 - 1] = \underline{x^2} + \underset{=}{4} \underline{- x^2} + \underset{=}{5} + \underline{3x^2} + \underset{=}{1}$
   $= 3x^2 + 10$
28. $6x^2 - 7$
29. $(5x^2 - 2x + 1) + (-4x^2 + 6x - 8) - (2x^2 - 4x + 3)$
   $= \underline{5x^2} - \underline{2x} + \underset{\sim}{1} - \underline{4x^2} + \underline{6x} - \underset{=}{8} - \underline{2x^2} + \underline{4x} - \underset{\sim}{3}$
   $= -x^2 + 8x - 10$
30. $-4y$
31. $[(5 + xy^2 + x^3y) + (-6 - 3xy^2 + 4x^3y)] - [(x^3y + 3xy^2 - 4)$
   $+ (2x^3y - xy^2 + 5)] = [\underline{5} + \underline{xy^2} + \underline{x^3y} - \underset{\sim}{6} - \underline{3xy^2} + \underline{4x^3y}]$
   $- [\underline{x^3y} + \underline{3xy^2} - \underset{\sim}{4} + \underline{2x^3y} - \underline{xy^2} + \underset{\sim}{5}]$
   $= \underline{5x^3y} - \underline{2xy^2} - \underset{\sim}{1} - \underline{3x^3y} - \underline{2xy^2} - \underset{\sim}{1} = 2x^3y - 4xy^2 - 2$
32. $6m^2n + 6$
33. $\underline{7.239x^2} - \underline{4.028x} + 6.205 - \underline{2.846x^2} + \underline{8.096x} + \underset{\sim}{5.307}$
   $= 4.393x^2 + 4.068x + 11.512$
34. $37.528x^2 + 6.15x - 52.64$

Exercises 6.3A (page 246)

1. $\quad x - 3(x + y)$
   $= x - 3x - 3y$
   $= -2x - 3y$
2. $-y - 2z$
3. $\quad 2a - 4(a - b)$
   $= 2a - 4a + 4b$
   $= \quad -2a + 4b$
4. $c + 2d$
5. $\quad u(u^2 + 2u + 4) - 2(u^2 + 2u + 4)$
   $= u^3 + 2u^2 + \underline{4u} - \underline{2u^2} - \underline{4u} - 8$
   $= u^3 - 8$
6. $x^3 + 27$
7. $\quad x^2(x^2 + y^2) - y^2(x^2 + y^2)$
   $= x^4 + \underline{x^2y^2} - \underline{x^2y^2} - y^4$
   $= x^4 - y^4$
8. $w^4 - 16$
9. $\quad 2x(3x^2 - 5x + 1) - 4x(2x^2 - 3x - 5)$
   $= \underline{6x^3} - \underline{10x^2} + \underline{2x} - \underline{8x^3} + \underline{12x^2} + \underline{20x}$
   $= -2x^3 + 2x^2 + 22x$
10. $6x^3 - 4x^2 - 11x$
11. $\quad -3(a - 2b) + 2(a - 3b)$
   $= -3a + 6b + 2a - 6b$
   $= -a$
12. $2m - 2n$
13. $\quad -5(2x - 3y) - 10(x + 5y)$
   $= -10x + 15y - 10x - 50y$
   $= -20x - 35y$
14. $-36s + 60t$ **15.** $-30x^2y^2$ (Product of monomials)
16. $-30abc^2$ **17.** $6x^2y - 15xy^2$ (Distributive property used)
18. $10abc - 6ac^2$

19. $-15xy^2 - 10xy$ (Distributive property used)

20. $-6ac^2 - 15bc$ **21.** $3xy + 2x - 5y$ (Parentheses dropped)

22. $2ac + 5b - 3c$

23. $x^2y(3xy^2 - y) - 2xy^2(4x - x^2y)$
$= 3x^3y^3 - \underset{\sim}{x^2y^2} - \underset{\sim}{8x^2y^2} + 2x^3y^3$
$= 5x^3y^3 - 9x^2y^2$

24. $-4a^2b^2 + 2a^2b^3$

25. $2h(3h^2 - k) - k(h - 3k^3)$
$= 6h^3 - 2hk - hk + 3k^4$
$= 6h^3 - 3hk + 3k^4$

26. $11xy^2 - 14x^2$

27. $3f(2f^2 - 4g) - g(2f - g^2)$
$= 6f^3 - 12fg - 2fg + g^3$
$= 6f^3 - 14fg + g^3$

28. $3ab^2 - 7ab$

29. $2x - [3a + 4x - 5a]$
$= 2x - [-2a + 4x]$
$= 2x + 2a - 4x$
$= 2a - 2x$

30. $-c - y$

31. $5x + [-(2x - 10) + 7]$
$= 5x + [-2x + 10 + 7]$
$= 5x - 2x + 10 + 7$
$= 3x + 17$

32. $x + 9$

33. $25 - 2[3g - 5(2g - 7)]$
$= 25 - 2[3g - 10g + 35]$
$= 25 - 2[-7g + 35]$
$= 25 + 14g - 70$
$= 14g - 45$

34. $66h - 200$

35. $-2\{-3[-5(-4 - 3z) - 2z] + 30z\}$
$= -2\{-3[20 + 15z - 2z] + 30z\}$
$= -2\{-60 - 45z + 6z + 30z\}$
$= -2\{-60 - 9z\}$
$= 120 + 18z$

36. $138z + 90$ **37.** $(4x)(3x)^2 = 4x(9x^2) = 36x^3$ **38.** $50y^3$

39. $-45x^2 + 10xy$ **40.** $-27cd + 63d^2$

41. $(2xy^2)^2(3y) = (4x^2y^4)(3y) = 12x^2y^5$ **42.** $45a^4b^3$

43. $(9x - 2y) - 5x = 9x - 2y - 5x = 4x - 2y$ **44.** $3c - 16d$

Exercises 6.3B (page 251)

1. $-56x^4y^3$ **2.** $-24a^5b^5$ **3.** $x^2 + x - 6$

4. $a^2 - a - 12$ **5.** $6x^3 + 14x$ **6.** $12y^4 - 6y$

7. $y^2 - y - 72$ **8.** $z^2 + 7z - 30$

9. $15x^3 + 5x^2y - 6xy - 2y^2$ **10.** $2s^3 + 3s^2t - 8st - 12t^2$

11. $10a^3 - 2ab^2$ **12.** $21c^3 - 3cd^2$ **13.** $x^2 + 5x + 4$

14. $x^2 + 4x + 3$ **15.** $a^2 + 7a + 10$ **16.** $a^2 + 8a + 7$

17. $m^2 - 2m - 8$ **18.** $n^2 + 4n - 21$ **19.** $y^2 - 3y - 108$

20. $z^2 - 16z + 55$ **21.** $-abx - aby$ **22.** $abx - aby$

23. $ax - bx + ay - by$ **24.** $ax + bx - ay - by$ **25.** $16x^2$

26. $9x^2$ **27.** $(4 + x)^2 = (4 + x)(4 + x) = 16 + 8x + x^2$

28. $9 + 6x + x^2$

29. $(b - 4)^2 = (b - 4)(b - 4) = b^2 - 8b + 16$

30. $b^2 - 12b + 36$ **31.** $3x^2 + 7x + 2$ **32.** $2x^2 + 7x + 6$

33. $8x^2 + 10x - 12$ **34.** $6x^2 + 7x - 5$

35. $20x^2 - 38x + 12$ **36.** $15x^2 - 32x + 16$

37. $(2x + 5)^2 = (2x + 5)(2x + 5) = 4x^2 + 20x + 25$

38. $9x^2 + 24x + 16$ **39.** $6x^3 + 3x^2 - 4x - 2$

40. $24a^3 - 16a^2 + 3a - 2$ **41.** $8x^2 + 26xy - 7y^2$

42. $12x^2 + 7xy - 10y^2$ **43.** $49x^2 - 140xy + 100y^2$

44. $16u^2 - 72uv + 81v^2$

45. $(3a + 2b)^2 = (3a + 2b)(3a + 2b)$
$= 9a^2 + 12ab + 4b^2$

46. $4x^2 - 24xy + 36y^2$ **47.** $16c^2 - 9d^2$ **48.** $25e^2 - 4f^2$

Exercises 6.3C (page 255)

1.
$$\begin{array}{r} 2x^2 + x - 1 \\ \underline{x - 3} \\ -6x^2 - 3x + 3 \\ \underline{2x^3 + x^2 - x} \\ 2x^3 - 5x^2 - 4x + 3 \end{array}$$

2. $3x^3 - 5x^2 - 3x + 2$ **3.** $6x^3y^2 - 2x^2y^3 + 8xy^4$

4. $12x^2y^3 + 3x^3y^2 - 9x^4y$

5.
$$\begin{array}{r} x^2 + x + 1 \\ \underline{x^2 + x + 1} \\ x^2 + x + 1 \\ x^3 + x^2 + x \\ \underline{x^4 + x^3 + x^2} \\ x^4 + 2x^3 + 3x^2 + 2x + 1 \end{array}$$

6. $x^4 - 2x^3 - x^2 + 2x + 1$ **7.** $4z^3 - 16z^2 + 64z$

8. $-5a^3 - 25a^2 - 125a$

9.
$$\begin{array}{r} z^2 - 4z + 16 \\ \underline{z + 4} \\ 4z^2 - 16z + 64 \\ \underline{z^3 - 4z^2 + 16z} \\ z^3 \qquad\quad + 64 \end{array}$$

10. $a^3 - 125$

11.
$$\begin{array}{r} -3z^3 + z^2 - 5z + 4 \\ \underline{-z + 4} \\ -12z^3 + 4z^2 - 20z + 16 \\ \underline{3z^4 - z^3 + 5z^2 - 4z} \\ 3z^4 - 13z^3 + 9z^2 - 24z + 16 \end{array}$$

12. $v^4 - 4v^3 + 5v + 6$

13.
$$\begin{array}{r} 2x^2 - x + 4 \\ \underline{x^2 - 5x - 3} \\ -6x^2 + 3x - 12 \\ -10x^3 + 5x^2 - 20x \\ \underline{2x^4 - x^3 + 4x^2} \\ 2x^4 - 11x^3 + 3x^2 - 17x - 12 \end{array}$$

14. $12a^4 - 14a^3 + 18a^2 + 19a - 15$

15.
$$3y^2 - 2y + 7$$
$$\underline{4y^2 + 8y - 3}$$
$$-9y^2 + 6y - 21$$
$$24y^3 - 16y^2 + 56y$$
$$\underline{12y^4 - 8y^3 + 28y^2}$$
$$12y^4 + 16y^3 + 3y^2 + 62y - 21$$

16. $6x^4 - 17x^3 + 47x^2 + 21x - 36$

17.
$$5x - 2$$
$$\underline{5x - 2}$$
$$-10x + 4$$
$$\underline{25x^2 - 10x}$$
$$25x^2 - 20x + 4$$

18. $4x^2 - 20x + 25$

19. $(x + y)^2 = (x + y)(x + y)$
$$= x^2 + 2xy + y^2 \quad \text{and}$$
$(x - y)^2 = (x - y)(x - y)$
$$= x^2 - 2xy + y^2$$
Then $(x + y)^2(x - y)^2 = (x^2 + 2xy + y^2)(x^2 - 2xy + y^2)$
$$x^2 + 2xy + y^2$$
$$\underline{x^2 - 2xy + y^2}$$
$$x^2y^2 + 2xy^3 + y^4$$
$$-2x^3y - 4x^2y^2 - 2xy^3$$
$$\underline{x^4 + 2x^3y + x^2y^2}$$
$$x^4 \qquad - 2x^2y^2 \qquad + y^4$$

20. $x^4 - 8x^2 + 16$

21. $(x + 2)^3 = \underbrace{(x + 2)(x + 2)}(x + 2)$

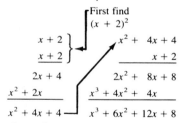

First find
$(x + 2)^2$

$$x + 2$$
$$\underline{x + 2}$$
$$2x + 4$$
$$\underline{x^2 + 2x}$$
$$x^2 + 4x + 4$$

$$x^2 + 4x + 4$$
$$\underline{x + 2}$$
$$2x^2 + 8x + 8$$
$$\underline{x^3 + 4x^2 + 4x}$$
$$x^3 + 6x^2 + 12x + 8$$

22. $x^3 + 9x^2 + 27x + 27$

23. $(x^2 + 2x - 3)^2 = (x^2 + 2x - 3)(x^2 + 2x - 3)$
$$x^2 + 2x - 3$$
$$\underline{x^2 + 2x - 3}$$
$$-3x^2 - 6x + 9$$
$$+2x^3 + 4x^2 - 6x$$
$$\underline{x^4 + 2x^3 - 3x^2}$$
$$x^4 + 4x^3 - 2x^2 - 12x + 9$$

24. $y^4 - 8y^3 + 6y^2 + 40y + 25$

25. $(x + 3)^2 = (x + 3)(x + 3) = x^2 + 6x + 9$
Then $(x + 3)^4 = (x + 3)^2(x + 3)^2$
$$= (x^2 + 6x + 9)(x^2 + 6x + 9)$$
$$x^2 + 6x + 9$$
$$\underline{x^2 + 6x + 9}$$
$$9x^2 + 54x + 81$$
$$6x^3 + 36x^2 + 54x$$
$$\underline{x^4 + 6x^3 + 9x^2}$$
$$x^4 + 12x^3 + 54x^2 + 108x + 81$$

26. $x^4 + 8x^3 + 24x^2 + 32x + 16$

Exercises 6.4A (page 256)

1. $(x + 3)(x - 3) = (x)^2 - (3)^2 = x^2 - 9$ **2.** $z^2 - 16$

3. $(w - 6)(w + 6) = (w)^2 - (6)^2 = w^2 - 36$ **4.** $y^2 - 25$

5. $(5a + 4)(5a - 4) = (5a)^2 - (4)^2 = 25a^2 - 16$

6. $36a^2 - 25$

7. $(2u + 5v)(2u - 5v) = (2u)^2 - (5v)^2 = 4u^2 - 25v^2$

8. $9m^2 - 49n^2$

9. $(4b - 9c)(4b + 9c) = (4b)^2 - (9c)^2 = 16b^2 - 81c^2$

10. $49a^2 - 64b^2$

11. $(2x^2 - 9)(2x^2 + 9) = (2x^2)^2 - (9)^2 = 4x^4 - 81$

12. $100y^4 - 9$

13. $(1 + 8z^3)(1 - 8z^3) = (1)^2 - (8z^3)^2 = 1 - 64z^6$

14. $81v^8 - 1$

15. $(5xy + z)(5xy - z) = (5xy)^2 - (z)^2 = 25x^2y^2 - z^2$

16. $100a^2b^2 - c^2$

17. $(7mn + 2rs)(7mn - 2rs) = (7mn)^2 - (2rs)^2 = 49m^2n^2 - 4r^2s^2$

18. $64h^2k^2 - 25e^2f^2$

Exercises 6.4B (page 258)

1. $(x - 1)^2 = (x)^2 - 2(x)(1) + (1)^2 = x^2 - 2x + 1$

2. $x^2 - 10x + 25$

3. $(x + 3)^2 = (x)^2 + 2(x)(3) + (3)^2 = x^2 + 6x + 9$

4. $x^2 + 8x + 16$

5. $(4x - 1)^2 = (4x)^2 - 2(4x)(1) + (1)^2 = 16x^2 - 8x + 1$

6. $49x^2 - 14x + 1$

7. $(5x - 3y)^2 = (5x)^2 - 2(5x)(3y) + (3y)^2 = 25x^2 - 30xy + 9y^2$

8. $16x^2 - 56xy + 49y^2$

9. $(3x + 2z)^2 = (3x)^2 + 2(3x)(2z) + (2z)^2 = 9x^2 + 12xz + 4z^2$

10. $4x^2 + 28xs + 49s^2$

Exercises 6.5A (page 260)

1. $\dfrac{3x + 6}{3} = \dfrac{3x}{3} + \dfrac{6}{3} = x + 2$ **2.** $2x + 3$

3. $\dfrac{4 + 8x}{4} = \dfrac{4}{4} + \dfrac{8x}{4} = 1 + 2x$ **4.** $1 - 2x$

5. $\dfrac{6x - 8y}{2} = \dfrac{6x}{2} + \dfrac{-8y}{2} = 3x - 4y$ **6.** $x - 2y$

7. $\dfrac{2x^2 + 3x}{x} = \dfrac{2x^2}{x} + \dfrac{3x}{x} = 2x + 3$ **8.** $4y - 3$

9. $\dfrac{15x^3 - 5x^2}{5x^2} = \dfrac{15x^3}{5x^2} + \dfrac{-5x^2}{5x^2} = 3x - 1$ **10.** $2y^2 - 1$

11. $\dfrac{3a^2b - ab}{ab} = \dfrac{3a^2b}{ab} + \dfrac{-ab}{ab} = 3a - 1$ **12.** $5n - 1$

13. $\dfrac{5x^5 - 4x^3 + 10x^2}{-5x^2} = \dfrac{5x^5}{-5x^2} + \dfrac{-4x^3}{-5x^2} + \dfrac{10x^2}{-5x^2} = -x^3 + \dfrac{4}{5}x - 2$

14. $-y^2 + \dfrac{5}{7}y - 2$

15. $\dfrac{-15x^2y^2z^2 - 30xyz}{-5xyz} = \dfrac{-15x^2y^2z^2}{-5xyz} + \dfrac{-30xyz}{-5xyz} = 3xyz + 6$

16. $3abc + 2$

17. $\dfrac{13x^4y^2 - 26x^2y^3 + 39x^2y^2}{13x^2y^2} = \dfrac{13x^4y^2}{13x^2y^2} + \dfrac{-26x^2y^3}{13x^2y^2} + \dfrac{39x^2y^2}{13x^2y^2}$
$$= x^2 - 2y + 3$$

18. $3n^3 - 5m - 2$

Exercises 6.5B (page 266)

NOTE: Sign changes for the subtractions are not shown here.

1.
$$
\begin{array}{r}
x + 3 \\
x + 2\overline{\smash{\big)}\,x^2 + 5x + 6} \\
\underline{x^2 + 2x} \\
3x + 6 \\
\underline{3x + 6} \\
0
\end{array}
$$

2. $x + 2$

3.
$$
\begin{array}{r}
x + 3 \\
x - 4\overline{\smash{\big)}\,x^2 - x - 12} \\
\underline{x^2 - 4x} \\
3x - 12 \\
\underline{3x - 12} \\
0
\end{array}
$$

4. $x - 4$

5.
$$
\begin{array}{r}
2x + 3 \\
3x - 2\overline{\smash{\big)}\,6x^2 + 5x - 6} \\
\underline{6x^2 - 4x} \\
9x - 6 \\
\underline{9x - 6} \\
0
\end{array}
$$

6. $4x + 5$

7.
$$
\begin{array}{r}
3v + 8 \quad \text{R } 66 \\
5v - 7\overline{\smash{\big)}\,15v^2 + 19v + 10} \\
\underline{15v^2 - 21v} \\
40v + 10 \\
\underline{40v - 56} \\
66
\end{array}
$$

8. $5v - 7$ R 52

9.
$$
\begin{array}{r}
3x + 5 \\
2x - 3\overline{\smash{\big)}\,6x^2 + x - 15} \\
\underline{6x^2 - 9x} \\
10x - 15 \\
\underline{10x - 15} \\
0
\end{array}
$$

10. $4x - 2$

11.
$$
\begin{array}{r}
4x^2 + 8x + 8 \quad \text{R} - 6 \\
-x + 2\overline{\smash{\big)}\,-4x^3 + 8x + 10} \\
\underline{-4x^3 + 8x^2} \\
-8x^2 + 8x \\
\underline{-8x^2 + 16x} \\
-8x + 10 \\
\underline{-8x + 16} \\
-6
\end{array}
$$

12. $x^2 + 3x - 3$ R -6

13.
$$
\begin{array}{r}
3a - 2b \quad \text{R } 7b^2 \\
2a + 3b\overline{\smash{\big)}\,6a^2 + 5ab + b^2} \\
\underline{6a^2 + 9ab} \\
-4ab + b^2 \\
\underline{-4ab - 6b^2} \\
7b^2
\end{array}
$$

14. $2a + 3b$ R $5b^2$

15.
$$
\begin{array}{r}
a^2 + 2a + 4 \\
a - 2\overline{\smash{\big)}\,a^3 + 0a^2 + 0a - 8} \\
\underline{a^3 - 2a^2} \\
2a^2 + 0a \\
\underline{2a^2 - 4a} \\
4a - 8 \\
\underline{4a - 8} \\
0
\end{array}
$$

16. $c^2 + 3c + 9$

17.
$$
\begin{array}{r}
x^2 + 4x + 8 \quad \text{R } 17 \\
x - 4\overline{\smash{\big)}\,x^3 + 0x^2 - 8x - 15} \\
\underline{x^3 - 4x^2} \\
4x^2 - 8x \\
\underline{4x^2 - 16x} \\
8x - 15 \\
\underline{8x - 32} \\
17
\end{array}
$$

18. $2x^2 - 3x + 6$ R -3

19.
$$
\begin{array}{r}
x^2 + x - 1 \quad \text{R } 3 \\
x^2 + x - 1\overline{\smash{\big)}\,x^4 + 2x^3 - x^2 - 2x + 4} \\
\underline{x^4 + x^3 - x^2} \\
x^3 + 0x^2 - 2x \\
\underline{x^3 + x^2 - x} \\
-x^2 - x + 4 \\
\underline{-x^2 - x + 1} \\
3
\end{array}
$$

20. $x^2 - x + 1$ R 6

21.
$$
\begin{array}{r}
x^2 + x + 1 \quad \text{R} - 8 \\
x^2 + 2x + 3\overline{\smash{\big)}\,x^4 + 3x^3 + 6x^2 + 5x - 5} \\
\underline{x^4 + 2x^3 + 3x^2} \\
x^3 + 3x^2 + 5x \\
\underline{x^3 + 2x^2 + 3x} \\
x^2 + 2x - 5 \\
\underline{x^2 + 2x + 3} \\
-8
\end{array}
$$

22. $x^2 + 2x + 3$ R 2

Review Exercises 6.6 (page 268)

1a. Third degree
b. Third degree

2. Not a polynomial   **3.** Not a polynomial

4a. Fourth degree
b. First degree

5a. $3x^4 + 7x^2 + x - 6$
b. 3

6a. $y^4 + 3xy^2 + 3x^2y + x^3$
b. 1

7. $\underline{5x^2y} + \underline{3xy^2} - \cancel{4y^3} + \underline{2xy^2} + \cancel{4y^3} + \underline{3x^2y} = 8x^2y + 5xy^2$

8. $-18x^4y^4$ **9.** $9x^3y^3 - 18x^2y^3$

10. $5a^2b + 9ab^2 - 2b^2 + 4a - 8$ **11.** $\dfrac{8x + 2}{2} = \dfrac{8x}{2} + \dfrac{2}{2} = 4x + 1$

12. $-20x^3y^3 + 12xy^3$ **13.** $x^2 + 2x - 35$ **14.** $x^3 - 8$

15.
$$
\begin{array}{r}
3x + 0 \quad \text{R} 10 \\
2x - 3 \overline{)6x^2 - 9x + 10} \\
\underline{6x^2 - 9x} \\
+ 10
\end{array}
$$

16. $x^2 - 16x + 64$

17. $(x^2 + 3 - 5x) - (4x^3 - x + 4x^2 - 1) =$
$x^2 + 3 - 5x - 4x^3 + x - 4x^2 + 1 = -4x^3 - 3x^2 - 4x + 4$

18. $a^2 - 1$

19. $(2m^2 - 5) - [(7 - m^2) - (4m^2 - 3)]$
$= (2m^2 - 5) - [7 - m^2 - 4m^2 + 3]$
$= (2m^2 - 5) - [10 - 5m^2]$
$= 2m^2 - 5 - 10 + 5m^2$
$= 7m^2 - 15$

20. $-14x^2 + 22x - 2$ **21.** $12x^3y^4 - 15x^2y^3 - 30xy^2$

22. $x^3 - 6x^2 + 12x - 8$ **23.** $9a^2 - 25$

24. $9 + 6x + 6y + x^2 + 2xy + y^2$

25.
$$
\begin{array}{r}
-3y + 2 \\
3y + 2 \overline{)-9y^2 + 0y + 4} \\
\underline{-9y^2 - 6y} \\
6y + 4 \\
\underline{6y + 4}
\end{array}
$$

26. $3ab^2 - \frac{4}{5}b + 2$ **27.** $-3xyz - z^2$ **28.** $3xy$

29.
$$
\begin{array}{r}
7a^2 - 3ab + 5 - b^2 \\
\underline{-(9a^2 + 2ab - 8 + b^2)} \\
\end{array} \Rightarrow
\begin{array}{r}
7a^2 - 3ab + 5 - b^2 \\
\underline{-9a^2 - 2ab + 8 - b^2} \\
-2a^2 - 5ab + 13 - 2b^2
\end{array}
$$

30. $6x^3 - 7x^2 + x - 4$

Chapter 6 Diagnostic Test (page 273)

Following each problem number is the textbook section number (in parentheses) where that kind of problem is discussed.

1. (6.1) **a.** Second degree
b. Third degree
c. -4
d. -4

2. (6.2)
$$
\begin{array}{r}
-7x^3 + 4x^2 + 3 \\
3x^3 + 6x - 5 \\
7x^2 - 4x + 8 \\
\hline
-4x^3 + 11x^2 + 2x + 6
\end{array}
$$

3. (6.2) $(6xy^2 - 5xy) + (17xy - 7x^2y) + (3xy^2 - y^3)$
$= 6xy^2 - 5xy + 17xy - 7x^2y + 3xy^2 - y^3$
$= 9xy^2 + 12xy - 7x^2y - y^3$

4. (6.2) $(-3x^2 - 6x + 9) - (8 - 2x + 5x^2)$
$= -3x^2 - 6x + 9 - 8 + 2x - 5x^2$
$= -8x^2 - 4x + 1$

5. (6.2)
$$
\begin{array}{r}
-6a^3 + 5a^2 + 4 \\
-(4a^3 + 6a - 7)
\end{array} \Rightarrow
\begin{array}{r}
-6a^3 + 5a^2 + 4 \\
-4a^3 - 6a + 7 \\
\hline
-10a^3 + 5a^2 - 6a + 11
\end{array}
$$

6. (6.2) $(8x^2y^2 - 5xy + 3) - (8xy - 6xy^2 + 4)$
$= 8x^2y^2 - \underline{5xy} + \underline{3} - \underline{8xy} + 6xy^2 - \underline{4}$
$= 8x^2y^2 - 13xy - 1 + 6xy^2$

7. (6.3) $(8x - 3) - (10x - 5) + (9 - 5x)$
$= 8x - 3 - 10x + 5 + 9 - 5x$
$= -7x + 11$

8. (6.3) $-6xy(3x^2 - 5xy^2 + 8y) = -18x^3y + 30x^2y^3 - 48xy^2$

9. (6.3) $(9x - 7)(8x + 9) = 72x^2 + 25x - 63$

10. (6.4) $(3x - 8)^2 = (3x)^2 - 2(3x)(8) + (8)^2 = 9x^2 - 48x + 64$

11. (6.3)
$$
\begin{array}{r}
x^2 - 3x + 9 \\
x + 3 \\
\hline
3x^2 - 9x + 27 \\
x^3 - 3x^2 + 9x \\
\hline
x^3 + 27
\end{array}
$$

12. (6.3)
$$
\begin{array}{r}
2x^2 - x - 4 \\
x^2 + 3x - 5 \\
\hline
-10x^2 + 5x + 20 \\
6x^3 - 3x^2 - 12x \\
2x^4 - x^3 - 4x^2 \\
\hline
2x^4 + 5x^3 - 17x^2 - 7x + 20
\end{array}
$$

13. (6.3) $(4abc^2)(3b)(-2a^2c) = -24a^3b^2c^3$

14. (6.3 and 6.4) $(x - 1)^4 = (x - 1)^2(x - 1)^2$
$(x - 1)^2 = (x)^2 - 2(x)(1) + (1)^2 = x^2 - 2x + 1$
Therefore, $(x - 1)^4 = (x^2 - 2x + 1)(x^2 - 2x + 1)$
$$
\begin{array}{r}
x^2 - 2x + 1 \\
x^2 - 2x + 1 \\
\hline
x^2 - 2x + 1 \\
-2x^3 + 4x^2 - 2x \\
x^4 - 2x^3 + x^2 \\
\hline
x^4 - 4x^3 + 6x^2 - 4x + 1
\end{array}
$$

15. (6.5) $\dfrac{8x^4 - 4x^3 + 12x^2}{4x^2} = \dfrac{8x^4}{4x^2} + \dfrac{-4x^3}{4x^2} + \dfrac{12x^2}{4x^2} = 2x^2 - x + 3$

16. (6.5)

$$\begin{array}{r} 5x + 2 \quad \text{R}3 \\ 3x - 1\overline{)15x^2 + \ x + 1} \\ \underline{15x^2 - 5x} \\ 6x + 1 \\ \underline{6x - 2} \\ 3 \end{array}$$

17. (6.5)

$$\begin{array}{r} x^2 - 2x - \ 8 \quad \text{R} - 20 \\ x - 4\overline{)x^3 - 6x^2 + 0x + 12} \\ \underline{x^3 - 4x^2} \\ -2x^2 + 0x \\ \underline{-2x^2 + 8x} \\ -8x + 12 \\ \underline{-8x + 32} \\ -20 \end{array}$$

$$or \ x^2 - 2x - 8 - \frac{20}{x - 4}$$

18. (6.5)

$$\begin{array}{r} 3x^2 - 2x - \ 3 \\ x - 4\overline{)3x^3 - 14x^2 + 5x + 12} \\ \underline{3x^3 - 12x^2} \\ -2x^2 + 5x \\ \underline{-2x^2 + 8x} \\ -3x + 12 \\ \underline{-3x + 12} \\ 0 \end{array}$$

Cumulative Review Exercises: Chapters 1–6 (page 274)

1. $F = \dfrac{9}{5}C + 32$

$F = \dfrac{9}{\cancel{5}}(\overset{-4}{\cancel{-20}}) + 32$
$\phantom{F = \dfrac{9}{5}}{}_1$

$F = -36 + 32 = -4$

2. -13

3. $5[11 - 2(6 - 9)] - 4(13 - 5)$
$= 5[11 - 2(-3)] - 4(13 - 5)$
$= 5[11 + 6] - 4(13 - 5)$
$= 5[17] - 4(8)$
$= 85 - 32 = 53$

4. $6x^3 - 4x + 24$

5. $-5^2 \cdot 4 - 15 \div 3\sqrt{25}$
$= -25 \cdot 4 - 15 \div 3(5)$
$= -25 \cdot 4 - 5(5)$
$= -100 - 25 = -125$

6. -22

7.
$$\begin{array}{r|r} 2 & 294 \\ 3 & 147 \\ 7 & 49 \\ & 7 \end{array}$$
Therefore, $294 = 2 \cdot 3 \cdot 7^2$

8. $6x^3 - 19x^2 + 16x - 15$

9. $(9x - 2)(9x + 2) = (9x)^2 - (2)^2$
$ = 81x^2 - 4$

10. $25x^2 - 30x + 9$

11.
$$\begin{array}{r} 5x^2 + 3x + 7 \quad \text{R}\,6 \\ x - 1\overline{)5x^3 - 2x^2 + 4x - 1} \\ \underline{5x^3 - 5x^2} \\ 3x^2 + 4x \\ \underline{3x^2 - 3x} \\ 7x - 1 \\ \underline{7x - 7} \\ 6 \end{array}$$

12. 15, 25, and 35

13. Twenty-two is what percent of 25?
Let x = percent
$25x = 22$
$\dfrac{25x}{25} = \dfrac{22}{25} = 0.88 = 88\%$ (her percent score)

14. \$175

15. Let $\quad x$ = number of adults' tickets
$\quad 9 - x$ = number of children's tickets
$2.50x + 1.25(9 - x) = 16.25$
$250x + 125(9 - x) = 1625$
$250x + 1125 - 125x = 1625$
$125x + 1125 = 1625$
$\underline{ -1125 \quad -1125}$
$125x = 500$
$\dfrac{125x}{125} = \dfrac{500}{125}$
$x = 4$ adults' tickets
$9 - x = 5$ children's tickets

16. -2

17. Let $\quad x$ = number of pounds of walnuts
$\quad 10 - x$ = number of pounds of almonds
$4.50x + 3.10(10 - x) = 3.52(10)$
$450x + 310(10 - x) = 3520$
$450x + 3100 - 310x = 3520$
$140x + 3100 = 3520$
$\underline{ -3100 \quad -3100}$
$140x = 420$
$\dfrac{140x}{140} = \dfrac{420}{140}$
$x = 3$ pounds of walnuts
$10 - x = 7$ pounds of almonds

Exercises 7.1A (page 283)

1. $4(3x + 2)$ **2.** $3(2x + 3)$ **3.** Not factorable

4. Not factorable **5.** $2(x + 4)$ **6.** $3(x + 3)$ **7.** $5(a - 2)$

8. $7(b - 2)$ **9.** $3(2y - 1)$ **10.** $5(3z - 1)$

11. $3x(3x + 1)$ **12.** $4y(2y - 1)$ **13.** $5a^2(2a - 5)$

14. $9b^2(3 - 2b^2)$ **15.** Not factorable **16.** Not factorable

17. $2ab(a + 2b)$ **18.** $3mn^2(1 + 2m)$ **19.** $6c^2d^2(2c - 3d)$

20. $15ab^3(1 - 3ab)$ **21.** $4x(x^2 - 3 - 6x)$

22. $6y(3 - y - 5y^2)$ **23.** Not factorable   **24.** Not factorable

25. $2x(4x^2 - 3x + 1)$ **26.** $3y^2(3y^2 + 2y - 1)$

27. $8(3a^4 + a^2 - 5)$ **28.** $15(3b^3 - b^4 - 2)$

29. $14xy^3(-x^7y^6 + 3x^4y - 2)$ **30.** $7uv^5(-3u^6v^3 - 9 + 5u)$

31. Not factorable   **32.** Not factorable

33. $11a^{10}b^4(-4a^4b^3 - 3b + 2a)$ **34.** $13e^8f^5(-2f + e^2f^3 - 3e^4)$

35. $2(9u^{10}v^5 + 12 - 7u^{10}v^6)$ **36.** $15(2a^3b^4 - 1 + 3a^8b^7)$

37. $6y^2(3x^3y^2 - 2z^3 - 8x^4y)$ **38.** $8m^3(4m^2n^7 - 3m^5p^9 - 5n^6)$

Exercises 7.1B (page 285)

1. $(s + t)(c + b)$ [GCF is $(s + t)$]   **2.** $(b + c)(a + d)$

3. $(a - b)(x + 5)$ [GCF is $(a - b)$]   **4.** $(s - t)(y + 7)$

5. $(u + v)(x - 3)$ [GCF is $(u + v)$]   **6.** $(t + u)(s - 2)$

7. $(x - y)(8 - a)$ [GCF is $(x - y)$]   **8.** $(a - b)(7 - c)$

9. $(s - t)(4 + u - v)$ [GCF is $(s - t)$]

10. $(x - y)(a + 5 - b)$

11. $3xy^2(a + b)(xy + 3)$ [GCF is $3xy^2(a + b)$]

12. $5ab(s + t^2)(2b^2 + 3a)$

Exercises 7.2A (page 286)

1. $\sqrt{64} = 8$ because $(8)^2 = 64$ **2.** 9

3. $\sqrt{4x^2} = 2x$ because $(2x)^2 = 4x^2$ **4.** $3y$

5. $\sqrt{100a^8} = 10a^4$ because $(10a^4)^2 = 100a^8$ **6.** $7b^3$

7. $\sqrt{m^4n^2} = m^2n$ because $(m^2n)^2 = m^4n^2$ **8.** u^5v^3

9. $\sqrt{x^{10}y^4} = x^{10/2}y^{4/2} = x^5y^2$ **10.** x^6y^4

11. $\sqrt{25a^4b^2} = 5a^{4/2}b^{2/2} = 5a^2b^1$ **12.** $10b^2c$

13. $\sqrt{36e^8f^2} = 6e^{8/2}f^{2/2} = 6e^4f^1$ **14.** $9h^6k^7$

15. $\sqrt{100a^{10}y^2} = 10a^{10/2}y^{2/2} = 10a^5y^1$ **16.** $11a^{12}b^2$

17. $\sqrt{9a^4b^2c^6} = 3a^{4/2}b^{2/2}c^{6/2} = 3a^2b^1c^3$ **18.** $12x^4y^1z^3$

Exercises 7.2B (page 288)

1. $m^2 - n^2 = (\sqrt{m^2} + \sqrt{n^2})(\sqrt{m^2} - \sqrt{n^2}) = (m + n)(m - n)$

2. $(u + v)(u - v)$

3. $x^2 - 9 = (\sqrt{x^2} + \sqrt{9})(\sqrt{x^2} - \sqrt{9}) = (x + 3)(x - 3)$

4. $(x + 5)(x - 5)$

5. $a^2 - 1 = (\sqrt{a^2} + \sqrt{1})(\sqrt{a^2} - \sqrt{1}) = (a + 1)(a - 1)$

6. $(1 + b)(1 - b)$

7. $4c^2 - 1 = (\sqrt{4c^2} + \sqrt{1})(\sqrt{4c^2} - \sqrt{1}) = (2c + 1)(2c - 1)$

8. $(4d + 1)(4d - 1)$

9. $16x^2 - 9y^2 = (\sqrt{16x^2} + \sqrt{9y^2})(\sqrt{16x^2} - \sqrt{9y^2})$
$= (4x + 3y)(4x - 3y)$

10. $(5a + 2b)(5a - 2b)$ **11.** Not factorable   **12.** Not factorable

13. $2x(2x^3 - 1)$ [GCF is $2x$]   **14.** $3a(3a^3 - 1)$

15. Not factorable   **16.** Not factorable

17. $49u^4 - 36v^4 = (\sqrt{49u^4} + \sqrt{36v^4})(\sqrt{49u^4} - \sqrt{36v^4})$
$= (7u^2 + 6v^2)(7u^2 - 6v^2)$

18. $(9m^3 + 10n^2)(9m^3 - 10n^2)$

19. $x^6 - a^4 = (\sqrt{x^6} + \sqrt{a^4})(\sqrt{x^6} - \sqrt{a^4}) = (x^3 + a^2)(x^3 - a^2)$

20. $(b + y^3)(b - y^3)$

21. $2x^2 - 18 = 2(x^2 - 9) = 2(x + 3)(x - 3)$

22. $3(x + 2)(x - 2)$ **23.** Not factorable   **24.** Not factorable

25. $a^2b^2 - c^2d^2 = (\sqrt{a^2b^2} + \sqrt{c^2d^2})(\sqrt{a^2b^2} - \sqrt{c^2d^2})$
$= (ab + cd)(ab - cd)$

26. $(mn + rs)(mn - rs)$

27. $49 - 25w^2z^2 = (\sqrt{49} + \sqrt{25w^2z^2})(\sqrt{49} - \sqrt{25w^2z^2})$
$= (7 + 5wz)(7 - 5wz)$

28. $(6 + 5uv)(6 - 5uv)$

29. $4h^4k^4 - 1 = (\sqrt{4h^4k^4} + \sqrt{1})(\sqrt{4h^4k^4} - \sqrt{1})$
$= (2h^2k^2 + 1)(2h^2k^2 - 1)$

30. $(3x^2y^2 + 1)(3x^2y^2 - 1)$

31. $81a^4b^6 - 16m^2n^8 = (\sqrt{81a^4b^6} + \sqrt{16m^2n^8})(\sqrt{81a^4b^6} - \sqrt{16m^2n^8})$
$= (9a^2b^3 + 4mn^4)(9a^2b^3 - 4mn^4)$

32. $(7c^4d^2 + 10e^3f)(7c^4d^2 - 10e^3f)$

Exercises 7.3A (page 295)

1. $(x + 2)(x + 4)$ **2.** $(x + 1)(x + 8)$ **3.** $(x + 1)(x + 4)$

4. $(x + 2)(x + 2)$ or $(x + 2)^2$ **5.** $(k + 1)(k + 6)$

6. $(k + 2)(k + 3)$ **7.** $u^2 + 7u + 10 = (u + 2)(u + 5)$

8. $(u + 1)(u + 10)$ **9.** Not factorable   **10.** Not factorable

11. $(b - 2)(b - 7)$ **12.** $(b - 1)(b - 14)$ **13.** $(z - 4)(z - 5)$

14. $(z - 2)(z - 10)$ **15.** $x^2 - 11x + 18 = (x - 2)(x - 9)$

16. $(x - 3)(x - 6)$ **17.** $(x + 10)(x - 1)$ **18.** $(y - 5)(y + 2)$

19. $(z - 3)(z + 2)$ **20.** $(m + 6)(m - 1)$

21. $5x(x + 2)$ [The GCF is $5x$]   **22.** $4y^2(2y + 1)$

23. $(x + 5)(x - 1)$ **24.** $(y + 7)(y - 1)$ **25.** Not factorable

26. Not factorable

27. $z^5 + 9z^4 - 10z^3 = z^3(z^2 + 9z - 10) = z^3(z + 10)(z - 1)$

28. $x^2(x + 8)(x - 1)$ **29.** Not factorable   **30.** Not factorable

31. $u^4 + 12u^2 - 64 = (u^2 - 4)(u^2 + 16)$
$= (u + 2)(u - 2)(u^2 + 16)$

32. $(v^2 + 2)(v^2 - 32)$ **33.** $(4 - v)(4 - v)$ or $(4 - v)^2$ or $(v - 4)^2$

34. $(2 - v)(8 - v)$ or $(v - 2)(v - 8)$ **35.** $(b + 4d)(b - 15d)$

36. $(c + 20x)(c - 3x)$ **37.** $(r + 3s)(r - 16s)$

38. $(s + 24t)(s - 2t)$

39. $x^4 + 2x^3 - 35x^2 = x^2(x^2 + 2x - 35) = x^2(x + 7)(x - 5)$

40. $x^2(x + 8)(x - 6)$

41. $14x^2 - 15x + x^3 = x(x^2 + 14x - 15) = x(x + 15)(x - 1)$

42. $x^2(x + 9)(x - 1)$ **43.** $x^4 + 6x^3 + x^2 = x^2(x^2 + 6x + 1)$

44. $y^2(y^2 + 5y + 1)$ **45.** $x^4 - 9 = (x^2 + 3)(x^2 - 3)$

46. $(a^2 + 5)(a^2 - 5)$

Exercises 7.3B (page 301)

1. $(x + 2)(3x + 1)$ **2.** $(x + 1)(3x + 2)$ **3.** $(x + 1)(5x + 2)$

4. $(x + 2)(5x + 1)$ **5.** $4x^2 + 7x + 3 = (x + 1)(4x + 3)$

6. $(x + 3)(4x + 1)$ **7.** Not factorable   **8.** Not factorable

9. $(a - 3)(5a - 1)$ **10.** $(m - 1)(5m - 3)$

11. $(b - 7)(3b - 1)$ **12.** $(u - 1)(3u - 7)$

13. $(z - 7)(5z - 1)$ **14.** $(z - 1)(5z - 7)$

15. $3n^2 + 14n - 5 = (n + 5)(3n - 1)$ **16.** $(k - 1)(5k + 7)$

17. $(3x + 7)(3x - 7)$ **18.** $(4y + 1)(4y - 1)$

19. $(x + 3y)(7x + 2y)$ **20.** $(a + 6b)(7a + b)$

21. $(h - k)(7h - 4k)$ **22.** $(h - 2k)(7h - 2k)$

23. Not factorable   **24.** Not factorable

25. $(7x - 3)(7x - 3)$ or $(7x - 3)^2$

26. $(5x - 2)(5x - 2)$ or $(5x - 2)^2$

27. $18u^3 + 39u^2 - 15u = 3u(6u^2 + 13u - 5)$
    $= 3u(2u + 5)(3u - 1)$

28. $2y(2y - 1)(3y + 5)$ **29.** $4(x^2 + x + 1)$

30. $6(x^2 + x + 3)$ **31.** $7(x^2 - 7)$ **32.** $5(x^2 - 5)$

33. $(3 - v)(2 - 5v)$ **34.** $(1 - v)(6 - 5v)$

35. $3x^2 + 20x + 12 = (3x + 2)(x + 6)$ **36.** $(3x + 4)(x + 3)$

37. $45x^2 - 120x + 80 = 5(9x^2 - 24x + 16) = 5(3x - 4)(3x - 4)$
    or $5(3x - 4)^2$

38. $4(4x - 3)(4x - 3)$ or $4(4x - 3)^2$ **39.** $(2x - 3)(4x + 5)$

40. $(8x + 5)(x - 3)$ **41.** $(2y - 5)(3y - 2)$

42. $(2y - 1)(3y - 10)$ **43.** $(3a - 2)(3a + 10)$

44. $(3b + 5)(3b - 4)$ **45.** $(2e^2 - 5)(3e^2 + 4)$

46. $(5f^2 + 3)(2f^2 - 7)$ **47.** Not factorable

48. Not factorable

Review Exercises 7.4 (page 302)

1. $4(2x - 1)$ **2.** Not factorable   **3.** $(m + 2)(m - 2)$

4. Not factorable   **5.** $(x + 3)(x + 7)$ **6.** Not factorable

7. $2u(u + 2)$ **8.** $3b(1 - 2b + 4b^2)$ **9.** $(z + 2)(z - 9)$

10. Not factorable   **11.** $(4x - 1)(x - 6)$

12. $(2x + 5)(2x + 5)$ or $(2x + 5)^2$

13. $9(k^2 - 16) = 9(k + 4)(k - 4)$ **14.** Not factorable

15. $2(4 - a^2) = 2(2 + a)(2 - a)$ **16.** $2(5c - 2)(c - 4)$

17. $3uv(5u - 1)$ **18.** $5ab(b - 2a - 1)$

19. $(2x + 3y)(5x - 8y)$ **20.** $4x^2y^2(2x^3 - 3y^2 - 4)$

Exercises 7.5 (page 307)

1. $am + bm + an + bn = m(a + b) + n(a + b)$
    $= (a + b)(m + n)$

2. $(u + v)(c + d)$

3. $mx - nx - my + ny = x(m - n) - y(m - n)$
    $= (m - n)(x - y)$

4. $(h - k)(a - b)$

5. $xy + x - y - 1 = x(y + 1) - 1(y + 1) = (y + 1)(x - 1)$

6. $(a - 1)(d + 1)$

7. $3a^2 - 6ab + 2a - 4b = 3a(a - 2b) + 2(a - 2b)$
    $= (a - 2b)(3a + 2)$

8. $(h - 3k)(2h + 5)$

9. $6e^2 - 2ef - 9e + 3f = 2e(3e - f) - 3(3e - f)$
    $= (3e - f)(2e - 3)$

10. $(2m - n)(4m - 3)$

11. $h^2 - k^2 + 2h + 2k = (h + k)(h - k) + 2(h + k)$
    $= (h + k)(h - k + 2)$

12. $(x + y)(x - y + 4)$

13. $x^3 + 3x^2 - 4x + 12 = x^2(x + 3) - 4(x - 3)$
    Therefore, not factorable

14. Not factorable

15. $a^3 - 2a^2 - 4a + 8 = a^2(a - 2) - 4(a - 2) = (a - 2)(a^2 - 4)$
    $= (a - 2)(a + 2)(a - 2) = (a - 2)^2(a + 2)$

16. $(x - 3)(x + 3)(x - 3) = (x - 3)^2(x + 3)$

17. $10xy - 15y + 8x - 12 = 5y(2x - 3) + 4(2x - 3)$
    $= (2x - 3)(5y + 4)$

18. $(7 + 3n)(5 - 6m)$

19. $a^2 - 4 + ab - 2b = (a + 2)(a - 2) + b(a - 2)$
    $= (a - 2)(a + 2 + b)$

20. $(x - 5)(x + 5 - y)$

Exercises 7.6 (page 309)

1. MP $= 3 \cdot 2 = 6$
    $6 = (-1)(-6) = (1)(6) \Rightarrow (1) + (6) = 7$
      $= (-2)(-3) = (2)(3)$
        $3x^2 + 1x + 6x + 2$
      $= x(3x + 1) + 2(3x + 1)$
      $= (3x + 1)(x + 2)$

2. $(x + 1)(3x + 2)$

3. MP $= 5 \cdot 2 = 10$
    $10 = (-1)(-10) = (1)(10)$
      $= (-2)(-5) = (2)(5) \Rightarrow (2) + (5) = 7$
        $5x^2 + 2x + 5x + 2$
      $= x(5x + 2) + 1(5x + 2)$
      $= (5x + 2)(x + 1)$

4. $(x + 2)(5x + 1)$

5. $4x^2 + 7x + 3$
    MP $= 4 \cdot 3 = 12$
    $12 = (-1)(-12) = (1)(12)$
      $= (-2)(-6) = (2)(6)$
      $= (-3)(-4) = (3)(4) \Rightarrow (3) + (4) = 7$
        $4x^2 + 3x + 4x + 3$
      $= x(4x + 3) + 1(4x + 3)$
      $= (4x + 3)(x + 1)$

6. $(x + 3)(4x + 1)$

7. MP $= 5 \cdot 4 = 20$
    $20 = (1)(20) = (-1)(-20)$ None of the
      $= (2)(10) = (-2)(-10)$ sums of these
      $= (4)(5) = (-4)(-5)$ pairs is $+20$
    Not factorable

8. Not factorable

9. MP $= 5 \cdot 3 = 15$
    $15 = (1)(15) = (-1)(-15) \Rightarrow (-1) + (-15) = -16$
      $= (3)(5) = (-3)(-5)$
        $5a^2 - 1a - 15a + 3$
      $= a(5a - 1) - 3(5a - 1)$
      $= (5a - 1)(a - 3)$

10. $(m - 1)(5m - 3)$

11. MP $= 3 \cdot 7 = 21$

$21 = (1)(21) = (-1)(-21) \Rightarrow (-1) + (-21) = -22$

$\quad = (3)(7) \ = (-3)(-7)$

$$\underbrace{3b^2 - 1b}_{} - \underbrace{21b + 7}_{}$$

$$= b(3b - 1) - 7(3b - 1)$$

$$= (3b - 1)(b - 7)$$

12. $(u - 1)(3u - 7)$

13. MP $= 5 \cdot 7 = 35$

$35 = (1)(35) = (-1)(-35) \Rightarrow (-1) + (-35) = -36$

$\quad = (5)(7) \ = (-5)(-7)$

$$\underbrace{5z^2 - 1z}_{} - \underbrace{35z + 7}_{}$$

$$= z(5z - 1) - 7(5z - 1)$$

$$= (5z - 1)(z - 7)$$

14. $(z - 1)(5z - 7)$

15. $3n^2 + 14n - 5$

MP $= 3 \cdot (-5) \ = -15$

$-15 = (1)(-15) = (-1)(15) \Rightarrow (-1) + (15) = 14$

$\quad = (3)(-5) \ = (-3)(5)$

$$\underbrace{3n^2 - 1n}_{} + \underbrace{15n - 5}_{}$$

$$= n(3n - 1) + 5(3n - 1)$$

$$= (3n - 1)(n + 5)$$

16. $(k - 1)(5k + 7)$

17. (MP method need not be used.)

$(3x + 7)(3x - 7)$

18. $(4y + 1)(4y - 1)$

19. MP $= 7 \cdot 6 = 42$

$42 = (-1)(-42) = (1)(42)$

$\quad = (-2)(-21) = (2)(21) \Rightarrow (2) + (21) = 23$

$\quad = (-3)(-14) = (3)(14)$

$\quad = (-6)(-7) \ = (6)(7)$

$$\underbrace{7x^2 + 2xy}_{} + \underbrace{21xy + 6y^2}_{}$$

$$= x(7x + 2y) + 3y(7x + 2y)$$

$$= (7x + 2y)(x + 3y)$$

20. $(a + 6b)(7a + b)$

21. MP $= 7 \cdot 4 = 28$

$28 = (1)(28) = (-1)(-28)$

$\quad = (2)(14) = (-2)(-14)$

$\quad = (4)(7) \ = (-4)(-7) \Rightarrow (-4) + (-7) = -11$

$$\underbrace{7h^2 - 4hk}_{} - \underbrace{7hk + 4k^2}_{}$$

$$= h(7h - 4k) - k(7h - 4k)$$

$$= (7h - 4k)(h - k)$$

22. $(h - 2k)(7h - 2k)$

23. MP $= 3(-6) \ = -18$

$-18 = (1)(-18) = (-1)(18)$ ⎫
$\quad = (2)(-9) \ = (-2)(9)$ ⎬ None of the sums of these pairs is -18
$\quad = (3)(-6) \ = (-3)(6)$ ⎭

Not factorable

24. Not factorable

25. MP $= 49 \cdot 9 = 441$

$441 = (1)(441) = (-1)(-441)$

$\quad = (3)(147) = (-3)(-147)$

$\quad = (7)(63) \ = (-7)(-63)$

$\quad = (9)(49) \ = (-9)(-49)$

$\quad = (21)(21) = (-21)(-21) \Rightarrow (-21) + (-21) = -42$

$$49x^2 - 21x - 21x + 9$$

$$= 7x(7x - 3) - 3(7x - 3)$$

$$= (7x - 3)(7x - 3) \text{ or } (7x - 3)^2$$

26. $(5x - 2)(5x - 2)$ or $(5x - 2)^2$

27. $3u(6u^2 + 13u - 5)$

To factor $6u^2 + 13u - 5$:

MP $= 6 \cdot (-5) = -30$

$-30 = (1)(-30) = (-1)(30)$

$\quad = (2)(-15) = (-2)(15) \Rightarrow (-2) + (15) = 13$

$\quad = (3)(-10) = (-3)(10)$

$\quad = (5)(-6) \ = (-5)(6)$

$$6u^2 - 2u + 15u - 5$$

$$= 2u(3u - 1) + 5(3u - 1)$$

$$= (3u - 1)(2u + 5)$$

The final answer is $3u(2u + 5)(3u - 1)$.

28. $2y(2y - 1)(3y + 5)$

29. $4(x^2 + x + 1)$

To factor $x^2 + x + 1$:

MP $= 1 \cdot 1 = 1$

$1 = (1)(1) = (-1)(-1)$

The sum of neither pair is 1. Therefore, $x^2 + x + 1$ is not factorable. The final answer is $4(x^2 + x + 1)$.

30. $6(x^2 + x + 3)$

31. $7(x^2 - 7)$ (MP method need not be used.)

32. $5(x^2 - 5)$

33. MP $= 6 \cdot (5) = 30$

$30 = (1)(30) = (-1)(-30)$

$\quad = (2)(15) = (-2)(-15) \Rightarrow (-2) + (-15) = -17$

$\quad = (3)(10) = (-3)(-10)$

$\quad = (5)(6) \ = (-5)(-6)$

$$6 - 2v - 15v + 5v^2$$

$$= 2(3 - v) - 5v(3 - v)$$

$$= (3 - v)(2 - 5v)$$

34. $(1 - v)(6 - 5v)$

35. $3x^2 + 20x + 12$

MP $= 3 \cdot 12 = 36$

$36 = (-1)(-36) = (1)(36)$

$\quad = (-2)(-18) = (2)(18) \Rightarrow (2) + (18) = 20$

$\quad = (-3)(-12) = (3)(12)$

$\quad = (-4)(-9) \ = (4)(9)$

$\quad = (-6)(-6) \ = (6)(6)$

$$3x^2 + 2x + 18x + 12$$

$$= x(3x + 2) + 6(3x + 2)$$

$$= (3x + 2)(x + 6)$$

36. $(3x + 4)(x + 3)$

37. $5(9x^2 - 24x + 16)$
To factor $9x^2 - 24x + 16$:
MP $= 9 \cdot 16 = 144$

$$144 = (1)(144) = (-1)(-144)$$
$$= (2)(72) \quad = (-2)(-72)$$
$$= (3)(48) \quad = (-3)(-48)$$
$$= (4)(36) \quad = (-4)(-36)$$
$$= (6)(24) \quad = (-6)(-24)$$
$$= (8)(18) \quad = (-8)(-18)$$
$$= (9)(16) \quad = (-9)(-16)$$
$$= (12)(12) = (-12)(-12) \Rightarrow (-12) + (-12) = -24$$

$$9x^2 - 12x - 12x + 16$$
$$= 3x(3x - 4) - 4(3x - 4)$$
$$= (3x - 4)(3x - 4) \text{ or } (3x - 4)^2$$

The final answer is $5(3x - 4)(3x - 4)$ or $5(3x - 4)^2$.

38. $4(4x - 3)(4x - 3)$ or $4(4x - 3)^2$

39. MP $= 8 \cdot (-15) = -120$

$$-120 = (-1)(120) = (1)(-120)$$
$$= (-2)(60) \quad = (2)(-60)$$
$$= (-3)(40) \quad = (3)(-40)$$
$$= (-4)(30) \quad = (4)(-30)$$
$$= (-5)(24) \quad = (5)(-24)$$
$$= (-6)(20) \quad = (6)(-20)$$
$$= (-8)(15) \quad = (8)(-15)$$
$$= (-10)(12) = (10)(-12) \Rightarrow (10) + (-12) = -2$$

$$= 8x^2 + 10x - 12x - 15$$
$$= 2x(4x + 5) - 3(4x + 5)$$
$$= (4x + 5)(2x - 3)$$

40. $(8x + 5)(x - 3)$

41. MP $= 6 \cdot 10 = 60$

$$60 = (1)(60) = (-1)(-60)$$
$$= (2)(30) = (-2)(-30)$$
$$= (3)(20) = (-3)(-20)$$
$$= (4)(15) = (-4)(-15) \Rightarrow (-4) + (-15) = -19$$
$$= (5)(12) = (-5)(-12)$$
$$= (6)(10) = (-6)(-10)$$

$$6y^2 - 4y - 15y + 10$$
$$= 2y(3y - 2) - 5(3y - 2)$$
$$= (3y - 2)(2y - 5)$$

42. $(2y - 1)(3y - 10)$

43. MP $= 9 \cdot (-20) = -180$

$$-180 = (1)(-180) = (-1)(180)$$
$$= (2)(-90) \quad = (-2)(90)$$
$$= (3)(-60) \quad = (-3)(60)$$
$$= (4)(-45) \quad = (-4)(45)$$
$$= (5)(-36) \quad = (-5)(36)$$
$$= (6)(-30) \quad = (-6)(30) \Rightarrow (-6) + (30) = 24$$
$$= (9)(-20) \quad = (-9)(20)$$
$$= (10)(-18) = (-10)(18)$$
$$= (12)(-15) = (-12)(15)$$

$$9a^2 - 6a + 30a - 20$$
$$= 3a(3a - 2) + 10(3a - 2)$$
$$= (3a - 2)(3a + 10)$$

44. $(3b + 5)(3b - 4)$

45. MP $= 6 \cdot (-20) = -120$

$$-120 = (-1)(120) = (1)(-120)$$
$$= (-2)(60) \quad = (2)(-60)$$
$$= (-3)(40) \quad = (3)(-40)$$
$$= (-4)(30) \quad = (4)(-30)$$
$$= (-5)(24) \quad = (5)(-24)$$
$$= (-6)(20) \quad = (6)(-20)$$
$$= (-8)(15) \quad = (8)(-15) \Rightarrow (8) + (-15) = -7$$
$$= (-10)(12) = (10)(-12)$$

$$6e^4 + 8e^2 - 15e^2 - 20$$
$$= 2e^2(3e^2 + 4) - 5(3e^2 + 4)$$
$$= (3e^2 + 4)(2e^2 - 5)$$

46. $(5f^2 + 3)(2f^2 - 7)$

47. Not factorable   (Sum of two squares; MP method need not be used.)

48. Not factorable

Exercises 7.7 (page 311)

1. $2(x^2 - 4y^2) = 2(\sqrt{x^2} + \sqrt{4y^2})(\sqrt{x^2} - \sqrt{4y^2}) = 2(x + 2y)(x - 2y)$

2. $3(x + 3y)(x - 3y)$

3. $5(a^4 - 4b^2) = 5(\sqrt{a^4} + \sqrt{4b^2})(\sqrt{a^4} - \sqrt{4b^2})$
$= 5(a^2 + 2b)(a^2 - 2b)$

4. $6(m + 3n^2)(m - 3n^2)$

5. $x^4 - y^4 = (\sqrt{x^4} + \sqrt{y^4})(\sqrt{x^4} - \sqrt{y^4}) = (x^2 + y^2)(x^2 - y^2)$
$= (x^2 + y^2)(x + y)(x - y)$

6. $(a^2 + 4)(a + 2)(a - 2)$

7. $2(2v^2 + 7v - 4) = 2(v + 4)(2v - 1)$

8. $3(v - 5)(2v + 1)$

9. $4(2z^2 - 3z - 2) = 4(z - 2)(2z + 1)$ **10.** $3(2z - 3)(3z + 1)$

11. $4(x^2 - 25) = 4(\sqrt{x^2} + \sqrt{25})(\sqrt{x^2} - \sqrt{25}) = 4(x + 5)(x - 5)$

12. $9(x + 2)(x - 2)$ **13.** $2(6x^2 + 5x - 4) = 2(2x - 1)(3x + 4)$

14. $3(3x + 2)(5x - 4)$

15. $a(b^2 - 2b + 1) = a(b - 1)(b - 1) = a(b - 1)^2$

16. $a(u - 1)^2$

17. $x^4 - 81 = (\sqrt{x^4} + \sqrt{81})(\sqrt{x^4} - \sqrt{81}) = (x^2 + 9)(x^2 - 9)$
$= (x^2 + 9)(x + 3)(x - 3)$

18. $(4y^4 + z^2)(2y^2 + z)(2y^2 - z)$ **19.** $16(x^2 + 1)$

20. $25(b^2 + 4)$ **21.** $2u(u^2 + uv - 6v^2) = 2u(u + 3v)(u - 2v)$

22. $3m(m + 3n)(m - 4n)$

23. $4h(2h^2 - 5hk + 3k^2) = 4h(2h - 3k)(h - k)$

24. $5k(h - 2k)(3h - k)$

25. $a^3b^2(a^2 - 4b^2) = a^3b^2(a + 2b)(a - 2b)$

26. $x^2y^2(y + 10x)(y - 10x)$

27. $2a(x^2 - 4a^2y^2) = 2a(\sqrt{x^2} + \sqrt{4a^2y^2})(\sqrt{x^2} - \sqrt{4a^2y^2})$
  $= 2a(x + 2ay)(x - 2ay)$

28. $3b^2(x^2 + 2y)(x^2 - 2y)$

29. $12 + 4x - 3x^2 - x^3 = 4(3 + x) - x^2(3 + x)$
  $= (3 + x)(4 - x^2) = (3 + x)(2 + x)(2 - x)$

30. $(5 - z)(3 + z)(3 - z)$

31. $6my - \underline{4nz} + 15mz - \underline{5zn} = 6my - 9nz + 15mz$
  $= 3(2my - 3nz + 5mz)$

32. $4(xy + mn)$

33. $x^4 - 8x^2 + 16 = (x^2 - 4)(x^2 - 4)$
  $= (x + 2)(x - 2)(x + 2)(x - 2) = (x + 2)^2(x - 2)^2$

34. $(y + 3)^2(y - 3)^2$

35. $x^8 - 1 = (x^4 + 1)(x^4 - 1) = (x^4 + 1)(x^2 + 1)(x^2 - 1)$
  $= (x^4 + 1)(x^2 + 1)(x + 1)(x - 1)$

36. $(a^4 + b^4)(a^2 + b^2)(a + b)(a - b)$

37. $x^2 - a^2 - 4x + 4a = (x + a)(x - a) - 4(x - a)$
  $= (x - a)(x + a - 4)$

38. $(m - 5)(m + 5 + n)$

39. $6ac - 6bd + 6bc - 6ad = 6[ac + bc - bd - ad]$
  $= 6[c(a + b) - d(b + a)] = 6(a + b)(c - d)$

40. $(2c + d)(5y - 3z)$

Exercises 7.8 (page 316)

1. $(x - 5)(x + 4) = 0$
  $x - 5 = 0 \quad | \quad x + 4 = 0$
  $x = 5 \quad | \quad x = -4$

2. $-7, 2$

3. $3x(x - 4) = 0$
  $3x = 0 \quad | \quad x - 4 = 0$
  $x = 0 \quad | \quad x = 4$

4. $0, -6$

5. $(x + 10)(2x - 3) = 0$
  $x + 10 = 0 \quad | \quad 2x - 3 = 0$
  $x = -10 \quad | \quad 2x = 3$
  $\quad\quad\quad\quad\quad | \quad x = \frac{3}{2} = 1\frac{1}{2}$

6. $8, -\frac{2}{3}$

7. $x^2 + 9x + 8 = 0$
  $(x + 1)(x + 8) = 0$
  $x + 1 = 0 \quad | \quad x + 8 = 0$
  $x = -1 \quad | \quad x = -8$

8. $-2, -4$

9. $x^2 - x - 12 = 0$
  $(x - 4)(x + 3) = 0$
  $x - 4 = 0 \quad | \quad x + 3 = 0$
  $x = 4 \quad | \quad x = -3$

10. $-4, 3$

11. $\quad\quad x^2 = 64$
  $\quad\quad x^2 - 64 = 0$
  $(x + 8)(x - 8) = 0$
  $x + 8 = 0 \quad | \quad x - 8 = 0$
  $x = -8 \quad | \quad x = 8$

12. $12, -12$

13. $6x^2 - 10x = 0$
  $2x(3x - 5) = 0$
  $2x = 0 \quad | \quad 3x - 5 = 0$
  $x = 0 \quad | \quad 3x = 5$
  $\quad\quad\quad\quad | \quad x = \frac{5}{3}$

14. $0, \frac{7}{2}$

15. $24w = 4w^2$
  $0 = 4w^2 - 24w$
  $0 = 4w(w - 6)$
  $4w = 0 \quad | \quad w - 6 = 0$
  $w = 0 \quad | \quad w = 6$

16. $0, 4$

17. $\quad\quad 5a^2 = 16a - 3$
  $\quad 5a^2 - 16a + 3 = 0$
  $(1a - 3)(5a - 1) = 0$
  $a - 3 = 0 \quad | \quad 5a - 1 = 0$
  $a = 3 \quad | \quad 5a = 1$
  $\quad\quad\quad\quad | \quad a = \frac{1}{5}$

18. $7, \frac{1}{3}$

19. $\quad\quad 3u^2 = 2u + 5$
  $\quad 3u^2 - 2u - 5 = 0$
  $(1u + 1)(3u - 5) = 0$
  $u + 1 = 0 \quad | \quad 3u - 5 = 0$
  $u = -1 \quad | \quad 3u = 5$
  $\quad\quad\quad\quad | \quad u = \frac{5}{3} = 1\frac{2}{3}$

20. $7, -\frac{1}{5}$

21. $(x - 2)(x - 3) = 2$
  $x^2 - 5x + 6 = 2$
  $x^2 - 5x + 4 = 0$
  $(x - 1)(x - 4) = 0$
  $x - 1 = 0 \quad | \quad x - 4 = 0$
  $x = 1 \quad | \quad x = 4$

22. $2, 6$

23.
$$x(x - 4) = 12$$
$$x^2 - 4x = 12$$
$$x^2 - 4x - 12 = 0$$
$$(x - 6)(x + 2) = 0$$

$x - 6 = 0$	$x + 2 = 0$
$x = 6$	$x = -2$

24. $-3, 5$

25. $4x(2x - 1)(3x + 7) = 0$

$4x = 0$	$2x - 1 = 0$	$3x + 7 = 0$
$x = 0$	$2x = 1$	$3x = -7$
	$x = \dfrac{1}{2}$	$x = -\dfrac{7}{3} = -2\dfrac{1}{3}$

26. $0, \dfrac{3}{4}, \dfrac{6}{7}$

27.
$$2x^3 + x^2 = 3x$$
$$2x^3 + x^2 - 3x = 0$$
$$x(2x^2 + x - 3) = 0$$
$$x(1x - 1)(2x + 3) = 0$$

$x = 0$	$x - 1 = 0$	$2x + 3 = 0$
	$x = 1$	$2x = -3$
		$x = -\dfrac{3}{2} = -1\dfrac{1}{2}$

28. $0, -5, \dfrac{1}{2}$

29.
$$2a^3 - 10a^2 = 0$$
$$2a^2(a - 5) = 0$$
$$2aa(a - 5) = 0$$

$2 \ne 0$	$a = 0$	$a = 0$	$a - 5 = 0$
		↑	$a = 5$

$a = 0$ need not be listed twice

30. $0, 6$

Exercises 7.9 (page 321)

1. Let $x = $ smaller number $\Big\}$ since their difference
$x + 5 = $ larger number $\Big\}$ is 5
$$x(x + 5) = 14$$
$$x^2 + 5x = 14$$
$$x^2 + 5x - 14 = 0$$
$$(x + 7)(x - 2) = 0$$

$x + 7 = 0$	$x - 2 = 0$
$x = -7$	$x = 2$
$x + 5 = -2$	$x + 5 = 7$

There are two answers: $\{2, 7\}$ and $\{-7, -2\}$

2. Two answers: $\{3, 9\}$ and $\{-9, -3\}$

3. Let $x = $ one number
$12 - x = $ other number
$$x(12 - x) = 35$$
$$12x - x^2 = 35$$
$$0 = x^2 - 12x + 35$$
$$0 = (x - 5)(x - 7)$$

$x - 5 = 0$	$x - 7 = 0$
$x = 5$	$x = 7$
$12 - x = 7$	$12 - x = 5$

One answer: $\{5, 7\}$

4. $2, -6$

5. Let $x = $ altitude (in inches)
$x + 3 = $ base (in inches)
$$\frac{1}{2}x(x + 3) = \text{area}$$
$$\frac{1}{2}x(x + 3) = 20$$
$$(2)\left(\frac{1}{2}x\right)(x + 3) = 20(2)$$
$$x(x + 3) = 40$$
$$x^2 + 3x - 40 = 0$$
$$(x - 5)(x + 8) = 0$$

$x - 5 = 0$	$x + 8 = 0$
$x = 5$ in., altitude	$x = -8$ Reject
$x + 3 = 8$ in., base	(A length cannot be negative.)

6. Altitude $= 12$ cm, base $= 15$ cm

7. Let $x = $ the first integer
$x + 1 = $ the second integer
$x + 2 = $ the third integer
$$x(x + 1) + (x + 1)(x + 2) = 8$$
$$x^2 + x + x^2 + 3x + 2 = 8$$
$$2x^2 + 4x - 6 = 0$$
$$2(x^2 + 2x - 3) = 0$$
$$2(x - 1)(x + 3) = 0$$

$2 \ne 0$	$x - 1 = 0$	$x + 3 = 0$
	$x = 1$	$x = -3$
	$x + 1 = 2$	$x + 1 = -2$
	$x + 2 = 3$	$x + 2 = -1$

Two answers: $\{1, 2, 3\}$ and $\{-3, -2, -1\}$

8. One answer: $\{2, 3, 4\}$

9. Let $x = $ the first even integer
$x + 2 = $ the second even integer
$x + 4 = $ the third even integer
$$2x(x + 2) = 16 + (x + 2)(x + 4)$$
$$2x^2 + 4x = 16 + x^2 + 6x + 8$$
$$x^2 - 2x - 24 = 0$$
$$(x + 4)(x - 6) = 0$$

$x + 4 = 0$	$x - 6 = 0$
$x = -4$	$x = 6$
$x + 2 = -2$	$x + 2 = 8$
$x + 4 = \quad 0$	$x + 4 = 10$

Two answers: $\{-4, -2, 0\}$ and $\{6, 8, 10\}$

10. Two answers: $\{5, 7, 9\}$ and $\{-15, -13, -11\}$

11. Let $W = $ width
Area $= (W + 5)W$
$$(W + 5)W = 84$$
$$W^2 + 5W = 84$$
$$W^2 + 5W - 84 = 0$$
$$(W + 12)(W - 7) = 0$$

$W + 12 = 0$	$W - 7 = 0$
$W = -12$	$W = 7$ ft, width
No meaning	$W + 5 = 12$ ft, length

12. Length $= 7$ ft, width $= 4$ ft

13. Let $\quad x =$ length of side of larger square

$\quad\quad x - 3 =$ length of side of smaller square

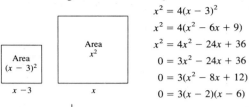

$$x^2 = 4(x - 3)^2$$
$$x^2 = 4(x^2 - 6x + 9)$$
$$x^2 = 4x^2 - 24x + 36$$
$$0 = 3x^2 - 24x + 36$$
$$0 = 3(x^2 - 8x + 12)$$
$$0 = 3(x - 2)(x - 6)$$

$x - 2 = 0$	$x - 6 = 0$
$x = 2$	$x = 6$ cm, side of larger square
$x - 3 = -1$	$x - 3 = 3$ cm, side of smaller square
Not possible	

14. Side of smaller square $= 2$ ft
Side of larger square $= 6$ ft

15.

Let $L =$ length

$L - 4 =$ width

Area $= L(L - 4)$

Perimeter $= 2L + 2(L - 4)$

$$L(L - 4) = 17 + 2L + 2(L - 4)$$
$$L^2 - 4L = 17 + 2L + 2L - 8$$
$$L^2 - 4L = 4L + 9$$
$$L^2 - 8L - 9 = 0$$
$$(L + 1)(L - 9) = 0$$

$L + 1 = 0$	$L - 9 = 0$
$L = -1$	$L = 9$ yd, length
No meaning	$L - 4 = 5$ yd, width

16. Side $= 8$

17. Let $\quad x =$ altitude

$\quad\quad x + 3 =$ base

Area $= \dfrac{1}{2}(x + 3)x$

$$\frac{1}{2}(x + 3)x = 35$$
$$2\left(\frac{1}{2}\right)(x + 3x)x = 2(35)$$
$$x^2 + 3x = 70$$
$$x^2 + 3x - 70 = 0$$
$$(x + 10)(x - 7) = 0$$

$x + 10 = 0$	$x - 7 = 0$
$x = -10$	$x = 7$ in., altitude
Reject	$x + 3 = 10$ in., base
(A length cannot be negative.)	

18. Altitude $= 4$ m, base $= 9$ m

19. Let $\quad x =$ altitude (in inches)

$\quad\quad 19 - x =$ base (in inches)

$$\frac{1}{2}x(19 - x) = \text{area}$$
$$\frac{1}{2}x(19 - x) = 42$$
$$(2)\left(\frac{1}{2}x\right)(19 - x) = 42(2)$$
$$x(19 - x) = 84$$
$$19x - x^2 = 84$$
$$0 = x^2 - 19x + 84$$
$$0 = (x - 7)(x - 12)$$

$x - 7 = 0$	$x - 12 = 0$
$x = 7$ in., altitude	$x = 12$ in., altitude
$19 - x = 12$ in., base	$19 - x = 7$ in., base

20. Altitude $= 6$ cm, base $= 9$ cm
or Base $= 6$ cm, altitude $= 9$ cm

21. Let $\quad x =$ height (in cm)

$\quad\quad x + 4 =$ length (in cm)

$$5x(x + 4) = \text{volume}$$
$$5x(x + 4) = 225$$
$$5x^2 + 20x = 225$$
$$5x^2 + 20x - 225 = 0$$
$$5(x^2 + 4x - 45) = 0$$
$$5(x - 5)(x + 9) = 0$$

$5 \neq 0$	$x - 5 = 0$	$x + 9 = 0$
	$x = 5$ cm, height	$x = -9$ Reject
	$x + 4 = 9$ cm, length	(A length cannot be negative.)

22. Width $= 7$ in., length $= 9$ in.

Review Exercises 7.10 (page 324)

1. $5(1 + 3a)$ **2.** $2(5 + n)(5 - n)$

3. $ab + 2b - a - 2 = b(a + 2) - 1(a + 2) = (a + 2)(b - 1)$

4. $y(y + 2)(y + 8)$ **5.** Not factorable **6.** $3b(1 - 2b)$

7. $(5c - 2)(c - 4)$ **8.** $(m - 5)(n - 1)$

9. $5(x^2 - 7x - 30) = 5(x + 3)(x - 10)$ **10.** Not factorable

11. $x^3 + 5x^2 + 3x + 15 = x^2(x + 5) + 3(x + 5) = (x + 5)(x^2 + 3)$

12. $(x + 3)^2(x - 3)$

13. $x^2 - y^2 + x - y = (x + y)(x - y) + 1(x - y)$
$$= (x - y)([x + y] + 1)$$
$$= (x - y)(x + y + 1)$$

14. $15(a + 2b)(a - b)$

15. $(x - 5)(x + 3) = 0$

$x - 5 = 0$	$x + 3 = 0$
$x = 5$	$x = -3$

16. $8, -3$

17. $m^2 - 3m - 18 = 0$

$(m - 6)(m + 3) = 0$

$m - 6 = 0$	$m + 3 = 0$
$m = 6$	$m = -3$

18. $6, -6$

19. $3z^2 - 12z = 0$

$3z(z - 4) = 0$

$\begin{array}{c|c|c} 3 \neq 0 & z = 0 & z - 4 = 0 \\ & & z = 4 \end{array}$

20. $\frac{5}{3}, \frac{9}{4}$

21. $3x^2 + 13x - 10 = 0$

$(3x - 2)(x + 5) = 0$

$\begin{array}{c|c} 3x - 2 = 0 & x + 5 = 0 \\ 3x = 2 & x = -5 \\ x = \dfrac{2}{3} & \end{array}$

22. $\frac{1}{5}, -\frac{3}{2}$

23. $2u(u + 6)(u - 2) = 0$

$\begin{array}{c|c|c|c} 2 \neq 0 & u = 0 & u + 6 = 0 & u - 2 = 0 \\ & & u = -6 & u = 2 \end{array}$

24. $0, \frac{3}{2}, \frac{2}{3}$

25. Let $\quad x =$ one number

$x + 3 =$ other number

$x(x + 3) = 28$

$x^2 + 3x = 28$

$x^2 + 3x - 28 = 0$

$(x + 7)(x - 4) = 0$

$\begin{array}{c|c} x + 7 = 0 & x - 4 = 0 \\ x = -7, \text{one number} & x = 4, \text{one number} \\ x + 3 = -4, \text{other number} & x + 3 = 7, \text{other number} \end{array}$

26. Small: 2 ft; large: 8 ft

27. Let $\quad x =$ width in meters

$x + 3 =$ length in meters

$x(x + 3) = 40$

$x^2 + 3x = 40$

$x^2 + 3x - 40 = 0$

$(x + 8)(x - 5) = 0$

$\begin{array}{c|c} x + 8 = 0 & x - 5 = 0 \\ x = -8 \quad \text{Reject} & x = 5\,\text{m, width} \\ \text{(A length cannot be negative.)} & x + 3 = 8\,\text{m, length} \end{array}$

28. Width: 4; length: 10

29. Let $\quad x =$ first integer

$x + 1 =$ second integer

$x(x + 1) = 1 + x + (x + 1)$

$x^2 + x = 2 + 2x$

$x^2 - x - 2 = 0$

$(x - 2)(x + 1) = 0$

$\begin{array}{c|c} x - 2 = 0 & x + 1 = 0 \\ x = 2 & x = -1 \\ x + 1 = 3 & x + 1 = 0 \end{array}$

Two answers: $\{2, 3\}$ and $\{0, -1\}$

30. 1, 3, and 5

Chapter 7 Diagnostic Test (page 327)

Following each problem number is the textbook section number (in parentheses) where that kind of problem is discussed.

1. (7.1) $\quad 8x + 12 = 4(2x + 3)$

2. (7.1) $\quad 5x^3 - 35x^2 = 5x^2(x - 7)$

3. (7.2) $\quad 25x^2 - 121y^2 = (5x + 11y)(5x - 11y)$

4. (7.1) $\quad 8x^2 + 7$ is not factorable

5. (7.1 and 7.2) $\quad 5a^2 - 180 = 5(a^2 - 36) = 5(a + 6)(a - 6)$

6. (7.3) $\quad z^2 + 9z + 8 = (z + 1)(z + 8)$

7. (7.3) $\quad m^2 + 5m - 6 = (m + 6)(m - 1)$

8. (7.3) $\quad 11x^2 - 18x + 7 = (11x - 7)(x - 1)$

9. (7.3) $\quad x^2 + 7x - 6$ is not factorable

10. (7.3) $\quad 4y^2 + 19y - 5 = (4y - 1)(y + 5)$

11. (7.5) $\quad 5n - mn - 5 + m = n(5 - m) - 1(5 - m)$

$= (5 - m)(n - 1)$

12. (7.3) $\quad 6h^2k - 8hk^2 + 2k^3 = 2k(3h^2 - 4hk + k^2)$

$= 2k(3h - k)(h - k)$

13. (7.8) $\quad 3x^2 - 12x = 0$

$3x(x - 4) = 0$

$\begin{array}{c|c|c} 3 \neq 0 & x = 0 & x - 4 = 0 \\ & & x = 4 \end{array}$

14. (7.8) $\quad x^2 + 20 = 12x$

$x^2 - 12x + 20 = 0$

$(x - 10)(x - 2) = 0$

$\begin{array}{c|c} x - 10 = 0 & x - 2 = 0 \\ x = 10 & x = 2 \end{array}$

15. (7.8) $\quad 3x^3 = x^2 + 10x$

$3x^3 - x^2 - 10x = 0$

$x(3x^2 - x - 10) = 0$

$x(3x + 5)(x - 2) = 0$

$\begin{array}{c|c|c} x = 0 & 3x + 5 = 0 & x - 2 = 0 \\ & 3x = -5 & x = 2 \\ & x = -\dfrac{5}{3} & \end{array}$

16. (7.8) $\quad (x - 7)(x + 6) = -22$

$x^2 - x - 42 = -22$

$x^2 - x - 20 = 0$

$(x - 5)(x + 4) = 0$

$\begin{array}{c|c} x - 5 = 0 & x + 4 = 0 \\ x = 5 & x = -4 \end{array}$

17. (7.9) Let $\quad x =$ one odd integer

$x + 2 =$ next odd integer

$x + 4 =$ third odd integer

$2x(x + 2) - x(x + 4) = 49$

$2x^2 + 4x - x^2 - 4x = 49$

$x^2 - 49 = 0$

$(x + 7)(x - 7) = 0$

$\begin{array}{c|c} x + 7 = 0 & x - 7 = 0 \\ x = -7, \text{one number} & x = 7, \text{one number} \\ x + 2 = -5, \text{next number} & x + 2 = 9, \text{next number} \\ x + 4 = -3, \text{third number} & x + 4 = 11, \text{third number} \end{array}$

18. (7.9) Let $\quad x = \text{width (in ft)}$

$x + 3 = \text{length (in ft)}$

$x(x + 3) = \text{area}$

$x(x + 3) = 28$

$x^2 + 3x - 28 = 0$

$(x - 4)(x + 7) = 0$

$x - 4 = 0 \quad | \quad x + 7 = 0$

$x = 4 \text{ ft, width} \quad | \quad x = -7 \quad \text{Reject}$

$x + 3 = 7 \text{ ft, length} \quad | \quad$ (A length cannot be negative.)

Cumulative Review Exercises: Chapters 1–7 (page 328)

1. $24 \div 2\sqrt{16} - 3^2 \cdot 5 = 24 \div 2(4) - 9 \cdot 5$

$\quad = 12(4) - 9 \cdot 5 = 48 - 45 = 3$

2. $\frac{12}{5}$ or $2\frac{2}{5}$

3. $C = \dfrac{a}{a + 12} \cdot A$

$C = \dfrac{8}{8 + 12} \cdot 35$

$C = \dfrac{\overset{2}{\cancel{8}}}{\underset{\cancel{5}}{\cancel{20}}} \cdot \dfrac{\overset{7}{35}}{1} = 14$

4. $\dfrac{y^6}{9z^4}$ **5.** $\left(\dfrac{\overset{3}{\cancel{15}}\overset{2}{x^2y}}{\underset{2}{\cancel{10}}\underset{x}{x^3}}\right)^3 = \left(\dfrac{3}{2x}\right)^3 = \dfrac{3^3}{2^3x^3} = \dfrac{27}{8x^3}$

6a. 5.73×10^7

  b. 3.51×10^{-3}

7. $(2x^2 + 5x - 3) - (-4x^2 + 8x + 10) + (6x^2 + 3x - 8)$

$\quad = 2x^2 + 5x - 3 + 4x^2 - 8x - 10 + 6x^2 + 3x - 8$

$\quad = 12x^2 - 21$

8. $3y^3 - 14y^2 + 13y - 20$

9. $\dfrac{12a^2 - 3a}{3a} = \dfrac{12a^2}{3a} - \dfrac{3a}{3a} = 4a - 1$

10. $4x + 5 \text{ R} -2$ or $4x + 5 - \dfrac{2}{2x - 3}$ **11.** $18x(2x - 1)$

12. $(1 + 6t)(1 - 6t)$ **13.** $(x - 3)(x - 5)$ **14.** Not factorable

15. $(5k + 1)(k - 7)$ **16.** $(3n - 5)(n + 1)$

17. Let $x = \text{number of pounds of cashews}$

(There will be $20 + x$ pounds in the mixture.)

$7.50x + 3.50(20) = 5.90(20 + x)$

$75x + 35(20) = 59(20 + x)$

$75x + 700 = 1180 + 59x$

$\underline{-59x - 700 \quad -700 - 59x}$

$16x \quad = \quad 480$

$x = 30 \text{ pounds of cashews}$

18. Let $\quad 3x = \text{length (in meters)}$

$2x = \text{width (in meters)}$

$(3x)(2x) = \text{area}$

$(3x)(2x) = 150$

$6x^2 = 150$

$6x^2 - 150 = 0$

$6(x^2 - 25) = 0$

$6(x + 5)(x - 5) = 0$

$6 \neq 0 \quad | \quad x + 5 = 0 \quad | \quad x - 5 = 0$

$\quad | \quad x = -5 \quad \text{Reject} \quad | \quad x = 5$

(A length cannot be negative.)

$3x = 3(5 \text{ m}) = 15 \text{ m, length}$

$2x = 2(5 \text{ m}) = 10 \text{ m, width}$

Exercises 8.1 (page 332)

1. 2, because 2 would make the denominator zero   **2.** -3

3. None   **4.** None

5. 0 and 2, because either of these would make the denominator zero

6. 0, 3

7. -1 and 2, because either of these would make the denominator zero

8. 3, -4

9. Yes, because if we multiply both x and $2y$ by 5, we get $5x$ and $10y$

10. Yes

11. No; we can't get the second fraction from the first by multiplying both x and $2y$ by the same number

12. No

13. Yes, because if we multiply both $(x + 1)$ and $2(3x - 2)$ by 6, we get $6(x + 1)$ and $12(3x - 2)$

14. Yes

15. -5. Signs of denominators are the same and signs of fractions are different; signs of numerators must be different.

16. -8

17. y. Signs of fractions are the same and signs of numerators are different; signs of denominators must be different.

18. -2

19. $x - 2$. Signs of fractions are the same and signs of denominators are different; signs of numerators must be different.

20. $y - 5$

21. -5. Signs of fractions are the same and signs of numerators are different; signs of denominators must be different.

22. -7

23. $5 - x$. Signs of numerators are the same and signs of fractions are different; signs of denominators must be different.

24. $4 - x$

25. $b - a$. Signs of fractions are the same and signs of denominators are different; signs of numerators must be different.

26. $y - x$

Exercises 8.2 (page 337)

1. $\dfrac{9}{12} = \dfrac{\overset{1}{\cancel{3}} \cdot 3}{\underset{1}{\cancel{3}} \cdot 4} = \dfrac{3}{4}$ **2.** $\dfrac{4}{7}$ **3.** $\dfrac{\overset{2}{\cancel{6}}ab^2}{\underset{1}{\cancel{3}}ab} = 2b$ **4.** $2m$

5. $\dfrac{\overset{2}{4x^2y}}{\underset{1}{2xy}} = 2x$ 6. $3x^2$ 7. $\dfrac{5x-10}{x-2} = \dfrac{5(\overset{1}{x-2})}{\underset{1}{x-2}} = 5$ 8. 3

9. $-\dfrac{5x-6}{6-5x} = +\dfrac{5x-6}{-(6-5x)} = \dfrac{5x-6}{5x-6} = 1$ 10. -1

11. $\dfrac{5x^2+30x}{10x^2-40x} = \dfrac{\overset{1}{5x}(x+6)}{\underset{2}{10x}(x-4)} = \dfrac{x+6}{2(x-4)}$ 12. $\dfrac{x}{3}$

13. $\dfrac{2+4}{4} = \dfrac{6}{4} = \dfrac{3}{2}$ Incorrect to cancel the 4's.   14. 4

15. Cannot be reduced   16. Cannot be reduced

17. $\dfrac{x^2-1}{x+1} = \dfrac{(\overset{1}{x+1})(x-1)}{\underset{1}{(x+1)}} = x-1$ 18. $x+2$

19. $\dfrac{6x^2-x-2}{10x^2+3x-1} = \dfrac{(3x-2)(\overset{1}{2x+1})}{(5x-1)(\underset{1}{2x+1})} = \dfrac{3x-2}{5x-1}$ 20. $\dfrac{2x-3}{3x+2}$

21. $\dfrac{x^2-y^2}{(x+y)^2} = \dfrac{(\overset{1}{x+y})(x-y)}{(\underset{1}{x+y})(x+y)} = \dfrac{x-y}{x+y}$ 22. $\dfrac{a+3b}{a-3b}$

23. $\dfrac{2y^2+xy-6x^2}{3x^2+xy-2y^2} = \dfrac{(2y-3x)(y+2x)}{(3x-2y)(x+y)}$

$= \dfrac{-(\overset{1}{3x-2y})(y+2x)}{(\underset{1}{3x-2y})(x+y)} = -\dfrac{2x+y}{x+y}$

24. $-\dfrac{3x+2y}{2x+3y}$

25. $\dfrac{8x^2-2y^2}{2ax-ay+2bx-by} = \dfrac{2(4x^2-y^2)}{a(2x-y)+b(2x-y)}$

$= \dfrac{2(2x+y)(\overset{1}{2x-y})}{(\underset{1}{2x-y})(a+b)} = \dfrac{2(2x+y)}{a+b}$

26. $\dfrac{3(x-2y)}{a+b}$ 27. $\dfrac{(-1)(z-8)}{8-z} = \dfrac{\overset{1}{8-z}}{\underset{1}{8-z}} = 1$ 28. 1

29. $\dfrac{(-1)(a-2b)(b-a)}{(2a+b)(a-b)}$

$= \dfrac{(a-2b)(\overset{1}{a-b})}{(2a+b)(\underset{1}{a-b})} = \dfrac{a-2b}{2a+b}$

30. $\dfrac{8}{3n+m}$

31. $\dfrac{9-16x^2}{16x^2-24x+9} = \dfrac{(3+4x)(\overset{-1}{3-4x})}{(4x-3)(\underset{1}{4x-3})}$

$= -\dfrac{3+4x}{4x-3}$ or $\dfrac{3+4x}{3-4x}$

32. $-\dfrac{5+3x}{3x-5}$ or $\dfrac{5+3x}{5-3x}$

33. $\dfrac{10+x-3x^2}{2x^2+x-10} = \dfrac{(5+3x)(\overset{-1}{2-x})}{(2x+5)(\underset{1}{x-2})} = -\dfrac{5+3x}{2x+5}$

34. $-\dfrac{4-5x}{3x+4}$ or $\dfrac{5x-4}{3x+4}$

35. $\dfrac{18-3x-3x^2}{6x^2+6x-36} = \dfrac{3(6-x-x^2)}{6(x^2+x-6)} = \dfrac{(\overset{1}{3+x})(\overset{-1}{2-x})}{2(\underset{1}{x+3})(\underset{1}{x-2})} = -\dfrac{1}{2}$

36. $-\dfrac{1}{4}$

Exercises 8.3 (page 342)

1. $\dfrac{5}{6} \div \dfrac{5}{3} = \dfrac{5}{6} \cdot \dfrac{3}{5} = \dfrac{1}{2}$ 2. $\dfrac{3}{14}$ 3. $\dfrac{\overset{1}{4a^3}}{\underset{1}{5b^2}} \cdot \dfrac{\overset{2}{10b}}{\underset{2}{8a^2}} = \dfrac{a}{b}$ 4. $\dfrac{c}{d}$

5. $\dfrac{3x^2}{16} \div \dfrac{x}{8} = \dfrac{3x^2}{\underset{2}{16}} \cdot \dfrac{\overset{1}{8}}{x} = \dfrac{3x}{2}$ 6. $\dfrac{3y}{2}$

7. $\dfrac{3x^4y^2z}{\underset{2}{18xy}} \cdot \dfrac{\overset{5}{15z}}{x^3yz^2} = \dfrac{5x^4y^2z^2}{2x^4y^2z^2} = \dfrac{5}{2}$ 8. $\dfrac{9}{2}$ 9. $\dfrac{x}{x+2} \cdot \dfrac{5(\overset{1}{x+2})}{x^3} = \dfrac{5}{x^2}$

10. $\dfrac{3}{b^3}$ 11. $\dfrac{\overset{1}{y-2}}{y} \cdot \dfrac{6}{3(\underset{1}{y-2})} = \dfrac{2}{y}$ 12. $\dfrac{2}{a}$

13. $\dfrac{b^3}{a+3} \div \dfrac{4b^2}{2a+6} = \dfrac{b^3}{a+3} \cdot \dfrac{2(\overset{1}{a+3})}{4b^2} = \dfrac{b}{2}$ 14. $\dfrac{m}{2}$

15. $\dfrac{5s-15}{30s} \div \dfrac{s-3}{45s^2} = \dfrac{5(\overset{1}{s-3})}{\underset{6}{30s}} \cdot \dfrac{\overset{15}{45s^2}}{s-3} = \dfrac{15s}{2}$ 16. $4n$

17. $\dfrac{a+4}{a-4} \div \dfrac{a^2+8a+16}{a^2-16} = \dfrac{\overset{1}{a+4}}{a-4} \cdot \dfrac{(a+4)(a-4)}{(a+4)(a+4)} = 1$ 18. 1

19. $\dfrac{5}{z+4} \cdot \dfrac{(\overset{1}{z+4})(z-4)}{(z-4)(z-4)} = \dfrac{5}{z-4}$ 20. $\dfrac{7}{x-3}$

21. $\dfrac{3(\overset{1}{a-b})}{4(c+d)} \cdot \dfrac{2(\overset{1}{c+d})}{b-a} = -\dfrac{3}{2}$ or $-1\dfrac{1}{2}$ 22. $-\dfrac{7}{2}$ or $-3\dfrac{1}{2}$

23. $\dfrac{4(\overset{1}{x-2})}{4} \cdot \dfrac{x+2}{(x+2)(x-2)} = 1$ 24. 1

25. $\dfrac{4a+4b}{ab^2} \div \dfrac{3a+3b}{a^2b} = \dfrac{4(a+b)}{ab^2} \cdot \dfrac{a^2b}{3(a+b)} = \dfrac{4a}{3b}$ 26. $\dfrac{5y}{4x}$

27. $\dfrac{a^2-9b^2}{a^2-6ab+9b^2} \div \dfrac{a+3b}{a-3b} = \dfrac{(\overset{1}{a+3b})(\overset{1}{a-3b})}{(\underset{1}{a-3b})(\underset{1}{a-3b})} \cdot \dfrac{a-3b}{a+3b} = 1$

28. 1

29. $\dfrac{x-y}{9x+9y} \div \dfrac{x^2-y^2}{3(x^2+2xy+y^2)}$

$= \dfrac{\overset{1}{x-y}}{\underset{3}{9(x+y)}} \cdot \dfrac{3(x+y)(x+y)}{(x+y)(x-y)} = \dfrac{1}{3}$

30. $\dfrac{1}{2}$

31. $\dfrac{a^2-9b^2}{6(a^2-6ab+9b^2)} \div \dfrac{a+3b}{2a-6b}$

$= \dfrac{(a+3b)(a-3b)}{6(a-3b)(a-3b)} \cdot \dfrac{2(a-3b)}{a+3b} = \dfrac{1}{3}$

32. $\dfrac{1}{2}$

33. $\dfrac{2x^3y+2x^2y^2}{6x} \div \dfrac{x^2y^2-xy^3}{y-x} = \dfrac{2x^2y(x+y)}{6x} \cdot \dfrac{y-x}{xy^2(x-y)} = -\dfrac{x+y}{3y}$

34. $-\dfrac{t+s}{2s^2t}$

Exercise 8.4 (page 346)

1. $\dfrac{7}{a} + \dfrac{2}{a} = \dfrac{7+2}{a} = \dfrac{9}{a}$ 2. $\dfrac{7}{b}$

3. $\dfrac{6}{x-y} - \dfrac{2}{x-y} = \dfrac{6-2}{x-y} = \dfrac{4}{x-y}$ **4.** $\dfrac{6}{m+n}$

5. $\dfrac{2}{3a} + \dfrac{4}{3a} = \dfrac{2+4}{3a} = \dfrac{\cancel{6}^{\,2}}{\cancel{3a}} = \dfrac{2}{a}$ **6.** $\dfrac{2}{z}$

7. $\dfrac{2y}{y+1} + \dfrac{2}{y+1} = \dfrac{2y+2}{y+1} = \dfrac{2\cancel{(y+1)}}{\cancel{y+1}} = 2$ **8.** 5

9. $\dfrac{3}{x+3} + \dfrac{x}{x+3} = \dfrac{3+x}{x+3} = 1$ **10.** 1

11. $\dfrac{3x}{x-4} - \dfrac{12}{x-4} = \dfrac{3x-12}{x-4} = \dfrac{3\cancel{(x-4)}}{\cancel{x-4}} = 3$

12. 7

13. $\dfrac{x-3}{y-2} - \dfrac{x+5}{y-2} = \dfrac{x-3-(x+5)}{y-2}$

    $= \dfrac{x-3-x-5}{y-2} = -\dfrac{8}{y-2}$

14. $-\dfrac{7}{a-b}$

15. $\dfrac{a+2}{2a+1} - \dfrac{1-a}{2a+1} = \dfrac{a+2-(1-a)}{2a+1}$

    $= \dfrac{a+2-1+a}{2a+1}$

    $= \dfrac{2a+1}{2a+1} = 1$

16. 1

17. $\dfrac{-x}{x-2} - \dfrac{2}{2-x} = \dfrac{-x}{x-2} + \dfrac{2}{x-2} = \dfrac{-x+2}{x-2} = \dfrac{-1\cancel{(x-2)}}{\cancel{x-2}} = -1$

18. 1

19. $\dfrac{-15w}{1-5w} - \dfrac{3}{5w-1} = \dfrac{-15w}{1-5w} + \dfrac{3}{1-5w}$

    $= \dfrac{-15w+3}{1-5w} = \dfrac{3\cancel{(1-5w)}}{\cancel{1-5w}} = 3$

20. 5

21. $\dfrac{7z}{8z-4} + \dfrac{6-5z}{4-8z} = \dfrac{7z}{8z-4} + \dfrac{5z-6}{8z-4} = \dfrac{7z+5z-6}{8z-4}$

    $= \dfrac{12z-6}{8z-4} = \dfrac{\cancel{6}^{\,3}\cancel{(2z-1)}}{\cancel{4}_{2}\cancel{(2z-1)}} = \dfrac{3}{2}$

22. $\dfrac{4}{3}$

23. $\dfrac{31-8x}{12-8x} - \dfrac{5-16x}{8x-12} = \dfrac{31-8x}{12-8x} + \dfrac{5-16x}{12-8x} = \dfrac{31-8x+5-16x}{12-8x}$

    $= \dfrac{36-24x}{12-8x} = \dfrac{\cancel{12}^{\,3}\cancel{(3-2x)}}{\cancel{4}_{1}\cancel{(3-2x)}} = 3$

24. 2

Exercises 8.5 (page 350)

1. (1) The denominators are already factored.

  (2) 3, x are all the different bases.

  (3) 3^1, x^1 are the highest powers of each base.

  (4) LCD $= 3x$

2. $2y$

3. (1) $2^2 \cdot 3$, 2^2 are the denominators in factored form.

  (2) 2, 3 are all the different bases.

  (3) 2^2, 3^1 are the highest powers of each base.

  (4) LCD $= 2^2 \cdot 3 = 12$

4. 9

5. (1) $5 \cdot x$, $2 \cdot x$ are the denominators in factored form.

  (2) 2, 5, x are all the different bases.

  (3) 2^1, 5^1, x^1 are the highest powers of each base.

  (4) LCD $= 2^1 \cdot 5^1 \cdot x^1 = 10x$

6. $12z$

7. (1) The denominators are already factored.

  (2) x is the only base.

  (3) x^3 is the highest power of x in any denominator.

  (4) LCD $= x^3$

8. a^4

9. (1) $2^2 \cdot 3 \cdot u^3 \cdot v^2$; $2 \cdot 3^2 \cdot u \cdot v^3$ are the denominators in factored form.

  (2) 2, 3, u, v are all the different bases.

  (3) 2^2, 3^2, u^3, v^3 are the highest powers of different bases.

  (4) LCD $= 2^2 \cdot 3^2 \cdot u^3 \cdot v^3 = 36u^3v^3$

10. $100x^3y^5$

11. (1) Denominators are already factored.

  (2) a, $(a+3)$ are the different bases.

  (3) a^1, $(a+3)^1$ are the highest powers of each base.

  (4) LCD $= a(a+3)$

12. $b(b-5)$

13. (1) $2(x+2)$, 2^2x are the denominators in factored form.

  (2) 2, x, $(x+2)$ are the different bases.

  (3) 2^2, x^1, $(x+2)^1$ are the highest powers of each base.

  (4) LCD $= 4x(x+2)$

14. $3x(x+2)$

15. (1) $2^2 \cdot z^2$, $(z+1)^2$, $z+1$ are the factored denominators.

  (2) 2, z, $(z+1)$ are the different bases.

  (3) 2^2, z^2, $(z+1)^2$ are the highest powers of each base.

  (4) LCD $= 4z^2(z+1)^2$

16. $12x^3(x-1)^2$

17. (1) $(x+1)(x+2)$, $(x+1)^2$

  (2) $(x+1)$, $(x+2)$

  (3) $(x+1)^2$, $(x+2)^1$

  (4) LCD $= (x+1)^2(x+2)$

18. $(x-4)(x+3)^2$

19. (1) $2^2 \cdot 3 \cdot x^2(x+2)$, $(x-2)^2$, $(x+2)(x-2)$

  (2) 2, 3, x, $(x+2)$, $(x-2)$

  (3) 2^2, 3^1, x^2, $(x+2)^1$, $(x-2)^2$

  (4) LCD $= 12x^2(x+2)(x-2)^2$

20. $8y(y+3)^2(y-3)$

21. (1) $(2x - 1)(3x + 2)$; $x(3x + 2)^2$; $(2x - 1)^2$

(2) x; $(2x - 1)$; $(3x + 2)$

(3) x; $(2x - 1)^2$; $(3x + 2)^2$

(4) LCD $= x(2x - 1)^2(3x + 2)^2$

22. $x(2x + 3)^2(5x - 1)^2$

Exercises 8.6 (page 355)

1. LCD $= a^3$

$$\frac{3}{a^2} + \frac{2}{a^3} = \frac{3}{a^2} \cdot \frac{a}{a} + \frac{2}{a^3} = \frac{3a}{a^3} + \frac{2}{a^3} = \frac{3a + 2}{a^3}$$

2. $\dfrac{5u^2 + 4}{u^3}$

3. LCD $= 2x^2$

$$\frac{1}{2} + \frac{3}{x} - \frac{5}{x^2} = \frac{1}{2} \cdot \frac{x^2}{x^2} + \frac{3}{x} \cdot \frac{2x}{2x} - \frac{5}{x^2} \cdot \frac{2}{2}$$

$$= \frac{x^2}{2x^2} + \frac{6x}{2x^2} - \frac{10}{2x^2} = \frac{x^2 + 6x - 10}{2x^2}$$

4. $\dfrac{2y^2 - 3y + 12}{3y^2}$

5. LCD $= xy$

$$\frac{2}{xy} - \frac{3}{y} = \frac{2}{xy} - \frac{3}{y} \cdot \frac{x}{x} = \frac{2}{xy} - \frac{3x}{xy} = \frac{2 - 3x}{xy}$$

6. $\dfrac{5 - 4b}{ab}$

7. LCD $= x$

$$\frac{5}{1} + \frac{2}{x} = \frac{5}{1} \cdot \frac{x}{x} + \frac{2}{x} = \frac{5x + 2}{x}$$

8. $\dfrac{3y + 4}{y}$

9. First reduce $\dfrac{9}{6y}$ to $\dfrac{3}{2y}$.

LCD $= 4y^2$

$$\frac{5}{4y^2} + \frac{3}{2y} \cdot \frac{2y}{2y} = \frac{5 + 6y}{4y^2}$$

10. $\dfrac{8x + 3}{4x^2}$

11. LCD $= x$

$$\frac{3x}{1} - \frac{3}{x} = \frac{3x}{1} \cdot \frac{x}{x} - \frac{3}{x} = \frac{3x^2 - 3}{x}$$

12. $\dfrac{4y^2 - 5}{y}$

13. LCD $= a(a + 3)$

$$\frac{3 (a + 3)}{a (a + 3)} + \frac{a (a)}{(a + 3) (a)} = \frac{3a + 9}{a(a + 3)} + \frac{a^2}{a(a + 3)}$$

$$= \frac{3a + 9 + a^2}{a(a + 3)} \text{ or } \frac{a^2 + 3a + 9}{a(a + 3)}$$

14. $\dfrac{b^2 + 5b - 25}{b(b - 5)}$

15. LCD $= 4x(x + 2)$

$$\frac{x}{2x + 4} + \frac{-5}{4x} = \frac{x \ (2x)}{2(x + 2) \ (2x)} + \frac{-5 \ (x + 2)}{4x \ (x + 2)}$$

$$= \frac{2x^2}{4x(x + 2)} + \frac{-5x - 10}{4x(x + 2)}$$

$$= \frac{2x^2 - 5x - 10}{4x(x + 2)}$$

16. $\dfrac{8 + 4x - 2x^2}{3x(x + 2)}$

17. LCD $= x(x - 2)$

$$\frac{x}{1} + \frac{2}{x} + \frac{-3}{x - 2} = \frac{x}{1} \cdot \frac{x(x - 2)}{x(x - 2)} + \frac{2}{x} \cdot \frac{(x - 2)}{(x - 2)} + \frac{-3}{x - 2} \cdot \frac{x}{x}$$

$$= \frac{x^2(x - 2)}{x(x - 2)} + \frac{2(x - 2)}{x(x - 2)} + \frac{-3x}{x(x - 2)}$$

$$= \frac{x^3 - 2x^2 + 2x - 4 - 3x}{x(x - 2)} = \frac{x^3 - 2x^2 - x - 4}{x(x - 2)}$$

18. $\dfrac{m^3 - 4m^2 + m - 12}{m(m - 4)}$

19. LCD $= b(a - b)$

$$\frac{a + b}{b} + \frac{b}{a - b} = \frac{(a + b)}{b} \cdot \frac{(a - b)}{(a - b)} + \frac{b}{(a - b)} \cdot \frac{b}{b}$$

$$= \frac{a^2 - b^2}{b(a - b)} + \frac{b^2}{b(a - b)} = \frac{a^2 - b^2 + b^2}{b(a - b)} = \frac{a^2}{b(a - b)}$$

20. $\dfrac{y^2}{x(y - x)}$

21. LCD $= 6(e - 1)$

$$\frac{3}{2e - 2} - \frac{2}{3e - 3} = \frac{3}{2(e - 1)} \cdot \frac{3}{3} + \frac{-2}{3(e - 1)} \cdot \frac{2}{2}$$

$$= \frac{9}{6(e - 1)} + \frac{-4}{6(e - 1)} = \frac{5}{6(e - 1)}$$

22. $\dfrac{7}{15(f + 2)}$

23. First change the signs of the last fraction and of its denominator. Then we have

$$\frac{3}{m - 2} + \frac{5}{m - 2} = \frac{8}{m - 2}$$

24. $\dfrac{9}{n - 5}$

25. LCD $= (x - 1)(x + 1)$

$$\frac{(x + 1) \ (x + 1)}{(x - 1) \ (x + 1)} + \frac{-(x - 1) \ (x - 1)}{(x + 1) \ (x - 1)}$$

$$= \frac{x^2 + 2x + 1}{(x - 1)(x + 1)} + \frac{-x^2 + 2x - 1}{(x - 1)(x + 1)} = \frac{4x}{x^2 - 1}$$

26. $\dfrac{16x}{x^2 - 16}$

27. LCD $= xy(x - y)$

$$\frac{y}{x(x-y)} \cdot \frac{y}{y} + \frac{-x}{y(x-y)} \cdot \frac{x}{x} = \frac{y^2 - x^2}{xy(x-y)}$$

$$= -\frac{x^2 - y^2}{xy(x-y)} = -\frac{(x+y)\cancel{(x-y)}}{xy\cancel{(x-y)}} = -\frac{x+y}{xy}$$

28. $\dfrac{a+b}{ab}$

29. LCD $= (x - 3)(x + 3)$

$$\frac{2x \ (x+3)}{(x-3) \ (x+3)} + \frac{-2x \ (x-3)}{x+3 \ (x-3)} + \frac{36}{(x+3)(x-3)}$$

$$= \frac{2x^2 + 6x - 2x^2 + 6x + 36}{(x-3)(x+3)} = \frac{12x + 36}{(x-3)(x+3)}$$

$$= \frac{12\cancel{(x+3)}}{(x-3)\cancel{(x+3)}} = \frac{12}{x-3}$$

30. $\dfrac{8}{4-x}$

31. LCD $= (x + 2)^2$

$$\frac{x}{x^2 + 4x + 4} + \frac{1 \ (x+2)}{(x+2) \ (x+2)} = \frac{x+x+2}{(x+2)^2} = \frac{2x+2}{(x+2)^2}$$

32. $\dfrac{5 - 3x}{(x-1)^2}$

33. LCD $= (x + 5)(x - 1)(2x + 3)$

$$\frac{(2x+3) \ (2x+3)}{(x+5)(x-1) \ (2x+3)} - \frac{(x+5) \ (x+5)}{(2x+3)(x-1) \ (x+5)}$$

$$= \frac{(2x+3)(2x+3) - (x+5)(x+5)}{(x+5)(x-1)(2x+3)}$$

$$= \frac{4x^2 + 12x + 9 - (x^2 + 10x + 25)}{(x+5)(x-1)(2x+3)}$$

$$= \frac{4x^2 + 12x + 9 - x^2 - 10x - 25}{(x+5)(x-1)(2x+3)} = \frac{3x^2 + 2x - 16}{(x+5)(x-1)(2x+3)}$$

34. $\dfrac{8x^2 - 2x - 45}{(x-1)(x+7)(3x+2)}$

35. LCD $= (x + 4)^2(x - 4)$

$$\frac{x^2 \ (x-4)}{(x+4)^2 \ (x-4)} - \frac{(x+1) \ (x+4)}{(x+4)(x-4) \ (x+4)}$$

$$= \frac{x^3 - 4x^2 - (x^2 + 5x + 4)}{(x+4)^2(x-4)} = \frac{x^3 - 4x^2 - x^2 - 5x - 4}{(x+4)^2(x-4)}$$

$$= \frac{x^3 - 5x^2 - 5x - 4}{(x+4)^2(x-4)}$$

36. $\dfrac{x^3 - 4x^2 - 5x - 6}{(x+3)^2(x-3)}$

37. LCD $= x(2x + 5)(x - 1)(x + 1)$

$$\frac{(x+1) \ (x+1)}{x(2x+5)(x-1) \ (x+1)} - \frac{(x-1) \ (x-1)}{x(2x+5)(x+1) \ (x-1)}$$

$$= \frac{x^2 + 2x + 1 - (x^2 - 2x + 1)}{x(2x+5)(x-1)(x+1)} = \frac{x^2 + 2x + 1 - x^2 + 2x - 1}{x(2x+5)(x-1)(x+1)}$$

$$= \frac{4\cancel{x}}{\cancel{x}(2x+5)(x-1)(x+1)} = \frac{4}{(2x+5)(x-1)(x+1)}$$

38. $\dfrac{4}{(3x+4)(x-1)(x+1)}$

Exercises 8.7 (page 360)

1. LCD $= 12$

$$\frac{12}{12} \cdot \frac{\dfrac{3}{4}}{\dfrac{5}{6}} = \frac{\dfrac{\cancel{12}}{1} \cdot \dfrac{3}{\cancel{4}}}{\dfrac{\cancel{12}}{1} \cdot \dfrac{5}{\cancel{6}}} = \frac{9}{10}$$

2. $\dfrac{4}{5}$

3. LCD $= 9$

$$\frac{9}{9} \cdot \frac{\dfrac{2}{3}}{\dfrac{4}{9}} = \frac{\dfrac{\cancel{9}}{1} \cdot \dfrac{2}{\cancel{3}}}{\dfrac{\cancel{9}}{1} \cdot \dfrac{4}{\cancel{9}}} = \frac{6}{4} = \frac{3}{2} = 1\frac{1}{2}$$

4. $1\frac{1}{2}$

5. LCD of secondary denominators is 8.

$$\frac{8}{8} \cdot \frac{\dfrac{3}{4} - \dfrac{1}{2}}{\dfrac{5}{8} + \dfrac{1}{4}} = \frac{\dfrac{8}{1}\left(\dfrac{3}{4}\right) + \dfrac{8}{1}\left(-\dfrac{1}{2}\right)}{\dfrac{8}{1}\left(\dfrac{5}{8}\right) + \dfrac{8}{1}\left(\dfrac{1}{4}\right)} = \frac{6-4}{5+2} = \frac{2}{7}$$

6. $\dfrac{9}{7} = 1\frac{2}{7}$

7. LCD of secondary denominators is 20.

$$\frac{20}{20} \cdot \frac{\dfrac{3}{5} + \dfrac{2}{1}}{\dfrac{2}{1} - \dfrac{3}{4}} = \frac{\overset{4}{\cancel{20}}\left(\dfrac{3}{\cancel{5}}\right) + 20\left(\dfrac{2}{1}\right)}{20(2) + \overset{5}{\cancel{20}}\left(-\dfrac{3}{\cancel{4}}\right)} = \frac{12 + 40}{40 - 15}$$

$$= \frac{52}{25} = 2\frac{2}{25}$$

8. $1\frac{1}{82}$

9. LCD of secondary denominators is $9y^4$.

$$\frac{9y^4}{9y^4} \cdot \frac{\dfrac{5x^3}{3y^4}}{\dfrac{10x}{9y}} = \frac{15x^3}{10xy^3} = \frac{3x^2}{2y^3}$$

10. $6ab$

11. LCD $= 15a^3b^2$

$$\frac{15a^3b^2}{15a^3b^2} \cdot \frac{\dfrac{18cd^2}{5a^3b}}{\dfrac{12cd^2}{15ab^2}} = \frac{\dfrac{15a^3b^2}{1} \cdot \dfrac{18cd^2}{5a^3b}}{\dfrac{15a^3b^2}{1} \cdot \dfrac{12cd^2}{15ab^2}} = \frac{\overset{9}{\cancel{54}}bcd^2}{\underset{2}{\cancel{12}}a^2cd^2} = \frac{9b}{2a^2}$$

12. $\dfrac{2xz^2}{y}$

13. LCD $= 10$

$$\frac{\overset{2}{\cancel{10}}}{\cancel{10}} \cdot \frac{\left(\dfrac{x+3}{\cancel{5}}\right)}{\left(\dfrac{2x+6}{\cancel{10}}\right)} = \frac{2x+6}{2x+6} = 1$$

14. $1\frac{1}{2}$

15. LCD $= 4x^2$

$$\frac{4x^2}{4x^2} \cdot \frac{\dfrac{x+2}{2x}}{\dfrac{x+1}{4x^2}} = \frac{\dfrac{\overset{2x}{\cancel{4x^2}}}{1} \cdot \dfrac{x+2}{\cancel{2x}}}{\dfrac{\cancel{4x^2}}{1} \cdot \dfrac{x+1}{\cancel{4x^2}}} = \frac{2x^2 + 4x}{x+1}$$

16. $\dfrac{3x-9}{x^2 - 9x}$

17. LCD = b

$$\frac{b}{b} \cdot \frac{\left(\dfrac{a}{b} + 1\right)}{\left(\dfrac{a}{b} - 1\right)} = \frac{\cancel{b}\left(\dfrac{a}{\cancel{b}}\right) + b(1)}{\cancel{b}\left(\dfrac{a}{\cancel{b}}\right) - b(1)} = \frac{a + b}{a - b}$$

18. $\dfrac{2y + x}{2y - x}$

19. LCD = x

$$\frac{x}{x} \cdot \frac{\dfrac{1}{x} + x}{\dfrac{1}{x} - x} = \frac{x\left(\dfrac{1}{x}\right) + x(x)}{x\left(\dfrac{1}{x}\right) - x(x)} = \frac{1 + x^2}{1 - x^2}$$

20. $\dfrac{a^2 - 4}{a^2 + 4}$

21. LCD = d^2

$$\frac{d^2}{d^2} \cdot \frac{\left(\dfrac{c}{d} + 2\right)}{\left(\dfrac{c^2}{d^2} - 4\right)} = \frac{d^2\left(\dfrac{c}{d}\right) + d^2(2)}{d^2\left(\dfrac{c^2}{d^2}\right) - d^2(4)} = \frac{cd + 2d^2}{c^2 - 4d^2}$$

$$= \frac{d(\cancel{c + 2d})}{(c - 2d)(\cancel{c + 2d})} = \frac{d}{c - 2d}$$

22. $\dfrac{x + y}{y}$

23. LCD = y

$$\frac{y}{y} \cdot \frac{x + \dfrac{x}{y}}{1 + \dfrac{1}{y}} = \frac{y \cdot x + \cancel{y}\left(\dfrac{x}{\cancel{y}}\right)}{y \cdot 1 + \cancel{y}\dfrac{1}{\cancel{y}}} = \frac{yx + x}{y + 1} = \frac{x(y + 1)}{y + 1} = x$$

24. $\frac{1}{3}$

25. LCD = $x^2 y^2$

$$\frac{x^2y^2}{x^2y^2} \cdot \frac{\dfrac{1}{x^2} - \dfrac{1}{y^2}}{\dfrac{1}{x} + \dfrac{1}{y}} = \frac{\cancel{x^2}y^2\left(\dfrac{1}{\cancel{x^2}}\right) - x^2\cancel{y^2}\left(\dfrac{1}{\cancel{y^2}}\right)}{\cancel{x}y^2\left(\dfrac{1}{\cancel{x}}\right) + x^2\cancel{y}\left(\dfrac{1}{\cancel{y}}\right)}$$

$$= \frac{y^2 - x^2}{xy^2 + x^2y} = \frac{(\cancel{y + x})(y - x)}{xy(\cancel{y + x})} = \frac{y - x}{xy}$$

26. $\dfrac{a + 2}{2a}$

27. LCD = x^2

$$\frac{x^2}{x^2} \cdot \frac{\dfrac{2}{x} - \dfrac{4}{x^2}}{\dfrac{1}{x} - \dfrac{2}{x^2}} = \frac{\cancel{x^2}\left(\dfrac{2}{\cancel{x}}\right) - \cancel{x^2}\left(\dfrac{4}{\cancel{x^2}}\right)}{\cancel{x^2}\left(\dfrac{1}{\cancel{x}}\right) - \cancel{x^2}\left(\dfrac{2}{\cancel{x^2}}\right)} = \frac{2x - 4}{x - 2} = \frac{2(\cancel{x - 2})}{\cancel{x - 2}} = 2$$

28. $\frac{1}{4}$

Review Exercises 8.8 (page 363)

1. −4 must be excluded, because −4 makes the denominator zero.

2. 0

3. 3 and −3 must be excluded, because either makes the denominator zero.

4. 0, 1, −2

5. Yes, because if we multiply both 1 and 2 by $(x + 1)$, we get $x + 1$ and $2x + 2$.

6. No

7. $x - 6$. The signs of fractions are different and signs of denominators are different. Therefore, the signs of numerators must be *the same*.

8. 2

9. -4. The signs of fractions are the same and signs of numerators are different. Therefore, the signs of denominators must be different.

10. $\dfrac{2b}{a^2}$ **11.** $\dfrac{\cancel{2}(1 + 2m)}{\cancel{2}} = 1 + 2m$ **12.** $1 - 2n$

13. $\dfrac{(a + 2)(a - 2)}{a + 2} = a - 2$ **14.** $\dfrac{1}{x + 2}$

15. $\dfrac{a - b}{a(x + y) - b(x + y)} = \dfrac{a - b}{(x + y)(a - b)} = \dfrac{1}{x + y}$ **16.** $\dfrac{10}{k}$

17. LCD = $2x$

$$\frac{5(2x)}{1(2x)} - \frac{3}{2x} = \frac{10x - 3}{2x}$$

18. 1 **19.** $\dfrac{-5a^2}{3b} \div \dfrac{10a}{9b^2} = \dfrac{-\cancel{5}a^2}{\cancel{3}b} \cdot \dfrac{\cancel{9}b^2}{\cancel{10}a} = -\dfrac{3ab}{2}$

20. $-6x^2y^2$ **21.** $\dfrac{3(x + 2)}{\cancel{6}} \cdot \dfrac{\cancel{2}x^2}{4(x + 2)} = \dfrac{x^2}{4}$ **22.** $\dfrac{y - 17}{6}$

23. LCD = $(a - 1)(a + 2)$

$$\frac{(a - 2)(a + 2)}{(a - 1)(a + 2)} + \frac{(a + 1)(a - 1)}{(a + 2)(a - 1)} = \frac{a^2 - 4 + a^2 - 1}{(a + 2)(a - 1)}$$

$$= \frac{2a^2 - 5}{(a + 2)(a - 1)}$$

24. $-\dfrac{5}{(k - 2)(k - 3)}$

25. $\dfrac{2x^2 - 6x}{x + 2} \div \dfrac{x}{4x + 8} = \dfrac{2\cancel{x}(x - 3)}{\cancel{x + 2}} \cdot \dfrac{4\cancel{(x + 2)}}{\cancel{x}} = 8(x - 3)$ **26.** 8z

27. LCD = $(2x + 3)(x + 2)(3x + 4)$

$$\frac{(3x + 4)(3x + 4)}{(2x + 3)(x + 2)(3x + 4)} - \frac{(x + 2)(2x + 3)}{(3x + 4)(x + 2)(2x + 3)}$$

$$= \frac{9x^2 + 24x + 16 - (2x^2 + 7x + 6)}{(2x + 3)(x + 2)(3x + 4)}$$

$$= \frac{9x^2 + 24x + 16 - 2x^2 - 7x - 6}{(2x + 3)(x + 2)(3x + 4)} = \frac{7x^2 + 17x + 10}{(2x + 3)(x + 2)(3x + 4)}$$

28. $\dfrac{-3x^3 + x^2 - 248x + 400}{15x(x + 4)^2(x - 4)}$

29. LCD of secondary denominators = $9m^2$

$$\frac{9m^2}{9m^2} \cdot \frac{\dfrac{5k^2}{3m^2}}{\dfrac{10k}{9m}} = \frac{\dfrac{9m^2}{1}\left(\dfrac{5k^2}{3m^2}\right)}{\dfrac{9m^2}{1}\left(\dfrac{10k}{9m}\right)} = \frac{15k^2}{10km} = \frac{3k}{2m}$$

30. $\dfrac{3b - a}{2b + a}$

31. LCD $= b^2$

$$\frac{b^2}{b^2} \cdot \frac{\dfrac{a^2}{b^2} - 1}{\dfrac{a}{b} - 1} = \frac{\cancel{b^2}\left(\dfrac{a^2}{\cancel{b^2}}\right) - b^2 \cdot 1}{\cancel{b^{\cancel{2}}}\left(\dfrac{a}{\cancel{b}}\right) - b^2 \cdot 1} = \frac{a^2 - b^2}{ab - b^2}$$

$$= \frac{(a+b)(a - b)}{b(a - b)} = \frac{a+b}{b}$$

32. $\dfrac{x+4}{x}$

Exercises 8.9A (page 371)

(The checks usually will not be shown.)

1. LCD $= 12$

Multiply each term on both sides of the equation by 12.

$$(12)\frac{x}{3} + (12)\frac{x}{4} = (12)\,7$$
$$4x + 3x = 84$$
$$7x = 84$$
$$x = 12$$

2. 15

3. LCD $= c$

$$\frac{10}{c}\,(c) = 2\,(c) \qquad \text{Multiply both sides of the equation by } c.$$
$$10 = 2c$$
$$5 = c, \text{ or } c = 5$$

4. $1\frac{1}{2}$

5. LCD $= 10$

$$(10)\frac{a}{2} + (10)\frac{-a}{5} = (10)\,6$$
$$5a - 2a = 60$$
$$3a = 60$$
$$a = 20$$

6. 63

7. LCD $= 2x$

$$\frac{9}{2x}\,(2x) = 3\,(2x)$$
$$9 = 6x$$
$$\frac{9}{6} = x$$
$$x = \frac{3}{2} \quad \text{or} \quad 1\frac{1}{2}$$

8. $\frac{2}{3}$

9. LCD $= 15$

$$(15)\frac{M-2}{5} + (15)\frac{M}{3} = (15)\frac{1}{5}$$
$$3(M-2) + 5M = 3$$
$$3M - 6 + 5M = 3$$
$$8M = 9$$
$$M = \frac{9}{8} = 1\frac{1}{8}$$

10. $-\frac{5}{9}$

11. LCD $= x(x+4)$

$$\frac{7}{x+4}\,(x)(x+4) = \frac{3}{x}\,(x)(x+4)$$
$$7x = 3(x+4)$$
$$7x = 3x + 12$$
$$4x = 12$$
$$x = 3$$

12. 4

13. LCD $= x - 2$

$$\frac{3x}{x-2}\,(x-2) = 5\,(x-2)$$
$$3x = 5(x-2)$$
$$3x = 5x - 10$$
$$-2x = -10$$
$$x = 5$$

14. 3

15. LCD $= x - 2$

$$\frac{3x-1}{x-2}\,(x-2) = 3\,(x-2) + \frac{2x+1}{x-2}\,(x-2)$$
$$3x - 1 = 3(x-2) + (2x+1)$$
$$3x - 1 = 3x - 6 + 2x + 1$$
$$3x - 1 = 5x - 5$$
$$4 = 2x$$
$$x = 2$$

Check: $\quad \dfrac{3x-1}{x-2} = 3 + \dfrac{2x+1}{x-2}$

$$\frac{3(2)-1}{(2)-2} \overset{?}{=} 3 + \frac{2(2)+1}{(x)-2}$$
$$\frac{5}{0} \overset{?}{=} 3 + \frac{5}{0}$$

Not a real number.
Therefore, 2 is not a solution; the equation has no solution.

16. No solution

17. LCD $= (x^2+1)(1+2x)$

$$\frac{x}{x^2+1}\,(x^2+1)(1+2x) = \frac{2}{1+2x}\,(x^2+1)(1+2x)$$
$$x(1+2x) = 2(x^2+1)$$
$$x + 2x^2 = 2x^2 + 2$$
$$\underline{\quad -2x^2 \qquad -2x^2 \quad}$$
$$x \qquad = \qquad 2$$

18. $\frac{2}{3}$

19. LCD $= 12$

$$(\cancel{12})\frac{2x-1}{\cancel{3}} + (\cancel{12})\frac{3x}{\cancel{4}} = (\cancel{12})\frac{5}{\cancel{6}}$$
$$4(2x-1) + 3(3x) = 10$$
$$8x - 4 + 9x = 10$$
$$17x = 14$$
$$x = \frac{14}{17}$$

20. $1\frac{1}{9}$

21. LCD = 10

$$(\overset{2}{\cancel{10}}) \frac{2(m-3)}{\cancel{5}} + (\overset{5}{\cancel{10}}) \frac{(-3)(m+2)}{\cancel{2}} = (10)\frac{7}{10}$$

$$4(m-3) - 15(m+2) = 7$$

$$4m - 12 - 15m - 30 = 7$$

$$-11m = 49$$

$$m = \frac{49}{-11} = -4\frac{5}{11}$$

22. $7\frac{4}{11}$

23. LCD = $x - 2$

$$\left(\frac{x-2}{1}\right)\frac{x}{x-2} = \left(\frac{x-2}{1}\right)\frac{2}{x-2} + \left(\frac{x-2}{1}\right)(5)$$

$$x = 2 + 5x - 10$$

$$8 = 4x$$

$$x = 2 \quad \text{Not a solution}$$

Check: $\quad \dfrac{x}{x-2} = \dfrac{2}{x-2} + 5$

$$\frac{2}{2-2} \overset{?}{=} \frac{2}{2-2} + 5$$

$$\frac{2}{0} \overset{?}{=} \frac{2}{0} + 5$$

↑   ↑ — Not a real number.

Therefore, 2 is not a solution.

24. No solution

Exercises 8.9B (page 376)

1a. First term $= 8$
b. Second term $= 14$
c. Third term $= 16$
d. Fourth term $= x$
e. Means $= 14, 16$
f. Extremes $= 8, x$

2a. 3
b. 5
c. x
d. 20
e. $5, x$
f. $3, 20$

3. $\dfrac{x}{4} = \dfrac{2}{3}$

$$3x = 2(4) = 8$$

$$x = \frac{8}{3} = 2\frac{2}{3}$$

4. $7\frac{1}{2}$

5. $\dfrac{8}{x} = \dfrac{4}{5}$

$$4x = 8(5) = 40$$

$$x = 10$$

6. $2\frac{2}{3}$

7. $\dfrac{4}{7} = \dfrac{x}{21}$

$$7x = 4(21)$$

$$x = \frac{4(21)}{7} = 12$$

8. $11\frac{1}{4}$

9. $\dfrac{100}{x} = \dfrac{4\cancel{0}}{3\cancel{0}} = \dfrac{4}{3}$

$$4x = 300$$

$$x = 75$$

10. 24

11. $\dfrac{x+1}{x-1} = \dfrac{3}{2}$

$$2(x+1) = 3(x-1)$$

$$2x + 2 = 3x - 3$$

$$5 = x; x = 5$$

12. 4

13. $\dfrac{2x+7}{9} = \dfrac{2x+3}{5}$

$$5(2x+7) = 9(2x+3)$$

$$10x + 35 = 18x + 27$$

$$8 = 8x$$

$$x = 1$$

14. -2

15. $\dfrac{5x-10}{10} = \dfrac{3x-5}{7}$

$$7(5x-10) = 10(3x-5)$$

$$35x - 70 = 30x - 50$$

$$5x = 20$$

$$x = 4$$

16. $2\frac{2}{7}$

17. $\dfrac{\frac{3}{4}}{6} = \dfrac{P}{16}$

$$6P = \frac{3}{\cancel{4}} \cdot \frac{\overset{4}{\cancel{16}}}{1} = 12$$

$$\frac{\cancel{6}P}{\cancel{6}} = \frac{12}{6} = 2$$

18. $2\frac{1}{2}$

19. $\dfrac{A}{9} = \dfrac{3\frac{1}{3}}{5}$

$$5A = 9 \cdot 3\frac{1}{3}$$

$$5A = \frac{\overset{3}{\cancel{9}}}{1} \cdot \frac{10}{\cancel{3}} = 30$$

$$\frac{\cancel{5}A}{\cancel{5}} = \frac{30}{5} = 6$$

20. 1

21. $\dfrac{7.7}{B} = \dfrac{\overset{0.7}{\cancel{3.5}}}{\cancel{5}}$

$$0.7B = 7.7$$

$$\frac{\cancel{0.7}B}{\cancel{0.7}} = \frac{7.7}{0.7} = 11$$

22. 22.96

23. $\dfrac{P}{100} = \dfrac{\frac{3}{2}}{15}$

$15P = \dfrac{\overset{50}{\cancel{100}}}{1} \cdot \dfrac{3}{\cancel{2}}$

$15P = 150$

$P = 10$

24. 4

25. $\dfrac{12\frac{1}{2}}{100} = \dfrac{A}{48}$

$100A = 12\frac{1}{2} \cdot 48$

$100A = \dfrac{25}{\cancel{2}} \cdot \dfrac{\overset{24}{\cancel{48}}}{1}$

$\dfrac{\cancel{100}A}{\cancel{100}} = \dfrac{\overset{1}{\cancel{25}}(\overset{6}{\cancel{24}})}{\underset{\underset{1}{\cancel{4}}}{\cancel{100}}} = 6$

26. 54

Exercises 8.9C (page 380)

(Checks will usually not be shown.)

1. LCD $= z$

$(z)z + (\cancel{z})\dfrac{1}{\cancel{z}} = (\cancel{z})\dfrac{17}{\cancel{z}}$

$z^2 + 1 = 17$

$z^2 + 1 - 17 = 0$

$z^2 - 16 = 0$

$(z + 4)(z - 4) = 0$

$z + 4 = 0 \quad | \quad z - 4 = 0$

$z = -4 \qquad\; z = 4$

2. $-3, 3$

3. LCD $= 2x^2$

$\left(\dfrac{2x^2}{1}\right)\dfrac{2}{x} + \left(\dfrac{2x^2}{1}\right)\left(-\dfrac{2}{x^2}\right) = \left(\dfrac{2x^2}{1}\right)\dfrac{1}{2}$

$4x - 4 = x^2$

$0 = x^2 - 4x + 4$

$0 = (x - 2)(x - 2)$

$x - 2 = 0 \quad | \quad x - 2 = 2$

$x = 2 \qquad\;\; x = 2$

4. 2, 4

5. $\dfrac{x}{x + 1} = \dfrac{4x}{3x + 2}$ A proportion

$4x(x + 1) = x(3x + 2)$

$4x^2 + 4x = 3x^2 + 2x$

$x^2 + 2x = 0$

$x(x + 2) = 0$

$x = 0 \quad | \quad x + 2 = 0$

$\qquad\qquad x = -2$

Check for $x = 0$: $\quad \dfrac{(0)}{(0) + 1} = \dfrac{4(0)}{3(0) + 2}$

$\dfrac{0}{1} = \dfrac{0}{2}$

$0 = 0 \quad$ True

Check for $x = -2$: $\quad \dfrac{(-2)}{(-2) + 1} = \dfrac{4(-2)}{3(-2) + 2}$

$\dfrac{-2}{-1} = \dfrac{-8}{-4}$

$2 = 2 \quad$ True

6. 0, 2

7. LCD $= x$

$\dfrac{5}{x}(x) - 1(x) = \dfrac{x + 11}{x}(x)$

$5 - x = x + 11$

$\underline{-11 + x \quad\; x - 11}$

$-6 = 2x$

$x = -3$

8. -4

9. LCD $= 3(x + 1)(x - 1)$

$3(x + 1)(\cancel{x - 1})\dfrac{1}{\cancel{x - 1}} + 3(\cancel{x + 1})(x - 1)\dfrac{2}{\cancel{x + 1}}$

$= \cancel{3}(x + 1)(x - 1)\dfrac{5}{\cancel{3}}$

$3(x + 1) + 6(x - 1) = 5(x^2 - 1)$

$3x + 3 + 6x - 6 = 5x^2 - 5$

$0 = 5x^2 - 9x - 2$

$0 = (5x + 1)(x - 2)$

$5x + 1 = 0 \quad | \quad x - 2 = 0$

$x = -\dfrac{1}{5} \qquad x = 2$

10. $3, \frac{1}{21}$

11. LCD $= 4(2x + 5)$

$\dfrac{4(\cancel{2x + 5})}{1} \cdot \dfrac{3}{\cancel{2x + 5}} + \dfrac{\cancel{4}(2x + 5)}{1} \cdot \dfrac{x}{\cancel{4}} = \dfrac{\cancel{4}(2x + 5)}{1} \cdot \dfrac{3}{\cancel{4}}$

$4(3) + x(2x + 5) = 3(2x + 5)$

$12 + 2x^2 + 5x = 6x + 15$

$2x^2 - x - 3 = 0$

$(1x + 1)(2x - 3) = 0$

$x + 1 = 0 \quad | \quad 2x - 3 = 0$

$x = -1 \qquad\quad 2x = 3$

$\qquad\qquad\qquad\quad x = \dfrac{3}{2}$

12. $2, -\dfrac{19}{2}$

Exercises 8.10 (page 383)

1. $2x + y = 4$

$\quad\quad 2x = 4 - y$

$\quad\quad\quad x = \dfrac{4 - y}{2}$

2. $y = \dfrac{6 - x}{3}$

3. $y - z = -8$

$\quad\quad -z = -8 - y$

$\quad\quad\quad z = 8 + y$

4. $n = m + 5$

5. $2x - y = -4$

$\quad\quad -y = -2x - 4$

$\quad\quad\quad y = 2x + 4$

6. $z = 3y + 5$

7. $2x - 3y = 6$

$\quad\quad 2x = 3y + 6$

$\quad\quad\quad x = \dfrac{3y + 6}{2}$

8. $x = \dfrac{2y + 6}{3}$

9. $2(x - 3y) = x + 4$

$\quad\quad 2x - 6y = x + 4$

$\quad\quad\quad\quad x = 6y + 4$

10. $x = \dfrac{2y + 14}{7}$

11. $PV = k$

$\quad\dfrac{\not{P}V}{\not{P}} = \dfrac{k}{P}$

$\quad\quad V = \dfrac{k}{P}$

12. $R = \dfrac{E}{I}$

13. $I = prt$

$\quad\dfrac{I}{rt} = \dfrac{p\not{r}\not{t}}{\not{r}\not{t}}$

$\quad\dfrac{I}{rt} = p$

14. $\ell = \dfrac{V}{wh}$

15. $p = 2\ell + 2w$

$\quad p - 2w = 2\ell$

$\quad\dfrac{p - 2w}{2} = \dfrac{\not{2}\ell}{\not{2}}$

$\quad\dfrac{p - 2w}{2} = \ell$

16. $w = \dfrac{P - 2\ell}{2}$

17. $y = mx + b$

$\quad y - b = mx$

$\quad\dfrac{y - b}{m} = \dfrac{\not{m}x}{\not{m}}$

$\quad\dfrac{y - b}{m} = x$

18. $t = \dfrac{V - k}{g}$

19. $\quad\dfrac{S}{1} = \dfrac{a}{1 - r}$ A proportion

$\quad S(1 - r) = a$

$\quad S - Sr = a$

$\quad\quad -Sr = a - S$

$\quad\dfrac{-Sr}{-S} = \dfrac{a - S}{-S}$

$\quad\quad\quad r = \dfrac{S - a}{S}$

20. $R = \dfrac{E - Ir}{I}$

21. LCD $= 9$

$\quad 9(C) = \dfrac{\not{9}}{1}\left(\dfrac{5(F - 32)}{\not{9}}\right)$

$\quad\quad 9C = 5F - 160$

$\quad 9C + 160 = 5F$

$\quad\dfrac{9C + 160}{5} = F$

22. $B = \dfrac{2A - hb}{h}$

23. $\quad\quad L = a + (n - 1)d$

$\quad\quad\quad L = a + nd - d$

$\quad L - a + d = nd$

$\quad\dfrac{L - a + d}{d} = \dfrac{n\not{d}}{\not{d}}$

$\quad\dfrac{L - a + d}{d} = n$

24. $h = \dfrac{A - 2\pi r^2}{2\pi r}$

25. $\quad\dfrac{z}{1} = \dfrac{Rr}{R + r}$ A proportion

$\quad z(R + r) = 1 \cdot Rr$

$\quad zR + zr = Rr$

$\quad\quad zr = Rr - zR$

$\quad\quad zr = R(r - z)$

$\quad\dfrac{zr}{r - z} = \dfrac{R(\not{r - z})}{\not{r - z}}$

$\quad\dfrac{zr}{r - z} = R$

26. $b = \dfrac{ca}{a - c}$

27. $LCD = Fuv$

$$\frac{Fuv}{1}\left(\frac{1}{F}\right) = \frac{Fuv}{1}\left(\frac{1}{u}\right) + \frac{Fuv}{1}\left(\frac{1}{v}\right)$$

$$uv = Fv + Fu$$

$$uv - Fu = Fv$$

$$u(v - F) = Fv$$

$$\frac{u(v - F)}{v - F} = \frac{Fv}{v - F}$$

$$u = \frac{Fv}{v - F}$$

28. $a = \dfrac{bc}{b - c}$

Exercises 8.11 (page 389)

(The checks will not be shown.)

1. Let n = numerator

$n + 6$ = denominator

original fraction = $\dfrac{n}{n + 6}$

$$\frac{n + 4}{n + 6 - 4} = \frac{11}{10} \quad \text{A proportion}$$

$$10(n + 4) = 11(n + 2)$$

$$10n + 40 = 11n + 22$$

$$18 = n$$

original fraction = $\dfrac{n}{n + 6} = \dfrac{18}{18 + 6} = \dfrac{18}{24}$

2. $\frac{16}{20}$

3. Let x = numerator

$2x$ = denominator

$$\frac{x + 5}{2x - 5} = \frac{4}{5} \quad \text{A proportion}$$

$$(x + 5)(5) = 4(2x - 5)$$

$$5x + 25 = 8x - 20$$

$$\underline{-5x + 20 \qquad -5x + 20}$$

$$45 = 3x$$

$$x = 15, \text{ numerator}$$

$$2x = 30, \text{ denominator}$$

The fraction is $\dfrac{15}{30}$.

4. $\frac{15}{45}$

5. Let x = the fraction

$\dfrac{1}{x}$ = its reciprocal

$$x + \frac{1}{x} = \frac{29}{10} \quad \text{LCD is } 10x.$$

$$x\,(10x) + \frac{1}{x}\,(10x) = \frac{29}{10}\,(10x)$$

$$10x^2 + 10 = 29x$$

$$10x^2 - 29x + 10 = 0$$

$$(2x - 5)(5x - 2) = 0$$

$2x - 5 = 0$	$5x - 2 = 0$
$2x = 5$	$5x = 2$
$x = \dfrac{5}{2}$	$x = \dfrac{2}{5}$
$\dfrac{1}{x} = \dfrac{2}{5}$	$\dfrac{1}{x} = \dfrac{5}{2}$

Therefore, there are two answers: The number is $\frac{2}{5}$ and its reciprocal is $\frac{5}{2}$, or the number is $\frac{5}{2}$ and its reciprocal is $\frac{2}{5}$.

6. $\frac{3}{11}$, or $\frac{11}{3}$

7. Let W = width

$$\frac{1 \text{ in.}}{8 \text{ ft}} = \frac{2\frac{1}{2}\text{ in.}}{W\text{ ft}} \Rightarrow \frac{1}{8} = \frac{\frac{5}{2}}{W}$$

$$W = 8\left(\frac{5}{2}\right) = 20$$

Let L = length

$$\frac{1 \text{ in.}}{8 \text{ ft}} = \frac{3 \text{ in.}}{L \text{ ft}} \Rightarrow \frac{1}{8} = \frac{3}{L}$$

$$L = 8(3) = 24$$

Therefore, the room is 20 ft by 24 ft.

8. 26 ft by 34 ft

9. Let x = woman's weight on moon (in lb)

$$\frac{6 \text{ earth}}{1 \text{ moon}} = \frac{150 \text{ earth}}{x \text{ moon}}$$

$$\frac{6}{1} = \frac{150}{x}$$

$$6x = 150$$

$$x = \frac{150}{6} = 25 \text{ lb}$$

10. 78.4 lb

11. Let x = weight of lead bar

$$\frac{21 \text{ lead}}{5 \text{ aluminum}} = \frac{x \text{ lead}}{150 \text{ aluminum}}$$

$$\frac{21}{5} = \frac{x}{150}$$

$$5x = 21(150)$$

$$x = \frac{21(150)}{5} = 630 \text{ lb}$$

12. 7.5 lb

13. Let x = number of chains of fire line

$$10\,\text{hr} - 6\frac{1}{2}\,\text{hr} = 3\frac{1}{2}\,\text{hr}$$

$$\frac{26\,\text{chains}}{6\frac{1}{2}\,\text{hr}} = \frac{x\,\text{chains}}{3\frac{1}{2}\,\text{hr}}$$

$$6\frac{1}{2}x = 26\left(3\frac{1}{2}\right)$$

$$\frac{13}{2}x = \overset{13}{\cancel{26}}\left(\frac{7}{\cancel{2}}\right)$$

$$\frac{13x}{2} = \frac{91}{1}$$

$$13x = 2(91) = 182$$

$$\frac{13x}{13} = \frac{182}{13}$$

$$x = 14\,\text{chains}$$

14. $1875

15. Let x = amount received for 16 steers

$$\frac{3420\,\text{dollars}}{9\,\text{steers}} = \frac{x\,\text{dollars}}{16\,\text{steers}}$$

$$9x = 16(3420)$$

$$\frac{9x}{9} = \frac{54720}{9}$$

$$x = \$6080$$

16. $45 per hog; 45¢ per kg

17. $\left(\text{Bill's rate is } \dfrac{1\,\text{job}}{8\,\text{hr}}, \text{ or } \dfrac{1}{8}\dfrac{\text{job}}{\text{hr}}. \text{ Mike's rate is } \dfrac{1\,\text{job}}{10\,\text{hr}}, \text{ or } \dfrac{1}{10}\dfrac{\text{job}}{\text{hr}}.\right)$

Let x = number of hr for Bill and Mike together to do the job

They work for x hr

$$\frac{1}{8}x + \frac{1}{10}x = 1 \quad \text{LCD is 40}$$

$$\frac{1}{8}x\,(40) + \frac{1}{10}x\,(40) = 1\,(40)$$

$$5x + 4x = 40$$

$$9x = 40$$

$$x = \frac{40}{9}\ \text{hr or }\ 4\frac{4}{9}\,\text{hr}$$

18. $2\frac{11}{12}$ days

19. Let x = numerator

$42 - x$ = denominator

$$\frac{x}{42 - x} = \text{fraction}$$

$$\frac{x + 2}{(42 - x) + 6} = \frac{2}{3}$$

$$\frac{x + 2}{48 - x} = \frac{2}{3}$$

$$3x + 6 = 96 - 2x$$

$$\underline{2x - 6 \quad -6 + 2x}$$

$$5x = 90$$

$$x = 18,\ \text{numerator}$$

$$42 - x = 24,\ \text{denominator}$$

The fraction is $\dfrac{18}{24}$.

20. $\frac{17}{23}$

21. Ruth's rate is $\dfrac{220\,\text{pages}}{4\,\text{hour}}$, or $55\dfrac{\text{pages}}{\text{hr}}$.

Let t = time for Sandra to proofread 210 pages

Then Sandra's rate is $\dfrac{210\,\text{pg}}{t\,\text{hr}}$.

They read for 3 hr

$$\frac{210}{t}\,(3) + 55\,(3) = 291 \quad \text{Number of pages read}$$

$$\frac{630}{t}\,(t) + 165\,(t) = 291\,(t)$$

$$630 + 165t = 291t$$

$$\underline{\quad -165t \qquad -165t\quad}$$

$$630 = 126t$$

$$t = 5\,\text{hr}$$

22. 5 hr

23. Let x = Jerry's speed

$$\frac{3}{4}x = \text{Jim's speed}$$

They each ride 3 hr

$$x\,(3) + \frac{3}{4}x\,(3) = 63 \quad \text{LCD is 4}$$

$$3x\,(4) + \frac{9}{4}x\,(4) = 63\,(4)$$

$$12x + 9x = 252$$

$$21x = 252$$

$$x = 12\,\text{mph}$$

$$\frac{3}{4}x = 9\,\text{mph}$$

24. Rebecca's average speed is 16 mph, and Jill's is 14 mph.

25. Let x = speed of the boat in still water

$x + 5$ = speed with current

$x - 5$ = speed against current

Note: since $rt = d$, $t = \dfrac{d}{r}$

$$\frac{999}{x + 5} = \text{time going with current}$$

$$\frac{729}{x - 5} = \text{time going against current}$$

The times are equal.

$$\frac{999}{x + 5} = \frac{729}{x - 5} \quad \text{A proportion}$$

$$999(x - 5) = 729(x + 5)$$

$$999x - 4995 = 729x + 3645$$

$$\underline{-729x + 4995 \qquad -729x + 4995}$$

$$270x = 8640$$

$$x = 32\,\text{mph}$$

26. 32 mph

27. Let x = larger number

$x - 12$ = smaller number

$$\frac{1}{6}x - \frac{1}{5}(x - 12) = 2 \quad \text{LCD} = 30$$

$$\frac{1}{6}x\,(30) - \frac{1}{5}(x - 12)\,(30) = 2\,(30)$$

$$5x - 6(x - 12) = 60$$

$$5x - 6x + 72 = 60$$

$$-x = -12$$

$$x = 12, \text{ larger number}$$

$$x - 12 = 0, \text{ smaller number}$$

28. 48 and 56

Review Exercises 8.12 (page 393)

(The checks will not be shown.)

1. LCD = m

$$(m)\,\frac{6}{m} = (m)\,5$$

$$6 = 5m$$

$$\frac{6}{5} = m$$

$$m = 1\frac{1}{5}$$

2. 5

3. LCD = 40

$$(40)\,\frac{z}{5} + (40)\,\frac{-z}{8} = (40)\,3$$

$$8z - 5z = 120$$

$$3z = 120$$

$$z = 40$$

4. $\frac{11}{9}$

5. LCD = $2z$

$$(2z)\,\frac{4}{2z} + (2z)\,\frac{2}{z} = (2z)\,1$$

$$4 + 4 = 2z$$

$$8 = 2z$$

$$z = 4$$

6. $-2, \frac{4}{5}$

7. LCD = $18(2x - 1)$

$$\frac{7}{2x - 1}\,(18)(2x - 1) + \frac{1}{18}\,(18)(2x - 1) = \frac{x}{6}\,(18)(2x - 1)$$

$$126 + 2x - 1 = 3x(2x - 1)$$

$$125 + 2x = 6x^2 - 3x$$

$$0 = 6x^2 - 5x - 125$$

$$0 = (6x + 25)(x - 5)$$

$6x + 25 = 0$	$x - 5 = 0$
$6x = -25$	$x = 5$
$x = -\dfrac{25}{6}$	

8. $y = \dfrac{2x - 14}{7}$

9. LCD = n

$$(n)\frac{2m}{n} = (n)P$$

$$2m = nP$$

$$\frac{2m}{P} = n$$

$$\text{or } n = \frac{2m}{P}$$

10. $B = \dfrac{3V}{h}$

11. $\dfrac{F - 32}{C} = \dfrac{9}{5}$

$$9C = 5(F - 32)$$

$$\frac{9C}{9} = \frac{5F - 160}{9}$$

$$C = \frac{5F - 160}{9} \quad \text{or} \quad C = \frac{5}{9}(F - 32)$$

12. $\frac{17}{13}$

13. Let x = speed returning from work

$$\frac{1}{2}(x + 10) = \frac{3}{4}(x)$$

$$\frac{1}{2}x + 5 = \frac{3}{4}x$$

$$5 = \frac{1}{4}x$$

$$x = 20 \text{ mph}$$

$$\text{Distance} = rt = 20\left(\frac{3}{4}\right) = 15 \text{ mi}$$

14. Two answers: The number is $\frac{2}{3}$ and its reciprocal is $\frac{3}{2}$, or the number is $\frac{3}{2}$ and its reciprocal is $\frac{2}{3}$.

15. Let x = number of hours for machine B to do the job

Machine A's rate is $\dfrac{1 \text{ job}}{6 \text{ hr}}$.

Machine B's rate is $\dfrac{1 \text{ job}}{x \text{ hr}}$.

$$\underset{\text{Both machines worked 4 hr}}{\frac{1 \text{ job}}{6 \text{ hr}}\,(4 \text{ hr}) + \frac{1 \text{ job}}{x \text{ hr}}\,(4 \text{ hr}) = 1 \text{ job}} \quad \text{One job was completed}$$

$$\frac{1}{6}(4)\,(6x) + \frac{1}{x}(4)\,(6x) = 1\,(6x) \quad \text{LCD is } 6x$$

$$4x + 24 = 6x$$

$$24 = 2x$$

$$x = 12 \text{ hr}$$

16. 810 mi

17. Let x = number of yards in $\frac{1}{2}$ day

$$\frac{2\frac{1}{2}\,\text{yd}}{\frac{5}{8}\,\text{day}} = \frac{x\,\text{yd}}{\frac{1}{2}\,\text{day}} \Rightarrow \frac{\frac{5}{2}}{\frac{5}{8}} = \frac{x}{\frac{1}{2}}$$

$$\frac{5}{8}x = \left(\frac{5}{2}\right)\left(\frac{1}{2}\right)$$

$$\frac{5}{8}x = \frac{5}{4}$$

$$\frac{5}{8}x(8) = \frac{5}{4}(8)$$

$$5x = 10$$

$$x = 2\,\text{yd}$$

Chapter 8 Diagnostic Test (page 397)

Following each problem number is the textbook section number (in parentheses) where that kind of problem is discussed.

1. (8.1) **a.** -4 must be excluded, because it makes the denominator zero.

b. 0 and -4 must be excluded, because they make the denominator zero.

2. (8.1) **a.** 3. The signs of fractions are different and signs of numerators are different. Therefore, the signs of denominators must be *the same*.

b. -3. The signs of denominators are different and signs of fractions are the same. Therefore, the signs of numerators must be different.

3. (8.2) $\dfrac{x^2 - 9}{x^2 - 6x + 9} = \dfrac{(x + 3)(x \!-\! 3)}{(x - 3)(x \!-\! 3)} = \dfrac{x + 3}{x - 3}$

4. (8.2) $\dfrac{4x^2 - 23xy + 15y^2}{20x^2y - 15xy^2} = \dfrac{(4x \!-\! 3y)(x - 5y)}{5xy(4x \!-\! 3y)} = \dfrac{x - 5y}{5xy}$

5. (8.3) $\dfrac{a}{a^2 + 5a + 6} \cdot \dfrac{4a + 8}{6a^3 + 18a^2} = \dfrac{a}{(a \!+\! 2)(a + 3)} \cdot \dfrac{4(a \!+\! 2)}{6a^2(a + 3)}$

$= \dfrac{2}{3a(a + 3)^2}$

6. (8.4) $\dfrac{6x}{x - 2} - \dfrac{3}{x - 2} = \dfrac{6x - 3}{x - 2}$ or $\dfrac{3(2x - 1)}{x - 2}$

7. (8.3) $\dfrac{2x}{x^2 - 9} \div \dfrac{4x^2}{x^2 - 6x + 9} = \dfrac{2x}{(x + 3)(x - 3)} \cdot \dfrac{(x - 3)(x - 3)}{4x^2}$

$= \dfrac{x - 3}{2x(x + 3)}$

8. (8.6) LCD is $b(b - 1)$

$\dfrac{b \cdot b}{(b - 1) \cdot b} - \dfrac{(b + 1)(b - 1)}{b(b - 1)} = \dfrac{b^2 - (b^2 - 1)}{b(b - 1)}$

$= \dfrac{b^2 - b^2 + 1}{b(b - 1)} = \dfrac{1}{b(b - 1)}$

9. (8.6) LCD is $(x - 5)^2(2x + 3)$

$\dfrac{(x + 3)(2x + 3)}{(x - 5)^2(2x + 3)} + \dfrac{x(x - 5)}{(x - 5)(2x + 3)(x - 5)}$

$= \dfrac{2x^2 + 9x + 9 + x^2 - 5x}{(x - 5)^2(2x + 3)} = \dfrac{3x^2 + 4x + 9}{(x - 5)^2(2x + 3)}$

10. (8.6) LCD is $(x + 5)(x - 5)$

$\dfrac{x(x - 5)}{(x + 5)(x - 5)} - \dfrac{x(x + 5)}{(x - 5)(x + 5)} - \dfrac{50}{(x + 5)(x - 5)}$

$= \dfrac{x^2 - 5x - (x^2 + 5x) - (50)}{(x + 5)(x - 5)} = \dfrac{x^2 - 5x - x^2 - 5x - 50}{(x + 5)(x - 5)}$

$= \dfrac{-10x - 50}{(x + 5)(x - 5)} = \dfrac{-10(x + 5)}{(x + 5)(x - 5)} = \dfrac{-10}{x - 5}$

11. (8.7) $\dfrac{\frac{6x^4}{11y^2}}{\frac{9x}{22y^4}} = \dfrac{6x^4}{11y^2} \div \dfrac{9x}{22y^4} = \dfrac{6x^4}{11y^2} \cdot \dfrac{22y^4}{9x} = \dfrac{4x^3y^2}{3}$

12. (8.7) LCD is x^2 $\dfrac{x^2\left(\frac{6}{x} + \frac{15}{x^2}\right)}{x^2\left(2 + \frac{5}{x}\right)} = \dfrac{6x + 15}{2x^2 + 5x} = \dfrac{3(2x + 5)}{x(2x + 5)} = \dfrac{3}{x}$

13. (8.9) $\dfrac{x + 7}{3} = \dfrac{2x - 1}{4}$

$4(x + 7) = 3(2x - 1)$

$4x + 28 = 6x - 3$

$\underline{-4x + 3 \quad\quad -4x + 3}$

$31 = 2x$

$x = \dfrac{31}{2}$

14. (8.9) LCD is x $\quad x - \dfrac{4}{x} = 3$

$x(x) - \left(\dfrac{4}{x}\right)(x) = 3(x)$

$x^2 - 3x - 4 = 0$

$(x - 4)(x + 1) = 0$

$x - 4 = 0 \quad | \quad x + 1 = 0$

$x = 4 \quad | \quad x = -1$

Check for $x = 4$: $\qquad$ *Check for $x = -1$:*

$x - \dfrac{4}{x} = 3 \qquad\qquad x - \dfrac{4}{x} = 3$

$4 - \dfrac{4}{4} \overset{?}{=} 3 \qquad\qquad -1 - \dfrac{4}{-1} \overset{?}{=} 3$

$4 - 1 \overset{?}{=} 3 \qquad\qquad -1 + 4 \overset{?}{=} 3$

$3 = 3 \qquad\qquad\qquad 3 = 3$

15. (8.9) LCD is $6(2x + 1)$

$$6(2x + 1)\left(\frac{3}{2(2x + 1)}\right) + 6(2x + 1)\frac{(x + 1)}{6} = 1\,(6)(2x + 1)$$

$$9 + 2x^2 + 3x + 1 = 12x + 6$$

$$2x^2 + 3x + 10 = 12x + 6$$

$$2x^2 - 9x + 4 = 0$$

$$(x - 4)(2x - 1) = 0$$

$$\begin{array}{l|l} x - 4 = 0 & 2x - 1 = 0 \\ \quad x = 4 & 2x = 1 \\ & \quad x = \dfrac{1}{2} \end{array}$$

Check for $x = 4$:

$$\frac{3}{2(2x + 1)} + \frac{x + 1}{6} = 1$$

$$\frac{3}{2(2[4] + 1)} + \frac{[4] + 1}{6} \overset{?}{=} 1$$

$$\frac{3}{2(8 + 1)} + \frac{5}{6} \overset{?}{=} 1$$

$$\frac{3}{2(9)} + \frac{5}{6} \overset{?}{=} 1$$

$$\frac{1}{6} + \frac{5}{6} \overset{?}{=} 1$$

$$1 = 1$$

Check for $x = \dfrac{1}{2}$:

$$\frac{3}{2(2x + 1)} + \frac{x + 1}{6} = 1$$

$$\frac{3}{2\left[2\left(\frac{1}{2}\right) + 1\right]} + \frac{\left(\frac{1}{2}\right) + 1}{6} \overset{?}{=} 1$$

$$\frac{3}{2[1 + 1]} + \frac{\frac{3}{2}}{6} \overset{?}{=} 1$$

$$\frac{3}{2[2]} + \frac{1}{4} \overset{?}{=} 1$$

$$\frac{3}{4} + \frac{1}{4} \overset{?}{=} 1$$

$$1 = 1$$

16. (8.9)

$$\frac{3x - 5}{x} = \frac{x + 1}{3}$$

$$3(3x - 5) = x(x + 1)$$

$$9x - 15 = x^2 + x$$

$$0 = x^2 - 8x + 15$$

$$0 = (x - 3)(x - 5)$$

$$\begin{array}{l|l} x - 3 = 0 & x - 5 = 0 \\ \quad x = 3 & \quad x = 5 \end{array}$$

Check for $x = 3$:

$$\frac{3x - 5}{x} = \frac{x + 1}{3}$$

$$\frac{3(3) - 5}{(3)} \overset{?}{=} \frac{(3) + 1}{3}$$

$$\frac{9 - 5}{3} \overset{?}{=} \frac{4}{3}$$

$$\frac{4}{3} = \frac{4}{3}$$

Check for $x = 5$:

$$\frac{3x - 5}{x} = \frac{x + 1}{3}$$

$$\frac{3(5) - 5}{(5)} \overset{?}{=} \frac{(5) + 1}{3}$$

$$\frac{15 - 5}{5} \overset{?}{=} \frac{6}{3}$$

$$\frac{10}{5} \overset{?}{=} \frac{6}{3}$$

$$2 = 2$$

17. (8.10) Solve for y:

$$5x + 7y = 18$$

$$7y = 18 - 5x$$

$$y = \frac{18 - 5x}{7}$$

18. (8.11) Let $\quad x = $ numerator

$$x + 6 = \text{denominator}$$

$$\frac{x - 1}{(x + 6) + 7} = \frac{1}{3}$$

$$\frac{x - 1}{x + 13} = \frac{1}{3}$$

$$3(x - 1) = 1(x + 13)$$

$$3x - 3 = x + 13$$

$$\underline{-x + 3 \qquad -x + 3}$$

$$2x = 16$$

$$x = 8, \text{ numerator}$$

$$x + 6 = 14, \text{ denominator}$$

The fraction is $\dfrac{8}{14}$.

Check: $\dfrac{8 - 1}{14 + 7} = \dfrac{7}{21} = \dfrac{1}{3}$

19. (8.11) Let $x = $ distance covered in $2\dfrac{2}{3}\text{hr}\left(\text{or } \dfrac{8}{3}\text{hr}\right)$

$$\frac{21\,\text{mi}}{\frac{2}{5}\,\text{hr}} = \frac{x\,\text{mi}}{\frac{8}{3}\,\text{hr}}$$

$$\frac{2}{5}x = 21\left(\frac{8}{3}\right)$$

$$\frac{2}{5}x = 56$$

$$(5)\left(\frac{2}{5}x\right) = 56\,(5)$$

$$2x = 280$$

$$x = 140\,\text{mi}$$

Check: Rate is $\dfrac{21\,\text{mi}}{\frac{2}{5}\,\text{hr}} = 52.5\,\dfrac{\text{mi}}{\text{hr}}$

$$52.5\,\frac{\text{mi}}{\text{hr}}\left(\frac{8}{3}\,\text{hr}\right) = 140\,\text{mi}$$

or Rate is $\dfrac{140\,\text{mi}}{\frac{8}{3}\,\text{hr}} = 52.5\,\dfrac{\text{mi}}{\text{hr}}$

20. (8.11) Let $t = $ the number of hours for Barbara to type 111 pages

Then Barbara's rate is $\dfrac{111\,\text{pages}}{t\,\text{hr}}$.

Sherry's rate is $\dfrac{128\,\text{pages}}{4\,\text{hour}}$, or $32\,\dfrac{\text{pages}}{\text{hr}}$.

They type for 5 hr

$$\frac{111}{t}\,(5) + 32\,(5) = 345 \quad \text{Number of pages typed}$$

$$\frac{555}{t}\,(t) + 160\,(t) = 345\,(t)$$

$$555 + 160t = 345t$$

$$\underline{\quad -160t \qquad -160t \quad}$$

$$555 = 185t$$

$$t = 3\,\text{hr}$$

Check:

$$\frac{111\,\text{pg}}{3\,\text{hr}} = 37\,\frac{\text{pg}}{\text{hr}}$$

$$37\,\frac{\text{pg}}{\text{hr}}(5\,\text{hr}) + 32\,\frac{\text{pg}}{\text{hr}}(5\,\text{hr}) = 185\,\text{pg} + 160\,\text{pg} = 345\,\text{pg}$$

Cumulative Review Exercises: Chapters 1–8 (page 399)

1. $10 - (3\sqrt{4} - 5^2) = 10 - (3 \cdot 2 - 25)$
$$= 10 - (6 - 25)$$
$$= 10 - (-19) = 29$$

2. $V = \frac{10}{9}$

3. $\left(\frac{24y^{-3}}{8y^{-1}}\right)^{-2} = (3y^{-2})^{-2} = 3^{-2}y^4 = \frac{y^4}{3^2} = \frac{y^4}{9}$ **4.** $3x - 1$

5.

$$
\begin{array}{r}
5z + 7 \ \ R\,18 \\
3z - 2\overline{)15z^2 + 11z + 4} \\
\underline{15z^2 - 10z} \\
21z + 4 \\
\underline{21z - 14} \\
18
\end{array}
$$

6. x^3

7. $\dfrac{2x}{x + 1} - \dfrac{2x - 1}{x + 1} = \dfrac{2x - (2x - 1)}{x + 1} = \dfrac{2x - 2x + 1}{x + 1} = \dfrac{1}{x + 1}$

8. $\dfrac{10x^3 + 3}{2x^3}$

9. LCD is $(x + 4)(x + 5)(2x - 1)$

$$\frac{(2x - 1)(2x - 1)}{(x + 4)(x + 5)(2x - 1)} - \frac{(x + 5)(x + 5)}{(2x - 1)(x + 4)(x + 5)}$$

$$= \frac{(4x^2 - 4x + 1) - (x^2 + 10x + 25)}{(x + 4)(x + 5)(2x - 1)}$$

$$= \frac{4x^2 - 4x + 1 - x^2 - 10x - 25}{(x + 4)(x + 5)(2x - 1)}$$

$$= \frac{3x^2 - 14x - 24}{(x + 4)(x + 5)(2x - 1)}$$

10. $x = \frac{140}{3}$

11. LCD is $4x^2$

$$(4x^2)\left(\frac{3}{x}\right) - (4x^2)\left(\frac{8}{x^2}\right) = \frac{1}{4}(4x^2)$$
$$12x - 32 = x^2$$
$$0 = x^2 - 12x + 32$$
$$0 = (x - 4)(x - 8)$$

$$
\begin{array}{c|c}
x - 4 = 0 & x - 8 = 0 \\
x = 4 & x = 8
\end{array}
$$

(The check is left to the student.)

12. $(x + 6)(x - 7)$

13. $3w^2 - 48 = 3(w^2 - 16) = 3(w + 4)(w - 4)$

14. $(5a - 3b)(4a + b)$

15. $3x^2 - 6x - 2xy + 4y = 3x(x - 2) - 2y(x - 2)$
$= (x - 2)(3x - 2y)$

16. 16 and 17

17. Let $\quad x = $ one number
$\quad 5 - x = $ other number
$$x(5 - x) = -24$$
$$5x - x^2 = -24$$
$$0 = x^2 - 5x - 24$$
$$0 = (x - 8)(x + 3)$$

$$
\begin{array}{c|c}
x - 8 = 0 & x + 3 = 0 \\
x = 8 & x = -3 \\
5 - x = -3 & 5 - x = 8
\end{array}
$$

The numbers are 8 and -3. (The check is left to the student.)

18. 5 lb of oranges at 99 cents per pound and 10 lb of oranges at 78 cents per pound

19. Let $\quad n = $ number of nickels
$\quad 20 - n = $ number of dimes
$$5n + 10(20 - n) = 165$$
$$5n + 200 - 10n = 165$$
$$-5n = -35$$
$$n = 7 \text{ nickels}$$
$$20 - n = 13 \text{ dimes}$$

20. 15 nickels, 6 dimes, 5 quarters

Exercises 9.1 (page 406)

1a. $(3, 1)$: Start at the origin, move right 3 units, then move up 1 unit.
b. $(-4, -2)$: Start at the origin, move left 4 units, then down 2 units.
c. $(0, 3)$: Start at the origin, move right 0 units, then up 3 units.
d. $(5, -4)$: Start at the origin, move right 5 units, then down 4 units.
e. $(4, 0)$: Start at the origin, move right 4 units, then up 0 units.
f. $(-2, 4)$: Start at the origin, move left 2 units, then up 4 units.

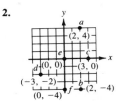

2.

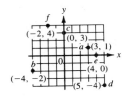

3a. $(3, 0)$ Not in any quadrant
b. $(0, 5)$ Not in any quadrant
c. $(-5, 2)$ II
d. $(-4, -3)$ III

4a. $(4, 3)$ I
b. $(-6, 0)$ Not in any quadrant
c. $(3, -4)$ IV
d. $(0, -3)$ Not in any quadrant

5a. 5 because A is 5 units to the right
b. -3 because C is 3 units to the left
c. 3 because E is 3 units to the right
d. 1 because F is 1 unit to the right

6a. 5
 b. −2
 c. 0
 d. −5

7a. 1 because *F* is 1 unit to the right. Abscissa is the number of units to the right or left of the vertical axis.

 b. 2 because *C* is 2 units above the horizontal axis

8a. 5
 b. −4

9. 0. The *y*-coordinate is the distance up or down from the horizontal axis.

10. 0

11.

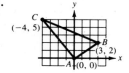

12.

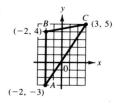

Exercises 9.2 (page 411)

1a. (0, 0) is not a solution, because 3(0) + 2(0) = 0 ≠ 6.
 b. (0, 3) is a solution, because 3(0) + 2(3) = 0 + 6 = 6.
 c. (3, 0) is not a solution, because 3(3) + 2(0) = 9 ≠ 6.
 d. (2, 0) is a solution, because 3(2) + 2(0) = 6 + 0 = 6.

2a. No
 b. No
 c. Yes
 d. Yes

3a. (0, 0) is a solution, because 3(0) + 4(0) = 0.
 b. (0, 3) is not a solution, because 3(0) + 4(3) = 0 + 12 ≠ 0.
 c. (4, −3) is a solution, because 3(4) + 4(−3) = 12 − 12 = 0.
 d. (−4, 3) is a solution, because 3(−4) + 4(3) = −12 + 12 = 0.

4a. Yes
 b. No
 c. No
 d. Yes

5a. $4(0) - 3y = 12$
 $-3y = 12$
 $y = -4$
 $(0, -4)$
 b. $4x - 3(0) = 12$
 $4x = 12$
 $x = 3$
 $(3, 0)$

c. $4(3) - 3y = 12$
 $12 - 3y = 12$
 $-3y = 0$
 $y = 0$
 $(3, 0)$
 d. $4x - 3(-4) = 12$
 $4x + 12 = 12$
 $4x = 0$
 $x = 0$
 $(0, -4)$

6a. (0, 3)   **c.** (3, −6)
 b. (1, 0)   **d.** (2, − 3)

7a. $2(0) - 5y = 0$
 $-5y = 0$
 $y = 0$
 $(0, 0)$
 b. $2x - 5(0) = 0$
 $2x = 0$
 $x = 0$
 $(0, 0)$
 c. $2(2) - 5y = 0$
 $4 - 5y = 0$
 $-5y = -4$
 $y = \frac{4}{5}$
 $\left(2, \frac{4}{5}\right)$
 d. $2x - 5(2) = 0$
 $2x - 10 = 0$
 $2x = 10$
 $x = 5$
 $(5, 2)$

8a. (0, 0)
 b. (0, 0)
 c. (3, −9)
 d. (1, −3)

9a. $x - 5 = 0$
 $x = 5$
 $(5, 0)$
 b. $x - 5 = 0$
 $x = 5$
 $(5, 5)$
 c. $x - 5 = 0$
 $x = 5$
 $(5, 2)$
 d. $x - 5 = 0$
 $x = 5$
 $(5, - 2)$

10a. (0, −3)
 b. (3, −3)
 c. (−3, −3)
 d. (5, −3)

Exercises 9.3 (page 419)

1. $x + y = 3$

$y = 3 - x$

Substitute three values for x and find the corresponding values for y.

If $x = 0$, $y = 3 - 0 = 3$

If $x = 1$, $y = 3 - 1 = 2$

If $x = 3$, $y = 3 - 3 = 0$

Table of values

x	y
0	3
1	2
3	0

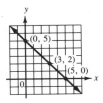

2.

3. $2x - 3y = 6$

Intercepts:

Set $y = 0$. Then $x = 3$.

Set $x = 0$. Then $y = -2$.

This gives the points $(3, 0)$ and $(0, -2)$.

Checkpoint: Set $x = 2$. Then $y = -\frac{2}{3}$.

4.

5. x can be any number but y is always 8.

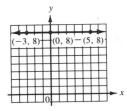

6.

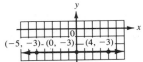

7. y can be any number, but x is always -2.

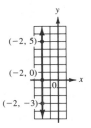

8.

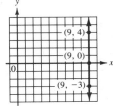

9. x is always -5. $x + 5 = 0 \Rightarrow x = -5$.

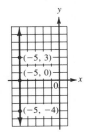

10.

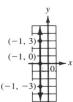

11. x can have any value, but y must be -5. (If $y + 5 = 0$, $y = -5$.)

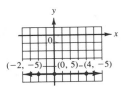

12.

13. $3x = 5y + 15$

Set $y = 0$.

Then $3x = 15$

$x = 5$.

This gives point $(5, 0)$.

Set $x = 0$.

Then $5y = -15$

$y = -3$.

This gives point $(0, -3)$.

Checkpoint: Set $x = 2$. Then $y = -1\frac{4}{5}$.

x	y
0	-3
5	0
2	$-1\frac{4}{5}$

14.

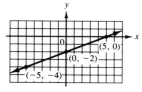

15. $x = -2y$

Set $y = 0$.
Then $x = 0$.

Set $y = 2$.
Then $x = -4$.

x	y
0	0
-4	2
2	-1

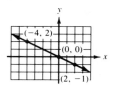

16.

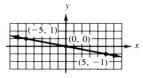

17.

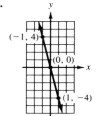

$y = -4x$
Set $x = 0$.
Then $y = 0$.
Set $x = 1$.
Then $y = -4(1) = -4$.
Checkpoint: Set $x = -1$.
Then $y = -4(-1) = 4$

18.

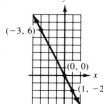

19.

$4 - y = x$
Set $x = 0$.
Then $y = 4$.
Set $y = 0$.
Then $x = 4$.
Checkpoint: Set $x = 2$.
Then $y = 2$.

20.

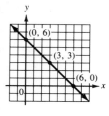

21. $y = x$

x	y
0	0
4	4
-3	-3

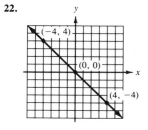

22.

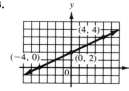

23.

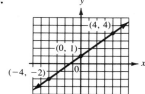

$y = \frac{1}{2}x + 2$
Set $x = 0$.
Then $y = 2$.
Set $y = 0$.
Then $x = -4$.
Checkpoint: Set $x = 4$.
Then $y = 4$.

24.

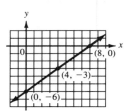

25. Find intercepts.
$3x = 24 + 4y$

x	y
0	-6
8	0
4	-3

26.

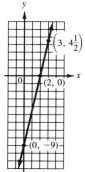

27. $x - y = 5$

a.

x	y
0	−5
5	0
−1	−6

$x + y = 1$

x	y
0	1
1	0
−1	2

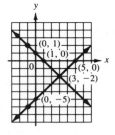

b. $(3, -2)$

28. Each line represents two units.

a.

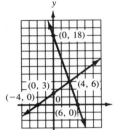

b. $(4, 6)$

Exercises 9.4 (page 426)

1. $m = \dfrac{y_2 - y_1}{x_2 - x_1} = \dfrac{3 - 1}{7 - (-2)} = \dfrac{2}{9}$　**2.** $\frac{5}{6}$

3. $m = \dfrac{y_2 - y_1}{x_2 - x_1} = \dfrac{-3 - (-1)}{5 - (-1)} = \dfrac{-2}{6} = -\dfrac{1}{3}$　**4.** $-\frac{1}{2}$

5. $m = \dfrac{y_2 - y_1}{x_2 - x_1} = \dfrac{-3 - (-3)}{4 - (-5)} = \dfrac{0}{9} = 0$　**6.** 0

7. $m = \dfrac{y_2 - y_1}{x_2 - x_1} = \dfrac{(-3) - (5)}{(8) - (-7)} = \dfrac{-8}{15}$　**8.** $-\frac{5}{17}$

9. $m = \dfrac{y_2 - y_1}{x_2 - x_1} = \dfrac{(-5) - (-15)}{(4) - (-6)} = \dfrac{10}{10} = 1$　**10.** -1

11. $m = \dfrac{y_2 - y_1}{x_2 - x_1} = \dfrac{(8) - (5)}{(-2) - (-2)} = \dfrac{3}{0}$

Slope does not exist

12. Does not exist

13. $m = \dfrac{y_2 - y_1}{x_2 - x_1} = \dfrac{-2 - 3}{-4 - (-4)} = \dfrac{-5}{0}$　Slope does not exist.

14. Does not exist

15.

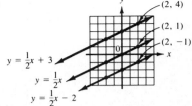

$y = \frac{1}{2}x$

x	y
0	0
2	1
−2	−1

$y = \frac{1}{2}x + 3$

x	y
0	3
2	4
−2	2

$y = \frac{1}{2}x - 2$

x	y
0	−2
2	−1
−2	−3

a. $m = \dfrac{1 - 0}{2 - 0} = \dfrac{1}{2}$

b. $m = \dfrac{4 - 3}{2 - 0} = \dfrac{1}{2}$

c. $m = \dfrac{-1 - (-2)}{2 - 0} = \dfrac{1}{2}$

16.

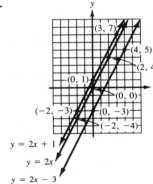

a. $m = 2$

b. $m = 2$

c. $m = 2$

17.

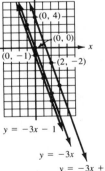

$y = -3x$

x	y
0	0
2	−6
−1	3

$y = -3x + 4$

x	y
0	4
2	−2
3	−5

$y = -3x - 1$

x	y
0	−1
2	−7
1	−4

a. $m = \dfrac{-6 - 0}{2 - 0} = -3$

b. $m = \dfrac{-2 - 4}{2 - 0} = -3$

c. $m = \dfrac{-7 - (-1)}{2 - 0} = -3$

18.

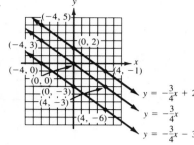

a. $m = -\dfrac{3}{4}$

b. $m = -\dfrac{3}{4}$

c. $m = -\dfrac{3}{4}$

$y = -\dfrac{3}{4}x + 2$

$y = -\dfrac{3}{4}x$

$y = -\dfrac{3}{4}x - 3$

Exercises 9.5 (page 430)

1. $\quad 3x = 2y - 4$

$3x - 2y + 4 = 0$

2. $2x - 3y - 7 = 0$

3. $\qquad y = -\dfrac{3}{4}x - 2$

$4y = -3x - 8$

$3x + 4y + 8 = 0$

4. $3x + 5y + 20 = 0$

5. $\quad 2(3x + y) = 5(x - y) + 4$

$6x + 2y = 5x - 5y + 4$

$x + 7y - 4 = 0$

6. $4x - 9y + 5 = 0$

7. $y - y_1 = m(x - x_1)$

$y - 4 = \dfrac{1}{2}(x - 3)$

$2y - 8 = x - 3$

$0 = x - 2y + 5 \ \text{ or } \ x - 2y + 5 = 0$

8. $x - 3y + 13 = 0$

9. $\qquad y - y_1 = m(x - x_1)$

$y - (-2) = -\dfrac{2}{3}[x - (-1)]$

$3(y + 2) = -2(x + 1)$

$3y + 6 = -2x - 2$

$2x + 3y + 8 = 0$

10. $5x + 4y + 22 = 0$

11. $\qquad y = mx + b$

$y = \dfrac{3}{4}x - 3$

$4y = 3x - 12$

$3x - 4y - 12 = 0$

12. $2x - 7y - 14 = 0$

13. $\qquad y = mx + b$

$y = -\dfrac{2}{5}x + \dfrac{1}{2}$

$\text{LCD} = 10$

$10y = -4x + 5$

$4x + 10y - 5 = 0$

14. $20x + 12y - 9 = 0$

15. Use the two points to find the slope, then use m and one point to find the equation of the line.

$m = \dfrac{y_2 - y_1}{x_2 - x_1} = \dfrac{4 - (-1)}{2 - 4} = -\dfrac{5}{2}$

$y - y_1 = m(x - x_1)$

$y - 4 = -\dfrac{5}{2}(x - 2)$

$2y - 8 = -5x + 10$

$5x + 2y - 18 = 0$

16. $3x + 2y - 11 = 0$

17. $m = \dfrac{4 - 0}{3 - 0} = \dfrac{4}{3}$

$y - y_1 = m(x - x_1)$

$y - 0 = \dfrac{4}{3}(x - 0)$

$3y = 4x$

$4x - 3y = 0$

18. $5x - 2y = 0$

19. Because the line is horizontal, every point on the line must have $y = 5$. Therefore, $y - 5 = 0$.

20. $y = -3 \ \text{ or } \ y + 3 = 0$

21. Because the line is vertical, every point on the line must have $x = -3$. Therefore, the equation is $x = -3$ or $x + 3 = 0$.

22. $x = -2$ or $x + 2 = 0$

23. $3x + 4y + 12 = 0$. Solve for y.

a. $4y = -3x - 12$

$y = -\dfrac{3}{4}x - 3$

b. $m = -\dfrac{3}{4}$

c. y-intercept $= -3$

24a. $y = -\dfrac{2}{5}x - 2$

b. $m = -\dfrac{2}{5}$

c. y-intercept $= -2$

25. $2x + 3y - 9 = 0$. Solve for y:

$3y = -2x + 9$

a. $y = -\dfrac{2}{3}x + 3$

b. $m = -\dfrac{2}{3}$

c. y-intercept $= 3$

26a. $y = -\dfrac{2}{5}x + 3$

b. $m = -\dfrac{2}{5}$

c. y-intercept $= 3$

Exercises 9.6 (page 436)

1. $y = x^2$
$y = (-2)^2 = 4$
$y = (-1)^2 = 1$
$y = 0^2 = 0$
$y = 1^2 = 1$
$y = 2^2 = 4$
x- and *y*- intercepts
are (0, 0).

x	y
-2	4
-1	1
0	0
1	1
2	4

2.

x	y
-3	$2\frac{1}{4}$
-2	1
-1	$\frac{1}{4}$
0	0
1	$\frac{1}{4}$
2	1
3	$2\frac{1}{4}$

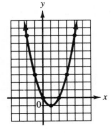

x- and *y*-intercepts are (0, 0).

3. $y = x^2 - 2x$
$y = (-2)^2 - 2(-2)$
$\quad = 4 + 4 = 8$
$y = (-1)^2 - 2(-1)$
$\quad = 1 + 2 = 3$
$y = 0^2 - 2(0) = 0$
$y = 1^2 - 2(1) = -1$
$y = 2^2 - 2(2) = 0$
$y = 3^2 - 2(3) = 3$
$y = 4^2 - 2(4) = 8$

x	y
-2	8
-1	3
0	0
1	-1
2	0
3	3
4	8

x-intercepts are (0, 0) and (2, 0).

y-intercept is (0, 0).

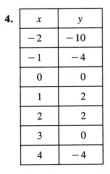

4.

x	y
-2	-10
-1	-4
0	0
1	2
2	2
3	0
4	-4

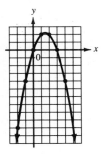

x-intercepts are (0, 0) and (3, 0).

y-intercept is (0, 0).

5.

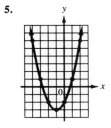

$y = x^2 + 2x - 2$
$y = (-4)^2 + 2(-4) - 2 = 6$
$y = (-3)^2 + 2(-3) - 2 = 1$
$y = (-2)^2 + 2(-2) - 2 = -2$
$y = (-1)^2 + 2(-1) - 2 = -3$
$y = (0)^2 + 2(0) - 2 = -2$
$y = (1)^2 + 2(1) - 2 = 1$
$y = (2)^2 + 2(2) - 2 = 6$

x	y
-4	6
-3	1
-2	-2
-1	-3
0	-2
1	1
2	6

6.

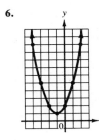

x	y
-4	10
-3	5
-2	2
-1	1
0	2
1	5
2	10

7. $y = 2x - x^2$
$y = 2(-2) - (-2)^2$
$\quad = -4 - 4 = -8$
$y = 2(-1) - (-1)^2$
$\quad = -2 - 1 = -3$
$y = 2(0) - 0^2 = 0$
$y = 2(1) - (1)^2$
$\quad = 2 - 1 = 1$
$y = 2(2) - (2)^2$
$\quad = 4 - 4 = 0$
$y = 2(3) - (3)^2$
$\quad = 6 - 9 = -3$
$y = 2(4) - (4)^2$
$\quad = 8 - 16 = -8$

x	y
-2	-8
-1	-3
0	0
1	1
2	0
3	-3
4	-8

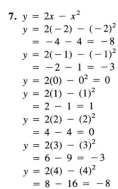

8.

x	y
-3	3
-2	0
-1	-1
0	0
1	3

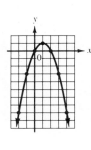

9. $y = x^3$
$y = (-2)^3 = -8$
$y = (-1)^3 = -1$
$y = 0^3 = 0$
$y = 1^3 = 1$
$y = 2^3 = 8$

x	y
-2	-8
-1	-1
0	0
1	1
2	8

10.

x	y
-2	2
-1	6
0	4
1	2
2	6

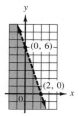

Exercises 9.7 (page 443)

1. $x + 2y < 4$
Change to $x + 2y = 4$ to find boundary line. Boundary line goes through:

x	y
0	2
4	0

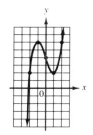

Boundary line is dashed because equality sign is not included with $<$.
Half-plane is the part that includes the origin because $(0, 0)$ makes $x + 2y < 4$ true.

2.

3. $2x - 3y > 6$
Boundary line:
$2x - 3y = 6$
$2(0) - 3y = 6$
$y = -2$
$2x - 3(0) = 6$
$x = 3$

Half-plane does not include the origin because
$2(0) - 3(0) \not> 6$
$0 \not> 6$

x	y
0	-2
3	0

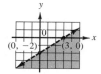

4.

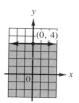

5. $x \geq -2$. All points to the right of and including the line $x = -2$

6.

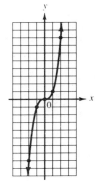

7. Boundary is $y = 4$, a horizontal line.

8.

The correct half-plane includes the origin.

9. $3y - 4x \geq 12$
Boundary line is $3y - 4x = 12$.
$x = 0, 3y = 12$
$y = 4$
$y = 0, -4x = 12$
$x = -3$

x	y
0	4
-3	0

The correct half-plane does not include the origin.

10.

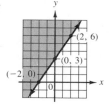

11. Boundary line is $3x + 2y = -6$.
Intercepts:
$x = 0, y = -3$
$y = 0, x = -2$

x	y	
0	-3	
-2	0	
-4	3	Checkpoint

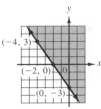

The correct half-plane includes the origin.

12.

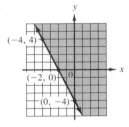

13. $3x - 4y \geq 10$
Boundary line:

$3x - 4y = 10$

$x = 0, y = \dfrac{10}{-4} = -2\dfrac{1}{2}$

$y = 0, x = \dfrac{10}{3} = 3\dfrac{1}{3}$

x	y
0	$-2\dfrac{1}{2}$
$3\dfrac{1}{3}$	0

The correct half-plane does not include the origin.

14.

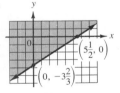

15. $2(x + 1) + 3 \leq 3(2x - 1)$

$2x + 2 + 3 \leq 6x - 3$

$2x - 6x \leq -2 - 3 - 3$

$-4x \leq -8$

$4x \geq 8$

$x \geq 2$

Boundary is $x = 2$, a vertical line.

16.

17. $4(2y + 3) - 1 \leq 5(2y - 1)$

$8y + 12 - 1 \leq 10y - 5$

$8y + 11 \leq 10y - 5$

$\underline{-8y + 5 \qquad -8y + 5}$

$16 \leq 2y$

$8 \leq y$

$y \geq 8$

Boundary is $y = 8$, a horizontal line.

18.

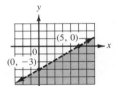

19. $\dfrac{x}{5} - \dfrac{y}{3} > 1$

$LCD = 15$

$15\left(\dfrac{x}{5}\right) + 15\left(-\dfrac{y}{3}\right) > 15(1)$

$3x - 5y > 15$

Boundary line: $3x - 5y = 15$
The point $(0, 0)$ does not make the inequality true. Therefore, the correct half-plane does not include the origin.

x	y
0	-3
5	0
3	$-1\dfrac{1}{5}$

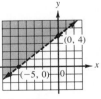

20.

21. $\dfrac{y}{4} - \dfrac{x}{5} > 1$ LCD is 20

$(20)\left(\dfrac{y}{4}\right) - (20)\left(\dfrac{x}{5}\right) > 1(20)$

$5y - 4x > 20$

Boundary is $5y - 4x = 20$.

x	y	
0	4	
-5	0	
-2	$2\dfrac{2}{5}$	(Checkpoint)

The correct half-plane does not include the origin.

22.

23. $\dfrac{2x + y}{3} - \dfrac{x - y}{2} \geq \dfrac{5}{6}$

LCD = 6

$\cancel{6}^{2}\left(\dfrac{2x + y}{\cancel{3}}\right) + \cancel{6}^{3}\left(\dfrac{-(x - y)}{\cancel{2}}\right) \geq \cancel{6}\left(\dfrac{5}{\cancel{6}}\right)$

$4x + 2y - 3x + 3y \geq 5$

$x + 5y \geq 5$

Boundary line: $x + 5y = 5$
The correct half-plane does not include the origin.

x	y
0	1
5	0
3	$\frac{2}{5}$

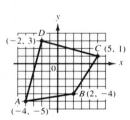

24.

Review Exercises 9.8 (page 445)

1.

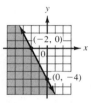

2a. y-coordinate = 4
b. abscissa = -2
c. x-coordinate = -2

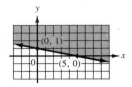

3. $x - y = 5$

x	y
0	-5
5	0
3	-2

4.

5. $x = -2$ is a vertical line.

6.

7. $4y - 5x = 20$

x	y
0	5
-4	0

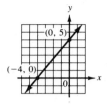

8.

9. $x + 2y = 0$

$x = -2y$

x	y
0	0
-4	2

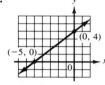

10. The slope does not exist.

11. $m = \dfrac{y_2 - y_1}{x_2 - x_1}$

$= \dfrac{5 - (-6)}{-3 - 2}$

$= \dfrac{11}{-5} = -\dfrac{11}{5}$

12. $-\dfrac{9}{5}$

13.
$y - y_1 = m(x - x_1)$

$y - (-5) = -\dfrac{2}{3}(x - 3)$

$3(y + 5) = -2(x - 3)$

$3y + 15 = -2x + 6$

$2x + 3y + 9 = 0$

14. $x + 3y - 27 = 0$

15. First find m; then use m and one point to find the equation.

$m = \dfrac{y_2 - y_1}{x_2 - x_1} = \dfrac{-2 - 4}{1 - (-3)} = \dfrac{-6}{4} = -\dfrac{3}{2}$

$y - y_1 = m(x - x_1)$

$y - (-2) = -\dfrac{3}{2}(x - 1)$

$2(y + 2) = -3(x - 1)$

$2y + 4 = -3x + 3$

$3x + 2y + 1 = 0$

16. $y = 5$ or $y - 5 = 0$

17. $x = -2$ (The x-coordinate of *every* point on the line is -2.)
$x + 2 = 0$

18.

x	y
-2	10
-1	4
0	0
1	-2
2	-2
3	0
4	4

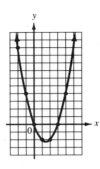

19. $y = x^3 - 3x$
$y = (-3)^3 - (3)(-3)$
$\quad = -27 + 9 = -18$
$y = (-2)^3 - 3(-2)$
$\quad = -8 + 6 = -2$
$y = (-1)^3 - 3(-1)$
$\quad = -1 + 3 = 2$
$y = 0^3 - 3(0) = 0$
$y = 1^3 - 3(1) = -2$
$y = 2^3 - 3(2) = 2$
$y = 3^3 - 3(3) = 18$

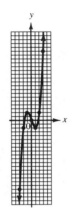

x	y
-3	-18
-2	-2
-1	2
0	0
1	-2
2	2
3	18

20.

x	y
-4	-18
-3	-2
-2	2
-1	0
0	-2
1	2
2	18

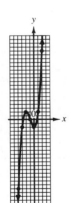

21. Boundary line is $3x - 7y = 21$.

x	y
0	-3
7	0
2	$-2\frac{1}{7}$ (Checkpoint)

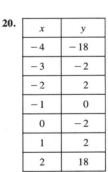

$(0, 0)$ makes the inequality true. Therefore, the correct half-plane includes the origin.

22. Boundary line: $x = -4y$

x	y
0	0
-4	1

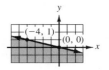

$(-1, -1)$ makes the inequality true. Therefore, the correct half-plane includes $(-1, -1)$.

23. The boundary is $y = 3$, which is a horizontal line. $(0, 0)$ makes the inequality false. Therefore, the correct half-plane does not include the origin.

24. LCD = 6

$$6\left(\frac{x}{2}\right) + 6\left(\frac{-y}{6}\right) \geq 6(1)$$

$$3x - y \geq 6$$

x	y
0	-6
2	0

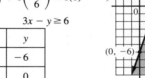

25.
$$\frac{y}{3} - \frac{x}{4} \leq 1 \quad \text{LCD is 12}$$

$$(12)\left(\frac{y}{3}\right) - (12)\left(\frac{x}{4}\right) \leq 1(12)$$

$$4y - 3x \leq 12$$

Boundary is $4y - 3x = 12$.

x	y
0	3
-4	0
4	6 (Checkpoint)

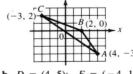

The correct half-plane includes the origin.

Chapter 9 Diagnostic Test (page 451)

Following each problem number is the textbook section number (in parentheses) where that kind of problem is discussed.

1. (9.1) **a.**

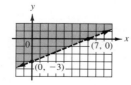

b. $D = (4, 5)$; $E = (-4, 1)$; $F = (-2, -6)$;
$G = (5, -4)$

2. (9.2) **a.** $(0, 0)$ is not a solution, because $5(0) - 3(0) = 0 \neq 15$.
b. $(3, 0)$ is a solution, because $5(3) - 3(0) = 15 - 0$
$= 15$.
c. $(0, 3)$ is not a solution, because $5(0) - 3(3) = 0 - 9 \neq$
15.
d. $(0, -5)$ is a solution, because $5(0) - 3(-5) = 0 + 15$
$= 15$.

3. (9.3) $y = -2$ is a horizontal line; the y-coordinate of every point on the line is -2.

4. (9.3) $3x + y = 3$

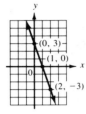

Set $x = 0$.
Then $y = 3$.
Set $y = 0$.
Then $x = 1$.
Checkpoint: Set $x = 2$.
Then $y = -3$.

5. (9.3) $2x - y = 4$

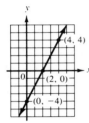

Set $x = 0$.
Then $y = -4$.
Set $y = 0$.
Then $x = 2$.
Checkpoint: Set $x = 4$.
Then $y = 4$.

6. (9.3) $4x + 9y = 18$

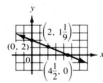

Set $x = 0$.
Then $y = 2$.
Set $y = 0$.
Then $x = 4\frac{1}{2}$.
Checkpoint: Set $x = 2$.
Then $y = 1\frac{1}{9}$.

7. (9.4) **a.** $m = \dfrac{y_2 - y_1}{x_2 - x_1} = \dfrac{1 - 2}{-3 - 5} = \dfrac{-1}{-8} = \dfrac{1}{8}$

b. $m = \dfrac{y_2 - y_1}{x_2 - x_1} = \dfrac{0 - 2}{-3 - [-3]} = \dfrac{-2}{0}$
The slope does not exist.

c. $m = \dfrac{y_2 - y_1}{x_2 - x_1} = \dfrac{-4 - [-4]}{0 - 2} = \dfrac{0}{-2} = 0$

8. (9.5) **a.** $x = -3$. (The x-coordinate of every point on the line is -3.) $x + 3 = 0$

b. $y = 4$. (The y-coordinate of every point on the line is 4.) $y - 4 = 0$.

9. (9.5) Using the slope-intercept form, we have

$$y = mx + b$$
$$y = 5x + 3$$
$$-5x + y - 3 = 0$$
$$5x - y + 3 = 0$$

10. (9.4 and 9.5) **a.** $m = \dfrac{y_2 - y_1}{x_2 - x_1} = \dfrac{-4 - 2}{1 - [-3]} = \dfrac{-6}{4} = -\dfrac{3}{2}$

b. Using the point-slope form, we have

$$y - y_1 = m(x - x_1)$$
$$y - 2 = -\frac{3}{2}(x - [-3]) \quad \text{LCD is 2}$$
$$2(y - 2) = \overset{1}{\cancel{2}}\left(-\frac{3}{\cancel{2}_{1}}\right)(x + 3)$$
$$2y - 4 = -3(x + 3)$$
$$2y - 4 = -3x - 9$$
$$3x + 2y + 5 = 0$$

11. (9.6)

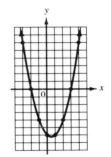

x	y	
-3	6	$y = x^2 - x - 6$
-3	6	$y = (-3)^2 - (-3) - 6 = 6$
-2	0	$y = (-2)^2 - (-2) - 6 = 0$
-1	-4	$y = (-1)^2 - (-1) - 6 = -4$
0	-6	$y = (0)^2 - (0) - 6 = -6$
1	-6	$y = (1)^2 - (1) - 6 = -6$
2	-4	$y = (2)^2 - (2) - 6 = -4$
3	0	$y = (3)^2 - (3) - 6 = 0$
4	6	$y = (4)^2 - (4) - 6 = 6$

12. (9.7) Graph $3x - 2y \le 6$. Boundary line is $3x - 2y = 6$.

x	y
0	-3
2	0
4	3

$(0, 0)$ makes the inequality true. Therefore, the half-plane includes the origin.

Cumulative Review Exercises: Chapters 1–9 (page 452)

1. $A = \dfrac{h}{2}(B + b)$

$= \dfrac{5}{2}(11 + 7)$

$= \dfrac{5}{2}(\overset{9}{\cancel{18}}) = 45$

2. $\dfrac{9b^2}{49a^6}$

3.

$$3z - 5 \overline{)\begin{array}{r} 2z + \quad 3\,R\,8 \\ 6z^2 - \quad z - \quad 7 \\ \underline{6z^2 - 10z} \\ 9z - \quad 7 \\ \underline{9z - \quad 15} \\ 8 \end{array}}$$

4. $\dfrac{2x^2 + 1}{(x - 2)(x + 1)}$

5. $\dfrac{b + 1}{a^2 + ab} \div \dfrac{b^2 - b - 2}{a^3 - ab^2}$

$\dfrac{b + 1}{a(a + b)} \cdot \dfrac{a(a + b)(a - b)}{(b + 1)(b - 2)} = \dfrac{a - b}{b - 2}$

6. $\dfrac{xy - 2}{y}$

7. LCD is $(x - 6)(x - 1)(2x + 3)$

$\dfrac{3(2x + 3)}{(x - 6)(x - 1)(2x + 3)} - \dfrac{4(x - 6)}{(2x + 3)(x - 1)(x - 6)}$

$= \dfrac{6x + 9 - (4x - 24)}{(x - 6)(x - 1)(2x + 3)}$

$= \dfrac{6x + 9 - 4x + 24}{(x - 6)(x - 1)(2x + 3)}$

$= \dfrac{2x + 33}{(x - 6)(x - 1)(2x + 3)}$

8. $\frac{8}{7}$, or $1\frac{1}{7}$

9. LCD is 18

$18\left(\dfrac{2x}{9}\right) - 18\left(\dfrac{11}{3}\right) = \dfrac{5x}{6}(18)$

$$4x - 66 = 15x$$
$$\underline{-4x \qquad\qquad -4x}$$
$$-66 = 11x$$
$$x = -6$$

10. $k = \dfrac{-2h^2}{2hC - 5}$ or $k = \dfrac{2h^2}{5 - 2hC}$

11. $y = 3x - 2$

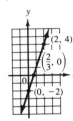

Set $x = 0$.
Then $y = -2$.
Set $y = 0$.
Then $x = \frac{2}{3}$.
Checkpoint: Set $x = 2$.
Then $y = 4$.

12.

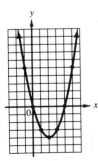

x	y
-1	5
0	0
1	-3
2	-4
3	-3
4	0
5	5

13. We first find the slope: $m = \dfrac{y_2 - y_1}{x_2 - x_1} = \dfrac{-4 - 1}{2 - 7} = \dfrac{-5}{-5} = 1$. Using the point-slope form, we have

$$y - y_1 = m(x - x_1)$$
$$y - 1 = 1(x - 7)$$
$$y - 1 = x - 7$$
$$-x + y + 6 = 0$$
$$x - y - 6 = 0$$

14. Two answers: $\left\{-\dfrac{1}{2} \text{ and } 3\dfrac{1}{2}\right\}$ or $\left\{\dfrac{1}{2} \text{ and } -3\dfrac{1}{2}\right\}$.

15. Let $\quad x = $ number of lb of Delicious apples

$50 - x = $ number of lb of Jonathan apples

$$98x + 85(50 - x) = 4458$$
$$98x + 4250 - 85x = 4458$$
$$13x = 208$$
$$x = 16 \text{ lb of Delicious apples}$$
$$50 - x = 34 \text{ lb of Jonathan apples}$$

16. Altitude is 8 cm, base is 11 cm

17. Let $x = $ the number

$\dfrac{1}{x} = $ its reciprocal

$$x + \dfrac{1}{x} = \dfrac{53}{14} \quad \text{LCD is } 14x$$
$$(14x)x + (14x)\left(\dfrac{1}{x}\right) = \left(\dfrac{53}{14}\right)(14x)$$
$$14x^2 + 14 = 53x$$
$$14x^2 - 53x + 14 = 0$$
$$(2x - 7)(7x - 2) = 0$$

$2x - 7 = 0$	$7x - 2 = 0$
$2x = 7$	$7x = 2$
$x = \dfrac{7}{2}$	$x = \dfrac{2}{7}$
$\dfrac{1}{x} = \dfrac{2}{7}$	$\dfrac{1}{x} = \dfrac{7}{2}$

Therefore, there are two answers: The number is $\frac{2}{7}$ and its reciprocal is $\frac{7}{2}$, or the number is $\frac{7}{2}$ and its reciprocal is $\frac{2}{7}$.

18. Two answers: -2 and -1, or 3 and 4.

19. Let $x = $ number of cc of water. (There will be $[10 + x]$ cc in the mixture.)

There is 0% of disinfectant in water

$$10(0.17) + x(0) = (10 + x)(0.002)$$
$$10(0.17) = (10 + x)(0.002)$$
$$10(170) = (10 + x)(2) \quad \text{Multiplying both sides by 1000}$$
$$1700 = 20 + 2x$$
$$1680 = 2x$$
$$840 = x$$
$$x = 840 \text{ cc of water}$$

20. 55 words per minute

21. Let x = cost per cheaper lens

     $x + 23$ = cost per better lens

$$\frac{\text{Total cost of}}{\text{cheap lenses}} = \frac{\text{Total cost of}}{\text{better lenses}}$$

$$\begin{aligned} 18x &= 15(x + 23) \\ 18x &= 15x + 345 \\ 3x &= 345 \\ x &= \$115 \text{ for cheaper lens} \\ x + 23 &= \$138 \text{ for better lens} \end{aligned}$$

22. $5.50, $8.25

Exercises 10.1 (page 457)

1a. $(0, 8)$ is not a solution, because in the second equation $2(0) - (8) = 0 - 8 = -8 \neq 2$.

b. $(1, 0)$ is not a solution, because in the first equation, $3(1) + (0) = 3 + 0 = 3 \neq 8$.

c. $(2, 2)$ is a solution, because $3(2) + (2) = 6 + 2 = 8$ and $2(2) - (2) = 4 - 2 = 2$.

2a. Not a solution   **b.** A solution   **c.** Not a solution

3a. $(3, 2)$ is not a solution, because in the first equation $2(3) + 3(2) = 6 + 6 = 12 \neq 6$.

b. $(0, 2)$ is a solution, because $2(0) + 3(2) = 0 + 6 = 6$ and $4(0) + 6(2) = 0 + 12 = 12$.

c. $(3, 0)$ is a solution, because $2(3) + 3(0) = 6 + 0 = 6$ and $4(3) + 6(0) = 12 + 0 = 12$.

4a. A solution   **b.** A solution   **c.** A solution

5a. $(0, 4)$ is not a solution, because in the first equation, $(0) + (4) = 0 + 4 = 4 \neq 2$.

b. $(0, 2)$ is not a solution, because in the second equation, $(0) + (2) = 0 + 2 = 2 \neq 4$.

c. $(4, 0)$ is not a solution, because in the first equation $(4) + (0) = 4 + 0 = 4 \neq 2$.

6a. Not a solution   **b.** Not a solution   **c.** Not a solution

Exercises 10.2 (page 461)

1. (1) $\begin{cases} 2x + y = 6 \\ 2x - y = -2 \end{cases}$ (2)

(1) Intercepts $(0, 6)$, $(3, 0)$

(2) Intercepts $(0, 2)$, $(-1, 0)$

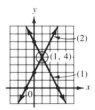

$(1, 4)$

2.

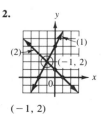

$(-1, 2)$

3. (1) $\begin{cases} x - 2y = -6 \\ 4x + 3y = 20 \end{cases}$ (2)

(1) Intercepts $(-6, 0)$, $(0, 3)$

(2) Intercepts $(5, 0)$, $\left(0, 6\frac{2}{3}\right)$

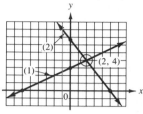

$(2, 4)$

4.

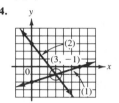

$(3, -1)$

5. (1) $\begin{cases} x + 2y = 0 \\ x - 2y = -2 \end{cases}$ (2)

(1)

x	y
0	0
2	-1

(2)

x	y
0	1
-2	0

$\left(-1, \frac{1}{2}\right)$

6.

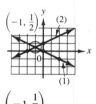

$\left(-\frac{3}{2}, 3\right)$

7. (1) $\begin{cases} 3x - 2y = -9 \\ x + y = 2 \end{cases}$ (2)

(1) Intercepts $(-3, 0)$, $\left(0, 4\frac{1}{2}\right)$

(2) Intercepts $(2, 0)$, $(0, 2)$

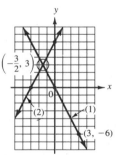

$(-1, 3)$

8.

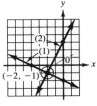

$(-2, -1)$

9. (1) $\begin{cases} 10x - 4y = 20 \\ 6y - 15x = -30 \end{cases}$ (2)

(1) Intercepts $(2, 0), (0, -5)$

(2) Intercepts $(2, 0), (0, -5)$

Since both lines have the same intercepts, they are the same line.

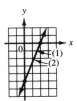

Dependent. Many solutions

10.

Dependent. Many solutions

11. (1) $\begin{cases} 2x - 4y = -2 \\ -3x + 6y = 12 \end{cases}$ (2)

(1)

x	y
-1	0
-3	-1
4	$\frac{5}{2}$

(2)

x	y
0	2
-4	0
3	$\frac{7}{2}$

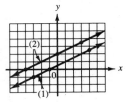

No solution (parallel lines)

12.

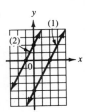

No solution (parallel lines)

Exercises 10.3 (page 466)

1. (1) $\begin{cases} 2x - y = -4 \\ x + y = -2 \end{cases}$ (2)

$$3x \quad\;\; = -6$$
$$x = -2$$

Substituting $x = -2$ into equation (2),

(2) $x + y = -2$
$$-2 + y = -2$$
$$y = 0$$

Solution: $(-2, 0)$

2. $(2, 2)$

3. $\begin{matrix} \\ -1] \end{matrix}\begin{cases} x - 2y = 10 \\ x + \;\; y = \;\; 4 \end{cases} \Rightarrow \begin{matrix} x - 2y = \;\; 10 \\ -x - \;\; y = -4 \end{matrix}$

$$-3y = \;\; 6$$
$$y = -2$$

Substituting $y = -2$ into equation (2),

(2) $x + y = 4$
$$x - 2 = 4$$
$$x = 6$$

Solution: $(6, -2)$

4. $(0, 1)$

5. (1) $\begin{cases} x - 3y = \;\; 6 \\ 4x + 3y = \;\; 9 \end{cases}$

$$5x \quad\;\; = 15$$
$$x = \;\; 3$$

Substituting $x = 3$ into equation (1),

(1) $x - 3y = 6$
$$3 - 3y = 6$$
$$-3y = 3$$
$$y = -1$$

Solution: $(3, -1)$

6. $(1, 0)$

7. $\begin{matrix} 2] \\ 1] \end{matrix}\begin{cases} 1x + 1y = \;\;\; 2 \\ 3x - 2y = -9 \end{cases} \Rightarrow \begin{matrix} 2x + 2y = \;\;\; 4 \\ 3x - 2y = -9 \end{matrix}$

$$5x \quad\quad = -5$$
$$x = -1$$

Substituting $x = -1$ into equation (1),

(1) $x + y = 2$
$$-1 + y = 2$$
$$y = 3$$

Solution: $(-1, 3)$

8. $(-2, -1)$

9. $\begin{matrix} 2] \\ 1] \end{matrix}\begin{cases} x + 2y = 0 \\ -2x + y = 0 \end{cases} \Rightarrow \begin{matrix} 2x + 4y = 0 \\ -2x + \;\; y = 0 \end{matrix}$

$$5y = 0$$
$$y = 0$$

Substituting $y = 0$ into equation (1).

$$x + 2y = 0$$
$$x + 2(0) = 0$$
$$x = 0$$

Solution: $(0, 0)$

10. (0, 0)

11.
$$\begin{array}{r} 3]\left\{4x + 3y = 2\right\} \Rightarrow 12x + 9y = 6 \\ -4]\left\{3x + 5y = -4\right\} \Rightarrow \underline{-12x - 20y = 16} \\ -11y = 22 \\ y = -2 \end{array}$$

Substituting $y = -2$ into equation (1),

(1) $\quad 4x + 3y = 2$

$4x + 3(-2) = 2$

$4x - 6 = 2$

$4x = 8$

$x = 2$

Solution: (2, −2)

12. (3, −2)

13.
$$\begin{array}{r} 9] \quad 3]\left\{6x - 10y = 6\right\} \Rightarrow 18x - 30y = 18 \\ -6]-2]\left\{9x - 15y = -4\right\} \Rightarrow \underline{-18x + 30y = 8} \\ 0 \neq 26 \end{array}$$

Inconsistent (no solution)

14. Inconsistent (no solution)

15.
$$\begin{array}{r} 2]\left\{3x - 5y = -2\right\} \Rightarrow 6x - 10y = -4 \\ 1]\left\{-6x + 10y = 4\right\} \Rightarrow \underline{-6x + 10y = 4} \\ 0 = 0 \end{array}$$

Dependent (many solutions)
Let $x = 0$ in equation (1).

(1) $\quad 3x - 5y = -2$

$3(0) - 5y = -2$

$-5y = -2$

$y = \dfrac{2}{5}$

One of the many solutions is $\left(0, \dfrac{2}{5}\right)$.

16. Dependent (many solutions)

Exercises 10.4 (page 472)

1. (1) $\left\{\begin{array}{l} 2x - 3y = 1 \\ x = y + 2 \end{array}\right\}$

Substitute $y + 2$ for x in equation (1):

(1) $\qquad 2x - 3y = 1$

$2(y + 2) - 3y = 1$

$2y + 4 - 3y = 1$

$-y = -3$

$y = 3$

Substitute $y = 3$ in
$x = y + 2$:

$x = 3 + 2 = 5.$

Solution: (5, 3)

2. (2, 7)

3. (1) $\left\{\begin{array}{l} 3x + 4y = 2 \\ y = x - 3 \end{array}\right\}$

Substitute $x - 3$ for y in equation (1):

$3x + 4(x - 3) = 2$

$3x + 4x - 12 = 2$

$7x = 14$

$x = 2$

Substitute $x = 2$ in (2):

$y = x - 3 = 2 - 3 = -1$

Solution: (2, −1)

4. (1, 3)

5. (1) $\left\{\begin{array}{l} 4x + y = 2 \\ 7x + 3y = 1 \end{array}\right\} \Rightarrow y = 2 - 4x$

Substitute $2 - 4x$ for y in equation (2):

(2) $\qquad 7x + 3y = 1$

$7x + 3(2 - 4x) = 1$

$7x + 6 - 12x = 1$

$-5x = -5$

$x = 1$

Substitute $x = 1$ in
$y = 2 - 4x$

$y = 2 - 4(1) = -2$

Solution: (1, −2)

6. (3, −2)

7. (1) $\left\{\begin{array}{l} 4x - y = 3 \\ 8x - 2y = 6 \end{array}\right\} \Rightarrow y = 4x - 3$

Substitute $4x - 3$ for y in (2):

(2) $\qquad 8x - 2y = 6$

$8x - 2(4x - 3) = 6$

$8x - 8x + 6 = 6$

$6 = 6 \quad$ True

Therefore, *any* ordered pair that satisfies (1) *or* (2) is a solution.
Dependent (many solutions)

8. *Any* ordered pair that satisfies (1) or (2) is a solution.
Dependent (many solutions)

9. (1) $\left\{\begin{array}{l} x + 3 = 0 \\ 3x - 2y = 6 \end{array}\right\} \Rightarrow x = -3$

Substitute $x = -3$ into (2):

(2) $\qquad 3x - 2y = 6$

$3(-3) - 2y = 6$

$-9 - 2y = 6$

$-2y = 15$

$y = \dfrac{15}{-2} = -7\dfrac{1}{2}$

Solution: $\left(-3, -7\dfrac{1}{2}\right)$

10. $\left(-\dfrac{3}{5}, 4\right)$

11. (1) $\begin{cases} 8x + 4y = 7 \\ 3x + 6y = 6 \end{cases} \Rightarrow 3x = 6 - 6y$

$$x = \frac{6 - 6y}{3} = \frac{3(2 - 2y)}{3} = 2 - 2y$$

Substitute $2 - 2y$ for x in (1):

(1) $\qquad 8x + 4y = 7$

$$8(2 - 2y) + 4y = 7$$

$$16 - 16y + 4y = 7$$

$$-12y = -9 \Rightarrow y = \frac{3}{4}$$

Substitute $y = \frac{3}{4}$ in $x = 2 - 2y$:

$$x = 2 - 2\left(\frac{3}{4}\right) = \frac{1}{2}$$

Solution: $\left(\frac{1}{2}, \frac{3}{4}\right)$

12. $\left(\frac{3}{5}, \frac{1}{4}\right)$

13. (1) $\begin{cases} 3x - 2y = 8 \\ 2y - 3x = 4 \end{cases} \Rightarrow 2y = 3x + 4$

$$y = \frac{3x + 4}{2}$$

Substitute $\frac{3x + 4}{2}$ for y in (1):

(1) $\qquad\qquad 3x - 2y = 8$

$$3x - 2\left(\frac{3x + 4}{2}\right) = 8$$

$$3x - 3x - 4 = 8$$

$$-4 \neq 8$$

Inconsistent (no solution)

14. Inconsistent (no solution)

15. (1) $\begin{cases} 8x + 5y = 2 \\ 7x + 4y = 1 \end{cases} \Rightarrow 4y = 1 - 7x$

$$y = \frac{1 - 7x}{4}$$

Substitute $\frac{1 - 7x}{4}$ for y in (1):

$$8x + 5\left(\frac{1 - 7x}{4}\right) = 2$$

$$\frac{4}{1} \cdot \frac{8x}{1} + \frac{\cancel{4}}{1} \cdot \frac{5}{1}\left(\frac{1 - 7x}{\cancel{4}}\right) = \frac{4}{1} \cdot \frac{2}{1}$$

$$32x + 5 - 35x = 8$$

$$-3x = 3$$

$$x = \frac{3}{-3} = -1$$

Substitute $x = -1$ in $y = \frac{1 - 7x}{4}$:

$$y = \frac{1 - 7(-1)}{4} = \frac{8}{4} = 2$$

Solution: $(-1, 2)$

16. $(4, 1)$

17. (1) $\begin{cases} 4x + 4y = 3 \\ 6x + 12y = -6 \end{cases} \Rightarrow 4x = 3 - 4y$

$$x = \frac{3 - 4y}{4}$$

Substitute $\frac{3 - 4y}{4}$ for x in (2):

(2) $\qquad\qquad 6x + 12y = -6$

$$\overset{3}{\cancel{6}}\left(\frac{3 - 4y}{\underset{2}{\cancel{4}}}\right) + 12y = -6 \quad \text{LCD} = 2$$

$$\frac{\cancel{2}}{1} \cdot 3\left(\frac{3 - 4y}{\cancel{2}}\right) + 2(12y) = 2(-6)$$

$$9 - 12y + 24y = -12$$

$$12y = -21$$

$$y = -\frac{21}{12} = -\frac{7}{4} = -1\frac{3}{4}$$

Substitute $y = -\frac{7}{4}$ in $x = \frac{3 - 4y}{4}$:

$$x = \frac{3 - 4\left(-\frac{7}{4}\right)}{4} = \frac{3 + 7}{4} = \frac{10}{4} = 2\frac{1}{2}$$

Solution: $\left(2\frac{1}{2}, -1\frac{3}{4}\right)$

18. $\left(2\frac{1}{2}, -2\frac{1}{3}\right)$

Exercises 10.5 (page 476)

(The checks will not be shown.)

1. Let x = one number

$\qquad y$ = other number

(1) $\begin{cases} x + y = 80 \\ x - y = 12 \end{cases}$
(2)

$\qquad\qquad 2x = 92 \qquad$ Adding the equations

$\qquad\qquad\ x = 46$

Substituting $x = 46$ in (1),

$\qquad\quad 46 + y = 80$

$\qquad\qquad\quad y = 34$

2. 47 and 43

3. Let x = measure of first angle

$\qquad y$ = measure of second angle

(1) $\begin{cases} x + y = 90 \\ x - y = 60 \end{cases}$
(2)

$\qquad\qquad 2x = 150 \qquad$ Adding the equations

$\qquad\qquad\ x = 75°$, measure of first angle

Substituting $x = 75°$ in (1),

$\ 75° + y = 90$

$\qquad\qquad y = 15°$, measure of second angle

4. 106° and 74°

5. Let x = smaller number

y = larger number

(1) $\begin{cases} 2x + 3y = 34 \\ 5x - 2y = 9 \end{cases}$ (2)

Using addition

(1) 2] $2x + 3y = 34 \Rightarrow 4x + 6y = 68$

(2) 3] $5x - 2y = 9 \Rightarrow \underline{15x - 6y = 27}$

$19x = 95$

$x = 5$ (smaller)

Substituting $x = 5$ in (1),

(1) $2x + 3y = 34$

$2(5) + 3y = 34$

$10 + 3y = 34$

$3y = 24$

$y = 8$ (larger)

6. 7 and 4

7. Let x = number of lb of apple chunks

y = number of lb of granola

(1) $\begin{cases} x + y = 6 \\ 4.24x + 2.10y = 16.88 \end{cases}$ (2)

(1) -210] $x + y = 6 \Rightarrow -210x - 210y = -1260$

(2) 100] $4.24x + 2.10y = 16.88 \Rightarrow \underline{424x + 210y = 1688}$

$214x = 428$

$x = 2$ lb of apple chunks

Substituting $x = 2$ in (1),

$2 + y = 6 \Rightarrow y = 4$ lb of granola

8. 43 lb Grade A and 57 lb Grade B

9. Let x = number of 18-cent stamps

y = number of 22-cent stamps

(1) $\begin{cases} x + y = 22 \\ 0.18x + 0.22y = 4.48 \end{cases}$ (2)

(1) -18] $x + y = 22 \Rightarrow -18x - 18y = -396$

(2) 100] $0.18x + 0.22y = 4.48 \Rightarrow \underline{18x + 22y = 448}$

$4y = 52$

$y = 13$ 22-cent stamps

Substituting $y = 13$ in (1),

$x + 13 = 22 \Rightarrow x = 9$ 18-cent stamps

10. 44 22-cent stamps and 6 45-cent stamps

11. Let W = width

L = length

$2W + 2L$ = perimeter

(1) $\begin{cases} 2W + 2L = 19 \\ L = W + 1.5 \end{cases}$ (2)

Substituting $W + 1.5$ for L in equation (1), we get

$2W + 2L = 19$

$2W + 2(W + 1.5) = 19$

$2W + 2W + 3 = 19$

$4W = 16$

$W = 4$

Substituting $W = 4$ in equation (2), we get

$L = W + 1.5$

$= (4) + 1.5$

$= 5.5 = 5$ ft 6 in.

12. Width = 5 ft; length = 7 ft 6 in.

13. Let n = numerator

d = denominator

Then

(1) $\dfrac{n}{d} = \dfrac{2}{3}$ and (2) $\dfrac{n + 4}{d - 2} = \dfrac{6}{7}$

$2d = 3n$ $6(d - 2) = 7(n + 4)$

$d = \dfrac{3n}{2}$ $6d - 12 = 7n + 28$

(3) $6d = 7n + 40$

Substitute $\dfrac{3n}{2}$ for d in (3) $\Rightarrow 6\left(\dfrac{3n}{2}\right) = 7n + 40$:

$9n = 7n + 40$

$2n = 40$

$n = 20$

Substitute $n = 20$ in $d = \dfrac{3n}{2}$:

$= \dfrac{3(20)}{2}$

$= 30$

Therefore, the original fraction is $\frac{20}{30}$.

14. $\frac{36}{48}$

15. Let x = speed of boat in still water

y = speed of current

Note: We multiply both sides of (1) by $\frac{1}{7}$ and both sides of (2) by $\frac{1}{9}$.

(1) $\begin{cases} 7(x + y) = 252 \\ 9(x - y) = 252 \end{cases}$ (2) $\Rightarrow \begin{matrix} x + y = 36 \\ \underline{x - y = 28} \end{matrix}$ Add

$2x = 64$

$x = 32$, speed of boat in still water

$32 + y = 36 \Rightarrow y = 4$, speed of current

16. The speed of the plane in still air is 120 mph. The speed of the wind is 30 mph.

17. Let t = cost of tie in cents

p = cost of pin in cents

(1) $\begin{cases} t + p = 110 \\ t = 100 + p \end{cases}$ (2)

Substitute $100 + p$ for t in (1):

(1) $t + p = 110$

$100 + p + p = 110$

$2p = 10$

$p = 5$

$t = 100 + p = 100 + 5 = 105$

Pin costs 5¢; tie costs $1.05

18. 7 on upper branch; 5 on lower branch

Review Exercises 10.6 (page 479)

1. (1) $\begin{cases} x + y = 6 \\ x - y = 4 \end{cases}$
  (2)

(1) Intercepts (6, 0), (0, 6); checkpoint (3, 3)
(2) Intercepts (4, 0), (0, −4); checkpoint (3, −1)

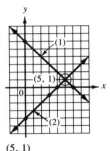

(5, 1)

2.

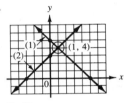

(1, 4)

3.

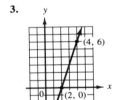

$3x - y = 6$	
x	y
0	−6
2	0
4	6

$6x - 2y = 12$	
x	y
0	−6
2	0
4	6

The system is dependent; there are many solutions.

4. (3, 2)

5.
$$\begin{array}{r} 3\ ][4x - 8y = 4 \Rightarrow\ \ 12x - 24y =\ \ \ 12 \\ -4\ ][3x - 6y = 3 \Rightarrow -12x + 24y = -12 \\ \hline 0 =\ \ \ 0 \end{array}$$

Dependent (many solutions)

6. Inconsistent (no solution)

7. (1) $\begin{cases} x = y + 2 \\ 4x - 5y = 3 \end{cases}$
  (2)

Substitute $y + 2$ for x in (2):

(2) $4x - 5y = 3$

$4(y + 2) - 5y = 3$

$4y + 8 - 5y = 3$

$-y = -5 \Rightarrow y = 5$

Substitute $y = 5$ in $x = y + 2$

$x = 5 + 2 = 7$

Solution: (7, 5)

8. (4, 9)

9. $\begin{cases} x + y = 4 \\ 2x - y = 2 \end{cases} \Rightarrow y = 4 - x$

Substitute $4 - x$ for y in (2):

$2x - (4 - x) = 2$

$2x - 4 + x = 2$

$3x = 6$

$x = 2$

Substitute 2 for x in $y = 4 - x$:

$y = 4 - 2 = 2$

Solution: (2, 2)

10. (−3, −2)

11. Let x = larger number

      y = smaller number

(1) $\begin{cases} x + y =\ \ 84 \\ x - y =\ \ 22 \end{cases}$ Add
(2)

$2x\ \ \ \ \ = 106$

$x =\ \ 53$ (larger)

Substitute $x = 53$ in (1):

(1) $x + y = 84$

$53 + y = 84$

$y = 31$ (smaller)

12. $\frac{24}{30}$

13. Let x = number of boxes of paper for laser printer

      y = number of boxes of paper for copy machine

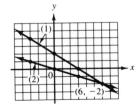

(1) $-425]$ $x +$ $y = 20$ $\Rightarrow -425x - 425y = -8500$
(2) $100]\ 7.50x + 4.25y = 107.75 \Rightarrow \underline{\ \ 750x + 425y =\ \ \ 10775}$

$325x\ \ \ \ \ \ \ \ =\ \ \ 2275$

$x = 7$ boxes of paper for laser printer

Substitute 7 for x in (1):

$7 + y = 20 \Rightarrow y = 13$ boxes of paper for copy machine

14. The speed of the boat in still water is 27 mph. The speed of the current is 6 mph.

Chapter 10 Diagnostic Test (page 483)

Following each problem is the textbook section number (in parentheses) where that kind of problem is discussed.

1. (10.2) (1) $\begin{cases} 2x + 3y = 6 \\ x + 3y = 0 \end{cases}$
      (2)

Solution: (6, −2)

$2x + 3y = 6$	
x	y
0	2
3	0
−3	4

$x + 3y = 0$	
x	y
0	0
−3	1
3	−1

2. (10.3) (1) $\left\{\begin{array}{l} x + 3y = 5 \\ -x + 4y = 2 \end{array}\right\}$ Add

$$7y = 7$$
$$y = 1$$

Substitute $y = 1$ in (1): $x + 3(1) = 5 \Rightarrow x = 2$

Solution: $(2, 1)$

3. (10.3) $\begin{array}{l} 2] \\ 3] \end{array} \left\{\begin{array}{l} 4x + 3y = 5 \\ 5x - 2y = 12 \end{array}\right\} \Rightarrow \begin{array}{l} 8x + 6y = 10 \\ \underline{15x - 6y = 36} \end{array}$

$$23x = 46$$
$$x = 2$$

Substitute $x = 2$ in (1): $\quad 4(2) + 3y = 5 \Rightarrow y = -1$

Solution: $(2, -1)$

4. (10.4) $\left\{\begin{array}{l} 2x - 5y = -20 \\ x = y - 7 \end{array}\right\}$

$$2(y - 7) - 5y = -20$$
$$2y - 14 - 5y = -20$$
$$-3y - 14 = -20$$
$$-3y = -6$$
$$y = 2$$

Substitute $y = 2$ in (2): $x = (2) - 7 = -5$
Solution: $(-5, 2)$

5. (10.3) $\begin{array}{l} 5] \\ 4] \end{array} \left\{\begin{array}{l} 3x + 4y = -1 \\ 2x - 5y = -16 \end{array}\right\} \Rightarrow \begin{array}{l} 15x + 20y = -5 \\ \underline{8x - 20y = -64} \end{array}$

$$23x = -69$$
$$x = -3$$

Substitute $x = -3$ in (1): $3(-3) + 4y = -1 \Rightarrow y = 2$

Solution: $(-3, 2)$

6. (10.3) $\begin{array}{l} -2] \\ 1] \end{array} \left\{\begin{array}{l} 5x + y = 10 \\ 10x + 2y = 9 \end{array}\right\} \Rightarrow \begin{array}{l} -10x - 2y = -20 \\ \underline{10x + 2y = 9} \end{array}$

$$0 = -11 \quad \text{False}$$

Inconsistent (no solution)

7. (10.3) $\begin{array}{l} 10] \, 5] \\ 6] \, 3] \end{array} \left\{\begin{array}{l} 6x - 9y = 3 \\ -10x + 15y = -5 \end{array}\right\} \Rightarrow \begin{array}{l} 30x - 45y = 15 \\ \underline{-30x + 45y = -15} \end{array}$

$$0 = 0 \quad \text{True}$$

Dependent (many solutions)

8. (10.5) Let $x =$ one number

$y =$ other number

(1) $\left.\begin{array}{l} x + y = 13 \\ (2) \; x - y = 37 \end{array}\right\}$ Add

$$2x = 50$$
$$x = 25$$

Substitute 25 for x in (1):

$$25 + y = 13 \Rightarrow y = -12$$

Check: $25 + (-12) = 13$

$$25 - (-12) = 25 + 12 = 37$$

The numbers are 25 and -12.

9. (10.5) Let $x =$ speed of plane in still air

$y =$ speed of wind

Then $x + y =$ speed with wind

and $x - y =$ speed against wind

Note: We will have small numbers to work with if we multiply both sides of (1) by $\frac{1}{11}$ and both sides of (2) by $\frac{1}{7.5}$.

(1) $\left\{\begin{array}{l} 11(x - y) = 1650 \\ (2) \; 7.5(x + y) = 1650 \end{array}\right\}$

(1) $\frac{1}{11}$] $11(x - y) = 1650 \Rightarrow x - y = 150$

(2) $\frac{1}{7.5}$] $7.5(x + y) = 1650 \Rightarrow \underline{x + y = 220}$

$$2x = 370$$
$$x = 185 \text{ mph, speed of plane in}$$
$$\text{still air}$$

Substitute 185 for x in (2): $185 + y = 220 \Rightarrow y = 35$ mph, speed of wind

Check: The speed of the plane against the wind is 185 mph

$- 35$ mph $= 150$ mph, and $150 \dfrac{\text{mi}}{\text{hr}}(11 \text{ hr}) = 1650$ mi. The speed

of the plane with the wind is 185 mph + 35 mph = 220 mph, and

$220 \dfrac{\text{mi}}{\text{hr}}(7.5 \text{ hrs}) = 1650$ mi.

10. (10.5) Let $x =$ number of cm in length

$y =$ number of cm in width

(1) $\left\{\begin{array}{l} x = y + 3 \\ (2) \; 2x + 2y = 98 \end{array}\right\}$

Substitute $y + 3$ for x in (2): $2(y + 3) + 2y = 98$

$$2y + 6 + 2y = 98$$
$$4y + 6 = 98$$
$$4y = 92$$
$$y = 23 \text{ cm}$$

Substitute 23 for y in (1): $x = 23$ cm $+ 3$ cm $= 26$ cm

Check: The length of the rectangle is 3 cm more than its width. Its perimeter is $2(23 \text{ cm}) + 2(26 \text{ cm}) = 46 \text{ cm} + 52 \text{ cm} = 98 \text{ cm}$.

Cumulative Review Exercises: Chapters 1–10 (page 484)

1. $\left(\dfrac{\overset{3}{\cancel{24}}w^{-2}z^4}{\underset{2}{\cancel{16}}w}\right)^{-3} = \left(\dfrac{3w^{-3}z^4}{2}\right)^{-3}$

$$= \dfrac{3^{-3}w^9z^{-12}}{2^{-3}} = \dfrac{2^3 w^9}{3^3 z^{12}} = \dfrac{8w^9}{27z^{12}}$$

2. 1322.50

3. $\dfrac{2 + 5x}{6x - 1} = \dfrac{3}{4}$ A proportion

$$4(2 + 5x) = 3(6x - 1)$$
$$8 + 20x = 18x - 3$$
$$2x = -11$$
$$x = -\dfrac{11}{2} \text{ or } -5\dfrac{1}{2}$$

4. -5

5. $\dfrac{1+a}{2-3a} + \dfrac{2+a}{4a}$ LCD $= 4a(2-3a)$

$$\dfrac{1+a}{2-3a} \cdot \dfrac{4a}{4a} + \dfrac{2+a}{4a} \cdot \dfrac{2-3a}{2-3a}$$

$$\dfrac{4a+4a^2}{4a(2-3a)} + \dfrac{4-4a-3a^2}{4a(2-3a)}$$

$$\dfrac{4a+4a^2+4-4a-3a^2}{4a(2-3a)}$$

$$\dfrac{a^2+4}{4a(2-3a)}$$

6.

7. $4y = x^2 \Rightarrow y = \frac{1}{4}x^2$

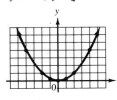

x	y
-4	4
-2	1
-1	$\frac{1}{4}$
0	0
1	$\frac{1}{4}$
2	1
4	4

8. $x \le 5$

9. $3x - 4y > 12$

Boundary line: $3x - 4y = 12$

Let $x = 0$: $3(0) - 4y = 12$

$y = -3$

$3x - 4y = 12$

Let $y = 0$: $3x - 4(0) = 12$

$x = 4$

The correct half-plane does not include the origin because:

$3(0) - 4(0) \not> 12$

$0 \not> 12$

x	y
0	-3
4	0

10. $3x + 7y - 5 = 0$

11. (1) $\begin{cases} 2x - 3y = 3 \\ -2x + 3y = 5 \end{cases}$

(2)

$0 = 8$ False

Inconsistent   (no solution)

12. Dependent   (many solutions)

13. $\begin{matrix} 3] \\ -1] \end{matrix} \begin{cases} 1x + 5y = 11 \\ 3x + 4y = 11 \end{cases} \Rightarrow \begin{matrix} 3x + 15y = 33 \\ -3x - 4y = -11 \end{matrix}$

$11y = 22$

$y = 2$

Substituting $y = 2$ in (1):

(1) $x + 5y = 11$

$x + 5(2) = 11$

$x + 10 = 11$

$x = 1$

Solution: (1, 2)

14. $\left(-\dfrac{7}{2}, \dfrac{9}{2}\right)$ or $\left(-3\dfrac{1}{2}, 4\dfrac{1}{2}\right)$

15. Not factorable   **16.** Not factorable

17. Let $x =$ cost of each cheaper camera

$x + 48 =$ cost of each more expensive camera

⌐The two costs are equal

$18x = 10(x + 48)$

$18x = 10x + 480$

$-10x \qquad -10x$

$8x = 480$

$x = 60$

$x + 48 = 108$

Check: $18(\$60) = \$1080$ and $10(\$108) = \$1080$

Therefore, the cheaper cameras cost $60 each, and the more expensive cameras would have cost $108 each.

18. $15,500 at 14% and $11,800 at 21%

19. Let $x =$ one number

$y =$ other number

(1) $\begin{cases} x + y = 6 \\ x - y = 40 \end{cases}$ Add

(2)

$2x = 46$

$x = 23$, one number

Substitute 23 for x in (1): $23 + y = 6$

$y = -17$, other number

Check: $23 + (-17) = 6$ and $23 - (-17) = 23 + 17 = 40$

20. The width is 12 m; the length is 17 m

Exercises 11.1 (page 489)

1. 17   **2.** 23   **3.** $4xy^2$ **4.** $9a^2b$ **5.** $x + 1$

6. $m - n$ **7.** 9   **8.** 100   **9.** 15   **10.** 29   **11.** $3a^2$

12. $7y^2$ **13.** $x + y$ **14.** $u + v$

Exercises 11.2A (page 492)

1. $\sqrt{25x^2} = \sqrt{25}\sqrt{x^2} = 5x$ **2.** $10y$

3. $\sqrt{81s^4} = \sqrt{81}\sqrt{s^4} = 9s^2$ **4.** $8t^3$ **5.** $\sqrt{4x^2} = \sqrt{4}\sqrt{x^2} = 2x$

6. $3y$ **7.** $\sqrt{16z^4} = \sqrt{16}\sqrt{z^4} = 4z^2$ **8.** $5b^3$

9. $\sqrt{12} = \sqrt{4 \cdot 3} = \sqrt{4}\sqrt{3} = 2\sqrt{3}$ **10.** $2\sqrt{5}$

11. $\sqrt{18} = \sqrt{9 \cdot 2} = \sqrt{9}\sqrt{2} = 3\sqrt{2}$ **12.** $3\sqrt{5}$

13. $\sqrt{8} = \sqrt{4 \cdot 2} = \sqrt{4}\sqrt{2} = 2\sqrt{2}$ **14.** $4\sqrt{2}$

15. $\sqrt{x^3} = \sqrt{x^2 \cdot x} = \sqrt{x^2}\sqrt{x} = x\sqrt{x}$ **16.** $y^2\sqrt{y}$

17. $\sqrt{m^7} = \sqrt{m^6 \cdot m} = m^3\sqrt{m}$ **18.** $n^4\sqrt{n}$

19. $\sqrt{24} = \sqrt{4 \cdot 6} = \sqrt{4}\sqrt{6} = 2\sqrt{6}$ **20.** $3\sqrt{6}$

21. $\sqrt{32x^2y^3} = \sqrt{2 \cdot 16x^2y^2y} = \sqrt{2}\sqrt{16}\sqrt{x^2}\sqrt{y^2}\sqrt{y} = 4xy\sqrt{2y}$

22. $7b\sqrt{2ab}$

23. $\sqrt{8s^4t^5} = \sqrt{2 \cdot 4s^4t^4t} = \sqrt{2}\sqrt{4}\sqrt{s^4}\sqrt{t^4}\sqrt{t} = 2s^2t^2\sqrt{2t}$

24. $2xy^3\sqrt{11x}$

25. $\sqrt{18a^2b^3} = \sqrt{2 \cdot 9a^2b^2b}$

$\qquad = \sqrt{2}\sqrt{9}\sqrt{a^2}\sqrt{b^2}\sqrt{b} = 3ab\sqrt{2b}$

26. $3c^2d\sqrt{3c}$

27. $\sqrt{40x^6y^2} = \sqrt{4 \cdot 10x^6y^2}$

$\qquad = \sqrt{4}\sqrt{10}\sqrt{x^6}\sqrt{y^2} = 2x^3y\sqrt{10}$

28. $3a^3b^4\sqrt{15}$

29. $\sqrt{60h^5k^4} = \sqrt{4 \cdot 15h^4hk^4} = \sqrt{4}\sqrt{15}\sqrt{h^4}\sqrt{h}\sqrt{k^4} = 2h^2k^2\sqrt{15h}$

30. $3m^3n^3\sqrt{10m}$

31. First express 280 in prime factored form:

$$\begin{array}{r|l} 2 & 280 \\ 2 & 140 \\ 2 & 70 \\ 5 & 35 \\ & 7 \end{array}$$

$\qquad 280 = 2^2 \cdot 2 \cdot 5 \cdot 7$

$\sqrt{280y^5z^6} = \sqrt{2^2 \cdot 70 \cdot y^4 \cdot y \cdot z^6} = 2y^2z^3\sqrt{70y}$

32. $3g^4h^4\sqrt{30h}$

33. $\sqrt{500x^7y^9} = \sqrt{100 \cdot 5 \cdot x^6 \cdot x \cdot y^8 \cdot y} = 10x^3y^4\sqrt{5xy}$

34. $6x^5y^6\sqrt{6xy}$

Exercises 11.2B (page 494)

1. $\sqrt{\dfrac{9}{25}} = \dfrac{\sqrt{9}}{\sqrt{25}} = \dfrac{3}{5}$ **2.** $\dfrac{6}{7}$ **3.** $\sqrt{\dfrac{16}{25}} = \dfrac{\sqrt{16}}{\sqrt{25}} = \dfrac{4}{5}$ **4.** $\dfrac{9}{10}$

5. $\sqrt{\dfrac{y^4}{x^2}} = \dfrac{\sqrt{y^4}}{\sqrt{x^2}} = \dfrac{y^2}{x}$ **6.** $\dfrac{x^3}{v^4}$ **7.** $\sqrt{\dfrac{4x^2}{9}} = \dfrac{\sqrt{4x^2}}{\sqrt{9}} = \dfrac{2x}{3}$

8. $\dfrac{4a}{7}$ **9.** $\sqrt{\dfrac{2}{8k^2}} = \sqrt{\dfrac{1}{4k^2}} = \dfrac{\sqrt{1}}{\sqrt{4k^2}} = \dfrac{1}{2k}$ **10.** $\dfrac{m}{3}$

11. $\sqrt{\dfrac{4x^3y}{xy^3}} = \sqrt{\dfrac{4x^2}{y^2}} = \dfrac{\sqrt{4x^2}}{\sqrt{y^2}} = \dfrac{2x}{y}$ **12.** $\dfrac{x^2}{3y}$

13. $\sqrt{\dfrac{1}{5}} = \sqrt{\dfrac{1 \cdot 5}{5 \cdot 5}} = \dfrac{\sqrt{5}}{\sqrt{5^2}} = \dfrac{\sqrt{5}}{5}$ **14.** $\dfrac{\sqrt{3}}{3}$

15. $\sqrt{\dfrac{7}{13}} = \sqrt{\dfrac{7 \cdot 13}{13 \cdot 13}} = \dfrac{\sqrt{91}}{\sqrt{13^2}} = \dfrac{\sqrt{91}}{13}$ **16.** $\dfrac{\sqrt{85}}{17}$

17. $\sqrt{\dfrac{1}{18}} = \sqrt{\dfrac{1 \cdot 2}{18 \cdot 2}} = \dfrac{\sqrt{2}}{\sqrt{36}} = \dfrac{\sqrt{2}}{6}$ **18.** $\dfrac{\sqrt{2}}{10}$

19. $\sqrt{\dfrac{7}{27}} = \sqrt{\dfrac{7 \cdot 3}{27 \cdot 3}} = \dfrac{\sqrt{21}}{\sqrt{81}} = \dfrac{\sqrt{21}}{9}$ **20.** $\dfrac{\sqrt{10}}{4}$

21. $\dfrac{3}{\sqrt{7}}\dfrac{\sqrt{7}}{\sqrt{7}} = \dfrac{3\sqrt{7}}{7}$ **22.** $\dfrac{2\sqrt{3}}{3}$ **23.** $\dfrac{10}{\sqrt{5}}\dfrac{\sqrt{5}}{\sqrt{5}} = \dfrac{\overset{2}{\cancel{10}}\sqrt{5}}{\underset{1}{\cancel{5}}} = 2\sqrt{5}$

24. $7\sqrt{2}$ **25.** $\sqrt{\dfrac{m^2}{3}} = \sqrt{\dfrac{m^2(3)}{3(3)}} = \dfrac{m\sqrt{3}}{3}$ **26.** $\dfrac{k\sqrt{5}}{5}$

27. $\sqrt{\dfrac{x^5z^4}{36y^2}} = \dfrac{\sqrt{x^4xz^4}}{\sqrt{36y^2}} = \dfrac{x^2z^2\sqrt{x}}{6y}$ **28.** $\dfrac{ac^3\sqrt{a}}{5b}$

29. $\sqrt{\dfrac{3a^2b}{4b^3}} = \sqrt{\dfrac{3a^2}{4b^2}} = \dfrac{a\sqrt{3}}{2b}$ **30.** $\dfrac{v\sqrt{5}}{2}$

31. $\sqrt{\dfrac{b^2c^4}{16d^3}} = \sqrt{\dfrac{b^2c^4d}{16d^3d}} = \dfrac{bc^2\sqrt{d}}{4d^2}$ **32.** $\dfrac{h^2k^4\sqrt{p}}{7p^3}$

33. $\sqrt{\dfrac{8m^2n}{2n^2}} = \sqrt{\dfrac{4m^2n}{n^2}} = \dfrac{2m\sqrt{n}}{n}$ **34.** $\dfrac{3y\sqrt{x}}{x}$

Exercises 11.3A (page 497)

1. $\sqrt{3}\sqrt{3} = 3$ **2.** 7 **3.** $\sqrt{4}\sqrt{4} = 4$ **4.** 9

5. $\sqrt{3}\sqrt{12} = \sqrt{36} = 6$ **6.** 8 **7.** $\sqrt{9x}\sqrt{x} = \sqrt{9x^2} = 3x$

8. $5y$ **9.** $\sqrt{5}\sqrt{10}\sqrt{2} = \sqrt{5 \cdot 10 \cdot 2} = \sqrt{100} = 10$ **10.** 12

11. $\sqrt{5ab^2}\sqrt{20ab} = \sqrt{100a^2b^2b} = 10ab\sqrt{b}$ **12.** $9xy\sqrt{x}$

13. $\sqrt{2a}\sqrt{6}\sqrt{3a} = \sqrt{2a \cdot 6 \cdot 3a} = \sqrt{36a^2} = 6a$ **14.** $4h^2$

15. $5\sqrt{2x} \cdot \sqrt{8x^3} \cdot 2\sqrt{3x^5} = (5 \cdot 2)\sqrt{2x \cdot 8x^3 \cdot 3x^5}$

$\qquad = 10\sqrt{48x^9}$

$\qquad = 10\sqrt{3 \cdot 16 \cdot x^8 \cdot x}$

$\qquad = 10 \cdot 4 \cdot x^4\sqrt{3x} = 40x^4\sqrt{3x}$

16. $72M^3\sqrt{2M}$

Exercises 11.3B (page 499)

1. $\dfrac{\sqrt{20}}{\sqrt{5}} = \sqrt{\dfrac{20}{5}} = \sqrt{\dfrac{4}{1}} = 2$ **2.** $\dfrac{1}{2}$

3. $\dfrac{\sqrt{32}}{\sqrt{2}} = \sqrt{\dfrac{32}{2}} = \sqrt{\dfrac{16}{1}} = 4$ **4.** 7 **5.** $\dfrac{\sqrt{4}}{\sqrt{5}} = \dfrac{2}{\sqrt{5}}\dfrac{\sqrt{5}}{\sqrt{5}} = \dfrac{2\sqrt{5}}{5}$

6. $\dfrac{3\sqrt{7}}{7}$ **7.** $\dfrac{\sqrt{15x}}{\sqrt{5x}} = \sqrt{\dfrac{15x}{5x}} = \sqrt{3}$ **8.** $\sqrt{6}$

9. $\dfrac{\sqrt{72x^3y^2}}{\sqrt{2xy^2}} = \sqrt{\dfrac{72x^3y^2}{2xy^2}} = \sqrt{36x^2} = 6x$ **10.** $3y$

11. $\dfrac{\sqrt{x^4y}}{\sqrt{5y}} = \sqrt{\dfrac{x^4y}{5y}} = \dfrac{\sqrt{x^4}}{\sqrt{5}}\dfrac{\sqrt{5}}{\sqrt{5}} = \dfrac{x^2\sqrt{5}}{5}$ **12.** $\dfrac{m^3\sqrt{3}}{3}$

13. $\dfrac{4\sqrt{45m^3}}{3\sqrt{10m}} = \dfrac{4}{3}\sqrt{\dfrac{\overset{9}{\cancel{45}m^3}}{\underset{2}{\cancel{10}m}}} = \dfrac{4}{3}\dfrac{\sqrt{9m^2}}{\sqrt{2}} = \dfrac{4 \cdot \overset{}{\cancel{3}}m}{\cancel{3}\sqrt{2}} \cdot \dfrac{\sqrt{2}}{\sqrt{2}} = \dfrac{\overset{2}{\cancel{4}}m\sqrt{2}}{\underset{1}{\cancel{2}}}$

$\qquad = 2m\sqrt{2}$

14. $4x\sqrt{6x}$

Exercises 11.4 (page 502)

1. $2\sqrt{3} + 5\sqrt{3} = (2 + 5)\sqrt{3} = 7\sqrt{3}$ **2.** $7\sqrt{2}$

3. $3\sqrt{x} - \sqrt{x} = (3 - 1)\sqrt{x} = 2\sqrt{x}$ **4.** $4\sqrt{a}$

5. $\dfrac{3}{2}\sqrt{2} - \dfrac{\sqrt{2}}{2} = \left(\dfrac{3}{2} - \dfrac{1}{2}\right)\sqrt{2} = \dfrac{2}{2}\sqrt{2} = \sqrt{2}$ **6.** $\sqrt{3}$

7. $5 \cdot 8\sqrt{5} + \sqrt{5} = 40\sqrt{5} + \sqrt{5} = (40 + 1)\sqrt{5} = 41\sqrt{5}$

8. $13\sqrt{7}$

9. $\sqrt{25} + \sqrt{5} = 5 + \sqrt{5}$ **10.** $4 + \sqrt{6}$

11. $\sqrt{50} + \sqrt{50} = 5\sqrt{2} + 5\sqrt{2} = (5 + 5)\sqrt{2} = 10\sqrt{2}$ **12.** $8\sqrt{2}$

13. $\sqrt{68} - \sqrt{17} = 2\sqrt{17} - \sqrt{17} = (2 - 1)\sqrt{17} = \sqrt{17}$ **14.** $\sqrt{13}$

15. $\sqrt{45} - \sqrt{28} = 3\sqrt{5} - 2\sqrt{7}$ **16.** $3\sqrt{10} - 3\sqrt{6}$

17. $2\sqrt{3} + \sqrt{12} = 2\sqrt{3} + \sqrt{4 \cdot 3} = 2\sqrt{3} + 2\sqrt{3} = 4\sqrt{3}$

18. $5\sqrt{2}$

19. $2\sqrt{50} - \sqrt{32} = 2\sqrt{25 \cdot 2} - \sqrt{16 \cdot 2}$
$$= 2 \cdot 5\sqrt{2} - 4\sqrt{2}$$
$$= 10\sqrt{2} - 4\sqrt{2} = 6\sqrt{2}$$

20. $3\sqrt{6}$

21. $3\sqrt{32} - \sqrt{8} = 3\sqrt{16 \cdot 2} - \sqrt{4 \cdot 2}$
$$= 3 \cdot 4\sqrt{2} - 2\sqrt{2}$$
$$= 12\sqrt{2} - 2\sqrt{2} = 10\sqrt{2}$$

22. $6\sqrt{3}$

23. $\sqrt{\dfrac{1}{2}} + \sqrt{8} = \sqrt{\dfrac{1 \cdot 2}{2 \cdot 2}} + \sqrt{4 \cdot 2}$
$$= \frac{1}{2}\sqrt{2} + 2\sqrt{2}$$
$$= \left(\frac{1}{2} + 2\right)\sqrt{2} = \frac{5\sqrt{2}}{2}$$

24. $\dfrac{7\sqrt{3}}{3}$

25. $\sqrt{24} - \sqrt{\dfrac{2}{3}} = \sqrt{4 \cdot 6} - \sqrt{\dfrac{2 \cdot 3}{3 \cdot 3}}$
$$= 2\sqrt{6} - \frac{\sqrt{6}}{3}$$
$$= \left(2 - \frac{1}{3}\right)\sqrt{6} = \frac{5\sqrt{6}}{3}$$

26. $\dfrac{13\sqrt{5}}{5}$

27. $10\sqrt{\dfrac{3}{5}} + \sqrt{60} = 10\sqrt{\dfrac{3 \cdot 5}{5 \cdot 5}} + \sqrt{4 \cdot 15}$
$$= \frac{10\sqrt{15}}{5} + 2\sqrt{15}$$
$$= 2\sqrt{15} + 2\sqrt{15} = 4\sqrt{15}$$

28. $6\sqrt{3}$

29. $\sqrt{\dfrac{25}{2}} - \dfrac{3}{\sqrt{2}} = \dfrac{\sqrt{25}}{\sqrt{2}} - \dfrac{3}{\sqrt{2}}$
$$= \frac{5 - 3}{\sqrt{2}} = \frac{2}{\sqrt{2}} \cdot \frac{\sqrt{2}}{\sqrt{2}}$$
$$= \frac{2\sqrt{2}}{2} = \sqrt{2}$$

30. $\dfrac{7\sqrt{5}}{10}$

31. $3\sqrt{\dfrac{1}{6}} + \sqrt{12} - 5\sqrt{\dfrac{3}{2}} = 3\sqrt{\dfrac{1 \cdot 6}{6 \cdot 6}} + \sqrt{4 \cdot 3} - 5\sqrt{\dfrac{3 \cdot 2}{2 \cdot 2}}$
$$= \frac{\overset{1}{\cancel{3}}}{1} \cdot \frac{\sqrt{6}}{\underset{2}{\cancel{6}}} + \sqrt{4}\sqrt{3} - \frac{5}{1} \cdot \frac{\sqrt{6}}{2}$$
$$= \frac{1}{2}\sqrt{6} + 2\sqrt{3} - \frac{5}{2}\sqrt{6}$$
$$= \left(\frac{1}{2} - \frac{5}{2}\right)\sqrt{6} + 2\sqrt{3}$$
$$= -2\sqrt{6} + 2\sqrt{3}$$

32. $\sqrt{10} + 2\sqrt{5}$

33. $\dfrac{1}{3}\sqrt{7} + 2\sqrt{3} \doteq \dfrac{1}{3}(2.646) + 2(1.732)$
$$= 0.882 + 3.464 = 4.346$$

34. 7.906

Exercises 11.5 (page 505)

1. $\sqrt{2}(\sqrt{2} + 1) = \sqrt{2}(\sqrt{2}) + \sqrt{2}(1) = 2 + \sqrt{2}$ **2.** $3 + \sqrt{3}$

3. $\sqrt{3}(2\sqrt{3} + 1) = \sqrt{3}(2\sqrt{3}) + \sqrt{3}(1) = 6 + \sqrt{3}$ **4.** $15 + \sqrt{5}$

5. $\sqrt{x}(\sqrt{x} - 3) = \sqrt{x}(\sqrt{x}) + \sqrt{x}(-3) = x - 3\sqrt{x}$ **6.** $4\sqrt{y} - y$

7. $(\sqrt{7} + 2)(\sqrt{7} + 3)$
$$= \sqrt{7}\sqrt{7} + 5\sqrt{7} + 2(3)$$
$$= 7 + 5\sqrt{7} + 6$$
$$= 13 + 5\sqrt{7}$$

8. $11 + 6\sqrt{3}$

9. $(\sqrt{8} - 3\sqrt{2})(\sqrt{8} + 2\sqrt{5}) = \sqrt{8}\sqrt{8} - 3\sqrt{2}\sqrt{8} + 2\sqrt{8}\sqrt{5} - 3\sqrt{2} \cdot 2\sqrt{5}$
$$= 8 - 3\sqrt{16} + 2\sqrt{40} - 6\sqrt{10}$$
$$= 8 - 3(4) + 2\sqrt{4 \cdot 10} - 6\sqrt{10}$$
$$= 8 - 12 + 2 \cdot 2\sqrt{10} - 6\sqrt{10}$$
$$= -4 + 4\sqrt{10} - 6\sqrt{10}$$
$$= -4 - 2\sqrt{10}$$

10. 0

11. $(\sqrt{2} + \sqrt{14})^2 = (\sqrt{2})^2 + 2\sqrt{2}\sqrt{14} + (\sqrt{14})^2$
$$= 2 + 2\sqrt{2^2 \cdot 7} + 14 = 16 + 2 \cdot 2\sqrt{7}$$
$$= 16 + 4\sqrt{7}$$

12. $16 + 2\sqrt{55}$

13. $(\sqrt{2x} + 3)^2 = (\sqrt{2x})^2 + 2\sqrt{2x}(3) + (3)^2$
$$= 2x + 6\sqrt{2x} + 9$$

14. $7x + 8\sqrt{7x} + 16$

15. $\dfrac{\sqrt{8} + \sqrt{18}}{\sqrt{2}} = \dfrac{\sqrt{8}}{\sqrt{2}} + \dfrac{\sqrt{18}}{\sqrt{2}} = \sqrt{\dfrac{8}{2}} + \sqrt{\dfrac{18}{2}}$
$$= \sqrt{4} + \sqrt{9} = 2 + 3 = 5$$

16. 5

17. $\dfrac{\sqrt{20} + 5\sqrt{10}}{\sqrt{5}} = \dfrac{\sqrt{20}}{\sqrt{5}} + \dfrac{5\sqrt{10}}{\sqrt{5}} = \sqrt{\dfrac{20}{5}} + 5\sqrt{\dfrac{10}{5}}$
$$= \sqrt{4} + 5\sqrt{2} = 2 + 5\sqrt{2}$$

18. $2 + \dfrac{\sqrt{21}}{3}$ **19a.** $2 - \sqrt{3}$ **b.** $2\sqrt{5} + 7$

20a. $3\sqrt{2} + 5$ **b.** $\sqrt{7} - 4$

21. $\dfrac{3}{\sqrt{2} - 1} = \dfrac{3}{(\sqrt{2} - 1)} \dfrac{(\sqrt{2} + 1)}{(\sqrt{2} + 1)}$
$$= \frac{3(\sqrt{2} + 1)}{(\sqrt{2})^2 - 1^2} = \frac{3(\sqrt{2} + 1)}{2 - 1} = 3(\sqrt{2} + 1)$$

22. $5(\sqrt{2} + 1)$

23. $\dfrac{6}{(\sqrt{3} - \sqrt{2})} \dfrac{(\sqrt{3} + \sqrt{2})}{(\sqrt{3} + \sqrt{2})} = \dfrac{6(\sqrt{3} + \sqrt{2})}{(\sqrt{3})^2 - (\sqrt{2})^2}$
$$= \frac{6(\sqrt{3} + \sqrt{2})}{3 - 2} = 6(\sqrt{3} + \sqrt{2})$$

24. $-8(\sqrt{2} + \sqrt{3})$

25. $\dfrac{6}{(\sqrt{5}+\sqrt{2})}\dfrac{(\sqrt{5}-\sqrt{2})}{(\sqrt{5}-\sqrt{2})} = \dfrac{6(\sqrt{5}-\sqrt{2})}{5-2}$

$= \dfrac{\overset{2}{\cancel{6}}(\sqrt{5}-\sqrt{2})}{\underset{1}{\cancel{3}}} = 2(\sqrt{5}-\sqrt{2})$

26. $2(\sqrt{7}-\sqrt{5})$

27. $\dfrac{(x-4)\,(\sqrt{x}-2)}{(\sqrt{x}+2)(\sqrt{x}-2)} = \dfrac{\cancel{(x-4)}(\sqrt{x}-2)}{\cancel{(x-4)}} = \sqrt{x}-2$

28. $\sqrt{y}+3$

29. $\dfrac{1}{3}\sqrt{7}-2\sqrt{3} \doteq \dfrac{1}{3}(2.646)-2(1.732)$

$= 0.882-3.464$

$= -2.582$

$\doteq -2.58$

30. $\doteq 6.79$

31. $\dfrac{3+2\sqrt{11}}{6} \doteq \dfrac{3+2(3.317)}{6} = \dfrac{3+6.634}{6} = \dfrac{9.634}{6} \doteq 1.61$

32. $\doteq -0.61$

Exercises 11.6 (page 511)

1. $\sqrt{x}=5$ $\quad$ *Check:* $\sqrt{x}=5$

$(\sqrt{x})^2 = 5^2$ $\quad\quad\quad\quad \sqrt{25} \overset{?}{=} 5$

$\quad\quad x = 25$ $\quad\quad\quad\quad\quad\quad 5 = 5$

2. 16

3. $\sqrt{2x}=4$ $\quad$ *Check:* $\sqrt{2x}=4$

$(\sqrt{2x})^2 = 4^2$ $\quad\quad\quad\quad \sqrt{2(8)} \overset{?}{=} 4$

$\quad\quad 2x = 16$ $\quad\quad\quad\quad\quad \sqrt{16} \overset{?}{=} 4$

$\quad\quad\quad x = 8$ $\quad\quad\quad\quad\quad\quad\quad 4 = 4$

4. 12

5. $\sqrt{x-3}=2$ $\quad$ *Check:* $\sqrt{x-3}=2$

$(\sqrt{x-3})^2 = 2^2$ $\quad\quad\quad\quad \sqrt{7-3} \overset{?}{=} 2$

$\quad\quad x-3 = 4$ $\quad\quad\quad\quad\quad \sqrt{4} \overset{?}{=} 2$

$\quad\quad\quad x = 7$ $\quad\quad\quad\quad\quad\quad 2 = 2$

6. 32

7. $\sqrt{2x+1}=9$ $\quad$ *Check:* $\sqrt{2x+1}=9$

$(\sqrt{2x+1})^2 = 9^2$ $\quad\quad\quad \sqrt{2\cdot 40+1} \overset{?}{=} 9$

$\quad\quad 2x+1 = 81$ $\quad\quad\quad\quad \sqrt{81} \overset{?}{=} 9$

$\quad\quad\quad 2x = 80$ $\quad\quad\quad\quad\quad\quad 9 = 9$

$\quad\quad\quad\quad x = 40$

8. 4

9. $\sqrt{3x+1}=5$ $\quad$ *Check:* $\sqrt{3x+1}=5$

$(\sqrt{3x+1})^2 = 5^2$ $\quad\quad\quad \sqrt{3\cdot 8+1} \overset{?}{=} 5$

$\quad\quad 3x+1 = 25$ $\quad\quad\quad\quad \sqrt{25} \overset{?}{=} 5$

$\quad\quad\quad 3x = 24$ $\quad\quad\quad\quad\quad\quad 5 = 5$

$\quad\quad\quad\quad x = 8$

10. 4

11. $(\sqrt{x+1})^2 = (\sqrt{2x-7})^2$ $\quad$ *Check:* $\sqrt{x+1}=\sqrt{2x-7}$

$\quad\quad x+1 = 2x-7$ $\quad\quad\quad\quad \sqrt{8+1} \overset{?}{=} \sqrt{2\cdot 8-7}$

$\quad\quad\quad\quad 8 = x$ $\quad\quad\quad\quad\quad\quad \sqrt{9} \overset{?}{=} \sqrt{9}$

$\quad\quad\quad\quad x = 8$ $\quad\quad\quad\quad\quad\quad\quad 3 = 3$

12. 3

13. $(\sqrt{3x-2})^2 = (x)^2$

$\quad\quad 3x-2 = x^2$

$\quad\quad\quad 0 = x^2-3x+2$

$\quad\quad\quad 0 = (x-2)(x-1)$

$x-2=0 \quad\Big|\quad x-1=0$

$\quad x=2 \quad\Big|\quad\quad x=1$

Check: For $x=2$: $\quad\Big|\quad$ *Check:* For $x=1$:

$\sqrt{3x-2}=x$ $\quad\quad\Big|\quad \sqrt{3x-2}=x$

$\sqrt{3(2)-2} \overset{?}{=} 2$ $\quad\Big|\quad \sqrt{3(1)-2} \overset{?}{=} 1$

$\sqrt{4} \overset{?}{=} 2$ $\quad\quad\quad\Big|\quad\quad \sqrt{1} \overset{?}{=} 1$

$2 = 2$ $\quad\quad\quad\quad\Big|\quad\quad\quad 1 = 1$

14. 2, 3

15. $(\sqrt{4x-1})^2 = (2x)^2$

$\quad\quad 4x-1 = 4x^2$

$\quad\quad\quad 0 = 4x^2-4x+1$

$\quad\quad\quad 0 = (2x-1)(2x-1)$

$\quad\quad 2x-1 = 0$

$\quad\quad\quad 2x = 1$

$\quad\quad\quad\quad x = \dfrac{1}{2}$

Check: $\quad\quad \sqrt{4x-1} = 2x$

$\sqrt{4\left(\dfrac{1}{2}\right)-1} \overset{?}{=} 2\left(\dfrac{1}{2}\right)$

$\sqrt{2-1} \overset{?}{=} 1$

$1 = 1$

16. $\dfrac{1}{3}$

17. $\sqrt{x-3}+5 = x$

$(\sqrt{x-3})^2 = (x-5)^2$

$\quad\quad x-3 = x^2-10x+25$

$\quad\quad\quad 0 = x^2-11x+28$

$\quad\quad\quad 0 = (x-4)(x-7)$

$x-4=0 \quad\Big|\quad x-7=0$

$\quad x=4 \quad\Big|\quad\quad x=7$

Check: For $x=7$: $\quad\Big|\quad$ For $x=4$:

$\sqrt{x-3}+5 = x$ $\quad\Big|\quad \sqrt{x-3}+5 = x$

$\sqrt{7-3}+5 \overset{?}{=} 7$ $\quad\Big|\quad \sqrt{4-3}+5 \overset{?}{=} 4$

$\sqrt{4}+5 \overset{?}{=} 7$ $\quad\quad\Big|\quad \sqrt{1}+5 \overset{?}{=} 4$

$2+5 \overset{?}{=} 7$ $\quad\quad\quad\Big|\quad 1+5 \overset{?}{=} 4$

$7 = 7$ $\quad\quad\quad\quad\quad\Big|\quad\quad 6 \neq 4$

Therefore, 7 is a $\quad\quad\Big|\quad$ Therefore, 4 is not

solution. $\quad\quad\quad\quad\quad\Big|\quad$ a solution.

18. 5

19.

$$\sqrt{2x + 2} = 1 + \sqrt{x + 2}$$

$$(\sqrt{2x + 2})^2 = (1 + \sqrt{x + 2})^2 \qquad \text{Squaring both sides}$$

$$2x + 2 = 1 + 2\sqrt{x + 2} + (x + 2)$$

$$2x + 2 = \quad 3 + 2\sqrt{x + 2} + x$$

$$\underline{-x - 3 \qquad -3 \qquad\qquad -x} \qquad \text{Adding } -x - 3 \text{ to both sides}$$

$$x - 1 = \qquad 2\sqrt{x + 2}$$

$$(x - 1)^2 = (2\sqrt{x + 2})^2 \qquad \text{Squaring both sides again}$$

$$x^2 - 2x + 1 = 4(x + 2)$$

$$x^2 - 2x + 1 = \quad 4x + 8$$

$$\underline{-4x - 8 \qquad -4x - 8} \qquad \text{Adding } -4x - 8 \text{ to both sides}$$

$$x^2 - 6x - 7 = 0$$

$$(x - 7)(x + 1) = 0$$

$$x - 7 = 0 \quad | \quad x + 1 = 0$$

$$x = 7 \quad | \quad\quad x = -1$$

Check for x = 7: $\quad \sqrt{2x + 2} = 1 + \sqrt{x + 2}$

$$\sqrt{2(7) + 2} \overset{?}{=} 1 + \sqrt{(7) + 2}$$

$$\sqrt{16} \overset{?}{=} 1 + \sqrt{9}$$

$$4 \overset{?}{=} 1 + 3$$

$$4 = 4$$

$x = 7$ is a solution.

Check for x = −1: $\quad \sqrt{2x + 2} = 1 + \sqrt{x + 2}$

$$\sqrt{2(-1) + 2} \overset{?}{=} 1 + \sqrt{(-1) + 2}$$

$$\sqrt{0} \overset{?}{=} 1 + \sqrt{1}$$

$$0 \overset{?}{=} 1 + 1$$

$$0 \neq 2$$

$x = -1$ does not check.

The only solution is $x = 7$.

20. 5

21.

$$\sqrt{3x - 5} = \sqrt{x + 6} - 1$$

$$(\sqrt{3x - 5})^2 = (\sqrt{x + 6} - 1)^2 \qquad \text{Squaring both sides}$$

$$3x - 5 = (x + 6) - 2\sqrt{x + 6} + 1$$

$$3x - 5 = \quad x + 7 - 2\sqrt{x + 6}$$

$$\underline{-x - 7 \qquad -x - 7} \qquad \text{Adding } -x - 7 \text{ to both sides}$$

$$2x - 12 = \qquad\quad -2\sqrt{x + 6}$$

$$(2x - 12)^2 = (-2\sqrt{x + 6})^2 \qquad \text{Squaring both sides again}$$

$$4x^2 - 48x + 144 = 4(x + 6)$$

$$4x^2 - 48x + 144 = \quad 4x + 24$$

$$\underline{-4x - 24 \qquad -4x - 24} \qquad \text{Adding } -4x - 24 \text{ to both sides}$$

$$4x^2 - 52x + 120 = 0$$

$$4(x^2 - 13 + 30) = 0$$

$$4(x - 3)(x - 10) = 0$$

$$4 \neq 0 \quad | \quad x - 3 = 0 \quad | \quad x - 10 = 0$$

$$\qquad\quad | \quad\quad x = 3 \quad | \quad\quad x = 10$$

Check for x = 3: $\quad \sqrt{3x - 5} = \sqrt{x + 6} - 1$

$$\sqrt{3(3) - 5} \overset{?}{=} \sqrt{(3) + 6} - 1$$

$$\sqrt{4} \overset{?}{=} \sqrt{9} - 1$$

$$2 \overset{?}{=} 3 - 1$$

$$2 = 2$$

$x = 3$ is a solution.

Check for x = 10: $\quad \sqrt{3x - 5} = \sqrt{x + 6} - 1$

$$\sqrt{3(10) - 5} \overset{?}{=} \sqrt{(10) + 6} - 1$$

$$\sqrt{25} \overset{?}{=} \sqrt{16} - 1$$

$$5 \overset{?}{=} 4 - 1$$

$$5 \neq 3$$

$x = 10$ does not check. The only solution is $x = 3$.

22. 2

23. $\quad I = \sqrt{\dfrac{P}{R}}; \quad R = 9, I = 7$

$$7 = \sqrt{\dfrac{P}{9}}$$

$$(7)^2 = \left(\sqrt{\dfrac{P}{9}}\right)^2$$

$$49 = \dfrac{P}{9}$$

$$441 = P$$

24. 100

25. $\quad I = \sqrt{\dfrac{P}{R}}; \quad I = 5, P = 500$

$$5 = \sqrt{\dfrac{500}{R}}$$

$$(5)^2 = \left(\sqrt{\dfrac{500}{R}}\right)^2$$

$$25 = \dfrac{500}{R}$$

$$25R = 500$$

$$R = 20$$

26. 24

27. $\quad \sigma = \sqrt{npq}; \sigma = 3, q = \dfrac{3}{4}, n = 48$

$$3 = \sqrt{48\left(\dfrac{3}{4}\right)p}$$

$$3 = \sqrt{36p}$$

$$(3)^2 = (\sqrt{36p})^2 \quad \text{Squaring both sides}$$

$$9 = 36p$$

$$p = \dfrac{1}{4}$$

28. $\frac{1}{5}$

29. $\quad c = \sqrt{a^2 + b^2}; \quad c = 5, b = 3$

$$5 = \sqrt{a^2 + 3^2}$$

$$(5)^2 = (\sqrt{a^2 + 9})^2$$

$$25 = a^2 + 9$$

$$0 = a^2 - 16$$

$$0 = (a + 4)(a - 4)$$

$$a + 4 = 0 \quad | \quad a - 4 = 0$$

$$a = -4 \quad | \quad\quad a = 4$$

Reject; a length cannot be negative

The length of one side is 4.

30. 12

Review Exercises 11.7 (page 514)

1a. $9x$ **b.** $\frac{1}{2}$ **c.** $x - 5$ **2.** 10 **3.** 3 **4.** $a + b$

5. $\sqrt{x^3} = \sqrt{x^2 \cdot x} = \sqrt{x^2}\sqrt{x} = x\sqrt{x}$ **6.** $6a^2b$

7. $\sqrt{a^3 b^3} = \sqrt{a^2 \cdot a \cdot b^2 \cdot b} = \sqrt{a^2}\sqrt{a}\sqrt{b^2}\sqrt{b} = ab\sqrt{ab}$ **8.** 11

9. $\sqrt{2}\sqrt{32} = \sqrt{64} = 8$ **10.** $5\sqrt{2}$

11. $\sqrt{180} = \sqrt{2^2 \cdot 3^2 \cdot 5} = \sqrt{2^2}\sqrt{3^2}\sqrt{5} = 2 \cdot 3\sqrt{5} = 6\sqrt{5}$

12. $3\sqrt{2}$ **13.** $\sqrt{18} - \sqrt{8} = 3\sqrt{2} - 2\sqrt{2} = (3 - 2)\sqrt{2} = \sqrt{2}$

14. $\sqrt{3}$ **15.** $\frac{1}{\sqrt{5}} = \frac{1 \cdot \sqrt{5}}{\sqrt{5} \cdot \sqrt{5}} = \frac{\sqrt{5}}{5}$ **16.** $\frac{\sqrt{7}}{7}$

17. $\frac{6}{\sqrt{3}} = \frac{6 \cdot \sqrt{3}}{\sqrt{3} \cdot \sqrt{3}} = \frac{\overset{2}{\cancel{6}}\sqrt{3}}{\cancel{3}_{1}} = 2\sqrt{3}$ **18.** $\frac{3\sqrt{2}}{2}$

19. $\sqrt{8} - \sqrt{\frac{1}{2}} = \sqrt{4 \cdot 2} - \sqrt{\frac{1 \cdot 2}{2 \cdot 2}} = 2\sqrt{2} - \frac{\sqrt{2}}{2} = 2\sqrt{2} - \frac{1}{2}\sqrt{2} = \frac{3\sqrt{2}}{2}$

20. $\frac{8\sqrt{6}}{3}$

21. $\sqrt{2}(\sqrt{8} + \sqrt{18}) = \sqrt{2}(\sqrt{8}) + \sqrt{2}(\sqrt{18}) = \sqrt{16} + \sqrt{36} = 4 + 6$
$= 10$

22. 23

23. $(2\sqrt{3} + 1)^2 = (2\sqrt{3})^2 + 2(2\sqrt{3})(1) + (1)^2$
$= 12 + 4\sqrt{3} + 1$
$= 13 + 4\sqrt{3}$

24. $19 - 6\sqrt{2}$

25. $\frac{8}{(\sqrt{3} - 2)} \frac{(\sqrt{3} + 2)}{(\sqrt{3} + 2)} = \frac{8(\sqrt{3} + 2)}{3 - 4} = -8(\sqrt{3} + 2)$

26. $6(\sqrt{5} - 2)$

27. $\sqrt{x} = 4$ *Check:* $\sqrt{x} = 4$
$(\sqrt{x})^2 = 4^2$ $\sqrt{16} \overset{?}{=} 4$
$x = 16$ $4 = 4$

28. 12

29. $(\sqrt{2x - 1})^2 = 5^2$ *Check:* $\sqrt{2x - 1} = 5$
$2x - 1 = 25$ $\sqrt{2(13) - 1} \overset{?}{=} 5$
$2x = 26$ $\sqrt{25} \overset{?}{=} 5$
$x = 13$ $5 = 5$

30. 3

31. $(\sqrt{7x - 6})^2 = (x)^2$ *Check* $\sqrt{7x - 6} = x$
$7x - 6 = x^2$ *for* $x = 6$: $\sqrt{7(6) - 6} \overset{?}{=} 6$
$0 = x^2 - 7x + 6$ $\sqrt{36} = 6$
$0 = (x - 6)(x - 1)$ *for* $x = 1$: $\sqrt{7(1) - 6} \overset{?}{=} 1$
                       $\sqrt{1} = 1$

$x - 6 = 0 \;\Big|\; x - 1 = 0$
$\quad x = 6 \;\Big|\; \quad x = 1$

32. $1, 8$

33. $(\sqrt{2x + 7})^2 = (\sqrt{x} + 2)^2$ *Check* $\sqrt{2x + 7} = \sqrt{x} + 2$
$2x + 7 = x + 4\sqrt{x} + 4$ *for* $x = 9$: $\sqrt{2(9) + 7} \overset{?}{=} \sqrt{9} + 2$
$x + 3 = 4\sqrt{x}$ $\sqrt{25} \overset{?}{=} 3 + 2$
$(x + 3)^2 = (4\sqrt{x})^2$ $5 = 5$
$x^2 + 6x + 9 = 16x$
$x^2 - 10x + 9 = 0$ *for* $x = 1$: $\sqrt{2(1) + 7} \overset{?}{=} \sqrt{1} + 2$
$(x - 9)(x - 1) = 0$ $\sqrt{9} \overset{?}{=} 1 + 2$
                         $3 = 3$

$x - 9 = 0 \;\Big|\; x - 1 = 0$
$\quad x = 9 \;\Big|\; \quad x = 1$

34. 1

35. $c = \sqrt{a^2 + b^2}; \quad c = 13, b = 12$
$13 = \sqrt{a^2 + 12^2}$
$(13)^2 = (\sqrt{a^2 + 144})^2$
$169 = a^2 + 144$
$0 = a^2 - 25$
$0 = (a + 5)(a - 5)$
$a + 5 = 0 \;\Big|\; a - 5 = 0$
$\quad a = -5 \;\Big|\; \quad a = 5$

       ⌐Reject; a length cannot be negative
The length of one side is 5.

Chapter 11 Diagnostic Test (page 517)

Following each problem number is the textbook section number (in parentheses) where that kind of problem is discussed.

1. (11.1)   **a.** $3x^3y$ **b.** $2x + 5$

2. (11.2) $\sqrt{16z^2} = \sqrt{16}\sqrt{z^2} = 4z$

3. (11.2) $\sqrt{50} = \sqrt{25 \cdot 2} = \sqrt{25}\sqrt{2} = 5\sqrt{2}$

4. (11.2) $\sqrt{x^6 y^5} = \sqrt{x^6 \cdot y^4 \cdot y} = \sqrt{x^6}\sqrt{y^4}\sqrt{y} = x^3 y^2 \sqrt{y}$

5. (11.2) $\sqrt{\frac{3}{14}} = \sqrt{\frac{3 \cdot 14}{14 \cdot 14}} = \frac{\sqrt{42}}{14}$

6. (11.2) $\sqrt{\frac{48}{3n^2}} = \sqrt{\frac{16}{n^2}} = \frac{4}{n}$

7. (11.2) $\sqrt{60} = \sqrt{4 \cdot 15} = \sqrt{4}\sqrt{15} = 2\sqrt{15}$

8. (11.3) $\sqrt{13}\sqrt{13} = 13$

9. (11.3) $\sqrt{3}\sqrt{75y^2} = \sqrt{225y^2} = \sqrt{225}\sqrt{y^2} = 15y$

10. (11.5) $\sqrt{5}(2\sqrt{5} - 3) = \sqrt{5}(2\sqrt{5}) - \sqrt{5}(3)$
$= 2 \cdot \sqrt{5 \cdot 5} - 3\sqrt{5}$
$= 2 \cdot 5 - 3\sqrt{5} = 10 - 3\sqrt{5}$

11. (11.5) $(\sqrt{13} + \sqrt{3})(\sqrt{13} - \sqrt{3}) = (\sqrt{13})^2 - (\sqrt{3})^2$
$= 13 - 3 = 10$

12. (11.5) $(4\sqrt{x} + 3)(3\sqrt{x} + 4) = (4\sqrt{x})(3\sqrt{x}) + 4(4\sqrt{x}) + 3(3\sqrt{x}) + (3)(4)$
$= 12\sqrt{x \cdot x} + \underline{16\sqrt{x} + 9\sqrt{x}} + 12$
$= 12x + 25\sqrt{x} + 12$

13. (11.5) $(8 + \sqrt{10})^2$
$= (8)^2 + 2(8)(\sqrt{10}) + (\sqrt{10})^2$
$= 64 + 16\sqrt{10} + 10$
$= 74 + 16\sqrt{10}$

14. (11.3) $\frac{\sqrt{5}}{\sqrt{80}} = \sqrt{\frac{5}{80}} = \sqrt{\frac{1}{16}} = \frac{1}{4}$

15. (11.3) $\frac{\sqrt{3}}{\sqrt{6}} = \sqrt{\frac{3}{6}} = \sqrt{\frac{1}{2}}$
$= \sqrt{\frac{1 \cdot 2}{2 \cdot 2}} = \frac{\sqrt{2}}{2}$

16. (11.5) $\frac{\sqrt{18} - \sqrt{36}}{\sqrt{3}} = \frac{\sqrt{18}}{\sqrt{3}} - \frac{\sqrt{36}}{\sqrt{3}}$
$= \sqrt{\frac{18}{3}} - \sqrt{\frac{36}{3}}$
$= \sqrt{6} - \sqrt{12} = \sqrt{6} - 2\sqrt{3}$

17. (11.3) $\dfrac{\sqrt{x^4 y}}{\sqrt{3y}} = \sqrt{\dfrac{x^4 y}{3y}} = \sqrt{\dfrac{x^4(3)}{3(3)}}$

$\qquad = \dfrac{\sqrt{x^4}\sqrt{3}}{\sqrt{9}} = \dfrac{x^2\sqrt{3}}{3} \text{ or } \dfrac{x^2}{3}\sqrt{3}$

18. (11.5) $\dfrac{6}{1+\sqrt{5}} = \dfrac{6}{(1+\sqrt{5})} \cdot \dfrac{(1-\sqrt{5})}{(1-\sqrt{5})} = \dfrac{6(1-\sqrt{5})}{1^2 - (\sqrt{5})^2}$

$\qquad = \dfrac{6(1-\sqrt{5})}{1-5} = \dfrac{\overset{3}{\cancel{6}}(1-\sqrt{5})}{\underset{2}{-\cancel{4}}}$

$\qquad = \dfrac{3(1-\sqrt{5})}{-2} \text{ or } -\dfrac{3}{2}(1-\sqrt{5})$

19. (11.4) $8\sqrt{t} - \sqrt{t} = (8-1)\sqrt{t} = 7\sqrt{t}$

20. (11.4) $7\sqrt{3} - \sqrt{27} = 7\sqrt{3} - 3\sqrt{3} = (7-3)\sqrt{3} = 4\sqrt{3}$

21. (11.4) $\sqrt{54} + 4\sqrt{24} = 3\sqrt{6} + 4 \cdot 2\sqrt{6}$

$\qquad = 3\sqrt{6} + 8\sqrt{6} = 11\sqrt{6}$

22. (11.4) $\sqrt{45} + 6\sqrt{\dfrac{4}{5}} = 3\sqrt{5} + 6\sqrt{\dfrac{4 \cdot 5}{5 \cdot 5}} = 3\sqrt{5} + 6\left(\dfrac{2}{5}\right)\sqrt{5}$

$\qquad = \left(3 + \dfrac{12}{5}\right)\sqrt{5} = \left(\dfrac{15}{5} + \dfrac{12}{5}\right)\sqrt{5} = \left(\dfrac{27}{5}\right)\sqrt{5} \text{ or } \dfrac{27\sqrt{5}}{5}$

23. (11.6)

$\sqrt{8x} = 12 \qquad\qquad$ *Check:* $\sqrt{8x} = 12$

$(\sqrt{8x})^2 = (12)^2 \qquad\qquad \sqrt{8(18)} \overset{?}{=} 12$

$8x = 144 \qquad\qquad\qquad \sqrt{144} \overset{?}{=} 12$

$x = 18 \qquad\qquad\qquad\quad 12 = 12$

24. (11.6)

$\sqrt{5x + 11} = 6 \qquad$ *Check:* $\sqrt{5x + 11} = 6$

$(\sqrt{5x+11})^2 = 6^2 \qquad\qquad \sqrt{5(5) + 11} \overset{?}{=} 6$

$5x + 11 = 36 \qquad\qquad\qquad \sqrt{36} \overset{?}{=} 6$

$5x = 25 \qquad\qquad\qquad\qquad 6 = 6$

$x = 5$

25. (11.6) $\sqrt{6x - 5} = x$

$\qquad (\sqrt{6x-5})^2 = x^2$

$\qquad\qquad 6x - 5 = x^2$

$\qquad\qquad\quad 0 = x^2 - 6x + 5$

$\qquad\qquad\quad 0 = (x-5)(x-1)$

$x - 5 = 0 \quad | \quad x - 1 = 0$

$x = 5 \qquad | \qquad x = 1$

Check for $x = 5$: $\qquad$ *Check for $x = 1$:*

$\sqrt{6x - 5} = x \qquad\qquad \sqrt{6x - 5} = x$

$\sqrt{6(5) - 5} \overset{?}{=} 5 \qquad\quad \sqrt{6(1) - 5} \overset{?}{=} 1$

$\sqrt{25} \overset{?}{=} 5 \qquad\qquad\qquad \sqrt{1} \overset{?}{=} 1$

$5 = 5 \qquad\qquad\qquad\qquad 1 = 1$

Cumulative Review Exercises: Chapters 1–11 (page 518)

1. $\left(\dfrac{\overset{5}{\cancel{30}}x^4 y^{-3}}{\underset{2}{\cancel{12}}y^{-1}}\right)^{-2} = \left(\dfrac{5x^4 y^{-2}}{2}\right)^{-2}$

$\qquad = \dfrac{5^{-2}x^{-8}y^4}{2^{-2}} = \dfrac{2^2 y^4}{5^2 x^8} = \dfrac{4y^4}{25x^8}$

2. $4(3 + h)(3 - h)$

3. $\dfrac{x^2 + x - 2}{x^2 - 1} \div \dfrac{x^2 - 2x - 8}{x^2 - 4x}$

$\qquad = \dfrac{(x-1)(x+2)}{(x-1)(x+1)} \cdot \dfrac{x(x-4)}{(x+2)(x-4)} = \dfrac{x}{x+1}$

4. $x \le 4$

$$\overset{\longleftarrow}{\underset{-7\ -6\ -5\ -4\ -3\ -2\ -1\ \ 0\ \ 1\ \ 2\ \ 3\ \ 4\ \ 5\ \ 6\ \ 7}{\rule{6cm}{0.4pt}}\longrightarrow$$

5. $\dfrac{2x - 3}{2} = \dfrac{5x + 4}{6} - \dfrac{5}{3} \quad$ LCD $= 6$

$\dfrac{\overset{3}{\cancel{6}}}{1} \cdot \dfrac{2x - 3}{\underset{1}{\cancel{2}}} = \dfrac{\overset{1}{\cancel{6}}}{1} \cdot \dfrac{5x + 4}{\underset{1}{\cancel{6}}} - \dfrac{\overset{2}{\cancel{6}}}{1} \cdot \dfrac{5}{\underset{1}{\cancel{3}}}$

$3(2x - 3) = 1(5x + 4) - 2(5)$

$6x - 9 = 5x + 4 - 10$

$x = 3$

6. $\dfrac{2x^2 - 7x - 10}{(x+2)(x-1)}$

7. (1) $\begin{cases} 4x - 3y = -9 \\ 3y + x = -6 \end{cases}$

(1) Intercepts: $\left(-2\dfrac{1}{4}, 0\right), (0, 3)$

(2) Intercepts: $(-6, 0), (0, -2)$

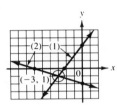

Solution: $(-3, -1)$

8. $(6, -5)$

9. (1) $\begin{cases} 2x - 5y = 2 \\ 3x - 7y = 2 \end{cases} \Rightarrow \begin{aligned} 2x &= 5y + 2 \\ x &= \dfrac{5y + 2}{2} \end{aligned}$

Substitute $\dfrac{5y + 2}{2}$ for x in (2):

(2) $\qquad\qquad\qquad\quad 3x - 7y = 2$

$\qquad\qquad 3\left(\dfrac{5y + 2}{2}\right) - 7y = 2 \quad$ LCD $= 2$

$\dfrac{\overset{1}{\cancel{2}}}{1} \cdot \dfrac{3}{1}\left(\dfrac{5y + 2}{\underset{1}{\cancel{2}}}\right) - \dfrac{2}{1} \cdot \dfrac{7y}{1} = \dfrac{2}{1} \cdot \dfrac{2}{1}$

$15y + 6 - 14y = 4$

$y = -2$

Substitute $y = -2$ in $x = \dfrac{5y + 2}{2}$:

$x = \dfrac{5(-2) + 2}{2} = \dfrac{-10 + 2}{2} = \dfrac{-8}{2} = -4$

Solution: $(-4, -2)$

10. $-\dfrac{7}{4}$ $\qquad$ **11.** $2x$ $\qquad$ **12.** $25 + 2\sqrt{154}$

13. $\sqrt{5}(3 + \sqrt{80}) = (\sqrt{5})(3) + (\sqrt{5})(\sqrt{80}) = 3\sqrt{5} + \sqrt{400}$

$\qquad\qquad\qquad\qquad\qquad\qquad\qquad\qquad = 3\sqrt{5} + 20$

14. $3\sqrt{7}$

15. $\dfrac{8}{(\sqrt{6} - 2)} \cdot \dfrac{(\sqrt{6} + 2)}{(\sqrt{6} + 2)} = \dfrac{8(\sqrt{6} + 2)}{(\sqrt{6})^2 - 2^2} = \dfrac{8(\sqrt{6} + 2)}{6 - 4} = \dfrac{\overset{4}{\cancel{8}}(\sqrt{6} + 2)}{\underset{1}{\cancel{2}}}$

$\qquad\qquad\qquad\qquad\qquad = 4(\sqrt{6} + 2)$

16. $\dfrac{14}{3}$ or $4\dfrac{2}{3}$

(In Exercises 17–20, the checks will not be shown.)

17. Let x = one number

y = other number

(1) $\begin{cases} x + y = \frac{1}{2} \\ x - y = 7\frac{1}{2} \end{cases}$ Add

$$2x = 8$$
$$x = 4$$

Substitute 4 for x in (1):

$$4 + y = \frac{1}{2} \Rightarrow y = \frac{1}{2} - 4 = -3\frac{1}{2}$$

18. 12 mi

19. Let x = number of dimes

$32 - x$ = number of nickels

$$10x + 5(32 - x) = 245$$
$$10x + 160 - 5x = 245$$
$$5x + 160 = 245$$
$$\underline{-160 \quad -160}$$
$$5x = 85$$
$$x = 17 \text{ dimes}$$
$$32 - x = 15 \text{ nickels}$$

20. 33, 35, and 37

Exercises 12.1 (page 521)

1. $\begin{aligned} 2x^2 &= 5x + 3 \\ 2x^2 - 5x - 3 &= 0 \end{aligned}$ $\begin{cases} a = 2 \\ b = -5 \\ c = -3 \end{cases}$

2. $3x^2 + 2x - 4 = 0$ $\begin{cases} a = 3 \\ b = 2 \\ c = -4 \end{cases}$

3. $\begin{aligned} 6x^2 &= x \\ 6x^2 - 1x + 0 &= 0 \end{aligned}$ $\begin{cases} a = 6 \\ b = -1 \\ c = 0 \end{cases}$

4. $3x^2 - 2x + 0 = 0$ $\begin{cases} a = 3 \\ b = -2 \\ c = 0 \end{cases}$

5. $\dfrac{3x}{2} + 5 = x^2$ LCD = 2

$$\frac{2}{1} \cdot \frac{3x}{2} + 2 \cdot 5 = 2 \cdot x^2$$

$\begin{aligned} 3x + 10 &= 2x^2 \\ 2x^2 - 3x - 10 &= 0 \end{aligned}$ $\begin{cases} a = 2 \\ b = -3 \\ c = -10 \end{cases}$

6. $3x^2 + 2x - 12 = 0$ $\begin{cases} a = 3 \\ b = 2 \\ c = -12 \end{cases}$

7. $x^2 - \dfrac{5x}{4} + \dfrac{2}{3} = 0$ LCD = 12

$$12(x^2) + \frac{12}{1}\left(\frac{-5x}{4}\right) + \frac{12}{1}\left(\frac{2}{3}\right) = 12 \cdot 0$$

$12x^2 - 15x + 8 = 0.$ $\begin{cases} a = 12 \\ b = -15 \\ c = 8 \end{cases}$

8. $30x^2 + 9x - 5 = 0$ $\begin{cases} a = 30 \\ b = 9 \\ c = -5 \end{cases}$

9. $\begin{aligned} x(x - 3) &= 4 \\ x^2 - 3x &= 4 \\ 1x^2 - 3x - 4 &= 0 \end{aligned}$ $\begin{cases} a = 1 \\ b = -3 \\ c = -4 \end{cases}$

10. $2x^2 + 2x - 12 = 0$ $\begin{cases} a = 2 \\ b = 2 \\ c = -12 \end{cases}$

11. $\begin{aligned} 3x(x + 1) &= (x + 1)(x + 2) \\ 3x^2 + 3x &= x^2 + 3x + 2 \\ 2x^2 + 0x - 2 &= 0 \end{aligned}$ $\begin{cases} a = 2 \\ b = 0 \\ c = -2 \end{cases}$

12. $3x^2 - 2x + 3 = 0$ $\begin{cases} a = 3 \\ b = -2 \\ c = 3 \end{cases}$

Exercises 12.2 (page 525)

The checks will not be shown.

1. $x^2 + x - 6 = 0$

$(x + 3)(x - 2) = 0$

$\begin{array}{c|c} x + 3 = 0 & x - 2 = 0 \\ x = -3 & x = 2 \end{array}$

2. 3, -2

3. $x^2 + x - 12 = 0$

$(x + 4)(x - 3) = 0$

$\begin{array}{c|c} x + 4 = 0 & x - 3 = 0 \\ x = -4 & x = 3 \end{array}$

4. 3, -5

5. $2x^2 - x - 1 = 0$

$(2x + 1)(x - 1) = 0$

$\begin{array}{c|c} 2x + 1 = 0 & x - 1 = 0 \\ x = -\frac{1}{2} & x = 1 \end{array}$

6. $\frac{1}{2}$, -1

7.
$$\frac{x}{8} = \frac{2}{x} \quad \text{A proportion}$$
$$x^2 = 16 \quad \text{Cross-multiplying}$$
$$x^2 - 16 = 0$$
$$(x - 4)(x + 4) = 0$$
$\begin{array}{c|c} x - 4 = 0 & x + 4 = 0 \\ x = 4 & x = -4 \end{array}$

8. 6, -6

9.
$$\frac{x + 2}{3} = \frac{-1}{x - 2} \quad \text{A proportion}$$
$$(x + 2)(x - 2) = 3(-1) \quad \text{Cross-multiplying}$$
$$x^2 - 4 = -3$$
$$x^2 - 1 = 0$$
$$(x + 1)(x - 1) = 0$$
$\begin{array}{c|c} x + 1 = 0 & x - 1 = 0 \\ x = -1 & x = 1 \end{array}$

10. 1, -1

11. $x^2 + 9x + 8 = 0$

$(x + 1)(x + 8) = 0$

$x + 1 = 0 \quad | \quad x + 8 = 0$

$x = -1 \quad | \quad x = -8$

12. $-1, -6$

13. $2x^2 + 4x = 0$

$2x(x + 2) = 0$

$2x = 0 \quad | \quad x + 2 = 0$

$x = 0 \quad | \quad x = -2$

14. $0, 3$

15. $x^2 = x + 2$

$x^2 - x - 2 = 0$

$(x - 2)(x + 1) = 0$

$x - 2 = 0 \quad | \quad x + 1 = 0$

$x = 2 \quad | \quad x = -1$

16. $3, -2$

17. $\dfrac{x}{2} + \dfrac{2}{x} = \dfrac{5}{2}$ $\quad$ LCD $= 2x$

$\dfrac{\cancel{2x}}{1} \cdot \dfrac{x}{\cancel{2}} + \dfrac{2\cancel{x}}{1} \cdot \dfrac{2}{\cancel{x}} = \dfrac{\cancel{2}x}{1} \cdot \dfrac{5}{\cancel{2}}$

$x^2 + 4 = 5x$

$x^2 - 5x + 4 = 0$

$(x - 1)(x - 4) = 0$

$x - 1 = 0 \quad | \quad x - 4 = 0$

$x = 1 \quad | \quad x = 4$

18. $1, 6$

19. $\dfrac{x - 1}{4} + \dfrac{6}{x + 1} = 2$ $\quad$ LCD $= 4(x + 1)$

$\dfrac{\cancel{4}(x + 1)}{1} \cdot \dfrac{x - 1}{\cancel{4}} + \dfrac{4\cancel{(x + 1)}}{1} \cdot \dfrac{6}{\cancel{(x + 1)}} = 4(x + 1)2$

$(x + 1)(x - 1) + 24 = 8(x + 1)$

$x^2 - 1 + 24 = 8x + 8$

$x^2 - 8x + 15 = 0$

$(x - 3)(x - 5) = 0$

$x - 3 = 0 \quad | \quad x - 5 = 0$

$x = 3 \quad | \quad x = 5$

20. $2, 3$

21. $\dfrac{2x^2}{1} = \dfrac{2 - x}{3}$ $\quad$ A proportion

$3(2x^2) = 1(2 - x)$ $\quad$ Cross-multiplying

$6x^2 = 2 - x$

$6x^2 + x - 2 = 0$

$(2x - 1)(3x + 2) = 0$

$2x - 1 = 0 \quad | \quad 3x + 2 = 0$

$2x = 1 \quad | \quad 3x = -2$

$x = \dfrac{1}{2} \quad | \quad x = -\dfrac{2}{3}$

22. $1\frac{1}{2}, \frac{1}{4}$

23. Let $W =$ width

$W(W + 5) = 24$

$W^2 + 5W = 24$

$W^2 + 5W - 24 = 0$

$(W + 8)(W - 3) = 0$

$W + 8 = 0 \quad | \quad W - 3 = 0$

$W = -8 \quad | \quad W = 3$ (width)

No meaning $\quad | \quad W + 5 = 3 + 5$

$= 8$ (length)

$W + 5 =$ length
Area $= W(W + 5)$

24. Width: 7; length: 11

25. Let $x =$ first even integer

$x + 2 =$ second even integer

$x(x + 2) + 4 = 84$

$x^2 + 2x + 4 = 84$

$x^2 + 2x - 80 = 0$

$(x + 10)(x - 8) = 0$

$x + 10 = 0 \quad | \quad x - 8 = 0$

$x = -10 \quad | \quad x = 8$

$x + 2 = -8 \quad | \quad x + 2 = 10$

Two answers: -10 and -8, or 8 and 10

26. Two answers: 6 and 7, or -7 and -6

27. Let $r =$ rate from LA to Mexico

	d	r	t
LA to Mexico	120	x	$\dfrac{120}{x}$
Mexico to LA	120	$x + 20$	$\dfrac{120}{x + 20}$

$t = \dfrac{d}{r}$

$t = \dfrac{120}{x}$

$t = \dfrac{120}{x + 20}$

Total time $\quad$ is $\quad$ 5 hr

$\dfrac{120}{x} + \dfrac{120}{x + 20} = 5$

LCD $= x(x + 20)$

$\cancel{x}(x + 20)\dfrac{120}{\cancel{x}} + x\cancel{(x + 20)}\dfrac{120}{\cancel{x + 20}} = 5 \cdot x(x + 20)$

$120x + 2400 + 120x = 5x^2 + 100x$

$240x + 2400 = 5x^2 + 100x$

$0 = 5x^2 - 140x - 2400$

$0 = 5(x^2 - 28x - 480)$

$0 = 5(x + 12)(x - 40)$

$5 \neq 0 \quad | \quad x + 12 = 0 \quad | \quad x - 40 = 0$

$\quad | \quad$ Reject $x = -12 \quad | \quad x = 40$

Therefore, average speed to Mexico is 40 mph.

28. 45 mph

Exercises 12.3 (page 531)

1. $8x^2 = 4x$

$8x^2 - 4x = 0$

$4x(2x - 1) = 0$

$4x = 0 \quad | \quad 2x - 1 = 0$

$x = 0 \quad | \quad 2x = 1$

$\quad | \quad x = \dfrac{1}{2}$

2. $0, \frac{1}{3}$

3. $x^2 - 9 = 0$

$x^2 = 9$

$x = \pm\sqrt{9}$

$x = \pm 3$

4. $6, -6$

5. $x^2 - 4x = 0$

$x(x - 4) = 0$

$x = 0 \mid x - 4 = 0$

$x = 4$

6. $0, 16$

7. $x^2 + 16 = 0$

$x^2 = -16$

$x = \pm\sqrt{-16}$ The solutions are not real numbers.

8. The solutions are not real numbers.

9. $x^2 - 4 = 0$

$x^2 = 4$

$x = \pm\sqrt{4}$

$x = \pm 2$

10. ± 4

11. $5x^2 = 4$

$x^2 = \frac{4}{5}$

$x = \pm\sqrt{\frac{4}{5}}$

$x = \pm\frac{2}{\sqrt{5}} = \pm\frac{2}{\sqrt{5}} \cdot \frac{\sqrt{5}}{\sqrt{5}}$

$x = \pm\frac{2\sqrt{5}}{5}$

12. $\pm\frac{5\sqrt{3}}{3}$

13. $8 - 2x^2 = 0$

$2x^2 = 8$

$x^2 = 4$

$x = \pm\sqrt{4}$

$x = \pm 2$

14. ± 3

15. $x(x + 3) = 3x - 4$

$x^2 + 3x = 3x - 4$

$x^2 = -4$

$x = \pm\sqrt{-4}$ The solutions are not real numbers.

16. The solutions are not real numbers.

17. $2(x + 3) = 6 + x(x + 2)$

$2x + 6 = 6 + x^2 + 2x$

$x^2 = 0$

$x = \pm\sqrt{0}$

$x = 0$

18. 0

19. $2x(3x - 4) = 2(3 - 4x)$

$6x^2 - 8x = 6 - 8x$

$6x^2 = 6$

$x^2 = 1$

$x = \pm\sqrt{1}$

$x = \pm 1$

20. $\pm\frac{1}{5}\sqrt{30}$

Exercises 12.4 (page 535)

1. $3x^2 - 1x - 2 = 0$ $\begin{cases} a = 3 \\ b = -1 \\ c = -2 \end{cases}$

$x = \dfrac{-(-1) \pm \sqrt{(-1)^2 - 4(3)(-2)}}{2(3)}$

$x = \dfrac{1 \pm \sqrt{1 + 24}}{6} = \dfrac{1 \pm \sqrt{25}}{6}$

$x = \dfrac{1 \pm 5}{6} = \begin{cases} \dfrac{1 + 5}{6} = \dfrac{6}{6} = 1 \\ \dfrac{1 - 5}{6} = \dfrac{-4}{6} = -\dfrac{2}{3} \end{cases}$

2. $\frac{1}{2}, -2$

3. $1x^2 - 4x + 1 = 0$ $\begin{cases} a = 1 \\ b = -4 \\ c = 1 \end{cases}$

$x = \dfrac{-(-4) \pm \sqrt{(-4)^2 - 4(1)(1)}}{2(1)}$

$x = \dfrac{4 \pm \sqrt{16 - 4}}{2(1)} = \dfrac{4 \pm \sqrt{12}}{2}$

$x = \dfrac{4 \pm 2\sqrt{3}}{2} = \dfrac{2(2 \pm \sqrt{3})}{2} = 2 \pm \sqrt{3}$

4. $2 \pm \sqrt{5}$

5. $\dfrac{x}{2} + \dfrac{2}{x} = \dfrac{5}{2}$ LCD = $2x$

$\dfrac{2x}{1} \cdot \dfrac{x}{2} + \dfrac{2x}{1} \cdot \dfrac{2}{x} = \dfrac{2x}{1} \cdot \dfrac{5}{2}$

$x^2 + 4 = 5x$

$1x^2 - 5x + 4 = 0$ $\begin{cases} a = 1 \\ b = -5 \\ c = 4 \end{cases}$

$x = \dfrac{-(-5) \pm \sqrt{(-5)^2 - 4(1)(4)}}{2(1)}$

$x = \dfrac{5 \pm \sqrt{25 - 16}}{2} = \dfrac{5 \pm \sqrt{9}}{2}$

$x = \dfrac{5 \pm 3}{2} = \begin{cases} \dfrac{5 + 3}{2} = \dfrac{8}{2} = 4 \\ \dfrac{5 - 3}{2} = \dfrac{2}{2} = 1 \end{cases}$

6. $1, 6$

7.
$$2x^2 = 8x - 5 \qquad \begin{cases} a = 2 \\ b = -8 \\ c = 5 \end{cases}$$
$$2x^2 - 8x + 5 = 0$$

$$x = \frac{-(-8) \pm \sqrt{(-8)^2 - 4(2)(5)}}{2(2)}$$

$$x = \frac{8 \pm \sqrt{64 - 40}}{4} = \frac{8 \pm \sqrt{24}}{4} = \frac{8 \pm 2\sqrt{6}}{4}$$

$$x = \frac{\overset{1}{\cancel{2}}(4 \pm \sqrt{6})}{\underset{2}{\cancel{4}}} = \frac{4 \pm \sqrt{6}}{2}$$

8. $\dfrac{3 \pm \sqrt{3}}{3}$

9. $\dfrac{1}{x} + \dfrac{x}{x-1} = 3 \quad \text{LCD} = x(x-1)$

$$\frac{\cancel{x}(x-1)}{1} \cdot \frac{1}{\cancel{x}} + \frac{x\cancel{(x-1)}}{1} \cdot \frac{x}{\cancel{x-1}} = \frac{x(x-1)}{1} \cdot \frac{3}{1}$$

$$(x-1) + x^2 = 3x(x-1)$$
$$x - 1 + x^2 = 3x^2 - 3x$$
$$0 = 2x^2 - 4x + 1$$

$$\left.\begin{array}{l} a = 2 \\ b = -4 \\ c = 1 \end{array}\right\}$$

$$x = \frac{-(-4) \pm \sqrt{(-4)^2 - 4(2)(1)}}{2(2)}$$

$$= \frac{4 \pm \sqrt{16 - 8}}{4} = \frac{4 \pm \sqrt{8}}{4}$$

$$= \frac{4 \pm 2\sqrt{2}}{4} = \frac{2(2 \pm \sqrt{2})}{4} = \frac{2 \pm \sqrt{2}}{2} \quad \text{The checks will not be shown.}$$

10. $\dfrac{-1 \pm \sqrt{11}}{5}$

11. $x^2 + x + 1 = 0 \quad \begin{cases} a = 1 \\ b = 1 \\ c = 1 \end{cases}$

$$x = \frac{-(1) \pm \sqrt{(1)^2 - 4(1)(1)}}{2(1)}$$

$$= \frac{-1 \pm \sqrt{1 - 4}}{2}$$

$$= \frac{-1 \pm \sqrt{-3}}{2} \quad \text{The answers are not real numbers.}$$

12. The answers are not real numbers.

13.
$$4x^2 + 4x = 1 \qquad \begin{cases} a = 4 \\ b = 4 \\ c = -1 \end{cases}$$
$$4x^2 + 4x - 1 = 0$$

$$x = \frac{-(4) \pm \sqrt{(4)^2 - 4(4)(-1)}}{2(4)}$$

$$= \frac{-4 \pm \sqrt{16 + 16}}{8}$$

$$= \frac{-4 \pm \sqrt{32}}{8} = \frac{-4 \pm 4\sqrt{2}}{8} = \frac{4(-1 \pm \sqrt{2})}{8} = \frac{-1 \pm \sqrt{2}}{2}$$

14. $\dfrac{1 \pm \sqrt{3}}{3}$

15. $3x^2 + 2x + 1 = 0 \quad \begin{cases} a = 3 \\ b = 2 \\ c = 1 \end{cases}$

$$x = \frac{-(2) \pm \sqrt{(2)^2 - 4(3)(1)}}{2(3)}$$

$$x = \frac{-2 \pm \sqrt{4 - 12}}{6} = \frac{-2 \pm \sqrt{-8}}{6}$$

Solutions are not real numbers because radicand is negative.

16. Solutions are not real numbers because radicand is negative.

In Exercises 17–24, the checks will not be shown.

17. Let x = number

$$\frac{1}{x} = \text{its reciprocal}$$

$$x - \frac{1}{x} = \frac{2}{3} \quad \text{LCD is } 3x$$

$$(3x)x - (3x)\left(\frac{1}{x}\right) = \left(\frac{2}{3}\right)(3x)$$

$$3x^2 - 3 = 2x$$

$$3x^2 - 2x - 3 = 0 \quad \begin{cases} a = 3 \\ b = -2 \\ c = -3 \end{cases}$$

$$x = \frac{-(-2) \pm \sqrt{(-2)^2 - 4(3)(-3)}}{2(3)}$$

$$= \frac{2 \pm \sqrt{4 + 36}}{6}$$

$$= \frac{2 \pm \sqrt{40}}{6} = \frac{2 \pm 2\sqrt{10}}{6} = \frac{2(1 \pm \sqrt{10})}{6} = \frac{1 \pm \sqrt{10}}{3}$$

Two answers: $\dfrac{1 + \sqrt{10}}{3}$ or $\dfrac{1 - \sqrt{10}}{3}$

18. Two answers: $\dfrac{-5 + 3\sqrt{17}}{8}$ or $\dfrac{-5 - 3\sqrt{17}}{8}$

19. $4x^2 + 4x - 1 = 0 \quad \begin{cases} a = 4 \\ b = 4 \\ c = -1 \end{cases}$

$$x = \frac{-(4) \pm \sqrt{(4)^2 - 4(4)(-1)}}{2(4)} = \frac{-4 \pm \sqrt{16 + 16}}{8}$$

$$= \frac{-4 \pm \sqrt{16 \cdot 2}}{8} = \frac{-4 \pm 4\sqrt{2}}{8} = \frac{4(-1 \pm \sqrt{2})}{8} = \frac{-1 \pm \sqrt{2}}{2}$$

$$\doteq \frac{-1 \pm 1.414}{2} = \begin{cases} \dfrac{0.414}{2} \doteq 0.21 \\[2mm] \dfrac{-2.414}{2} \doteq -1.21 \end{cases}$$

20. $\dfrac{1 + \sqrt{3}}{3} \doteq 0.91$

$$\frac{1 - \sqrt{3}}{3} \doteq -0.24$$

21. Let W = width

Area = $(W + 2)W = 2$

$W^2 + 2W = 2$

$1W^2 + 2W - 2 = 0$ $\begin{cases} a = 1 \\ b = 2 \\ c = -2 \end{cases}$ $W + 2$ = length

$W = \dfrac{-(2) \pm \sqrt{(2)^2 - 4(1)(-2)}}{2(1)} = \dfrac{-2 \pm \sqrt{4 + 8}}{2}$

$W = \dfrac{-2 \pm \sqrt{12}}{2} = \dfrac{-2 \pm 2\sqrt{3}}{2} = \dfrac{2(-1 \pm \sqrt{3})}{2} = -1 \pm \sqrt{3}$

$W = -1 \pm \sqrt{3} = \begin{cases} -1 + \sqrt{3} \doteq -1 + 1.732 \doteq 0.73 \\ -1 - \sqrt{3} \doteq -1 - 1.732 \doteq -2.73 \end{cases}$

When $W = 0.73$	When $W = -2.73$
Width = $W = 0.73$	Not possible
Length = $W + 2 = 2.73$	

22. Width = $-2 + \sqrt{10} \doteq 1.16$; length = $2 + \sqrt{10} \doteq 5.16$

23. Let x = length of a side of square

$4x = x^2 - 6$

$x^2 - 4x - 6 = 0$ $\begin{cases} a = 1 \\ b = -4 \\ c = -6 \end{cases}$

Area = x^2

Perimeter = $4x$

$x = \dfrac{-(-4) \pm \sqrt{(-4)^2 - 4(1)(-6)}}{2(1)}$

$= \dfrac{4 \pm \sqrt{16 + 24}}{2} = \dfrac{4 \pm \sqrt{40}}{2}$

$= \dfrac{4 \pm 2\sqrt{10}}{2} = \dfrac{2(2 \pm \sqrt{10})}{2} = 2 \pm \sqrt{10}$

$x = \begin{cases} 2 + \sqrt{10} \doteq 2 + 3.162 \doteq 5.16 \quad \text{Side of square} \\ 2 - \sqrt{10} \doteq 2 - 3.162 \doteq -1.16 \quad \text{Not possible} \end{cases}$

24. $2 + \sqrt{6} \doteq 4.45$

Exercises 12.5 (page 539)

1. $x^2 = (16)^2 + (12)^2$

$x^2 = 256 + 144$

$x^2 = 400$

$x = \sqrt{400} = 20$

$x = 20$

2. 15

3. $x^2 = 6^2 + 2^2$

$x^2 = 36 + 4$

$x^2 = 40$

$x = \sqrt{40} = \sqrt{4 \cdot 10} = 2\sqrt{10}$

$x = 2\sqrt{10}$

4. $2\sqrt{5}$

5. $x^2 = (\sqrt{5})^2 + 2^2$

$x^2 = 5 + 4$

$x^2 = 9$

$x = \sqrt{9} = 3$

6. 4

7. $5^2 = x^2 + 3^2$

$25 = x^2 + 9$

$16 = x^2$

$\sqrt{16} = x$

$4 = x$

$x = 4$

8. 5

9. $(3\sqrt{3})^2 = 3^2 + x^2$

$9 \cdot 3 = 9 + x^2$

$27 = 9 + x^2$

$x^2 = 18$

$x = \sqrt{18} = \sqrt{9 \cdot 2} = 3\sqrt{2}$

$x = 3\sqrt{2}$

10. $2\sqrt{2}$

11. $x^2 = 10^2 + 6^2$

$x^2 = 100 + 36$

$x^2 = 136$

$x = \sqrt{136} = \sqrt{4 \cdot 34} = 2\sqrt{34}$

$x = 2\sqrt{34}$

12. $4\sqrt{13}$

13. $(x + 1)^2 + (\sqrt{20})^2 = (x + 3)^2$

$x^2 + 2x + 1 + 20 = x^2 + 6x + 9$

$12 = 4x$

$x = 3$

14. 7

15. Let W = width

$(24)^2 + W^2 = (25)^2$

$576 + W^2 = 625$

$W^2 = 49$

$W = \sqrt{49}$

$W = 7\,\text{cm}$

16. 9 in.

17. Let x = length of one leg (in meters)

$2x - 4$ = length of other leg (in meters)

$(10)^2 = (2x - 4)^2 + x^2$

$100 = 4x^2 - 16x + 16 + x^2$

$0 = 5x^2 - 16x - 84$

$0 = (5x + 14)(x - 6)$

$5x + 14 = 0$	$x - 6 = 0$
$x = -\dfrac{14}{5}$	$x = 6$ m, one leg
	$2x - 4 = 12 - 4$
Not a solution	$= 8$ m, other leg

18. Width = 9 yd; length = 12 yd

19. Let x = length of diagonal (in in.)

$x^2 = (\sqrt{6})^2 + (\sqrt{6})^2$

$x^2 = 6 + 6$

$x^2 = 12$

$x = \sqrt{12}$

$x = 2\sqrt{3}$ in.

20. $2\sqrt{7}$ m

21. Let x = side

$$(4\sqrt{2})^2 = x^2 + x^2$$
$$32 = 2x^2$$
$$16 = x^2$$
$$\sqrt{16} = x$$
$$x = 4\,\text{cm}$$

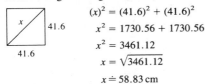

22. 5 in.

23. Let x = length of diagonal (in cm)

$$(x)^2 = (41.6)^2 + (41.6)^2$$
$$x^2 = 1730.56 + 1730.56$$
$$x^2 = 3461.12$$
$$x = \sqrt{3461.12}$$
$$x \doteq 58.83\,\text{cm}$$

24. $\doteq 2.66$ m

Review Exercises 12.6 (page 543)

The checks will not be shown.

1. $x^2 + x = 6$

$$x^2 + x - 6 = 0$$
$$(x + 3)(x - 2) = 0$$
$$x + 3 = 0 \quad | \quad x - 2 = 0$$
$$x = -3 \quad | \quad x = 2$$

2. 0, 49

3. $x^2 - 2x - 4 = 0 \quad \begin{cases} a = 1 \\ b = -2 \\ c = -4 \end{cases}$

$$x = \frac{-(-2) \pm \sqrt{(-2)^2 - 4(1)(-4)}}{2(1)}$$
$$x = \frac{2 \pm \sqrt{4 + 16}}{2} = \frac{2 \pm \sqrt{20}}{2}$$
$$x = \frac{2 \pm 2\sqrt{5}}{2} = \frac{2(1 \pm \sqrt{5})}{2} = 1 \pm \sqrt{5}$$

4. $2 \pm \sqrt{3}$

5.
$$x^2 = 5x$$
$$x^2 - 5x = 0$$
$$x(x - 5) = 0$$
$$x = 0 \quad | \quad x - 5 = 0$$
$$\qquad \qquad x = 5$$

6. $\pm\dfrac{5}{6}$

7. $\dfrac{x + 2}{3} = \dfrac{1}{x - 2} + \dfrac{2}{3}$ LCD $= 3(x - 2)$

$$\frac{3(x - 2)}{1} \cdot \frac{(x + 2)}{3} = \frac{3(x - 2)}{1} \cdot \frac{1}{(x - 2)} + \frac{3(x - 2)}{1} \cdot \frac{2}{3}$$
$$(x - 2)(x + 2) = 3 + 2(x - 2)$$
$$x^2 - 4 = 3 + 2x - 4$$
$$x^2 - 2x - 3 = 0$$
$$(x - 3)(x + 1) = 0$$
$$x - 3 = 0 \quad | \quad x + 1 = 0$$
$$x = 3 \quad | \quad x = -1$$

8. The solutions are not real numbers.

9. $5(x + 2) = x(x + 5)$

$$5x + 10 = x^2 + 5x$$
$$x^2 = 10$$
$$x = \pm\sqrt{10}$$

10. $\pm 2\sqrt{3}$

11. $\dfrac{2}{x} + \dfrac{x}{x + 1} = 5$ LCD $= x(x + 1)$

$$\frac{x(x + 1)}{1} \cdot \frac{2}{x} + \frac{x(x + 1)}{1} \cdot \frac{x}{(x + 1)} = \frac{x(x + 1)}{1} \cdot \frac{5}{1}$$
$$2(x + 1) + x^2 = 5x(x + 1)$$
$$2x + 2 + x^2 = 5x^2 + 5x$$
$$4x^2 + 3x - 2 = 0 \quad \begin{cases} a = 4 \\ b = 3 \\ c = -2 \end{cases}$$
$$x = \frac{-(3) \pm \sqrt{(3)^2 - 4(4)(-2)}}{2(4)} = \frac{-3 \pm \sqrt{9 + 32}}{8} = \frac{-3 \pm \sqrt{41}}{8}$$

12. $\dfrac{-1 \pm \sqrt{73}}{6}$

13. $3x^2 + 2x + 1 = 0 \quad \begin{cases} a = 3 \\ b = 2 \\ c = 1 \end{cases}$

$$x = \frac{-(2) \pm \sqrt{(2)^2 - 4(3)(1)}}{2(3)}$$
$$x = \frac{-2 \pm \sqrt{4 - 12}}{6} = \frac{-2 \pm \sqrt{-8}}{6} \quad \text{Not real numbers}$$

14. $\pm\dfrac{3\sqrt{5}}{5}$

15.
$$15^2 = 12^2 + x^2$$
$$225 = 144 + x^2$$
$$81 = x^2$$
$$\sqrt{81} = x$$
$$x = 9$$

16. 12 in.

17. Let x = length of diagonal (in meters)

$$x^2 = (\sqrt{10})^2 + (\sqrt{10})^2$$
$$x^2 = 10 + 10$$
$$x^2 = 20$$
$$x = \sqrt{20}$$
$$x = 2\sqrt{5}\,\text{m}$$

18. $8\sqrt{2}$ yd

19. Let x = length of diagonal

$$x^2 = 8^2 + 4^2$$
$$x^2 = 64 + 16$$
$$x^2 = 80$$
$$x = \sqrt{80}$$
$$x = \sqrt{16 \cdot 5} = 4\sqrt{5}$$
$$\doteq 4(2.236) \doteq 8.94\,\text{in.}$$

20. The width is $(-3 + \sqrt{15})$ ft $\doteq 0.87$ ft, and the length is $(3 + \sqrt{15})$ ft $\doteq 6.87$ ft.

Chapter 12 Diagnostic Test (page 547)

Following each problem number is the textbook section number (in parentheses) where that kind of problem is discussed.

1. (12.1)　　$5 + 3x^2 = 7x$　$\begin{cases} a = 3 \\ b = -7 \\ c = 5 \end{cases}$

$3x^2 - 7x + 5 = 0$

2. (12.1)　$3(x + 5) = 3(4 - x^2)$

$3x + 15 = 12 - 3x^2$

$3x^2 + 3x + 3 = 0$　$\begin{cases} a = 3 \\ b = 3 \\ c = 3 \end{cases}$

3. (12.4)　　$x^2 = 8x + 15$

$x^2 - 8x - 15 = 0$

$x^2 - 8x - 15 = 0$　$\begin{cases} a = 1 \\ b = -8 \\ c = -15 \end{cases}$

$x = \dfrac{-(-8) \pm \sqrt{(-8)^2 - 4(1)(-15)}}{2(1)}$

$= \dfrac{8 \pm \sqrt{64 + 60}}{2}$

$= \dfrac{8 \pm \sqrt{124}}{2} = \dfrac{8 \pm 2\sqrt{31}}{2} = \dfrac{2(4 \pm \sqrt{31})}{2}$

$= 4 \pm \sqrt{31}$

4. (12.2)　$3x^2 + 2x = 8$

$3x^2 + 2x - 8 = 0$

$(3x - 4)(x + 2) = 0$

$3x - 4 = 0 \quad | \quad x + 2 = 0$

$3x = 4 \qquad\quad x = -2$

$x = \dfrac{4}{3}$

5. (12.3)　　$5x^2 = 17x$

$5x^2 - 17x = 0$

$x(5x - 17) = 0$

$x = 0 \quad | \quad 5x - 17 = 0$

$\qquad\qquad 5x = 17$

$\qquad\qquad x = \dfrac{17}{5}$

6. (12.2)　$2x^2 - 9x - 5 = 0$

$(2x + 1)(x - 5) = 0$

$2x + 1 = 0 \quad | \quad x - 5 = 0$

$2x = -1 \qquad\quad x = 5$

$x = -\dfrac{1}{2}$

7. (12.4)　$x^2 + 2x + 6 = 0$　$\begin{cases} a = 1 \\ b = 2 \\ c = 6 \end{cases}$

$x = \dfrac{-(2) \pm \sqrt{(2)^2 - 4(1)(6)}}{2(1)}$

$= \dfrac{-2 \pm \sqrt{4 - 24}}{2}$

$= \dfrac{-2 \pm \sqrt{-20}}{2}$ The solutions are not real numbers.

8. (12.3)　$6x^2 = 54$

$x^2 = \dfrac{54}{6} = 9$

$x = \pm 3$

9. (12.4)　$3(x + 1) = x^2$

$3x + 3 = x^2$

$0 = x^2 - 3x - 3$　$\begin{cases} a = 1 \\ b = -3 \\ c = -3 \end{cases}$

$x = \dfrac{-(-3) \pm \sqrt{(-3)^2 - 4(1)(-3)}}{2(1)}$

$= \dfrac{3 \pm \sqrt{9 + 12}}{2}$

$= \dfrac{3 \pm \sqrt{21}}{2}$

10. (12.3)　$x(x + 3) = x$

$x^2 + 3x = x$

$x^2 + 2x = 0$

$x(x + 2) = 0$

$x = 0 \quad | \quad x + 2 = 0$

$\qquad\qquad x = -2$

11. (12.5)　$(\sqrt{11})^2 + x^2 = (4)^2$

$11 + x^2 = 16$

$x^2 = 5$

$x = \sqrt{5}$

12. (12.5)　Let $x =$ length of diagonal

$x^2 = (7)^2 + (7)^2$

$x^2 = 49 + 49$

$x^2 = 98$

$x = \sqrt{98} = 7\sqrt{2}\,\text{m}$

13. (12.5)　Let　$x =$ width of rectangle

$x + 6 =$ length of rectangle

$x(x + 6) = 78$

$x^2 + 6x - 78 = 0$　$\begin{cases} a = 1 \\ b = 6 \\ c = -78 \end{cases}$

$x = \dfrac{-(6) \pm \sqrt{(6)^2 - 4(1)(-78)}}{2(1)}$

$= \dfrac{-6 \pm \sqrt{36 + 312}}{2}$

$= \dfrac{-6 \pm \sqrt{348}}{2} = \dfrac{-6 \pm 2\sqrt{87}}{2} = \dfrac{2(-3 \pm \sqrt{87})}{2} = -3 \pm \sqrt{87}$

We reject $-3 - \sqrt{87}$ because a length cannot be negative. The width is $(-3 + \sqrt{87})$ ft and the length is $[(-3 + \sqrt{87}) + 6]$ ft $= (3 + \sqrt{87})$ ft.

Cumulative Review Exercises: Chapters 1–12 (page 548)

1. No　**2.** No　**3.** 0　**4.** $\dfrac{4s^4 t^6}{9}$　**5.** Not possible

6. $9 - 2\sqrt{3}$

7.
$$4x^2 + 3x - 2$$
$$5x - 3$$
$$\overline{}$$
$$-12x^2 - 9x + 6$$
$$20x^3 + 15x^2 - 10x$$
$$\overline{}$$
$$20x^3 + 3x^2 - 19x + 6$$

8. $26 - 2\sqrt{105}$

9. $\dfrac{x^2 + 3x}{2x^2 + 7x + 5} \div \dfrac{x^2 - 9}{x^2 - 2x - 3} = \dfrac{x(x + 3)}{(2x + 5)(x + 1)} \cdot \dfrac{(x + 1)(x - 3)}{(x + 3)(x - 3)}$

$$= \dfrac{x}{2x + 5}$$

10. $x - 11 - \dfrac{38}{x - 3}$

11. $(x^2 + 2x - 8) \div \dfrac{x^2 - 4}{x^2 + 2x} = \dfrac{(x + 4)(x - 2)}{1} \cdot \dfrac{x(x + 2)}{(x + 2)(x - 2)}$

$$= x(x + 4)$$

12. $\dfrac{6x}{x^2 - 9}$

13.
$$\dfrac{x^2 + 3x - 2}{(x + 2)(x + 3)} - \dfrac{x - 1}{(2x + 1)(x + 3)}$$

$$= \dfrac{(x^2 + 3x - 2)(2x + 1)}{(x + 2)(x + 3)(2x + 1)} - \dfrac{(x - 1)(x + 2)}{(2x + 1)(x + 3)(x + 2)}$$

$$= \dfrac{2x^3 + 7x^2 - x - 2}{(x + 2)(x + 3)(2x + 1)} - \dfrac{x^2 + x - 2}{(2x + 1)(x + 3)(x + 2)}$$

$$= \dfrac{(2x^3 + 7x^2 - x - 2) - (x^2 + x - 2)}{(x + 2)(x + 3)(2x + 1)}$$

$$= \dfrac{2x^3 + 7x^2 - x - 2 - x^2 - x + 2}{(x + 2)(x + 3)(2x + 1)} = \dfrac{2x^3 + 6x^2 - 2x}{(x + 2)(x + 3)(2x + 1)}$$

14. $\dfrac{27\sqrt{2}}{2}$

15. $7 - (2x + 3) = 3x - 2 - (4x - 3)$

$$4 - 2x = -x + 1$$
$$\underline{-1 + 2x \qquad 2x - 1}$$
$$3 \quad = \quad x$$
$$x = 3$$

16. $\left(\dfrac{29}{2}, -6\right)$ or $\left(14\dfrac{1}{2}, -6\right)$

17. (1) $\quad 2]\begin{cases} 3x + 4y = 3 \\ 2x - 5y = 25 \end{cases} \Rightarrow \begin{aligned} 6x + 8y = 6 \\ -6x + 15y = -75 \end{aligned}$
(2) $-3]$
$$\overline{}$$
$$23y = -69$$
$$y = -3$$

Substitute $y = -3$ into (1):

(1) $\qquad 3x + 4y = 3$
$$3x + 4(-3) = 3$$
$$3x - 12 = 3$$
$$3x = 15$$
$$x = 5$$

Solution: $(5, -3)$

18. $-\dfrac{2}{3}$

19. $(\sqrt{5x + 3})^2 = (10x)^2$

$$5x + 3 = 100x^2$$
$$0 = 100x^2 - 5x - 3$$
$$0 = (20x + 3)(5x - 1)$$

$20x + 3 = 0$	$5x - 1 = 0$
$20x = -3$	$5x = 1$
$x = -\dfrac{3}{20}$	$x = \dfrac{1}{5}$

Check for $x = \dfrac{1}{5}$:

$$\sqrt{5\left(\dfrac{1}{5}\right) + 3} = 10\left(\dfrac{1}{5}\right)$$
$$\sqrt{1 + 3} \overset{?}{=} 2$$
$$\sqrt{4} = 2 \quad \text{True}$$

Check for $x = -\dfrac{3}{20}$:

$$\sqrt{5\left(-\dfrac{3}{20}\right) + 3} = 10\left(-\dfrac{3}{20}\right)$$
$$\sqrt{-\dfrac{3}{4} + 3} \overset{?}{=} -\dfrac{3}{2}$$
$$\sqrt{\dfrac{9}{4}} = -\dfrac{3}{2} \quad \text{False}$$

Only one answer: $\dfrac{1}{5}$

20. $-\dfrac{3}{7}$

21. $\quad 12x - 3 \le 7x - 13$
$$\underline{-7x + 3 \qquad -7x + 3}$$
$$5x \quad \le \quad -10$$
$$x \le -2$$

22. $-7, 2$

23. $3 + 4x^2 - 12x = 0$

$$4x^2 - 12x + 3 = 0 \quad \begin{cases} a = 4 \\ b = -12 \\ c = 3 \end{cases}$$

$$x = \dfrac{-(-12) \pm \sqrt{(-12)^2 - 4(4)(3)}}{2(4)}$$

$$= \dfrac{12 \pm \sqrt{144 - 48}}{8}$$

$$= \dfrac{12 \pm \sqrt{96}}{8} = \dfrac{12 \pm 4\sqrt{6}}{8} = \dfrac{4(3 \pm \sqrt{6})}{8} = \dfrac{3 \pm \sqrt{6}}{2}$$

24. $2, -1$

25. $7x^2 - 1 = 0$
$$7x^2 = 1$$
$$x^2 = \dfrac{1}{7}$$
$$x = \pm\sqrt{\dfrac{1}{7}} = \pm\sqrt{\dfrac{1 \cdot 7}{7 \cdot 7}} = \pm\dfrac{\sqrt{7}}{7}$$

26. $6 + 2\sqrt{2}$

27. $3x^2 - 12x - 15 = 3(x^2 - 4x - 5) = 3(x - 5)(x + 1)$

28. $5(x + 3)(x - 3)$ **29.** Not factorable **30.** $(7x + 1)(x - 2)$

In Exercises 31–34, the checks will not be shown.

31. Let x = number of inches corresponding to 12 ft
y = number of inches corresponding to 18 ft

$$\frac{\frac{1}{4}\text{in.}}{1\,\text{ft}} = \frac{x\,\text{in.}}{12\,\text{ft}}$$

$$x = \frac{1}{4}(12) = 3 \text{ inches}$$

$$\frac{\frac{1}{4}\text{in.}}{1\,\text{ft}} = \frac{y\,\text{in.}}{18\,\text{ft}}$$

$$x = \frac{1}{4}(18) = \frac{9}{2} \text{ inches}$$

32. Two answers: -3 and -2, or 4 and 5

33. Let x = number of lb of stew meat
$5 - x$ = number of lb of back ribs

$$2.10x + 1.20(5 - x) = 7.35$$
$$210x + 120(5 - x) = 735$$
$$210x + 600 - 120x = 735$$
$$90x = 135$$
$$x = \frac{135}{90} = \frac{3}{2} \text{ or } 1\frac{1}{2}$$
$$5 - x = 5 - \frac{3}{2} = \frac{10}{2} - \frac{3}{2} = \frac{7}{2} \text{ or } 3\frac{1}{2}$$

Therefore, there are $1\frac{1}{2}$ lb stew meat and $3\frac{1}{2}$ lb back ribs.

34. The shorter leg is 5 cm and the longer one is 12 cm.

Appendix A Exercises A.1 (page 554)

1. Yes, because it is a collection of objects or things   **2.** Yes

3. Yes, because they have exactly the same members. Writing a member more than once tells you no more than when you write it once—namely, that that element is a member of the set.

4. $\{0, 1, 2\}$

5. { }, since there are no digits greater than 9

6. $\{0, 1, 2\}$

7. $\{10, 11, 12, \ldots\}$. This is the way we show the set of whole numbers greater than 9. It is an infinite set.

8. $\{5\}$

9. { }, since there are no whole numbers greater than 4 and at the same time less than 5

10. $\{0, 1, 2, 3\}$

11. 2, a, 3. Because of the way the set is written, we know that these are its elements.

12. 0, 1, 2, 3, 4, 5, 6, 7, 8, 9

13a. $n(\{1, 1, 3, 5, 5, 5\}) = n(\{1, 3, 5\}) = 3$. This set has three elements: 1, 3, and 5.

b. $n(\{0\}) = 1$. This set has one element: 0.

c. $n(\{a, b, g, x\}) = 4$. This set has four elements: a, b, g, and x.

d. $n(\{0, 1, 2, 3, 4, 5, 6, 7\}) = 8$. You can count its elements and see that there are eight of them.

e. $n(\varnothing) = 0$. The empty set has no elements.

14. $\varnothing$ has no elements; $\{0\}$ has one element, namely 0. Since the sets do not have exactly the same elements, they are not equal.

15a. The set of digits is finite because when we count its elements the counting comes to an end. In this case the counting ends at 10.

b. The set of whole numbers is infinite because when we attempt to count its elements the counting never comes to an end.

c. Finite. The counting ends at 7.

d. Finite. If we started counting the books in the ELAC library, we would eventually finish counting them.

16a. True   **b.** True   **c.** False   **d.** False

17a. True   **b.** False   **c.** False   **d.** True

18a. False   **b.** True   **c.** True   **d.** False

Exercises A.2 (page 556)

1a. $\{3, 5\}$ is a proper subset of M because each of its elements, 3 and 5, is an element of M; and M has at least one element, such as 2, that is not an element of $\{3, 5\}$.

b. $\{0, 1, 7\}$ is not a subset of M because elements 0 and 7 are not elements of M.

c. $\varnothing$ is a proper subset of M because $\varnothing$ is a proper subset of every set except itself.

d. $\{2, 4, 1, 3, 5\}$ is an improper subset of M because each element of C is an element of M, and M has no element that is not an element of C.

2a. Proper   **b.** Improper   **c.** Proper   **d.** Not a subset of P

3. $\{R, G, Y\}$ All the subsets with three elements
$\{R, G\}, \{R, Y\}, \{G, Y\}$ All the subsets with two elements
$\{R\}, \{G\}, \{Y\}$ All the subsets with one element
$\{\ \}$ All the subsets with no elements

4. $\{\square, \triangle\}, \{\square\}, \{\triangle\}, \{\ \}$

Exercises A.3 (page 558)

1a. $\{1, 5, 7\} \cup \{2, 4\} = \{1, 5, 7, 2, 4\}$
$\{1, 5, 7\} \cap \{2, 4\} = \{\ \}$

b. $\{a, b\} \cup \{x, y, z, a\} = \{a, b, x, y, z\}$
$\{a, b\} \cap \{x, y, z, a\} = \{a\}$

c. $\{\ \} \cup \{k, 2\} = \{k, 2\}$
$\{\ \} \cap \{k, 2\} = \{\ \}$

d. {river, boat} $\cup$ {boat, streams, down} = {river, boat, streams, down}
{river, boat} $\cap$ {boat, streams, down} = {boat}

2a. $\{1, 3, 5, 7, 2, 4, 6\}$

b. $\{1, 2, 3, 4, 5, 6, 7\}$

c. $\{\ \}$

d. $\{6\}$

3. C and D are disjoint because they have no member in common, that is, $C \cap D = \varnothing$. All other pairs of sets have at least one member in common.

4a. $P = \{c, d, k\}$

b. $Q = \{k, j, f\}$

c. $P \cup Q = \{c, d, k, j, f\}$

d. $P \cap Q = \{k\}$

e. $U = \{a, e, g, h, c, d, k, j, f\}$

5a. $X \cap Y = \{5, 11\}$ because these are the only elements in both X and Y.

b. $Y \cap X = \{5, 11\}$ for the same reason as (a).

c. Yes. $X \cap Y = Y \cap X$ because they have exactly the same elements.

6a. $K \cap L = \{4, b\}$

  b. $n(K \cap L) = 2$

  c. $L \cup M = \{m, 4, 6, b, n, 7, t\}$

  d. $n(L \cup M) = 7$

7.

$A \cup B$

8.

$P \cap Q$

9. $R \cap S$ because the shaded area is in both R and S.

10. $Y \cup Z$

Appendix B Exercises B.1 (page 568)

1. $\dfrac{\overset{2}{\cancel{6}}}{\underset{3}{\cancel{9}}} = \dfrac{2}{3}$

2. $\dfrac{3}{5}$ **3.** $\dfrac{49}{24}$ Already in lowest terms

4. $\dfrac{7}{8}$

5. $\dfrac{5}{2} = 2\overline{)5}^{\,2\,R\,1\,=\,2\frac{1}{2}}$

6. $1\dfrac{3}{8}$

7. $\dfrac{47}{25} = 25\overline{)47}^{\,1\,R\,22\,=\,1\frac{22}{25}}$

        $\dfrac{25}{22}$

8. $2\dfrac{4}{11}$ **9.** $3\dfrac{7}{8} = \dfrac{3 \cdot 8 + 7}{8} = \dfrac{24 + 7}{8} = \dfrac{31}{8}$

10. $\dfrac{25}{9}$

11. $2\dfrac{5}{6} = \dfrac{2 \cdot 6 + 5}{6} = \dfrac{12 + 5}{6} = \dfrac{17}{6}$

12. $\dfrac{43}{5}$

13. $\quad \dfrac{2}{3} = \dfrac{8}{12} \quad$ LCD = 12

    $+\dfrac{1}{4} = \dfrac{3}{12}$

        $\dfrac{11}{12}$

14. $\dfrac{1}{2}$

15. $\quad \dfrac{7}{10} = \dfrac{7}{10} \quad$ LCD = 10

    $-\dfrac{2}{5} = -\dfrac{4}{10}$

        $\dfrac{3}{10}$

16. $\dfrac{7}{20}$

17. $\dfrac{\cancel{3}^{1}}{\underset{4}{\cancel{16}}} \cdot \dfrac{\cancel{20}^{5}}{\underset{3}{\cancel{9}}} = \dfrac{5}{12}$

18. $\dfrac{9}{28}$

19. $\dfrac{4}{3} \div \dfrac{8}{9} = \dfrac{\cancel{4}^{1}}{\underset{1}{\cancel{3}}} \cdot \dfrac{\cancel{9}^{3}}{\underset{2}{\cancel{8}}} = \dfrac{3}{2}$

20. 4

21. $\quad 2\dfrac{2}{3} = 2\dfrac{10}{15}$

    $+\,1\dfrac{3}{5} = 1\dfrac{9}{15}$

        $3\dfrac{19}{15} = 4\dfrac{4}{15}$

22. $7\dfrac{3}{4}$

23. $\quad 5\dfrac{4}{5} = \quad 5\dfrac{8}{10}$

    $-\,3\dfrac{7}{10} = -3\dfrac{7}{10}$

        $2\dfrac{1}{10}$

24. $2\dfrac{1}{6}$

25. $4\dfrac{1}{5} \cdot 2\dfrac{1}{7} = \dfrac{\cancel{21}^{3}}{\underset{1}{\cancel{5}}} \cdot \dfrac{\cancel{15}^{3}}{\underset{1}{\cancel{7}}} = 9$

26. $10\dfrac{1}{2}$

27. $2\dfrac{2}{5} \div 1\dfrac{1}{15} = \dfrac{12}{5} \div \dfrac{16}{15} = \dfrac{\cancel{12}^{3}}{\underset{1}{\cancel{5}}} \cdot \dfrac{\cancel{15}^{3}}{\underset{4}{\cancel{16}}} = \dfrac{9}{4} = 2\dfrac{1}{4}$

28. $\dfrac{2}{3}$

29. $\dfrac{\frac{5}{8}}{\frac{5}{6}} = \dfrac{5}{8} \div \dfrac{5}{6} = \dfrac{\cancel{5}^{1}}{\underset{4}{\cancel{8}}} \cdot \dfrac{\cancel{6}^{3}}{\underset{1}{\cancel{5}}} = \dfrac{3}{4}$

30. $\dfrac{1}{2}$

31. 34.5

     1.74

     18.

     $\underline{0.016}$

    54.256

32. 94.745

33. 356.40

    $\underline{34.67}$

    321.73

34. 336.56   **35.** $100 \times 7.\underset{\longrightarrow}{45} = 745$ **36.** 3540

37. $\dfrac{46.8}{100} = 0.468$

38. 0.0895

39.
$$9\ 4.7\ 8\ (2\text{ decimal places})$$
$$\underline{7\ 0.9\ (1\text{ decimal place})}$$
$$8\ 5\ 3\ 0\ 2$$
$$\underline{6\ 6\ 3\ 4\ 6}$$
$$6\ 7\ 1\ 9.9\ 0\ 2 \doteq 6719.9$$

40. 2798.5

41.
$$
\begin{array}{r}
0.8\,2\,8 \doteq 0.83 \\
7.2\,5\,\overline{)6.0\,0\,0.7\,0\,0} \\
\underline{5\ 8\ 0\ 0} \\
2\ 0\ 7\ 0 \\
\underline{1\ 4\ 5\ 0} \\
6\ 2\ 0\ 0 \\
\underline{5\ 8\ 0\ 0} \\
4\ 0\ 0
\end{array}
$$

42. 1.71

43. $5\frac{3}{4} = \frac{23}{4} = 4\overline{)2\ 3^30^20}$ with $5.7\,5$ quotient

44. 4.6

45. $0.65 = \dfrac{\cancel{65}^{13}}{\cancel{100}_{20}} = \dfrac{13}{20}$

46. $\dfrac{1}{4}$

47. $5.9 = 5 + .9 = 5 + \dfrac{9}{10} = 5\dfrac{9}{10}$

48. $4\dfrac{3}{10}$

49. $4.70_\rightarrow = 470\%$

50. 0.1%   **51.** $0.25_\rightarrow 8 = 25.8\%$ **52.** 1200%

53. $._\leftarrow 03.5\% = 0.035$ **54.** 0.0002   **55.** $1._\leftarrow 57\% = 1.57$

56. 0.178   **57.** $\dfrac{1}{8} = 0.12_\rightarrow 5 = 12.5\%$ **58.** 60%

59. $\dfrac{7}{12} \doteq 0.58_\rightarrow 33 = 58.33\%$ **60.** 14.29%

61. $35\% = \dfrac{35}{100} = \dfrac{7}{20}$ **62.** $\dfrac{4}{5}$ **63.** $12\% = \dfrac{12}{100} = \dfrac{3}{25}$

64. $\dfrac{9}{50}$

Appendix C Exercises C.1 (page 573)

1. The set is a function, because no two ordered pairs have the same first element but different second elements. The domain is {4, 2, 3, 0}.

2. The set is a function. The domain is {6, −2, 4, 0}.

3. The set is not a function, because the pairs (2, −5) and (2, 0) have the same first element but different second elements.

4. The set is not a function.

5. The set is a function, because no two ordered pairs have the same first element but different second elements. The domain is {−8, 3, 6, 9}.

6. The set is a function. The domain is {−3, −1, 0, 3}.

7. The domain is the set of all real numbers except 12. (Twelve is an excluded value.)

8. The domain is the set of all real numbers except 2.

9. The domain is the set of all real numbers.

10. The domain is the set of all real numbers.

11. The radicand must be ≥ 0.
$$
\begin{array}{rcl}
x - 5 & \geq & 0 \\
\underline{+\ 5} & & \underline{+\ 5} \\
x & \geq & 5
\end{array}
$$
The domain is the set of all real numbers greater than or equal to 5.

12. The domain is the set of all real numbers greater than or equal to 10.

13. There is a restriction: $x \geq 0$. Therefore, the domain is the set of all real numbers greater than or equal to 0.

14. The domain is the set of all real numbers greater than or equal to −3.

15. b is the graph of a function; a and c are not graphs of functions.

16. c is the graph of a function; a and b are not graphs of functions.

Exercises C.2 (page 575)

1. $f(x) = (x + 2)^2$
  a. $f(0) = ([0] + 2)^2 = (2)^2 = 4$
  b. $f(2) = ([2] + 2)^2 = (4)^2 = 16$
  c. $f(-3) = ([-3] + 2)^2 = (-1)^2 = 1$
  d. $f(1) = ([1] + 2)^2 = (3)^2 = 9$

2a. 9   **b.** 25   **c.** 1   **d.** 49

3. $g(x) = x^2 + 4x + 4$
  a. $g(0) = (0)^2 + 4(0) + 4 = 0 + 0 + 4 = 4$
  b. $g(2) = (2)^2 + 4(2) + 4 = 4 + 8 + 4 = 16$
  c. $g(-3) = (-3)^2 + 4(-3) + 4 = 9 - 12 + 4 = 1$
  d. $g(1) = (1)^2 + 4(1) + 4 = 1 + 4 + 4 = 9$

4a. 9   **b.** 25   **c.** 1   **d.** 49

5. $F(x) = x^2 + 4$
  a. $F(0) = (0)^2 + 4 = 0 + 4 = 4$
  b. $F(2) = (2)^2 + 4 = 4 + 4 = 8$
  c. $F(-3) = (-3)^2 + 4 = 9 + 4 = 13$
  d. $F(1) = (1)^2 + 4 = 1 + 4 = 5$

6a. 9   **b.** 13   **c.** 25   **d.** 25

7. $f(x) = 3x^2 + x - 1$
  a. $f(0) = 3(0)^2 + (0) - 1 = 0 + 0 - 1 = -1$
  b. $f(-2) = 3(-2)^2 + (-2) - 1 = 12 - 2 - 1 = 9$
  c. $f(5) = 3(5)^2 + (5) - 1 = 75 + 5 - 1 = 79$
  d. $f(1) = 3(1)^2 + (1) - 1 = 3 + 1 - 1 = 3$

8a. −5   **b.** 4   **c.** 39   **d.** 0

9. $f(x) = x^3 - 1$
  a. $f(0) = (0)^3 - 1 = 0 - 1 = -1$
  b. $f(-1) = (-1)^3 - 1 = -1 - 1 = -2$
  c. $f(2) = (2)^3 - 1 = 8 - 1 = 7$
  d. $f(1) = (1)^3 - 1 = 1 - 1 = 0$

10a. 1   **b.** −7   **c.** 2   **d.** 0

11. $g(t) = \dfrac{4t^2 + 5t - 1}{t + 2}$
  a. $g(0) = \dfrac{4(0)^2 + 5(0) - 1}{(0) + 2} = \dfrac{0 + 0 - 1}{2} = -\dfrac{1}{2}$
  b. $g(-1) = \dfrac{4(-1)^2 + 5(-1) - 1}{(-1) + 2} = \dfrac{4 - 5 - 1}{1} = -2$
  c. $g(2) = \dfrac{4(2)^2 + 5(2) - 1}{(2) + 2} = \dfrac{16 + 10 - 1}{4} = \dfrac{25}{4} \text{ or } 6\dfrac{1}{4}$
  d. $g(4) = \dfrac{4(4)^2 + 5(4) - 1}{(4) + 2} = \dfrac{64 + 20 - 1}{6} = \dfrac{83}{6} \text{ or } 13\dfrac{5}{6}$

12a. $\dfrac{1}{3}$ **b.** −1   **c.** −1   **d.** 1

Index

Abscissa, 402
Absolute value, 13–14
Addition
  associative property of, 36–37
  commutative property of, 35
  of decimals, 564
  and distributive property, 93
  of fractions, 343–344, 351–357, 566
  of polynomials, 239–240
  of signed numbers, 11–17
  of square roots, 499–503
  systems of equations, addition method for,
    461–467
  of unlike fractions, 351–357
Addition property
  of equality, 135–138
  for inequalities, 161
Additive identity, 15–16
Additive inverse, 18
Algebra
  defined, 2
  reducing fractions in, 335
Algebraic expressions. *See also* Simplification
    of algebraic expressions
  defined, 68
  key word expressions, 176–178
  word expressions, changing of, 176–183
Algebraic fractions, 330–335
Approximately equal symbol, 46, 564
Arithmetic, 2
  reducing fractions in, 335
  review of, 563–569
Ascending powers, 238
Associative properties, 36–41
  of addition, 36–37
  determining use of, 38–39
  of multiplication, 37

Bars. *See* Grouping symbols
Base, defined, 41
Binomials, 236
  factoring expressions with common binomial
    factor, 284–285
  multiplication of, 248–252
  squares of, 257–259
Boundary line of half-plane, 438–439
Braces. *See* Grouping symbols
Brackets. *See* Grouping symbols
Business formula evaluation, 75, 76, 77

Calculators
  for solutions of quadratic equations, 537
  and scientific notation, 122
  finding square roots with, 46–47
Cardinal number of set, 555
Chemical solution problems, 221–225
Chemistry formula evaluation, 75, 76
Circles, 180
Coefficients, 87
Combining like terms, 100–104
Commutative properties, 35–36
  of addition, 35
  of multiplication, 36
Completing the square, 534
Complex fractions, 357–361, 567

Composite numbers, 52
Conditional equations, 134, 155–158
Conditional inequality, 161
Cones, 180
Conjugates, 504–506
Consecutive integers, 179
Consecutive numbers, 10
Consistent systems, 459, 462, 468
Constants, 68
Coordinates, 402
Counting numbers, 2–3
Cubes, 180
Cubic roots, 48
Curves, graphing of, 432–437
Cylinders, 180

Decimal fractions, 4
Decimal notation, 4, 121
Decimal places, 4–5, 564
Decimals
  fractions to, 567
  to fractions, 567
  as solutions of equations, 150
  to percents, 567
  review of, 564–565
Degree of polynomials, 237–238
Denominators, 4. *See also* Rationalizing the
    denominator
  of algebraic fractions, 330
Dependent systems, 460, 462, 468
Dependent variables, 572
Descending powers, 238
Diagonal of rectangle, 539
Difference. *See also* Subtraction
  defined, 17
  product of sum and difference of two terms,
    256–257
Digits, set of, 3
Disjoint sets, 560
Distance-rate-time problems, 211–217
  with fractions, 391–392
Distributive property, 93–97
  for binomial multiplication, 248
  extensions of, 94–95
Dividends, 26
  for fractions, 340
  for polynomials, 261–267
Divisibility, tests for, 50–51
Division
  associative property, absence of, 38
  commutative property, absence of, 36
  of decimals, 565
  of exponential numbers, 106–111
  of fractions, 340–341, 567
  of polynomials, 259–267
  of signed numbers, 26–30
  of square roots, 498–499, 503–507
  with zero, 27
Division property
  of equality, 138–143
  for inequalities, 161
Divisors, 26–27
  for fractions, 340
  for polynomials, 261–267
Domain, 572
  rules for finding, 573

Electricity formula evaluation, 75
Elements of set, 554
  defined, 2
Empty set, 555
Equal sets, 555
Equal sign, 2
Equations, 72–73, 133–173. *See also*
    Quadratic equations; Solutions to
    equations; Systems of equations
  changing signs in, 147
  defined, 72–73
  of lines, 427–432
  literal equations, 381–384
  with no solution, 156–157
  radical equations, 507–512
  word problems changed to, 184–187
Equivalent equations, 134–135
Equivalent fractions, 331, 566
Equivalent systems of equations, 461–462
Evaluating expressions with variables and
    numbers, 68–72
Evaluating functions, 576–577
Even index, 48
Even integers, 10
Even powers, 41
Excluded values, 330–331
Exponential expressions, 41
Exponents
  defined, 41, 68
  dividing exponential numbers, 106–111
  first rule of, 89–90
  general rule of, 108–109
  multiplying exponential numbers, 89–90
  negative exponents, 112–115
  and order of operations, 63–64
  powers of exponential numbers, 104–106
  scientific notation, 120–123
  simplified form of terms with, 116–117
  zero exponent, 111
Expressions
  algebraic, 68
  exponential expressions, 41
  and negative exponents, 113–114
  with numerical bases, 115–116
Extraneous roots, 369
  in radical equations, 507
Extremes of proportion, 373

Factoring, 51, 277–328
  binomial factor, expressions with common,
    284–285
  equations solved by, 312–317
  by grouping, 305–307
  and incomplete quadratic equations,
    530–531
  Master Product method, 308–309
  monomial factor, expressions with common,
    278–284
  and quadratic equations, 312–317, 524–529
  rule of signs for factoring trinomials, 290
  selecting method for, 309–312
  squares, difference of two, 285–289
  of trinomials, 289–302
  uniqueness of, 52, 280–283
  and word problems, 317–323
Factors, 86–87. *See also* Factoring
  defined, 22